14th Conference on Research & Development in Power Engineering (RDPE 2019)

E3S Web of Conferences Volume 137 (2019)

Warsaw, Poland
3 – 6 December 2019

Editors:

Rafal Laskowski
Adam Smyk
Slawomir Bielecki
Łukasz Szabłowski

ISBN: 978-1-7138-0029-3

Printed from e-media with permission by:

Curran Associates, Inc.
57 Morehouse Lane
Red Hook, NY 12571

Some format issues inherent in the e-media version may also appear in this print version.

No changes have been made to the content of these proceedings. There may be changes to pagination and minor adjustments for aesthetics.

Printed with permission by Curran Associates, Inc. (2020)

For additional information, please contact EDP Sciences – Web of Conferences
at the address below.

EDP Sciences – Web of Conferences
17, Avenue du Hoggar
Parc d'Activité de Courtabœuf
BP 112
F-91944 Les Ulis Cedex A
France

Phone: +33 (0) 1 69 18 75 75
Fax: +33 (0) 1 69 28 84 91

contact-edps@webofconferences.org

Additional copies of this publication are available from:

Curran Associates, Inc.
57 Morehouse Lane
Red Hook, NY 12571 USA
Phone: 845-758-0400
Fax: 845-758-2633
Email: curran@proceedings.com
Web: www.proceedings.com

TABLE OF CONTENTS

Preface

The 14th Conference on Research & Development in Power Engineering (RDPE 2019) will take place on 3-6 December 2019 in Warsaw, Poland, with the Honorary Patronage of the Rector of Warsaw University of Technology (WUT).

The conference is organized jointly by the Institute of Heat Engineering, Warsaw University of Technology and the Applied Research Institute, Warsaw University of Technology, in collaboration with the Committee on Energy, Thermodynamics and Combustion, Polish Academy of Sciences and the Polish Chamber of Power Industry and Environmental Protection. Over the years, this biannual event has become a leading discussion forum related to the sector of power engineering; it is here where science meets industry and representatives of both the former and the latter are given an opportunity to exchange knowledge, raise questions, find answers on the state-of-the-art power engineering and to offer the expertise for the benefit of each-another.

The topics presented during the Conference include:

- operational tests and experience from completed retrofitting
- technical and financial aspects of retrofitting and reconstruction in the power sector
- emission control technologies
- rational lifetime extensions for power installations
- flexibility of power unit
- operational optimisation for power and CHP plants
- energy market
- new power technologies
- adaptation of the power and heating sector to EU regulations
- mathematical modelling of physical processes in the power sector, process machinery, power and process systems
- nuclear energy issues.

Please visit the Conference website for detailed information: http://www.pbec.itc.pw.edu.pl.

The organizers of the RDPE Conference would like to thank all conference participants for preparing interesting articles.

On behalf of the Organizers
Chairman of the Organizing Committee

Prof. Krzysztof Badyda

Honorary Committee of XIV RDPE Conference

- prof. Jan Szmidt, Rector of Warsaw University of Technology
- Grzegorz Należyty, Energy Sector SIEMENS Sp. z o.o.
- prof. Janusz Lewandowski, Warsaw University of Technology
- prof. Jarosław Mikielewicz, Institute of Fluid-Flow Machinery, Polish Academy of Sciences
- dr inż. Jerzy Trzeszczyński, Przedsiębiorstwo Usług Naukowo-Technicznych „Pro Novum" Sp. z o.o.
- dr inż. Krzysztof Sadowski, The Nacional Centre for Research and Development
- Wiesław Różacki, Mitsubishi Hitachi Power Systems Europe
- dr inż. Tomasz Wilczak, PGNiG Termika
- Adam Ogrodnik, Urząd Dozoru Technicznego - Office of Technical Inspection (UDT)

Scientific Committee of XIV RDPE Conference

- prof. Henryk Anglart, KTH Royal Institute of Technology
- prof. Janusz Badur, Institute of Fluid-Flow Machinery, Polish Academy of Sciences
- prof. Krzysztof Badyda, Warsaw University of Technology – CHAIRMAN
- prof. Jerzy Banaszek, Warsaw University of Technology
- prof. Asfaw Beyene, San Diego State University
- prof. Tadeusz Bohdal, Koszalin University of Technology
- prof. Wojciech Bujalski, Warsaw University of Technology – VICE-CHAIRMAN
- prof. Tadeusz Chmielniak, Silesian University of Technology
- prof. Dorota Chwieduk, Warsaw University of Technology
- prof. Umberto Desideri, Università degli Studi di Perugia
- prof. Janusz Frączek, Warsaw University of Technology
- prof. Piotr Furmański, Warsaw University of Technology
- prof. Norihiko Iki, Energy Technology Research Institute, Tsukuba
- prof. Maciej Jaworski, Warsaw University of Technology
- prof. Krzysztof Jesionek, Wrocław University of Science and Technology
- dr hab. inż. Hanna Jędrzejuk, Warsaw University of Technology
- prof. Soteris Kalogirou, Cyprus University of Technology
- prof. Henryk Kaproń, Warsaw University of Technology
- prof. Sotirios Karellas, National Technical University of Athens
- prof. Janusz Kotowicz, Silesian University of Technology
- prof. Władysław Kryłłowicz, Lódź University of Technology
- prof. Janusz Lewandowski, Warsaw University of Technology
- prof. Henrik Lund, Aalborg University
- prof. Dariusz Mikielewicz, Gdańsk University of Technology
- prof. Jarosław Milewski, Warsaw University of Technology
- prof. Tetyana Morozyuk, Technical University of Berlin
- prof. Wojciech Nowak, AGH University of Science and Technology

Organizing Team of XIV RDPE Conference

- Krzysztof Badyda
- Sławomir Bielecki
- Janusz Buchowski
- Wojciech Bujalski
- Aleksandra Dzido
- Kamil Futyma
- Andrzej Grzebielec
- Karolina Jazowska
- Michalina Kurkus-Gruszecka
- Jerzy Kuta
- Katarzyna Langiewicz
- Rafał Laskowski
- Piotr Łapka
- Jarosław Milewski
- Grzegorz Niewiński
- Adam Smyk
- Michał Stępień
- Ewa Stefaniak
- Łukasz Szabłowski
- Arkadiusz Szczęśniak
- Marcin Wołowicz

RESEARCH & DEVELOPMENT IN POWER ENGINEERING 2019

INSTITUTE OF HEAT ENGINEERING
FACULTY OF POWER AND AERONAUTICAL ENGINEERING
WARSAW UNIVERSITY OF TECHNOLOGY
Warsaw, Poland 2019

Statement of Peer review

In submitting conference proceedings to *Web of Conferences*, the editors of the proceedings certify to the Publisher that

1. They adhere to its **Policy on Publishing Integrity** in order to safeguard good scientific practice in publishing.
2. All articles have been subjected to peer review administered by the proceedings editors.
3. Reviews have been conducted by expert referees, who have been requested to provide unbiased and constructive comments aimed, whenever possible, at improving the work.
4. Proceedings editors have taken all reasonable steps to ensure the quality of the materials they publish and their decision to accept or reject a paper for publication has been based only on the merits of the work and the relevance to the journal.

Title, date and place of the conference

XIV RESEARCH & DEVELOPMENT IN POWER ENGINEERING Conference,
03.12.2019-06.12.2019,
Warsaw, Poland

Proceedings editor(s):

Dr Eng. Rafał Laskowski
Dr Eng. Adam Smyk
Dr Eng. Sławomir Bielecki
Dr Eng. Łukasz Szabłowski

Date and editor's signature

29.11.2019

E3S Web of Conferences **137**, 01001 (2019) https://doi.org/10.1051/e3sconf/201913701001
RDPE 2019

RESEARCH & DEVELOPMENT IN POWER ENGINEERING, 2019

The analysis of a Velox-type CHP plant for various gaseous fuels

Krystian Smołka[1,*], and *Sławomir* Dykas[1]

[1] Silesian University of Technology, Institute of Power Engineering and Turbomachinery, Konarskiego 18, 44-100 Gliwice, krystian.smolka@polsl.pl, slawomir.dykas@polsl.pl, Poland

Abstract. For many years, the Institute of Power Engineering and Turbomachinery of the Silesian University of Technology has been using a small-capacity (about 500 kW$_e$) steam-gas power plant. Based on many years of experience in operation of this power plant utilizing the Velox-type gas-steam system, an idea arose to modify this type of thermal power cycle to create a combined heat and power (CHP) plant of small capacity, dedicated for distributed heat and power production of the range up to 500 kW or production process steam with high temperature. Previous thermodynamic and economic analysis of that type of CHP plant were conducted for natural gas as a fuel. The new idea is use the alternative gas fuels or waste heat for Velox-type CHP plant. An adaptation of the Velox-type CHP plant for various fuels can be done in simple way by moving the combustion chamber out from the set of heat exchangers, in similar way as it is done for HRSG. This paper presents a thermodynamic analysis of the Velox-type steam-gas cycle fired with various alternative gas fuels such as coke gas, blast furnace gas, biogas or gas from gasification. The systems are modelled in the EBSILON® Professional 13 program.

1 Introduction

Historically, the first idea for combining gas and steam turbine cycles was to install a steam generator in place of the gas turbine combustion chamber. This concept was put forward by the Brown Boveri company, which patented the Velox-type gas-steam system in 1932. In 1939, a gas turbine was used for the first time in the world to produce electricity. It was also the first gas-steam system used for electricity generation [1, 2].

Considering the amount of steam generated, the Velox-type steam generator is characterized by a compact structure and relatively small heat transfer surface area, especially in the evaporator. This is due to the fact, that the heat exchange process in the combustion chamber takes place in overpressure, which intensifies the heat transfer between exhaust gases and water and steam in the tubes of the evaporator or steam superheaters. The high velocity of the exhaust gas enabled effective use of the combustion chamber surface area [2]. The implementation of this type of steam generator in power engineering made it possible to create highly efficient plants generating electricity and heat, i.e., the first combined heat and power plant (CHP plant) [1, 3].

For many years, the Institute of Power Engineering and Turbomachinery of the Silesian University of Technology has been using a small-capacity (about 500 kW$_e$) steam-gas power plant of a Velox-type cycle for purposes related to both teaching and experimental testing of steam condensing flows [4]. The experience gained so far from the operation of this power plant based on a Velox-type gas-steam system confirms very good flexibility of the system operation, the short time needed to prepare live steam with required parameters, the parameters stability and the fast rate of load change.

Based on this an idea arose to modify this type of thermal power cycle to create a combined heat and power (CHP) plant of small capacity, dedicated for distributed heat and power production of the range up to 500 kW or production process steam with high temperature. Previous thermodynamic and economic analysis [5, 6] of that type of CHP plant fired with natural gas, and compared with classical gas-steam cycle [7, 8] led to conclusions:

- Velox-type plant can operate for much less load than classical CHP plant,
- Velox-type steam-gas unit creates an opportunity to use higher temperatures of live steam,
- Velox-type plant can operate for much less load than classical CHP plant,
- Velox-type plant can be much cheaper than classical CHP plant.

The new idea is use the alternative gas fuels or waste heat for Velox-type CHP plant. An adaptation of Velox-type CHP plant for various fuels can be done in simple way by moving on combustion chamber out from the set of heat exchangers.

This paper presents a thermodynamic analysis obtained by modelling the Velox-type steam-gas cycle fired with various alternative gas fuels in the EBSILON® Professional 13 program.

The considerations presented herein concern small-capacity power plant with the power output of the order of a few megawatts of electric power, intended for distributed generation of electricity and heat based on the various gas fuel.

https://doi.org/10.1051/e3sconf/201913701001

2 Thermal power cycle and parameters

The Velox-type CHP plant (Fig. 1) is characterized by a much higher ratio of electricity obtained from the steam turbine and the gas turbine, respectively, compared to classical gas-steam cycles which are now in use. From the same amount of fuel more steam with higher enthalpy is produced [6]. The Velox-type CHP plant systems was modelled in the EBSILON® Professional 13 program according to technological structure depicted in the Fig. 1.

In Table 1 the basic parameters are presented for nominal load of the Velox-type CHP plant under consideration. It sets out all important parameters for cycle presented above, as well as efficiency of main and auxiliary machines.

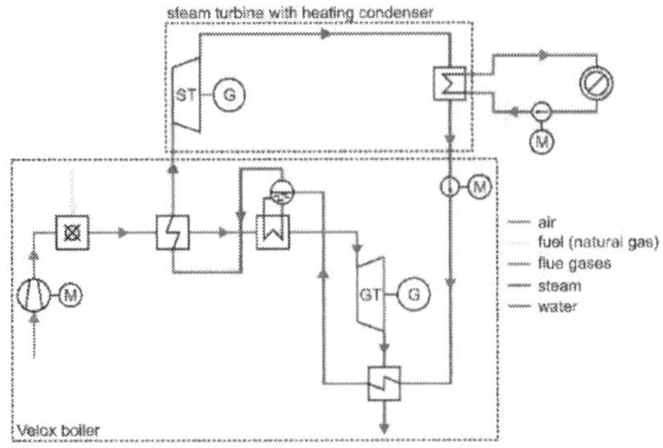

Fig. 1. Velox-type CHP plant scheme.

Table 1. Basic operating parameters for the Velox-type CHP plant under consideration (nominal load).

Parameter	Value	Unit
Air pressure upstream the compressor	0.1	MPa(a)
Air temperature upstream the compressor	20	°C
Air mass flow	0.1	kg/s
Air pressure downstream the compressor	0.3	MPa(a)
Fuel pressure upstream the combustion chamber	0.3	MPa(a)
Fuel temperature upstream the combustion chamber	20	°C
Exhaust gas temperature downstream the combustion chamber	700	°C
Exhaust gas pressure downstream the gas turbine	0.1	MPa(a)
Live steam temperature	680	°C
Live steam pressure	2	MPa(a)
Steam temperature downstream of the steam turbine	265	°C
Steam pressure downstream the steam turbine	0.1	MPa(a)
Inlet temperature of the heating water circuit	30	°C
Outlet temperature of the heating water circuit	94.6	°C
Gas turbine internal efficiency	90	%
Steam turbine internal efficiency	90	%
Efficiency of generators	98.6	%
Compressor efficiency	85	%
Efficiency of pumps	80	%
Electric efficiency of motors driving the pumps and compressor	95	%
Mechanical efficiency of motors (for pumps and compressor)	99.8	%

3 Considered gas fuels

Four types of alternative gas fuels, already successfully used in the production of electricity and heat, were selected for the analysis of the Velox-type CHP plant work. These are biogas, blast furnace gas, coke gas and syngas from coal gasification. In addition, two different syngases obtained from the most popular gasification processes were selected to analyse the Velox-type CHP plant work with syngas. Thus, a total of five selected gaseous fuels were analysed.

Brief characteristics of each of these fuels together with volume fractions and calculated calorific value are presented below. The calorific values of each of the gaseous fuels discussed below were calculated in the EBSILON® Professional 13 program automatically after entering presented volume fractions of individual components of the gas mixture, and these values were adopted for further analysis.

https://doi.org/10.1051/e3sconf/201913701001

3.1 Biogas

Biogas is the mixture of gases produced by the breakdown of organic matter in the absence of oxygen. It can be produced from raw materials such as agricultural waste, manure, municipal waste, plant material, sewage, green waste or food waste. Biogas is primarily methane and carbon dioxide and may have small amounts of hydrogen sulfide, moisture and siloxanes [9].

Because of low calorific value and flame speed biogas is found to be worse fuel than the conventional fuels, therefore it is often enriched by hydrogen or methane.

For our analysis we have decided to use biogas from work [10]. The composition of this biogas is presented in Table 2.

Table 2. Composition of biogas used for analysis [10].

Component	Value	Unit
CH_4	49.777	Volume %
CO_2	43.535	Volume %
CO	1.197	Volume %
H_2	0.489	Volume %
N_2	4.759	Volume %
O_2	0.243	Volume %
Calorific value	13977.464	kJ/kg

3.2 Blast furnace gas

Blast furnace gas is a by-product of blast furnaces that is generated when the iron ore is reduced with coke to metallic iron. It has a very low heating value because it consists of about 60 percent nitrogen and 18-20% carbon dioxide, which are not flammable. The rest is mostly carbon monoxide, which has a fairly low heating value already and some hydrogen [11].

It is commonly used as a fuel within the steel works, but it can be used in boilers and power plants equipped to burn it. It may be combined with natural gas or coke oven gas before combustion or a flame support with richer gas or oil is provided to sustain combustion. Because of the high concentration of carbon monoxide blast furnace gas is hazardous.

Composition of blast furnace gas used in our analysis is presented in Table 3 and the source of the volume fractions values is paper [12].

Table 3. Composition of blast furnace gas used for analysis [12].

Component	Value	Unit
CO_2	20.09	Volume %
CO	23.74	Volume %
H_2	2.05	Volume %
N_2	53.41	Volume %
O_2	0.71	Volume %
Calorific value	2348.173	kJ/kg

3.3 Coke gas

Coal gas is a flammable gaseous fuel made from coal. It is a by-product of the coking process and it is produced when coal is heated strongly in the absence of air. Coal gas contains a variety of calorific gases including hydrogen, carbon monoxide, methane, ethylene and volatile hydrocarbons together with small quantities of non-calorific gases such as CO_2 and N_2 [13].

Because of high amount of CH_4 and H_2 coke gas has a higher calorific value than other alternative gas fuels. Due to this fact, coal gas can be supplied to the user via a piped distribution system in the same way as natural gas.

The coke gas used in our analysis is produced in steel plant placed in Krakow, Poland [14]. Composition of this gas is presented in Table 4.

Table 4. Composition of coke gas used for analysis [14].

Component	Value	Unit
CH_4	23.7	Volume %
CO_2	2.6	Volume %
CO	6.6	Volume %
H_2	57.3	Volume %
N_2	7.2	Volume %
O_2	0.2	Volume %
C_nH_m	2.4	Volume %
Calorific value	35498.049	kJ/kg

3.4 Syngas

Syngas is a fuel gas mixture consisting primarily of hydrogen, carbon monoxide, and very often some carbon dioxide. Syngas is usually a product of coal gasification and the main application is electricity generation. Syngas is combustible and can be used as a fuel of internal combustion engines. Syngas has less than half the energy density of natural gas [15].

Syngas can be produced from many sources, including natural gas, coal, biomass, or virtually any hydrocarbon feedstock, by reaction with steam (steam reforming), carbon dioxide (dry reforming) or oxygen (partial oxidation).

Due to the fact, that in Poland we still have large coal reserves, which can be used in electricity and heat production, we have decided to use in our analysis only syngas from two most popular coal gasification processes. It means from Koppers–Totzek gasification system and Lurgi gasification system.

3.4.1 Koppers-Totzek gasification

The Koppers-Totzek gas generator is based on a turbulent flux. Coal is supplied to direct-flow reactors together with gasifying agents – steam and oxygen. The reactor is at elevated pressure and temperature. In contrast to processes with a compact bed, the use of turbulent flux permits the gasification of different types of coal and produces gas at high temperature; the gas is free of tars and phenol [15].

https://doi.org/10.1051/e3sconf/201913701001

Composition of syngas obtained from Koppers-Totzek gasification process used in the analysis is presented in Table 5.

Table 5. Composition of syngas Koppers-Totzek used for analysis [15].

Component	Value	Unit
CO_2	65.2	Volume %
CO	0.8	Volume %
H_2	25.5	Volume %
N_2	8.2	Volume %
H_2S	0.3	Volume %
Calorific value	11507.881	kJ/kg

3.4.2 Lurgi gasification

One of the commonest and oldest-established technologies is gasification in a steady bed by the Lurgi method. Gasification in a dense fuel bed at atmospheric pressure extracts practically all of the valuable components. Gasification of a dense bed of coal pieces at elevated pressure is also employed [15].

Composition of syngas obtained from Lurgi gasification process used in the analysis is presented in Table 6.

Table 6. Composition of syngas Lurgi used for analysis [15].

Component	Value	Unit
CH_4	7.6	Volume %
CO_2	2.6	Volume %
CO	60.6	Volume %
H_2	27.8	Volume %
N_2	1.0	Volume %
C_nH_m	0.4	Volume %
Calorific value	15245.344	kJ/kg

4 Thermodynamic analysis of a Velox-type CHP plant for considered gas fuels

Figures 2, 3 and 4 and Table 7 shows a comparison of the calculated, selected thermodynamic parameters for the nominal load of the analysed Velox-type CHP plant working with the gaseous fuels presented above.

Fig. 2 shows a comparison of cycle efficiency with a breakdown into the efficiency of electricity and heat generation. The total cycle efficiency, depending on the fuel used, remains basically unchanged and fluctuates slightly within 77%. Larger differences can be seen in the efficiency of generating electricity or heat alone. It can be seen here that the cycle fired with blast furnace gas demonstrate the highest efficiency of electricity generation. This efficiency is 23.8%, compared to approximately 20.5% for other cases. However, the heat generation efficiency is reduced for this case. For a cycle fired with blast furnace gas it is 53.8%, compared to approximately 56% for other cases.

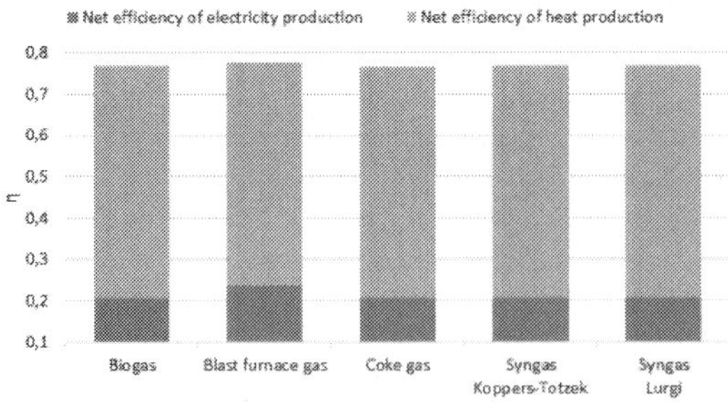

Fig. 2. Efficiency for nominal load of the thermal cycle fired with gas fuels under consideration.

For the cycle fired with blast furnace gas compared to other cases, higher production of both electricity and heat was noticed (see also Fig. 6). At the same operating parameters, such as pressure and temperature of steam and exhaust gases, for each case, these differences are obviously associated with both exhaust gas and steam mass flow rates. As can be seen in Fig. 3 primarily the exhaust gas mass flow rate for a cycle fired with blast furnace gas is much larger compared to other cases, hence more exhaust gas energy can be converted into useful work here.

However, the production of larger amounts of exhaust gas is associated with largest demand for fuel, the more that blast furnace gas has the lowest calorific value of all fuels analysed here. Additionally, in Fig. 3 we can see that for the nominal load of the analysed Velox-type CHP plant, the least fuel should be supplied to the cycle fired with coke gas (0.0018 kg/s) and this is not much more than for previously conducted analyses for natural gas [6]. The next in order are cases fired with fuels: syngas Lurgi (0.0042 kg/s), biogas (0.0047 kg/s), syngas Koppers-Totzek (0.0057 kg/s) and at the very end the already mentioned blast furnace gas (0.0376 kg/s). Considering the calorific values of individual considered gaseous fuels, this order should have been expected.

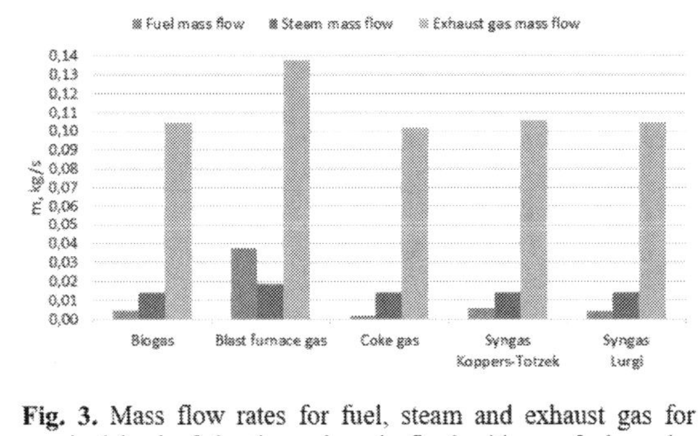

Fig. 3. Mass flow rates for fuel, steam and exhaust gas for nominal load of the thermal cycle fired with gas fuels under consideration.

It should also be noted that the larger mass flow rates of exhaust gases or steam, the heat exchange surfaces also increase (Fig. 4 and Table 7). In Fig. 4 and Table 7 we can also see that the largest heat exchange area should be provided for the heat exchanger in the heating water circuit for the cycle fired with blast furnace gas. This cycle, compared to other cases, would also need the largest heat exchange areas in the economizer, evaporator and superheater. Other cases have heat exchange surfaces very similar, therefore a change of fuel for the Velox-type CHP plant under consideration would not require major structural changes, except for the use of blast furnace gas as a fuel.

Table 7. Surface area of the heat exchangers (kA parameter) of the thermal cycle fired with various gas fuels (nominal load).

Parameter	Biogas	Blast furnace gas	Coke gas	Syngas Koppers-Totzek	Syngas Lurgi	Unit
Economizer	0.241	0.322	0.232	0.239	0.236	kW/K
Evaporator	0.131	0.168	0.127	0.130	0.128	kW/K
Superheater	0.129	0.168	0.126	0.128	0.127	kW/K
Heat exchanger in the heating water circuit	1.503	1.951	1.462	1.493	1.478	kW/K
Sum	2.003	2.610	1.947	1.990	1.969	kW/K

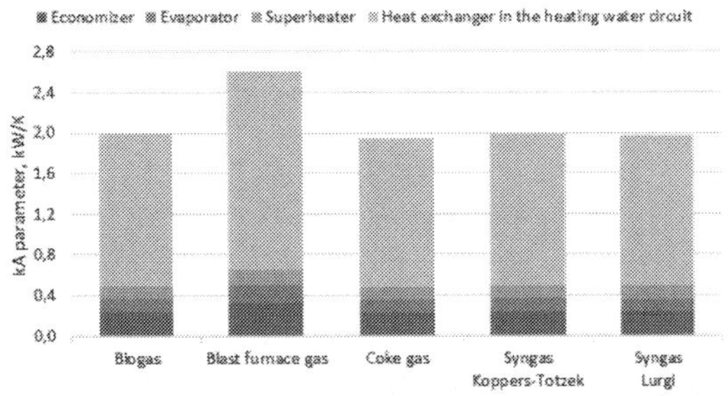

Fig. 4. Surface area of the heat exchangers (kA parameter) for nominal load of the thermal cycle fired with gas fuels under consideration.

Figures 5 – 9 presents the operational characteristics of Velox-type CHP plant working with the gaseous fuels under consideration for various loads. From the analysis of these figures, it can be concluded that for each fuel the operating range of the considered cases is very similar. The flexibility of the Velox-type CHP plant fired with various gaseous fuels is again confirmed, as it was for natural gas in earlier analysis presented in [6].

In addition, in Fig. 6 it can be also seen a slightly larger operating range for the cycle fired with blast furnace gas, compared to other cases, as well as greater production of both electric power and thermal power, and slightly greater efficiency, which has already been shown for nominal load and is associated with energy flow for this case.

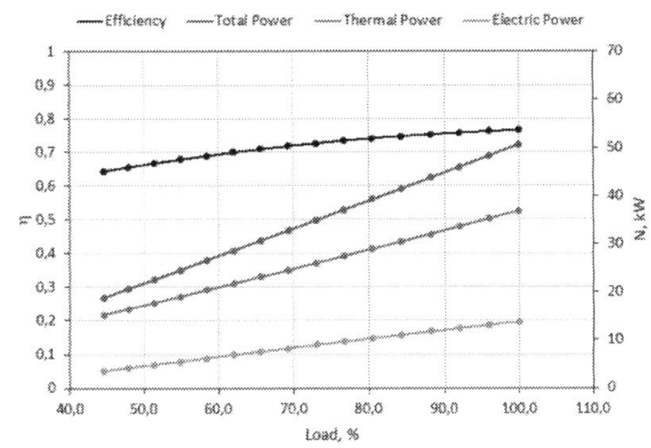

Fig. 5. Operational characteristics of the Velox-type CHP plant fired with biogas for different loads.

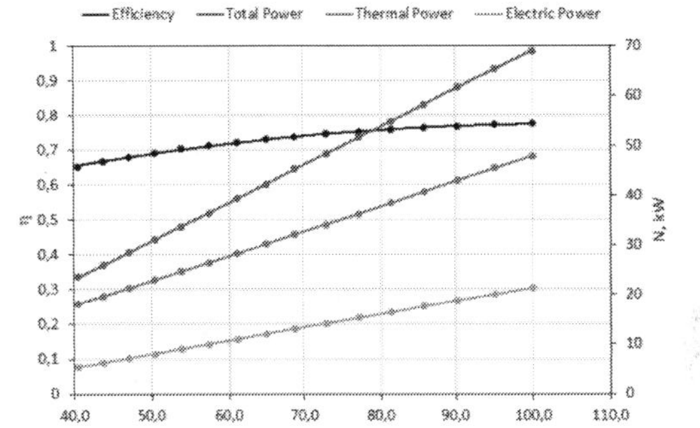

Fig. 6. Operational characteristics of the Velox-type CHP plant fired with blast furnace gas for different loads.

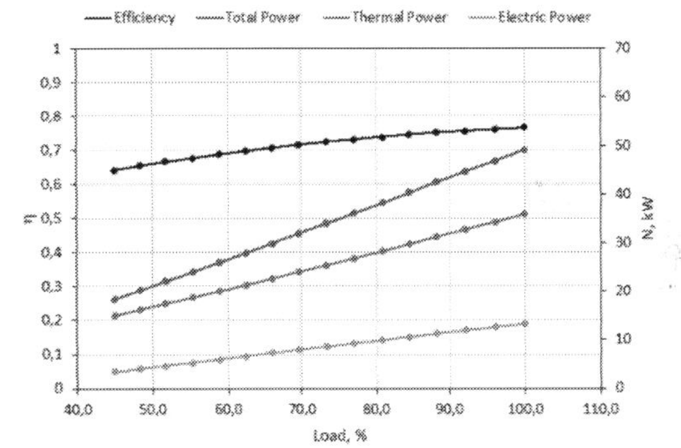

Fig. 7. Operational characteristics of the Velox-type CHP plant fired with coke gas for different loads.

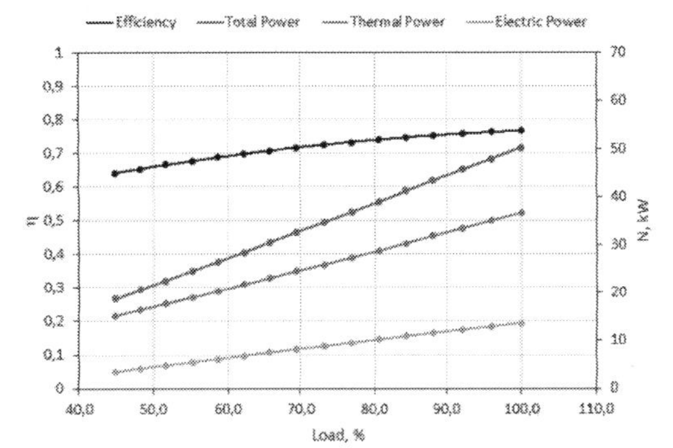

Fig. 8. Operational characteristics of the Velox-type CHP plant fired with syngas Koppers-Totzek for different loads.

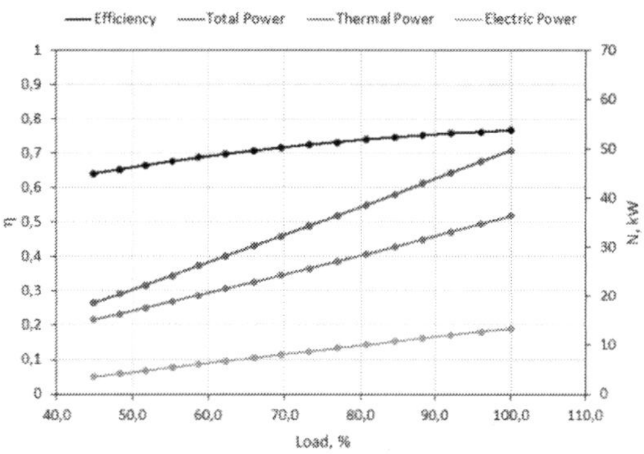

Fig. 9. Operational characteristics of the Velox-type CHP plant fired with syngas Lurgi for different loads.

5 Conclusions

Obtained results from the analysis of the work of Velox-type CHP plant fired with alternative gaseous fuels led to the following conclusions:

- the cycle under consideration can work successfully with any of the proposed unconventional gaseous fuels,
- for the same operating parameters assumed for each case, the largest demand for fuel is shown for the cycle fired with blast furnace gas,
- also the cycle fired with blast furnace gas would be the largest CHP plant due to the largest heat exchange areas required for each of the heat exchanger used,
- the cycles fired with biogas, coke gas and both types of syngas could work with any of these fuels without major structural changes.

Thus, Velox-type CHP plant creates great opportunities for distributed generation of electricity and heat with the power output of the order of a few megawatts. This cycle can be adapted at a low cost to work with each of the alternative gas fuels under consideration.

The presented work was supported by the Silesian University of Technology within Statutory Research Funds.

References

1. Gaffert G.A., *Steam Power Stations*, McGraw-Hill Book Company, New York and London, (1940).
2. *The Brown Boveri Review Issued by Brown, Boveri & Company*, Baden (Switzerland), Vol. XXXI, Nos. 1/2, (1944).
3. Kotowicz J., *Elektrownie gazowo-parowe [Gas-steam power plants]*, KAPRINT, Lublin, (2008).
4. Dykas S., Kotowicz J., Smołka K., Niekomercyjna instalacja parowej mini-siłowni kondensacyjnej w Politechnice Śląskiej [Non-Commercial Installation of a Mini Steam Power Plant in the Silesian University of Technology], *Rynek Energii*, No. 4(95), pp. 47-52, (2011).
5. Smołka K., Dykas S., Rulik S., A comparative analysis of the performance of a steam-gas system of a Velox-type cycle and a gas-steam power station using a waste-heat boiler, *Proceedings of 4th International Conference on Contemporary Problems of Thermal Engineering*, (2016).
6. Smołka K., Dykas S., The performance of a steam-gas power unit of a Velox-type cycle, *Journal of Power Technologies*, Vol. 97, No 5, pp. 343-348, (2017).
7. Bartela Ł., *Optymalizacja termodynamiczna oraz ekonomiczna pracy elektrociepłowni gazowo-parowej [Thermodynamic and economic optimization of the gas-steam power plant]*, PhD dissertation, Gliwice, (2009).
8. Kotowicz J., Bartela Ł., The influence of economic parameters on the optimal values of the design variables of a combined cycle plant, *Energy*, Vol. 35 Issue 2, pp. 911-919, (2010).
9. Wellinger A., Murphy J.D., Baxter D., *The Biogas Handbook: Science, Production and Applications*, Woodhead Publishing, (2013).
10. Kruczek G., *Experimental and numerical investigation of combustion process of conventional and alternative fuels in Internal Combustion Engine*, PhD dissertation, Gliwice, (2018).
11. Liu Y., Wang Y., Wang H., Li P., Experimental Investigations on Combustion Characteristics of the Blast Furnace Gas, *Proceedings of 3rd International Conference on Bioinformatics and Biomedical Engineering*, (2009).
12. Bhattacharya A., Muthusamy S., Static Heat Energy Balance Mathematical Model for an Iron Blast Furnace, *International Journal of Mineral Processing and Extractive Metallurgy*, Vol. 2, Issue 5, pp. 57-67, (2017).
13. Razzaq R., Li C., Zhang S., Coke oven gas: Availability, properties, purification, and utilization in China, *Fuel*, Vol. 113, pp. 287-299, (2013).
14. Bieda B., Grzesik K., Sala D., Gaweł B., Life cycle inventory processes of the integrated steel plant (ISP) in Krakow, Poland – coke production, a case study, *The International Journal of Life Cycle Assesment*, Vol. 20, Issue 8, pp. 1089-1101, (2015).
15. Podgorodetskii G.S., Yusfin Yu.S., Sazhin A.Yu., Gorbunov V.B., Polulyakh L.A., Production of Generator Gas from Solid Fuels, *Steel in Translation*, Vol. 45, Issue 6, pp. 395-402, (2015).

RESEARCH & DEVELOPMENT IN POWER ENGINEERING, 2019

Reverse engineering methodology as a way of steam turbine blades designing for Loviisa Nuclear Power

Radosław Bondyra[1,*], Krzysztof Dominiczak[2], Jacek Matuszak[3]

[1] GE Power LTD, ul. Stoczniowa 2, 82-300 Elbląg, Instytut Maszyn Przepływowych PAN, ul. Generała Stanisława Fiszera 14, 80-231 Gdańsk, radoslaw.bondyra@ge.com, Poland
[2] GE Power LTD, ul. Stoczniowa 2, 82-300 Elbląg, Instytut Maszyn Przepływowych PAN, ul. Generała Stanisława Fiszera 14, 80-231 Gdańsk, krzysztof.dominiczak@ge.com, Poland
[3] GE Power LTD, ul. Stoczniowa 2, 82-300 Elbląg, jacek.matuszak@ge.com, Poland

Abstract. This article concerns a reverse engineering-based design process of last stage blade (LSB) for other original equipment manufacturer (oOEM). For Loviisa Power Plant (Finland) GE designed and delivered a set of oOEM LSBs to be fit into existing low pressure (LP) turbine module steam path. Although cost competitiveness is a one of major selection criterion for steam turbine spare parts components supplier, diversification of suppliers is also a strategic for power plant owner. Considered here is a process of reengineering of oOEM LSB and all relevant challenges related to this process especially management of geometry deviations between reverse-engineered and oOEM blade. In this article, there are a design steps described taken to qualify reverse-engineered design. Moreover, a manufacturing process of the LSB is shown.

1 Introduction

To meet constantly incising customer requirements, GE has launched a reverse engineering process to compete on oOEM market, which is a design and manufacturing process based on a projection of already existing object. Reverse engineering capabilities significantly increase a steam turbine manufacturer competitiveness and increase its product portfolio. This article describes one of the GE projects, which delivered a reversed-engineered LSB for LP module of oOEM nuclear steam turbine.

Described here reverse engineering capabilities, allowed GE Power to offer, design, manufacture and install oOEM LSB. However, the reverse engineering process is characterized by a high-risk level, which needs to be carefully managed. A main risk is related to a creation of a virtual 3D model of an object as well as the validation of newly designed elements based on it. Regardless all challenges, a risk management of reverse engineering process allowed to mitigate all potential risks on a level accepted by a nuclear safety and enabled a delivery of LSB to the Loviisa Nuclear Power Plant, located in Finland.

2 Loviisa – Nuclear Power Plant

Loviisa - a nuclear power plant, is located in Finland on the small Hästholmen peninsula, near the town of Loviisa, about 100km from Helsinki. It produces 7.8TWh of electricity, which is more than 11% of the energy produced in Finland. In addition, it produces district heating energy. The power plant, which consists of two units, has been commissioned in 1977. Power generation is based on two Russian pressurized water reactors (PWR) with a rated power of 507 MW per each. An interesting fact is that plant plants owns a training facility. There is a reactor installed which was purchased from Poland, where originally was supposed to be installed in Żarnowiec.

Each unit has a two 200MW class steam turbine installed. These units were modernized by OEM in late 90's [1]. The turbine shaft line consists of a high-pressure part and two low-pressure parts made in impulse technology (Fig. 1). Currently, one unit (i.e. two turbines) is able to produce 522-524MW of electricity and about 1500MW of thermal energy.

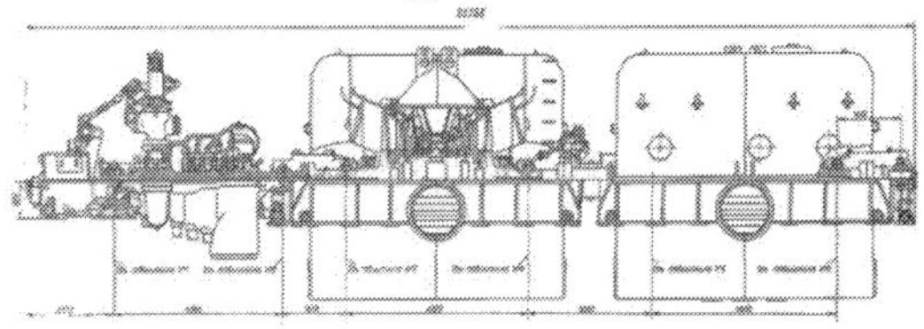

Fig. 1. Loviisa unit 1 steam turbine cross-section.

3 OEM last stage blades

Despite nearly 40 years of cooperation between the Loviisa power plant and OEM, in 2016 GE Power signed a contract for the supply of three sets of last stage blades for the low-pressure part of steam turbines installed in Loviisa. Power plant wanted to diversify a LSB supply source due to make LSB prices more competitive and ensure an alternative supply source to mitigate risks related to unstable political situation in oOEM country.

3.1 Design description

All relevant to LSB data was obtain from reverse engineering process. For a purpose of the process based on contract agreement, GE was handed four OEM blades from the power plant warehouse. Two of them were used and two were new. Additionally, an on-site measurement campaign was conducted to capture a relevant rotor geometry i.e. interface between LP rotor and LSB. Figure 2 shows a sketch of the reverse-engineered last stage blade.

LSB has a shroud and one dumping wire located in vicinity of a mean diameter. The total blade length is 1103,5mm, whereas airfoil high is 1030mm. LSB root has fir-tree design. The material of OEM was identified as high alloyed steel.

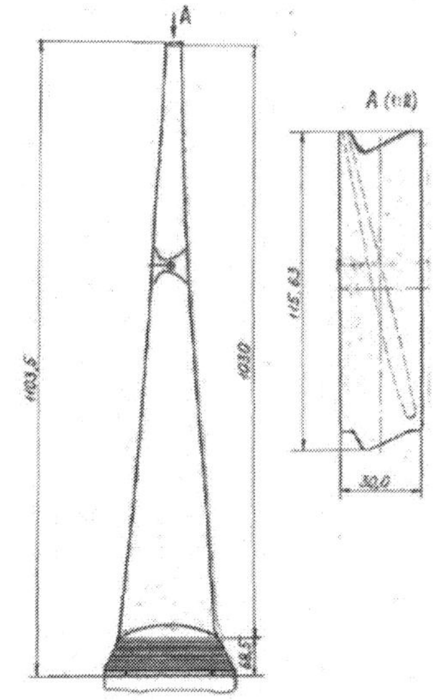

Fig. 2. Draft of the 5th stage blade of low-pressure steam turbine module.

3.2 Requirements for LSB

The main power plant owner requirement was to provide a blade which would not be geometrically different than oOEM LSB, which allow assembly on existing LP rotors. The plant owner required LSB strength characteristics at the similar level to oOEM design. Additionally, there was a requirement for LSB material Charpy impact test not to be lower than 50J. Moreover, reverse-engineered LSB was required to have similar dynamic characteristic than oOEM blade not to interfere with dynamic behaviour of the steam turbine. As a part of contractual obligation, GE delivered a few dummy blades, which allowed power plant verification of LSB geometry and feasibility of assembly process.

3.3 Description of reverse engineering methodology

The LSB was reconstructed only based on a physically existing element. For this purpose, a series of tests and measurements have been conducted to design LSB on required similarity level to oOEM design.

3.3.1 Blade 3D model reconstruction method

The reverse engineering process began with a selection of appropriate blades to be scanned with blue light scanner. The selection took placed in power plant storage, where 14 LSB (10 new and 4 used) were fully inspected by GE. The inspection revealed a poor quality of available for measurement LSBs, especially in leading and trailing edges, which were not straight (Fig. 3). The poor quality of available samples significantly hindered a reconstruction process. After further investigation and measurements, origins for reversed engineered LSB were selected and followed by detail scanning.

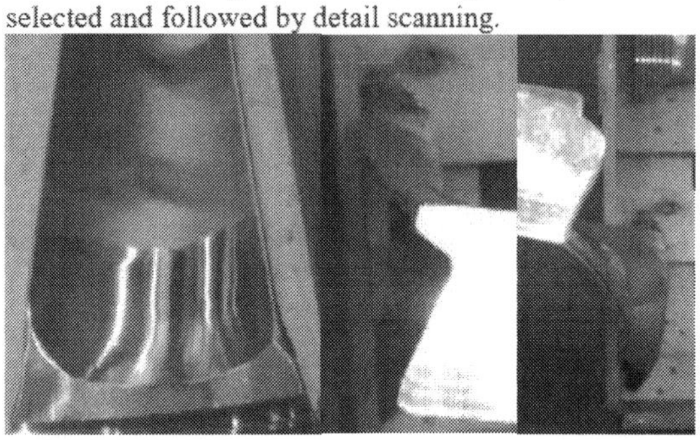

Fig. 3. Poor condition of a new oOEM LSB used for reverse engineering.

Due to the fact, that blades selected for scanning had significant geometrical discrepancies, each of the selected blades was scanned three times using blue light scanner. As a result, 18 points cloud models were obtained and next postprocessed to choose the most representative one. The state-of-the art engineering software were used for postprocessing where scanned points were appropriately corrected and LSB geometry verified. In the next stage of the reverse engineering process, LSB geometry model surfaces were created in engineering software from verified cloud points. In the next step model were imported to Catia V5, which was a target software to create 3D models and manufacturing documentation. As a result of the process, 18 3D models of LSB were obtained. The largest and smallest profiles were selected and compared (Fig. 4). The shape averaged in size finally served as nominal and values from extremes were treated as production deviations

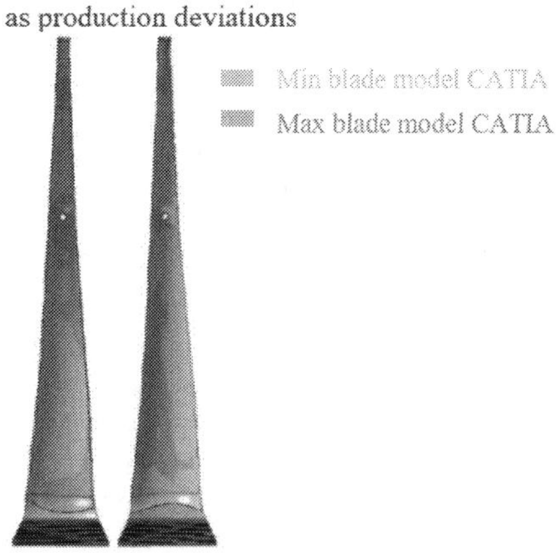

Fig. 4. Comparison of min, max models CATIA.

An important aspect of steam turbine LSB design is its root fastening. In this particular case, it is an interface that must have been properly adapted to the existing LP rotor groove and oOEM LSB. To obtain rotor groove geometry and assembly tolerances, a similar approach was taken as for LSB. There was an onsite measurement campaign to capture this aspect of the reverse engineering process. After visual inspection, five grooves and related to them LSB from each flow were selected as representative. Further roots models were compared with groove models and their fitting was verified [Fig.5]. Some geometry corrections were required but in general this part of the reverse engineering process was ended successfully.

Fig. 5. The process of scanning the rotor grooves and verification by analysing the groove and root fit.

3.3.2 Technical specification and manufacture technology of the last stage blade

Parallel to the creation of 3d models, other investigation relevant for LSB technical specification were carried out, i.e. selection of equivalent material, corrosion protection technology for the leading edge, etc. Based on available samples of oOEM LSB, a series of tests were conducted to determine a baseline for LSB acceptance criteria. For instance, to find the state-of-the-art solution to protect the leading edge of LSB, GE Power conducted a following tests:

1. visual inspection
2. verification of the blade surface under the microscope
3. hardness test on leading edge & shroud (HV5 & HV1 method)
4. comparative chemical analysis of leading Edge
5. blade natural frequencies test stress calculations

Further in this section, these tests are described in detail.

All blades provided by the power plant were subjected to visual assessment (1). First, the test was carried out on new oOEM LSB. Surfaces without wear caused by operational factors, were used to determine requirements for a manufacturing technology. It was observed that the protection of the leading edge was made by spraying a special coating. The protected area covers a larger surface than required by GE Power design standards. The presence of this coating was also visible on the radius between the shroud and the airfoil.

LSB after 120000 operating hours of service life, were also been inspected to determine a weak points of oOEM to improve the design. In general, technical condition of used LSB was very good. However, erosion of the leading edge was observed over a section of about 300mm in the upper part of the blade and on the transition radius of the shroud and airfoil. This clearly indicates a need for a larger protective coating zone.

The main purpose of the microscope examination (2) was to check whether the blade surface bears signs of shot peening. This kind of treatment is performed to increase the hardness of the treated surface. As a result of this test, no features of such a procedure were noticed, so the theory of the use of a special coating instead of shot peening was confirmed.

The next stage of investigation was hardness tests performed (3) on the leading edge of the airfoil and on a shroud. The HV5 and HV1 methods were applied and the Zwick3212 tool was used to perform them. The analysis has shown that coated zone has about 200HV better results than the parent material. The obtained results confirmed the presence of a special coating based on the difference in results for the surface of the airfoil (where visual inspection stand that coating appears) and the remaining part of the blade. Differences were about 200HV.

The last study was a comparative analysis of the chemical composition (4). The main purpose was to check what kind of coating was used to protect the leading edge in the original blades. Based on this test results, it was concluded that the most likely a sprayed titanium tungsten coating (cemented carbide) was used to protect oOEM LSB. Base on this investigation results, a leading-edge protection technology for new LSB was selected not to be worse than for oOEM LSB.

3.4 Verification of the LSB manufacturing process

Based on the requirements of the Loviisa Power Plant, GE Power manufactured and delivered on site dummy blades at the early stage of the project to allow amendments in the LSB design if required. Six dummy blades per flow were delivered to the power plant. The blades were made of high alloyed steel and were not subjected to any treatments as per final manufacturing process of reverse engineered blades

The main purpose of this procedure was to check the three main interfaces:

- fitting of the blade root to the rotor groove
- position of the damping wire hole
- blade shroud fit

As a result of the dummy blades assembly on site in a spare LP rotor, there were a few amendments to the LSB design. The design of the blade root and dumping wire were slightly adjusted. It was also spotted that the shape of the shroud corresponds to oOEM but there is a slight difference in shroud thickness. Additionally, the shape of leading and trailing edges of the aerofoil was not satisfactory and had to be corrected. However, the dummy blades trial installation was successful and GE gained an experience which allowed to adjust LSB design accordingly.

An important aspect of dummy blades installation was a verification of rotating row assembly procedure, especially assembly of the last blade in the row. Dummy blades installation proved that oOEM assembly and disassembly procedures are feasible for reverse engineered LSBs. This was an important project

 https://doi.org/10.1051/e3sconf/201913701002

milestone, which gave a green light of manufacturing of 624 LSBs, which were a final deliverable of the project.

4 Last stage blades strength & dynamic assessment

In order to prove that LSB design meets all contractual requirements as well as all internal GE design criteria, mechanical integrity and dynamic analysis were performed. The design was verified by means of:

- static and dynamic stress
- dynamic assessment
- fatigue assessment

All relevant analyses were made using finite element method in ABAQUS, which is a commercial FE software.

4.1 Description of 3D model used for calculation

For the sake of assessment, a geometry model was prepared, and appropriate boundary conditions were applied. 3D approach was used to accurately capture all relevant aspects of the design [3]. Analysis covers maximum and minimum aerofoil geometry to mitigate any risk related to accuracy of the reverse engineering. It was decided to proceed with maximum and minimum aerofoil geometry despite the small differences between both geometries. A dynamic behaviour of the LSB was a concern. However, analysis shown that differences between two considered models are insignificant but nuclear safety always go first.

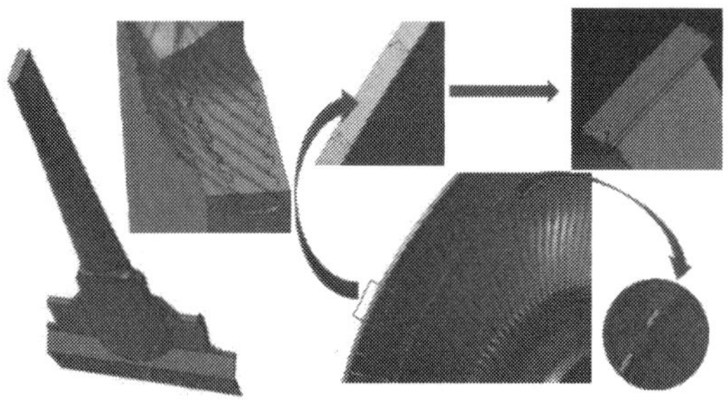

Fig. 6. 3D Geometry model for calculation purpose overview.

Figure 6 shows the model prepared for the analysis. Only one sector of the rotor circumference was modelled, which corresponded to one LSB. A cyclic symmetry was used in the model. Figure 6 also presents the contact between LSB root and LP rotor blade groove. Other interfaces are also visible, i.e. the interaction between shrouds of subsequent blades and the interface with the damping wire. This modelling strategy allowed to significantly reduce computation time but without influencing analysis accuracy. Next, discreet model was prepared based on quadratic hexahedron elements. Later, relevant loads were applied to the model.

4.2 Strength analysis of reverse engineered LSB

In order to verify the strength of the reverse engineered LSB according to GE Power design rules, stress in all relevant LSB feature were compared to permissible values. One of the major design criteria concerns stresses in the aerofoil. The maximum stress was observed in the transition radius between root and airfoil (Figure 7). Stresses in blade root, shroud as well as in the dumping wire have been checked. Based on a stress level in the LSB and based on positive service history, a good operation behaviour of the LSB was foreseen. Nevertheless, all stress related design criteria were met, and blade were qualified as per GE internal design criteria.

The leading-edge protection zone was also verified. This area of blade was described in section 3.3.2. The surface is located on the blade airfoil i.e. from the shroud up to damming wire. Based on stress field analysis, it was concluded that protection can be done by means of coating process.

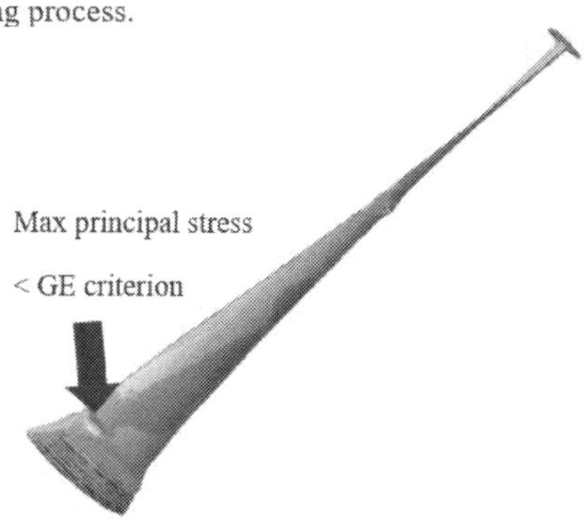

Fig. 7. Results of blade airfoil strength analysis.

4.3 Dynamic assessment

Natural vibration frequencies analysis was conducted to verify dynamic behaviour of the reverse engineered LSB. The analysis covers all relevant loads [4] related to LSB operation regime as well as loads for rotor unbalance.

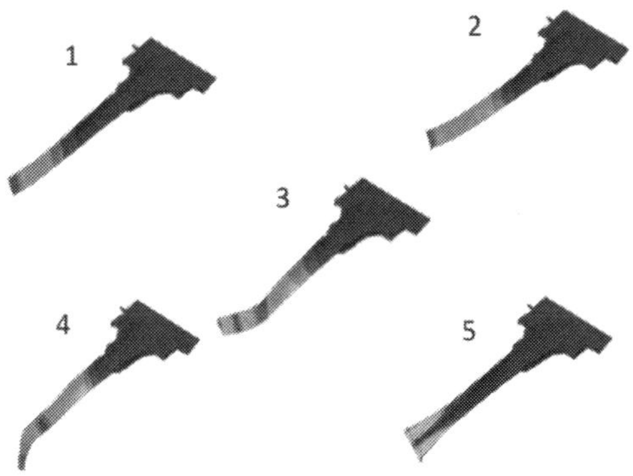

Fig. 8. Visualization of modes of frequencies.

One of the main customer requirements was to compare the frequencies of oOEM blade with reverse engineered one. Additionally, the GE Power project team decided to compare the calculated natural frequencies values for the maximum and the minimum airfoil geometry. This was to verify the impact of slight geometrical changes on a dynamic of LSB row. Table 1 shows the results for five modes of vibration at 0rpm. Figure 8 shows vibration modes of the LSB. Most of mode were bending or torsional-bending. Comparing the maximum and the minimum airfoil models, it can be seen, that the maximum deviation between frequencies values reaches 8.5% at 5 mode. Based on all results it was stated that differences had no significant impact on the dynamics of the entire LSB row. To fully compare both models an interference diagram was prepared (Figure 9). There are no discrepancies up to 7 nodal diameters for all relevant engine orders. There are slight differences between two models for higher nodal diameters. However, the deviation is small up to 4-5%.

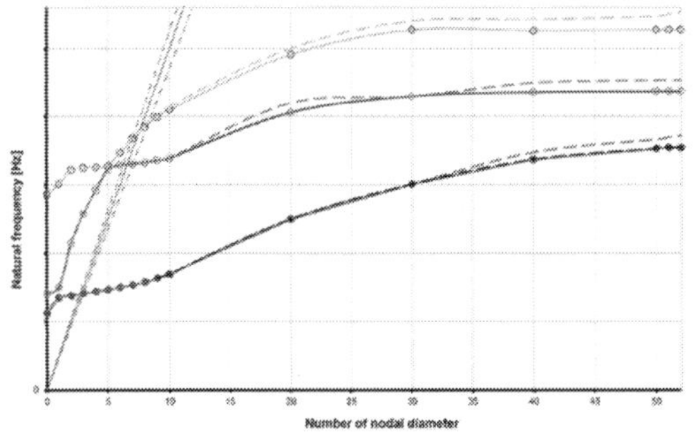

Fig. 9. Related frequencies comparison of max and min model. All frequencies related to the first mode frequency form oOEM test certificate.

Table 1. Formatting sections, subsections and subsubsections

Mode No	model min [%]	model max [%]	Difference %
1	100,53	105,67	5,11
2	193,67	194,93	0,65
3	493,2	480,13	2,66
4	741,27	742,17	0,12
5	801,1	869,1	8,49

The Campbell chart was a final verification of LSB dynamic analyses. As it can be seen in the Figure 10, the first mode is in proximity to the forbidden zone as per GE Power rules. An additional stress and strain analysis were made for these conditions to verify additional requirements applicable for this kind of cases (Figure 10). In a summary, there are no concerns regarding good dynamic behaviour of the reverse engineered LSB.

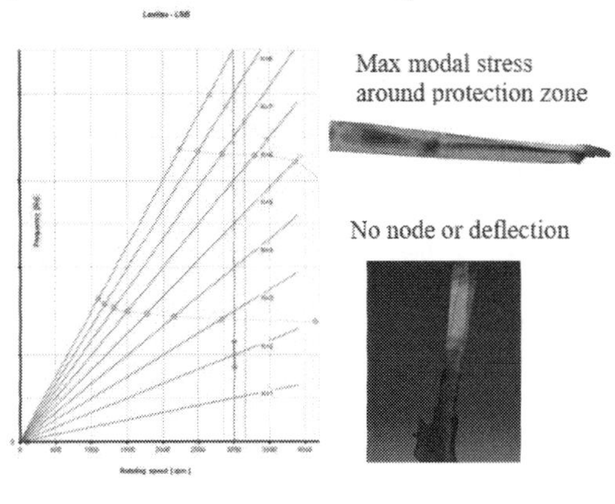

Fig. 10. Campbell diagram and point of concern analysis.

4.4 LSB fatigue analysis summary

A fatigue analysis was performed to reversed engineered design. As in other design criteria, fatigue analysis didn't rise any concerns. The critical location was identified in vicinity of damping wire. Lifetime consumption in this location is twice higher in comparison to average LSB lifetime consumption but yet below permissible limit.

4.5 Summary of stress analysis

It can be concluded that all internal GE design rules are fulfilled for reverse engineered LSB for Loviisa power plant. Moreover, all contractual requirements related to the appropriate level of similarity to oOEM design are also met. It was demonstrated that residual deviations from oOEM design for maximum and minimum models don't affect the dynamics of the blade row as a whole.

It is also wort to empesize usage of cutting edge nanoHVOF (High Velocity Oxygen Fuel) technology method to protect LSB against erosion. It was a pilot application of this method [5].

5 Conclusion

The reverse engineering process on an example of LSB for Loviisa power plant was presented in this article. The article describes also GE Power competences and capabilities to deliver spare parts for oOEM turbine. Described process can be used to deliver any component of a steam turbine, based on a physical element. As it was described all related risk to the reverse engineering process can me mitigate to meet even nuclear safety standard.

References

1. Shvetsova V. L., Kozheshkurta I. I., Koneva V. A., Boikob A. V., Usatyib A. P., Solodovc V. G., and Khandrimailov A. A., Improving the HighPressure Cylinder of the K220442M Turbine at the Loviisa Nuclear Power Station *Thermal Engineering*, Vol. 60, No. 2, pp. 98–105. (2013)
2. Murzynowski W., Grzesiczek E.: Wpływ geometrii i modeli obliczeniowych MES na rezultaty analizy stanu naprężeń w obrzeżach tarcz wirnikowych, *VIII Sympozjum Informacyjno-Szkoleniowe*, (2006)
3. Lipka J., Wytrzymałość Maszyn Wirnikowych, Wydawnictwa Naukowo-Techniczne, Warszawa (1967).
4. Rządkowski R., Sokołowski J.: Drgania własne ułopatkowanych tarcz wirnikowych osadzonych na wale, Akademia Marynarki Wojennej (2005)
5. Rehmus-Forc A., Topolski J., Matuszak J., Kucharski W., Ochrona antyerozyjna a konstrukcja na przykładzie inżynierii odwrotnej łopatki ostatniego stopnia części NP. turbin parowych, *Energetyka*, Vol. 27, pp. 36-38, (2016)

RESEARCH & DEVELOPMENT IN POWER ENGINEERING, 2019

Identification of the liquid mass fraction content in the wet steam

Mirosław Majkut[1,*], *Sławomir* Dykas[1], *Krystian* Smołka[1] and *Michał* Strozik[1]

[1] Silesian University of Technology (SUT), Institute of Power Engineering and Turbomachinery, Konarskiego 18, 44-100 Gliwice, miroslaw.majkut@polsl.pl, Poland

Abstract. The experimental research, initiated at the Institute of Power Engineering and Turbomachinery of the Silesian University of Technology at the end of the last century, created an opportunity to supplement knowledge about the phenomenon of steam flow with condensation, tested by numerical fluid mechanics by IMiUE employees. In parallel with the numerical work, significant progress has been made in experimental studies on the flow of wet steam. Recently, measuring techniques based on recording the phenomenon of extinction of light have been elaborated or improved. The basic value determined in the final stage in the experimental way was the content of the liquid phase defined as the wetness fraction. The methodology of tests and experimental investigations was presented, as well as the applied and developed measurement systems. Next, some developed designs of X probe and their modifications are described. Article present also some examples of applications of the developed measurement techniques in application to experimental research conducted on wet steam in the IMiUE. Examples of comparison between experimental and numerical tests are also provided.

1 Introduction

In recent years, thermal power plants have requirements to minimize their impact on the natural environment, regardless of the load and parameters of the system. Thanks to the operation experience gained over decades and progress in the field of numerical modelling of fluid flow, stress and strain states modelling in solid bodies, the efficiency of electricity generation in such power units has more than doubled. The current requirements regarding the reduction of pollutant emissions and the flexibility of steam power units, force them to continuously improve the operation of the machines and power equipment included in their structure. In the case of steam condensing turbines, which are a key element of, for example, coal-fired, steam-gas or nuclear blocks, an important element from the point of view of maintenance is the measurement of wetness mass fraction $y[\%]$ (or dryness $x = 100\%-y$) at the turbine outlet. There are many experimental methods used to determine the volume of the liquid phase and, as a result, also wet steam enthalpy in the turbine. These methods have been created and have been partly used since the early 1960s. Simple calorimetric methods [19] or early optical methods can be mentioned here. The introduction of various types of analyzes of the obtained flow image has now enabled not only the determination of the total value of wet steam wetness, but also enabled an accurate analysis of its composition. There are currently many measuring methods available for two-phase mixtures. They are: impedance [1]), optical, using electromagnetic or ultrasonic radiation. In the case of tests conducted for opaque factors, there are methods based on Doppler shift of reflected light (LDV), or analysis of shifted images (PIV). Numerous measurement techniques using a laser beam have also been developed. These include Rayleigh scattering [31], Raman spectroscopy [31], laser induced fluorescence (LIF) [27], spectral spectroscopy [21], as well as techniques using Fraunhofer diffraction of laser radiation by droplets [26], phase Doppler anemometry (PDA) [2] or the use of laser light extinction [6]. Each technique has its advantages and disadvantages and is useful in a specific field of application. They have different temporal and spatial resolution, and some of them may require wide optical access, so practical implementation of such measurement in industrial conditions becomes difficult. Among the methods tested so far, techniques based on the phenomenon of light extinction are of particular interest, attractive due to their relative simplicity of implementation and data interpretation, the possibility of continuous measurement with a quick time response, as well as very limited optical requirements when it comes to access to the measurement space. Optical methods use the phenomenon of wave weakening (in this case light over a wide range of wavelengths) by the associated scattering. One of the main techniques currently being developed in application for measurements in steam turbine channels by several significant research centers from Germany [17, 24], China [4, 5] or the Czech Republic [20] is the method based on light absorption, which uses in its basic form the theory of scattering Mie on spherical particles.

2 Condensation phenomenon and elements of light scattering theory

In two-phase flow systems, the typical droplet formation process mainly begins through homogeneous or heterogeneous condensation. Phenomena lead to the creation of so-called nucleation nuclei, which are a set of molecules that grow under the conditions of the surrounding gas phase. The resulting droplets (primary and secondary) have sizes from 1 μm to 100 μm and high concentrations. The measured moisture content of water vapor as a two-phase mixture y is usually determined from the volume fraction of droplets (liquid phase) C_{vl} and the density of the gas phase ρ_v and liquid phase ρ_l in the form:

$$y = \frac{m_l}{m_m} = \frac{m_l}{m_l + m_v} = \frac{C_{vl}\rho_l}{C_{vl}\rho_l + (1 - C_{vl})\rho_v} \quad (1)$$

$$C_{vl} = \frac{\pi}{6} D_{32}{}^3 N \quad (2)$$

where:
D_{32} –Sauter diameter
N – number of droplets

In the last several years, numerical modelling of steam flow with condensation in the final stages of turbines has achieved significant progress [7, 8, 22, 23, 29] and now these methods enable turbine designers and scientists to fully simulate the processes of supercooling, nucleation and droplet growth in the considered area research. However, the current advancement of these methods and the relatively small effort, needed to obtain the results of numerical simulations of condensation of steam flow, often overshadows the fact that the calculations are essentially based on mathematical models, e.g. nucleation and drop growth. As a consequence, at least some of the results of computational fluid dynamics must be validated against experimental studies to verify their accuracy.

Such tests are often carried out in laboratory channels with measuring elements in the form of various types of nozzles, in which the steam is continuously expanded from the conditions in the expansion tank installed at the inlet to a constant pressure in the outlet chamber, and the condensation process occurs during expansion in the channel. Apart from the instability of the flow, which may result from the occurring heat losses, there are usually no additional disturbances caused, for example, by the presence of blade channels in the actual machine. Also taking into account the fact that, even for nozzle flows, the results of numerical analyzes often differ from the results of experimental tests, especially with respect to droplet sizes and their concentration, it is necessary to further confirm the numerical results for flows with steam condensation in turbines by means of experimental tests. In this regard, it should be mentioned that although the previously determined vapor wetness level is an important parameter for assessing the quality of flow calculations, the droplet size distribution and diameter is the main criterion, since the wetness level itself is usually determined by the thermodynamic conditions at the beginning of the condensation process.

Depending on the geometrical relationships and the wavelength of light, two types of scattering are mainly distinguished: Rayleigh and Mie [18]. There are also other specific types, such as scattering on very large particles, called geometric (Fig. 1).

The main scattering parameters are:

- wavelength of the incident radiation λ,
- scattering particle size, usually expressed as dimensionless parameter $X = \frac{2\pi r}{\lambda}$,
  where: r - radius of the spherical scattering particle,
- quantity taking into account optical properties, i.e. the refractive index.

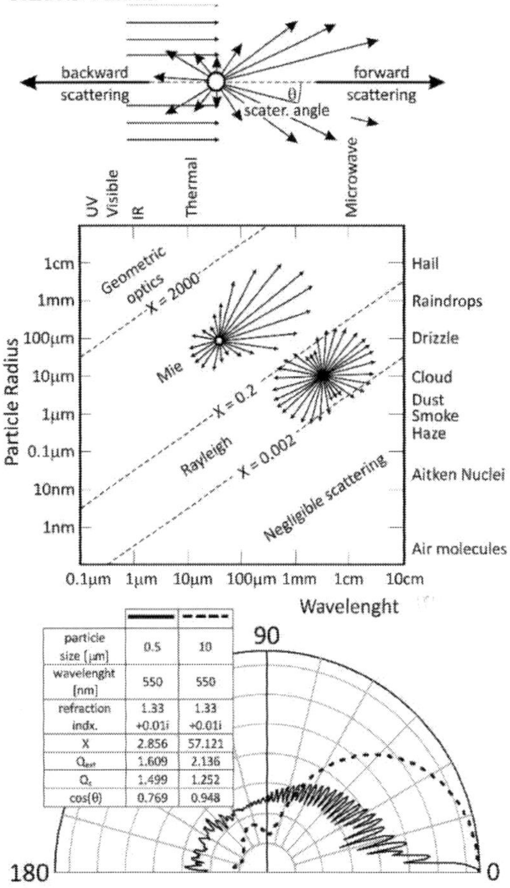

Fig. 1. Light scattering regimes and scattering geometry as well as exemplary scattering curves for two different particle diameters (obtained according to the IMiUE procedure)

As can be seen from the (Fig. 1), in the case of increasing the diameter of the scattering particles, we are dealing with a change in the scattering symmetry to an increased forward scattering. We are dealing here with the Mie effect. Its main essence is the assumption of particle sphericity and its homogeneity, therefore a single refractive index at a given wavelength can be assumed. However, it requires the adoption of a relative refractive index, which is the refractive index of the particle divided by the refractive index of the medium.

For water, the value is usually taken as n = 1.333, but this value also depends on the wavelength of the incident light and temperature.

3 Dependencies regarding the extinction method

In the method of light absorption (extinction), the wavelength λ and the original intensity I0(λ), after passing the distance L through the medium of homogeneous particles (drops), is reduced to the intensity I(λ) (Fig. 2).

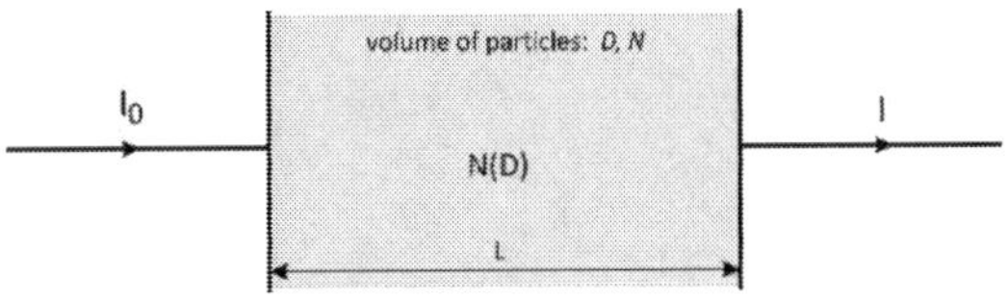

Fig. 2. Scheme of attenuation of the primary intensity Io after crossing the distance L in the light extinction method

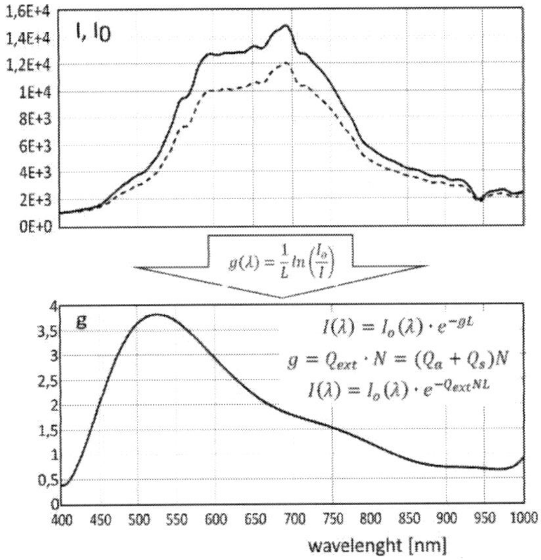

Fig. 3. Light intensity curves I and Io (experimental measurement) and the curve of the extinction coefficient g as a function of the wavelength of light

In the case of the considered method, considering the dependence of both intensities on the wavelength and marking the optical thickness as ξ, it can be written that:

$$I(\lambda) = I_0(\lambda)\exp[-\xi(\lambda)] \tag{3}$$

For the homogeneous distribution of particles, the modified Bouger-Lambert-Beer equation (which is used in absorption spectroscopy) takes the form [32]:

$$\frac{1}{t}\ln\left(\frac{I(\lambda)}{I_0(\lambda)}\right) = \frac{\pi}{4}\int_{D_{min}}^{D_{max}} Q_{ext}(n,D,\lambda)N(D)D^2\,dD \tag{4}$$

In most cases, we use the Sauter diameter in comparisons, which we define as follows for the above considerations:

$$D_{32} = \frac{\int_{Dmin}^{Dmax} D^3\,N(D)\,dD}{\int_{Dmin}^{Dmax} D^2\,N(D)\,dD} \tag{5}$$

The determination procedure for the case adopted in IMIUE consists in the first phase of determining the dependencies resulting from the Bouguer-Lambert-Beer law for the entire measured wavelength range of light. Fig. 3 presents this principle graphically, obtaining a curve of the dependence of the extinction coefficient from the measurement depending on the wavelength.
Then, using the dependencies according to Mie®law, the Q_{ext} coefficient and the final form of the equation were determined according to

$$g(\lambda) = \frac{1}{L}\ln\left(\frac{I_o}{I}\right) \tag{6}$$

and

$$g(\lambda) = \int_{D_{min}}^{D_{max}} \frac{\pi}{4}D^2 \cdot Q_{ext}(D,\lambda,n)\cdot N(D)\,dD \tag{7}$$

marking as $K_{j,i}$ the part of the above dependence taking into account the droplet dimensions:

$$K_{j,i} = \frac{\pi}{4}D_j^{\,2}\cdot Q_{ext}(D_j,\lambda_i,n_i)\Delta D \tag{8}$$

and as f_j quantities regarding their numbers:

$$f_j = N(D_j) \tag{9}$$

we can save the form of the pattern as:

$$\begin{pmatrix} g(\lambda_1) \\ \vdots \\ g(\lambda_m) \end{pmatrix} = \begin{bmatrix} K_{1,1} & \cdots & K_{1,n} \\ \vdots & \ddots & \vdots \\ K_{m,1} & \cdots & K_{m,n} \end{bmatrix} \cdot \begin{pmatrix} f(D_1) \\ \vdots \\ f(D_n) \end{pmatrix} \tag{10}$$

in other words: $g_i = K_{j,i}\cdot f_j$

in equation (10) g_i is a vector determined in the course of direct measurement on a spectrometer, $K_{j,i}$ is a matrix calculated using Mie procedures, and function f_j is a vector that requires determination. In this case, we have to solve the inverse problem, in which it is necessary to determine the inverse matrix $K_{j,i}^{\ -1}$, i.e. solve the equation:

$$f_j = K_{j,i}^{\ -1}\cdot g_i \tag{11}$$

Considering the curve in the drawing (Fig. 4), showing an example of the course of Q_{ext} variability depending on the Mie - X parameter, plotted in logarithmic coordinates, it can be seen that for the range 1 <X <2.5 the curve runs almost linearly, so that it is possible to determination of the inverse matrix. However, for larger particles, the larger the number of populations and the more dispersed their sizes are, the more it is impossible to determine by standard methods the inverse matrix [25]. The main problem in this case is the stability of the vector g_j for the measured wavelength range of light.

For this reason, it is proposed to use the smoothing curve g_j or to determine the size of I_o and I in a way that allows obtaining the vector g_j with the smallest possible standard deviation for the entire spectrum.

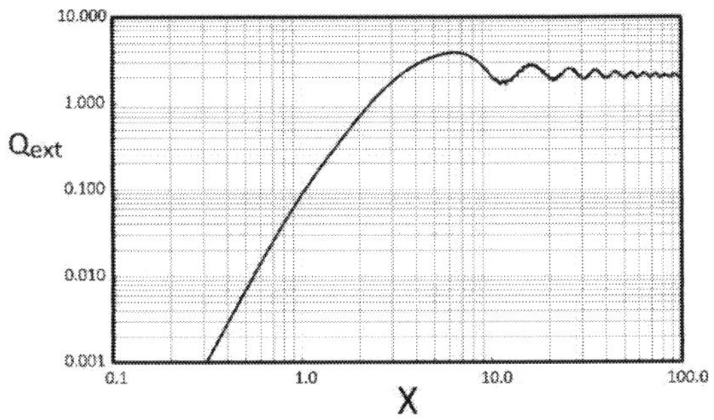

Fig. 4. Extinction coefficient as a function of the Mie parameter X for m =1.33+0.0i

In the literature [25], several methods can be found to solve this problem. The most common is the method presented by Philips and Twomey [28] as well as Lawson and Hanson [68], using the classic NNLS (non-negative least squares) approach. The NNLS method is a regression method that guarantees obtaining only positive (or zero) coefficients and solves equation (11) as:

$$\min_f \|K \cdot f - g\|^2 , \quad \text{where} \quad f \geq 0 \tag{12}$$

4 Construction of X probe

Recently, optical methods, based on the technique of light absorption, together with the development of technology, give increasingly accurate and reproducible results. The most important of the advantages of this method can be formulated as follows:

- it is possible to obtain not only information about the wetness level of the steam, but also about the size and concentration of liquid phase droplets,
- fiber optic probes are treated rather as non-invasive (or minimally invasive) due to the transparency of the measuring space in which only light is located,
- scattered light data analysis is usually conducted for the visible wave range (0.45-0.80 µm), so you can use simple and relatively cheap optics and spectroscopic systems,
- measurements can be made directly inside the device. No sampling is necessary. Local thermodynamic parameters are not disturbed,
- the measurement is fast, possible to do in a few seconds or less,
- optical probes can be built into long holders with a circular cross section with a small outer diameter. Therefore, they can be used to make measurements not only outside the last turbine stage, but also between subsequent stages. Such a handle also allows for quite simple integration with machine structure.

The X probe measuring system at IMiUE includes elements of the lighting system in the form of a deuterium-halogen illuminator and a spectrometer, as well as a traverse system. In addition, probe software was created to automate the measurement process and conduct on-line signal analysis.

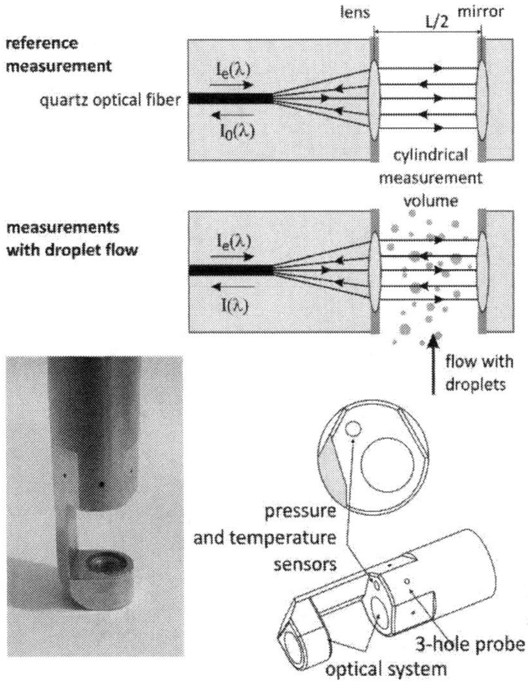

Fig. 5. Determination of light extinction coefficient by measurement of reference intensity and attenuated intensity and view of the current version of the X probe (IMiUE)

For the correct determination of the size and number of drops, as well as for a full description of the stream parameters, it is necessary to know the basic thermodynamic parameters of the tested flow. The probe design has implemented a 3-hole probe for measuring several temperatures and pressures, as well as to determine the angle of the flow in relation to the measuring space. The part of the probe responsible for determining the flow angle was placed in the upper part of the sensor, on the element having the triangle cross-section. The probe was aerodynamically optimized, reducing the sensor diameter to 10 mm, making it less invasive (Figure 5).

5 Description of the X probe measuring system

The X probe measuring system (Fig. 6) consists of two main blocks. The first of them is responsible for the production, registration and analysis of the light spectrum, while the second for measurements of thermodynamic parameters. The illuminator used for testing was equipped with two types of light sources enabling the production of a collimated beam both in the visible range (400 nm to 2000 nm) and extended by the ultraviolet range (200 nm to 400 nm) (Fig. 6).

Pressure registration was carried out using high-precision 0.16% transducers from Applisens. The figure 6 presents a view of the X probe measurement system during tests in the steam tunnel and the spectrum of the lamps.

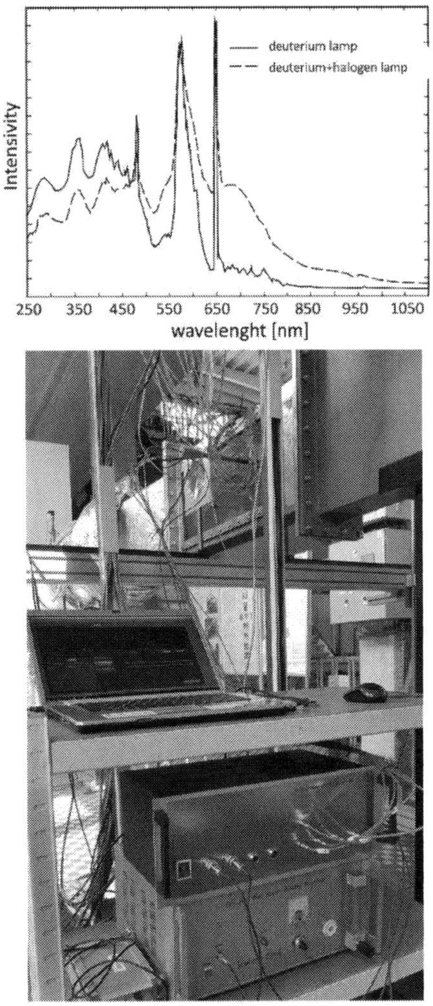

Fig. 6. View of the X probe measurement system during tests in the steam tunnel and the spectrum of the deuterium and deuterium + halogen lamp

The test system, prepared and checked for correct operation, was subjected to test and calibration measurements in the medium stream with known droplets diameter and its concentration, as well as in a specially prepared vacuum test chamber. The pressure recording system has been calibrated before being placed in the tunnel in accordance with the procedure enabling determination of appropriate calibration factors.

6 Static and comparative tests of the proposed light extinction system

In the first phase, the values of droplet diameter and their concentration were determined using static tests. For this purpose, polystyrene particles with a diameter $D \cong 2\,\mu m \pm 0.05\,\mu m,\ N = 1.2\ 10^{11} 1/m^3$, suspended in an aqueous medium, were used. The figure (Fig. 7) presents the view of the probe during the tests, as well as the light spectrum I and I_o along with the graph of particle diameter distribution and their numbers.

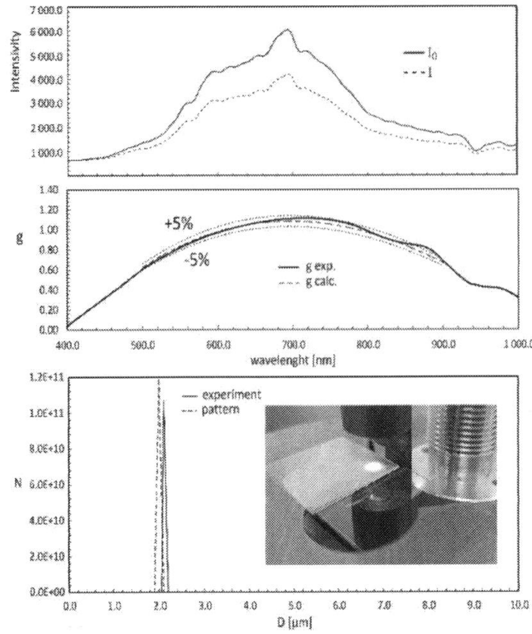

Fig. 7. I and I_o light spectrum, distribution of extinction coeff. g from measurement and calculations, distribution of particle diameter and their number, view of X probe during tests

The figure 7 compares the distribution of the extinction coefficient $g(\lambda)$ determined during test measurements with the coefficient obtained from the relationship (7). In addition, curves showing the relative error (± 5%) of the calculated value were marked on the graph. Satisfactory test results for polystyrene particles were achieved, for which the error was 5.75% (2.115 μm – value determined during the test) and the number of particles was 12.1% ($1.06\text{x}10^{11}$ $1/m^3$). The larger error in the number of determined particles could have been caused by the difficulty associated with the correct estimation of their number in the measurement volume. In addition, the test results obtained and the calculation procedure itself were subjected to additional verification, checking them against the procedure developed at the Technical University in Stuttgart. Researchers at this University have also developed their own measurement procedure, based on the method of light extinction, proven many times in a large number of different type of tests, as well as in tests carried out in a real turbine or using numerical analysis [17, 24].

The comparative test measurement was based on data obtained from the actual steam flow during tests in the nozzle installed in the IMiUE steam tunnel. Extensive research on this stand, conducted with the use of wet steam and humid air, has been described in co-author monographs [16] and in numerous publications [9-13]. For critical analysis of results, a case was selected for which the results of the comparison of procedures were presented in the figure 8.

 https://doi.org/10.1051/e3sconf/201913701003

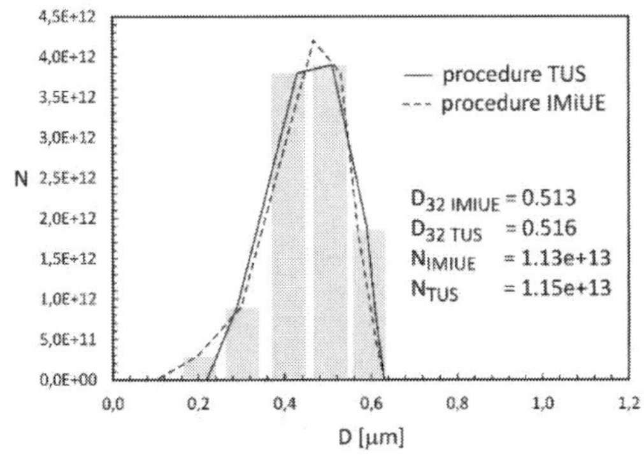

Fig. 8. Distributions of droplet diameter and their numbers presented for the IMiUE and TUS procedure

Comparative studies have shown high compliance of the proposed calculation procedures. The size deviations determined for the interesting wavelength range of light after optimization of the IMiUE system did not exceed 0.3% for the droplet diameter, and 1.7% for their number, using identical input data.

7 Determination of the wetness and the diameter of the droplets for the real wet steam flow at the outlet from the steam tunnel nozzle

From the point of view of the practical application of the presented test method, tests carried out under real conditions of wet steam flow in a steam tunnel give the largest view on its measuring capabilities. The possibility of comparing experimental measurements with advanced numerical fluid mechanics methods carried out in parallel is also valuable. For this purpose, the X probe was used to tests carried out in a research steam tunnel located in the IMiUE Thermal Machinery Hall, used for a wide spectrum of steam flow tests, using various types of nozzles and blade channels. Research using the installation has been extensively described in publications over the last several years [14-16]. To carry out the tests, a measuring nozzle with the geometry shown in the figure 9, [14], and the Ansys CFX software were used. Solver© criteria and other relevant code settings have been set in accordance with Ansys recommendation for wet steam flow calculations, and detailed description of the model used, can be found in the literature [30]. The obtained results of numerical analyzes were compared with the results obtained at the laboratory, i.e. with the distribution of static pressure along the nozzle, the image of the flow field obtained using the Schlieren technique and, most importantly, with the measured wetness and the Sauter diameter, determined at the outlet of the measuring nozzle. A more detailed description of the experimental research object and the complementary measuring techniques used in the experiment were presented in publications [3, 10, 12].

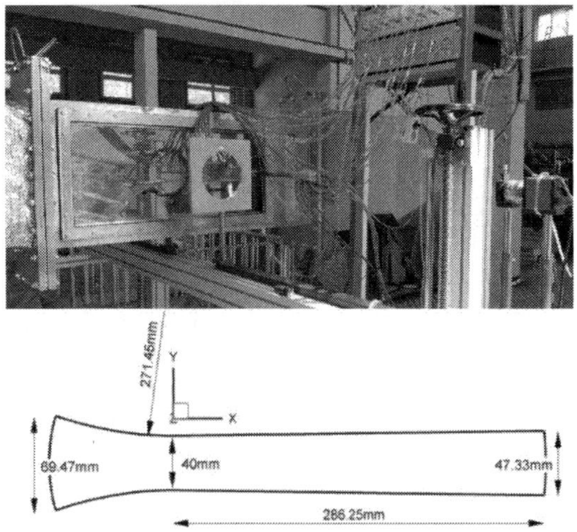

Fig. 9. The geometry of the measured nozzle

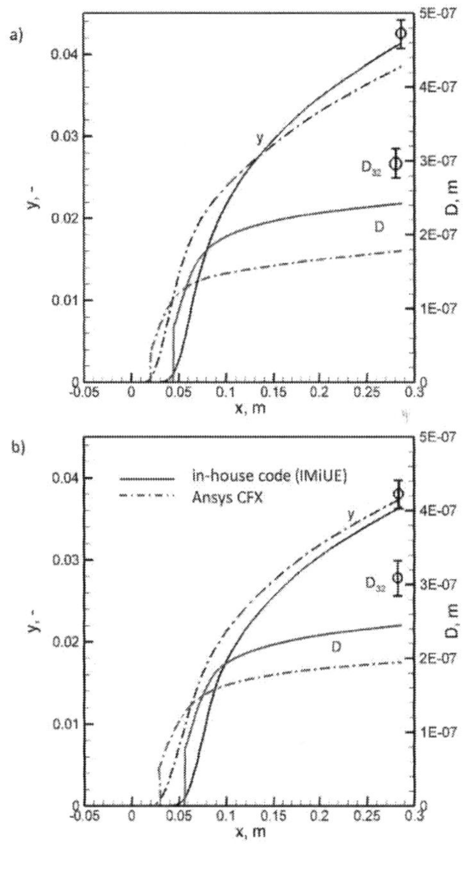

Fig. 10. Liquid phase parameter distribution along the nozzle for (a) p_0 = 124 kPa, t_0 = 118°C and (b) p_0 = 128 kPa, t_0 = 121°C with Sauter mean diameter D_{32} and y

Two measurement series were selected for analysis from many sessions of conducted tests. They differ in the total temperature at the inlet of the nozzle, at almost the same total pressure. In the first case, the values at the nozzle inlet are p_0 = 124 kPa and t_0 = 118°C, in the second case p_0 = 128 kPa and t_0 = 121°C. For the first case (Fig. 10(a)), comparing the wetness mass fraction and the average droplet size from measurements, in relation to both numerical codes, it can be observed that both determined values are higher than the values determined by Ansys CFX program and calculated by own numerical code developed at IMiUE. Similar relationships can be seen for the second selected case. Although for the degree of humidity the results of numerical tests and the experiment are quite similar, in the case of the Sauter diameter, the results of the experiment are about 50% higher, compared to the results of numerical calculations. However, even the results of two numerical analyzes alone in an interesting area show significant differences in the values of the obtained quantities. Although the tests performed in real flow conditions did not give significant results of quantitative tests, they showed additional technical and measuring imperfections of the constructed probe. Other obstacles having a significant impact on the quality of the results obtained were, for example, problems with determining the I_o value during long-term tests and changing optical conditions during the flow of wet steam (dirt in the optics in the field of the light beam) or narrow water jets that pass through the optics, resulting from surface imperfections and the probe structure itself. This effect was also intensified by the pressure measuring impulse tube cleaning system, which allowed a temporary flow of atmospheric air into the tunnel.

8 Conclusions

The aim of the work was to develop a proprietary measurement procedure for a comprehensive method for determining the parameters of two-phase flow occurring in power machines mainly at the flow of wet steam in steam turbines.

The basic value, determined experimentally in the final phase, was the content of the liquid phase, defined as the wetness mass fraction of the medium. The design and testing procedure was described, as well as the tests in real conditions of the own probe X. For the probe, both static and dynamic tests were carried out at specially designed test stands. A comparative analysis of the correctness of the calculation method was carried out. Test results show that this method has good repeatability and reliability of obtained results under static conditions. Errors in determining the diameter size D_{32} are in the range of 5.75% and the number of particles 12.1%, which was considered good values for this type of measuring instruments, comparable to the error size of similar systems developed in other research centres. A calibration and functional check of the other temperature and pressure sensors included in the probe head were also carried out. In this case, it was also found that the flow velocity and angles were determined correctly, as well as the temperature values. During tests in real flow, the experiments showed the possibility of using the constructed probe for wet steam flow tests, the need to refine the mechanical design of the probe was noticed, and a slight discrepancy was obtained in the values obtained from the numerical analysis. However, it should be stated that both the results of real and static tests show that the developed measuring system is capable of measuring online the moisture content of steam in the IMiUE laboratory steam tunnel as well as the low-pressure steam turbine.

The use of determining y using the light extinction technique with additional determined full thermodynamic parameters and flow visualization, e.g. with the schlieren technique, will allow the creation of a comprehensive measuring device enabling quick and simultaneous determination of the necessary amounts of wet steam flow.

In addition, to be able to carry out a more accurate analysis of losses in condensation steam flows, in further test tunnels it is desirable to further miniaturize the design of the probe X. Using probe X at the outlet of the blade palisade will allow more accurate determination of losses in the experiment and will also be very helpful in further work on the numerical model developed at IMiUE, including its validation.

The presented results were performed within the project TANGO 3 08/050/TAN19/0200 of The National Centre for Research and Development and Statutory Research Funds of the Silesian University of Technology.

References

1. Andreussi P., Di Donfrancesco A, and Messia M.(1988) An impedance method for the measurement of liquid hold-up in two-phase flow, *Int. 1. Multiphase Flow*, Vol. 14. p 777-785
2. Bachalo, W.D. and Houser, M.J. (1984) Phase Doppler spray analyzer for simultaneous measurements of drop size and velocity distributions, *Optical Engineering*, p.583-90
3. Blaszczuk A (2013) Experimental investigation of natural convection inside a upper part of vertical converging air channel using the Schlieren technique. *Exp Thermal Fluid Sci*, vol.50, p.178–186.
4. Cai X. S., Wang N. N., Wei J. M, Zheng G.(1992), Experimental Investigation of the Light Extinction Method For Measuring Aerosol Size Distributions, *J. Aerosol Sci.*, Vol.23, No. 7 p. 749-757
5. Cai X., Wanc L., Ouyang X., Pan Y. (1992), A Novel Integrated Probe System for Measuring the Two Phase Wet Steam Flow in Steam Turbine, *Journal of Engineering Thermophisics*, 22 (6), p. 52–57.
6. Dobbins, R.A. and Jizmagia, G.S. (1966): Particle size measurements based on use of mean scattering cross sections, *Journal of the Optical Society of America*
7. Doerffer P., Dykas S. (2005), Numerical analysis of shock induced separation delay by air humidity, *Journal of Thermal Science*, 14 (2), p. 120–125.
8. Dykas S. (2001), Numerical Calculation of the Steam Condensing Flow, *TASK QUARTERLY* 5, no. 4, p. 519–535.

9. Dykas S., Majkut M., Smołka K., Strozik M. (2013), Research on Steam Condensing Flows in Nozzles with Shock Wave, *Journal of Power Technologies*, vol. 93, issue 5, p. 288–294.
10. Dykas S., Majkut M., Smołka K., Strozik M. (2015), Experimental Research on Wet Steam Flow With Shock Wave, *Experimental Heat Transfer*, vol. 28, issue 5, p. 417–429.
11. Dykas S., Majkut M., Smołka K., Strozik M. (2018), Analysis of the steam condensing flow in a linear blade cascade, *Proc. Inst. Mech. Eng., A J. Power Energy*, vol. 232 iss. 5, p. 501-514.
12. Dykas S., Majkut M., Strozik M., Smołka K. (2015), Experimental Study of Condensing Steam Flow in Nozzles and Linear Blade Cascade, *International Journal of Heat and Mass Transfer*, vol. 80, p. 50–57.
13. Dykas S., Majkut M., Strozik M., Smołka K. (2015), Non-Equilibrium Spontaneous Condensation in Transonic Steam Flow Through Linear Cascade, *Proceedings of the 11th European Conference on Turbomachinery*.
14. Dykas S., Majkut M., Strozik M., Smołka K. (2018), An attempt to make a reliable assessment of the wet steam flow field in the de Laval nozzle., *Heat Mass Transfer*, p. 1-7
15. Dykas S., Majkut M., Strozik M., Smołka K. (2018), Comprehensive investigations into thermal and flow phenomena occurring in the atmospheric air two-phase flow through nozzles, *Int. J. Heat Mass Transf.* 2017 vol. 114, p. 1072-1085
16. Dykas S., Smołka K., Majkut M., Strozik M. (2018), Identyfikacja pola przepływu pary mokrej w dyszach i płaskich palisadach łopatkowych, *Wydawnictwo Politechniki Śląskiej*, Gliwice.
17. Eberle T., Schatz M. (2011), Experimental Study of Droplet Size and Wetness in a Model Steam Turbine Using Light Extinction Measurements, 25th *Turbomachinery Workshop*, Gdańsk/Krzeszna,
18. Horvath H. (2009), Gustav Mie and the scattering and absorption of light by particles: Historic developments and basics, *Journal of Quantitative Spectroscopy & Radiative Transfer*, no. 110, p. 787–799.
19. Kasprzyk, S. (1964): Ein Selbstansaugendes Kalorimeter zur Bestimmung der Dampffeucht, *Brennstuff-Werme-Kraftwerk BWK*, Vol. 16, p. 350-355.
20. Kolovratník M., Bartoš O. (2015):CTU Optical probes for liquid phase detection in the 1000 MW steam turbine, *EPJ Web of Conferences* 92
21. Ma, L., Sanders, S.T., Jeffries, J.B., Hanson, R.K. (2003): Monitoring and control of a pulse detonation engine using a diode-laser fuel concentration and temperature sensor, *Proceedings of the Combustion Institute*, vol. 29, p.161-166
22. Puzyrewski R. (1969), Theoretical and experimental studies on formation and growth of water drops in LP steam turbines, *Transactions IFFM*, 42–44, p. 289–303.
23. Puzyrewski R., Gardzilewicz A., Baginska M. (1973), Shock Waves in Condensing Steam Flowing Through a Laval Nozzle, *Archives of Mechanics*, 25, no. 3, p. 393–409
24. Schatz M., Eberle T. (2014),Experimental study of steam wetness in a model steam turbine rig: presentation of results and comparison with computational fluid dynamics data, *Proc IMechE Part A: J Power and Energy*, Vol. 228(2) p. 129–142
25. Su M.,, Xu F., Cai X., Ren K., Shen J.,(2007): Optimization of regularization parameter of inversion in particle sizing using light extinction method, *China Particuology*, Vol. 5, Iss. 4, p. 295-299
26. Swithenbank, J., Beer, J.M., Taylor, D.S., Abbot, D., McCreath, G.C. (1977), A laser diagnostic technique for the measurement of droplet and particle size distribution, *Progress in Aeronautics and Astronautics*, vol. 53, p.421-447
27. Thurber, M.C. and Hanson, R.K. (2001), Simultaneous imaging of temperature and mole fraction using acetone planar laser-induced fluorescence, *Experiments in Fluids*, vol. 30, p. 93-101
28. Twomney S. (1963): On the Numerical Solution of Fredholm Integral Equations of the First Kind by the Inversion of the Linear System Produced by Quadrature, *Journal of the ACM (JACM)*, Vol. 10, Iss. 1, p. 97-101
29. White A.J. (1992), Condensation in steam turbine cascades, *Department of Engineering, University of Cambridge*. Praca doktorska.
30. Wróblewski W, Dykas S, Gepert A (2009): Steam Condensing Flow in Turbine Channels. *Int J Multiphase Flow* vol. 35, Iss.6
31. Zhao, H. and Ladommatos, N.A. (1998): Optical diagnostics for in-cylinder mixture formation measurements in IC engines, *Progress in Energy and Combustion Science*, vol. 24, p. 297-336
32. Majkut, M. (2019): Zaawansowane techniki badań eksperymentalnych i wizualizacji przepływu gazów ściśliwych w maszynach energetycznych, *Wydawnictwo Politechniki Śląskiej*, Gliwice 2019, ISBN 978-83-7880-623-3

RESEARCH & DEVELOPMENT IN POWER ENGINEERING, 2019

Experimental and numerical analysis of the biomass innovative solar pyrolysis process

Sebastian Werle[1,*], *Szymon* Sobek[1], *Zuzanna* Kaczor[1], *Łukasz* Ziółkowski[1], *Zbigniew* Buliński[1] and *Mariusz* Dudziak[2]

[1] Silesian University of Technology, Institute of Thermal Technology, 22 Konarskiego St, 44-100 Gliwice, Sebastian.werle@polsl.pl, Poland

[2] Silesian University of Technology, Institute of Water and Wastewater Engineering, 18 Konarskiego St, 44-100 Gliwice, Poland

Abstract. Paper present the experimental and numerical analysis of biomass photopyrolysis process. The experimental tests is performed on the solar pyrolysis installation, designed in Institute of Thermal Technology, Gliwice. It consist of the copper reactor powered by artificial light simulating sun. The paper shows the result of the solar pyrolysis of wood. The yield of the main fraction as a function of the process temperature is presented. Additionally the gas composition is determined. The numerical model is prepared in the Ansys Fluent 18.2 software, which allow at the same time for capturing geometry of the real system and easy change of input data. The results indicate that both the product yields (liquid, solid and gaseous) and gas components shares are strongly influenced by pyrolysis parameters and feedstock composition.

1 Introduction

The European Environment Agency (EEA) has estimated that around 235 Mtoe/year of biomass could be made available in the European Union (EU) by 2020 without harming the environment [1]. Agriculture (95 Mtoe), waste (100 Mtoe) and wood industry (30 Mtoe) are the key suppliers [2]. Due to this fact, in EU countries, biomass is currently one of the main renewable energy sources used to heat and electricity production and for transportation purposes [3]. Moreover, its use is rapidly increasing. Nevertheless, there are a number of restrictions related to the production of biomass, in particular legal solutions concerning the environmental protection and the principles of biodiversity crops. Additionally, the low energy density biomass is distributed in a wide range of remote areas. Therefore, for energy purposes locally available waste products from agriculture, agricultural - food industry, spatial and other biodegradable waste, like the sludge should be used [4]. The annual biomass energy potential in Poland is equal to [5]:

- More than 20 million Mg waste straw,
- Approx. 4 million Mg of waste wood,
- Approx. 6 million Mg of sewage sludge.

Thermal methods of the waste biomass utilization are gaining importance for many years. The main reason for this fact are the EU requirements because the aim of the European Commission (EC) is to increase the share of the renewable energy sources in overall energy consumption to 32% by 2030 [6, 7].

The thermal processes include combustion, co-combustion, gasification and pyrolysis [8]. The most common is direct combustion of biomass material. The biggest problems with biomass-fired plants are in handling and pre-processing the fuel. This is the case with both small grate-fired plants and large pulverized coal power plants. Drying the biomass before combustion improves the overall process efficiency, but may not be economically motivated in many cases. Exhaust systems are used to vent combustion by-products to the environment. Emission controls might include a cyclone or multi-cyclone, a baghouse, or an electrostatic precipitator. The primary function of all of the equipment listed is particulate matter control, and is listed in order of increasing capital cost and effectiveness. Cyclones and multi-cyclones can be used as pre-collectors to remove larger particles upstream of a baghouse (fabric filter) or electrostatic precipitator. In addition, emission controls for unburned hydrocarbons, nitrogen oxides, and sulfur might be required, depending on fuel properties and the Law regulations. Despite the observed ecological, economic and social benefits, the use of biomass creates many technical problems: (1) a wide range of humidity causing difficulty in combustion stabilization, (2) biomass fuels have ash that is more alkaline in nature, which may exaggerate the fouling problems, (3) low density makes it difficult for transportation and storage, (4) high volatile content causing combustion process to be rapid and difficult to control, (5) the relatively low the lower heating value (6) heterogeneity and diversity of composition, including the content of chlorines lead to process during which hydrogen chloride, dioxins and furans are formed [9]. These problems cause that co-combustion of biomass with fossil fuels (mainly hard coal) gained importance. This process can be advantageous with regard to substitution of fossil fuels, reducing fuel cost and emissions of NO_x and CO_2 minimizing waste and reduce soil and water pollution and increasing boiler efficiency.

However, attention must be taken to increase deposit formation in the boiler and limitations in ash use due to compositions in biomass, especially alkali metals, which may disable the use of ash in building materials. Due to undesired changes of ash compositions, the share of biomass is usually limited to approximately 20% of the fuel input [10]. Additionally, there are other barriers of co-combustion biomass with coal. These barriers include: (1) biomass procurement practices to obtain low-cost fuels in a long term reliable manner; the impact of co-combustion on ash composition and ability to sell fly ash; (2) the trade-off between the impact of biomass on emissions and fuel cost [11]. In case of the Polish market the new Act of the renewable energy sources should be also emphasized [12]. Based on this document, co-financial assistance of the co-combustion installation will be limited.

All presented facts cause that new pioneer and innovative thermal solutions for biomass conversion are needed. An example of such technology is pyrolysis.

Biomass pyrolysis is defined as a thermal degradation of the biopolymers present in the organic matter under an inert oxygen-free atmosphere [13]. Three products are always produced (solid, liquid and gaseous), but the proportions can be varied over a wide range by adjustment of the process parameters [14]. Pyrolysis is a process carried out at a lower temperature in comparison to combustion or co-combustion [15]. Thanks to it, the formation of toxic substances is limited and the complex gas cleaning system is not necessary. There are many other important advantages of pyrolysis in comparison to combustion, as follows: (1) Pyrolysis can be performed at relatively small scale and at remote locations which enhance energy density of the biomass resource and reduce transport and handling costs; (2) Due to lower temperature of the process corrosion problem is reduced – the maintenance costs of the installation are lower; (3) Due to the lower temperature of the process, recovery of the selected elements (mainly non-ferrous metals) from solid products is possible; (4) Due to the endothermic nature of the pyrolysis, control of the process is easier in comparison to combustion; (5) Pyrolysis is characterized by high level of the fuel flexibility; (6) The pyrolysis products can be stored and later used for energy purposes [16].

Despite its advantages, pyrolysis has the disadvantage of requiring an external energy input to reach the operating temperature [17]. This external energy input is generally derived from a non-renewable source that has a negative impact on the environment. A possible solution to this problem is to use thermo-solar energy to heat the reactor to achieve conditions appropriate for solar pyrolysis initialization. In such process the concentrated solar radiation supplies high temperature heat for biomass pyrolysis reaction [18]. Then biomass and solar energy can be converted into transportable and dispatchable solar fuel [19]. Solar processes has the potential to produce higher calorific value products with lower CO_2 emission compared with conventional process [20, 21]. The biomass energy is upgraded through solar energy providing pyrolysis reaction enthalpy transferred into products. Solar pyrolysis is an endothermic process of converting a biomass in an inert atmosphere in which the required heat is provided by concentrated solar energy. The direct solar radiation is concentrated and redirected to the pyrolytic reactor and the biomass to reach pyrolytic temperatures.

Paper present experimental and numerical analysis of wood photopyrolysis process. The yield of the main fraction as a function of the process temperature is presented. Additionally the gas composition is determined. The numerical model is prepared in the Ansys Fluent 18.2 software, which allow at the same time for capturing geometry of the real system and easy change of input data. The results indicate that both the product yields (liquid, solid and gaseous) and gas components shares are strongly influenced by pyrolysis parameters and feedstock composition.

2 Experiment

2.1 Fuel properties

Table 1 present properties of wood which was used as a feedstock in the solar pyrolysis experiment.

Table 1. Wood analysed properties.

Property	Status	Unit	Value
Moisture, total	ar	%mass	9.00
Ash	dry	%mass	0.75
Volatiles	dry	%mass	83.3
Higher heating value	ar	kJ/kg	18,760
Lower heating value	ar	kJ/kg	16,870
Carbon	dry	%vol	49.61
Hydrogen	dry	%vol	5.94
Nitrogen	dry	%vol	0.06
Sulphur	dry	%vol	0.02
Oxygen	dry	%vol	44.36
Lignin	ar	%mass	27.90
Cellulose	ar	%mass	54.80
Hemicellulose	ar	%mass	11.80

ar – as received, dry – dry basis

Ultimate and proximate analysis has been determined using plasma spectrometer Thermo iCAP 6500 Duo ICP according to Polish norms: PN-G-04528-02:1977, PN-EN 196-2:2013-11, PN-EN 15936:2013-02 and PN-EN 15936:2013-02. Chemical analysis has been carried out using extraction system Ankom A200 and FOSS Fibertec 2010 and 2011 according to application note FOSS 3434 – AOAC 2002:04/ISO 16472:2005

2.2 Experimental stand

Figure 1 presents the scheme and the photo of the main elements of the solar installation and Figure 2 shows the view of the xenon arc lamp.

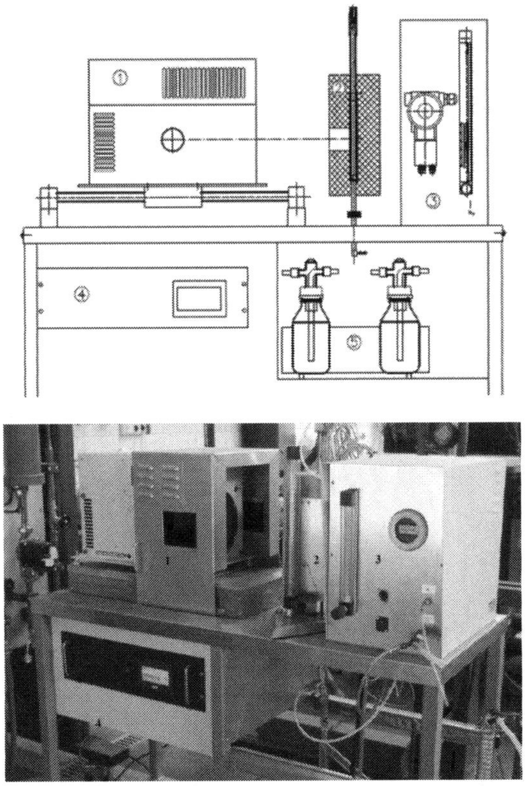

Fig. 1. Experimental setup scheme (upper) and actual photo (lower) of the research stand: 1, xenon arc lamp, 2, insulated reactor, 3, PLC and measurements devices container, 4, lamp power supply, 5, bio-oil condenser.

Fig. 2. Xenon arc lamp view.

The heart of the laboratory station was artificial sun, the high-power xenon arc lamp. The 1.6 kW Sciencetech© lamp produces stable heat flux in a form of concentrated radiation, with spectral characteristics close to the natural sunlight. Lamp radiation is directed onto reactor surface, containing biomass samples. Reactor is a copper block with four ducted channels, with 169 mm of total length. Volatiles released during pyrolysis are moved out from reaction zone with flow of inert gas (nitrogen with 1.5 l/min flow rate) and directed to liquid fraction (bio-oil) condenser. Bio-oil condenser consists of a set of water- cooled laboratory condensers with assigned Dreshl scrubbers in order to trap condensed bio-oil and pass non-condensable gases to ABB© gas analyser through Bronkhorst® El-Flow Prestige mass-flow meter.

2.3 Experimental procedure

Table 2. The experimental procedure.

Fuel form	Output power* of the lamp, %	Inert** gas flow, l/min	Measured parameters
Wood pellets Dxh=6x20mm	80.0 85.5 90.0	1.5	Sample temperature Product yield Gas composition shares

D- outer dimeter
h – height
**-maximum power is equal to 1.6kW*
***-nitrogen was used*

Table 2 presents the experimental procedure. The total mass of wood feed was 25 grams distributed in four reactor channels in linear form wrapped in copper net. The pyrolysis process was realized in a wide range of temperatures, from approximately 500 to 800 °C. Six K-type thermocouples with the outer diameter 0.5 mm were located in precisely drilled holes in 3 pellets from 2 channels. Control of the process temperature was executed by the variable output power (80-90%) of the lamp controlled by the power supply of the lamp. After being loaded with wood pellets, the reactor was exposed to xenon lamp radiation for 90 min duration time. Measurements were carried out for 3 pre-selected lamp power values 80, 85.5 and 90%. During pyrolysis, the temperature and pyrolysis gas composition were recorded using WAGO© PFC 100 programmable logic controller (PLC). After each test products were separated. The volatiles continuously flew out from the reaction zone with the flow of nitrogen, while bio-oil condensed in the water-cooled condenser, passing dry pyrogas to gas analyser. Solid residue remained in the reactor was removed from the reactor and weighted. Liquid phase was collected from scrubbers and extracted with acetone, then weighted.

2.4 Experimental results

Figure 3 presents final results of the study, listing impact of lamp power on following product yields with dry gas composition. Observed trends in product formation are typical for slow pyrolysis, the higher the final temperature and heating rate, the higher the bio-oil yields with lowering amount of dry gas and char. Char yield is strongly correlated to lignin content of the biomass samples (Table 1), bio-oil to cellulose content, and dry-gas to hemicellulose. Lower yield of bio-oil than hemicellulose content can be explained by cracking of tar components to lighter molecules increasing dry gas yield.

Dry gas quality is typical for slow pyrolysis. Low temperatures favor formation of CO_2 and CO, with lower content of CH_4 and H_2. Finally, the Figure 4 presents indication of gas analyser related to sample temperature for experiment with 80.0% of lamp power. Peaks of CO and CO_2 at around 320 °C correspond to decomposition of cellulose, while later CH_4 and H_2 release can be assigned to slow char formation reactions, with wide peaks at 380 °C and 560 °C respectively.

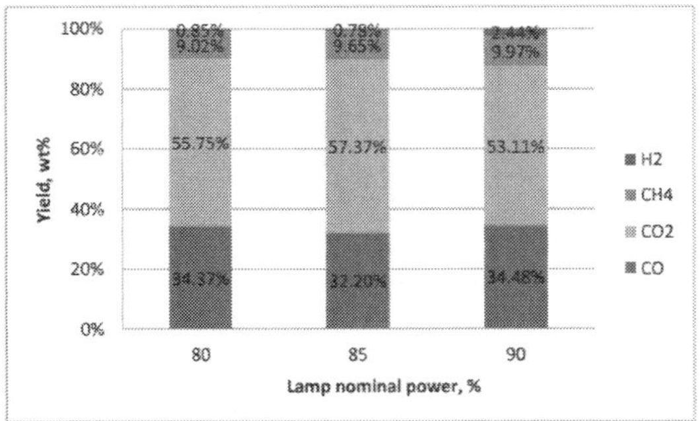

Fig. 3. Effect on lamp power on product yield and pyrogas composition.

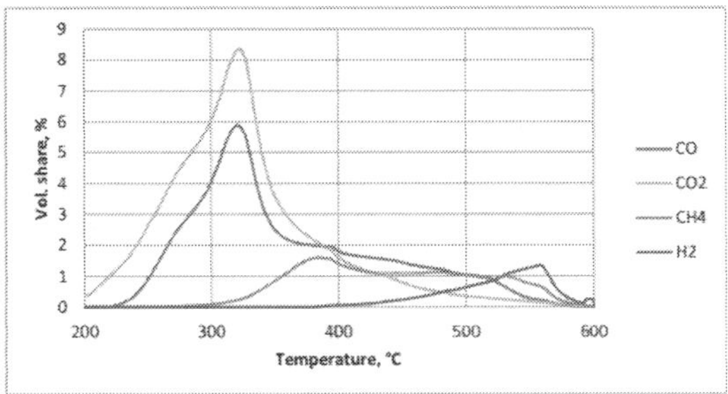

Fig. 4. Gas analyser indications as a function of sample temperature during experiment with 80% of lamp power.

4 Numerical investigation

The main aim of the realization this part of the investigation was definition how to model solar pyrolysis process and what phenomenon must be included in such analysis. The most important element of the experimental system is the pyrolytic reactor (Fig. 5), in which four pellet packages are placed in a fixed position which is additionally secured with a copper gauze. The gauze holds the pellets and at the same time allows for free flow of gaseous pyrolysis products out of the reacting bed and for nitrogen flow to provide an inert atmosphere for the process.

The reactor is made of copper. Its outer surface is insulated, except the spot in the middle of its upper surface, where the focused xenon lamp rays fall. This light incidence spot will ultimately be covered with a substance of a high, known absorption capacity.

Nitrogen supplied to the rector will be split to feed all 4 reactor channels. At the outlet of the channels, nitrogen with a mixture of gaseous products (*tar* and *gas*) will get through again in a common collector and flow further to the cooling zone, to separate the *tar* off the *gas*, which then goes to the analyser.

Such a concept causes that the composition analysis of the process products will average the results for 4 packages of pellets.

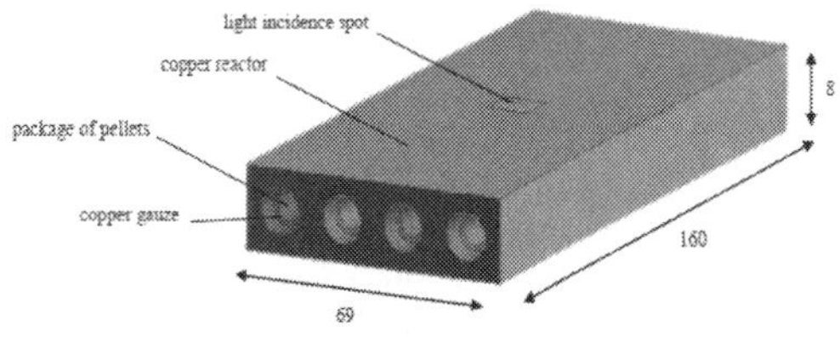

Fig. 5. Pyrolytic reactor scheme for CFD investigation – key elements and dimensions (in mm).

Based on the the literature review [22 - 26] it can be concluded that the model being prepared within this work can definitely be included into the group of large particles pyrolysis models, where heat and mass transport through the reacting bed requires special attention and obligatory including the changes of physical properties. Moreover, due to the fact that many pellets are simultaneously subjected to pyrolysis, being placed at different distance from the heat source, thus in different thermal conditions, it will be necessary to accurately reflect the geometry of all elements of the reactor, to appropriate capture thermal conditions. The model have to include chemical reactions, heat transfer, physical properties change and moisture evaporation, shrinkage.

5 Conclusions

The paper present both theoretical (introduction to CFD modelling) and practise (experiment) aspects of the solar pyrolysis process.

Solar pyrolysis of waste wood can be promising way to utilize waste biomass from wood industry, allowing to produce high-quality, solar enriched, bio-oil and bio-char.

Changing the heat transfer conditions is a key element in the mathematical description of the process, especially if the scale of the system is as large as in the presented case.

Evaporation of moisture and deposit shrinkage are essential elements of the pyrolysis process and certainly at further stages of the model development would be a valuable complement to it.

Within the experimental part of the paper, solar pyrolysis of waste wood study results has been presented. Experiments have been conducted on solar pyrolysis reactor own-designed by authors. Impact of investigated biomass chemical composition on actual products distribution and dry gas composition have been presented and discussed. It was confirmed that during wood solar pyrolysis chemical components cellulose, hemicellulose and lignin favours decomposition course leading towards dry gas, bio-oil and bio-char formation respectively. Product shares and dry pyrolysis gas quality followed results published in the literature with well-known trends of increasing bio-oil yield with increase of final temperature and average heating rate. Xenon-arc lamp provided stable radiant flux and plenty of power. Precise temperature monitor within pellet bed gave insight into samples behaviour during pyrolysis process.

ACKNOWLEDGMENT

This paper has been prepared within the frame of the project "Study on the solar pyrolysis process of the Waste biomass", financed by the National Science Centre, Poland (registration number 2016/23/B/ST8/02101).

References

1. European Environment Agency, How much bioenergy can Europe produce without harming the environment, 2006, p. 6.
2. Eurostat main tables, IEA, WEO 2011, Annexes, p. 26.
3. Marculescu C., Comparative Analysis on Waste to Energy Conversion Chains Using Thermal-Chemical Processes, *Energy Procedia*, 18, pp. 604-611, (2012).
4. Magdziarz A., Wilk M., Thermal characteristics of the combustion process of biomass and sewage sludge, *Journal of Thermal Analysis and Calorimetry*, 14, pp. 519-529, (2013).
5. Werle S. Wilk R.K., A review of methods for the thermal utilization of sewage sludge: the Polish perspective, *Renewable Energy*, 35, pp. 1914–1919 (2010).
6. Refuse derived fuel, current practice and perspectives – Final Report, European Commission – Directorate General Environment 2003.
7. Report from the Commission to the European Parliament and the Council on the implementation of the EU Strategy on adaptation to climate change, COM(2018)0738, 12 November 2018.
8. Communication from the Commission of 7 December 2005 – Biomass Action Plan
9. Werle S., Impact of feedstock properties and operating conditions on sewage sludge gasification in a fixed bed gasifier, *Waste Management and Research*, 32, pp. 954-960, (2014)
10. Werther J. Saenger M. Hartge E.-U, Ogada T, Siagi Z, Combustion of agricultural residues, *Progress in Energy and Combustion Science*, 26, pp. 1-27, (2000).
11. Khan A.A., de Jong W., Jansens P.J., Spliethoff H., Biomass combustion in fluidized bed boilers: potential problems and remedies, *Fuel Processing Technology*, 90, pp. 21–50, (2009).
12. Smoot L.D., International research centers' activities in coal combustion. *Progress in Energy and Combustion Science*, 24, pp. 409–501, (1998).
13. Act of the renewable energy sources (Dz. U. 2015 poz. 478), Poland, 20 February 2015r.
14. Rutkowski P., Kubacki A., Influence of polystyrene addition to cellulose on chemical structure and properties of bio-oil obtained during pyrolysis, *Energy Conversion and Management*, 47, pp. 716-731, (2006).
15. Yaman S., Pyrolysis of biomass to produce fuels and chemical feedstock, *Energy Conversion and Management*, 45, pp. 651-671, (2004).
16. Basu P., Biomass gasification and pyrolysis. Practical design and theory, Elsevier, 2010.
17. Zeng K., Minh D.P., Gauthier D., Weiss-Hortala E., Nzihou A., Flamant G., The effect of temperature and heating rate on char properties obtained from solar pyrolysis of beech wood, *Bioresource Technology*, 182, pp. 114-119, (2015).
18. Morales S., Miranda R., Bustos D. Cazares T., Tran H., Solar biomass pyrolysis for the production of bio-fuels and chemical commodities, *Journal of Analytical and Applied Pyrolysis*, 109, pp. 65-78, (2014).
19. Piatkowski N., Wieckert C., Weimer AW., Steinfeld A., Solar-driven gasification of carbonaceous feedstock—a review, *Energy and Environmental Science*, 4, pp. 73–82, (2011).
20. Chueh W.C., Falter C., Abbott M., Scipio D., Furler P., Haile S.M., Steinfeld A., Highflux solar-driven thermochemical dissociation of CO_2 and H_2O using nonstoichiometric ceria, Science, 330, pp. 1797–801, (2010).
21. Nzihou A., Flamant G., Stanmore B., Synthetic fuels from biomass using concentrated solar energy - a review, Energy, 42, pp. 121–31, (2011).
22. Grønli M.G., Melaaen M.C., Mathematical model for wood pyrolysis - Comparison of experimental measurements with model predictions, *Energy & Fuels*, vol. 14, no. 4, pp. 791–800, (2000).
23. Blondeau J., Jeanmart H., Biomass pyrolysis at high temperatures : Prediction of gaseous species yields from an anisotropic particle, *Biomass and Bioenergy*, vol. 41, pp. 107–121, (2012).
24. Xue Q., Dalluge D., Heindel T.J., Fox R.O., Brown R.C., Experimental validation and CFD modeling study of biomass fast pyrolysis in fluidized-bed reactors, *Fuel*, vol. 97, pp. 757–769, (2012).
25. Tolvanen H., Kokko L., Raiko R., Fast pyrolysis of coal , peat , and torrefied wood : Mass loss study with a drop-tube reactor , particle geometry analysis , and kinetics modeling, *Fuel*, vol. 111, pp. 148–156, (2013).

https://doi.org/10.1051/e3sconf/201913701004

26. Yu X., Hassan M., Ocone R., Y. Makkawi Y., A CFD study of biomass pyrolysis in a downer reactor equipped with a novel gas-solid separator-II thermochemical performance and products *Fuel Processing Technoogy.*, vol. 133, pp. 51–63, (2015).

E3S Web of Conferences 137, 01005 (2019) https://doi.org/10.1051/e3sconf/201913701005
RDPE 2019

RESEARCH & DEVELOPMENT IN POWER ENGINEERING, 2019

Analysis of a micro oxy gas turbine for variable oxidizer and fuel parameters

Wojciech Nowak[1], *Mikołaj* Złotkowski[1],
[1]AGH University of Science and Technology, Krakow, wnowak@agh.edu.pl, Poland
Ahmed A. Alharbi[2]
[2]King Abdul Aziz City for Science & Technology – KACST, ahalharbi@kacst.edu.sa, Saudi Arabia

Abstract. The article describes the behavior of a micro oxy gas turbine after replacing the nitrogen with carbon dioxide. The accurate simulation of the oxy turbine behavior is of great importance to understand its performance under a variety of operational parameters. This work presents combustion analysis for the comprehensive simulation of a micro oxy gas turbine.

1 Introduction

Carbon capture and storage (CCS) of CO_2 from large energy sources can be the effective solution to decrease the global release of greenhouse gases. CCS has the possibility of storing the large amount of produced CO_2 in geological formations.
Three main technology options of CO_2 reduction are:

- separation CO_2 after the combustion process (post-combustion capture),
- separation CO_2 before the process (pre-combustion capture),
- combustion in oxygen atmosphere (oxy-fuel combustion) [1].

One cannot show on the present stage of the technology development, which among these technologies has the largest potential to reduce costs and energy needed to run the process of carbon capture and storage. Therefore, there is a need for intensive R&D since each of these technologies has the different approach to capture CO_2. Although the CCS technology causes increasing fuel consumption because of the inevitable decrease of the process efficiency, the achievement of the significant reduction of greenhouse gases is until now a better solution than the potential which is possible through the increase of power efficiency.
The gas turbine allows the conversion of gas fuels into mechanical energy with efficiency of up to 25-45%, it is a simple construction having as few as possible movable or friction elements, unlike the piston engine is a flow machine, it does not have connecting rods, oil sumps, valves or camshaft, which makes with the failure rate of individual elements may be lower. In order for the gas turbine to work properly, the fuel must be supplied in a gas form in the purest possible form. Fuel contamination negatively affects the lifetime of the turbine. It was assumed that fuel would be methane. This gives the chance that in the future it may be the only hydrocarbon fuel regarded as renewable and clean in combustion.
In conventional air combustion, the capture of CO_2 from flue gas is a complicated process due to the low partial pressure of CO_2 and large amount of nitrogen in the flue gases. Since air contains 79% of nitrogen, all of nitrogen in different chemical forms is emitted to atmosphere during the combustion process, this the CO_2 concentration in the flue gases is only in a range of 13-16% vol.
Oxy-fuel combustion is a process whereby oxygen is separated from the air and used directly to burn the fuel in order to increase the CO_2 concentration in the flue gases to 90% vol., or greater. The volume of flue gases is significantly reduced by elimination of nitrogen and a CO_2-enriched gas after removing water is ready for sequestration without CO_2 capture. Oxy-combustion power production involves three major components: oxygen production (air separation unit ASU), the oxy-combustion gas turbine (combustion unit), and CO_2 purification and compression [2-5].
These components, along with different design options, are shown in Figure 1.

https://doi.org/10.1051/e3sconf/201913701005

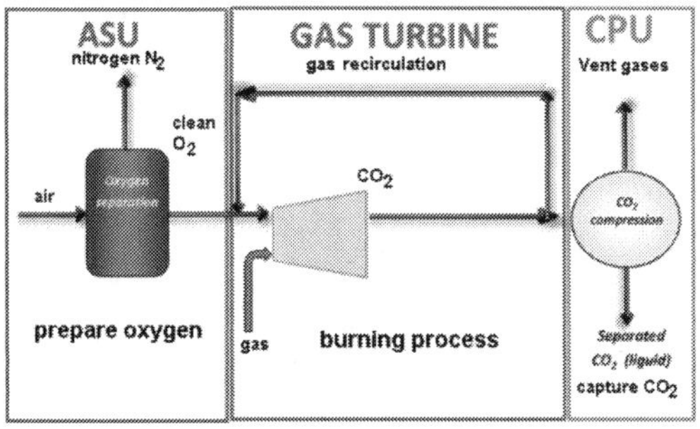

Fig. 1. Components of the oxy gas combustion unit.

Oxy-combustion gas turbine can be configured differently with these components, resulting in different energetic and economic performances. Systems can be configured in either low- or high-temperature designs. In low-temperature designs, flame temperatures are similar to that of air-fired combustion(~1600°C), while flame temperatures exceed 2500 °C in the advanced high-temperature design. Low-temperature designs for new or retrofit applications recycle combustion products to lower the flame temperature to approximate the heat transfer characteristics of air-fired turbines. High-temperature designs use increased radiant heat transfer in new construction applications to reduce the size and capital cost of the gas turbine.

Micro Turbine Natural Gas C30 kW has been selected in the present studies. The parameters of the turbine are as follows:

Electric power output	30 kW
Frequency	50/60 Hz
Fuel	8 – 13 MJ/m^3
Consumption fuel max	444 MJ/h
Heat emission	14,4MJ/kWh
NOx emission	18 mg/ m^3
NOx emission	0,23g/hp(h)
Exhaust gas flow	0,31kg/s
Exhaust gas temperature	275 °C

Gas power system achieves ultra-low emissions and reliable electricity from natural gas and the system can be transferred to oxy combustion after modifications. Microturbines use a lean premix combustion system to achieve low emissions levels at a full power range. Microturbine control and power electronic systems allow for different operation modes, such as: grid connect, stand-alone, dual mode and multiple units for potentially enhanced reliability. Micro oxy turbine achieves almost zero NO_x emissions and reliable electricity from natural gas. In addition, oxy-gas combustion is emerging as an attractive option to facilitate capture and sequestration of CO_2 from gas-fired power plants.

The accurate simulation of the oxy turbine behavior is of great importance to understand its performance under a variety of operational parameters. The paper discusses the oxy-fuel combustion development addressing the problem of oxy-fuel concept in a gas turbine. In particular, this work presents combustion analysis for the comprehensive simulation of a micro oxy gas turbine.

2 Method

Calculations of turbine operation parameters are carried out using the IPSEpro. It is a highly flexible environment for modelling, simulation, analysis and design of components and processes for energy and chemical engineering. Initial conditions of the turbine performance were taken from the documentation of the gas unit. Then the air atmosphere was changed to oxy one. Carbon dioxide, having a lower heat capacity, needs less fuel and oxygen to be supplied to the combustion chamber, resulting in similar turbine performance as with air. In case of oxy combustion mass share of exhaust gases is over 90% CO_2, 8% O_2 and 2% methane.

Controlled CO_2 recirculation allows to keep the temperature in the combustion chamber same as during air combustion. CO_2-rich flue gases with gas recirculation leaving the gas turbine contains 96% CO_2 and 4% H_2O. A drying unit may be used, depending on requirements for the final CO_2 stream applications. The NOx formation, when applying O_2/CO_2 recycle combustion, is expected to be very small since the nitrogen supplied by the air is removed, thus eliminating the thermal NOx formation. In case of air combustion, the exhaust gas contains only 14% CO_2 and 12% H_2O – the rest is nitrogen.

Figure 2 shows simulations of a gas turbine working on oxy atmosphere mode. Air is compressed in a 1:4 ratio and the use of heat exchangers between stages of compression, reduces energy demand. The water removes heat from the air from 28 to 55 °C, taking 1 kW of thermal power from the compressors. Air compressed to 4 bar is heated to 130°C and then mixed with methane to form a lean mixture, $\lambda = 1.05$. Behind the combustion chamber, the temperature rises to 1600 °C and hot gases drive the turbine. Around 40% of the energy transferred to the turbine is used to drive the compressors, the rest is converted into electricity generating 23 kW. The flue gases behind the turbine have high temperature and low pressure. The last exchanger absorbs this heat, converting water into superheated steam of 740 °C and 1.8 bar; the temperature of flue gases behind the exchanger is approx. in a range of 120 - 150 °C.

For oxy combustion compressing gas needs more power than oxygen and nitrogen. Temperature gases going to turbine are 950 °C, the turbine generates 6 kW electric power, gas outlet turbine have 785 °C; the entrophy lose on the turbine is only 220 kJ/kg. The water with cooling cycle heated to 60 °C after two heat exchanger; after third cheat exchanger the temperature is 107 °C. It is near evaporation (2000 kJ/kg). The exhaust gas after third heat exchanger have temperature less than 100 °C, the water can by condensing; its pH is about 7 and condensation is safety.

https://doi.org/10.1051/e3sconf/201913701005

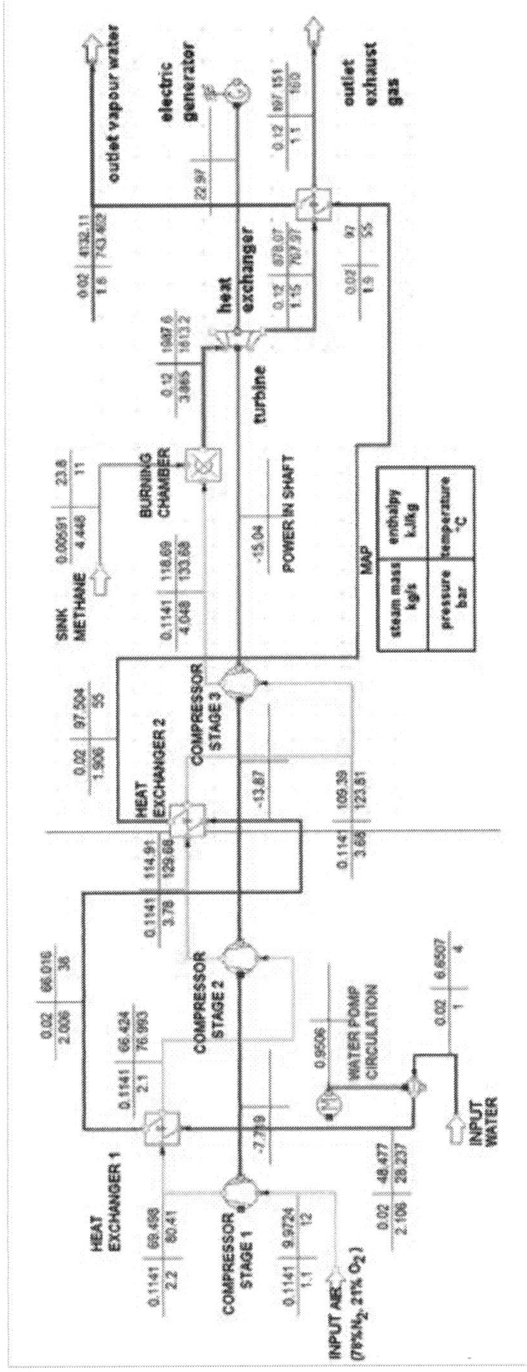

Fig. 2. Simulations in IPSEpro.

3 Results

Figure 3 shows the exhaust gas temperature at the turbine outlet as a function of the oxygen mass flow at a variable CO_2 flow. For an oxygen flow of 8 g/s the mixture is rich (λ = 0.18), so the temperature of the exhaust gas tends to increase. The higher the CO_2 content, the temperature gradient is lower since the more heat can be absorbed by the higher the CO_2 flow. The maximum temperature of 1120 °C for a of 24 g/s O_2 flow was reached (λ = 1) at the lowest CO_2 flow (55g/s).

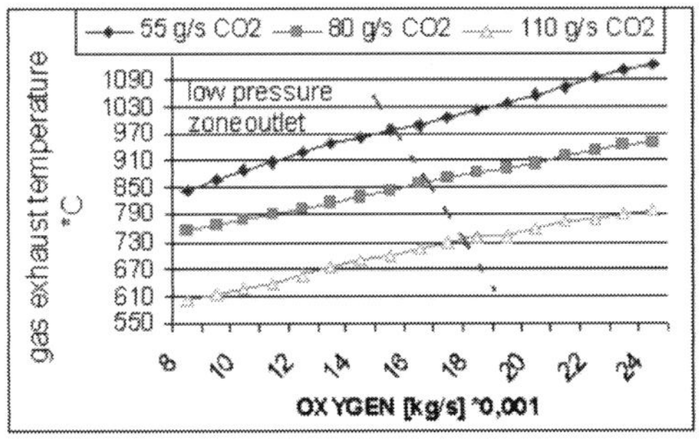

Fig. 3. Exhaust gas temperature at the turbine outlet as a function of the oxygen mass flow for different CO2 flows and 30 kW of power generation.

The higher the CO_2 recycled flow, the less the temperature profile since the more heat can be absorbed by a higher gas flow. The maximum temperature of 770 °C was reached for the 9 g/s O_2 when the mixture becomes lean.

Figure 4 shows the flue gas temperature as a function of CO_2 mass flow. Four curves are shown for fuel mass composition 24:6 (λ = 1), 48:6 (λ = 2), 72:6 (λ = 3) and 96:6 (λ = 4) g/s and for the constant load of the generator 26 kW. Hyperbolic course of temperature drop can be observed when the CO_2 share increases. In this way, the combustion process is controlled so that the power generated by the turbine can be smoothly adjusted. The response to the change in the exhaust gas temperature is a quick correction of the CO_2 flow at the entrance to the combustion chamber. Keeping the stoichiometric mixture (blue line), the temperature of the exhaust gases can be controlled over a wide range from 300 to 1150°C. In order for the turbine to be able to generate 26 kW it is necessary to have an outlet pressure lower than atmospheric pressure. The turbine will not be able to generate 26 kW for CO_2 flow larger than 140 g/s.

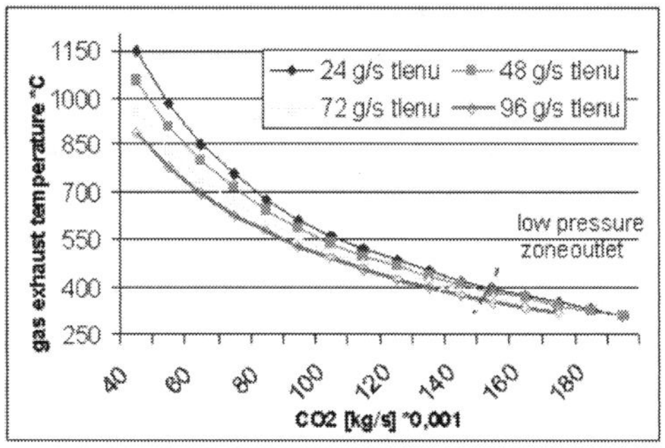

Fig. 4. The flue gas temperature as a function of CO_2 mass flow.

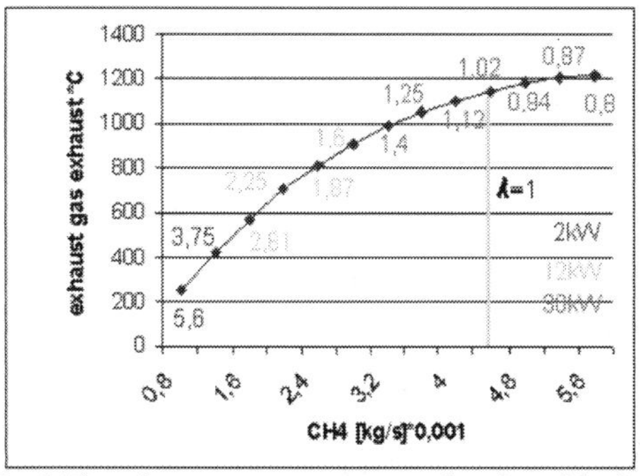

Fig. 6. The exhaust gas temperature as a function of the fuel flow.

Figure 5 presents a comparison of maximum temperatures for 40 g/s of CO_2 with different oxygen flows. As can be seen, raising the oxygen content above the stoichiometric ratio results in a decrease in the temperature measured at the turbine outlet.

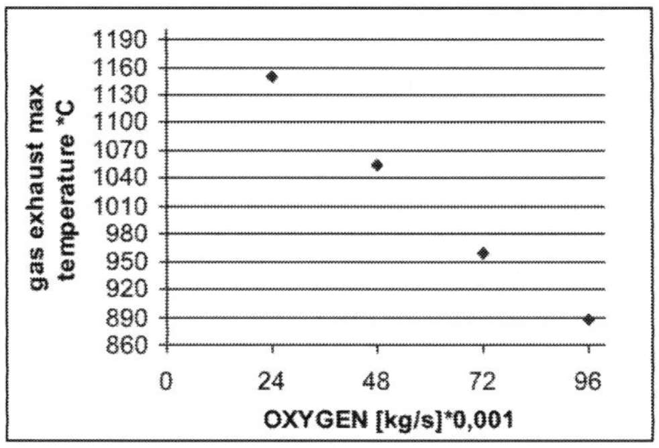

Fig.5. Comparison of average temperatures with the oxygen flow.

Figure 6 shows the exhaust gas temperature as a function of the fuel flow for different turbine loads of 2, 12 and 30 kW, respectively.

By increasing the concentration of CH_4 in the mixture, the power generated by the turbine increases. For a small fuel flow the turbine generates a low power of only 1 kW and then the power generated increases rapidly. It is possible to generate even 12 kW for a fuel flow of 2 g/s. If λ is less than 1.4 the temperature increase and it is possible to generate 30 kW of power. For stoichiometric mixtures, the temperature of the exhaust gases reaches 1150 °C. When the mixture becomes rich, the temperature increase is no longer significant and for $\lambda = 0.8$ reaches 1200 °C (CO_2/oxygen ratio is 70:18).

Figure 7 shows the gas exhaust temperature as a function of the fuel flow for the turbine loads of 1 and 10 kW, and constant the CO_2 mass flow.

By increasing the oxygen concentration in the mixture, the power generated by the turbine increases; for low fuel content the turbine generates less power, but by increasing the fuel content the power increases rapidly to a level of 10 kW. The temperature reaches 700 °C and stabilizes for richer mixtures.

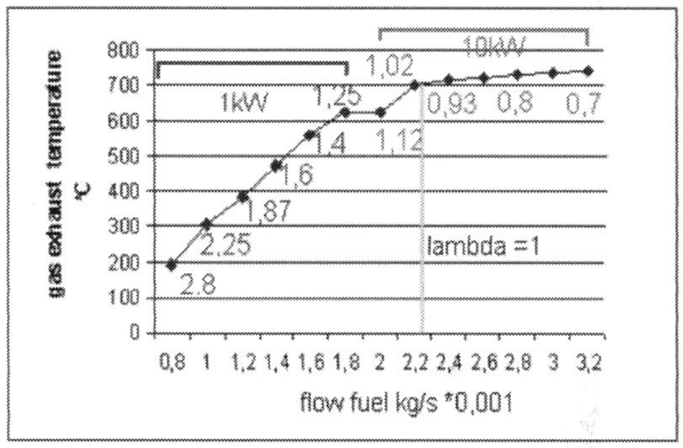

Fig. 7. The gas exhaust temperature as a function of the fuel flow for the turbine loads of 1 and 10 kW.

Figure 8 presents simulation in a case of lean combustion using the same parameters as in Fig. 7.

As the fuel flow increases to 4.8 g/s, the exhaust temperature linearly raises. Further the load is increased to 28 kW and the temperature in the area of 4.8 - 5 g/s is at a near constant level. It then rises again linearly to reach 980 °C at a stoichiometric mixture with a CO_2 to oxygen ratio of 70:24.

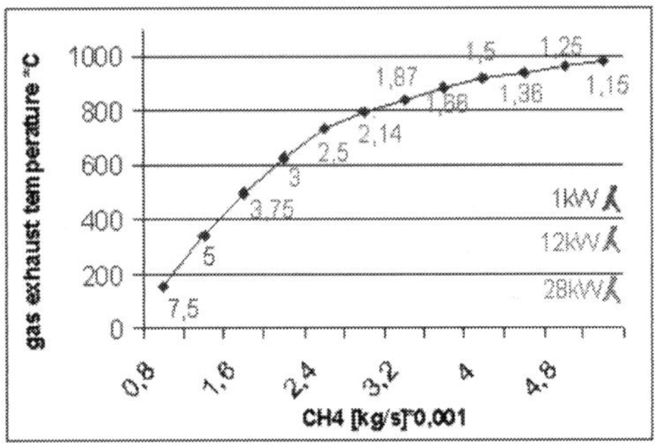

Fig. 8. Simulation in a case of lean combustion using more fuel.

Figure 9 shows a family of characteristics of changes in the mixture flow in which the ratio of the amount of fuel to oxygen and CO_2 does not change; this was determined as the ratio of the amount of oxygen and CO_2 to the amount of fuel $(CO_2 + O_2)/F$. Four cases were tested for ratio $(CO_2 + O_2)/F = 16, 28, 36, 45$.

The power generated increases linearly from 1 to 30 kW. It should also be added that changes in the ratio $(CO_2 + O_2)/F$ cause changes in λ of the mixture. For $(CO_2 + O_2)/F = 16$, the turbine runs on a very rich mixture (λ = 0.32) over the entire range.

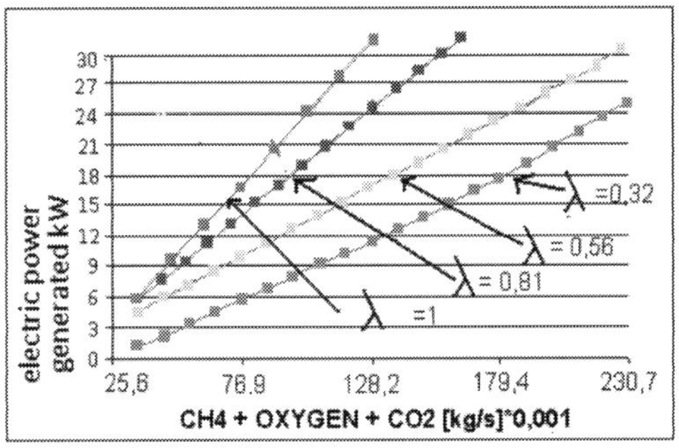

Fig. 9. Comparison of turbine power generated by the turbine for 4 different mixtures from rich to stoichiometric.

Figure 10 presents a similar situation to Fig. 9 showing in addition thermal power that can be used. One can see the heat transferred to the secondary circuit and the generated electric power as a function of the amount of the mixture for the ratio $(CO_2 + O_2)/F = 36$, i.e. per 1 gram CH_4 there is 36 g of $CO_2 + O_2$ (in mass ratio 10 % O_2 and 90 % CO_2) in the range from 40 to 250 g/s (λ = 1). Increasing the oxygen content to 10% results in the temperature increase in the combustion chamber and thus the power generated by the turbine, the heat measured has a constant proportion to the power generated, it is 3.6 times higher. The turbine achieves its nominal power at a mixture flow of 190 g/s and it produces 110 kW of heat.

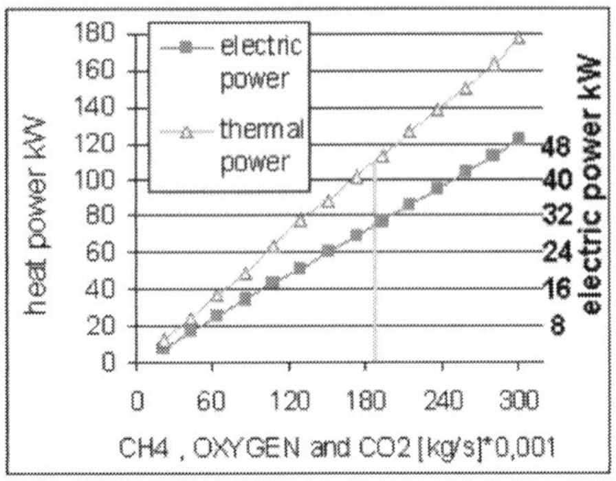

Fig. 10. Available power generated as a function of increasing the amount of mixture in ratio $(CO_2 + O_2)/F = 36$

Figure 11 presents the exhaust gas temperature behind the turbine and oxygen demand as a function of the percentage of oxygen in the mixture for a turbine load of 30 kW. The composition is controlled so that the combustion is stoichiometric and the generated electric power is at the level of 30 kW, the oxy composition changes in the range of 5 to 16% of oxygen resulting in the inlet gas flow changes.

As the oxygen concentration increases, the mass demand decreases, the smaller amount of CO_2 causes a gradual increase in the gas temperature at the turbine outlet to the level of about 930 °C for a 16% of oxygen and the input gas flow does not exceed 70 g/s.

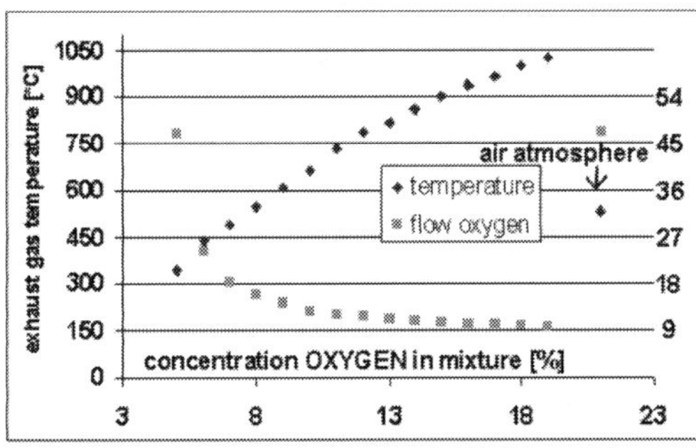

Fig. 11. Exhaust gas temperature behind the turbine and oxygen demand as a function of the percentage of oxygen in the gas mixture for a load of 30 kW.

4 Conclusions

Providing the turbine with a stoichiometric mixture is essential to ensure that the flue gases are highly clean and can be processed with little energy. Such parameters are for the ratio $(CO_2 + O_2)/F$ in the range of 35 – 55.

Depending on the oxygen content in the oxidant, adjustable parameters such as : amount of injected fuel, the amount of oxygen supplied and the amount of CO_2 returned make it possible to smoothly change $(CO_2 + O_2)/F$ ratio.

References

1.Czakiert T., W. Nowak. Polish National Strategic Program on Oxy-Fuel Combustion – 5 Years of Studies and Investigations. Proc. of the 22nd International Conference on Fluidized Bed Conversion, June 14-17, Turku, Finland, 301-313, (2015).
2.Dennis R. ARPA-E Workshop, NETL, (2017).
3.Ricardo Llorente Manso. Master's Thesis. Norwegian University of Science and Technology, (2013).
4. Snarheim S.. PhD Thesis "Control issues in oxy-fuel combustion", The Norwegian University of Science and Technology, (2009).
5. Ditaranto M., Inge Saanum, Petter E. Røkke, Janczewski J., Testing of a gas turbine oxy-fuel burner. 5th Meeting of the IEAGHG InternationalOxyfuel Combustion Research NetworkWuhan, Hubei, China26 - 29 October,(2015).

The authors would like to thank King AbdulazizCity forScience and Technology (KACST) of Saudi Arabia for their support.

RESEARCH & DEVELOPMENT IN POWER ENGINEERING, 2019

Estimation of maximum loads of residential electricity users

Sławomir Bielecki.[1]

[1] Warsaw University of Technology, Faculty of Power and Aeronautical Engineering – Institute of Heat Engineering, Nowowiejska Str. 21/25 00-665 Warsaw, Poland, e-mail: slawomir.bielecki@pw.edu.pl

Abstract. The subject matter of coefficient factors is not a frequent object of analysis in the scientific literature. However, because of the use of this term in engineering tasks and because of the constantly changing manner of energy use (resulting from the dissemination of new appliances), there is a need for current analysis and possible revision of the approach based on coefficient factors. The aim of the paper is to compare currently recommended values of this coefficient in the task of estimating the peak output of installations on the example of standards applied in Poland. On the basis of smart metering data and statistical methods, coefficient factors for residential consumers supplied from the municipal distribution network were determined. The results were compared with the currently applied values in accordance with the industry standard issued by the Association of Polish Electrical Engineers (standard N SEP-E-002).

1 Introduction

Due to the non-monotonous and usually characterised by more than one extreme of the individual users (consumers load), the peak (maximum) of load profile of the feeder does not have to occur at the time of peak at any of supplied users. There is a non-linear or even entangled superposition of individual peaks. A simplified model of these dependencies is the coincidence factor which takes different values for different numbers and types of loads forming the analysed structure.

The basic definition of *the coincidence factor* is the ratio of the peak power of the feeder to the sum of the peak power of the loads:

$$k_j = \frac{P_{j-group\ peak}}{\sum_{i=1}^{n} P_{i-peak}} \tag{1}$$

where: $P_{j\text{-group peak}}$ – peak power in the feeder which the *n* loads; $P_{i\text{-peak}}$ – peak load of the *i*-th load.

It takes values from the range:

$$k_j \in [\tfrac{1}{n}; 1] \tag{2}$$

where *n* is the number of loads supplied from the feeder.

The concept of the *kj* factor is based on the finding that the behavioural patterns of different types of consumers overlap. It is unlikely that all consumers will be used at full capacity at the same time and, therefore, smaller load capacity sections of installations or networks can be designed than the simple sum of the expected peak loads. The value of the *kj* factor depends on the number of loads (consumers) and decreases with the increase in this number.

In literature there are reversals of coincidence factor, called a *diversity factor*. The simplicity of the idea of the coincidence factor as a parameter, characterising the load of the network or installation, forces significant limitations of its accuracy. In particular, the definition does not specify the relationship between the moments of the highest loads and does not directly specify the conditions, ranges and certainty of its occurrence (values may change different time horizons – replacement of equipment, change of habits and use of energy among tenants).

The *kj* coefficient binds the peaks of loads of single loads supplied from a feeder. If a too low value is adopted during designing, it may lead to a reduction of capacity reserve or even overload during peak periods, while a too high value in turn leads to unnecessary oversizing of the installation and the increase in investment costs.

2 Literature review

The idea and method of the coincidence factor has been known since the beginning of the 20th century [1], [2], [3]. Its values are determined primarily for household consumers [4], [5].

Assuming that the power demand in the adopted time intervals of consumers has a normal distribution and using the statistical approach, the *kj* coefficients can be determined on the basis of formulas that directly depend their value on the number of consumers and parameters describing the distribution. In fact, the use of active power and energy by users does not have to create normal distributions [6]. This means that formulas based on this assumption (e.g. Rusck, Velander, Strand-Axelsson) are currently being questioned [7].

Assumptions simplifying and unifying loads adopted to determine the coefficients characterising the use of electricity such as *kj* were needed in the conditions of limited availability of measurement data. With the development of AMI technology, the possibilities of using measurement data in load modelling are increasing. In [8], the calculated values of diversity factors with their dispersion (presenting the average and maximum values and the percentiles: 99th, 95th, 5th and 1st) for different numbers of consumers on the basis of the measurement data of 1898 consumers were presented.

In [9] the concept of a perceptron-type multilayer neural network for the estimation of coincidence factors values was presented. The network requires information about 10 issues influencing the factor value, concerning the load (monthly peak demand, difference between peak and valley demand), environmental (weather) conditions, users and urbanization parameters. The network requires a data-learning process.

In the paper [10] it was pointed out that engineering estimation of maximum power demand in low voltage networks, based on peak demand and corrected by coincidence factors of simultaneity, does not take into account the stochastic character of the phenomenon and does not contribute to obtaining a proper outlook of aspects related to energy quality. Therefore, the authors proposed a model of the Monte Carlo simulation of energy demand by the consumer, using data from the gamma distribution. The result of the simulation is corrected, among others, by temperature data.

In the paper [11], presenting the results of calculations of average values of diversity factors, obtained on the basis of the method of bootstrap for household consumers (in the number from 1 to 60), it was shown that these values for a selected number of consumers (with 1000 bootstrapping samples) do not form normal distributions. The authors suggest that a more appropriate distribution for these values is the gamma distribution. Moreover, comparing the obtained results with [5], the authors indicate that these factors for each assumed number of household appliances, obtained as a result of calculations according to the bootstrap method, are independent values. In the paper [12] the methods of calculation of losses based on fuzzy grouping of monthly energy consumption by consumers, without determining the peak demand, are presented.

The paper [13] attempts to introduce and determine coincidence factors for charging electric vehicles from the network (mutual coincidence factor), defined as the ratio of the peak active power in the system after connecting the vehicle to the network to the sum of the peak charging power of the vehicle and the peak power of other loads in the network. On the basis of data from the urban network area of Beijing and the Monte Carlo method, the ranges of these coefficients were estimated, depending on the charging power. A normal coincidence factor for charging electric vehicles, defined as the ratio of the peak charging power of a vehicle to the product of the number of vehicles charged and the charging power of a single vehicle, has also been proposed. These values, therefore, depend on the number of vehicles and the charging power. The concept of coincidence factors in the analysis of the impact of V2G services on the distribution network was also used in the paper [14].

In the Polish papers [15], [16] a proposal of determination of coefficients of simultaneity based on, among others, the estimated degree of filling of the load diagram and correlations between the consumers was presented. The calculations of coincidence factor or peak power made in [15], [16] on the examples turned out to be even greater in comparison to the values given in the SEP (the Association of Polish Electrical Engineers) standard [17].

3 Methodology

Having at its disposal the measurement data of active energy consumed by 7671 single residential consumers during one year in 1-hour intervals, it was possible to estimate the peak power demand by consumer. The 1-hour values of the single consumer©active energy input in kWh are numerically equal to the values of the active power load in kW averaged over 1-hour intervals.

Based on a random selection it was possible to create sets simulating the system of consumers supplied from one installation and to determine the relation between the peak load of a hypothetical feeder supplying such a system of consumers and the sum of peaks of individual consumers. This relation is represented by the value of *kj* factor. The trend of *kj* value was searched for as a function of the number of residential consumers in the system.

Table 1. Results of statistical tests on the values of active power loads of consumers.

Test	Jarque-Bera	Kolmogorv-Smirnov	Lilliefors
Significance level	0,05	0,05	0,05
Maximum of probability-value	0,001	0	0,001
Minimum of difference between test statistic and critical value *)	136,19	0,486	0,0435
Percent of cases where the null hypothesis**) can be rejected	100%	100%	100%

*) In case of normally distributed values the difference should be negative.

**) The null hypothesis: measured values are normally distributed with unspecified mean and standard deviation.

Distributions of 1-hour loads with the active power of the consumers do not have to form normal distributions. Kurtosis as a dimensionless statistical coefficient for normal distributions assumes the value of 0. In the analysed data of the consumers©loads the lowest value of kurtosis was 1,089 (Fig. 1). Statistical tests of the measurement data provided reasons to reject the hypothesis of the normal distribution of values (Table 1).

Therefore, the application of formulas for calculating coincidence factor based on the assumption of equality of standard deviation of standard loads of consumers will not be appropriate. A better approach is to determine the values of these coefficients directly from their definition in a sufficiently large number of random samples of data sets and to determine statistical measures describing them.

The values of coincidence factor *kj* differ depending on the selected consumers making up the set. In order to determine the representative size of the factor sought, the bootstrapping approach was used.

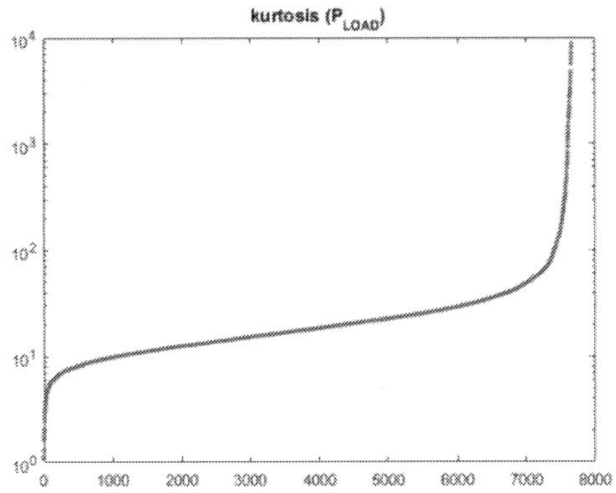

Fig. 1. Diagram of the sorted value of kurtosis of active power loads for each of the 7671 consumers.

The bootstrap is a simulation method of statistical inference developed in 1979 [18] and can be applied to both independent and dependent variables, also with unequal distribution. The method is predestined for situations when the distribution of statistics or a random variable describing a feature is unknown. Bootstrapping may be a better method of estimating the distribution of an estimated parameter than classical methods based on a central limit statement [19]. In a non-parametric approach, this method is free from model assumptions. The method allows to estimate the parameter characterising a given population on the basis of a randomly selected output sample. The approximation of statistics values on the basis of bootstrap samples is based on the Monte Carlo method. It consists in drawing independent samples of no more than the size of the original sample [20], calculating their statistics, where their average value is an estimation of the sought-after parameter describing the community. The method using bootstrapping to calculation the *kj* coefficients was applied in the paper [11].

The possessed measurement data of loads of consumers were considered as the primary sample. Bootstraps are sets of data on loads drawn from set of users. The key issue is to select a sufficient number of independent secondary samples. There are many rules for selecting the number of draws. A correct approach, although not very practical (a large number of sets), is to select the number of draws as equal to all possible variants of the bootstrap set elements of the assumed number [21]. Apart from complicated algorithms of optimal number of samples (e.g. [22]), often to determine the bootstrap distribution estimator, fixed values are taken as the element from the number of the output sample [19] or a simple recommendation that at least 1000 samples of the bootstrap type should be used [23].

In the task of determining the value of *kj* factors, the own algorithm of random selection of a set of consumers was developed. The built algorithm allowed for one-time use of data of each consumer in a number of sets. A single set concerned the selected number of consumers forming the system of loads supplied from one feeder. Therefore, the number of secondary samples was determined by the number of consumers n. The creation of sets was repeated 250 times, obtaining an integer number from $250 \times \left[\frac{L}{n}\right]$ samples for each number n, where $L=7671$ is the number of all the measured consumers. Thus, numerous samples from 958 750 (at $n=2$) to 9 500 (at $n=200$) were obtained.

4 Results

Fig. 2 presents the statistical measurements of population of coincidence factors values calculated on the basis of the algorithm as a function of the number of residential consumers. For comparison, the values recommended by the SEP standard [17] were imposed in two extreme cases, namely for older modernised installations (*SEP1*) and installations with electric heating of water (SEP2), as well as the theoretically minimum value (*min*) determined according to formula (2). The average value and the median *kj* were approximated by the power function. The graph of standard deviations in relation to the average of *kj* value depending on the number of flats presents Fig. 3.

Fig. 4. presents a comparison of peak power estimates in relation to the number of consumers (flats) obtained in three ways, namely on the basis of:

- *kj* coefficients according to the SEP standard,
- *kj* coefficients calculated in this paper,
- direct determination of peak power on the basis of analysed measurement data.

Peak capacities according to the SEP standard (*SEP1* and *SEP2*) are the products of the number of flats and given in this standard: relevant *kj* coefficients and the power demanded by one flat (7 kVA for modernised installations or 30 kVA for flats with electric heating of usable water). The mean value of the apparent peak power of 7671 consumers analysed in this paper was 3.4 kVA and 2.3 kVA median and 99th percentile was 5.2 kVA.

For precautionary purposes, the 99th percentile (5.2 kVA) was used in the peak power calculations on the basis of the previously determined *kj* coefficients. The estimated peak capacity is, therefore, the product of the number of consumers, the 5.2 kVA value and the relevant *kj* value calculated at the previous stage of this analysis. Two variants of estimations were presented – on the basis of appropriate average values of *kj* (*S calc kj mean*) and 95th percentile (*S calc kj 95-perc*).

In the second stage of the bootstrapping analyses, peak loads in absolute units for one feeder with residential consumers were estimated directly. The algorithm looked similar to the previous one, but this time it did not calculate the value of *kj*. The graph (Fig. 4). shows the average values of the samples obtained from the peak values depending on the number of consumers (*Pmean*) and the 95th percentile of peak power of the feeder (*P 95-perc*). According to the SEP standard, it is assumed that in residential buildings the power factor cosφ≈1, so these two values in kW would be numerically equal to the values in kVA. The average value *(Pmean)* was converted into apparent power (*Pmean/cos(phi)*) on the basis of the determined median power factor of the analysed consumers cosφ, which in this case was 0.3.

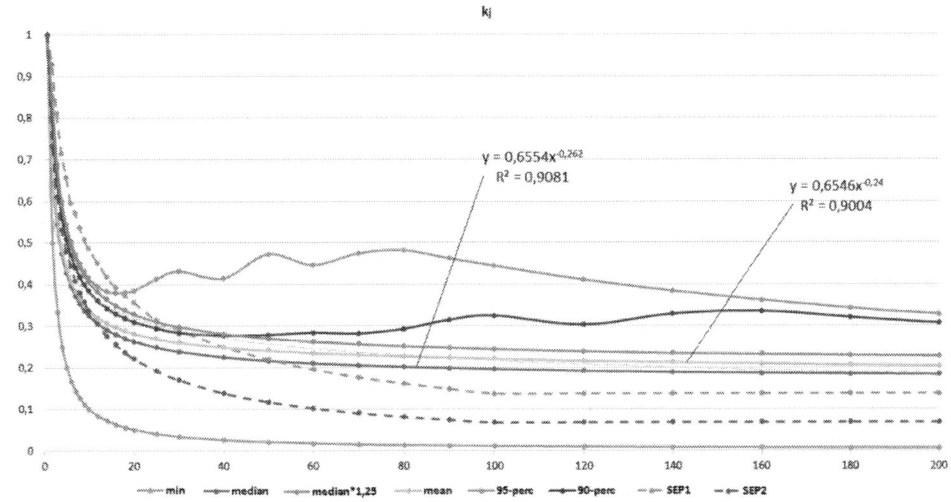

Fig. 2. The coincidence factor values obtained from calculations based on sets of randomly selected consumers as a function of their number in the installation – mean, median, 95th and 90th percentile, median increased by 25% (*median*1.25*), coefficients according to SEP standard (*SEP1, SEP2*) and theoretically minimal value of the coefficient (*min*) according to formula (2).

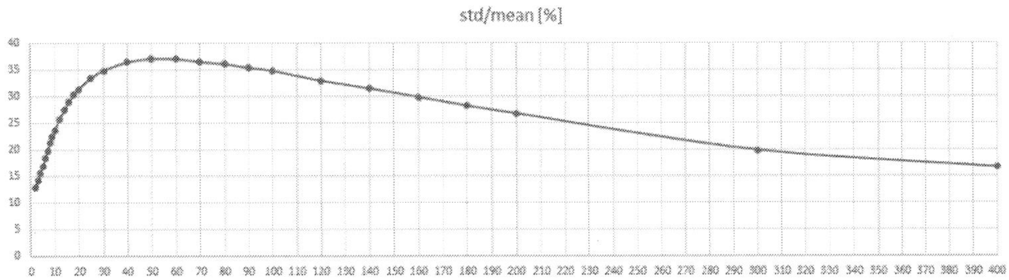

Fig. 3. Percentage of standard deviations to the mean value of the *kj* coefficients of the samples as a function of the number of flats supplied from one feeder.

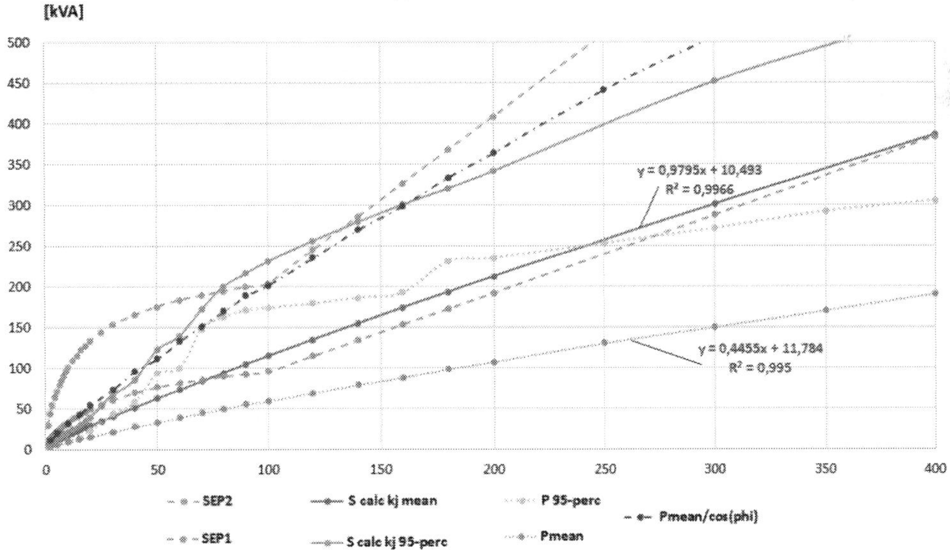

Fig. 4. Comparison of the determined peak capacities based on: SEP standards (*SEP1* and *SEP2*); determined coincidence factors *kj* - mean (*S calc kj mean*) and 95th percentile (*S calc kj 95-perc*); directly determined peak capacities in absolute values - mean (*P mean*), 95th percentile (*P 95-perc*) and mean taking into account the average value of tangent power of consumers. Values are presented as a function of the number of flats.

5 Discussion

The values of statistical measures, apart from the higher percentile, are monotonously decreasing in relation to the increasing number of flats supplied from the common feeder. The average value for more than 20 consumers is located in the middle between the median values: calculated and increased by 25%. The highest dynamics of changes in *kj* value is observed for small numbers of consumers (below 20) supplied from one feeder (Fig. 2). The highest values of standard deviations from the samples in relation to their mean value reach 40% (Fig. 3).

The determined *kj* coefficients in comparison to the SEP standard take smaller values for smaller numbers of consumers. For a number of flats below 10, the smallest of the *kj* values (according to the SEP standard with electric water heating) are between the average value and 90th percentile of the obtained *kj* values from random samples. The highest values of *kj* coefficients (for modernised installations) according to SEP are higher than 125% of obtained median of samples for the number of flats below 30. For larger numbers of flats the SEP coefficients turn out to be lower than those obtained as a result of analyses (Fig. 2). The values of median and average are similar, the biggest differences do not exceed 12% (maximum 0.025 in absolute values).

The median and mean values as a function of the number of flats are monotonously decreasing in the whole analysed range. To 14 flats the 95th percentile practically coincides with the median value increased by 25%. The 90th percentile does not behave monotonously as a function of the number of flats in the range between 40 and 160, while the 95th percentile in the range 20…90. In these ranges (number of flats 20...160) the highest standard deviation values are observed among the samples of *kj* in relation to their mean values (30-40%) – Fig. 3. For this number of flats, more samples with higher *kj* values were observed, despite the fact that the monotonous course of averages and median *kj* were maintained. In the data set there were series of peak load points of relatively high values, which were equally probable to be drawn at the mentioned numbers of users and increased the extreme *kj* values.

The characteristic feature is the occurrence of a linear relationship between the number of loads and the average value of directly determined peak power values of the feeder – Fig. 4. The linear relationship is confirmed by the matching of a straight trend line with a high determination factor R^2. The lowest peak power values were obtained for mean values under the assumption $\cos\varphi \approx 1$. These values are significantly lower than according to the SEP standard.

A separate issue is the demand for reactive power of residential consumers. Nowadays, the assumption adopted in accordance with the SEP standard with a significant advantage of resistive devices in residential installations does not seem appropriate. An additional aspect complicating the issue is the controversy around the power theory [24]. Apart from such detailed issues, on the basis of the measurement data from the meters of 7671 residential users, the median power factor was determined, which reached only $\cos\varphi=0.3$. Taking this value into account for the determined average value of peak power resulted in the fact that for a smaller number of users this value is similar to the estimation made on the basis of 95th percentile of calculated *kj*. For a large number of flats, this value is between the extreme variant according to the SEP standard and the values determined on the basis of 95th percentile *kj*. The assumption of such a low $\cos\varphi$ value for all users does not have to be justified. Reactive power is compensable. It is therefore possible to compensate each other for reactive power flows in the installation. On the one hand, it requires an additional flow analysis and, on the other hand, it enables the specific use and optimisation of this capacity within the framework of the extended energy management concept [25].

Peak apparent power values estimated on the basis of average *kj* values and assumed peak power of a single flat were also linear in relation to the number of residential users. For a smaller number of flats (up to 60), the values are lower than the least demanding variant of the SEP standard. The values of 95th percentiles of the directly estimated active peak power to approximately 250 flats vary between the values corresponding to the peak power estimated on the basis of the average of the determined *kj* and the directly estimated average peak power, corrected by the power factor value. Above 250 flats, 95th percentile of directly estimated peak power are quasi-linear, below the value based on the SEP standard.

6 Conclusions

The method of peak power estimation based on coincidence factors *kj* is simple to apply in engineering practice. The key step is to adopt the correct values for these factors. The distributions of the power used values do not have to form normal distributions, so it will not be appropriate to determine them on the basis of the assumption of similar values of standard deviations. Estimating *kj* directly from the definition requires access to large sets of measurement data, which in the era of smart metering should not be a problem. In this case, a simpler and more effective method may be to determine directly the values in absolute peak power units, without calculating the coincidence factor. Especially since the *kj*-method additionally requires estimation of the peak power of a single consumer.

The analyses carried out indicate that the peak power may be overestimated by the applicable standards giving the *kj*-values This overestimation results mainly from assuming a too high value of the peak power of a single user. The calculations of the peak power of the feeder conducted directly for the average values from random samples indicate a linear dependence on the number of residential consumers. The results obtained on the basis of *kj* averaged from random samples are characterised by a similar dependence. The results obtained on the basis of the *kj*-value from the standards do not have a general linear trend. The impact of reactive power demand complicates the issue due to the complexity and small recognition of reactive power using at the level of household consumers.

References

1. W. Lackie, The Influence of Load and Diversity Factors on Methods of Charging for Electrical Energy, *Journal of the Institution of Electrical Engineers*, vol. 42, pp. 100-114, (1909).
2. H. Gear, Diversity factor, *the Chicago Section of the American Institute of Electrical Engineers*, (1910).
3. C. Bary, Coincidence-Factor Relationships of Electric-Service-Load Characteristics, *AIEE Transactions*, vol.64, pp. 623-629, (1945).
4. A. Sargent, R. Broadwater, J. Thompson and J. Nazarko, Estimation of Diversity and kWh-to-Peak-kW Factors from Load Research Data, *IEEE Trans. on Power Systems*, vol.9, pp. 1450-1456, (1994).
5. J. Nazarko, R. Broadwater and N. Tawalbeh, Identification of Statistical Properties of Diversity and Conversion Factors from Load Research Data, *Electrotechnical Conference*, vol.1, (1998).
6. K. Billewicz, Tests normally distributed population of the values energy (in Polish), *Przegląd Elektrotechniczny*, no. 4, pp. 78-79, (2008).
7. J. Dickert and P. Schegner, Residential Load Models for Network Planning Purposes, *IEEE 2010 Modern Electric Power Systems*, Wrocław, (2010).
8. J. Peppanen and J. Taylor, Enhanced Load Modeling with Expanded System Monitoring, *IEEE Clemson University Power Systems Conference (PSC)*, (2018).
9. X. Liu, H. Zheng, S. Song and G. Fu, Selection Method of Community Load Coincidence Factor Based on BP Neural Network, *IEEE 3rd International Conference on Computer Science and Network Technology*, (2013).
10. D. H. McQueen, P. Hyland and S. Watson, Monte Carlo Simulation of Residential Electricity Demand for Forecasting Maximum Demand on Distribution Networks, *IEEE Trans. on Power Systems*, vol. 19, no. 3, pp. 1685-1689, (2004).
11. V. Chatlani, D. Tylavsky, D. Montgomery and M. Dyer, Statistical Properties of Diversity Factors for Probabilistic Loading of Distribution Transformers, *IEEE 39th North American Power Symposium*, (2007).
12. S. Kaykahie and S. Kowsari Movahed, A New Approach for Calculating Load and Loss Factor Base on a New Approach for Calculating Load and Loss Factor Base on Consumer Data with Fuzzy Modelling, *CIRED 22nd International Conference on Electricity Distribution*, (2013).
13. X. Tong, C. Guo, X. Yang and C. Chen, Research on Characteristics of Electric Vehicle Charging Load and Distribution Network Supportability, *IEEE PES Asia-Pacific Power and Energy Conference*, (2016).
14. U. C. Chukwu, O. A. Nworgu and D. O. Dike, Impact of V2G penetration on Distribution System Components Using Diversity Factor, *IEEE SOUTHEASTCON*, (2014).
15. S. Ciura and H. Kocot, Rzeczywiste i obliczeniowe obciążenia poszczególnych elementów sieci, zasilających odbiorców bytowych, przyłączonych do sieci niskiego napięcia., *Rynek energii elektrycznej. Aktualne problemy energetyki.*, P. Pijarski, Z. Połecki, Eds., Politechnika Lubelska, pp. 33-42, (2018).
16. S. Ciura and P. Kubek, Wyznaczanie obciążeń maksymalnych w sieciach nn – teoria i praktyka, *Rynek energii elektrycznej. Rozwój i funkcjonowanie rynków energii.*, Z. Połecki, P. Pijarski, Eds., Politechnika Lubelska, pp. 7-18, (2016).
17. N SEP-E-002. Instalacje elektryczne w obiektach budowlanych. Instalacje elektryczne w budynkach mieszkalnych., COSiW SEP, (2003).
18. B. Efron, Bootstrap Methods: Another Look at the Jackknife, *The Annals of Statistics*, vol.7, pp.1-26, (1979).
19. K. Singh and M. Xie, Bootstrap: A Statistical Method, Rutgers University, (2008).
20. B. Efron, Estimating the Error Rate of a Prediction Rule: Improvement on Cross-validation, *Journal of the American Statistical Association*, vol.78, (1983).
21. J. Fox, Applied Regression Analysis and Generalized Linear Models, *SAGE Publications Inc.*, (2008).
22. R. Davidson and J. MacKinnon, Bootstrap tests: how many bootstraps?, *Econometric Reviews – Taylor&Francis Journals*, vol. 19(1), (2000).
23. B. Efron and R. Tibshirani, An Introdution to the Bootstrap, Chapman&Hall, (1993).
24. L. Czarnecki, Meta-theory of electric powers and present state of power theory of circuits with periodic voltages and currents, *Przegląd Elektrotechniczny* no.6, pp. 26-31, (2013).
25. S. Bielecki and T. Skoczkowski, An enhanced concept of Q-power management, *Energy* no.162, pp. 335-353, (2018).

RESEARCH & DEVELOPMENT IN POWER ENGINEERING, 2019

The cost-effectiveness of an energy storage system using hydrogen and underground gas storage

Sebastian Lepszy[1]

[1] Silesian University of Technology, ul. Konarskiego 18, 44-100 Gliwice, sebastian.lepszy@polsl.pl, Poland

Abstract. Due to the random nature of the production, the use of renewable energy sources requires the use of technologies that allow adjustment of electricity production to demand. One of the ways that enable this task is the use of energy storage systems. The article focuses on the analysis of the cost-effectiveness of energy storage from the grid. In particular, the technology was evaluated using underground hydrogen storage generated in electrolysers. Economic analyzes use historical data from the Polish energy market. The obtained results illustrate, among other things, the proportions between the main technology modules selected optimally in technical and economic terms.

1 Introduction

Energy storage technologies make it possible to change the way electricity networks operate. Currently, the amount of energy produced and consumed is balanced in short periods, due to the low storage capacity. Increasing energy storage capacity can help in some cases reduce costs and pollutant emissions. Storage systems can also provide additional services for power networks such as: frequency regulation, rotating reserve, voltage regulation, reactive power compensation, network stabilization [1].

In Poland, there is little use of storage systems. The installed capacity of storage systems in the form of pumped storage power plants is 1413 MW, which represents 3.2% of the installed capacity of the power plant in Poland. The electricity produced in these systems represents only 0.27% of the electricity produced in Poland, and their capacity is about 10GWh.

The main reason for considering energy storage should be making a profit for the energy storage company. This purpose of running a business also guarantees rational use of resources. Making a profit by the storage company means that it is possible to operate conventional power plants with greater power during periods of low demand for electricity and still with full power during periods of peak demand. The effect of such work may be an increase in the efficiency of electricity production of a conventional power plant, which may led to reduction in energy costs and a decrease in the production of electricity in peak systems.

Systems of this type can also contribute to reducing emissions by enabling greater use of overproduction of energy from renewable sources.

The current weak development of this technologies suggests that this are uneconomical solutions [2]. However, the rapidly changing economic environment should be taken into account, which is largely influenced by legal regulations, concerning the reduction of pollutant emissions and the use of renewable energy sources. The same regulations may also cause a high risk of introducing new technologies.

One of the significant factors affecting the economic efficiency of energy storage is the cost of energy transmission. In economic analyzes, for simplicity, transmission costs have been omitted, which corresponds to the situation of location of the storage system together with the power plant generating electricity.

The analyzes also are based on the assumption that the introduction of the storage system will not significantly affect price levels.

In the literature, it is difficult to find informations on optimal technology parameters in terms of power and storage capacity. The available analyzes mainly concern storage options in a short time. For example, in [3] an economic analysis of storage systems based on price arbitrage was conducted. The compared technologies did not contain, unfortunately, energy storage technology in the form of hydrogen. Economic profitability was demonstrated by CAES technologies and pumped storage power stations. All technologies had a storage capacity of less than one day, which made it impossible to take advantage of seasonal fluctuations in energy prices. In the literature, you can find descriptions of algorithms for assessing energy efficiency or economic energy storage systems cooperating with renewable energy sources. In [4] an algorithm for energy and economic calculations for a distributed system was presented, in which the energy store was located near a wind farm. Determined economic and energy parameters partly based on average values taking into account the normal distribution of probability. Although the results were obtained for the whole year, it was based on average prices of wind energy and production, which did not allow for the assessment of solutions in the case of long-term storage.

This article presents the methodology for selecting the size of machines and devices in the installation, and sets out the main economic parameters for estimating the profitability conditions of this type of installation

2 Electricity prices

To simplify the analysis, in order to estimate the economic effects, no additional costs of electricity transmission were assumed in further calculations.
The costs of purchasing energy can be determined by analysing prices on energy markets. Their analysis reveals the possibility of using price fluctuations to generate revenue in an energy storage company.
Figure 1 presents an ordered price chart on the Polish Power Exchange for the day-ahead market in 2017. The arithmetic average price is 158.37 PLN, the maximum to average ratio is 6.2 and the minimum to average ratio is 0.47.

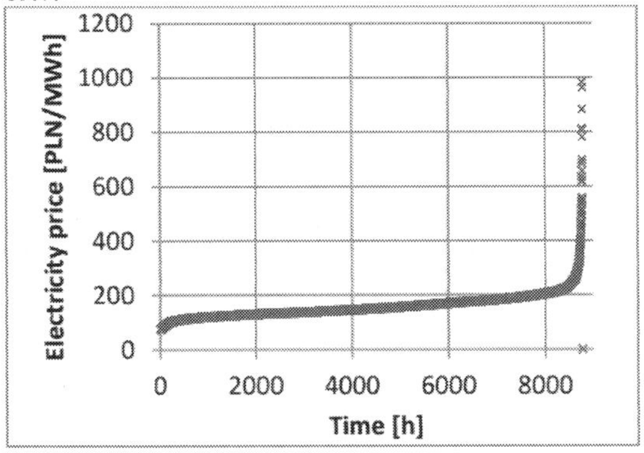

Fig. 1. An ordered chart of electricity prices (day-ahead market) on the Polish Power Exchange in 2017

Based on the data, the relationship between the average price of energy production and energy consumption for different levels of storage efficiency and the ratio of charging and generating power can be analyzed. The choice of the listed variables makes it easy to determine the difference in revenues and costs of electricity in the enterprise, e.g. to the amount of energy taken from the grid. For selected variables, revenues from the sale of electricity S_{el} can be specified:

$$S_{el} = C_{srS} E_S \quad (1)$$

where: C_{srS} - average sale price of electricity for its production in the τ_S period characterized by the highest prices on the energy market; E_s - the amount of energy sold.
Electricity purchase costs are calculated using following equation:

$$K_{el} = C_{srK} E_K \quad (2)$$

where: C_{srK} - average purchase price of electricity for its collection in the period τ_K characterized by the lowest prices on the energy market; E_k – the amount of energy purchased.

There is a relationship between the amount of purchased and generated energy:

$$E_S = \eta_m E_K \quad (3)$$

where: η_m - energy storage energy efficiency.
By specifying the ratio of storage loading power (power taken from the network) and storage discharge power (produced power, fed into the grid), it can be written:

$$\tau_S = \eta_m \frac{N_K}{N_S} \tau_K \quad (4)$$

where: τ_s - storage discharge time with constant N_S power; τ_k - storage loading time with constant N_K power. For analyses carried out for the year, it can be seen that the maximum time of energy consumption should be less than 8,760 hours, and the production time is its function. Using the data cited in Fig. 1, it is possible to determine the dependence of the price difference C_{srS}-C_{srk} as a function of energy consumption time τ_k, for different power ratios, for a given storage efficiency. Figures 2,3,4 show these functions for a storage efficiency of 0.2, 0.5 and 0.8 respectively.

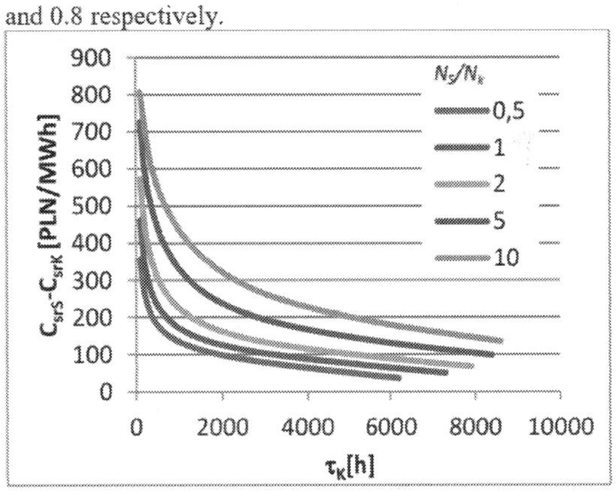

Fig. 2. Relationship between the average sale and purchase price difference C_{srS}-C_{srk} for different value of working time τ_k and the ratio of production and consumption power N_S/N_k for storage efficiency η_m=0.2

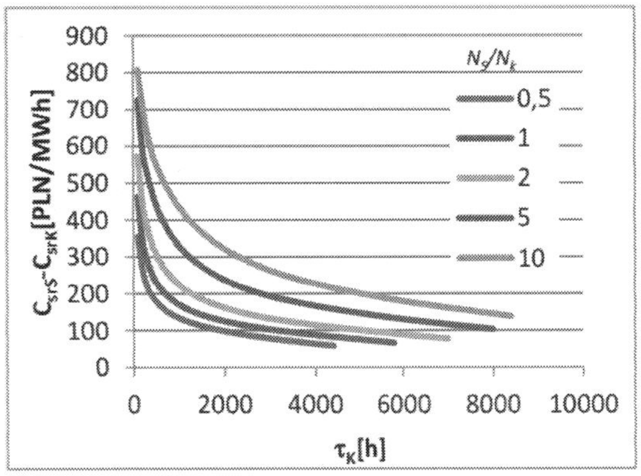

Fig. 3. Relationship between the average sale and purchase price difference C_{srS}-C_{srk} for different value of working time τ_k and the ratio of production and consumption power N_S/N_k for storage efficiency η_m=0.5

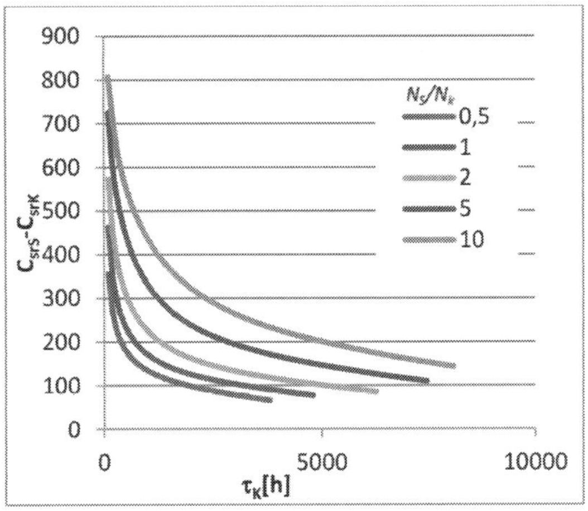

Fig. 4. Relationship between the average sale and purchase price difference C_{srS}-C_{srk} for different value of working time τ_k and the ratio of production and consumption power N_S/N_k for storage efficiency η_m=0.8

Based on the analysis, it can be concluded that the highest prices differences can be obtained only for a very short time of energy consumption, and thus for the production of energy from storage in periods of highest prices. Increasing the ratio of generation capacity to consumption capacity increases the difference in average prices due to the possibility of producing energy from the storage with greater power during high prices.
Increasing storage efficiency reduces price differences for individual values of energy consumption times, due to the extension of the time of energy production from the storage at lower prices.

3 Electricity costs and revenues

The price difference, although it is a factor enabling economic efficiency, does not directly illustrate the scale of the project. For small amounts of time consuming electricity, the price difference is large, while the amount of stored energy is small, which results in a small difference in revenues and costs of electricity. On the other hand, for a large time of energy consumption, the production time is also long, while the price difference is small, which means that the difference in revenues and costs will also be small. To illustrate this issue, based on the data, the dependence of the difference in revenues and costs was developed for installations with 1MW input power for different values of N_S/N_k power and storage efficiency (Fig. 5, 6, 7). For a relatively low storage efficiency, it can be determined that only for a relatively short storage operation time and a large N_S/N_k power ratio the difference in revenues and costs is positive. The main reason is the low value of storage efficiency. The increase in storage efficiency increases the value of the difference in revenues and costs as well as the optimal loading time of the storage.

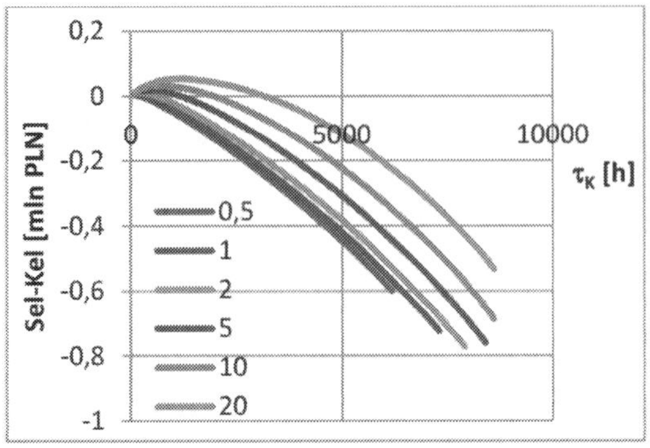

Fig. 5. Dependence of the difference in revenues and costs of electricity S_{el}-K_{el} for different values of working time τ_k and the ratio of production and consumption power N_S/N_k for storage efficiency η_m=0.2

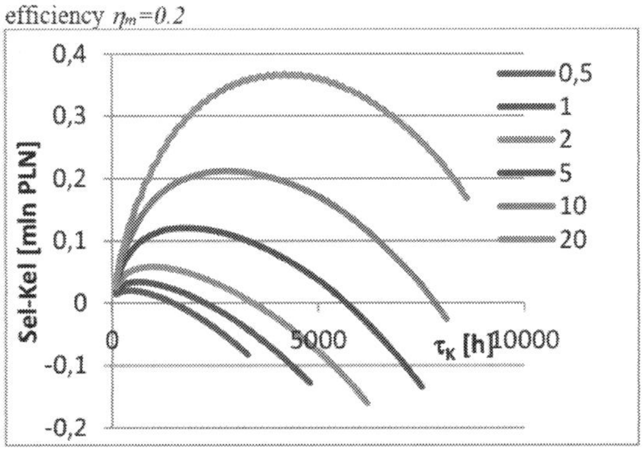

Fig. 6. Dependence of the difference in revenues and costs of electricity S_{el}-K_{el} for different values of working time τ_k and the ratio of production and consumption power N_S/N_k for storage efficiency η_m=0.5

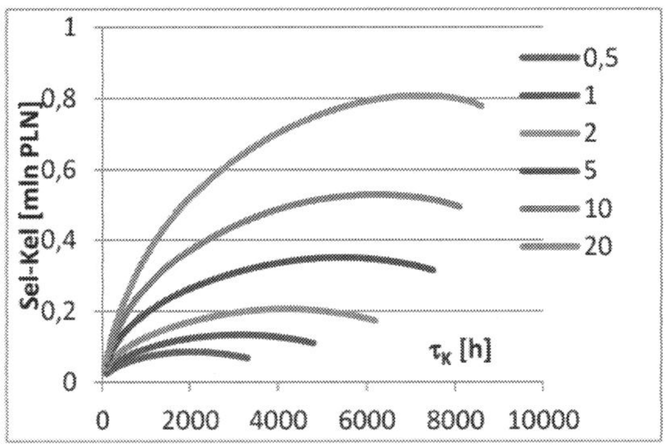

Fig. 7. Dependence of the difference in revenues and costs of electricity S_{el}-K_{el} for different values of working time τ_k and the ratio of production and consumption power N_S/N_k for storage efficiency η_m=0.8

4 Relative profit

Increasing the difference in revenues and costs for large power ratios tends to analyze the ratio between the said difference and the sum of the system power in the storage loading and unloading mode. This analysis is interesting because the sum of power is a parameter that significantly affects the investment expenditure related to the construction of the system. Figures 8, 9, 10 illustrate these relationships for different storage efficiency values. The highest revenue values in relation to the total power can be obtained for the highest storage efficiency. Presented charts, also take into account the impact of the power ratio and working time on the optimal size of the storage and related investment outlays. The presented analysis can lead to rational conclusions when investment outlays for gas storage would be proportional to the sum of charging and discharging power.

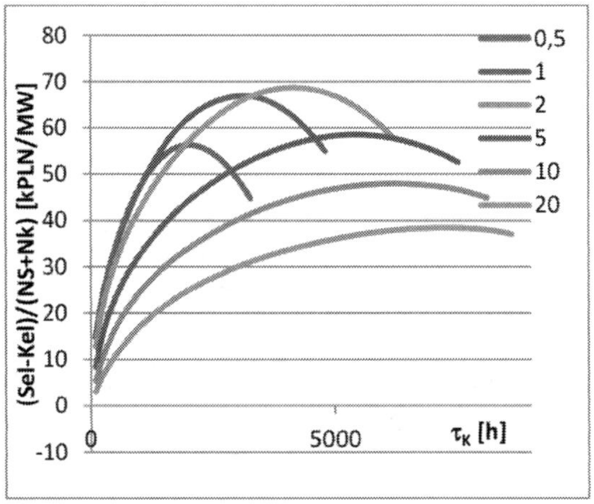

Fig. 8. Dependence of the profit to power ratio for different values of working time τ_k and the ratio of production and consumption power N_S/N_k for storage efficiency $\eta_m=0.8$

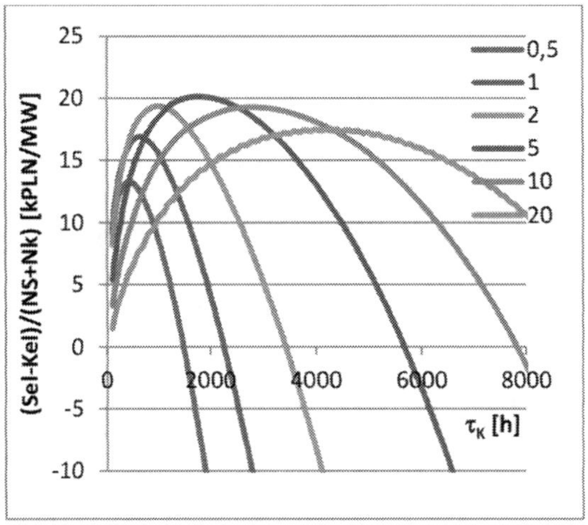

Fig. 9. Dependence of the profit to power ratio ratio for different values of working time τ_k and the ratio of production and consumption power N_S/N_k for storage efficiency $\eta_m=0.5$

The presented methodology illustrates the possibilities of obtaining revenues for energy storage installations assuming no transmission costs. Including them further reduces the price difference, which can significantly affect the company's energy efficiency.

The presented analysis allows only to estimate the basic parameters of the storage system operating based on one selected market and based on the price level of the selected time. Significant changes in the energy market that we observe nowadays can cause real changes in the results of analyzes.

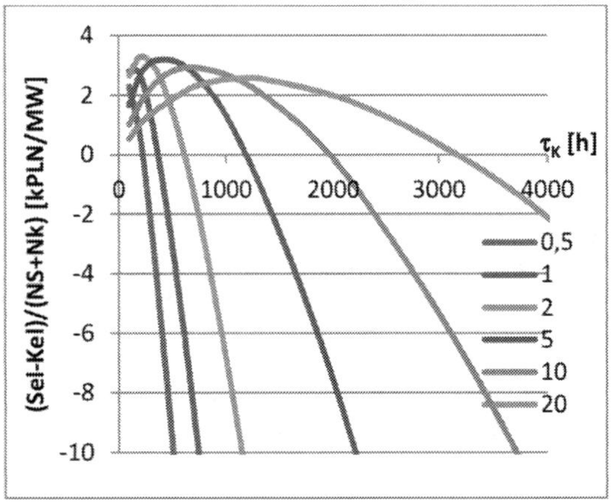

Fig. 10. Dependence of the profit to power ratio for different values of working time τ_k and the ratio of production and consumption power N_S/N_k for storage efficiency $\eta_m=0.2$

5 Gas storage capacity

To conduct a preliminary economic analysis of the energy storage system, in addition to information about the energy consumption of energy production and, therefore, the power of devices, information on the size of the energy storage is necessary, which, inter alia, affects investment expenditure related to the construction and operating costs of the system.

Based on the time course of energy prices in a given market, the time intervals in which energy is stored or taken from the storage can be specified. Assuming constant energy flows, it is possible to specify the time variability of the amount of gas in the storage system. Assuming the value of gas losses from the storage system, it is possible to determine its working capacities as the difference from the maximum and minimum amount of medium in the cavern

The analysis of the energy storage capacity value can be carried out based on such parameters as:

- ratio of gas charging and discharging flow;
- time of charging the storage in a year.

For each time interval it is possible to determine if the price that can be obtained when buying is less than the limit price, which is related to the time the system works in the loading process. If the above inequality occurs, we assume that in this time interval the storage is loaded with the assumed capacity.

Similarly, for each time interval, it is possible to determine, if the energy price is greater than the cut-off price, related with the system operation time in the unloading process. If this inequality occurs, the gas is unloaded with capacity depending on the assumed flow ratio. There may be periods of time for which the gas storage is not unloaded or loaded. Also in the long term (e.g. year), the quantity introduced into the storage must be balanced with the amount of gas released. On this basis, it can be written:

$$\tau_S = (1 - s_m)\frac{V_K}{V_S}\tau_K \quad (5)$$

where: V_k – gas storage loading flow; Vs – gas storage discharge flow; s_m - relative gas losses related to charging efficiency.

An example of the dependence of the amount of gas in the storage over time is illustrated in Fig. 11. It was assumed that the capacity ratio is equal to 4 and the loading time is 7000h. The amount of stored gas is expressed using the E_{m_h} value:

$$E_{m\_h} = \sum_{t=1}^{8760} V(t)\Delta t \quad (6)$$

where: $V(t)$ – time function of charging flow (in mass or mol units charging flow per h); Δt – time interval (1h). From the chart you can read that for a selected year of operation the gas storage should be emptied in about 4000 hour of operation from the beginning of the year and filled in about 1000 hour of work per year, respectively.

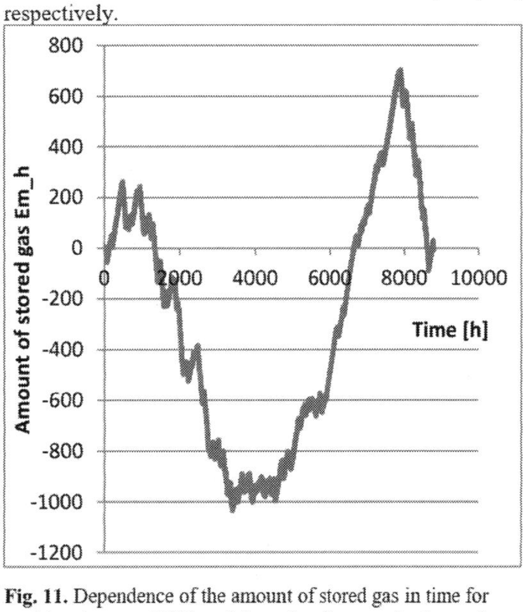

Fig. 11. Dependence of the amount of stored gas in time for working time $\tau_{k=}$ 7000h and the ratio of gas production and consumption flow $V_k/V_s=4$

The dependence of the required storage capacity can be determined as a function of the running time of the energy consuming system. The results for different values of storage loading and unloading efficiency are shown in Fig. 12. It can be seen that for low values of the charging time the storage capacity is almost linearly dependent on the said time. This is directly related to high energy prices far exceeding the average value and the need to store this part of energy until the low price. For larger values of the charging time, changes in capacity are no longer always increasing.

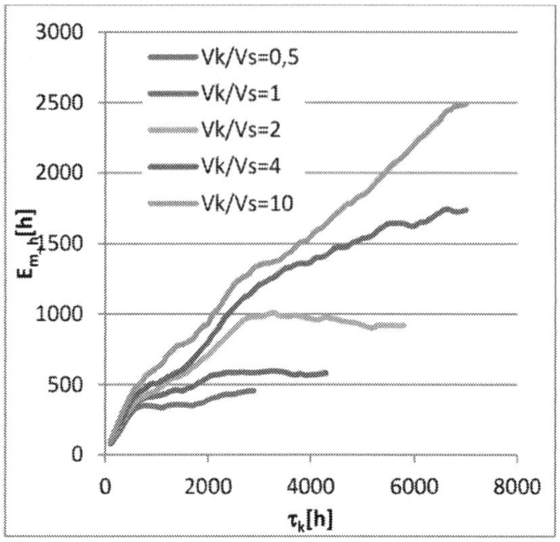

Fig. 12. Dependence of the necessary working capacity in ours of gas production as a function of working time $\tau_{k=}$ and the ratio of gas production and consumption flow $V_k/V_s=4$

For the example, installation with the V_k/V_s ratio of 4 and loading time of the storage 1000 we get 277 beginnings of the charging system operation and a similar number of 273 beginnings of the discharge system operation. The maximum continuous operation time of the charging system is 168 hours, and for the discharge - 16 hours.

6 Storage of energy using hydrogen - case study

The profitability analysis of storage theology was conducted based on the system block diagram presented in Fig. 13.

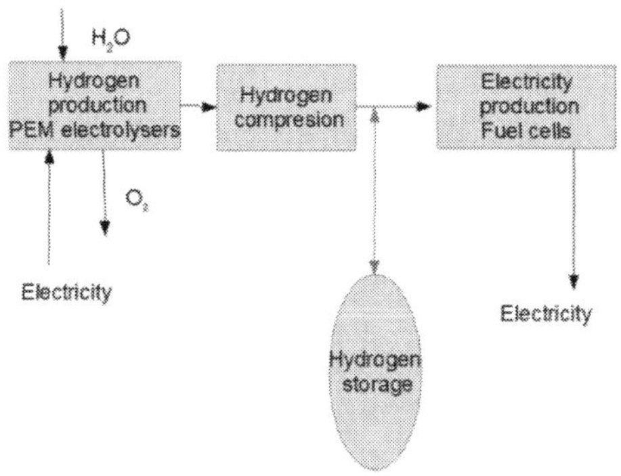

Fig. 13. Block diagram of system

In the presented system, electricity is converted into chemical energy of hydrogen in PEM type cells. Unlike alkaline electrolysers, these devices have a better development perspective, in particular in terms of electricity conversion efficiency or capital expenditure. It was assumed that their energy efficiency related to LHV is at the level of 68%, which is the value possible to achieve in the near future [5].

Hydrogen from the cells is directed to the compressor system increasing the pressure to the level required in the gas storage. To simplify the analysis, electricity consumption in this module was omitted. This consumption can be from 0% (pressure in the gas storage below 30 bar) to 2% (pressure in the gas storage approx. 200bar) of energy consumed by electrolysers.

The compressed hydrogen goes to the gas storage in the form of e.g. salt caverns

Production of electricity takes place in PEM type fuel cells with 58% energy conversion efficiency (LHV based).

These values allow to determine the maximum storage efficiency at the level of 39.4%. It was assumed that the ratio of electricity production power to input power is 4.7, which is the optimal value for charging time of 1100 hours a year.

Assuming the power of the energy intake system at 50MW, it is necessary to use fuel cell modules with 235MW electric power. These power values correspond to 1.02Mg/h hydrogen production and 12.15Mg/h utilization rate. Based on the performance ratio, the gas storage working capacity of 673Mg can be estimated.

Based on the day-ahead market data of the 2017 energy exchange, it can be determined that the cost of purchasing electricity would be 6 022 000PLN while revenues from the sale of energy 9 730 000PLN, which gives a difference of 3 707 000PLN. The difference obtained can be compared to capital expenditure on main devices, i.e. electrolysers and fuel cells. Assuming the unit price of electrolysers at the level of 1000 Euro/kWe and fuel cells at the level of 3000 Euro/kWe, we obtain capital expenditure worth 216 000 000 + 3 046 000 000 = 3 262 million PLN, which is about 900 times greater than the difference calculated earlier

7 Conclusions

The day-ahead market of the commodity energy exchange was characterized in 2017. a small number of hours in which the price was much lower than the average and a small number of hours with the price significantly exceeding the average price. This price distribution theoretically enables the operation of the storage system using price arbitrage, with particular emphasis on the operation of the system during high prices.

Taking into account energy balancing in the storage system, it is possible to determine the difference in purchase costs and revenues from the sale of electricity. The highest values of this indicator can be obtained for a relatively high power of the electric energy production system in relation to the power of the electric energy consuming system. The increase in the ratio of the production system's power to the consumed power caused the increase of the optimal value of the time of energy consumption.

The day-ahead market of the commodity energy exchange was characterized in 2017 by a very small number of hours in which the price was much lower than the average and a small number of hours with the price significantly exceeding the average price. This price distribution theoretically enables the operation of the storage system using price arbitrage, with particular emphasis on the operation of the system during high prices.

Taking into account energy balancing in the storage system, it is possible to determine the difference in purchase costs and revenues from the sale of electricity. The highest values of this indicator can be obtained for a relatively high power of the electric energy production system in relation to the power of the electric energy consuming system. The increase in the ratio of the production system's power to the consumed power caused the increase of the optimal value of the time of energy consumption.

Due to significant changes in the value of revenues from the purchase and sale of electricity depending on the power ratio, the ratio of revenues in relation to the power of main elements was determined. Optimal working times of the collection system were on the same level like optimal values taking into account only revenues.

A case study for a system using PEM electrolysers and fuel cells indicates the lack of economic efficiency of the considered solution for the market in 2017.

References

[1] M. Arbabzadeh, J.X. Johnson, G.A. Keoleian: Parameters driving environmental performance of energy storage systems across grid applications. *J Energy Storage,* 12 (2017), pp. 11-28

[2] E. Fertig, J. Apt: Economics of compressed air energy storage to integrate wind power: a case study in ERCOT. *Energy Policy*, 39 (5) (May 2011), pp. 2330-23422

[3] Bradbury K., Pratson L., Patiño-Echeverri D.: Economic viability of energy storage systems based on price arbitrage potential in real-time U.S. electricity markets. *Applied Energy* 114 (2014) 512–519

[4] M. Korpaas, A. T. Holen, R. Hildrum, Operation and sizing of energy storage for wind power plants in a market system, *International Journal of Electrical Power & Energy Systems* 25 (8) (2003) 599–606.

[5] Bertuccioli L., Chan A., Hart D., Lehner F., Madden B., Standen E.: Study on development of water electrolysis in the EU. Fuel Cells and Hydrogen Joint Undertaking, Final Report; February 2014

RESEARCH & DEVELOPMENT IN POWER ENGINEERING, 2019

Experimental and numerical tip leakage flow visualization in the LP turbine labyrinth seal

Krzysztof Bochon[1,*], *Włodzimierz* Wróblewski[1], *Artur* Szymański[1], *Mirosław* Majkut[1], *Michał* Strozik[1] and *Krzysztof* Marugi[2]

[1] Silesian University of Technology - Institute of Power Engineering and Turbomachinery, Konarskiego18, 44-100 Gliwice, Poland, krzysztof.bochon@polsl.pl
[2] Avio Aero, Grażyńskiego 141 43-300 Bielsko-Biała, Poland

Abstract. The subject of this publication is the identification of basic flow parameters and flow structures in the seal experimentally and compare them with CFD results. A straight-through seal with two leaning fins and smooth or honeycomb land was analysed. The sealing concept is characteristic for the tip seal of the last stage of an aircraft low-pressure turbine. Due to the limitations of the test rig the analyses presented here were conducted on a highly simplified, stationary model of the seal itself, with an axial inflow and no curvature in the circumferential direction. The characteristics of the discharge coefficient as a function of the pressure ratio for different clearances and the pressure distribution along the seal, for different pressure ratios are presented. In addition, an attempt was made to visualize the flow using the schlieren technique. The main idea of application schlieren photography was to observe the vortex and separation structures occurring during the flow through the labyrinth seal, which is the major source of pressure losses. CFD calculations were carried out using the Ansys CFX commercial code.

1 Introduction

Ever increasing demand for high efficiency and low emission solutions in aerospace enforces significant efforts on engineers and scientists. Perfecting the aero-engine compartments is one of the areas with a significant impact on turbomachinery performance. Research by Rosic & Denton [1] shows 2% of turbine efficiency drop due to internal leakages. The labyrinth seals are the most popular solution to eliminate or regulate the leakage in aero-engines. Despite the fact that state-of-the-art sealing solutions such as brush seals or finger seals reduce leakage much better [2], labyrinth seals are favoured mostly due to resistance to high temperature, high-pressure ratios, high speed, rubbing and solid particles contamination [3]. Historically, a significant effort in labyrinth seals evaluation, including experiments [4-6], CFD calculations [7-10] and analytical methods [11-13] have been presented in the literature. The research to evaluate and assess labyrinth seals works has been going on since the 1970s when Detroit Diesel Allison performed some of the first experimental campaigns [14]. The first CFD calculations were performed [15, 16], using 2D models. Later on, 3D models were applied for various sealing solutions, with recently deployed LES methods [17, 18]. Recently labyrinth seals are often applied with hole-patterns against it, for instance, honeycomb. Hole-patterns cause some difficulties in mesh generation, hence the ideas for simplifying these structures during the numerical modelling. In [19] the honeycomb structure was replaced by square holes, to simplify the geometry, and lower the computational cost during optimization.

The most popular parameter used for the seal performance evaluation is the discharge coefficient C_D:

$$C_D = \frac{\dot{m}}{\dot{m}_{ideal}} \tag{1}$$

defined as the ratio of mass flow occurring in the seal to m_{ideal}, a theoretical isentropic flow through the clearance described by the relation for an isentropic flow through a nozzle:

$$\dot{m}_{ideal} = \frac{p_0 A_{nom}}{\sqrt{T_0}} \sqrt{\frac{2\kappa}{R(\kappa-1)} \left[\left(\frac{1}{\pi}\right)^{2/\kappa} - \left(\frac{1}{\pi}\right)^{(\kappa+1)/\kappa} \right]} \tag{2}$$

The leakage highly depends on clearance size, and flow direction.

The studies presented here were a part of the research of the rotor tip seal of the low-pressure part of an aircraft engine turbine. Some parts of the research were described in [20-23]. The main goal of the research was to improve the efficiency of the analysed seal. Works within the research included numerous experimental studies and CFD analyses. In the part of the research, which is the subject of this publication, the focus was on the identification of the main flow parameters and flow structures in the seal on the test rig using the schlieren technique and comparison of the obtained results with the results of CFD analyses. The flow field and discharge characteristics were evaluated utilising commercial CFD tools, such as ANSYS CFX and the in-house experiment.

2 Methodology

2.1 Investigated geometries

The seal analysed in this paper is a straight-through configuration with two fins. The labyrinth with smooth land (Fig. 1a) and honeycomb land (Fig. 1b) was investigated. The fins are inclined towards the inflow. The dimensions of the labyrinth as well as of the land structure can be found in [22].

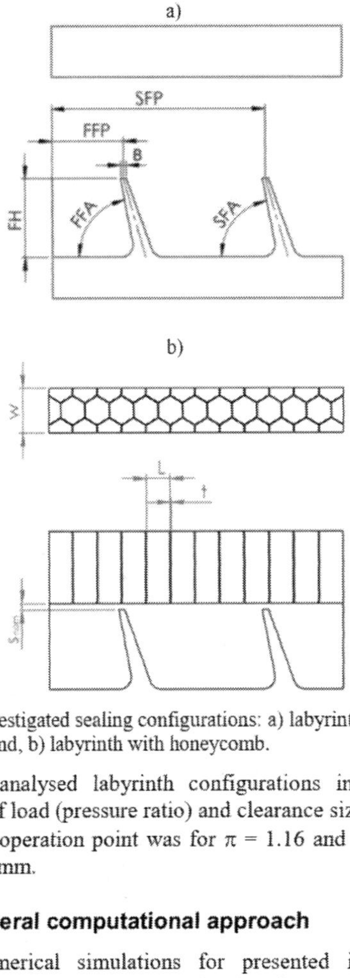

Fig. 1 Investigated sealing configurations: a) labyrinth with smooth land, b) labyrinth with honeycomb.

The analysed labyrinth configurations included a variety of load (pressure ratio) and clearance size, but the nominal operation point was for π = 1.16 and clearance s = 0.75 mm.

2.2 General computational approach

The numerical simulations for presented in Fig. 1 configurations were performed using the ANSYS CFX code. The numerical study presented in the paper focuses on the sealing structure only. The other structures commonly present together with the seals, such as inlet and outlet cavities are not evaluated here. The reason to simplify the computational domain was to keep it the same as in the experimental setup, where the inlet and outlet cavities are not taken into account.

An exemplary model of the labyrinth seal with facing is presented in Fig. 2. The width of the domain model is determined by land structure periodicity. For example, in the honeycomb land, the flow domain width is equal to two honeycomb cells. Therefore, the translational periodicity condition is assumed in the lateral areas of the computational domain. To obtain a uniform velocity profile at the inlet, an extra inlet channel is added to the seal area. Because a strong jet is formed downstream of the last fin clearance, causing a substantial disturbance to the outlet zone and creating large vortex structures, the outlet channel is also lengthened. At the ends of the channel extensions, the inlet and outlet boundary conditions are specified. Generally, the land area and the main channel (sealing, inlet and outlet channel) are prepared as separate domains because of the meshing strategy. Appropriate surfaces of these domains are joined using the General Grid Interface method, which permits non-matching of the element node location. On the domain remaining surfaces (top and bottom ones and on the walls of the honeycomb cells) the adiabatic wall conditions are assumed. The mesh has been checked with respect to its sensitivity to the refinement. More details on the CFD approach applied to this problem was described in the paper [24].

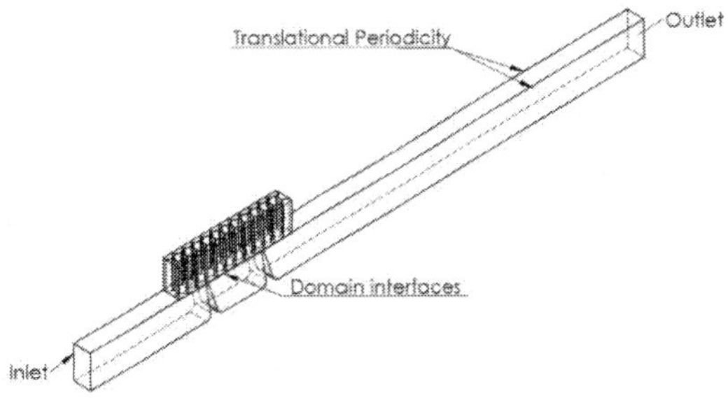

Fig. 2 Computational domain.

The flow is steady and adiabatic. The two-equation k-ω SST turbulence model is chosen. The energy equation includes the viscous work term. The higher-order upwind advection scheme is selected to solve the governing equation system. Following boundary conditions were applied: inlet parameters were specified as total pressure and temperature (101.325 kPa, 288.2 K), at a direction normal to the boundary. Walls were treated as adiabatic and smooth. Static pressure defined the outlet boundary condition, with the blend profile of 0.05. It varied from 50 to 95 kPa. Air treated as an ideal gas, with Sutherland formula for viscosity change was the working fluid. The boundary condition at the outlet is the mean value of static pressure, which depends on the conditions of the seal operation. The labyrinth seal loading is evaluated by pressure ratio π which is defined as:

$$\pi = \frac{p_0}{p_s}, \tag{3}$$

where p_s is outlet static pressure and p_0 is the inlet total pressure.

2.3 Experimental approach

The experiment was realised using a test section with a vacuum installation. A testing approach based on atmospheric air supply guarantees precise determination of stagnation parameters at the inlet section and no thermal deformation of a specimen. The static temperature during the test was in the range of 5 – 25°C, which guaranteed unchangeable sizes of the clearance and constant shape of the fins. The vacuum in the system is generated using a Roots blower with the output of 12.4 Nm3/min. The pressure vessel with the 3 m^3 capacity compensates the pressure fluctuations and stabilises parameters downstream of the test section. On the test section side, the air is sucked in through a 4 m long inlet pipeline. In this point, first flow measurement is performed using the hot wire anemometry probe (HWA). Next, the air flows through the test section. After the air leaves the testing section, it expands and gets into the outlet pipeline, where the second HWA probe and an ISA orifice plate are installed. The mass flow measurement applied in three points guarantees high accuracy of seal leakage determination. It also makes it possible to detect potential leaks in the installation.

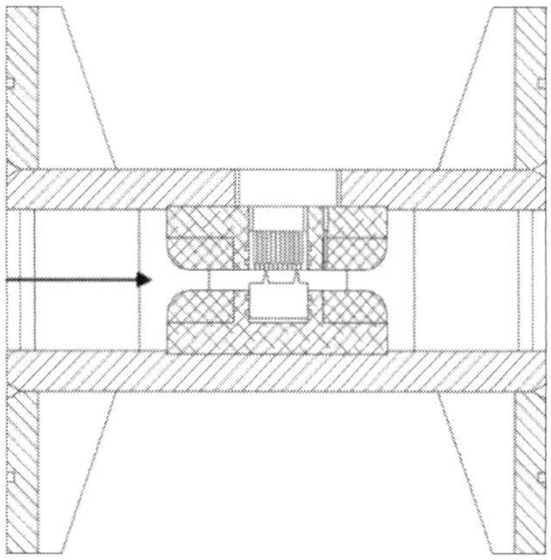

Fig. 3 Test rig cross-section.

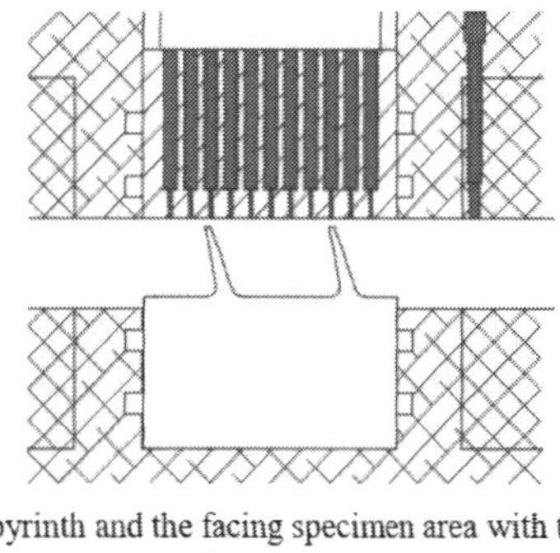

Fig. 4 Labyrinth and the facing specimen area with the holes for pressure measurements.

The cross-section of the rig shows its components, as presented in Fig. 3. A quasi-2D approach of the labyrinth seal was applied – stationary specimens were used without keeping the circumferential curvature. As proved in previous works [19, 25], as well as from own experience, this approach ensures adequate accuracy in determining flow parameters. It can also be related successfully to axially symmetrical geometries. The results of the experimental works [26, 27] proved that a stationary configuration can be used with great accuracy if the flow velocity over the labyrinth fin is higher than the rotor tangential velocity in the labyrinth area ($u/c_{ax} <$ 1 condition is satisfied) which in practice concerns the aero-engine low-pressure turbines as well as heavy-duty turbines because of a relatively low rotational speed.

Apart from determining leakage for a given structure, the test rig also enables the measurement of the air pressure distribution along the labyrinth seal, as presented in Fig. 4. In every case, there are 10 – 12 points in every labyrinth-facing specimen where the static pressure value is measured. The pressure ratio π, defined by the formula (3), is calculated as ambient pressure divided by static pressure measured in the last pressure tap, downstream the labyrinth. The pressure ratio during CFD calculations was established in the same manner i.e. using the static pressure value from the location corresponding with the last pressure tap location. The width of the test rig is 240 mm, what with clearances of 1 mm and less causes that the effect of the side walls boundary layer is negligible. The model adopted for the CFD study differs insignificantly when compared with the geometry deployed in the experiment. There is a small step before the labyrinth Fig. 4, which is a result of the need to regulate the clearance and is very small in relation to the height of the entire channel. The step is located in the bottom part of the channel, which, as it appeared from previous experience, has a negligible effect on the flow in the seal, because the jet flowing on the first fin is formed continuously before the clearance. The influence of the inlet nozzle location and its shape was numerically examined at the design stage of the test rig. The change of C_D value was lower than 0.2% and therefore neglected.

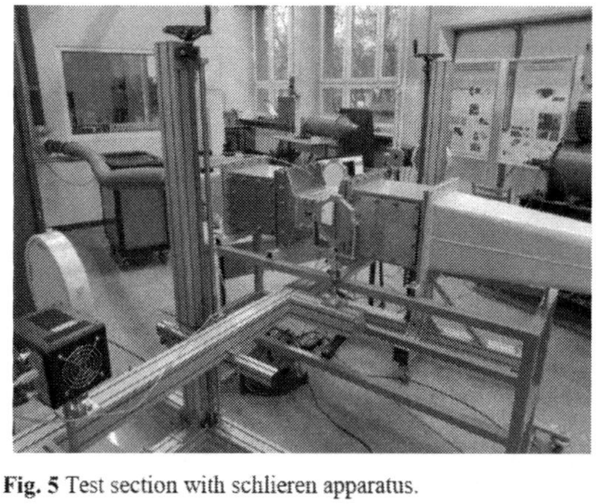

Fig. 5 Test section with schlieren apparatus.

2.4 Schlieren optical flow field measurement

Apart from the mass flow characteristics and pressure distribution along the seal, the test section brings about the flow field visualisation with schlieren technique. The standard implementation of a schlieren system uses light from a single collimated source shining on, or from behind, a target object, lens, schlieren knife and receiver - camera. Variations in refractive index caused by density gradients in the flowing fluid distort the collimated light beam. This distortion creates a spatial variation in the intensity of the light, which can be visualized directly with a shadowgraph system. The main advantage of this method is its non-invasive idea, short time of measurement and simplicity. Fast Speed Camera Phantom

Miro C110 has been deployed to capture the flow field. It achieves up to 800 frames-per-second (fps) at full 1.3 Mpx resolution of 1280x1024, and up to over 29,800 fps at smaller resolutions. In this case, the light source was NANOLITE Nanosecond-Flash lamps. The main idea of application schlieren photography was to observe the vortex and separation structures occurring during the flow through the labyrinth seal, which is the major source of pressure losses. Fig. 5 presents an overview of the schlieren installation in the laboratory and Fig. 6 shows the schlieren setup used for the presented study.

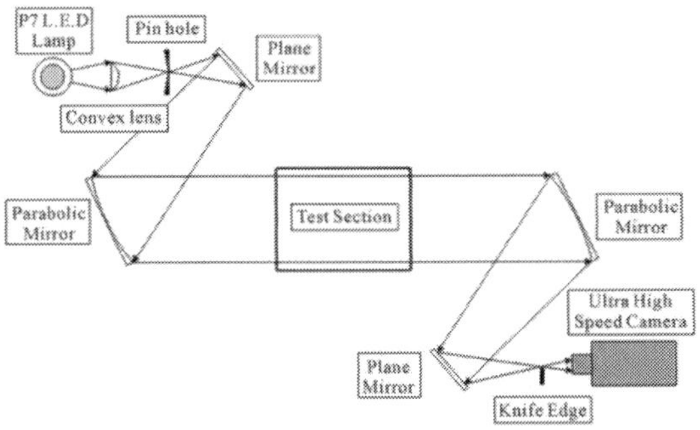

Fig. 6 Schlieren setup scheme.

For the sake of comparison, so-called 'numerical schlieren' results based on the CFD simulations were compared with the experiment. The numerical schlieren is defined as the absolute value of the density gradient on the planar cross section in the flow domain.

Comparison of results for experimental and numerical schlieren for honeycomb land raises some problems. The point is that with this type of land, the flow structure is not uniform in the circumferential direction and changes with the change in the relationship of the honeycomb cell's position and the sealing fin. It means that the jets behind the seal fins and the swirls in the seal chambers will look differently in different cross-sections in the circumferential direction. This will also apply to the density gradient and thus what we see in the images presenting for the schlieren results. In the case of numerical schlieren, the presented results show the distribution of the density gradient in a specific cross-section. In the case of an experiment, the presented image is the result of the refraction of light in subsequent layers of fluid along the entire width of the measuring chamber.

Detailed comparison of obtained images and extracting specific conclusions from them is very difficult and unjustified in this situation. Therefore, the schlieren results presented in the paper are aimed at a very general analysis of emerging structures, observing certain trends and relating them to the numerical results presented in selected axial cross-sections of the seal.

3 Results

3.1 Seal with Smooth land

The C_D value for the nominal conditions obtained from CFD simulations amounts to C_D =0.589. The agreement of CFD with experimental results is substantial (Fig. 7). The result indicates the linear character of changes of the C_D with pressure ratio and clearance.

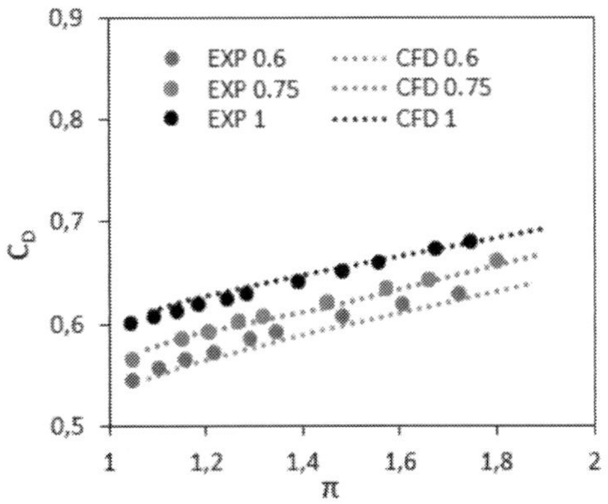

Fig. 7 Labyrinth seal with smooth land – experimental and numerical discharge coefficient in function of pressure ratio.

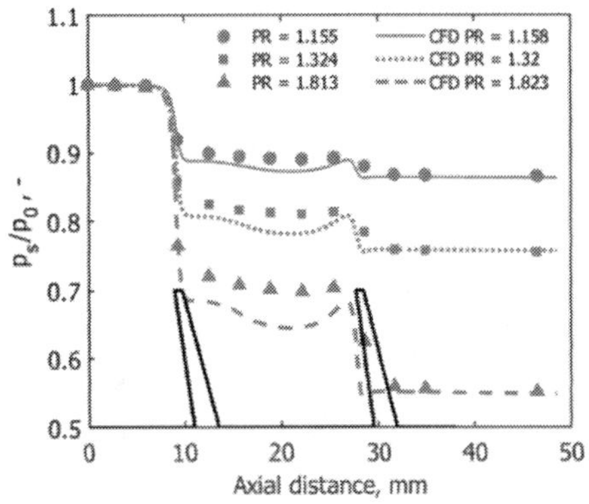

Fig. 8 Experimental and numerical static pressure distribution – labyrinth against smooth land, clearance s = 0.75mm

The pressure distributions for different pressure ratios along the seal are depicted in Fig. 8. The distributions indicate the regions responsible for pressure loss, which is the major mechanism responsible for leakage limitation. They also suggest possible velocities in the cavity and carryover effect. Presented results combine computed and measured pressure distributions. The pressure distributions are shown on the example of a 0.75 mm clearance. The highest pressure drop is always observed at the first fin. When the pressure ratio rises, the second fin contribution in pressure loss generation is more significant. For clearances of 0.6 and 1 mm the distribution looks very similar, the main difference is that as the clearance decreases, the pressure drop on the first fin increases and decreases on the second fin. Overall, the contribution of the second fin is lower than 10% of total pressure drop for low pressure ratios (π < 1.15), and reaches up to 33% (s = 0.75mm) and 18% (s = 1mm) at high pressure ratios $\pi \approx 1.8$, respectively. This means that the carry-over effect increases with the increase of the clearance and Reynolds number.

The experimental and CFD flow visualisation of reference and optimised labyrinth seal is presented in Fig. 9. The figures indicate the density gradient maps at two values of pressure ratio – low (1.16) and high (1.8), for the clearance of 0.75 mm. The same range of parameter was applied to the results presented in the next case. Therefore, some difference may be spotted, i.e. the experiments do not show the circular vortexes between the fins or downstream the second one. The CFD calculations predict the circular one, while the experiment shows significantly stochastic behaviour. Both methods predicted strong bottom vortex in front of the first fin, having a similar shape for both pressure ratios. Both methods predicted expanding jet angle downstream the labyrinth fin tips – which shows the place of higher velocity and density change. The effect is stronger for high-pressure ratios, and the mentioned angle is wider.

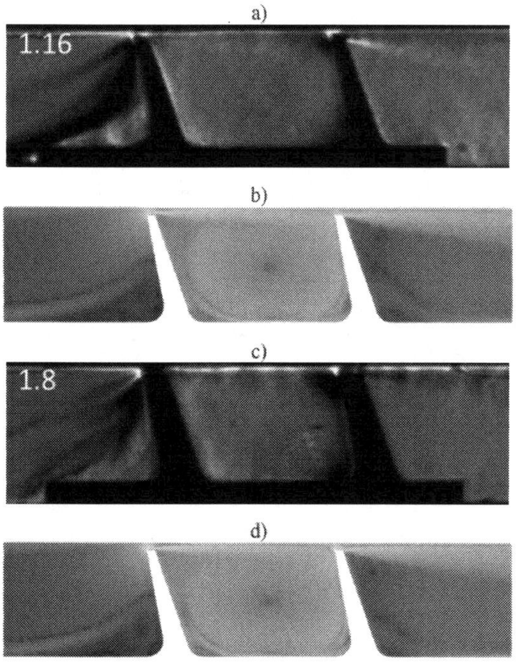

Fig. 9 Experimental a), c) and numerical b), d) schlieren pictures for labyrinth with smooth land, at low (1.16) a), b) and high (1.8) c), d) pressure ratios, clearance $s = 0.75$ mm

3.2 Seal with honeycomb land

The C_D value for the nominal conditions for the honeycomb case amounts to 0.590. The experimental results of the discharge coefficient distributions with pressure ratio match well with the numerically predicted data (Fig. 10). Slightly larger difference can be observed only for the clearance of 0.75. The honeycomb case results revealed behaviour specific for labyrinth seals with cell structures – the discharge coefficient depends highly on clearance size. It is higher for low clearances and reduces with the clearance rise. This behaviour is opposite to trends observed in the configuration with smooth land. Similar behaviour was previously indicated by Alizadeh [9], where the C_D raised by 40-50% after application of honeycomb on the labyrinth seal with three straight fins. The mechanism was also extensively described by Schramm [28], where the experimental results presented even 65% increase of leakage after the honeycomb was applied. This experience shows that using honeycomb may have a negative effect on the mass flow rate.

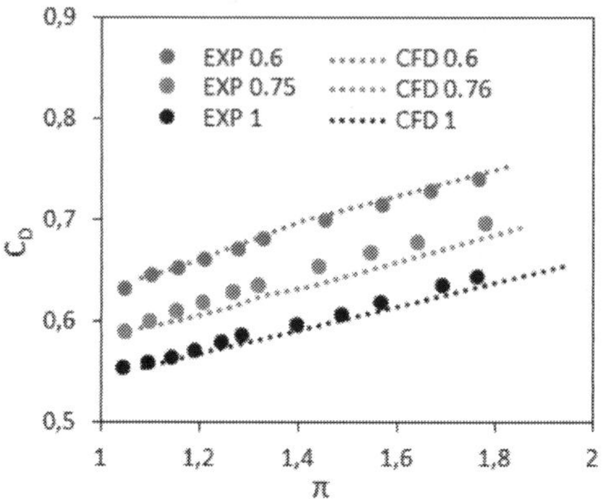

Fig. 10 Labyrinth seal with honeycomb land – experimental and numerical discharge coefficient in function of pressure ratio.

The comparison of experimental and CFD results for pressure distribution in the seal for clearance 0.75 mm is depicted in Fig. 11 for different pressure ratios. The computed results are collected from the top part of the honeycomb land to be in agreement with the positions of the pressure probes. Therefore, the lines of the pressure distribution have characteristic step-like shape, which reflects different pressure levels in the honeycomb cells. In contrast to seal with the smooth land, here with honeycomb land, the second fin plays a more important role in pressure loss mechanism and leakage limitations. The pressure drop on the second fin is around of 50% of the overall pressure drop in the seal.

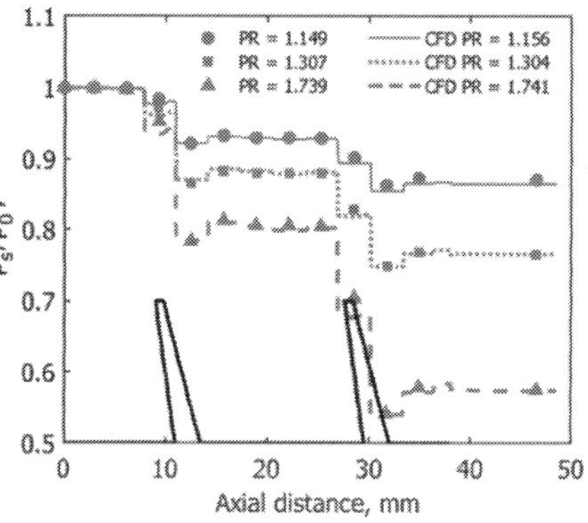

Fig. 11 Experimental and numerical static pressure distribution – labyrinth against honeycomb land, clearance s = 0.75mm.

The shape of pressure distribution in the case of honeycomb land differs from the smooth case. The damping effects of the honeycomb are clearly visible. Moreover, the pressure drop in the area just behind fin tips indicates locally lower pressure values than in area downstream fin tips. The mentioned area corresponds to the honeycomb cell, in which the pressure prevails. The pressure above the fin can be substantially lower than in area downstream of the labyrinth, especially for the higher pressure ratios. The agreement of CFD results with pressure ratio experiment was substantial (Fig. 11). Due to much lower jet velocity, the RANS model predictions were more appropriate than in the structure with smooth land.

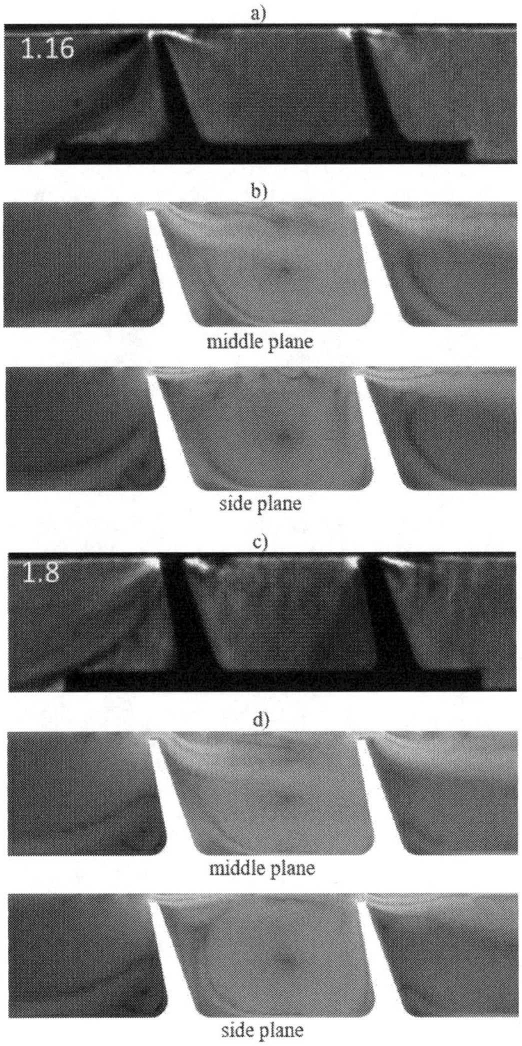

Fig. 12 Experimental a), c) and numerical b), d) schlieren pictures for labyrinth with honeycomb land, at low (1.16) a), b) and high (1.8) c), d) pressure ratios, clearance s = 0.75 mm.

Fig. 12 shows the schlieren and CFD flow visualisation of the labyrinth seal with honeycomb land. Similarly, as in the smooth land case the schlieren photograph presented instantaneous flow field, therefore, some of the time-averaged flow field features may not be visible. As it was explained earlier the schlieren photographs reflect the changes in the flow structure extended along the width of the labyrinth specimen (240mm), in contrast, the numerical schlieren presents the values calculated on the plane. Therefore, numerical schlieren views for every case are represented on two planes. One is located in the middle of the computational domain and the second on the side of the domain. Generally, both planes intersect the honeycomb structure in the middle of cells but the cells are shifted by a half of the cell size in both planes.

Also in the area after the second fin, the flow is highly turbulent. The eddy's shape after the second fin depends on the pressure ratio and reverses at high loads. This reversal flow has an adverse effect on the leakage amount, which is noticeable due to relatively low leakage at high loads, previously indicated in Fig. 10. The experimental schlieren presents much stronger turbulent structures in front of the first fin. In addition, the effects of high value of pressure ratio can be noted, with the air density shadows originating from the trailing edge of the fins. It is noticeable, that due to the honeycomb structure the inclination of the jets just behind fins take place. The jets are headed towards the central part of the cavity, which enables stronger dissipation of kinetic energy.

4 Discussions and conclusions

The characteristics of the discharge coefficient as a function of the pressure ratio for different clearances and the pressure distribution along the seal, for different pressure ratios were presented. Important part was also the flow visualization using schlieren technique. Numerical and experimental investigations were performed for straight-through seal with two leaning fins and smooth or honeycomb land.

The agreement of CFD with experimental results is good concerning discharge coefficient parameter. The result indicates the linear character of changes of the C_D with pressure ratio and clearance. The value of the discharge coefficient increases as the pressure ratio increases. In addition, in the case of a smooth land, the C_D value increases as the gap increases. However, in the case of the honeycomb land, the situation is reversed, the smaller the clearance, the larger the C_D.

When it comes to pressure distributions, better agreement of the numerical simulations with the experiment was obtained for the honeycomb. In the case of smooth land, a small share of the second fin was observed in the overall pressure drop in the seal, which indicates the importance of the carry over effect characteristic of straight-through seals. Honeycomb seems to eliminate this effect because the pressure drop is similar on both fins. Similar conclusions regarding the carry over effect can be drawn by analysing schlieren results. Both in the case of experimental and numerical tests, the jets behind the fins for a smooth land are arranged more horizontally, which enhance the carry over effect, while in the case of honeycomb they deflect downwards, which gives a positive effect.

Schlieren pictures helped to understand the flow behaviour better. The major difference between the schlieren imaging and the CFD schlieren pictures is the fact that schlieren spatially averages the flow field in the direction perpendicular to the flow direction, and shows instantaneous results. The computed by RANS model numerical schlieren shows time-averaged results in selected planar location, therefore some differences can be seen, but the overall flow structures ale similar.

This work was supported by the Polish National Centre for Research and Development and by Avio Polska within the Innolot Programme, Coopernik Project.

References

1. Rosic B., Denton J.D. Control of Shroud Leakage Loss by Reducing Circumferential Mixing, *J Turbomach*, Vol. 130(2), (2008).
2. Chupp R.E., Hendricks R.C., Lattime S.B., Steinetz B.M., Sealing in Turbomachinery, *NASA Technical Report*, NASA/TM-2006-214341, (2006).
3. Steinetz B.M., Hendricks R.C., Braun M.J., Turbomachine Sealing and Secondary Flows, Part 1—Review of Sealing Performance, Customer, Engine Designer, and Research Issues, *NASA Technical Report*, NASA/TM-2004-211991/Part1, (2004).
4. Kong X., Liu G., Liu Y., Zheng L., Experimental testing for the influences of rotation and tip clearance on the labyrinth seal in a compressor stator well, *Aerosp Sci Technol*, Vol. 71 pp. 556–567, (2017).
5. Zhang X., Li W., Zhang X., Wang X., Zhu Y., Chen H., Experimental and Numerical Investigations of Closed Radial Inflow Turbine With Labyrinth Seals, *J Eng Gas Turbines Power*, Vol. 140(10), (2018).
6. Murata K., Kamishita M., Matsumoto K., Kawashita R., Iwasaki M., Tokimasa Y., Validation of CFD Analysis Method for Seal Dynamic Coefficients With Various Labyrinth Seal Types, *Proc ASME Turbo Expo 2018 Power Land, Sea Air*, GT2018-75251, pp. 1-10, (2018).
7. Elebiary K., Jin H., Untaroiu A., Hayrapetiau V., Fu G., The Effects of Fluid Preswirl and Swirl Brakes Design on the Performance of Labyrinth Seals, *J Eng Gas Turbines Power*, Vol. 140(8), (2018).
8. Wasilczuk F., Kaczynski P., Szwaba R., Marugi K., Flaszynski P., Doerffer P., Leakage flow analysis in the gas turbine shroud gap, *Aircr Eng Aerosp Technol*, (2018).
9. Alizadeh M., Nikkhahi B., Farahani A.S., Fathi A., Numerical study on the effect of geometrical parameters on the labyrinth–honeycomb seal performance, *Proc Inst Mech Eng Part G J Aerosp Eng*, pp. 1–12, (2017).
10. Lin Z., Wang X., Yuan X., Shibukawa N., Noguchi T., Investigation and improvement of the staggered labyrinth seal, *Chinese J Mech Eng*, Vol. 28, pp. 402–408, (2015).
11. Cangioli F., Pennacchi P., Nettis L., Ciuchicchi L., Design and Analysis of CFD Experiments for the Development of Bulk-Flow Model for Staggered Labyrinth Seal, *Int J Rotating Mach*, Vol. 2018, pp. 1–16 (2018).
12. Zimmermann H., Wolff K.H., Comparison Between Empirical and Numerical Labyrinth Flow Correlations, *Proc ASME 1987 International Gas Turbine Conference and Exhibition*, 1987, pp. 1–6, (1987).
13. Zimmermann H., Wolff K.H., Air System Correlations Part 1: Labyrinth Seals. *Proc Int Gas Turbine Aeroengine Congr Exhib*, 98-GT-206, (1998).
14. Stocker H.L., Cox D.M., Holle G.F., Aerodynamic Performance of Conventional And Advanced Design Labyrinth Seals With Solid-Smooth Abradable And Honeycomb Lands, *NASA Technical Report*, (1977).
15. Choi D-C., Rhode D.L., Development of a 2-D CFD Approach for Computing 3-D Honeycomb Labyrinth Leakage, *Proc ASME Turbo Expo 2003 Power Land, Sea Air*, GT2003-38238, pp. 965–975. (2003).
16. Wittig S., Schelling U., Kim S., Jacobsen K., Numerical Predictions and Measurements of Discharge Coefficients in Labyrinth Seals, *Proc ASME 1987 International Gas Turbine Conference and Exhibition*, 87-GT-188, (1987).
17. Tyacke J.C., Jefferson-Loveday R., Tucker P.G., Application of LES to labyrinth seals, *Proc 20th AIAA Comput Fluid Dyn Conf*, pp. 1–24, (2011).
18. Tyacke J., Jefferson-Loveday R., Tucker P.G., On the application of LES to seal geometries, *Flow, Turbul Combust*, Vol. 91(4), pp. 827–848, (2013).
19. Wróblewski W., Dykas S., Bochon K., Rulik S., Optimization of the rotor tip seal with honeycomb land in a gas turbine, *9th Eur Conf Turbomach Fluid Dyn Thermodyn ETC 2011 - Conf Proc*, pp. 1–13 (2011).
20. Frączek D., Bochon K., Wróblewski W., Influence of Honeycomb Land Geometry on Seal Performance, *Proc ASME Turbo Expo 2016 Power Land, Sea Air*, GT2016-57569, pp. 1–11, (2016).
21. Rulik S., Wróblewski W., Frączek D., Metamodel-Based Optimization of the Labyrinth Seal, *Arch Mech Eng*, Vol. 64, pp. 75–91, (2017).
22. Wróblewski W., Frączek D., Marugi K., Leakage reduction by optimisation of the straight – through labyrinth seal with a honeycomb and alternative land configurations, *Int J Heat Mass Transf*, Vol. 126, pp. 725–739, (2018).
23. Szymanski A., Bochon K., Wróblewski W., Marugi K., Dykas S., Fraczek D., Optimization of the Straight-Through Labyrinth Seal With a Smooth Land, *J Eng Gas Turbines Power*, Vol. 140(12), pp. 122503-122511, (2018).
24. Frączek D., Wróblewski W., Validation of numerical models for flow simulation in labyrinth seals, *J Phys Conf Ser*, Vol 760, (2016).
25. Morgan N.R., Wood H.G., Untaroiu A., Numerical Optimization of Leakage By Multifactor Regression of Trapezoidal, *Proc ASME Turbo Expo 2015 Power Land, Sea Air*, GT2015-43794, pp. 1–10, (2015).
26. Waschka W., Wittig S., Kim S., Influence of High Rotational Speeds on the Heat Transfer and Discharge Coefficients in Labyrinth Seals, *J Turbomach*, Vol. 114(2), pp. 462-468, (1992).
27. Paolillo R., Moore S., Cloud D., Glahn JA, Impact of Rotational Speed on the Discharge Characteristic of Stepped Labyrinth Seals, *Proc ASME Turbo Expo 2007 Power Land, Sea Air*, GT2007-28248, pp. 1291–1298, (2007).
28. Schramm V., Willenborg K., Kim S., Wittig S., Influence of a Honeycomb Facing on the Flow Through a Stepped Labyrinth Seal, *J Eng Gas Turbines Power* Vol. 124(1), pp140-146, (2002).

RESEARCH & DEVELOPMENT IN POWER ENGINEERING, 2019

Speciation of inorganic gaseous species and condensed phases during coconut husk combustion based on thermodynamic equilibrium calculations

Wojciech Jerzak[1,*], and *Monika* Kuźnia[2]

[1] AGH University of Science and Technology, Mickiewicza 30 Av., 30-059 Krakow, wjerzak@agh.edu.pl, Poland
[2] AGH University of Science and Technology, Mickiewicza 30 Av., 30-059 Krakow, kuznia@agh.edu.pl, Poland

Abstract. Thermodynamic equilibrium calculations to predict coconut husks (CH) combustion products have been carried out in this work. The selected type of biomass belongs to problematic fuels due to the fibrous structure preventing its grinding, and high chlorine content. The calculations results showed, that the combustion temperature for the tested range of 600-1000°C clearly affects the concentrations of chlorine species in the flue gas. When the temperature was below 820°C, the highest concentration had HCl(g), and above 820°C KCl(g). The chlorine was also present in ash, as KCl-NaCl-RbCl solid solution, when the combustion temperature $T \leq 700$°C, and KCl-NaCl-K_2SO_4-Na_2SO_4 liquid solution, when $600 \leq T \leq 960$ °C. High content of chlorine in ash from CH combustion at T = 650°C has been confirmed experimentally. Speciations of inorganic gaseous species and condensed phases we investigated also during flue gas cooling from 1000 to 400°C. Major condensed phase composition were dominated by alkali metal salts in both solid and liquid phase states. Finally, we presented sixteen eutectic points for different binary systems calculated in the FactSage software.

1 Introduction

The long-term European Union policy related to the reduction of CO_2 emissions from the fossil fuels combustion is an incentive for the development of technologies that use waste biomass for energy production. Biomass is considered as a renewable carbon-neutral fuel. However, there are many problems associated with the use of lignocellulosic biomass as a fuel, e.g.: high moisture content, low heating value, high chlorine, sulfur and alkali metal elements contents responsible for fouling, corrosion and slagging, and its fibrous structure makes biomass difficult to grind. The behaviour of the inorganic part of biomass during the combustion process [1-3], and cooling of products gas [4-7], can be determined by thermodynamic equilibrium prediction. The trends of the main solid and liquid phases formation as a function of combustion temperature under equilibrium conditions were investigated by Kaknics et al. [1]. They confirmed, among others that the low temperature of biomass combustion (below 800 °C) favors the formation of the alkali salts: KCl(s), NaCl(s), K_2SO_4(s), Na_2SO_4(s) in ash. Part of the potassium reacts with the biomass-ash in the combustion chamber forming silicates: $K_2Si_4O_6$(s), $Ca_3K_2Si_6O_{16}$(s) or $K_2Si_3MgO_8$(s). In turn, Wei et al. [2] indicated: KCl(g), KOH(g) and K_2SO_4(g) as the main constituents of the flue gas containing alkali. Furthermore, the air excess coefficient has a limited effect on the release of chlorine, potassium and sodium, whereas the increase in pressure can enhances the release of HCl(g) and reduces the formation of KCl(g), NaCl(g), KOH(g) and NaOH(g) at high temperatures. The release of inorganic gaseous species is controlled by the trace elements that compete for chlorine and sulfur [3]. Increased concentration of HCl(g) in the combustion chamber reduces the retention of Cd, Cu, Mn, Zn, Cr and Ni in the biomass-ash. During the cooling of the flue gas, the condensation of alkali salts vapors occurs. The vapors condensed, and can easily form the eutectics e.g. NaCl-KCl, NaCl-Na_2SO_4, with lower melting point than the pure salts [4, 5]. As reported by Froment et al. [6] condensation of K and Cl is favoured, when the total pressure is increased. The condensed phase formed when the gas is cooled contains not only salts, but also metal oxides and hydroxides such as: FeO(s), Fe_2O_3(s), Na_2O(s), NaOH(s,l) an KOH(s,l) [7].

2 Materials and Methods

2.1 Coconut husk

This work focuses on the fate of inorganics during coconut husk (CH) combustion. CH is an agricultural waste from coconut processing. The world leaders in coconut production are: Indonesia, Philippines, India, and Brazil [8]. CH is a high-calorie energy source containing a lot of lignin and cellulose. It has a large potential for use in power plants, due to availability and low price. At present, CH is used as domestic fuel, a fiber source for ropes, for biochar production, and as potential silica source [9, 10]. Table 1 presents proximate, ultimate and trace elements analysis of CH. All analyses were performed with three replicates in the Central Laboratory of Energopomiar in Gliwice, and Table 1 shows the mean values. CH were tested in accordance with PN-EN ISO standards: 18134-2:2017-03 — Moisture, 18122:2016-01 — Ash, 18123:2016-01 — Volatile matter, 16948:2015-07 — C, H, and N, 16994:2016-10 — S, and Cl, 18125:2017-07 — Lower Heating Value (LHV), 16968:2015:07 — Trace elements. Whereas Hg determination was carried out in accordance with EPA Method 7473: 2007. CH contain as much as 1.9% chlorine. The chlorine accumulated in the husk comes from the fertilization system including sodium chloride or potassium chloride [8]. According to CH analysis, the molar ratio 2S/(K + Na) is 0.08. Values below 5 indicate a high propensity to create a deposit consisting of alkali chlorides [11]. In this case, the sulphation process reducing the tackiness of the deposit is very limited. There are also TEs present in CH among others: Rb, Sn, Pb, B, with concentrations above 10 mg/kg of dry fuel.

Table 1. Coconut husk analysis.

Proximate analysis (as received)	**%**
Moisture	8.5
Ash	5.39
Volatile matter	61.5
Calorific Value (dry)	**MJ/kg**
HHV	19.31
LHV	18.22
Ultimate analysis (dry)	**%**
C	49.59
H	5.3
N	0.38
S	0.06
Cl	1.9
O^{diff}	36.87
Trace elements (dry)	**mg/kg**
Zn	7.46
Cu	4.24
Pb	17.1
Ni	1.04
Cr	1.14
Cd	<0.06
Co	<0.1
As	<1
V	<0.5
Sb	<1.5
Ba	9.24
Sr	8.19
Mo	<3.2
Sn	17.8
Rb	47.1
Li	<0.15
B	11.8
Hg	0.008

TEs may react with chlorine and sulfur during combustion, so their identification in biomass plays an important role. As many as eighteen TEs listed in Table 1 were identified in CH, and Cd, Co, As, V, Sb, Mo, and Li were below the measurement range of the apparatus.

2.2 Ash

Ash from CH was obtained by biomass combustion at 650°C in an electric heated tube furnace. Chemical composition of ash included in Table 2 presents the average values from the three samples, and has been determined with the method of inductively coupled plasma optical emission spectrometry (ICP−OES) with the use of a plasma spectrometer Thermo iCAP 6500 Duo ICP. Chlorine and carbonates were determined according to the following standards: PN-EN 196-2: 2013-11, and PN-EN 15936: 2013-02. As can be seen from Table 2, silicon and potassium oxides predominate in the ash composition. Chlorine contained in ash accounts for 10.8 % of mass ash, which is in line with the value obtained by other researchers [12].

Table 2. Ash composition from CH at 650 °C

Component	**%**
SiO_2	31.6
Fe_2O_3	11.9
Al_2O_3	3.0
Mn_3O_4	0.10
TiO_2	0.30
CaO	2.33
MgO	2.19
P_2O_5	1.60
SO_3	1.12
Na_2O	4.82
K_2O	27.5
CO_2	1.5
Cl	10.8

2.3 Software and thermochemical database

Thermodynamic equilibrium calculations were performed in the Equilib module of the FactSage 6.3 commercial software. Thermodynamic equilibrium predictions are based on the minimizing of the total Gibbs free energy:

$$G = \sum_{\substack{ideal\\gas}} n_i\left(g_i^{\circ} + RT\ln p_i\right) + \sum_{\substack{pure\\condensed\\phases}} n_i\, g_i^{\circ} + \sum_{solution\ 1} n_i\left(g_i^{\circ} + RT\ln X_i + RT\ln \gamma_i\right) + \sum_{solution\ 2} n_i\left(g_i^{\circ} + RT\ln X_i + RT\ln \gamma_i\right) + \cdots \quad (1)$$

where:

$n_i, g_i^\circ, X_i, \gamma_i$ – successively moles, standard molar Gibbs energy, mole fraction, and activity coefficient of component "i",

p_i – partial pressure of the gas component "i",

R – universal gas constant,

T – temperature.

The system achieves thermodynamic equilibrium state, when:

$$G = min., \quad \text{or} \tag{2}$$

$$dG = 0 \tag{3}$$

The equilibrium criterion in equation (2) is mainly valid in multi-phase multicomponent systems.
In Equilib module, the following thermochemical databases were selected for pure components and solutions: FactPS, FToxide, ASalt-liq, FTpulp, and FTsalt. In addition, the Phase Diagram module was used to determine eutectic points in binary systems.

2.4 Calculation procedure

The calculation procedure covers two cases. The first is the combustion of 1 kg dry biomass, in the temperature range from 600 to 1 000 °C with a 20 °C steps, and λ=1.2. In the second case, the flue gas are cooled from 1 000 to 400°C, also with a 20°C step.

3 Result and discussion

3.1 Speciation of minor inorganic gaseous species calculated at different combustion temperature

The main flue gas components from CH combustion in the studied temperature range are: N_2>CO_2>H_2O>O_2>Ar. In addition, minor inorganic gaseous species are also present in the flue gas, the concentration of which varies with the combustion temperature. Figure 1 shows flue gas components present in minor quantities, i.e. ≥1 mg/m^3 wet flue gas. The fuel-bond chlorine during CH combustion reacts with potassium, hydrogen, sodium, and rubidium. When the combustion temperature is in the range of 600-780 °C, the flue gas are dominated by HCl. Increased concentration of HCl in the combustion chamber reduces the retention of Cd, Cu, Mn, Zn, Cr and Ni in the biomass-ash [3]. HCl have an inhibiting effect on biomass burning and CO → CO_2 oxidation [13]. HCl can easily be removed from the flue gas by a scrubbing process [13], or using an alkaline-carbonate sorbent [14], therefore, it belongs to the desired products containing chlorine. The following chlorides are also present in the flue gas: KCl, NaCl, RbCl, as dimers K_2Cl_2 and Na_2Cl_2. An increase in the combustion temperature above 780 °C, clearly promotes release of KCl, K_2Cl_2, NaCl, and reduction of HCl. As a result, the KCl concentration is fivefold as high as HCl at 1000 °C. Due to its high propensity to form aerosols in the cooling process, KCl is an undesirable constituent of flue gas, that contributes to deposition and corrosion. Apart from NO, Na_2Cl_2 and SO_2 in flue gas two trace elements (Rb, and B) can also be observed as RbCl, H_3BO_3 and KBO_2. The positive correlation of the increase in the volume of flue gases with temperature (resulting from the decrease in density) is the reason for the reduction of RbCl concentration in Fig. 1.

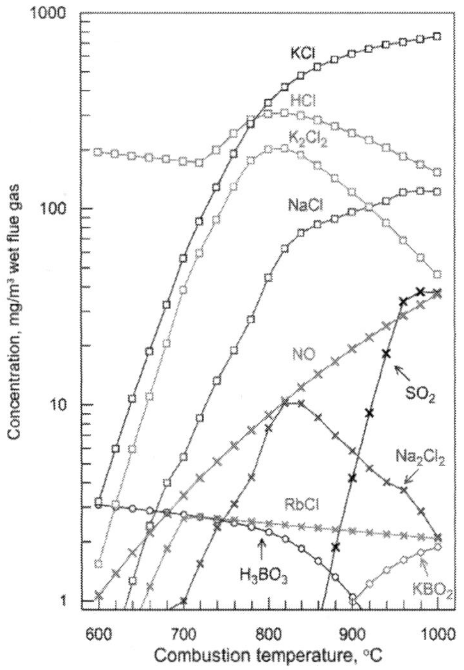

Fig. 1. Speciation of minor inorganic gaseous species (≥1 mg/m^3) during CH combustion in air.

3.2 Speciation of the condensed phase during CH combustion

The condensed phases shown in Fig. 2 are divided into three groups: a) alkali metal salts (solid and liquid), b) other solid salts, and c) solid oxides. Caring for the readability of Fig. 2, ash components with concentrations below 0.1 g/kg ash, i.e. $BaCrO_4$ and $NiTiO_3$, were not included. Although the low combustion temperature (<700 °C) inhibits potassium and sodium evaporation, however, these alkali are susceptible to forming the liquid solution KCl-NaCl-K_2SO_4-Na_2SO_4 already at a temperature of 600 °C. As known, chlorides and sulphates present in the ash, they interact with each other, and the melting temperature of the mixture is lower than the melting temperature of the individual salts [15]. The maximum total mass of the liquid salt solution was achieved at 720 °C, and the disappearance of the liquid phase at 980 °C. Potassium, sodium and rubidium chlorides in their solid state are formed only in the temperature range of 600 – 700 °C. According to Fig. 2b), the masses of individual solid ash components are conditioned by the combustion temperature. Only three solid salt compounds are present in the ash in the entire temperature range tested, ie. KAl_2SiO_6, $Na_2Ca_3Si_6O_{16}$, and $SrTiO_3$. Other salts are contained in the ash only in specific temperature ranges, with noticeable

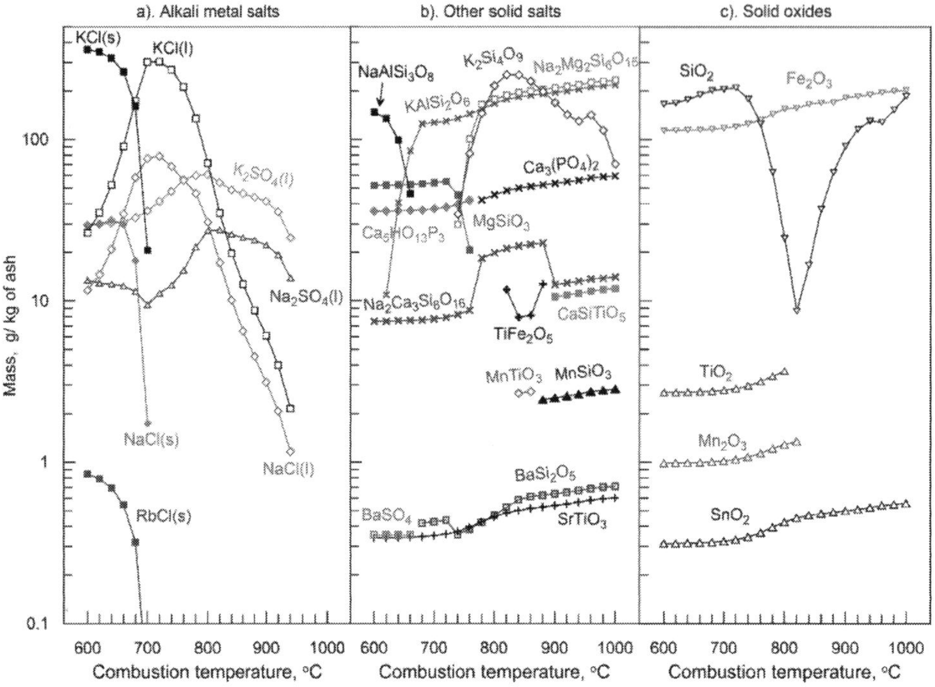

Fig. 2. Speciation of the condensed phase during CH combustion: a). alkali metal salts, b). other solid salts, and c). solid oxides.

transformations, e.g. $Ca_5HO_{13}P_3$ (T ≤ 860 °C) → $Ca_3(PO_4)_2$ (T ≥ 880 °C). In addition to salt, solid oxides are also thermodynamic predicted in ash: SiO_2, Fe_2O_3, TiO_2, Mn_2O_3, and SnO_2. The minimum SiO_2 concentration in Fig. 2c) occurs at the same temperature (820 °C) as the maximum $K_2Si_4O_9$ in Fig. 2b). The formation of potassium silicates depends on the molar ratios between the reactants and the gas atmosphere. When SiO_2:K_2CO_3 or SiO_2:KCl molar ratios is 4 or more, the formation of $K_2Si_4O_9$ occurs according to the reactions [16,17]:

$$4SiO_{2(s)} + K_2CO_{3(s)} \rightleftharpoons K_2Si_4O_{9(g)} + CO_{2(g)} \qquad (4)$$

$$4SiO_{2(s)} + KCl_{(l)} + H_2O_{(g)} \rightleftharpoons K_2Si_4O_{9(g)} + 2HCl_{(g)} \qquad (5)$$

3.3 Speciation of the condensed phase during the flue gas cooling

Condensable components during the flue gas cooling, was shown in Figs. 3a-b). Major condensed phase composition is dominated by alkali metal salts in both solid and liquid phase states. The liquid KCl-NaCl-K_2SO_4-Na_2SO_4 salt solution begins to form after the flue gas has been cooled to 960 °C. The flue gas cooling promotes an increase in the mass of liquid salt solution up to a temperature of 700 °C. The equilibrium between alkali (Me = K, or Na) chlorides and sulphates is determined on the basis of a series of reactions in which reactions (6) and (7) plays a leading role [18]:

$$2MeCl_{(g)} + H_2O_{(g)} + SO_{2(g)} + \tfrac{1}{2}O_{2(g)} \rightleftharpoons Me_2SO_{4(l)} + 2HCl_{(g)} \qquad (6)$$

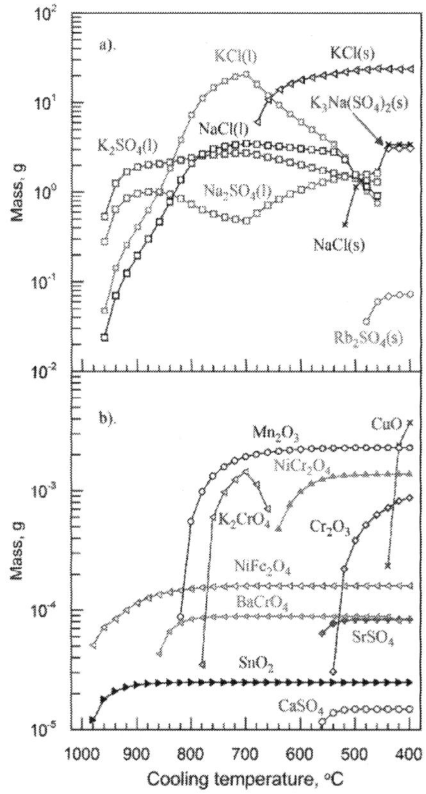

Fig. 3. Speciation of the condensed phase during the flue gas cooling: a). major condensed species, b). minor solid species.

$$2MeCl_{(g)} + SO_{2(g)} + O_{2(g)} \rightleftharpoons Me_2SO_{4(l)} + Cl_{2(g)} \quad (7)$$

As a result, the cooled flue gas contains less NaCl(g) and KCl(g), while more HCl(g) and $Cl_{2(g)}$. Further HCl(g) reacts with oxygen, according to the Deacon reaction:

$$2HCl_{(g)} + \tfrac{1}{2}O_{2(g)} \rightleftharpoons Cl_{2(g)} + H_2O_{(g)} \quad (8)$$

and the content of Cl_2 increases in the flue gas.
According to Fig. 3b), minor solid species (< 0.01g) are also forming in the flue gas cooling process. The cooler the flue gas, the more species it condenses (for example, at 980°C there are only two species, while at 420 °C as many as nine. Interestingly, based on thermodynamic prediction, condensed species form both salts and metal oxides. In the entire studied temperature range, only $NiFe_2O_4$ and SnO_2 have condensed. The remaining condensed components were only present in specific temperature ranges.

3.4 Eutectic points

In order to supplement the analysis of the occurrence of liquid species in the process of CH combustion and exhaust gas cooling, additional calculations were performed using the Phase Diagram module, determining eutectic points in binary systems. All calculations were carried out for a pressure of 1 atm, and the shares of components in binary systems were given in mass %. Calculations were made for sixteen binary systems, and the system with the lowest liquid temperature is presented in Fig. 4. The eutectic point for the $FeCl_2$–KCl phase diagram is determined by the composition of 48% mass KCl and 52% mass. $FeCl_2$ at 351°C.

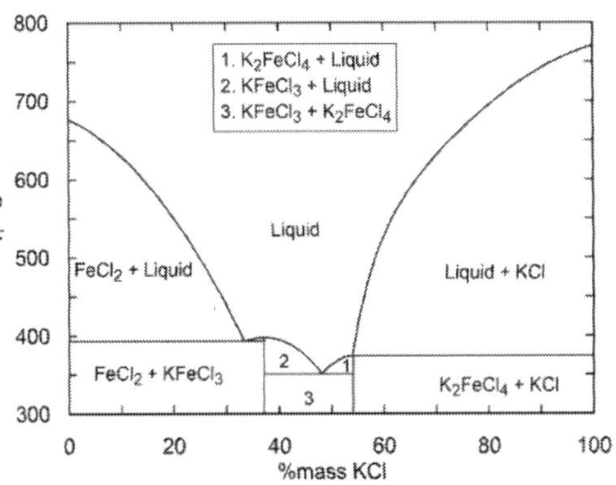

Fig. 4. Binary phase diagram of $FeCl_2$–KCl – calculated in the FactSage software.

The calculation results for selected binary systems are summarized in Table 3. The values are ordered according to lowest melting temperature, corresponding to the eutectic point. At this point, it should be emphasized that only binary phase systems are listed in Table 3, for which thermodynamic databases have already been developed. The authors of this article are aware of the need for further research in order to create databases for other systems, eg KCl-Na_2SO_4 (515 °C, 57% mass KCl) [19].

Table 3. Eutectic points for different binary systems calculated in the FactSage software.

System	Eutectic temperature, °C	Composition at eutectic point, mass %
KCl–$FeCl_2$	351	48
K_2CO_3–KOH	365	20
K_2SO_4–KOH	385	17
KCl–KOH	392	1→20
KCl–$PbCl_2$	408	20
K_2SO_4–Li_2SO_4	549	26
NaCl–RbCl	550	27.6
K_2O–SiO_2	584	61
KCl-$CaCl_2$	601	67
K_2O–Na_2O	602	84
KCl–K_2CO_3	617	47
NaCl-Na_2SO_4	625	32
KCl–NaCl	657	58.6
KCl–K_2SO_4	691	56
Na_2O –Na_2SO_4	781	11
K_2SO_4–Na_2SO_4	824	30

4 Conclusions

In this work, thermodynamic equilibrium calculations were performed in the FactSage 6.3 software, with the customized thermodynamic database for predicting the elemental composition of gaseous species and condensed phases under different temperature. Based on thermodynamic prediction, it has been shown that the dominant inorganic gaseous species of flue gas are: KCl(g), HCl(g), K_2Cl_2(g), and NaCl(g). Interpreting the calculation results of the combustion process, three groups were distinguished among condensed phases: a) alkali metal salts (solid and liquid), b) other solid salts, and c) solid oxides. KCl(s), NaCl(s) and RbCl(s) are formed only in the temperature range of 600 – 700 °C. It has been shown, that solid and liquid ash components are dependent by the combustion temperature. The consequence of flue gas cooling was the formation of a condensed phase. The phase composition was dominated by alkali metal salts in both solid and liquid states. The more strongly the exhaust gases were cooled, the more species of condensed phases therein were revealed. Finally, we analyzed sixteen binary systems for which eutectic points were determined.

This work was financially supported by the AGH University of Science and Technology (D.S. no. 16.16.110.663).

References

1. Kaknics J., Defoort F., Poirier J., *Energy Fuels*, Vol. 29, pp. 6433–6442, (2015).

2. Wei X., Schnell U., Hein K.R.G., *Fuel*, Vol. 84, pp. 841–848, (2005).
3. Miller B., Dugwell D.R., Kandiyoti R., *Energy Fuels*, Vol. 17, pp. 1382-1391, (2003).
4. Wang Y., Tan H., Wang X., Cao R., Wei B., *Fuel*, Vol. 187, pp. 33–42, (2017).
5. Lindberg D., Niemi J., Engblom M., Yrjas P., Laurén T., Hupa M., *Fuel Processing Technology*, Vol. 141, pp. 285–298, (2016).
6. Froment K., Defoort F., Bertrand C., Seiler J.M., Berjonneau J., Poirier J., *Fuel*, Vol. 107, pp. 269–281, (2013).
7. Wan W., Engvall K., Yang W., Fredriksson Möller B., *Energy*, Vol. 153, pp. 35–44, (2018).
8. Krishnakumar V., Thampan P.K., Achuthan Nair M., *The Coconut Palm (Cocos nucifera L.) - Research and Development Perspectives*, Springer Nature Singapore, (2019).
9. Suman S., Gautam S., *Energy Sources, Part A: Recovery, Utilization, and Environmental Effects*, Vol. 39(8), pp. 761–767, (2017).
10. Anuar M.F., Fen Y.W., Mohd Zaid M.H., Matori, K. A., Mohamed Khaidir R. E., *Results in Physics*, Vol. 11, pp. 1–4, (2018).
11. Robinson A.L., Junker H., Baxter L.L., Energy Fuels, Vol. 16, pp. 343-355, (2002).
12. Bonneau X., Haryanto I., Karsiwan T., Experimental Agriculture, Vol. 46 (3), pp. 401–414, (2010).
13. Glarborg P., *Computer Aided Chemical Engineering*, Vol. 45, pp. 603-645, (2019).
14. Ren X., Rokni E., Liu Y., Levendis Y.A., *Journal of Energy Engineering*, Vol. 144(4), 04018045, (2018).
15. Billen P., Van Caneghem J., Vandecasteele C., *Waste and Biomass Valorization*, Vol. 5, pp. 879–892, (2014).
16. Anicic B., Agglomeration Mechanisms during Fluidized Bed Combustion of Biomass. Kgs. Lyngby: Technical University of Denmark (DTU), (2018).
17. Defoort F., Dupont C., Durruty J., Guillaudeau J., Bedel L., Ravel S., Campargue M., Labalette F., Da Silva Perez D., Energy Fuels, Vol. 29, pp. 7242-7253, (2015).
18. Scandrett L.A., Clift R., *Journal of the Institute of Energy*, Vol. 57(433), pp. 391–397, (1984).
19. Rowe J.J., Morey G.W., Zen C.S., The Quinary Reciprocal Salt System Na, K, Mg, Ca/Cl, SO_4 – A Review of the Literature with New Data, Geological survey professional paper 741, US Govt. Print. Off., (1972).

 https://doi.org/10.1051/e3sconf/201913701010

RESEARCH & DEVELOPMENT IN POWER ENGINEERING, 2019

Pump cavitation in feed water systems - remedies

Krzysztof Karaśkiewicz[1,*]

[1] Warsaw University of Technology, 00-665 Warsaw, Nowowiejska Str. 21/25, krzysztof.karaskiewicz@itc.pw.edu.pl, Poland

Abstract. The Net Positive Suction Head incipient is a multiple of the Net Positive Suction Head Required in centrifugal pumps. With limited static heads of the feedwater systems, the suction impellers of the feed pumps work in cavitation. The article discusses the ways to deal with the problem of cavitation in such systems.

1 Introduction

Cavitation in industrial pumps reduces the head, efficiency, leads to excessive vibration and finally cuts the impeller® operating time off due to damage caused by cavitation erosion. This phenomenon is caused by excessive static pressure drop in the pump suction nozzle. Fig. 1 shows the stages of cavitation and their effect on the pump head while maintaining a constant flow.

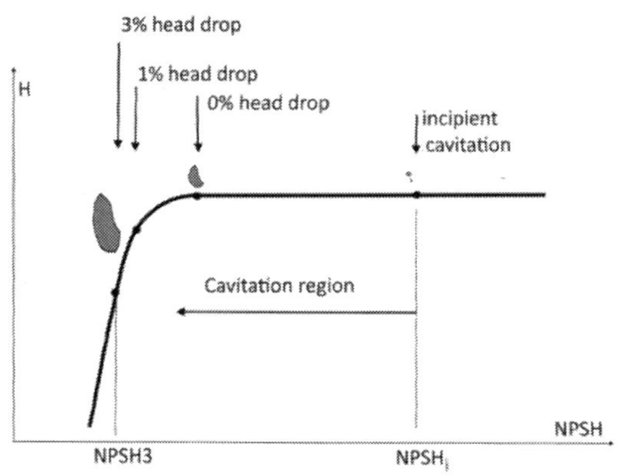

Fig. 1. Stages of cavitation at constant flow rate.

The static pressure before the pump is determined by the characteristics of the suction side of the pumping system determined by the Net Positive Suction Head available.

2 Net Positive Suction Head Available of a feedwater system

A diagram of a typical steam generation system in power plants and hot water generation system in an industrial plant is shown in Figure 2.

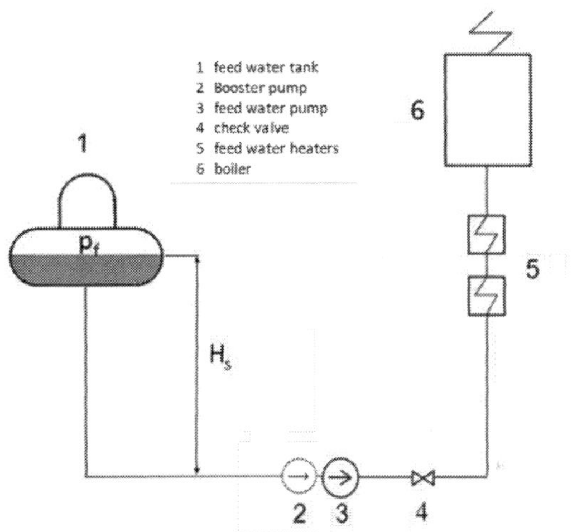

Fig. 2. Feedwater system

The low-pressure side of the system consists of a feedwater tank with a degasser, a suction pipeline with fittings (filter, elbows, etc.), feed water pump with or without a booster pump.

The water in the feed pump is of the temperature above 120°C. This affects the vapor head, which for such temperatures is $H_v > 22$m.

The disposable excess of the system is determined by the difference between the total amount of energy at the end of the suction system (inlet to the feed pump) and the evaporation height. It can be converted to the form:

$$NPSHA = (H_f - H_v + H_s) - H_{str} \quad (1)$$

wherein: $H_{str} \cong a(Q/Q_n)^2$

Taking into account that the pressure in the feed water tank nearly equals vapor pressure $H_f \cong H_v$

$$NPSHA \cong H_s - a(Q/Q_n)^2 \quad (2)$$

Typical static heads in industrial plants are $H_S = 25{\div}30$m. The losses for optimal flow usually do not

exceed a few meters, hence the typical values of hydraulic resistance in equation (2) $a = 2 \div 3$m.
The *NPSHA* characteristic limits the pump operation. To avoid cavitation, the Net Positive Suction Head Required must be smaller.

3 Net Positive Suction Head Required margin

A commonly used cavitation characteristics in industry is a 3% drop in head, i.e. *NPSHR* = *NPSH3*. Manufacturers provide the *NPSH3* characteristic for rated speed. For a suction impeller of a suction specific speed $n_{ss} = 230$ (single-suction $i = 1$ commercially available pump with a specific speed $n_q = 20$), rated rotational speed $n = 5000$ 1 / min, and flow rate $Q_{BEP} = 520\text{m}^3/\text{h}$, *NPSH3* can be expected:

$$NPSH3 = \left(\frac{n}{n_{ss}}\right)^{4/3}\left(\frac{Q_{BEP}}{i}\right) = 17m \quad (3)$$

This is a relatively large value. Such *NPSH* can be reduced in three ways:
1. Use of impeller of larger suction specific speed. For the suction specific speed $n_{ss} = 280$ and the same rated pump parameters, *NPSH3* ≅ 13m can be obtained.
2. Use double suction impeller with the same parameters n, n_{ss}, Q_{BEP} . It allows to obtain *NPSH3* 20% smaller.
3. Reducing the rotational speed of the pump while increasing the number of stages and impellers diameter to maintain the required head. For example, reducing the speed down to $n = 3500$ 1 / min reduces the Net Positive Suction Head to *NPSH3* ≅ 10m.
To avoid cavitation some margin is recommended. The NPSH margin is different depending on source literature, for example [1] gives:

$$\text{NPSH margin} = \begin{cases} 25\% \\ 0{,}6m \end{cases} \quad \text{which value is greater}$$

The Hydraulic Institute [2] recommends:

Table 1. Examples of NPSH margin by HI.

Industry	Application	NPSH Margin
Petroleum / hydrocarbon process	Typical, except vertical canned pumps	10% or 1.0m
Chemical process	Typical	10%÷20% OR 0,6m÷1,0m
Electric power generation	Circulating / cooling water	1,0 m
Electric power generation	Boiler feed < 250 kW/stage	30%
Water	Typical, stainless steel or aluminum-bronze impeller, < 75 kW/stage	10% or 1.5m
Building services	Typical for pumps in open systems (not pressurized)	Up to 10% or 0.6m
General	Often a standard catalog pump	10% or 1,0m

However, they are larger for high energy pumps.
McGuire [3] reports:

Table 2. Examples of NPSH margin according to [3].

Application	NPSH Margin
Water, Cold	10÷35%(1),(2)
Hydrocarbon	10%(2)
Boiler feed, small	50%(3)
High energy	200÷300%(4)

(1) Higher margin for larger pumps
(2) minimum 1m
(3) up to 1900kW at n=3600 rpm
(4) U_1 greater than 30 m/s

The correct pump operation is defined by *NPSHA* ≥ *NPSH3* + NPSH margin but avoiding cavitation imposes a stricter condition $NPSHA \geq NPSH_i$.

4 Incipient cavitation and NPSHi/NPSH3 ratio

To determine the safe pump operation area without cavitation, the ratio of *NPSH* for incipient cavitation and 3% cavitation is needed.
Figure 3 shows such ratio for pumps with a specific speed from $n_q = 16$ to $n_q = 90$.

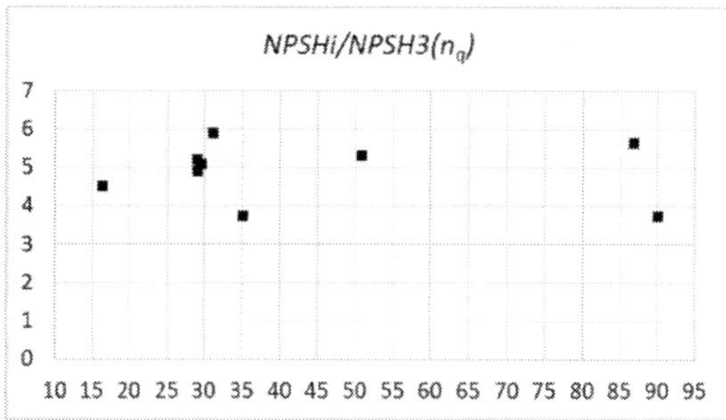

Fig. 3. NPSHi/NPSH3 ratio for the data from [4,5,6,7] as a function of specific speed.

The ratios range between about 4 and 6. For very good suction impellers, they are lower and range from 2 to 3 [1,4]. Figure 4 shows *NPSH* ratios for a typical feedwater system with the static head $H_S = 25$m, a feedwater pump with a double-suction impeller and 25% margin, and $NPSH_i$ for incipient cavitation. All is related to $NPSH3_{BEP} = 10$m. $NPSH_i$ values as a function of flow rate were estimated based on the studies in [4,5,6,7].

 https://doi.org/10.1051/e3sconf/201913701010

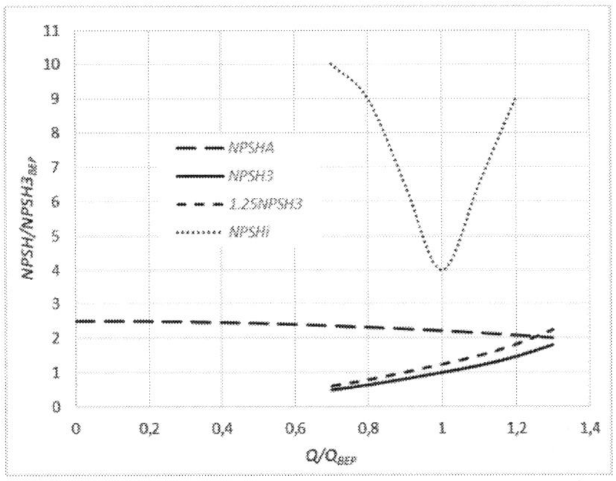

Fig. 4. *NPSH* curves for feed water pump system in small power plant

The field below the $NPSH_i$ curve is the cavitation area so feedwater pump operates in cavitation because of $NPSHA \leq NPSH_i$.

5 Feedwater system with booster pump

The use of a low-rotational speed booster pump significantly improves the suction conditions of the feed water pump. The booster pump power is often around 5% of the feed water pump power. With efficiency of over 80%, additional losses associated with booster pump operation reduce the efficiency of pumping into the boiler by less than a percent. With a booster pump's head of 6 to 8 times the *NPSH3* of feed water pump the latter can achieve a wide range of operation without cavitation. The simple structure of the booster pump allows for its long trouble-free operation.

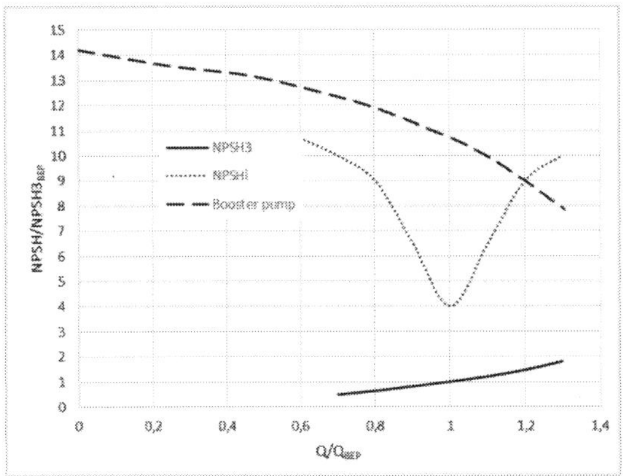

Fig. 5. Impact of the booster pump on a feed water pump operation

Figure 5 shows *NPSH* of the system with a booster pump where non-cavitation operation of the feed pump is guaranteed for flow rates $Q < 1.2Q_{BEP}$. The presence of the booster pump slightly increases system complication and decreases efficiency, but extends trouble-free operation time by over 100,000 hours. Despite the tendency to eliminate booster pumps, the final decision should be preceded by LCC analysis.

6 New materials for suction impeller

At existing static heads of the system, without the booster pump the suction impeller of the feed pump operates in cavitation. The conditions for suction impellers material is operation for 40,000 hours. Typical chromium cast steels traditionally used, e.g. GX20Cr14, does not meet this condition but alloys are offered whose cavitation erosion resistance is at least ten times larger. Table 3 contains materials ordered by increasing cavitation resistance according to KSB Lexicon

Table 3. Cavitation erosion: Materials graded by increasing cavitation resistance; weight loss index for typical cast metals (based on grey cast iron JL 1040 with an index of 1.0)

Name	Type	indeks
Cast steel	GP240GH+QT	0.8
Tin bronze	CC480K-GS	0.1
Cast chrome steel	GX20Cr14	0.2
Aluminium multi-alloy bronze	CC333G-GC	0.1
Cast chrome nickel steel	GX5CrNi19-10	0.05
Noridur	GX3CrNiMoCuN24-6-2-3	0.02

Leading producers offer cast steels with increased resistance to cavitation erosion e.g. Noridur and Noriclor (KSB) or X-Cavalloy (Flowserve). They are compared in table 4.

Table 4. Commercial cast steels resistant to cavitation erosion.

	Noridur	Noriclor	X-Cavalloy
C	0.04	0.04	0,1
Si	1.5	1.0	0,5
Mn	1.5	1.0	15,5
Cr	23.0–26.0	22.0–25.0	18
Ni	5.0–8.0	4.5–6.5	0,5
Mo	2.0–3.0	4.5–6.0	-
Cu	2.75–3.5	1.5–2.5	-
N	0.10–0.2	0.15–0.25	0,25

Noridur and Noriclor are based on chrome, nickle and molybdenum whereas X-Cavalloy distinguishes by a large share of 15.5% Manganese.

7 Conclusion

The feed pumps operation in power plant systems takes place at static heads of 25÷30m resulting from typical elevation heights of the feed water tank.

Even for suction impellers with large suction specific speed and relatively large NPSH margin it leads to cavitation operation. One way is to avoid cavitation by

using a booster pump. The reliability of single-stage, double-suction pumps working usually as booster pumps is very high due to the lack of balancing system, low rotational speed, and simple structure. The disadvantage of the system with a booster pump is slightly lower efficiency reaching a fraction of a percent. Therefore, the assessment of the system operation with or without a booster pump should be preceded by LCC analysis.
Without a booster pump, an acceptable impeller lifetime of 40,000 hours requires materials with high resistance to cavitation erosion. New cast steels with erosion resistance more than ten times higher than typical chromium cast steels are the answer to these needs. However, further material advancement extending impeller lifetime is required.

References

1. Gulich J.F. - Centrifugal Pump - *Springer-Verlag*, Berlin Heidelberg, (2008)
2. *ANSI/HI 9.6.1–2012* Rotodynamic Pumps Guideline for NPSH Margin
3. McGuire, J. T., So You Need Pumps for a Revamp!, *Pumps & Systems Magazine*, August (1996).
4. Schiavello B., Visser F.C. - Pump cavitation – various NPSHR criteria, NPSH margins, and impeller life expectancy - *Proceedings of the twenty-fifth 138 International Pump Users Symposium*, (2009)
5. Md Rakibuzzaman,·Sang-Ho Suh, Hyoung-Ho Kim, Young-Hoon J - Relationship between Cavitation Incipient and NPSH Characteristic for Inverter Drive Centrifugal Pumps - *The KSFM Journal of Fluid Machinery* - Vol. 18, No. 6, pp.76-80, (2015)
6. Sedlar M., Sputa O., Komarek M - CFD Analysis of Cavitation Phenomena in Mixed-Flow Pump, *International Journal of Fluid Machinery and Systems*, Vol. 5, No. 1, January-March (2012)
7. Sedlár M., Komárek M., Vyroubal M., Doubrava V., Varchola M., Hlbocan P. - Experimental and numerical analysis of unsteady behaviour of high efficiency mixed-flow pump - , *EPJ Web of Conferences 67*, 02104 (2014)
8. KSB Lexicon - https://www.ksb.com/centrifugal-pump-lexicon/cavitation-erosion/328182/
9. McCaul C. - A cavitation-resistant casting alloy for pumps - status after ten years' commercial use – in *Advanced Materials for Fluid Machinery*, Wiley, (2004)

E3S Web of Conferences **137**, 01011 (2019) https://doi.org/10.1051/e3sconf/201913701011
RDPE 2019

RESEARCH & DEVELOPMENT IN POWER ENGINEERING, 2019

Experimental and numerical studies of sound generated by cavities

Sebastian Rulik[1,*], Włodzimierz Wróblewski[2], Mirosław Majkut[3], Michał Strozik[4], Krzysztof Rusin[5]

[1,2,3,4,5] Silesian University of Technology, Poland, sebastian.rulik@polsl.pl, wlodzimierz.wroblewski@polsl.pl, miroslaw.majkut@polsl.pl, michal.strozik@polsl.pl, krzysztof.rusin@polsl.pl

Abstract. Cavities and gaps are an important element in the construction of many devices and machines, including energy sector applications. This type of flow is usually coupled with strong pressure fluctuations inside the cavity, which are emitted into the far field in the form of a sound wave responsible for the noise generation. This applies to both subsonic and supersonic flows. Pressure fluctuations often have the character of single tones of a specific frequency and high amplitude and their generation is associated with a vortex shedding formed directly above the inlet and its interaction with the walls of the cavity. The presented work include description of developed test stand and applied measurement techniques dedicated to the analysis of high frequency phenomena. In addition, the adopted numerical model will be described, including conducted two-dimensional and three-dimensional analysis. The developed models will be validated based on experimental measurements concerning wide range of flow conditions.

1 Introduction

Analysis of the flow in the cavities is a subject of a numerous experimental and numerical investigations. This type of flow is usually accompanied by strong pressure fluctuations inside the chamber itself, which are emitted into a far field in the form of a sound wave responsible for the generation of noise. It concerns both subsonic and supersonic flows. Pressure fluctuations are often characterized by a single mod with a certain frequency and high amplitude. Their generation is associated with the vortex path created directly above the cavity inlet and its interaction with cavity walls.

This type of flows occurs in many machines and devices. The most popular examples include: cavities in a landing gear of plane, bomb hatches, open car windows, gaps in car bodies. These examples pertain open flows. Besides, this type of flow occurs in many parts of a gas or steam turbine blade system such as labyrinth seals or rim seals applied between stator and rotor [4, 5] and spaces between guiding vanes and turbine casing resulting from its modular construction [12]. Another example is flame stabilizer applied in the ramjet engines. In all aforementioned cases, there are closed flows in which the influence of the flow channel walls on the mechanism of generation and propagation of disturbances should be additionally considered.

References [9,10,11] supplement research of this type of phenomenon with the relation between the parameters of the generated acoustic wave and the conditions of heat exchange inside the chamber. The paper presented herein is a part of the validation process of the applied numerical models based on conducted experimental investigations.

2 Investigated phenomenon

Mechanism of tonal noise generation inside the cavity, which is flown over by gas was described by Rossitier in the sixties of the past century [7]. It is based on the feedback loop between generated acoustic wave and the flow of subsequent vortices from the leading edge of the chamber.

Nowadays, this type of flow is a frequent subject of investigations. The main motivation is understanding of the phenomena responsible for the generation of the strong pressure pulsations, which subsequently will make it possible to reduce noise and may contribute to the improvement of unfavourable working conditions of machines, in which this phenomenon occurs.

Rossiter [8] was first to discover that the frequency of pulsation was connected with feedback loop mechanism. His observations with the use of shadowgraph technique showed the presence of large vortex structures responsible for pressure pulsations inside the cavity.

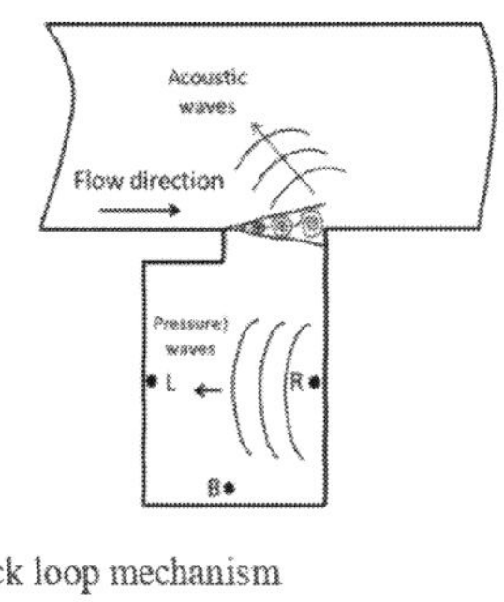

Fig. 1 Feedback loop mechanism

Their interaction with the rear cavity wall causes generation of pressure pulses which, moving towards the front wall of the cavity, intensify the flow of subsequent vortices. Part of the generated waves escapes the cavity and is audible as noise. This mechanism is presented for a cavity with additional cover in Fig. 1. Rossitier introduced also half-empirical formula which makes it possible to determine frequency of generated tonal noise:

$$St = \frac{fL}{U_\infty} = \frac{m-\alpha}{M_\infty + \frac{1}{k}} \qquad m=1,2,3\ldots \qquad (1)$$

Quantity k is a convective velocity of the vortices related to the main flow velocity and α is a delay time between the vortex impingement onto the cavity wall and moment of acoustic wave generation. In most cases k is within range 0.5-0.75 and a 0-0.25 [13]. Both quantities are chosen empirically. It is worth noting that in the presented equation Strouhal number, that is the dimensionless frequency of pressure pulsations for a given mod, is a function of only Mach number and two constant a and k.

Apart from the mechanism of sound wave generation based on resonance associated with feedback loop between frequency of generated vortices and frequency of acoustic waves, other types of resonance may occur, which significantly increase the level of generated sound. An example of such a resonance, which may occur in discussed flow, is a cavity working as Helmholtz resonator [2]. In this case air column inside the cavity is subjected to periodic compression and expansion which is associated with the compressibility of air. The resonating frequency of generated acoustic wave should be independent of free stream velocity [1]. This type of resonance occurs with relatively low Mach numbers (M<0.2), particularly for deep cavities (Length/Depth<2) and the ones with covered inlet. Frequency of acoustic wave generated by the Helmholtz resonator can be expressed as:

$$f = \frac{a}{2\pi}\sqrt{\frac{C}{V b_{ef}}}. \qquad (2)$$

The frequency of Helmholtz resonator can be also determined from equation (3) which is a modified version of the equation proposed by Rayleigh [5,6].

$$f = \frac{a}{2\pi}\sqrt{\frac{C}{V\left(b + \frac{1}{2}\sqrt{S\pi}\right)}}. \qquad (3)$$

In equations 2 and 3 a is speed of sound, C is area of inlet cross section, V is cavity volume, b is height of cavity neck and b_{ef} is effective height of the cavity neck which depends on b and cavity geometry.

For cavities with a partially covered inlet, pressure pulsations resulting from the feedback mechanism and pulsations associated with the operation of the cavity as a Helmholtz resonator can occur simultaneously, leading to its intensification [3].

3 Test stand

Presented experimental investigations were performer in the Fluid Flow Machines Laboratory of Institute of Power Engineering and Turbomachinery. The laboratory is equipped with the air system operating in underpressure. The diagram of air system is presented in Fig. 2. The main part of the installation is Roots blower DR124T with maximum volumetric flow rate equal to 12.4 m^3/min. It is propelled by 13.2 kW synchronous electric engine and maximum under pressure obtained is 50 kPa. The Roots blower is directly connected with the 3.2 m^3 air tank. This tank is designed to level pressure pulsations associated with the operation of the blower. Moreover, the blower is placed inside of the soundshield. The other parts of air installation consist of aluminum piping with 105 mm in diameter.

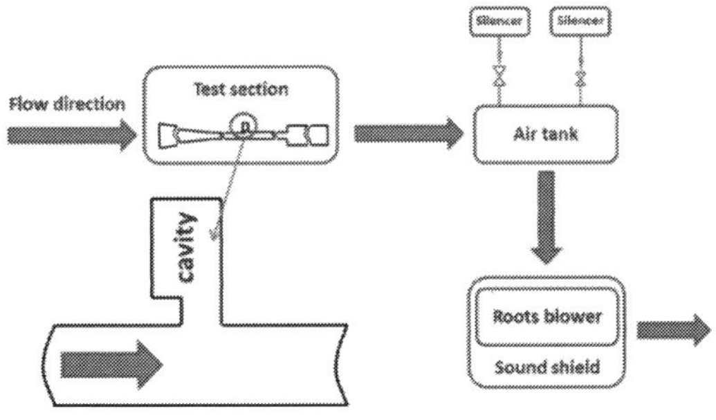

Fig. 2 Air system

The test rig constitutes a first element of the presented air installation operating in underpressure. It consists of 3 main parts (Fig. 3). The first one is converging inlet with an opening angle equal to 6^0. The next part is a rectangular channel with the dimensions of cross-section equal to 50x40mm, in which the investigated cavity is installed in the middle. The final part of the installation is the transition from a rectangular cross-section to a circular one and the connection with air supplied system.

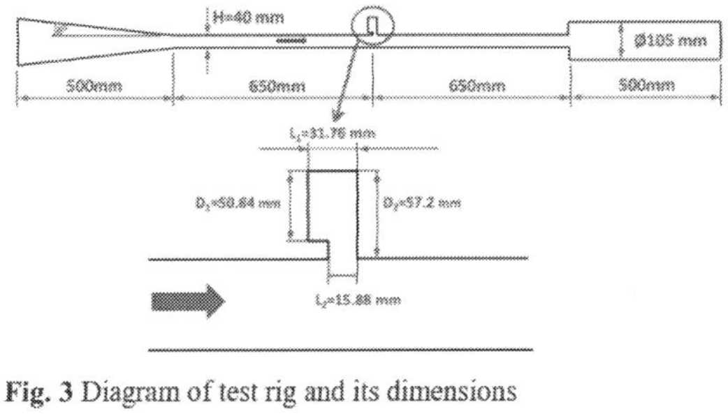

Fig. 3 Diagram of test rig and its dimensions

Fig. 4 Investigated cavity

The test stand is made out of polycarbonate and sidewalls of the cavity are demountable. The deep cavity with partially covered inlet is placed in the upper part of the main channel (Fig. 4).

A set of 6 mm microphones was prepared for measuring acoustic signals. Each of the microphones works with a separate microphone amplifier with adjustment. The calibration of the microphone measurement track was carried out on a specially prepared stand with a high-power pistonphone. The set of eight microphone amplifiers works with the National Instruments NI USB-6216 measuring module. It is a universal system containing a 16-channel, 16-bit AC converter with 400 kS / s sampling and communication with a master computer via a USB serial port. The mass flow rate measurement was carried out using an orifice and a calibrated thermoanemometer. In addition, a Pitot tube was installed in the flow channel to measure velocity in the channel axis.

4 Numerical model

Numerical analyses were carried out using various models differing in size and geometry of the analyzed computational area. The analyses were performed on both simplified two-dimensional models and three-dimensional models. The basic configuration of the computational area for two-dimensional analyses is presented in Fig. 5. The chamber in this case is located in the middle of the $40D_1$ main channel.

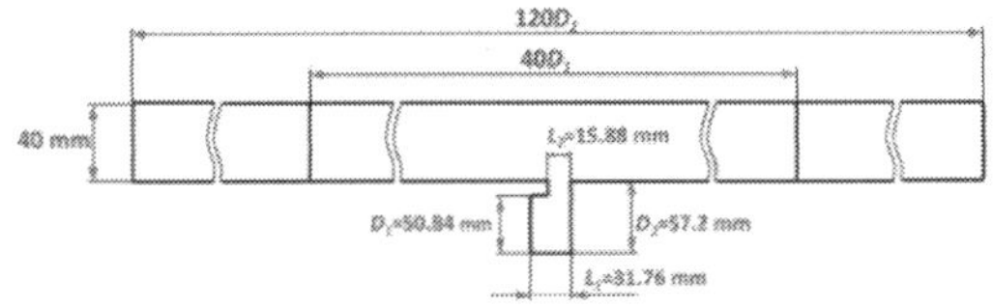

Fig. 5 Computational domain

Two-dimensional calculations were carried out based on two models (Fig. 6), aimed at eliminating the reflection of the acoustic wave from the edge of the computational domain. One method is to use non-reflecting boundary conditions (model 1, Fig. 6). This concept for high-amplitude waves is often unreliable, and its application can lead to instability of the calculation process, hence it was proposed to replace non-reflective boundary conditions with an additional buffer domain at the inlet and outlet with a relatively rare numerical mesh. Acoustic waves after passing through this area are further attenuated due to numerical dissipation, which prevents strong reflections from the inlet and outlet areas of the main channel. Such a solution requires, however, both an increase in the number of computational nodes and an extension of the entire computational area to the length of $120D_1$ (Fig. 6). The additional domain has been connected to the main channel with the use of General Grid Interface. In the buffer domain for the bottom and upper surface, the boundary condition of the free slip wall has been adopted, hence this area does not cause additional losses in the flow channel that result from fluid viscosity. Both analysed variants of the computational domain were compared and the obtained results are very similar. The second solution, however, is more stable for all considered flow conditions and configuration of boundary conditions. Due to the inability to conduct purely two-dimensional analyses, the calculation area for models 1 and 2 has a span of 1 mm. The boundary conditions adopted for analysis are presented in Table 1.

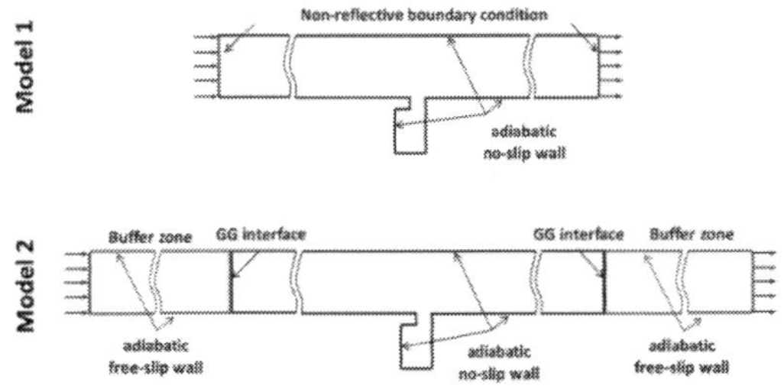

Fig. 6 Boundary conditions adopted to CFD analyses

The boundary condition of symmetry was assumed for the lateral walls of the channel and no-slip wall was adopted for the upper and bottom surfaces.

The numerical mesh is shown in Fig. 7. It is a structural, orthogonal mesh with 231,000 nodes. The span of the flow channel is discretized with a single element. The mesh is significantly finer at the cavity inlet. This area is particularly important for correct modelling of generation and propagation of acoustic waves. For all walls of both the main channel and the cavity, the dimensionless value of $y_+ \approx 1$.

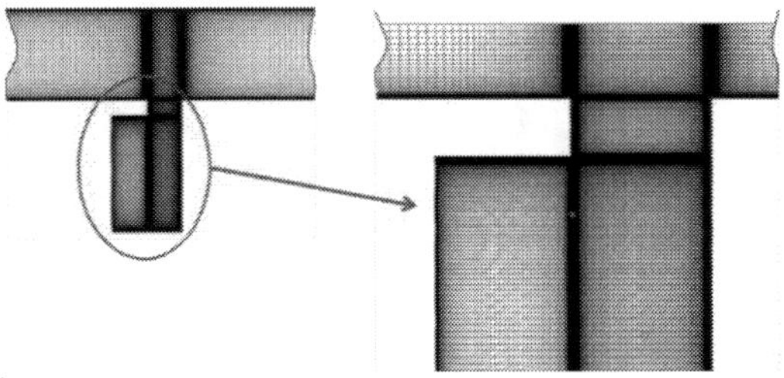

Fig. 7 Numerical mesh for CFD analyses

Tab 1 Basic boundary conditions and assumptions of the numerical analyses

Inlet	Velocity: 50 m/s Static temperature: 15°C Turbulence intensity: 1%
Outlet	Averaged static pressure: 1 bar
Channel and cavity walls	Adiabatic
Lateral walls	Symmetry
Fluid	Air ideal gas
Heat transfer	Energy equation based on total enthalpy with taking account viscous work associated with tangential stresses
Turbulence model	SST or DES

The numerical mesh is gradually coarsened in the direction of inlet and outlet from the main channel. For

 https://doi.org/10.1051/e3sconf/201913701011

model 2 (Fig. 6) in the interface area, the element size of the numerical mesh is abruptly changed. The mesh in the buffer domain is not refined along the channel edge due to the accepted boundary condition of free slip wall.

The third investigated numerical model (model 3) relates to calculations using methods based on filtered Navier-Stokes equations based on the DES method. It is assumed that the configuration of the computational domain and boundary conditions correspond to model 1 (Fig. 6), while the span of the computational area is extended to 20 mm, which allow for the three-dimensional character of the flow. In this case, the span of the flow channel and the cavity is discretized using 50 elements of the numerical mesh, which together gives a grid of 5.8 million nodes.

Models 4a and 4b relate to three-dimensional calculations using the averaged Navier-Stokes equations. The model k-ω SST was used to model the Reynolds stress tensor. In this case, the geometry of the computational domain exactly matches the dimensions of the test stand (Fig. 3). In the 4a model, the free slip boundary condition was adopted for the lateral walls of the computational domain, which allowed limiting the size of the numerical mesh. For model 4b, the numerical mesh was refined near the lateral walls of the domain and the no-slip boundary condition was assumed. This made it possible to correctly determine the velocity distribution along the span of the flow channel.

5 Results

Computations were carried out for a flow velocity in the channel equal to 50 m/s. For the simulation of non-stationary flow, a time step of 10^{-5}s was adopted, which gave 87 time steps for one acoustic wave period. The exception is time discretization used for the DES method. In this case, a time step of $2.5 \cdot 10^{-6}$ s was adopted, which provided 348 time steps for one period of pressure pulsation in the cavity. The process of determining the amplitude of the generated acoustic wave required approximately 14,000 iterations, which considerably increases computational time.

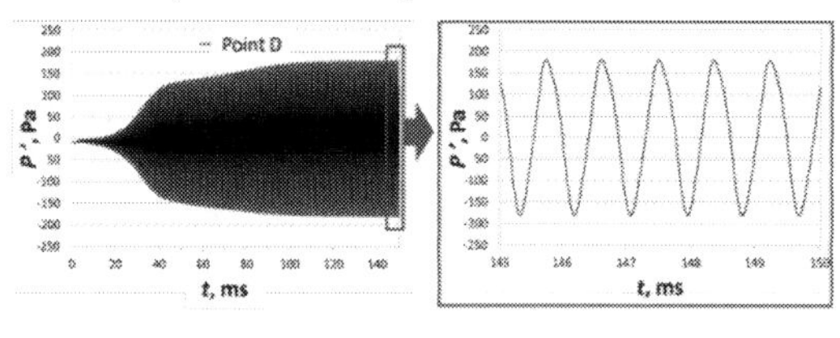

Fig. 8 Generation of sound pressure fluctuation at a 50 m/s flow velocity - Model 1

Fig. 8 shows the process of determination of acoustic wave amplitude. The obtained results concern two-dimensional analysis. Fig. 9 and 10 show the course of changes in pressure fluctuations for three control points located in the middle of the left, right and bottom walls (Fig. 1). These are respectively points L, R, D

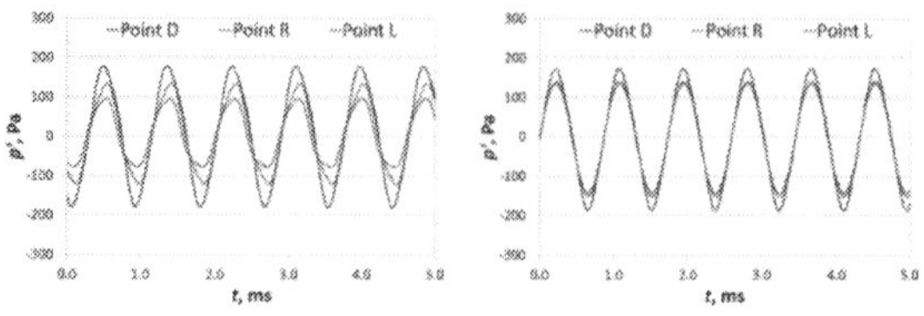

a) 2D k-ω SST (model 1) b) 3D k-ω SST (model 4b)

Fig. 9 Acoustic pressure fluctuations for flow velocity in the channel equal to 50 m/s – k-ω SST

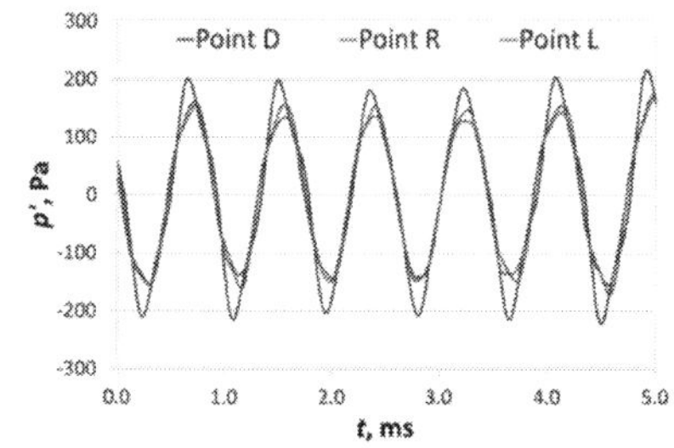

Fig. 10 Acoustic pressure fluctuations for flow velocity in the channel equal to 50 m/s – DES

The results of two and three-dimensional analysis based on the k-ω SST turbulence model and on the DES method are compared. In addition, the corresponding results of experimental studies are presented in Fig. 11. A fragment of the pressure pulsation course in the case of experiment covers a wider period of time, equal to 20 ms. In this case, the course of fast-changing signals was recorded at a frequency of 44 kHz. It can be seen in Fig. 11 that the sound pressure oscillations at 50 m/s are strongly periodical. The pressure course for all three analysed measuring points are sinusoidal. The analysis of measurement data also shows that the amplitude of acoustic pulsation for individual points remains almost unchanged. The obtained sound pressure fluctuation courses for all analysed numerical models are very similar and all are also characterized by ideal periodicity of the analysed phenomenon, which is also reflected, as mentioned earlier, in the obtained experimental data.

It is also important to note that for all three measuring points, the course of sound pressure fluctuations are coincidental in phase in spite of the slight differences in the pulsation amplitude. This indicates that this type of cavity acts as a Helmholtz resonator, and pressure pulsations occur uniformly throughout the cavity volume. However, this does not exclude the existence of an additional type of resonance resulting from the feedback mechanism described by Rossiter [8].

It is also worth noting that the results obtained from 2D simplified model and 3D model are in good agreement. The highest sound pressure level was obtained for point D, located on the bottom wall of the cavity, while the lowest for point R, located on the right wall. The relatively low sound pressure level for this point may be due to the fact that it is located much below the place where the vortices formed in the cavity inlet hit, hence the higher pressure pulsations on the right wall of the cavity will be observed rather in the vicinity of the

https://doi.org/10.1051/e3sconf/201913701011

main flow channel. The biggest difference between the sound pressure levels for the analyzed measuring points was obtained in the case of 2D analysis. The sound pressure fluctuation course for the DES are slightly sharper than in the case of other analyzed numerical models. It is also worth mentioning that the sound pressure fluctuations for the 4a and 4b models are very similar, hence only one of the results obtained is presented.

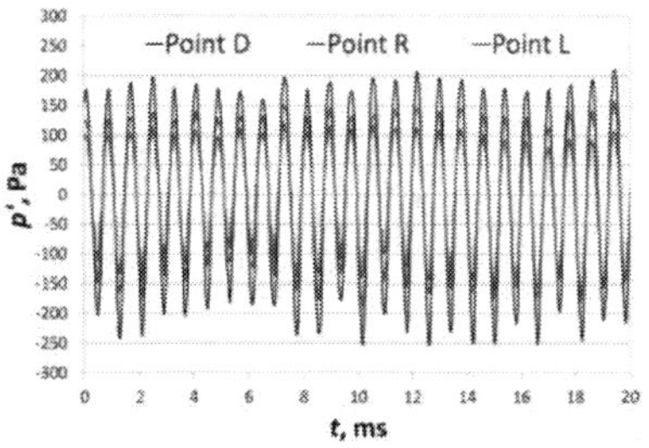

Fig. 11 Acoustic pressure fluctuations for flow velocity 50 m/s – experimental investigations

Non-stationary changes in sound pressure fluctuations obtained in experimental studies and from numerical calculations were subjected to FFT (Fast Fourier Transformation) spectral analysis. The sound pressure level spectra determined for each case calculated on the basis of numerical analyses are presented in Fig. 12, while in Fig. 13 the spectrum obtained on the basis of experiment is shown.

4096 samples after the stabilization of the acoustic pulsation level were taken into account for the FFT analysis. The FFT analysis of the sound pressure level in the case of experimental studies was additionally averaged. 10 separately performed spectral analyses of one continuous time period were taken into account, which were then averaged. Due to the relatively small changes in the level of pressure pulsation in time, the average FFT results do not differ significantly from the instantaneous course. Averaging, however, allows the elimination of individual sound pressure pulsations that deviate significantly from the mean values.

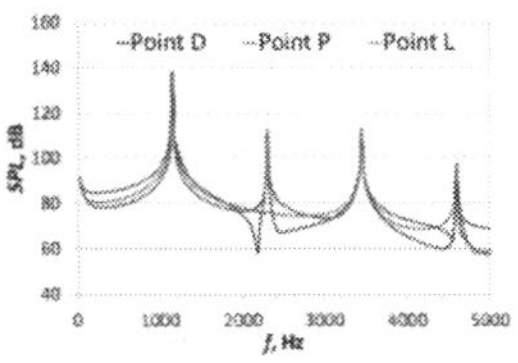

a) 2D k-ω SST (model 1)

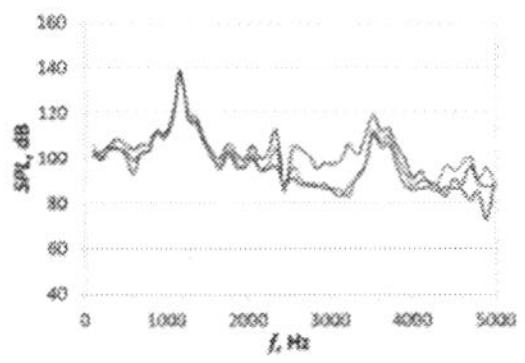

b) 3D DES (model 3)

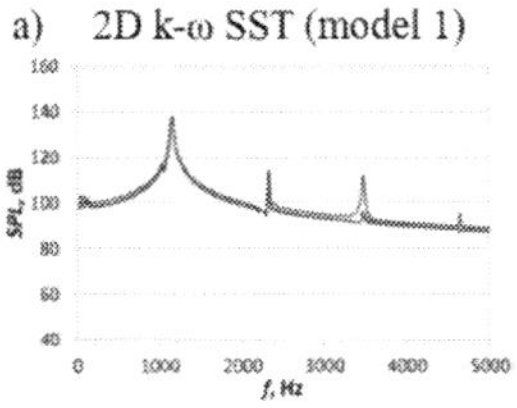

c) 3D k-ω SST (model 4a)

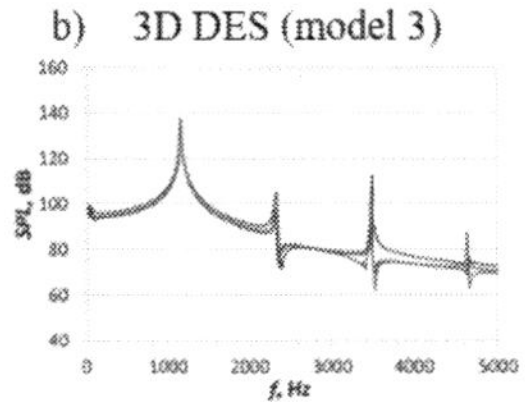

d) 3D k-ω SST (model 4b)

Fig. 12 Spectral analysis of the sound pressure level for fluid velocity 50 m/s - numerical investigations

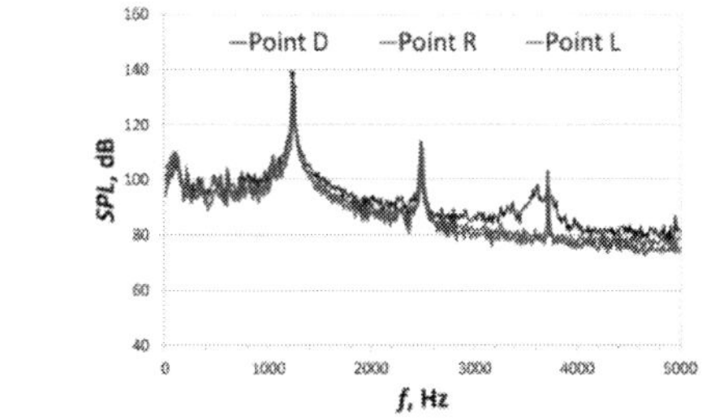

Fig. 13 Spectral analysis of the sound pressure level for flow velocity 50 m/s – experimental investigations

All applied models based on averaged Navier-Stokes equations give a very smooth picture of the sound pressure level spectrum, while in the case of the DES method, the spectral characteristics include many smaller peaks associated with the modelling of smaller turbulence scales. In this case, it was also possible to capture small differences in the course of spectral characteristics near the third harmonic, which are also visible in experimental studies (Fig. 13). Nevertheless, the correct prediction of the amplitude and the frequency of the first mod is crucial due to the almost sinusoidal nature of the generated acoustic wave. A detailed comparison of the parameters of the first three mods is presented in Table 2.

Tab. 2 Frequency and acoustic pressure level for selected mods

f, Hz					
Mod	Experiment	DES model 3 3D	k-ω SST model 4a 3D	k-ω SST model 4b 3D	k-ω SST model 1 2D
1	1235	1171	1147	1156	1148
2	2492	2344	2319	2311	2295
3	3717	3516	3467	3483	3442

SPL, dB					
Mod.	Experiment	DES model 3 3D	k-ω SST model 4a 3D	k-ω SST model 4b 3D	k-ω SST model 1 2D
1	138.5	137.7	137.9	137.4	138.8
2	111.4	112.4	114.7	104	112.5
3	103.3	119.0	112.0	111.9	113.6

Analysis of the data presented in Tab . 2 shows that all the numerical models under predict the frequency of mods. These differences for the first mod are in the range of 5.2-7.0% and are the smallest for the DES method, while for the third mod they are in the range of 5.4-7.4%. Nevertheless, the difference between obtained frequency and experimental data increases for subsequent mods. Very good agreement was obtained between all numerical models and experimental investigations in case of sound pressure level. For the first mod the maximum differences are 1.1 dB. The highest discrepancies were obtained for the third mod and they are equal to a maximum of 15.2%, but the amplitude of the sound pressure level is small in comparison to the first mod.

6 Conclusions

Performed investigations concerning analysis of the noise generation by the cavity showed good agreement between numerical models and experiment. In both cases the generated by the cavity acoustic wave is almost purely sinusoidal with one main mod.

The course of pressure fluctuations are coincidental in phase for three investigated measuring point located at the centre of each cavity wall. This indicates Helmholtz type resonance which can coincide with feedback mechanism described by Rossiter.

The comparison between CFD and experimental data shows that all investigated numerical models slightly under predicted frequency of the first mod of pressure fluctuations and the differences were within range 5-7% depending on the model. The discrepancies in sound pressure level between numerical models and experiment were maximum 1 dB.

Presented analyses made it possible to develop an optimal numerical model to shorten computational time. Further investigations will concern wider range fluid velocities and experimental investigations will also take into account heat transfer conditions in case of heated cavity. This will allow the validation of numerical studies on the heat transfer conditions in the cavities.

Acknowledgements

The presented research was conducted within the the Silesian University of Technology statutory research funds for young scientists.

References

1. De Jong, A.T., Bijl H., *Investigation of higher spanwise Helmholtz resonance modes in slender covered cavities*, The Journal of the Acoustical Society of America, vol. 128 (4), 2010, p. 1668-1678.
2. Durgin W., Graf H., *Flow-excited acoustic resonance in a deep cavity: an analytical model, Proceedings of the Third International Symposium on Flow-Induced Vibrations and Noise, vol. 7, 1992, p. 81-91*
3. *Loh Ching Y., Computation of Low Speed Cavity Noise, NASA/CR-2004-212892, 2004.*
4. *Patinios M., Ong I.L., Scobie J.A., Lock G.D., Sangan C.M., Influence of Leakage Flows on Hot Gas Ingress, Journal of Engineering for Gas Turbines and Power, Vol. 141/ 021010-1, 2019*
5. Patinios M., Scobie J.A., Sangan C.M., Owe J.M., Lock G.D., Measurements and Modeling of Ingress in a New 1.5-Stage Turbine Research Facility, Journal of Engineering for Gas Turbines and Power, Vol. 139/ 012603-1, 2017
6. Rayleigh L., *On the application of the principle of reciprocity*, Proceedings of the Royal Society of London, vol. 25, 1877, p. 118-122.
7. Rayleigh L., *The theory of sound*, Dover Publications, New York, revised edition, 1945.
8. Rossiter J.E., *Wind Tunnel Experiments on the Flow over Rectangular Cavities at Subsonic and Transonic Speeds*, Royal Aircraft Establishment, Technical Report 64037, 1964.
9. Rulik S., Wróblewski W., *A Numerical study of the heat transfer intensification using high amplitude acoustic waves*, Archives of Acoustics, vol. 43, 2018, p. 31-47
10. Rulik S., Wróblewski W., Nowak G., Szwedowicz., *Heat transfer intensification using acoustic waves in a cavity*, Energy, vol. 87, 2015, p. 21-30.
11. Rulik S., Wróblewski W., Rusin K., *A numerical study of the relation between the acoustic generator geometry and the heat transfer conditions*, Archives of Mechanical Engineering, vol. 65(1), 2018, p. 5-26.
12. Spura D., Lueckert J., Schoene S., Gampe U., *Concept development for the experimental investigation of forced convection heat transfer in circumferential cavities with variable geometry*, International Journal of Thermal Sciences, vol. 96, 2014, p. 277-289.
13. Ünalmis Ö.H., Clemens N.T., Dolling D.S., *Cavity Oscillation Mechanisms in High –Speed Flows*, AIAA Journal, vol. 42 (10), 2004.

RESEARCH & DEVELOPMENT IN POWER ENGINEERING, 2019

Evaluation of long-term start up costs impact on short-term price based operational optimization of a CCGT using MILP

Sylwia Gotzman[1,*], *Paweł* Ziółkowski[2], *Janusz* Badur[1]

[1] Szewalski Institute of Fluid-Flow Machinery, Polish Academy of Sciences, Fiszera 14, 80-231 Gdańsk, *corresponding author sylwia.gotzman@imp.gda.pl, jb@imp.gda.pl, Poland

[2] Gdansk University of Technology, Faculty of Mechanical Engineering, Narutowicza 11/12, 80-233 Gdańsk, e-mail: pawel.ziolkowski1@pg.edu.pl, Poland

Abstract: An increasing share of the weather-dependent RES generation in the power system leads to the growing importance of flexibility of conventional power plants. They were usually designed for base load operation and it is a challenge to determine the actual long-term cycling costs, which account for an increase in maintenance and overhaul expenditures, increased forced outage rates and shortened life expectancy of the plant and components. In this paper, the overall impact of start up costs is evaluated by formulating and solving price based unit commitment problem (PBUC). The electricity spot market is considered as a measure for remunerating flexibility. This approach is applied to a real-life case study based on the 70 MWe PGE Gorzów CCGT power plant. Different operation modes are calculated and results are used to derive a mixed integer linear programming (MILP) model to optimize the operation of the plant. The developed mathematical model is implemented in Python within the frame of the PuLP library and solved using GUROBI. Results of the application of the method to a numerical example are presented.

1 Introduction

The installed capacity growth of fluctuating renewable sources puts a strain on the remaining dispatchable power plants. Namely, the volatile character of the residual load forces conventional power plants to meet ever higher flexibility requirements, as more frequent start-ups and shut-downs, part-load operation and consequently falling operating hours are required from them. This effect can already be seen in the present and is expected to increase in the future. According to [1], as many as 200 start-ups and as few as 1500-4500 operating hours per year should be the target flexibility parameters of conventional power plants.

From the perspective of the Transmission System Operator (TSO), flexible power plants help ensure the stability of the power system. From the perspective of a power plant operator, enhancing the flexibility parameters results in a higher profitability of the power plant. As deregulation and an increasing share of volatile generation leads to more variability in electricity prices on the spot market [2], price signals give incentives for cyclic operation of power plants. If a power plant may follow daily, weekly and annual market trends, it may avoid electricity production when spot prices drop below marginal costs, and on the other hand, maximize electricity production in the high spread periods.

The last two decades have witnessed a growing interest in gas fueled power plants across Poland. These are usually combined cycle gas turbines (CCGTs). The total installed capacity of CCGTs in Poland today is around 2 GW and should approach 3 GW by 2020, when EC Stalowa Wola and EC Żerań power plants start operating [1]. CCGTs are much more capable of providing sufficient flexibility to the power system compared to coal-fired units, which today account for 70% of total installed capacity in Poland [3]. Nevertheless, these plants are usually built for base-load operation, which means they do incur high costs when operated cyclically. The subject of actual costs of cycling is not entirely obvious. Besides direct costs (mainly fuel costs during the start-up), power plant cycling entails long-term costs, like increased maintenance and repair expenditures, costs of unplanned outages and component life expectancy shortening. The long-term costs occur in the future and often can only be predicted with a degree of uncertainty. Consequently, they are prone to being underestimated. As the authors of [4] point out, the long-term costs often exceed the short-term profits if not taken into consideration when submitting a bid price. Therefore, it is essential to have a good understanding of the actual costs of cycling the plant and use the insights to determine how to respond to the time-sensitive electricity prices when making short-term scheduling decisions.

For this purpose, we address the issue of an optimal CCGT plant operation schedule according to 2019 Polish electricity spot market prices. Typical flexibility constraints to thermal generation units include start-up costs, minimum/maximum generation limits, ramping costs and ramp rates. As neither the heat demand constraints nor the TSO requirements are taken into consideration, ramping parameters are not included in the analysis. We derive a mixed integer linear programming (MILP) model of unit commitment (UC) with rolling horizon and implement it in PuLP Python. Then, PGE Gorzów power plant is analyzed and the model is run over a range of different flexibility parameters to evaluate the impact of start-up costs on optimal scheduling decisions.

2 Literature review

In the literature, the problem of scheduling of generation units over a daily or weekly time horizon in order to accomplish some objective while meeting some physical and operational constraints, vary considerably in size and scope. Consequently, there is no uniform terminology. We decided to stick to the term "unit commitment" as it is specific to the energy sector as opposed to "scheduling" and as it is more commonly used by solution techniques researches than "economic dispatch". The meaning of the term "unit commitment" is ambiguous [5]. There are authors using the term "unit commitment" in a narrow sense, where it refers to the on/off status of generators and then economic dispatch is performed to specify the generation level of the units [e.g. 6-9]. In a broad sense, as it is used in this paper, UC refers to determining both the on/off status and the generation level. At the power plant or one owner level, this problem is also called self commitment [e.g. 10], self-scheduling problem [e.g. 11] or simply scheduling problem [e.g.12].

Under the traditional regulated monopoly paradigm, unit commitment model seeks to determine the optimum operating point of the available generation units in order to satisfy the forecasted level of demand with the minimum operating cost. In a competitive setting, electricity and reserve prices are important factors in the decision process for unit commitment scheduling. Power plants have freedom to schedule their generators to produce less electricity than predicted load in order to maximize their profit. In the literature the problem is called the profit based unit commitment problem (PBUCP) [e.g. 13].

There are many different modeling approaches to the unit commitment problem. In many cases they were applied to combined cycle power plants [e.g. 5, 8, 10, 12, 14, 15]. Models proposed to simulate thermal unit operations differ considerably in the degree of detail. The least technologically detailed unit commitment models consider the units at a plant level, which means that the interactions of components within the plant are not modeled [e.g. 16]. More realistic models in terms of modeling operation of a CCGT accurately can be achieved either through introducing different power plant operation modes (configurations) [e.g. 14] or base the model on physical components [e.g. 12]. In the current work, we implemented the operation modes based model.

In the case of real-life case studies, typically in the first step the power plant is modeled [e.g. 8]. For components not included in the standard library, programmable components are used. After a validation of the plant model, an optimization tool is implemented in some numerical computing environment and simulations are run based on the plant technical data and technical and economic assumptions.

There has also been a proliferation of research on solution techniques for the unit commitment problem due to its complexity and practical importance. The most common solution techniques include MILP, dynamic programming (DP), Lagrange relaxation and priority-list. Readers can refer to [17] for a comprehensive overview of different methods. The solution approach that has become very popular is Mixed-Integer Linear Programming (MILP) approach due to significant improvements in off-the-shelf MILP solvers, based on the branch-and-cut algorithm [18].

3 Model formulation

Objective function

The objective function of the optimization states the profit from the operation, which shall be maximized. Revenues originate from selling the electricity on the wholesale market. The costs included in the model are composed of variable and fixed costs and will be discussed later in this section. The objective function is formulated as follows:

$$\sum_{t=0}^{T}\sum_{m=0}^{M} \begin{matrix} \left(s_{t,m} * P^{Futures} * c^{Futures}\right) + \\ +(P_t^{Spot} * c_t^{Spot}) + \\ -(P_t^{Generation} * varCost_m) + \\ -(s_{t,m} * fixCost_m) \end{matrix} \tag{1}$$

where T is the time horizon with a time step of one hour indexed with t; m and M are an operation mode index and the number of operation modes respectively; P is the amount of electricity in MWh; c is the price of electricity in PLN/MWh, $varCost_m$ is the variable unit cost of electricity generation for operation mode m in PLN/MWh, $fixCost_m$ are the fixed costs of electricity generation for operating mode m and as such are expressed in PLN per hour; $s_{t,m}$ is a status (binary) variable indicating which mode m is the plant on at hour t. The following constraint ensures that only one operation mode at a time is selected:

$$\sum_{m=0}^{M} s_{t,m} = 1 \tag{2}$$

We assumed that a constant amount of electricity for each hour of the analyzed period has been sold on the futures market at a constant price and the remaining output can be sold on the spot market. Also buying electricity on the spot market is made possible in order to avoid electricity generation when prices drop below marginal costs. $P_t^{Generation}$is the amount of electricity generated during hour t, $P^{Futures}$ is the amount of electricity sold on the futures market and it is constant throughtout the whole period, and P_t^{Spot} is the electricity sold (>0) or bought (<0) on the spot market so that at each hour:

$$P^{Futures} + P_t^{Spot} = P_t^{Generation} \tag{3}$$

Technical constraints

A typical combined cycle gas fueled plant consist of one or more gas turbine generators (GT), each one of them with its own heat recovery steam generator (HRSG), and one conventional steam turbine generator (ST). Lager CCGT plants frequently generate steam at multiple pressure levels, so that STs include dual or triple pressure sections in order to operate at a higher thermal efficiency level. The more components the system has, the larger number of possible operation modes (configurations of the components) should be taken into consideration to allow accurate optimization of the plant scheduling. In our model there are 6 operation modes:

1) off (*OFF*),
2) cold start-up (*coldSU*),
3) warm start-up (*warmSU*),
4) minimum complaint load operation (*MCL*),
5) part-load operation (*PART*),
6) nominal load operation (*NOM*).

The generation capacity, overall efficiency of the plant and CO_2 emissions change depending on the operation mode. They all are modeled as constant values characterizing each operation mode. Better accuracy of the model can be achieved through higher resolution of operation modes.

For each mode, a feasible operation area is defined by its minimum and maximum electricity generation. A continuous operation range between the minimum and maximum electricity generation can be formulated using the binary operating variable as:

$$\sum_{m=0}^{M} s_{t,m} * P_m^{min} \leq P_t^{Generation} \leq \sum_{m=0}^{M} s_{t,m} * P_m^{max} \qquad (4)$$

where P_m^{min} and P_m^{max} are the minimum and maximum electricity generation at mode m respectively.

These are all the technical constraints of the optimization problem. Constraints forcing the optimum to lay in the part-load operation area (e.g. heat demand or TSO constraints) are not considered in this model. Also, due to the hourly resolution of the model, ramping between different load points is considered fast and thus not included in the modeling.

Start-up modes

In order to reflect the higher cost of starting up a cold power plant, two start-up modes are introduced: cold start-up mode and warm start-up mode. Both the cold and the warm start-up modes are represented by hourly operation modes in the same manner as the offline mode or the different generation modes. Start-up duration is not taken into consideration, as the optimal offline time proceeding start-ups is long enough to assume the process had been initialized sufficiently in advance.

A start-up in the hour before normal operation can be forced by the following logical constraints:

$$-s_{t-1,OFF} + s_{t,OFF} + s_{t,\text{coldSU}} + s_{t,\text{warmSU}} \geq 0 \qquad (5)$$

Selection of the correct start-up type of the plant, depending on the prior offline time, is achieved by adding another inequality constraining the warm start-up status variable $s_{t,warmSU}$ to be zero at hour t unless the offline time prior to that hour was shorter or equal to L hours:

$$offTime_t - L \leq 1000 * (1 - s_{t,\text{warm}SU}) \qquad (6)$$

where $offTime_t$ denotes the number of offline hours within the period of length L preceding the hour t.

Production costs

We focus on the short run costs of electricity generation and capital expenditures do not appear in our model. It is common to divide generation costs into two categories: fixed costs, which are the costs independent of the power plant output, and variable costs that do vary with the power plant output. The fixed production costs consist of fixed O&M costs, labour costs, external services, and other general costs which do not vary with the output of the power plant. The variable production costs in our model consist of the expenses for fuel and CO_2 allowances and variable operations and maintenance (O&M) costs. We assumed that the expenses for fuel and CO_2 allowances differ depending on the operation mode, but the variable O&M costs are constant per 1 MWh of generated electricity and are common for all the operation modes. The variable O&M costs and fixed costs are strongly dependent on cyclicality of operation [19, p. 599]. In the model they should be set at a level appropriate for base-load operation, as any increase in the production costs due to start-ups, especially in the O&M costs, should be reflected in the start-up costs.

In the literature production costs are usually formulated via quadratic [e.g. 16] or cubic [e.g. 15] functions, which are further linearized into piecewise linear forms. The slope of each segment is then used in the objective function as it represents the incremental cost of a given generation range. In our optimization model production costs are represented by a step function, so that the marginal and average variable costs are constant for a given generation range. In particular, the heat rate is constant for a given operation mode.

Start-up costs

It is a broadly accepted approach in the literature to refer to the start-up costs as the transition costs between different operation modes. In our formulation of the problem the start-up costs are assigned to the start-up modes as their fixed costs, as they do not depend on the operation mode following the start-up mode.

Start-up costs are composed of [20]:

• costs of additional fuel due to heat rate losses during start-up; auxiliary power usage, chemicals and additional manpower required during start-up;

• depreciation of the components exposed to wearing and additional fuel due to heat rate losses stemming from the wear of the equipment; increased maintenance, repair and overhaul expenditures; increased forced outages; critical equipment life shortening.

 https://doi.org/10.1051/e3sconf/201913701012

The former are direct costs and are easily accounted for. The latter occur in the future and a great part of them can only be predicted with a degree of uncertainty.

4 Case study

Technical characteristics of the plant

The proposed formulation has been applied to solve a unit commitment problem for Gorzów CCGT plant, the first CCGT plant in Poland. The plant consists of one gas turbine of 55 MWe, one heat recovery steam generator (HRSG) and two steam-turbines of 5 and 6 MWe respectively. The layout of the plant is presented in Figure 1.

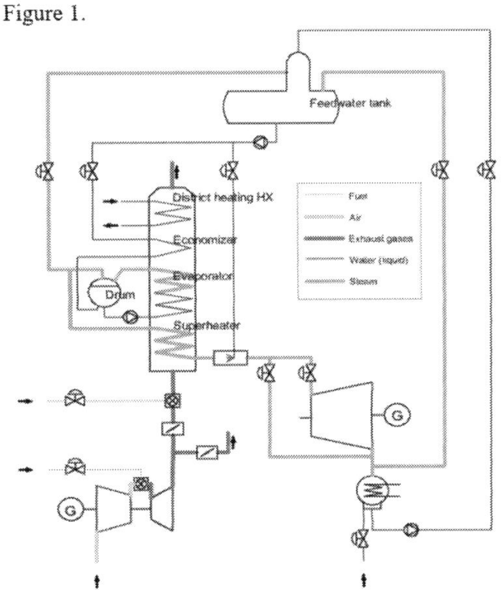

Fig. 1. The layout of the Gorzów CCGT plant.

A detailed thermodynamic cycle analysis of the steady-state design operating point of the system was conducted using an in-house code COM-GAS in [21]. The electrical efficiencies at the nominal load of the GT and of the two STs together were estimated at 34% and 15% respectively. However, the overall electrical efficiency of the CCGT power plant was estimated to be 41%. The overall efficiencies of the plant at *MCL* and *PART* operation modes were estimated according to data from [22]. Historical data were used to estimate the overall efficiencies of the plant at *coldSU* and *warmSU* operation modes and the CO_2 specific emission factor. The output at *MCL* was assumed to be 50% of the nominal GT output – the upper bound for one GT systems according to [23]. The maximum number of offline hours preceding the warm start-up mode is assumed at 30.

Costs and market assumptions

The assumptions about costs are based on the literature (Table 1). Market data assumptions are based on TGE data (Table 2). It is assumed that 1 € = 4.3 PLN.

Table 1. Costs assumptions

	Value	Source
Variable O&M costs	2.4 €/MWh	[24]
Fixed costs	15k €/MW/y	[20]
Warm start-up reference scenario costs	2x51.0 €/MW capacity	[25]
Cold start-up reference scenario costs	2x73.2 €/MW capacity	[25]

Table 2. Market data assumptions

	Value
Electricity spot prices	TGE DAM prices, first fixing
Electricity futures price	250 PLN/MWh
CO2 emission allowance cost	20 €/tCO_2
Fuel price	70 PLN/MWh

A scheduling problem is solved on a 4380 hour time horizon based on historical TGE day ahead market (DAM) electricity prices of the first half of the year 2019 (Figure 2). A rolling horizon method as described in [26] is used, i.e. the overall result is gained by optimizing successive 48h planning intervals, each shifted 24h forward, and concatenating the results.

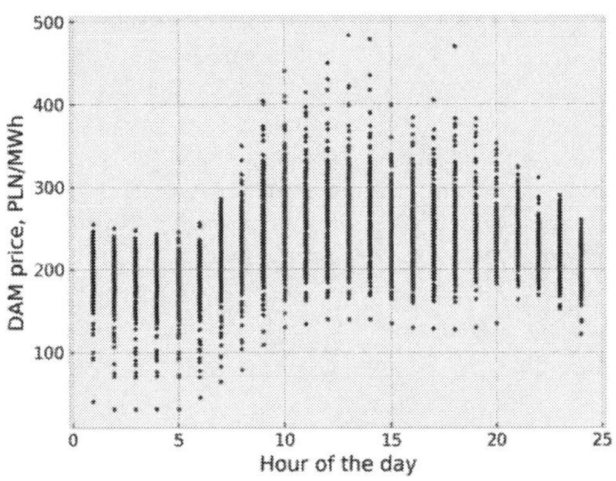

Fig. 2. TGE DAM prices in the first half of 2019.

5 Results and discussion

Four case study scenarios are provided to demonstrate the impact of start-up costs on optimal scheduling of the plant, in particular on the number of start-ups.

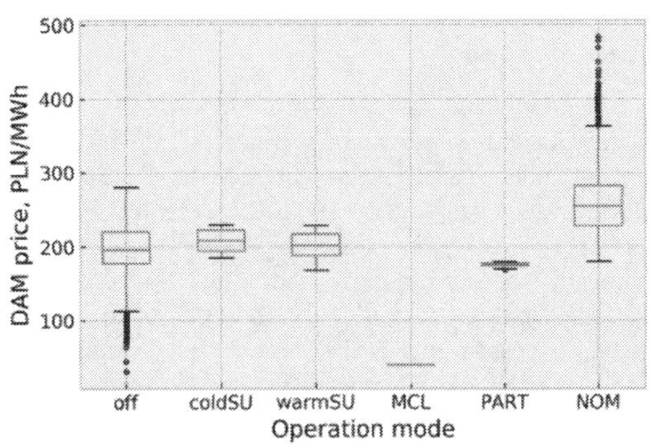

Fig. 3. Distribution of TGE DAM electricity prices at each operation mode for the reference scenario.

Figure 3 shows data on TGE DAM 2019 prices grouped into operation modes that resulted from solving the UC problem for the reference scenario (#0). The optimal number of start-ups within the time horizon amounts to 24 (12 cold and 12 warm, see Figure 4 and Table 3), which is circa one start-up a week.

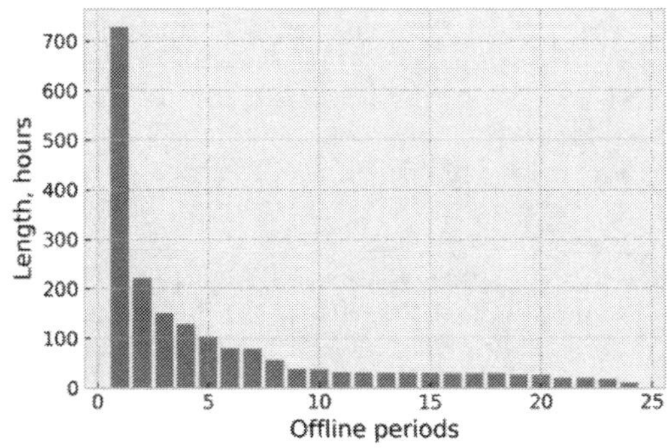

Fig. 4. Offline periods and their length for scenario #0.

Scenario #0 uses all the assumptions described in the Case study section. In particular, the start-up costs are twice higher than what are the best case assumptions according to [25], the maximum number of offline hours preceding the *warmSU* mode is 30 and the maximum output at MCL is 50% of the nominal GT output. In the scenarios #1, #2 and #3 we lift the assumptions one at a time.

Table 3. Counts of each operation mode hours

	off	*cold SU**	*warm SU**	*MCL*	*PART*	*NOM*
Scenario #0	1 983	**12**	**12**	1	22	2 338
Scenario #1	1 808	**23**	**33**	1	1	2 526
Scenario #2	2 030	**8**	**16**	1	18	2 319
Scenario #3	1760	**19**	**32**	76	0	2 505

* The meaning of the number of hours is the number of start-ups.

Scenario #1 assumes the best case start-up costs from [25], which are 51 and 73 € per 1 MW of capacity for warm and cold start-up respectively. If the plant operator underestimated the start-up costs, they would schedule the generation according to scenario #1 while incurring costs of scenario #0. As a result, there would be 23 cold start-ups and 33 warm start-ups throughout the whole time horizon of 26 weeks (Table 3, Figure 5), i.e. more than 2 start-ups a week. The profit seems to have increased by almost PLN 0.6 million (Table 4), but that is roughly the same amount as the overlooked difference in the overall costs of start-ups. It should be noted, though, that the number of start-ups with a prior offline period of no more than 8 hours has increased from 0 in scenario #0 to 12 in scenario #1 (Table 4), which means the start-up costs would have been lower, had we taken in consideration so called hot start-ups in the model.

In scenarios #2 and #3 we aim at estimating the revenues gained from flexibilization of the power plant, as they provide a target framework for possible investment costs. We want to answer the question: Can flexibility retrofitting costs be justified by the resultant reduction in cycling costs?

Table 4. The value of the objective function and the number and length of offline periods

	Profit [PLN / half a year]	**Number or offline periods**		
		Total	**≤8h (hot start-ups)**	**> 48h (very cold start-ups)**
Scenario #0	5 338 789	24	0	8
Scenario #1	5 932 543	56	12	14
Scenario #2	5 450 843	24	0	8
Scenario #3	5 956 448	51	8	14

In scenario #2 the maximum number of offline hours preceding the warm-start up mode is increased from the reference 30h to 48h, which could be achieved by keeping the components hot. The optimization results show an increase in the plant's profit of 112 k PLN (Table 4), which translates to about 0.8 €/kW/y. That certainly does not justify the investment costs required for flexibility retrofitting, which are roughly estimated to be 100 to 500 €/kW [28].

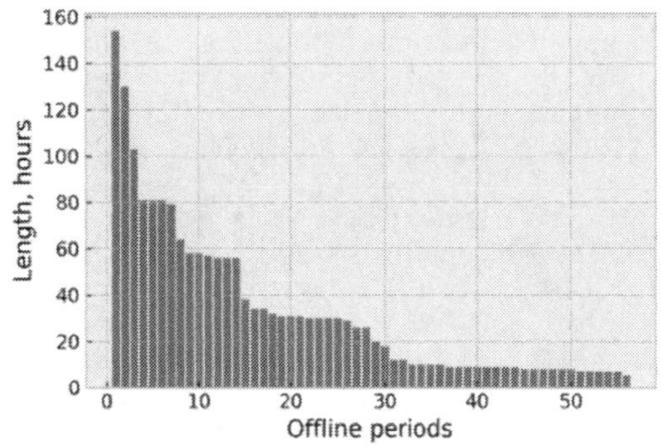

Fig. 5. Offline periods and their length for scenario #1.

In scenario #3 the maximum output at MCL is decreased from the reference 50% to 25% of the nominal GT output, so that the cost of staying online at minimum load is lowered. This flexibility improvement results in PLN 0.6 million higher profit in the optimized period (Table 4), which gives quite a significant amount of 1.2 million PLN/y. It is worth noting that lowering the minimum load – considered a measure to avoid cycling costs [4] – caused more frequent shut-downs and start-ups.

6 Conclusions and perspectives

In the paper we developed an optimizing tool for electricity generation scheduling and applied it to Gorzów CCGT plant data and historical DAM prices of the first half of 2019. The impact of underestimated start-up costs and flexibility improvements on operation of the power plant and the financial results were presented.

In future work we would like to investigate a case of a CCGT with an energy storage and a case of a CCGT with heat demand constraints. We expect the former to be very sensitive to misjudged long-term start-up costs and the latter to benefit much more than it was showed in this paper from improving flexibility parameters.

Acknowledgements

The authors would like to thank PGE Energia Ciepła SA Gorzów Wielkopolski Subsidiary for providing necessary data.

References

1. Bronk L., Czarnecki B., Magulski R., *Elastyczność krajowego systemu elektroenergetycznego*, Raport Forum Energii, (2019).
2. Green R., Vasilakos N. Market behaviour with large amounts of intermittent generation, *Energy Policy* 38(7):3211–3220, (2010).
3. Macuk R., Maćkowiak-Pandera J., Gawlikowska-Fyk A., Rubczyński A., *Transformacja energetyczna w Polsce*, Raport Forum Energii, (2019).
4. Lefton S. A., Hilleman D., Make Your Plant Ready for Cycling Operations, Intertek-Aptech, *www.powermag.com*, (2011).
5. Dai H., Zhang N., Su W, A Literature Review of Stochastic Programming and Unit Commitment, *Journal of Power and Energy Engineering*, Vol. 3, pp. 206-214, (2015).
6. Biggar D., Hesamzadeh R., *The Economics of Electricity Markets*, John Wiley & Sons Ltd., (2014).
7. Arabas J., Mościcki M., Świrski K., Rozszerzenie zadania ekonomicznego rozdziału obciążeń w warunkach rynku energii, Transition Technologies SA, *www.cire.pl*, (2001).
8. Angerer M., Djukow M., Riedl K., Gleis S., Hartmut S, Simulation of Cogeneration-Combined Cycle Plant Flexibilization by Thermochemical Energy Storage, *Journal of Energy Resources Technology*, Vol. 140(2), (2017).
9. Rist J., Dias M., Palman M., Zelazo D., Cukurel B., Economic dispatch of a single micro-gas turbine under CHP operation, *Applied Energy*, Vol. 200, pp. 1-18, (2017).
10. Papavasiliou A., He Y., Svoboda A., Self-Commitment of Combined Cycle Units Under Electricity Price Uncertainty. *IEEE Transactions on Power Systems*, Vol. 30, pp. 1690-1701, (2015).
11. Simoglou C., Biskas P., Bakirtzis A., Optimal Self-Scheduling of a Thermal Producer in Short-Term Electricity Markets by MILP, *IEEE Transactions on Power Systems*, Vol. 25, pp. 1965-1977, (2010).
12. Mitra, S., Sun L., Grossmann I., Optimal Scheduling of Industrial Combined Heat and Power Plants under Time-Sensitive Electricity Prices, *Energy*, Vol. 54, pp. 194-211. (2012)
13. Li, T., Shahidehpour, M., Price-Based Unit Commitment: A Case of Lagrangian Relaxation Versus Mixed Integer Programming. Power Systems, *IEEE Transactions on Power Systems*, Vol. 20, pp. 2015-2025, (2005).
14. López J., Gómez R., Guillén I., Practical commitment of combined cycle plants using dynamic programming. *Proceedings of the 2010 Electrical Power and Energy Conference*, (2010).
15. Alemany J., Moitre D., Pinto H., Magnago F., Short-Term Scheduling of Combined Cycle Units Using Mixed Integer Linear Programming Solution, *Energy and Power Engineering*, Vol. 5, pp. 161-170, (2013).
16. Carrion M., Arroyo J. M., A Computationally Efficient Mixed-Integer Linear Formulation for the Thermal Unit Commitment Problem, *IEEE Transactions on Power Systems*, Vol. 21(3), (2006).
17. Ackooij W., Lopez I., Frangioni A., Lacalandra F., Tahanan M., Large-scale unit commitment under uncertainty: an updated literature survey, *Annals of Operations Research*, Springer, Vol. 271(1), pp. 11-85, (2018).
18. Morales-España G., Latorre J., Ramos A., Tight and Compact MILP Formulation for the Thermal Unit Commitment Problem, *IEEE Transactions on Power Systems*, Vol. 28(4), (2013)..
19. Gülen S., *Gas Turbines for Electric Power Generation*, Cambridge University Press, (2019).
20. Schröder A., Kunz F., Meiss J., Mendelevitch R., von Hirschhausen C., Current and Prospective Costs of Electricity Generation until 2050, DIW Data Documentation, No. 68, Deutsches Institut für Wirtschaftsforschung (DIW), Berlin, (2013).
21. Ziółkowski P., Lemański M., Badur J., Nastałek L., Power augmentation of PGE Gorzow's gas turbine by steam injection – thermodynamic overview. *Rynek Energii*, Vol. 98, pp. 161–167, (2012).
22. Kehlhofer R., Rukes B., Hannemann F., Stirnimann F., *Combined-Cycle Gas & Steam Turbine Power Plants*, PennWell Books (2009)..
23. Chmielniak T., Lepszy S., Czaja D., Economic assessment of gas-steam systems taking account of variable loads, *Journal of Power of Technologies*, Vol. 95, pp. 54-62, (2015).
24. Mikołajuk H., Duda M., Radović U., Skwierz S., Lewarski M., Kowal I., Aktualizacja analizy porównawczej kosztów wytwarzania energii elektrycznej w elektrowniach jądrowych, węglowych i gazowych oraz odnawialnych źródłach energii, Agencja Rynku Energii, Warszawa, (2016).
25. Prina M., Fanali L., Manzolini G., Moser D., Sparber W., Incorporating combined cycle gas turbine flexibility constraints and additional costs into the EPLANopt model: The Italian case study, *Energy*, Vol. 160, pp. 33-43, (2018).
26. Prina M., Fanali L., Manzolini G., Moser D., Sparber W., Incorporating combined cycle gas turbine flexibility constraints and additional costs into the EPLANopt model: The Italian case study, *Energy*, Vol. 160, pp. 33-43, (2018).
27. Marquant J., Evins R., Carmeliet J., Reducing Computation Time with a Rolling Horizon Approach Applied to a MILP Formulation of Multiple Urban Energy Hub System, *Procedia Computer Science*, Vol. 51, pp. 2137-2146, (2015).
28. Agora Energiewende, Flexibility in thermal power plants – With a focus on existing coal-fired power plants, (2017).

https://doi.org/10.1051/e3sconf/201913701013

RESEARCH & DEVELOPMENT IN POWER ENGINEERING, 2019

Pilot installation for thermal utilization of meat-and-bone meal using the rotary kiln pyrolyzer and the fluidised bed boiler

Marcin Kantorek[1], Krzysztof Jesionek[2], Sylwia Polesek-Karczewska[3], Paweł Ziółkowski[4,*], Michał Stajnke[3] and Janusz Badur[3]

[1] Endress+Hauser, Wołowska 11, 51-116 Wrocław, marcin.kantorek@gmail.com, Poland
[2] Wrocław University of Science and Technology, Wybrzeże Wyspiańskiego 27, 50-370 Wrocław, krzysztof.jesionek@pwr.edu.pl, Poland
[3] Institute of Fluid-Flow Machinery, Polish Academy of Sciences, Fiszera 14, 80-231 Gdańsk, sylwia.polesek-karczewska@imp.gda.pl, michal.stajnke@imp.gda.pl, jb@imp.gda.pl, Poland
[4] Gdańsk University of Technology, Faculty of Mechanical Engineering, Department of Energy and Industrial Apparatus, Narutowicza 11/12, 80-233 Gdańsk, e-mail: pawel.ziolkowski1@pg.edu.pl, Poland.

Abstract. The technology of thermal utilization of meat-and-bone meal (MBM) is still at the research state. However, there are already a number of existing pilot-scale installations that provide the complete, effective and environmentally safe process conducting. The universal technology was developed that allows to combust various types of waste, including animal waste, municipal waste and sludge, mixed at any ratio with different types of biomass. It provides the possibility to utilize the waste-and-biomass fuel mixtures of up to 90%wt of moisture content, while maintaining the allowable pollutant emission to the atmosphere and soil. This regards mainly NOx, SO_2, HCl and VOC. Contrary to the typical large scale grate boilers used for waste incineration, the developed operating pilot-scale plant with a capacity of 12MW offers the complete combustion of animal meal, resulting in a flue gas which is proved to be free of flammable gaseous components and sooty particles in slag and fly ash. The thermal decomposition and combustion of waste using this technology ensures thermal conversion of chemical energy contained in waste and biomass into the heat and electricity, while preserving the optimum thermodynamic efficiency of technological system. The efficiency of the prototype installation varied between 88.36 and 84.84% depending on the load.

1 Introduction

An increasing worldwide production of meat leads to an increased amounts of animal waste that require management. Due to the risk of the Bovine Spongiform Encephalopathy (BSE) that arised in 1980s and 1990s, this type of waste started being considered as hazardous [1-3]. Since then, implemented regulations, restricted its use as an additive to cattle feed and direct disposal in landfilling as natural fertilizer [4-5].

Addressing the need for environmentally safe destruction of this type of waste, including meat-and bone meal (MBM), the efforts have been undertaken to develop efficient thermochemical technologies that would allow their energy utilization with low emissions [6-10]. Since MBM is classified as hazardous, its thermal utilization requires special processing conditions, namely, relatively high temperatures (above 800°C) and appropriate residence time. Therefore it used to be incinerated in cement kilns [11,12]. MBM is also considered as an alternative fuel in combustion and co-combustion with other solid fuels. This include the combustion in grate boilers [13], as well as in fluidized bed combustors [1, 5, 11,14], which offer high process temperatures and long residence times [15]. Combustion of MBM as the save disposal method is also justified due to valuable properties of its ashes, which being free from harmful pathogens may serve as fertilizer [16].

To carry out environmentally sound and efficient thermal conversion of meat industry by-products that would meet the regulations, the in-depth recognition of sub-processes and phenomena occurring during the process is needed. This also refers to the material properties and behavior as the basis for the determination of the potential use of this kind of waste for energy purposes [3, 18]. This has contributed to the development of technology, which proved to be thermodynamically efficient and to provide low levels of contamination and dust emissions. The paper presents the detailed characteristic of the system and shows the key performance parameters of a pilot-scale facility of 12MW capacity.

2 Description of a system

The described herein technology provides the possibility to incinerate not only the animal waste, but also other types of organic waste such as sewage sludge, municipal waste and biomass of different kind. Its application allows to utilize the biodegradable waste of the overall moisture content up to 90%wt and whilst preserving the maximum allowable levels of atmosphere and soil contamination.

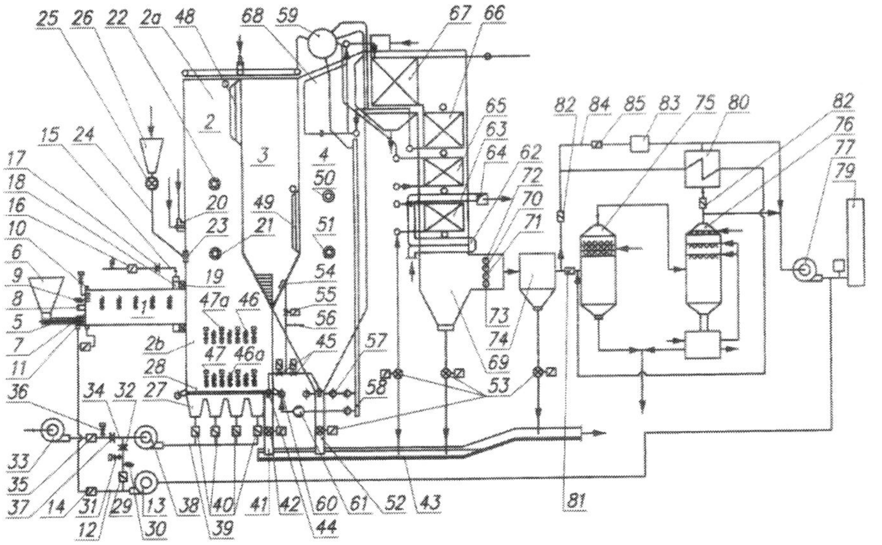

Fig. 1. The schematic of the installation for thermal utilization of meat-and-bone meal.

The energy recycling of hazardous and non-hazardous waste implementing the two-stage technology that use the rotary reactor to pyrolyze the feedstock and the fluidized bed chamber to combust the pyrolysis gas and the remaining char, ensures thermal conversion of waste with flue gases free from any environmentally harmful dioxins and furans. The combustion process is supplemented with the liquid or gas fuel, which is added in an amount ranging from 0.015 to 0.070 kg per kilogram of a feedstock, depending on its moisture content. The physical structure of waste to be utilized may be solid, as well as in form of a pulp or a dense slurry. In each case, the technology provides the optimum thermal effect. In view of this, the presented waste/biomass-to-energy technology allows to obtain the maximum levels of thermal efficiency and protection to the environment and human health and, additionally, the optimum economical effect.

The primary fuel in this case is an animal waste, biomass of different kind and alternative fuels. As aforementioned, the auxiliary fuel used to initiate the process is either gas or liquid fuel. However, it is preferable to use the renewables as supporting fuels since all the energy generated in the system would then come from the renewable energy sources. The minimum amount of feedstock material for which the installation can be operated is 20% of its nominal efficiency, whereas maximum reaches 150% at a moisture content of 80%.
The combustion takes place under conditions that comply with the relevant requirements of the directives and regulations concerning the safe utilization of animal waste. Namely, it is carried out at temperatures higher than 850°C, with oxygen content exceeding 8% and the residence time longer than 2 seconds.

The reported method of thermal utilization is a continuous-type process that proceeds in a system of integrated devices. These are equipped with measuring and control instruments for reading local parameters and remote transmission of signals to control units. Automatic control is performed according to the prescribed operation algorithm. The control system covers the course of a technological process, its visualization, archiving of operation parameters and occurring events, and monitoring of parameters of substances leaving the installation. And thereby, it enables direct and immediate interference with the process course in order to preserve the required parameters.

The schematic of the installation for thermal utilization of animal waste, including the rotary pyrolyzer and the fluidized bed boiler is shown in Fig. 1. The MBM to be burnt is fed into the boiler storage tank (6) together with the limestone in a quantity corresponding with a stoichiometric ratios needed to neutralise the sulphur and chlorine content in a fuel. They are mixed there and then transported by the fuel feeder (5) into the rotary kiln pyrolyzer (1). The mixture falls into the pyrolyzer chamber via the chute of a feeding unit. In order to ensure an even fuel feed into the chamber, the waste is supplied in a stream of flue gases that flow in a gas-box (11) surrounding the feeder. The exhaust gas is pumped through the pipeline (12) into the gas-box using the re-circulation fan (13), and taken out using the suction fan (77). The amount of flue gas required for the process is adjusted by opening or closing the flap (14).

On the side of fuel charging system, the rotary pyrolysis chamber is closed by a flat front panel (7), which is connected to the chamber through the channel with a labyrinth sealing (10) that is filled with recirculating exhaust gas. Its quantity is controlled via an adjustable slide. In a central axis of the flat front panel is located a supporting burner (gas or oil) (8), used to provide the appropriate thermal conditions for drying and pyrolysis of a fuel. The panel is also equipped with a thermocouple (9) and a pressure plug. The other end of a rotary chamber is connected to the fluidized bed chamber (2). This connection is additionally sealed through an air canal (18) equipped with a swirl unit (19) and the compensating slides of vertical elongations. The canal is supplied with air through the pipeline (15) fitted with the measurement orifice (17), the regulation flap (16) and a thermometer. In order to ensure the correct course of drying and devolatilization in a rotary chamber, the fuel feedstock is raised up to over 70% of the chamber diameter by the specially designed lifting flights mounted inside. This limits the possibility of sintering and agglomeration of a feedstock, and thereby intensifies the heat transfer within the material and promote the moisture and gas release. Additionally, along the rotary chamber centreline the thermometers are located to control the processing temperature. This, in case of thermal utilization of meat meal, should remain between 900 and 1200°C. The thermometers enable the proper control of a supplying burner operation so as to ensure the relevant temperature distribution. The rotation speed of a pyrolyzer within a range between 0.5 and 5 rpm provide the residence time of a char in a chamber at the level ranging from 5 to 10 minutes.

Gas mixture and solid carbon residue resulted from the drying and pyrolysis of a meat meal are directed to the fluidized bed chamber. It is built of a sealed membrane-type walls bearing a heating medium. Pyrolysis gas is supplied to the upper part of a chamber and combusted there, whereas char is burned in a fluidized bed in a hopper-shaped lower part of chamber. At the bottom, the chamber is closed by the orifice plate (28), which is covered with a refractory concrete on the fluidized bed side. The orifice plate is closed by a wind box (27), which is divided into a number of sections to allow the adjustment of pressure and the gas flow in each bed zone. The amount of gas supplied to each zone is controlled by opening and closing the flaps (40). It flows in from the collector (39), which is fed by the fluidizing fan (38) with adjustable settings of gas efficiency and pressure. The fan draws in the flue gas from the damper located downstream of the sucking fan (13). The flue gas is supplied to the fan through the duct equipped with the shut-off and/or regulatory slide (29), the thermometer (30), the manometer (31) and the measurement flange (32). To generate the fluidizing gas, the blower (33) provides additional air supply to the fluidizing fan. Air is transported via the pipeline (34), similarly having the shut-off and/or regulatory slide (35), the manometer (36) and the orifice (37). The exhaust and air fans are equipped with the systems for adjusting the quantity and the pressure of supplied media.

The residual ash is removed from the fluidized bed using an ash discharging system, comprising the discharge channel (41) and the remotely operated rotary feeder (42). The channel is situated next to the vertical rear wall of a fluidized chamber hopper. Ash is then directed to the bucket feeder (43), which is placed in a water tank that serves as a water trap for a fluidized bed chamber. The ash discharge channel is also equipped with a compressed air impulse nozzle (44) operated by a remote shut-off valve (45), which supports the ash removal efficiency.

The lower zone of a chamber, namely the fluidized hopper is formed by a vertical front and rear walls, and the side walls inclined toward the chamber centre with an inclination angle less than 45°. The hopper is insulated with a refractory concrete to protect metal elements of the walls from erosion.

To ensure the proper operation of a fluidized unit, the hopper is equipped with a relevant measuring system allowing to monitor the bed parameters, and to determine the fluidizing gas composition and the amount of inert material. The system include the temperature sensors (46) and (46a), as well as the manometers (47) and (47a), evenly spaced across the left and right inclined walls of a hopper. Manometers and thermocouples are spaced alternately in a horizontal plane at two heights, namely at the distance of 200÷500 mm and 2÷4 m above the orifice plate. Both, the lower and the upper measuring points in a vertical plane are positioned in pairs to enable the differential pressure measurement in a bed.

As aforementioned, a fluidized bed is generated in a chamber hopper as a result of a fluidizing gas flow i.e., the air/flue gas mixtures with the volumetric ratios ranging from 10%/90% to 90%/10%, through the inert bed material. This is composed of a mixture of silica sand and properly grinded slag, which is blended with limestone in its mass fraction ranging from 2 to 80 %, depending on the sulphur, chlorine and the fixed carbon contents in a fuel. The material is fed into the fluidized bed chamber periodically through the inlet window (23) from the storage tank (26) and through the duct (24) with a rotary feeder (25).

The flow velocity of a fluidizing gas through the bed, lies in a range between 1 and 4 m/s, regardless of the heat load. The fluidization process must be conducted in such a way that the temperature in a bed does not exceed the characteristic ash softening temperature. The process temperature ranges within 750 and 900°C, and the volumetric fraction of oxygen in fluidizing gas varies between 2 and 20 %.

As already stated, the pyrolysis gas is combusted in the upper zone of a fluidized bed chamber. The process is carried out with the use of a burner (20) mounted in a front chamber wall. The burner is powered by liquid or gas fuel and is needed to ignite and stabilize the process of gas combustion. This is conducted at temperatures of 1200÷1300°C and must be controlled so as not to exceed the maximum of 1300°C that fosters rapid formation of nitrogen oxides. To provide the process control the combustion air is distributed. The primary air is supplied through the sealing channel (18) equipped with a swirl vane, which connects the rotary pyrolyzer to a fluidized bed chamber. This air amounts to 0.2–0.4 of the stoichiometric air needed to burn the meal. Of the same quantity are also supplied the secondary and the third air streams, by the nozzles (21) and (22), respectively. Such air distribution helps to extend the combustion zone, leading in result to volumetric heat load of a chamber that allows to prevent from exceeding 1300°C in a burning zone.

Exhaust gases are directed through the upper festoon (48) to the separation chamber (3). The festoon is made of three rows of wall tubes of a separation chamber, to which the U-profiles are attached. These, with their open parts face the flue gas inflow, allowing the precipitation of particulates and the condensation of alkali and heavy metals' vapours. When hitting the U-profile, the exhaust particles decelerate and fall back along the section down to the bed. The heavy metals' vapours condense on a profile wall forming a glassy-like deposit, which is blown away downwards back to the bed with soot blower installed on a top of a chamber, above the festoon tubes. The separation chamber which constitutes the rear chamber wall, is closed by a baffle placed in front of rotary pyrolyzer and the chamber walls. The baffle, being bended in its lower part towards the chamber rear wall at a maximum angle of 45° serves as a tightly closing heated membrane-type surface. The bended baffle surface has the discharge hoppers along the chamber width to drive back the separated particles to the bed using compressed air supplied by the impulse nozzles (54). Rear wall of a fluidized bed chamber is of similar shape, however bended towards opposite direction it closes the separation chamber. The bendings are covered with refractory cement protecting them against erosion. Exhaust gases, partially purified of solids and vapours, change direction when passing upper festoon and flow through the separation chamber. There, due to change in direction and reducing the flow velocity, they are further de-dusted. Next, the treated flue gas flows through the lower festoon (49) to the afterburner chamber (4). The burn out of combustible gases remaining in a flue gas is carried out using oil or gas burner (50), ensuring ignition and the process stabilization. The complete burnout is provided by mixing the exhaust gases with an additional (fourth) air dosed by a nozzle (51) at the amount ranging from 0.1 to 0.3 of stoichiometric quantity.

There is a bulkhead superheater (68) mounted in an upper part of the afterburning chamber, and the second stage superheater (67) with the vapour temperature controller placed in a crossover duct. There are the first stage superheater (66), the water heaters (65) and (63), and the air heater (64) placed in a convection channel. Fumes, free from gaseous and solid combustible components, give off the heat by convection when flowing through the system. They are analysed at the outlet of the convection part. i.e., in an exhaust duct, in terms of their chemical composition, ash content, pressure and temperature, using spigots (70)-(73), respectively.
Flue gas flowing out of a boiler goes next through the fabric filter (74) to remove particulates. Purified gas is next fed to the water sprinkler (75) and the scrubber (76), where the sub-micron solid particles, as well as alkali and heavy metals' vapours are eventually separated, and gaseous molecules of sulphur and chlorine oxides are chemically bonded by a lime slurry sprayed in a scrubber. The parameters of exhaust gases entering the stack are monitored with the use of analyser (78). These include the temperature and the contents of ash and soot, moisture, carbon oxide, sulphur and nitrogen oxides, and hydrocarbons. In case when temperature exceeds the dew point temperature, the fumes flow through an exhaust fan (77) directly to the stack (79). Otherwise, a flue gas is first directed to the heat exchanger (80) and then to the chimney.

Ash removed from flue gas in the hopper of afterburning chamber, in the discharge hopper of firing zone and convection zones, and in the fabric filter, similarly as in the case of fluidized bed chamber, is conveyed through the channels equipped with rotary feeders. It further goes to a bucket feeder and finally to a disposal site. As regards the water circulation system, the medium is transferred from the heater to the drum (59), from which it is supplied through the down-comers (58) to the wall tube manifold of afterburning chamber (57) and the wall tube manifold of fluidized bed chamber (60). The steam generated in a boiler goes back to the drum, and then is directed to the superheaters and the turbine. The medium flow in a fluidized bed chamber is forced by a water circulation pump (61).

3 Pilot-scale installation of 12 MW capacity

An experimental facility was built in the town of Ostrowite, the Lniano municipality, in Kujawy and Pomerania Province. It was developed to serve as a source of thermal energy in the form of steam for technological use in an animal waste treatment plant. The installation continuous to operate. The system consists of the rotary pyrolyzing reactor, the fluidized bed chamber and the waste heat recovery boiler. As described previously, drying and devolatilization are carried out in a pyrolyzer, whereas the fluidized bed chamber is used to combust the pyrolysis products, namely the gas components in its upper zone and the remaining solid (char) in its lower zone. Burning of volatiles takes place in a controlled manner with the multistage air supply. Char combustion in a fluidized bed is controlled by an oxygen concentration in a fluidizing gas, which is a mixture of air and recirculating flue gases. The exhaust gas generated in a fluidized bed chamber, in the first place, undergoes cleaning in a separation chamber aimed to extract the fly ash. After partially purified, the flue gas is directed to the shell-type heat recovery boiler that produces steam at pressure of 5 bars and temperature of 250°C. The view of the installation is displayed in Fig. 2.

Fig. 2. The general view of the pilot-scale installation.

The presented technology system provides the complete combustion of animal meal and the elimination of combustible components from slag and the fly ash, as well as flammable gases from fumes. As confirmed by the measurements, the installation ensures the emissions of harmful substances below the permissible standards, including NO_x, SO_2, HCl and VOC. The implemented two-stage technology that utilize the rotary reactor for drying and degasifying waste fuel, and the fluidized bed chamber for separate combustion of pyrolysis gas and char, allows to control the individual stages of fuel combustion. These refer to the chemistry of a process. i.e., an elementary composition of a reactive atmosphere, and to its thermal regime, i.e., characteristic process temperature range.

The commonly used installations for waste incineration based on grate-type boilers do not provide the complete combustion and they generate a slag that contains substantial quantities of combustible components in form of char, exceeding mass percentage of 5%. It consists of carbon in nearly 100%. This slag is thus a waste, which has to be further disposed. In the case of studied prototype system for thermal conversion of waste, particularly the MBM, the remaining ash constitutes a valuable raw material for the production of phosphorous-, potassium- and calcium-magnesium-based fertilizers. The content of phosphates in ash reaches 25-28% and is about 5 times higher than typical contents in natural minerals. The obtained ash is a valuable resource, basically due to high levels of P_2O_5 and the trace levels of combustible parts remaining at 0.5% by weight.

In the case of considered prototype installation in Ostrowite, the amount of auxiliary (liquid) fuel ranges between 0.01 and 0.08 kilograms per 1 kg of a meal, depending on a moisture content in a target waste fuel. In any case, the system provides the positive thermal effect. The heat generated in a furnace is used in a heat recovery boiler that produces steam at the pressure of 4-6 bars and temperature T=250-300°C, depending on the technological needs. At the start-up stage, namely at the stage needed to establish the temperature balance in the system, the light heating oil is used. When the assumed temperature in a rotary pyrolysis reactor is achieved, the ignition fuelling system is switched from heating oil to heated animal fat of temperature T=80°C. This type of kindling fuel is classified as an alternative fuel from renewable resources, and thereby the total thermal energy generated by the system is regarded as renewable energy. The installation provide high availability and reliability. The minimum quantity of meal required for operating the system amounts to 20% of a nominal capacity, whereas the maximum corresponds to the nominal capacity of 120%.

As aforementioned, the process of MBM utilization is of continuous-type, and the facility equipped with automatic monitoring and control system based on the prescribed control algorithms allows an on-line adjustment of operating settings so as to meet the required output parameters.

3.1. Meal supply system

Meat meal is fed into the hopper of a unit of external conveyors, which transfer the fuel into an interim bunker hopper. From there the meal is fed into a loading hopper of a rotary pyrolyzer. To avoid dust lift-off and the spread of odours, the fuel supply unit is hermetically sealed. Additionally, it is placed in a room with continuous exhaust system of contaminated air, which is further used for fuel combustion. The installation is also fitted with a grate to separate bones exceeding the dimension of 50 mm. The amount of supplied meal is regulated with the use of inverter setting the travelling speed of a feeder.

3.2. Drying and thermal decomposition

The processes of drying and devolatilization take place in a rotary kiln reactor with a diameter of 1200 mm. Its chamber is equipped with specially designed material lifters, which are distributed alternately in three sectors along the chamber length. The chamber rotates at 0.5-5 rpm and is inclined by 2-3° towards an outlet. It is made from an uncooled pipe with a refractory lining inside and closed by a sealed faceplate from the front. In a faceplate there is installed an oil burner, equipped with the systems of ignition and flame control. There is also mounted the charging device of meal and sorbent. The quantity of liquid fuel to be combusted is automatically adjusted, so as the temperature inside the chamber remains within range between 850 and 1100°C, according to the type of animal meal being utilized. The amount of air supplied to the rotary kiln pyrolyzer depends on a supplementary fuel quantity and is set to ensure the complete combustion and nearly zero content of oxygen in the drying and pyrolysis zones. The process is carried out in a reductive atmosphere to avoid burning and heat release. The rotary chamber is connected with a fluidized bed chamber via a swirling element for primary air, which is necessary to partially combust evolved pyrolysis gas.

3.3. Combustion

Burning of pyrolysis products takes place in several stages in a fluidized bed boiler, which is directly connected to an outlet of the rotary kiln reactor. The boiler chamber is made of tight wall tubes that constitute the evaporator unit. At the chamber wall in a fluidized bed zone there is installed a burner for igniting and stabilizing combustion. There is an orifice plate at the chamber bottom, which provokes fluidization. It is composed of several sections, inclined towards an ash discharge hopper. The post-pyrolysis char is combusted in a fluidized bed, whereas pyrolysis gas is burnt in the upper part of a chamber in its consecutive sections being supplied with primary, secondary and the third air, respectively. Temperature in a flame zone is within 1100 and 1300°C, and an oxygen content in a flue gas amounts to 7-8%. Temperature inside a fluidized bed ranges between 700 and 950°C and is being set according to ash softening characteristics. An excess-air ratio is within 1.1 and 1.25, whereas an oxygen content in a fluidizing gas is 6%. The residence time of pyrolysis gas in a fluidized bed chamber at the required temperature range of 1200-1300°C is 4-6 seconds. The residence time of the solid pyrolysis residue in a fluidized bed is about 10 minutes. The exhaust gases leaving the chamber, being next treated when passing through the separation chamber, flow to the waste heat recovery boiler.

3.4. Tests of the facility

Experimental study of combustion of animal meal blends in the prototype installation was performed for four different heat loads, corresponding to the meal quantities of 500, 1000, 1500 and 2000 kilograms per hour. The chemical composition of examined blend, as well as post-combustion ash were analysed.

The experimental analysis covered mainly the measurement of: i) fuel quantity, ii) temperature inside a pyrolysis reactor and inlet/outlet pressure, iii) pyrolysis gas composition, iv) flue gas temperature and composition, v) steam parameters (output, temperature and pressure) and vi) carbon content in ash.

The performed measurements indicated low levels of CO (~0.15%), relatively low content of SO_2 (between 115 and 233 mg/Nm^3) and constant level of NOx emission (180-221 mg/Nm^3) independent on the waste fuel quantity. Also the content of unburned combustibles wass very low, particularly in case of a bottom ash (up to 0.4%), regardless of the heat load. In the case of fly ash this content fluctuated between 1.0 and 1.5 %.

Simultaneously, the thermodynamic efficiency of a system varied between 88.36 and 84.84%, depending on a thermal input.

4 Conclusions

In the paper we present the two-stage technology for waste utilization, which offers the fuel incineration with a maximum available thermal efficiency and the minimum levels of emitted gaseous components and unburned carbon in fly ash. This was confirmed by the tests of animal meal utilization carried out in a pilot-scale installation with an output of 12MW.

The developed technology is universal, allowing to incinerate any waste, as well as any mixtures of waste and biomass.

Acknowledgments: The part of work has been prepared within the frame of 3 institution mainly: IMP PAN, PG and PW.

References

1. Cascarosa E., Boldrin A., Astrup T., Pyrolysis and gasification of meat-and-bone-meal: Energy balance and GHG accounting, *Waste Management*, Vol. 33, pp. 2501-2508, (2013).
2. Staroń P., Kowalski Z., Staroń A., Banach M., Thermal treatment of waste from the meat industry in high scale rotary kiln, *International Journal of Environmental Science and Technology*, Vol. 14, pp. 1157-1168, (2017).
3. Kantorek M., Jesionek K., Polesek-Karczewska S., Ziółkowski P., Thermal utilization of meat and bone meals. Performance analysis in terms of drying process, pyrolysis and kinetics of volatiles combustion, *Fuel*, Vol. 254, 115548, (2019).
4. Campbell W.A., Fonstad T., Pugsley T., Gerspacher R., MBM fuel feeding system design and evaluation for FBG pilot plant, *Waste Management*, Vol. 32, pp. 1138-1147, (2012).
5. Gulyurtlu I., Boavida D., Abelha P., Lopes M.H., Cabrita I., Co-combustion of coal and meat and bone meal, *Fuel*, Vol. 84, pp. 2137-2148, (2005).
6. Poskrobko S., Identification and stabilization of combusting animal waste with active participation of bone material – Emission of SO_2 and HCl, *Fuel Processing Technology*, Vol. 113, pp. 20-27, (2013).
7. Lopes H., Gulyurtlu I., Abelha P., Crujeira T., Salema D., Freire M., Pereira R., Cabrita I., Particulate and PCDD/F emissions from coal co-firing with solid biofuels in a bubbling fluidised bed reactor, *Fuel*, Vol. 88, pp. 2373-2384, (2009).
8. Kantorek M. i inni: Możliwości technologiczne energetycznego wykorzystania mączki mięsno-kostnej, *Archiwum Spalania* 2009. Vol. 9. pp. 135-149, (2009).
9. Kantorek M.. i inni: Możliwości wykorzystania mączki mięsno-kostnej jako paliwa w instalacjach energetycznych, *Energetyka*, Vol. 1, pp. 39-47, (2009).
10. Karcz H.. Kozakiewicz A., Sposób termicznej utylizacji odpadów zwierzęcych, *Energetyka*, Vol. 11, pp. 823-831, (2007).
11. Fryda L., Panopoulos K., Vourliotis P., Pavlidou E., Kakaras E., Experimental investigation of fluidised bed co-combustion of meat and bone meal with coals and olive bagasse, *Fuel*, Vol. 85, pp. 1685-1699, (2006).
12. Vermeulen I., Van Caneghem J., Block C., Dewulf W., Vandecasteele C., Environmental impact of incineration of calorific industrial waste: Rotary kiln vs. cement kiln, *Waste Management*, Vol. 32, pp. 1853-1863, (2012).
13. Orszulik E.. Lenkiewicz D., Współspalanie węgla kamiennego z mączką mięsno-kostną w kotłach rusztowych. Energetyka, Vol. 11, pp. 831-836, (2007).
14. McDonnell K., Desmond J., Leahy J.J., Howard-Holdige R., Ward S., Behaviour of meat and bonemeal/peat pellets in a bench scale fluidized bed combustor, Energy, Vol. 26, pp.81-90, (2001).
15. Bis Z., *Kotły fluidalne: teoria i praktyka*, Wydaw. Politechniki Częstochowskiej, Częstochowa, (2010).
16. Simeon O., *Fluidized Bed Combustion*, Taylor&Francis, (2003).
17. Staroń P., Kowalski Z., Staroń A., Seidlerova J., Banach M., Residues from the thermal conversion of waste from the meat industry as a source of valuable macro- and micronutrients, *Waste Management*, Vol. 49, pp. 337-345, (2016).
18. Kantorek M., Energetyczny recykling mączki zwierzęcej, PhD Thesis [in Polish], Politechnika Wrocławska, (2015).

E3S Web of Conferences **137**, 01014 (2019) https://doi.org/10.1051/e3sconf/201913701014
RDPE 2019

RESEARCH & DEVELOPMENT IN POWER ENGINEERING, 2019

Heat transfer characteristics of an external tube heat exchanger in a 1296t/h circulating fluidized bed boiler

Artur Błaszczuk[1,*] and *Szymon* Jagodzik[2]

[1] Czestochowa University of Technology, Institute of Advances Energy Technologies, Dabrowskiego 71, 42-200 Czestochowa, ablaszczuk@is.pcz.pl, Poland
[2] Tauron Wytwarzanie S.A., Lagisza Power Plant, Pokoju 14, 42-500 Bedzin, Poland

Abstract. In this research article, experimental study was carried out to obtain the heat transfer characteristics between a submerged horizontal tube bundle and a fluidized bed in a large-scale circulating fluidized bed (CFB) boiler with an external heat exchanger (EHE). The operational parameters in the tube EHE were measured during performance tests at variable load conditions. The average heat transfer coefficient (HTC) was calculated using a mechanistic heat transfer model based on packed renewal theory. The heat transfer characteristics are considered in terms of heat transfer mechanisms such as emulsion phase convection, gas convection and also thermal radiation. The obtained heat transfer data exhibit a maximum value with variation mean bed particle size irrespective of pressure. The results showed that the average HTC increases with a decrease of the Sauter mean particle diameter and with the increase of the fluidizing number as a result of good mixing dynamics in emulsion phase (i.e. emulsion wall contact time, bubble fraction in the bed). Based on the heat transfer data, empirical correlations are proposed for predicting a heat transfer coefficient from fluidized bed to horizontal tube bundle. The mechanistic heat transfer model predicted the average HTC in sufficiently good agreement with CFB boiler data accessible in the literature.

1 Introduction

Knowledge about the heat transfer coefficient from bubbling fluidized bed particles to the tube heat exchanger is necessary for an optimal/more reliable design and scaling-up of heat transfer surfaces in processes where the control of temperature plays an essential role from the point of view of proper operation and optimum performance of CFB boilers. This is especially important in the case of CFB combustors fired with various fuel types, at turn down boiler loads and also SO_2 and NO_x emission control levels.

The heat transfer between the tubular surface and bed particles depends on the local properties of the emulsion phase (i.e. heat conductivity, emulsion density, emulsion heat capacity) and also the properties of the bubbling bed (i.e. bubble fraction and emulsion-wall contact time with horizontal tube bundles). The above mentioned properties affect the dynamics of the emulsion phase mixing process, thus ensuring a uniform bed temperature, a high bed-to-tube heat transfer coefficient, high combustion, maintenance of proper steam temperature and flexible operation across a wide range of CFB boiler loads [1]. In addition, the flow of gas bubbles is a factor that determines the intensity of the heat transfer process from the bubbling fluidized bed to the surface of the tubular heat exchanger.

Not many researchers have studied heat flow to a horizontal tube bundle immersed in a large-scale bubbling fluidized bed so as to better understand the effect of operational parameters on the bed-to-tube heat transfer coefficient. In the last decade, several experiments have been conducted in a laboratory-scale FBHE with submerged horizontal tube bundles in BFB to evaluate the impact of superficial gas velocity [2] and also placement of the immersed surface [3] on the value of bed-to-tube heat transfer coefficient. Di Natale, Bareschino and Nigro [4] also studied the effect of angular position around a horizontal cylinder with the fluidization velocity on the heat transfer coefficient inside a bubbling fluidized bed. Stenberg et al. [5] investigated the influence of the particle size and the superficial velocity on the bed-to-tube surface heat transfer coefficient in a horizontal tube at a high level of bed temperature (400-950°C) in a laboratory scale BFB system. In the literature, most studies focus on laboratory scale. Therefore, it is reasonable to conduct research studies focused on heat transfer characteristics of horizontal tubes submerged in large-scale gas fluidized beds.

Hence, it has been attempted in this research study to investigate the heat transfer characteristics in a large-scale bubbling fluidized bed and obtain more information about reliable predictions of the bed-to-tube heat transfer coefficient. In the present study, we have used a mechanistic heat transfer model based on the packed renewal theory [6, 7], which correlates the bed-to-surface heat transfer to the residence time of bed particles at the surface of horizontal tube bundles. In order to consider the impact of bubble fraction, emulsion-wall contact time, Sauter mean particle diameter, suspension density, and fluidization number on the enhancement of heat transfer in EHE, performance tests at variable CFB boiler loads were carried out. This research paper presents unique data on heat transfer between the bubbling fluidized bed and the horizontal tube bundles in a 1296t/h CFB boiler. It should be noted that the experimental investigations for commercial CFB systems are rarely published in the accessible scientific literature. Therefore, predicting the heat transfer characteristics will be very useful for the regulation of fluidized bed heat exchangers. Besides, the obtained heat transfer data can also be used to validate empirical correlations for the design and scale-up of heat transfer surfaces for commercial CFB boilers and bridge a gap in the accessible CFB & BFB (bubbling fluidized bed) literature.

2 Mechanistic heat transfer model

The methodology to predict the bed-to-tube heat transfer coefficient was adopted from the packed renewal model proposed by Mickley and Fairbanks [7]. According to the packed renewal approach, the active heat transfer surface will be alternatively in contact with the gas bubbles and the packed particles (i.e. parcels of emulsion phase). The packed particles sweep of the active heat transfer surface in the result of the gas bubbles flow. The occurrence of gas bubbles is purely randomized.

In the mechanistic heat transfer model presented, the heat transfer process between the bubbling fluidized bed and the submerged surface for the bottom tube bundles h_{bottom} in FBHE (fluidized bed heat exchanger) contains three varieties of heat transfer mechanisms, including emulsion phase convection during time of solid contact, gas convection during time of bubble contact and a radiative component in the case of high temperature operation [8, 9].

$$h_{bottom} = h_c + h_r = (1 - \delta_b)h_e + \delta_b h_g + h_r, \quad (1)$$

where h_c denotes convective heat transfer coefficient, h_e is packet particles convective heat transfer, h_g represents gas convection heat transfer coefficient, δ_b denotes bubble fraction and h_r means radiative heat transfer coefficient. A description of the individual heat transfer components in equation (1) is provided below.

The convective heat transfer coefficient due to particle packets is obtained according to the following relationship:

$$h_e = \frac{2}{\pi}\sqrt{k_e \rho_e c_e}\left(1/\sqrt{t_e}\right), \quad (2)$$

where c_e denotes emulsion phase heat capacity, k_e represents thermal conductivity of emulsion phase, t_e is emulsion-wall contact time and ρ_e means density of packets (emulsion phase). Specific information on formulas on physical and thermal properties of emulsion phase is summarized in earlier work [6].

The gas convection heat transfer coefficient for the tube surface uncovered by the particle packets is estimated from the relation given by Baskakov et al. [10],

$$h_g = 0.009\left(k_g/d_p\right)Ar^{0.5}Pr^{0.33}. \quad (3)$$

In the above equation (3), Ar denotes Archimedes number, d_p is Sauter mean particle diameter, k_g means thermal conductivity of gas phase and Pr refers to Prandtl number. The relation (3) is useable for a wide range of fluidizing conditions [11].

Thermal radiation contributes significantly to the bed-to-tube heat transfer coefficient especially for bubbling fluidized beds operated at high bed temperature ($T_b >$ 600°C). The empirical expression for the radiative heat transfer coefficient h_r is defined as:

$$h_r = \sigma(T_{eb}^4 - T_w^4)/\left(\frac{1}{e_b} + \frac{1}{e_w} - 1\right)(T_{eb} - T_w), \quad (4)$$

where e_b denotes effective emissivity of particles in the bed, e_w is effective emissivity of horizontal tube surface, T_{eb} means effective bed temperature, T_w represents metal (wall) temperature and σ refers to Stefan Boltzmann constant.

Equation (1) may not be used to predict the bed-to-tube heat transfer coefficient due to different particle packing behavior around the horizontal tube surface. There is a defluidized cap called a "*lee stack*" formed on the top section of tubes bundles at the velocity near the minimum fluidization velocity U_{mf}. For the lower portion of the tube bundles, there is an alternating character to the bubble flow around the tubes which results in a constant replacement of the emulsion phase in the bottom section. Therefore, the average heat transfer coefficient depends upon the placement of the active heat transfer surface in an external tube heat exchanger. To take into account of the influence of the tube bundle placement in the fluidized bed heat exchanger, it is necessary to calculate the bed-to-tube heat transfer coefficient for the top section of the tube. In this work, a semi-empirical correlation (5) based on Reynolds number Re, Prandtl number Pr, tube diameter D, Sauter mean particle diameter d_p and also heat capacity of packets c_e and gas phase c_g is used to predict the heat transfer coefficient at the top section of the tube bundles [2, 12].

$$Nu_{top} = 47.56Re^{0.43}Pr^{0.33}\left(D/d_p\right)^{-0.74}\left(c_e/c_9\right)^{-1.69}. (5)$$

Thus, the final correlation regarding the average heat transfer coefficient is estimated using the following equation (6):

$$h_{avg} = 8/12\, h_{bottom} + 4/12\, h_{top}, \quad (6)$$

where, h_{bottom} refers to bed-to-tube heat transfer coefficient at bottom tube section and h_{top} is a bed-to-tube heat transfer coefficient at top tube section. From the BFB & CFB literature data, the relationship (6) for the average heat transfer coefficient is valid, whereas the Sauter mean particle diameter is less than 0.535mm.

3 Input data to heat transfer modelling

The packed renewal model for the heat transfer flow in the bubbling fluidized bed with submerged horizontal superheater tube bundles is based on formulas (1)-(6). Under this approach, the following assumptions are used in formulating the heat transfer flow from the fluidized bed to the horizontal bundles of smooth tubes in an external heat exchanger: (i) the packed particles temperature and the gas temperature are the same, (ii) the gas bubbles and emulsion phase are treated as a continuous flow fluidized bed, (iii) the heat transfer surface is covered by the emulsion phase or gas phase

 https://doi.org/10.1051/e3sconf/201913701014

(bubbles), (iv) the bubble fraction in the fluidized bed is estimated in agreement with the classical two-phase theory for aggregative fluidization [12], (v) heat capacity of packets equal to the heat capacity of the solid [13].

In order to perform heat transfer modelling in accordance with the packed renewal theory, twelve performance tests were carried out under steady-state operating conditions of the CFB system. All performance tests were carried out in a 1296t/h circulating fluidized bed boiler with an external heat exchanger (2.7 × 2.3 × 1.3m – depth × width × height). For all performed measurements, the test procedures are previously described by Blaszczuk and Nowak [14]. Table 1 summarizes the operating parameters used in this study. Some operating data as confidential commercial information will be not presented here.

Table 1. Test conditions in the INTREX™ heat exchanger.

Item	Unit	Overall range
Superficial gas velocity, U_g	m/s	0.14-0.26
Minimum fluidization velocity, U_{mf}	m/s	0.0228-0.0979
Pressure drop, Δp	kPa	12.34-14.72
Bed temperature, T_b	K	929-1083
Sauter particle diameter, d_p	mm	0.219-0.444
Particle density, ρ_p	kg/m3	2650-2750

The fluidized bed heat exchanger was operated with Group B particles according to Geldart® classification (see Fig. 1).

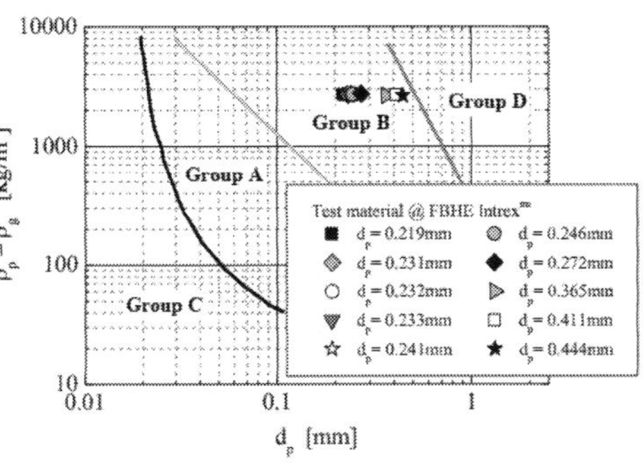

Fig. 1. Geldart powder classification of bed particles used in the FBHE at 1296t/h CFB boiler.

As can be shown in Fig. 1, the mean particle diameter of the bed material used in the FHBE is covered in a range of 0.219-0.444mm. Mean bed particle size is provided from a sieve analysis (i.e. direct dimensional measurement method). The device and the laboratory equipment used for the standard sieve analysis of bed material have been fully described in our other published works [6, 15].

In the external tube heat transfer exchanger, the horizontal tube bundles are generally arranged in staggered arrays (i.e. 32 columns × 12 rows). The detailed data on the apparatus in a fluidized bed heat exchanger have been described in [6, 16].

4 Results and discussions

In this work, the predicted values of the average heat transfer coefficient for horizontal bundles of smooth tubes are obtained from the mechanistic heat transfer model based on packed renewal theory. The empirical model of heat transfer coefficient used the parameters of solids and gas in a fluidized bed heat exchanger, which were recorded at different CFB boiler loads. The findings presented in this work are given in a dimensionless scale and presented in four subsections. Some measurement data are normalized by the maximum value of the recorded parameters during all performance tests. The influence of operating parameters (mean particle size, suspension density, superficial gas velocity, bed temperature) on the heat absorption characteristics of an external tube heat exchanger in a 1296t/h circulating fluidized bed boiler are studied by means of heat transfer correlations developed by us. The computational results obtained were correlated by the authors using regression analysis. In the current heat transfer study, average heat transfer data are compared with those in the available literature.

4.1 The correlation between average heat transfer data and Sauter mean bed particle size

In figure 2, a calculated average bed-to-tube heat transfer coefficient is plotted against the Sauter mean particle diameter. The computational results from the packet renewal model are shown for a wide range of bed particle size (i.e. from 0.219mm to 0.444mm). Each mean bed particle size was marked by an individual symbol for easier perception of findings. The average heat transfer data are heavily dependent on the mean bed particle size. During tests with the different Sauter mean diameters of bed particles, the average bed-to-tube heat transfer coefficient increased from 255-305[W/(m^{2}K)] for coarse particle sizes (i.e. d_p>0.365mm) to 317-391[W/(m^{2}K)] in the case of fine particles (i.e. d_p<0.272mm). The variation of the heat transfer coefficient is smaller for fine particles as opposed to coarser particles. The reason for such heat transfer behavior results from smaller voidage of the emulsion (packet) phase and in consequence the low emulsion packet conductivity for coarse bed particles. Fine particles have a larger number of contact points with the superheater tube wall than coarse particles. From that, therefore, it can be concluded that fine particles are more efficient in heat flow with a superheater tube surface. Wang et al. [17], Parmar et al. [18] and Pidwerbecki et al. [19] reported a similar trend in meaningful increases in the heat transfer coefficient with decreases in mean bed particle size. Heat transfer data for a bed particle size of 0.246m are characterized by a high scatter of results obtained. This fact is due to different physical properties of bed particles of about the same diameter. In the packet renewal approach, the packed convective heat transfer coefficient is directly proportional to packet conductivity and inversely proportional to the square root of emulsion-wall contact time. So, the differences in the values of average heat transfer coefficient can come from the fluidization characteristic of these particles. Another explanation for this fact is that the superficial gas velocity had a significant effect on the average bed-to tube heat transfer coefficient. When the bed particle size equals 0.246mm, the predicted values of heat transfer rate were obtained at a superficial gas velocity covering a range of 0.14-0.22m/s.

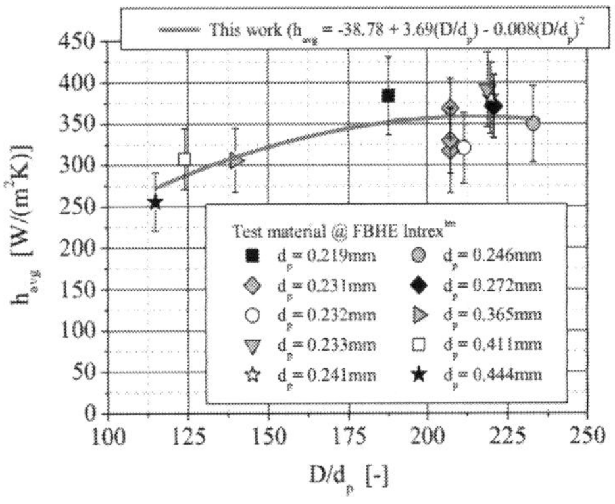

Fig. 2. Estimated average heat transfer coefficient from bed to tube at different Sauter mean particle sizes.

As can be shown in Fig. 2, at bed particle size equalling 0.233mm optimal heat transfer conditions (i.e. $h_{avg\ max}$ = 391[W/(m^{2}K)]) around the horizontal superheater tube are observed. In the present work, the heat transfer data are correlated with the ratio of diameter tube-to-mean particle size using regression analysis. The proposed empirical correlation is plotted as a red curve. The predicted values of the heat transfer coefficient have a good agreement with the proposed empirical correlation (R^2=0.5, moderate positive correlation). The uncertainty analysis of heat transfer data is carried out on the basis of the root-sum-square (RSS) approach. In Figs 2, 3, 4 and 5, the error bars represent uncertainty in calculated values of average heat transfer coefficient. Based on equations (1), (5) and (6), the uncertainty in the heat transfer data varied between ±35 and ±51[W/(m^{2}K)].

4.2 The correlation between average heat transfer data and suspension density for horizontal superheater tube

Fig. 3 depicts the heat transfer results of varying the suspension density for ten different mean bed particle sizes. The maximum value of the average heat transfer coefficient between the fluidized bed and the submerged superheater tube is recorded for normalized suspension density equalling 0.96 (@ ρ_b = 1006kg/m^3). All bed particle sizes show an almost linear/monotonic increase in the average heat transfer coefficient with increased suspension density. This tendency of heat transfer coefficient variation versus suspension density is in accordance with other published work [6]. The predicted heat transfer data are correlated with a normalized suspension density using non-linear regression. The calculated curve represented by the red line in Fig. 3 shows the same trend with the predicted values of HTC (i.e. heat transfer coefficient). The maximum deviation of heat transfer data from the obtained correlation function does not exceed +40[W/(m^{2}K)] in the current study. Plot 3 presents a high positive correlation for heat transfer data values at varying levels of normalized suspension density (R^2=0.8).

Figure 3 also reveals that the average bed-to-tube heat transfer coefficient for the fine particles appears to be somewhat higher at a high normalized suspension density (i.e. 0.91 < $\rho_b/\rho_{b\ max}$ < 1.0). The exception to the above mentioned rule is the value of the average heat transfer coefficient, which has been estimated for a mean bed particle size of 0.241mm. This fact can be explained by a very high bubble fraction (i.e. δ_b = 0.41 @ d_p=0.241 mm) around the horizontal tube bundles immersed in a large-scale bubbling fluidized bed. As a result, the bed particle had a short residence time on the tube surface (i.e. t_e = 0.13s @ d_p = 0.241mm). It should be noted that the emulsion packed conductivity is several times bigger than the conductivity of the gas phase (i.e. k_e/k_g = 5 @ d_p=0.241mm). Thus, the high heat transfer rate is dependent on the interaction of the emulsion phase and gas bubbles.

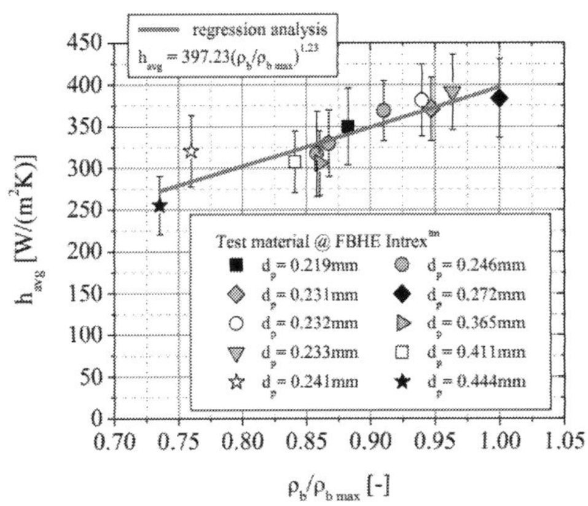

Fig. 3. Estimated average heat transfer coefficient from bed to tube at different normalized suspension densities.

Particle motion at the tube surface or particle exchange frequency (i.e. the replacement rate of packets particles by gas bubbles) depends upon excess gas velocity (U_g-U_{mf}) or fluidization number (U_g/U_{mf}). This issue will be discussed in the next subsection 4.3

4.3 The correlation between average heat transfer data and fluidization number for horizontal superheater tube

Figure 4 shows the calculated h_{avg} plotted against fluidization number for bed materials with Sauter mean diameters 0.219, 0.231, 0.232, 0233, 0.241, 0.246, 0.272, 0.365, 0.411 and 0.444mm respectively. The results presented in Fig. 4 are computed from the model described in section 2. Heat transfer data obtained for fluidization number varied between 1.75 and 10.18. It can be seen that for a horizontal tube submerged in a fluidized bed, there is an increase of average heat transfer

https://doi.org/10.1051/e3sconf/201913701014

coefficient with the fluidization number. The trend variation of average heat transfer coefficient as a function of fluidization number in figure 4 has been confirmed by other researchers [2, 5, 20, 21].
The lowest heat transfer coefficient is obtained for the bed particle diameter of 0.444mm with the fluidization velocity close to the minimum fluidization velocity. Meanwhile, the highest average heat transfer coefficient is recorded for U_g/U_{mf} = 10.18. It is evident that higher superficial gas velocity leads to faster heat flow between the fluidized bed and the superheater tube surface, as the average heat transfer coefficient showed an upward trend. At a sufficiently high fluidization velocity (i.e. 6.66<U_g/U_{mf}<10.18) coalescence takes place resulting in large gas bubbles which tend to cause a higher frequency of packet replacement, and good heat transfer is obtained. Besides, heat transfer characteristics in an external heat exchanger with horizontal tube bundles can be explained by means of amended bed fluidization quality (i.e. decrease of gas bubble size and increase of bubble frequency) and dynamic gas bubble flow at the tube surface with the variation of system pressure. Thus, the variation of h_{avg} is essentially affected by gas bubble action (i.e. growing bubble motion) at the heat transfer surface together with the downward sliding motion of bed particles or packets.

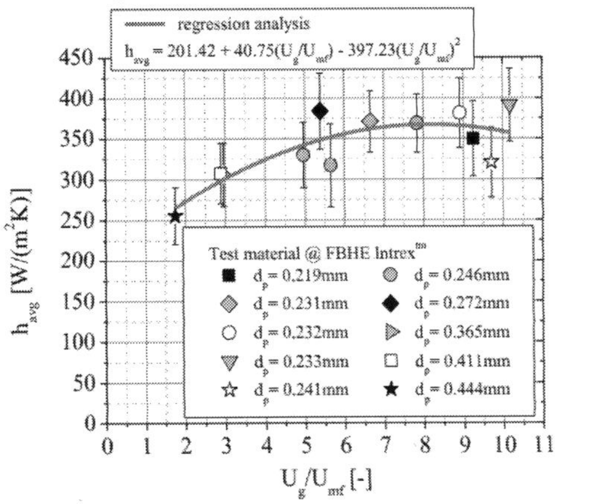

Fig. 4. Estimated average heat transfer coefficient from bed to tube at different fluidization numbers.

A polynomial of order two was fitted to plot the average heat transfer coefficient versus the fluidization number. The red curve in Fig. 4 is almost smooth with a medium gradient for coarse particles (i.e. d_p>0.365mm). Moreover, at the fluidization number covering a range of 6.66-8.91, the values of the average heat transfer coefficient slightly increase at a certain superficial velocity and then remain almost constant for 0.232mm<d_p<0233mm. Nevertheless, the preliminary increase of the average bed-to-tube heat transfer coefficient with fluidization number is caused by reduction in packet particles (emulsion) and residence time (i.e. from 0.25s to 0.17s), and further decrease is due to an increase in the bed porosity in vicinity of the tube wall at a higher fluidization velocity (i.e. 0.22<U_g<0.25m/s), as is depicted by determining the correlation function (red curve). In the current heat transfer study, the optimum fluidizing velocity equalled 0.25m/s for an enhancement of heat transfer rate in FBHE. As is seen from figure 4, the agreement between the correlation function and the predicted heat transfer data points is good (R^2=0.6, moderate positive correlation) for all particles (i.e. coarse, middle and fine particles).

4.4 The correlation between average heat transfer data and bed temperature for the horizontal superheater tube

Figure 5 illustrates the variation of the calculated average heat transfer coefficient at different bed temperatures. The error bars displayed designate the variation of the average bed-to-tube heat transfer coefficient in multiple performance tests in a large-scale 1296t/h CFB boiler. As the bed temperature varied between 929K and 950K, a high variation of average heat transfer coefficient for coarse particles is observed. For middle and fine bed particles, the normalized bed temperature had little effect on heat transfer rate, which follows from the red curve trend in Figure 5. The computed heat transfer coefficients slightly increased for bed temperatures in a range of 1071-1083K, except fine & middle particle sizes: 0.219mm, 0.231mm and 0.246mm, respectively.

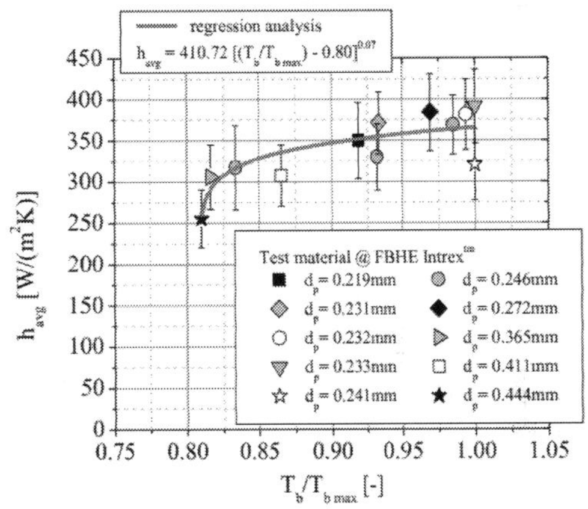

Fig. 5. Estimated average heat transfer coefficient from bed to tube at different normalized bed temperatures.

The higher values of heat transfer coefficient are probably a consequence of enough high suspension density and also a dominant role of the particle-convection component in the heat transfer mechanism. At the same time, the gas convection heat transfer coefficient decreased as a result of the decrease in gas phase density in a high temperature bed.
The average heat transfer coefficient to horizontal tube immersed in a fluidized bed is correlated to the normalized bed temperature $T_b/T_{b\,max}$, by this potential function:

$$h_{avg} = a[(T_b/T_{b\,max}) - b]^c. \qquad (7)$$

Here, T_b denotes bed temperature, $T_{b\,max}$ represents the maximum value of bed temperature during tests, c means exponent, a and b are regression coefficients. The values of the parameters in the above mentioned correlation a, b, *c* were found by regression analysis and they are displayed in Fig. 5. Graph 5 presents a moderate positive correlation for heat transfer data values at varying normalized bed temperatures (R^2=0.64). A similar trend of heat transfer coefficient increasing with bed temperature variation was observed by Stenberg el al [5], Botterill et al. [22] and reported by Prins [23].

5 Conclusions

The averaged heat transfer coefficients were computed from the raw data coming from 12 performance tests at an industrial CFB unit. The packed renewal model offers a convenient way to investigate the impact of key variables on heat transfer in bubbling fluidized beds with Geldart B powder. It was found that the model was satisfactory for all mean bed particle sizes. The average heat transfer coefficient heavily depended upon mean particle size, particle concentration in vicinity of tube surface, bed temperature and also fluidization quality.

An effort is made here to develop a correlation for large scale fluidized bed heat exchangers which can be used to predict h_{avg} with success. Moreover, a few empirical correlations with satisfactory good fit computational results obtained from a mechanistic model were developed. The obtained correlation functions are applicable to a wide range of Sauter mean particle diameters, similar fluidization conditions and also staggered tube bundles. These empirical correlations can be useful for engineering design or heat transfer characteristics in fluidized bed heat exchangers with horizontal tube bundles.

Acknowledgments

This work was financially supported by scientific research no BS-PB-400/301/19.

References

1. Wey M.Y., Lin Ch.L., You S.D., Fluidized behaviour and heat transfer in a bubbling fluidized bed incinerator, *J. Environ. Eng. Manage.*, **Vol.17**(3), pp. 169-175, (2007).
2. Kim S.W., Kim S.D., Heat transfer characteristics in a pressurized fluidized bed of fine particles with immersed horizontal tube bundle, *Int. J. Heat Mass Tran.*, **Vol. 64**, pp. 269-277, (2013).
3. Merzsch M., Lechner S., Krautz H.J., Heat transfer from single horizontal tubes in fluidized beds: Influence of tube diameter, moisture and diameter definition by Geldart C fines content, *Power Technol.*, **Vol.235**, pp. 1038-1046, (2013).
4. Di Natale F., Bareschino P., Nigro R., Heat transfer and void fraction profiles around a horizontal cylinder immersed in a bubbling fluidised bed, *Int. J. Heat Mass Tran.*, **Vol.53**, pp. 3525-3532, (2010).
5. Stenberg V., Sköldberg V., Öhryby L., Rydén M., Evaluation of bed-to-tube surface heat transfer coefficient for a horizontal tube in bubbling fluidized bed at high temperature, *Powder Technol.*, **Vol. 352**, pp. 488-500, (2019).
6. Blaszczuk A., Pogorzelec M., Shimizu T., Heat transfer characteristics in a large-scale bubbling fluidized bed with immersed horizontal tube bundles, *Energy*, **Vol. 162**, pp. 10-19, (2018).
7. Mickley H.S., Fairbanks D.F., Mechanism of heat transfer of fluidized beds, *AIChE J.*, **Vol.1**, pp. 374-384, (1955).
8. Yang W.Ch. (ed.), Fluidization, solids, handling, and processing. Industrial Applications, *Noyes Publications*, Westwood, New Jersey, USA, (1998).
9. Yang W.Ch. (ed.), Handbook of fluidization and fluid-particle systems, *Marcel Dekker Inc.*, New York, (2003).
10. Baskakov A.P., Berg B.V., Vitt O.K., Fillippovsky N.F., Kirakosyan V.A., Goldobin J.M., Maskaev V.K., Heat transfer to objects immersed in fluidized bed, *Powder Technol.*, **Vol.8**, pp. 273-282, (1973).
11. Glickeman L.R., Decker N.A., Design relationship for predicting heat transfer to tube bundles in fluidized bed combustion. In: *Proceedings of the sixth annual international conference on fluidized bed combustors*, Atlanta, GA, **Vol. III**, pp. 1152-1158, (1980).
12. Kunii D., Levenspielen O., Fluidization engineering. *Butterworth-Heinemann*, London,(1991).
13. Ozkaynak T.F., Chen J.C., Emulsion phase residence time and its use in heat transfer models in fluidized beds, *AIChE Journal*, **Vol.26**, pp. 544-550, (1980).
14. Blaszczuk A., Nowak W., Heat transfer behavior inside a furnace chamber of large-scale supercritical CFB boiler, *Int. J. Heat Mass Tran.*, **Vol.87**, pp. 464-480, (2015).
15. Blaszczuk A., Zylka A., Leszczynski J., Simulation of mass balance behaviour in a large-scale circulating fluidized bed reactor, *Particuology*, **Vol. 25**, pp. 51-58, (2016).
16. US Patent No US 6962676B1, Method and apparatus in a fluidized bed heat exchanger, (2005).
17. Wang L., Wu P., Ni X., Surface-particle-emulsion model of heat transfer between a fluidized bed and a immersed surface, *Powder Technol.*, **Vol.149**, pp. 127-138, (2005).
18. Parmar M.S., Hayhurst A.N., The heat transfer coefficient for a freely moving sphere in a bubbling fluidised bed, *Chem. Eng. Sci.*, **Vol.57**, pp. 3485-3494, (2002).
19. Pidwerbecki D., Welty J.R., Heat transfer to a horizontal tube in the splash zone of a bubbling fluidized bed, an experimental study of particle size effects, *Exp. Therm. Fluid Sci.*, **Vol.10(3)**, pp. 307-317, (1995).
20. Masoumifarad N, Mostoufi N, Hamidi AA, Sotudeh-Gharebagh R., Investigation of heat transfer between a horizontal tube and gas-solid fluidized bed, *Int. J Heat Fluid Flow*, **Vol.29**, pp. 1504-1511, (2008).
21. Lu P, Cao Y., Pan W.P., Ma Ch., Heat transfer characteristics in a horizontal swirling fluidized bed, *Exp. Therm. Fluid Sci.*, **Vol.35**, pp. 1127-1134, (2011).
22. Botterill J.S.M., Teoman Y., Yüregir K.R., Temperature effects on the heat transfer behaviour of

https://doi.org/10.1051/e3sconf/201913701014

gas fluidised beds, *AIChE Symposium Series*, 77, pp.330-340, (1981).

23. Prins W., Fluidised bed combustion of a single carbon particle, *PhD Thesis*, Twente University, Netherlands, (1987).

RESEARCH & DEVELOPMENT IN POWER ENGINEERING, 2019

Two-Phase Ferrofluid Systems: Surfactants and Thermal Properties of Iron Oxide Nanofluids for Condensation

Daniel Zabek[1,*], *Andrzej* Grzebielec[1], *Luca* Cmok[2], *Patricija* Hribar Boštjančič[2,3] and *Alenka* Mertelj[2]

[1] Institute of Heat Engineering, Warsaw University of Technology, ul. Nowowiejska 21/25, 00-665 Warsaw, *Daniel.Zabek@itc.pw.edu.pl, Poland.

[2]Department of Complex Matter, Jožef Stefan Institute, Jamova Cesta, 39, 1000 Ljubljana, Slovenia.

[3]Jožef Stefan International Postgraduate School, Jamova Cesta 39, 1000 Ljubljana, Slovenia.

Abstract. Evaporation heat transfer with ferrofluids has the potential to improve heat and mass transfer compared to conventional two-phase systems. The ferromagnetic nanoparticles which control the rheology and heat transfer properties of the ferrofluid, must be coated with a dispersing agent, also referred to as a surfactant, to prevent particle agglomeration. This paper compares ferromagnetic iron-oxide particles coated with fatty acid, polymeric and hydrophobic surfactants after undergoing evaporation and condensation. The outcome was measured by assessing the dispersion of the ferrofluid after condensation. The fatty acid surfactants demonstrated the best dispersion after condensation proving temperature stability.

1 Introduction

Heat and mass transfer is becoming the limiting factor of modern electrical components (e.g. CPUs, solid state memory), energy transformation (e.g. rectifiers and inverters), energy generation (e.g. fuel cells) and energy storage technologies (e.g. batteries). In these fields, extreme thermal loads of up to 1000 $W.cm^{-2}$ are expected and therefore, a need for an immediate breakthrough in cooling technology research has been identified [1]. One promising approach in resolving these limitations is the use of ferrofluids. Ferrofluids have, compared to pure solvents or water, solid ferromagnetic particles dispersed in a liquid carrier fluid. The dispersed particles can change the rheology and heat transfer properties of the fluid and can be used for various applications such as e. g. in magnetic transformer oil for electrical insulation and cooling [2], in lubrication and establishing thermal contact in shaft bearings in turbines, [3] as seals and for damping [4]. One particular advantage with ferrofluids is that the solid ferromagnetic particles dispersed in thecarrier fluid usually have a high thermal conductivity compared to the base liquid carrier fluid [5]. However, in ferrofluids the interaction between the solid particles dispersed in the liquid carrier fluids critical to the ferrofluid mixture. In order to prevent particle agglomeration and to maintain a repelling force between the particles, the particles must be coated with a stabilizing dispersing agent, also referred to as a surfactant. The surfactant are molecules which modify the surface properties of the ferromagnetic particles in order to obtain a stable suspension or colloid [6]. The repelling force between the particles is maintained despite strong magnetic fields or high temperatures. However, under high temperatures the liquid carrier fluid experiences boiling. This boiling phase transition is complex because it separates the solid particles from the gaseous. On the other hand, this phase transition can be favourable because the evaporation process has the potential to be much more effective than conventional thermal conductivity, due to the high level of thermal energy needed for evaporation. The possibility to cycle the ferromagnetic particles through an evaporation and condensation phase is primarily dependent on the surfactant maintaining the repelling force between the solid particles.

Three different classes of Fe_3O_4 iron-oxide(magnetite) ferrofluids with different surfactants were sourced from Ferrotech (US) and compared as follows: EMG 1200 - Fatty acid, EMG 1300 - Polymeric and EMG 1400 - Hydrophobic [7].When the ferrofluid is heated, the carrier fluid evaporates and a sludge of solid particles and surfactant remains. When the ferrofluid is cooled, the carrier fluid condenses and ideally mixes with the particle sludge. The aim of this experiment was to demonstrate and visualise the mixing behaviour of the condensed carrier fluid with the ferrofluid particles. Hence, this paper presents preliminary experimental results on ferrofluid evaporation and condensation demonstrating the possibility of two-phase ferrofluid applications by comparing the effects a fatty acid, a polymeric and a hydrophobic-surfactant have oniron oxide nanoparticles in a n-Heptane carrier fluid. These preliminary results enable further investigation into powerful ferrofluid cooling systems capable of dissipating extreme thermal loads and reducing the cooling device size.

2 Governing equations and experimental procedures

Thermal conductivity, evaporation and condensation has been extensively studied with solids and fluids providing sufficient theoretical and experimental evidence for most engineering applications. The proposed ferrofluid evaporation experiment takes place under constant thermal boundary conditions where the heat flow Q [W] is defined as [8,9]:

$$Q = -kA\frac{\Delta T}{\Delta x} = \frac{\Delta U}{\Delta t} + \frac{W}{\Delta t} \tag{1}$$

and where k [W.m^{-1}K^{-1}] is the thermal conductivity of the material, A [m^2] the surface area, ΔT [K] the available temperature difference, Δx [m] the thickness, ΔU [J] change in internal energy and W [J] work done by the thermodynamic system. When maximising the heat flow in equation 1, it is crucial to improve the thermal conductivity $k$of the material. The effective thermal conductivity determining the heat transfer in a ferrofluid suspension $k_{suspension}$ [W.m^{-1}K^{-1}] is based in the thermal conductivity of the carrier fluid k_{liquid}[W.m^{-1}K^{-1}] and the thermal conductivity of the suspended ferromagnetic particles k_{solid} [W.m^{-1}K^{-1}], and is generally correlated by the following expression for a two component ferrofluid mixture as follows[10]:

$$k_{suspension} = k_{liquid}\frac{1+0.5\left(\frac{k_{solid}}{k_{liqud}}\right)-\Phi(1-\left(\frac{k_{solid}}{k_{liqud}}\right)}{1+0.5\left(\frac{k_{solid}}{k_{liqud}}\right)+0.5\,\Phi(1-\left(\frac{k_{solid}}{k_{liqud}}\right)} \tag{2}$$

where Φ[-] is the volume phase fraction of the solid. Assuming that the solid particles are coated in a surfactant and heated in series under uniaxial heating conditions, the thermal conductivity for the solid component in equation 2 is defined as [11]:

$$\frac{1}{k_{solid}} = \frac{1}{k_{surfactant}} + \frac{1}{k_{particles}} \tag{3}$$

with proportional components for particles content and surfactant in dry particles. Most common ferrofluids have solid phase fractions between 1 – 10 % including the surfactant component. Hence, based on equation 2 the thermal conductivity of most ferrofluids is only slightly higher with the suspended particles than the thermal conductivity of the carrier fluid, and only moderately improves with increasing particle concentration, and overall does not improve the thermal conductivity of the fluid significantly [12]. According to equation 1, in order to improve the heat transfer of a ferrofluid, evaporating the liquid carrier fluid requires a substantial amount of thermal energy to change from the liquid phase to the gas phase. The corresponding thermal evaporation enthalpy ΔH_{vap} [J] is defined as [8]:

$$\Delta H_{vap} = \Delta U_{vap} + W \tag{4}$$

and the evaporation change in energy is ΔU_{vap} [J]. Substituting equation 4 into equation 1 enables to maximise the heat flow with ferrofluids using thermal evaporation. For the proposed evaporation experiment, dry ferrite particles (Ferrotech - US) were disperse in n-Heptane at a concentration of 1 % by weight and sonicated for 10 minutes [7]. The iron-oxide particles have an average particles size of 10 nm and a thermal conductivity of 2.4 W.m^{-1}.K^{-1} [13].The surfactants (fatty acid-, polymeric- and hydrophobic) are initially assumed to have an approximate thermal conductivity of 1 W.m^{-1}.K^{-1} [14]. For the n-Heptane carrier fluid a thermal conductivity of 0.14 W.m^{-1}.K^{-1}is assumed[15]. Essentially, the effective thermal conductivity increases linearly to the change in the weight phase fraction of the iron-oxide particles added [16]. Subsequently, the sonicated 1 % ferrofluid suspension was transferred into a 500µm x 100 µm square and 70 mm long sealed closed volume glass capillary.

3 Results and discussion

According to equation 2, the thermal conductivity was calculated for particle weight fractions between 1 – 10 % for the EMG 1200, EMG 1300 and EMG 1400 ferrofluids.

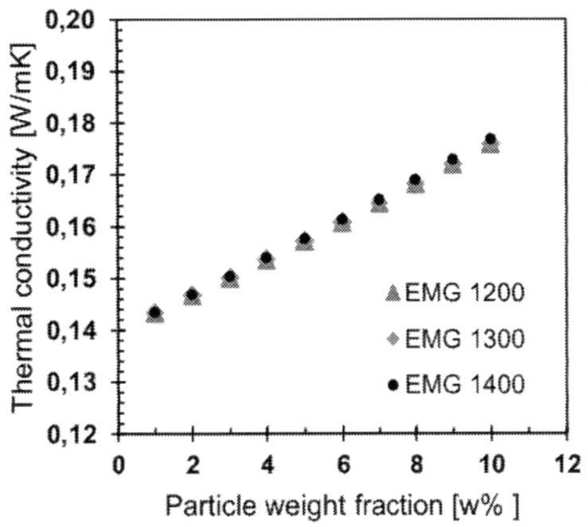

***Figure 1**: Calculated thermal conductivity for EMG 1200, EMG 1300 and EMG 1400 as a function ofiron-oxide particle weight concentration in percent.*

Based on equation 2, Figure 1 shows a linear increase in calculated thermal conductivity for the three ferrofluids with only a slightly higher thermal conductivity of 0.144 W.m^{-1}.K^{-1}for EMG1400 compared to 0.143W.m^{-1}.K^{-1} for EMG 1200 and EMG 1300 at 1 % ofIron-oxide particle weight phase fraction. These results are in good agreement with experimental measurements reporting thermal conductivity for iron-oxide ferrofluids with up to 25 % particle weight fraction [17].

***Table 1**: Properties of EMG 1200, EMG 1300 and EMG 1400 ferrofluids.*

Properties	**EMG 1200**	**EMG 1300**	**EMG 1400**
Iron-oxide content in dry particles [%]	70	70	80
Particle weight phase fraction [%]	1	1	1
Calculated thermal conductivity [$W.m^{-1}.K^{-1}$]	0.143	0.143	0.144
Surfactant [-]	Fatty acid	Poly-meric	Hydro-phobic

As can be seen from the properties of EMG 1200, EMG 1300 and EMG 1400 ferrofluids in Table 1, the slightly higher thermal conductivity with EMG 1400 stems from higher Iron-oxidecontent of 80 % compared to 70 % with EMG 1200 and EMG 1300. Due to the lack of information on the exact molecule structure of the surfactants employed, it is assumed that the fatty acid-, polymeric- and hydrophobic-surfactants have an identical thermal conductivity.

For the evaporation and condensation, the glass capillaries where filled with the EMG 1200, EMG 1300 and EMG 1400 ferrofluid and heated with a hot plate to 115 °C together with a pure heptane filled capillary as reference.

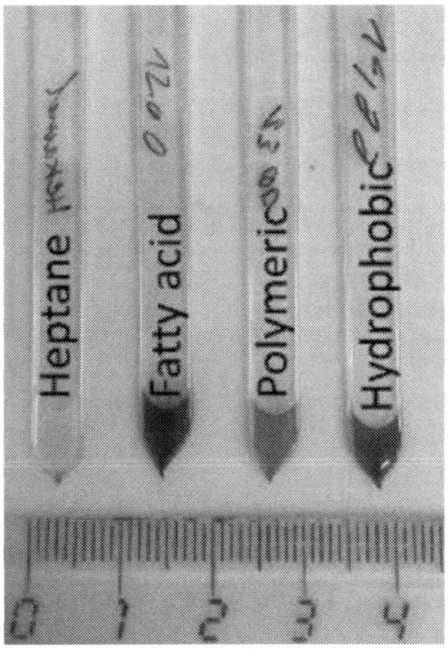

***Figure 2a**: (l-r) Pure n-Heptane capillary as reference, sonicated ferrofluid EMG 1200 in capillary, -ferrofluid EMG 1300 in capillary, and -ferrofluid EMG 1400 in capillary at room temperature,*

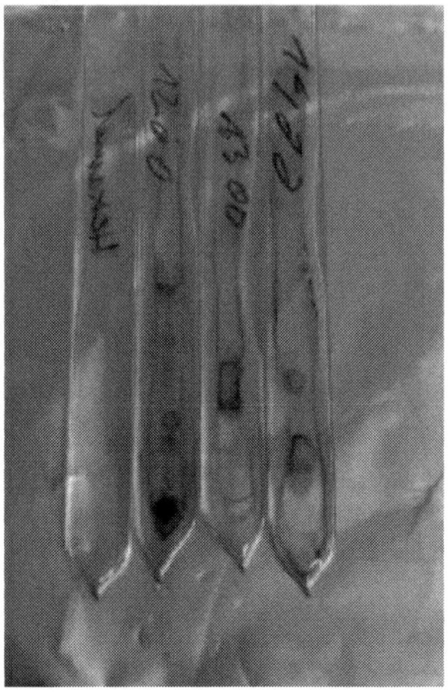

***Figure 2b**: Ferrofluids at 115 °C*

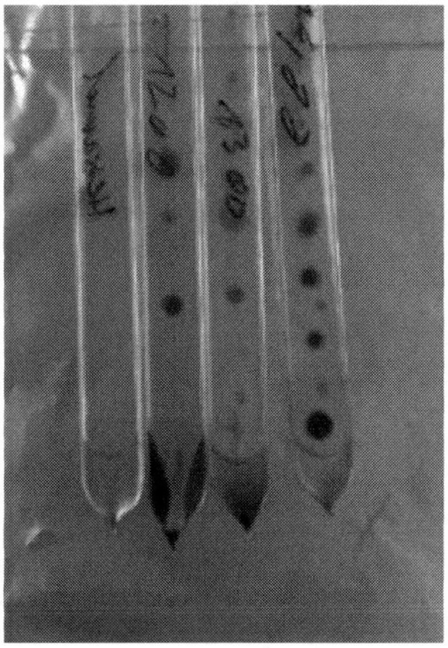

***Figure 2c**: Ferrofluids after condensation at room temperature at 1 w% in capillaries.*

Figure 2 (a) shows from left (l) to right (r)the capillaries filled with the sonicated ferrofluid EMG 1200, ferrofluid EMG 1300, ferrofluid EMG 1400 and pure Heptane for reference. Figure 2 (b) shows the capillaries heated to a temperature of 115 °C. At 115 °C, the Heptane carrier fluid has evaporated leaving a particle sludge along the capillary. Figure 2 (c) shows the ferrofluid at room temperature after the Heptane carrier fluid condensed. According to Figure 2 (c), the ferrofluid EMG 1200 with the fatty acid surfactant shows good dispersion and therefore good mixing after condensation. Compared to the EMG 1200, the ferrofluid EMG1300 with a polymeric surfactant and the condensed EMG 1400 with a hydrophobic surfactant shows adistinct separation between the heptane carrier fluid and the particle sludge compared to Figure 2 (a). As an example, oleic acid is a fatty acid compound and is commonly used in iron-oxide nanoparticles creating a chelate bond with the iron atoms, which is stronger than a simple physisorption and subsequently survives higher temperatures [18]. For the demonstrated polymeric and hydrophobic surfactants in EMG 1300 and EMG 1400 aggregation took place during evaporation. If the surfactant molecules are weakly bonded to the particles, their desorption from the particles is already apparent at slightly elevated temperatures thus causing aggregation during evaporation.

4 Conclusions

Ferrofluid surfactants are crucial to the interaction between the particles in the fluids. This paper investigated the different surfactants for iron-oxide ferromagnetic particles and shows different surfactants impact upon the thermal properties of the ferrofluid. With respect to thermal evaporation a fatty acid surfactant shows dispersion after condensation indicating good mixing of ferrofluid particles. In contrast, polymeric and hydrophobic surfactant ferrofluids show agglomeration of the particles and separation between the condensate and the particle sludge. Future work requires the chemical composition of the surfactant molecules to be known so that the bonding mechanism between the surfactant and the particles can be ascertained.

References

1. Kandlikar SG, Colin S, Peles Y, Garimella S, Pease RF, Brandner JJ, Tuckerman DB. Heat transfer in microchannels—2012 status and research needs. *Journal of Heat Transfer*. 1;135(9):091001,(2013).
2. Kosterec M, Kurimský J, Cimbala R, Čonka Z, Kruželák L, Rajňák M, Timko M, Kopčanský P, Vargová B. Analysis of Thermal Field in Mineral Transformer Oil Based Magnetic Fluids. *ActaPhysicaPolonica A*. 1;131(4):937-9, (2017).
3. Raj K, Moskowitz R. Commercial applications of ferrofluids. *Journal of Magnetism and Magnetic Materials*. 1990 1;85(1-3):233-45, (1990).
4. Rosensweig RE. Fluid dynamics and science of magnetic liquids. InAdvances in electronics and electron physics, Vol. 48, pp. 103-199*). Academic Press*. (1979).
5. Popplewell J, Al-Qenaie A, Charles SW, Moskowitz R, Raj K. Thermal conductivity measurements on ferrofluids. *Colloid and Polymer Science*, 260(3):333-8, (1982).
6. Raffa P, Wever DA, Picchioni F, Broekhuis AA. Polymeric surfactants: synthesis, properties, and links to applications. *Chemical reviews*, 16;115(16):8504-63 (2015).
7. www.ferrofluid.ferrotec.com/technology [Accessed: 30.08.2010].
8. Incropera FP, Lavine AS, Bergman TL, DeWitt DP. Fundamentals of heat and mass transfer. Wiley; (2007).
9. Jaworski M, Rusowicz A, Grzebielec A. Temperature stabilization in buildings by the use of heat/cold storage units with PCM integrated with the ventilation system. *Inżynieria Bezpieczeństwa Obiektów Antropogenicznych*, 3:31-37 (2019)
10. Chhabra RP, Richardson JF. Non-Newtonian flow in the process industries: fundamentals and engineering applications. Butterworth-Heinemann; (1999).
11. Pietrak K, Wiśniewski TS. A review of models for effective thermal conductivity of composite materials. *Journal of Power Technologies*, 23;95(1):14-24, (2014
12. Hong TK, Yang HS, Choi CJ. Study of the enhanced thermal conductivity of Fe nanofluids. *Journal of Applied Physics*, 97(6):064311, (2005).
13. Li M, Endo R, Akoshima M, Tanei H, Okada H, Susa M. Thermal Conductivity of Oxide Scale Thermally Grown on Iron Substrate Corrected by Temperature-dependent Interfacial Thermal Resistance in Laser Flash Measurement. *ISIJ International*.;59(3):398-403.(2019)
14. Lenin R, Joy PA. Role of base fluid on the thermal conductivity of oleic acid coated magnetite nanofluids. Colloids and Surfaces A: *Physicochemical and Engineering Aspects*;529:922-9, (2017).
15. Assael MJ, Bogdanou I, Mylona SK, Huber ML, Perkins RA, Vesovic V. Reference Correlation of the Thermal Conductivity of n-Heptane from the Triple Point to 600 K and up to 250 MPa. *Journal of Physical and Chemical Reference*;42(2):023101, (2013).
16. Altan CL, Elkatmis A, Yüksel M, Aslan N, Bucak S. Enhancement of thermal conductivity upon application of magnetic field to Fe3O4 nanofluids. *Journal of Applied Physics*.;110(9):093917, (2011).
17. Fertman VE, Golovicher LE, Matusevich NP. Thermal conductivity of magnetite magnetic fluids. Journal of Magnetism and Magnetic Materials.;65(2-3):211-4, (1987).
18. Soares PI, Alves AM, Pereira LC, Coutinho JT, Ferreira IM, Novo CM, Borges JP. Effects of surfactants on the magnetic properties of iron oxide colloids. *Journal of colloid and interface science*.;419:46-51, (2014).

RESEARCH & DEVELOPMENT IN POWER ENGINEERING, 2019

Experimental investigation of PWR accident scenarios at the PKL test facility

Rafał Bryk[1,2*], *Lars* Dennhardt[2] and *Simon* Schollenberger[2]

[1] Institute of Heat Engineering, Warsaw University of Technology, Nowowiejska 21/25, 00-665 Warsaw, Poland
[2] Framatome GmbH, Paul-Gossen-Straße 100, 91052 Erlangen, rafal.bryk@framatome.com, Germany

Abstract. PKL is the only test facility in Europe that replicates the entire primary side and the most important parts of the secondary side of western-type Pressurized Water Reactors (PWR) in the scale of 1:1 in heights. It is also worldwide the only test facility with 4 identical reactor coolant loops arranged symmetrically around the Reactor Pressure Vessel (RPV) for simulation of non-symmetrical boundary conditions between the reactor loops. Thermal-hydraulic phenomena observed in PWRs are simulated in the PKL test facility for over 40 years. The analyses carried out in these years encompass a large spectrum of accident scenario simulations and corresponding cool-down procedures. The overall goal of the PKL experiments is to show that under accident conditions – even for extreme and highly unlikely assumptions as additional loss of safety systems – the core cooling can be maintained or re-established by automatic or operator-performed procedures and that a severe accident e.g. a core melt-down can be avoided under all circumstances. Another goal of the tests performed in the PKL facility is the provision of data for validation of thermal-hydraulic system codes. This paper presents recent modifications of the PKL facility, applied in order to adapt the facility to the latest western-type designs currently built in the world. The paper discusses also important results obtained in the last years.

1 Introduction

PKL is the German abbreviation for Primary Side Circuit. For over 40 years, since the construction of the PKL test facility, various test scenarios have always been reflecting safety issues addressed in the debate of the nuclear community at the international level. There are four general objectives of all campaigns conducted thus far:

- Investigation of thermal-hydraulic system behavior under accident situations
- Demonstration of safety margins
- Training of nuclear power plant (NPP) operation personnel
- Provision of data for thermal-hydraulic codes validation

During the years the focus of the nuclear community was evolving, so were the issues investigated at the PKL facility. Over 200 integral tests performed to date have contributed to a deeper understanding of sometimes highly complex phenomena occurring in different accident scenarios and to a better assessment of countermeasures implemented for emergency control.

The first test at the PKL facility was performed in 1977. Since then specific objectives of various tests were changing and so was the test facility itself. The test rig has been refitted several times to suite additional and ongoing tasks and to follow the latest developments in the fields of measuring instrumentation, data processing and new reactor designs. Such a modification of the PKL facility has been introduced in the course of the currently conducted test campaign – PKL III i. The main target of the implemented changes was a better replication of currently built American- and French-type PWRs. Before the modification PKL facility was strictly replicating German KONVOI PWR. Nowadays, all configurations are possible.

2 PKL test programs

At the very beginning of the PKL test facility operation i.e. in the seventies, the Large Break (LB) Loss of Coolant Accident (LOCA) was the design-basis accident for the emergency core cooling systems based on the Maximum Credible Accident (MCA) concept. Thus, the first test campaign was focused on verification of the effectiveness of the emergency core cooling systems (ECCS) adapted for controlling such events. Applying conservative assumptions, the very first test was conducted in 1977.

The first LB-LOCA test campaign was interrupted in 1979 by the accident in Three Mile Island unit 2 (TMI-2). The small break (SB) LOCA which occurred in TMI-2 directed the interest of the nuclear community to this kind of events. Therefore, shortly after the accident the PKL facility was adapted for simulation of the small breaks. Consequently, the TMI-2 scenario was replicated with the main focus on the safety margins demonstration of the operating units. As a result the LB- and SB-LOCA scenarios were covered within the test programs PKL I and PKL II [1, 2].

After the completion of the PKL II, the next PKL III program was initiated in 1986. The goal of the subsequent project was the investigation of so-called accident transients with and without LOCA. The first test campaign of PKL III was focused on design-basis accidents and cool-down procedures specified in the operating manual. Progress in the field of probabilistic safety analyses then yielded values for the so-called integral Core Damage Frequency (CDF) for beyond-design-basis accidents. To reduce the CDF for the beyond-design-basis accidents, preventive accident management (AM) procedures were designed and tested at the PKL facility [3]. Typical topics of investigation studied within the test campaigns PKL III A to PKL III D are shown in the figure 1 [4].

In 1998 the Senior Group of Experts on Nuclear Safety Research Facilities and Programmes (SESAR/FAP) of the OECD/NEA Committee on Safety of Nuclear Installations (CSNI) presented a first set of results which not only comprised general and strategic recommendations but also proposed actions for immediate or near-term implementation. The safeguarding of integral test facilities for studying thermal-hydraulic issues was one of the actions assigned top priorities, provided test programs with unquestionable scientific interest are proposed. At its annual meeting in December 1999, the CSNI issued the recommendation to set up an international collaborative project in the field of thermal-hydraulics to implement the instruction made to this effect by SESAR/FAP. Based on programs, time schedules and cost offered by the various companies operating the test facilities, a proposal for a project called OECD/SETH (SESAR Thermal-Hydraulics) was elaborated. The project formulation was performed in consultation with the NEA Secretariat and it was submitted to the member countries in 2000. The proposal, which was based on experiments to be conducted in the PANDA and PKL test facilities, was approved by the project partners within OECD/NEA.

From the beginning of 2001 within the context of the OECD/SETH [6, 7] and OECD/PKL-programs [8] questions were examined in connection with inherent boron dilution [9], both after the SB-LOCA and accidents in cold shut-down condition (failure of residual heat removal system (RHRS) [10]), in particular with regard to the general sequence of events under various primary- and secondary-side boundary conditions.

Meanwhile, operational aspects of the scenario accidents in cold shut-down condition such as e.g. an optimization of procedures for troubleshooting in cold shut-down condition with regard to system use and return to safe RHR-operation were also focused on in the examinations. This scenario formed a main target of the OECD/PKL-2 program [11] that began in 2008. In the frames of the program the PKL III G tests campaign was conducted which additionally covered the topics indicated in the figure 1 [5].

Following the PKL III G tests campaign, the PKL III H test series was initiated in the framework of the OECD/PKL-3 program in April 2012. As the answer to open questions in the field of reactor safety – in particular as a contribution to the discussion related to the safety assessment of PWR in connection with the events in Fukushima 2011 – experimental investigations on the system behavior of PWR under beyond-design-basis accident scenarios (Station Blackout (SBO), LOCA with additional system failures) with a significant core heat-up were the central topics of the OECD/PKL-3 project [12]. The whole PKL III H tests campaign in the frames of the OECD/PKL-3 project covered the subjects depicted in the figure 1.

After completion of the PKL III H tests campaign, PKL III i program was proposed in the framework of OECD/PKL-4 project. The project was started in June 2016 and its goal is to address still open questions/topics that could not be covered within the previous investigations. It focuses on the parametric studies on the thermal-hydraulic procedures for the development and validation of the system codes and on the experimental

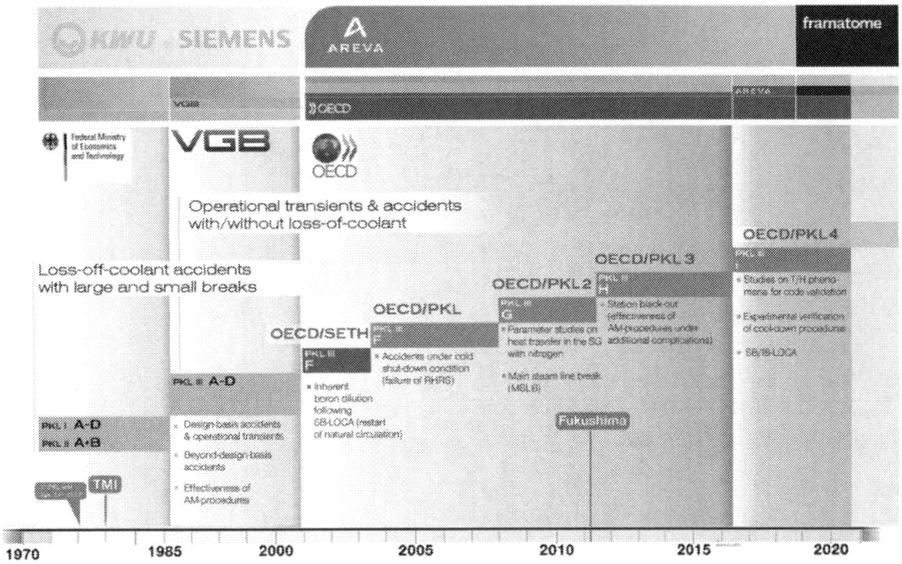

Fig. 1. PKL test programs

 https://doi.org/10.1051/e3sconf/201913701016

verification of the cool-down procedures and operation modes for different incidents and accidents (figure 1) [13].

The OECD/PKL-4 project will be conducted until June 2020.

3 PKL description

The PKL test facility models the nuclear steam supply system of a 1300 MWe nuclear power plant in a scale of 1:145. Detailed design was based to the largest possible extent on the Vorkonvoi type plant and on the specific data of Philippsburg nuclear power plant, unit 2. The facility replicates the entire primary side and the most relevant parts (excluding turbine and condenser) of the secondary side. As for the other test facilities of this size, the scaling concept aims to simulate overall thermal-hydraulic behavior of the full-scale power plant. For this reason all geodetic heights are represented in 1:1 ratio. Power as well as volume and cross-sectional areas are replicated in 1:145 scales, which correspond to a hydraulic diameter reduction of 1:12. The single-phase pressure losses correspond to a large extent to the values in the PWR. Together with the heat losses they have been determined in detail for single components and for both, entire load and temperature ranges i.e. during RCPs in operation, NC-conditions and under hot and cold conditions [5].

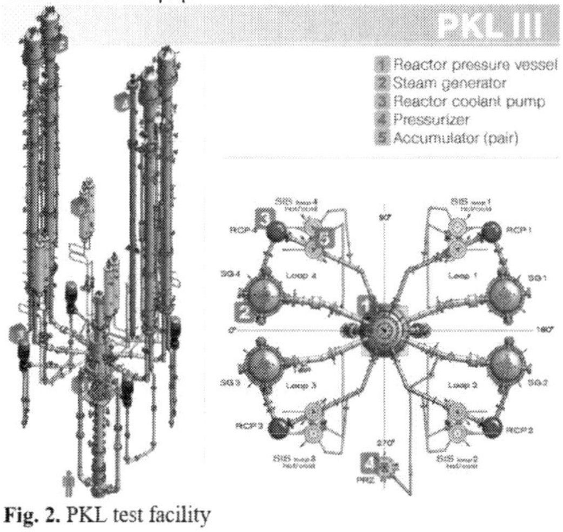

Fig. 2. PKL test facility

The Reactor Cooling System (RCS) comprises a rod bundle vessel containing a heater bundle (simulation of the reactor core), four loops with the reactor cooling pumps, steam generators, a downcomer model and a pressurizer. The pressurizer is connected to the RCS with a surge line.

The rod bundle vessel is a pressure vessel assembled from several elements and instrumentation rings. The whole vessel models the upper head (UH), upper plenum (UP), reactor core sections, reflector gap and the lower plenum.

The core is replicated as a section of the actual system. This means that the full-scale rod dimensions and spacers are used. However, the number of rods in the core is scaled down. Thus, the core is represented by 314 electrically heated rods with a total power of 2.5 MW corresponding to 10 % of the scaled-down nominal power. The PKL heater rod bundle has a uniform axial power profile and the rods are arranged in three concentric zones which are heated independently from one another. This enables the simulation of radial power profiles across the test bundle. The reflector gap between the rod bundle vessel and the bundle wrapper is modelled volumetrically to scale.

The RPV downcomer is replicated as an annulus in the upper region and as two stand pipes connected to the lower plenum. This configuration enables a symmetrical connection of the 4 cold legs (CL) to the RPV, a reliable determination of flow rates, a preservation of frictional pressure losses, and it does not unacceptably distort the volume/surface ratio. This is of importance with respect of the removal of stored heat from the walls during a cool-down procedure [5].

The upper head bypass is modelled by four parallel bypass lines associated with the respective loops to enable the detection of asymmetric flow phenomena in the RCS (e.g. single-loop operation). The replication of the bypass loss is provided by the installation of an appropriately scaled orifice in each bypass line.

The four SGs of the PKL facility are vertical U-tube bundle heat exchangers like their counterparts in the modelled nuclear power plant. The tube bundle is welded into the tube sheet. The cylindrical primary channel head located at the bottom is extended downward by two pipes, split into an inlet and an outlet plenum and flanged to the tube sheet above. The heat exchange surface in modelled to scale. Similarly to the reactor core rods, the SG U-tubes were modelled in the full-scale geometry with a reduced number of tubes according to the scaling principle of 1:145 with respect to the reference plant. Each of the four SGs is equipped with a separate main steam, feedwater, emergency feedwater and steam generator blowdown system.

The symmetrical arrangement of the 4 loops around the RPV also allows investigating individual effects of multiple system failures. Tests on the behavior of a 3-loop or a 2-loop plant can also be conducted by isolating one or two loops of the PKL facility. Each of the primary side loops contains an active reactor coolant pump (RCP), which is equipped with a speed controller to enable the simulation of various pump characteristics [5].

The PKL facility is also equipped with all relevant safety and operational systems on the primary and secondary side. On the primary side, four independent high and low pressure safety injection (HPSI and LPSI) systems are connected to both, the hot and cold legs. Furthermore, the RHRS, 8 accumulators (ACC), the pressurizer (PRZ) pressure control system and the chemical and volume control system (CVCS) are replicated at the facility.

On the secondary side the feedwater system, the emergency feedwater system and the main steam lines with all control features of the original designs are reproduced. For the realistic illustration of the events during secondary side bleed-and-feed operations, the complex geometry of the feedwater system was modelled according to the scaling principle with the possibility of setting the corresponding temperature distributions. All these characteristics allow for the simulation of a wide spectrum of accident scenarios including the interaction between the primary and secondary side in combination with various safety and operational systems [5].

The PKL facility is extensively instrumented with around 2000 measurement points. This enables the accurate analysis and correct interpretation of the phenomena occurring during the tests. Apart from conventional measurements (temperature, pressure), two-phase flow and boron concentration measurement instrumentation are installed at the facility.

4 PKL recent modifications

In principle all western type PWRs have similar Reactor Vessel Internals (RVI) and they differ only in several aspects. Some of these design differences can have, however, a significant influence on both, the reactor normal and accidental operating states. An example of important design diversity is the Upper Support Plate (USP) which is installed in some reactor designs and it is not in others. The USP is a rigid base that positions and supports the control rod guide tubes and the upper support columns that in turn position and uphold the Upper Core Plate (UCP). The USP also directs the thermocouple columns and guides.

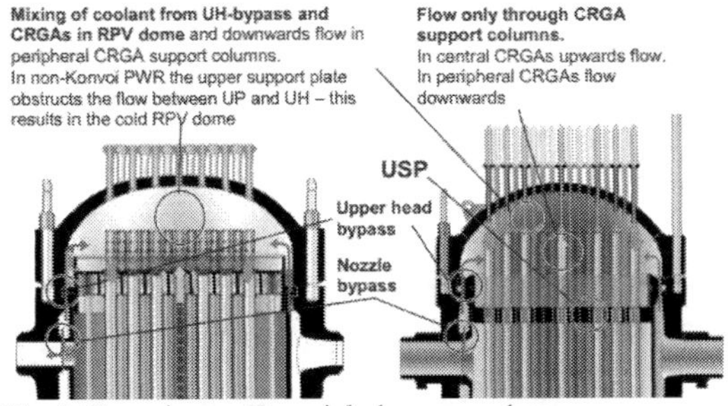

Fig. 3. Konvoi - non-Konvoi design comparison

The USP embedded between the UP and UH in the non-Konvoi design precludes free flow between the chambers and this has a significant influence on the operation of the reactor regardless of the operation mode. Figure 3 shows for instance liquid flows in the RPV during normal operation of the reactor. Under these conditions in the Konvoi reactor the water can freely flow between the UH and UP across practically the whole cross sectional area of the reactor (excluding the top plate area). In case of the non-Konvoi design the free flow is precluded and it is only possible through the control rod guide assemblies (CRGA), which in total strongly limit the flow area. Due to the limited direct flow from the UP to the UH the part of the indirect cold flow from the downcomer to the UH through the UH bypass is relatively bigger. This in turn results in lower temperature in the UH (so called cold dome configuration).

In case of the anomalous operation, for instance due to Loss of Offsite Power (LOOP) and resulting Natural Circulation (NC) the flow through the UH bypass would be reversed as a result of reversion of the pressure difference between the UH and the downcomer vessel. A so called sweeping flow would appear in the bypass due to the pressure difference. Further proceeding of the incident and application of the cool-down procedure i.e. depressurization of the RPV leads to flash boiling in the RPV being a reason for void formation. Flash boiling starts first in the hottest fluid volumes in the RCS which, for the Konvoi design, is usually the RPV dome. Consequently, the void would be collected in the UH of the RPV. In case of the non-Konvoi design it is likely that a void bubble would also appear below the USP resulting in the formation of two bubbles – one below the dome and one below the USP. The differences between the Konvoi and non-Konvoi designs are hence significant and they lead to meaningful discrepancies for certain operational and accident procedures.

In order to enable a simulation of the phenomena in the reactors of all types, the PKL facility was modified correspondingly to the non-Konvoi design. Figure 4 shows a comparison between the old and the new configuration of the facility.

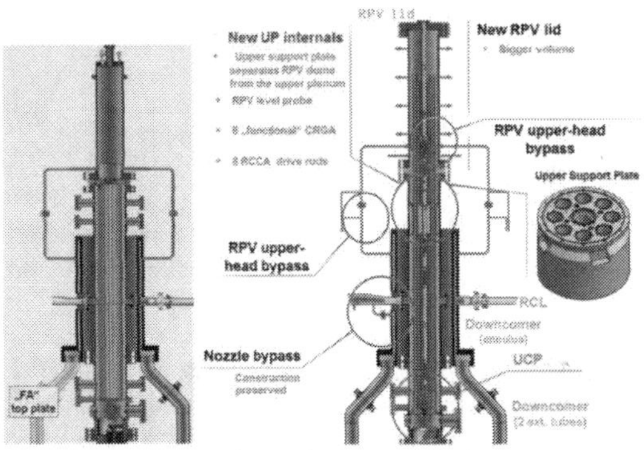

Fig. 4. PKL facility before and after modification

Due to the significant differences between the reference plants, replication of the non-Konvoi design by the PKL facility required substantial modifications. On the left side of the figure 4 there is a cross section through the PKL test facility before the modification. The UP and UH are basically two vessels separated by the top plate. The UH is practically empty while the UH houses filler tubes which serve for the correct replication of the volume/surface ratio of the facility. The pressure drop of the flow between the vessels is mainly simulated by the top plate which replicates the Konvoi design.

On the right side of the figure 4 the modified PKL facility is illustrated. Similarly to the non-Konvoi reference plant (EPR) the UH and the UP are separated by the USP. In the UP apart from the filler tubes there are also CRGA tubes with Rod Cluster Control Assemblies (RCCA) inside. The CRGAs are bolted to the UCP. They get through the USP and their top ends are in the UH. In the PKL facility, the pressure drop of the flow between the UP and UP is mainly simulated by the top caps of the CRGAs. The illustration on the right side of the figure 4 shows also a 3D picture of the USP. The component has thus 8 concentric holes for the CRGAs and one in the middle for the RPV water level measurement instrumentation.

An important feature of conducted modification is the possibility of exchanging the configurations. Therefore, after the adaptation of the facility to the non-Konvoi design, the original reference plant (and hence all the other designs which do not employ the USP) can still be simulated in the PKL test facility.

5 Test results

The total number of integral tests conducted at the PKL exceeds 200. Therefore, only a representative group of various tests can be presented.

5.1 Heat transfer in the SG with failure of RHRS under ¾ loop operation

Loss of residual heat removal system under shutdown conditions i.e. during mid-loop operation has occurred several times world-wide [14]. Increasing attention has been devoted to such incidents as various probabilistic safety assessments have shown that the contribution of these accidents accounts for 18% of the integral core damage frequency of 1.1×10^{-5} per reactor per year and it is much larger than originally anticipated.

Various tests on loss of residual heat removal in mid-loop operation have already been included in the previous PKL test programs, some with the reactor coolant system closed and some with it (partially) open. Tests performed with the reactor coolant system closed showed that residual heat removal is assured if at least one steam generator is still operable. However, these test results also show that under certain boundary conditions, especially with only one SG operable, the cold-leg boron concentration can be significantly reduced in the long term. Furthermore, the test results suggest that boron dilution in the cold leg can be prevented if more than one steam generator is operable.

The test results also demonstrate, that under certain circumstances, the additional injection of water into the reactor coolant system (e.g. by the accumulators) can lead to a degradation of the heat transfer in the steam generators and thereby even hinder the reactivation of the RHR system (assuming that operability has been restored). The experiments revealed that the flow phenomena connected with heat removal and boron dilution are highly sensitive to slight changes in the boundary conditions. Possible scenarios and their expected influence on the system are presented in the figure 5.

For further investigation of the phenomena occurring during RHRS failure a succeeding tests series was conducted. The following campaign was intended to provide a systematic analysis of the highly complex thermal hydraulic processes that occur in the SGs in the presence of nitrogen, steam and water, and to analyze the phenomena related to boron dilution and heat transfer to the secondary side. In the framework of the test campaign two tests were performed: one with a single SG under operation and the second one with two SGs available. Furthermore, variable primary side coolant inventory and resulting different swell level heights in U-tubes was investigated for provision of data for different heat transfer mechanisms and transition between the different flow patterns:

- Heat transfer in the SGs in presence of nitrogen, steam and liquid in the U-tubes
- Blocking of U-tubes by subcooled coolant slug with N_2 enclosed
- Continuous und discontinuous single- or two-phase coolant transport in U-tubes (prospect of boron dilution)

Initial conditions of the test are illustrated in the figure 6.

The test series comprises several changes of primary coolant inventory (PCI) with phases of steady-state conditions maintained by stable heat flow from primary to secondary side at constant primary pressure. Inventory was drained and replenished via lower plenum drain

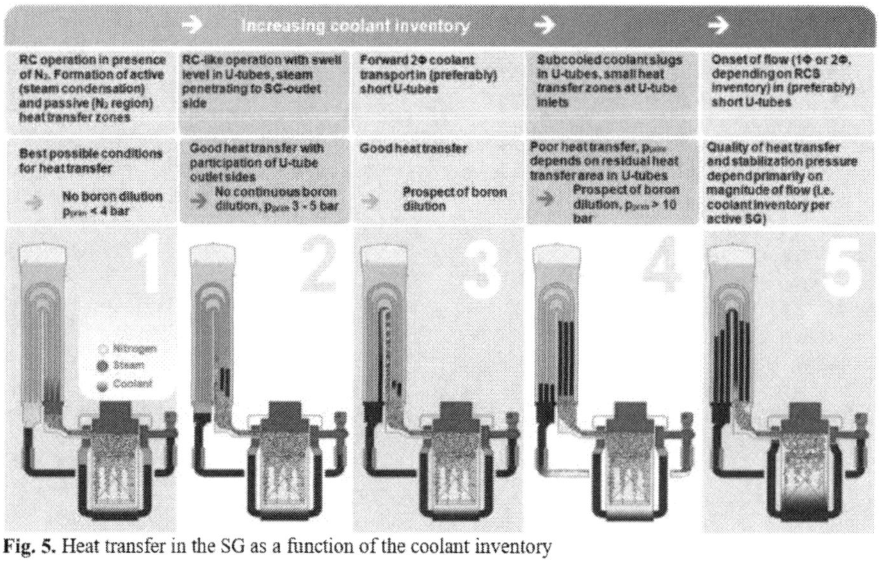

Fig. 5. Heat transfer in the SG as a function of the coolant inventory

valve and a modified CVCS piping. In order to avoid a disturbance of the steam formation in the core as a consequence of exterior influence the liquid injection was directed to lower sections of downcomer tubes. The boundary conditions were adjusted to enable an in-depth analysis of the thermal hydraulic phenomena occurring in the SGs.

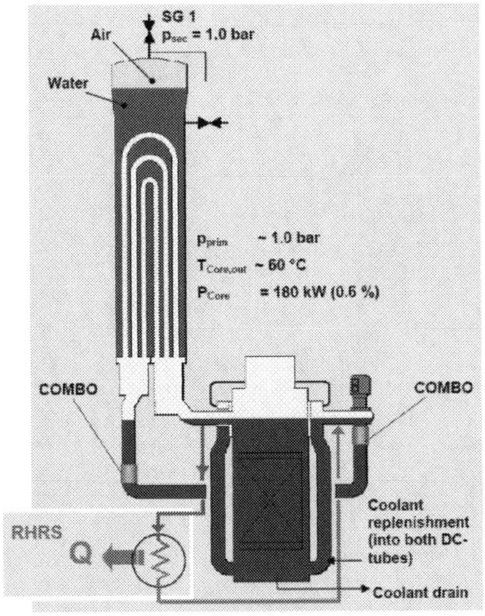

Fig. 6. Initial conditions of RHRS failure test

The whole run of single test lasted around 14 hours. During this time several steady state conditions were maintained as the coolant inventory was discharged and replenished (figure 5). Figure 7 shows conditions in the SG during overspilling of coolant in the shortest SG U-tubes.

Overspilling of the coolant in the shortest U-tubes supports the natural circulation in the RCS and it increases the heat transfer. The temperature distribution in the long tubes indicates that in these conditions, the tubes are practically blocked.

The performed test campaign supported basic principles observed before and helped to formulate further conclusions related to heat transfer mechanisms. The obtained results are useful for the assessment of a full-size PWR thermal-hydraulic behavior. The main conclusions from the test campaign are:

- In case of RHRS failure and at least one SG operational, heat removal from the core to the secondary side is maintained, i.e. core cooling is assured in either case
- Pressure at the primary side stabilizes without the employment of any actions

In the scope of development of various heat transfer modes in the SG U-tubes, the following findings are applicable to full-scale PWR:

- Displacement of liquid water into the active SGs results in hindering the heat transfer as long as no overspilling and forward coolant transport into the outlet part of the U-tubes takes place
- Two active SGs provide a larger heat transfer area and reduce the pressure level required for stabilization
- Presence of two active SGs enables homogenous distribution of the coolant and it reduces the impact of injected liquid on the pressure gradient
- The expulsion of nitrogen from SG U-tubes is necessary for the onset of coolant transport, that supports heat removal to the secondary side
- Commencement of coolant transport – overspilling and regular natural circulation (figure 7) improves heat removal and consequently decreases the pressure at the primary side

The obtained test results were employed for the support of conclusions regarding NPP operation after loss of RHRS in the scope of primary side pressure and temperature stabilizing at moderate levels and prevention of boron dilution. Furthermore, the results are a valuable data base for code validation in the scope of various heat transfer mechanisms.

5.2 Cool-down procedures under natural circulation and asymmetric boundary conditions (with isolated SGs)

Cool-down under asymmetric boundary conditions can occur in the PWR after a feedwater or main steam line

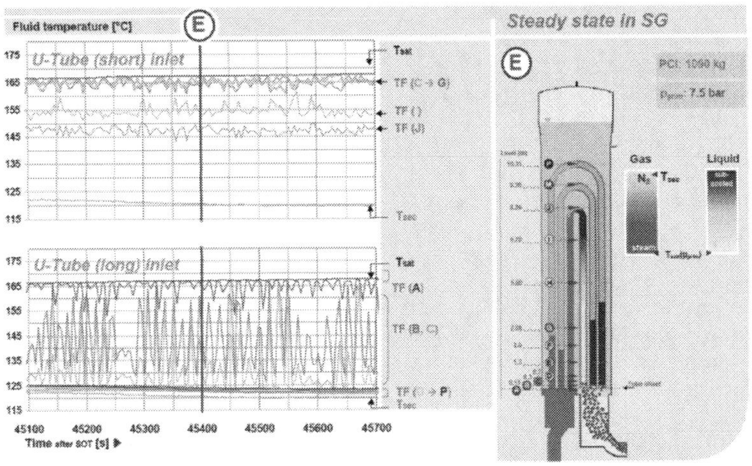

Fig. 7. Temperature profiles and flow conditions during overspilling in the shortest U-tubes

break as well as in case of a SG U-tube leakage. During the test additional loss of offsite power was assumed, so the core was cooled under natural circulation conditions. After the postulated break or leakage at the secondary side the defective SG is heated up, liquid water inventory evaporates and the U-tubes are no longer covered with water which leads to hindering of the heat transfer. After isolating the feedwater and main steam lines, the temperatures at the inlet and outlet of broken SG U-tubes equalize and the driving force for natural circulation diminishes in the affected loop.

In the course of further cooling of the primary side through the intact SGs, the temperature in the primary circle can decrease below the temperature in the isolated SGs. As a result an opposing driving force occurs in the loops with isolated SGs leading to flow stagnation.

The stagnation can appear in two different ways. In the first one, the opposing driving force in the affected SG can become larger than the driving force resulting from heating of the fluid in the core. In the second case, saturation conditions and evaporation in the U-tubes can be reached as a result of pressure decrease at the primary side and simultaneous heating of the fluid in the U-tubes of the isolated SG.

In order to investigate the potential consequences and course of events under asymmetric plant cool-down and after LOOP several runs of the test were performed. Figure 8 illustrates the initial conditions of the test.

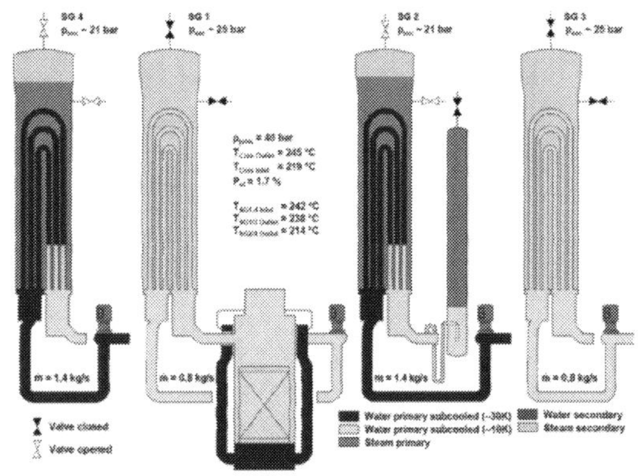

Fig. 8. Initial conditions of the cool-down procedure test

The particular emphasize was put on determination, whether NC can be maintained in the affected loops at continuous cool-down with 50 K/h and after evaporation of water in the secondary side of isolated SGs assuming different accident management activities. In this context the NC behavior in the function of RPV outlet temperature and temperature difference between the inlet and outlet sides of the SG was of special interest. Furthermore, the opposing driving force in the isolated SGs as well as its consequences was investigated.

The first run of the test assumed additional injection of water from CVCS for pressure reduction. The second run was performed without this injection. The third run was dedicated for investigation, whether the NC can be maintained by stepwise cool-down.

The conclusions from all three runs can be summarized as follows:

- After evaporation of liquid water and isolation of two SGs (figure 9A), NC establishes in all loops. In the affected loops the flow is approximately two times lower than in the intact loops, where the NC can be maintained without cooling-down of the secondary side
- In the longer time perspective – ca. 1.5 h after LOOP – and for continuous cooling of the secondary side with 50 K/h a complete stagnation of NC in the affected loops was observed (figure 10)

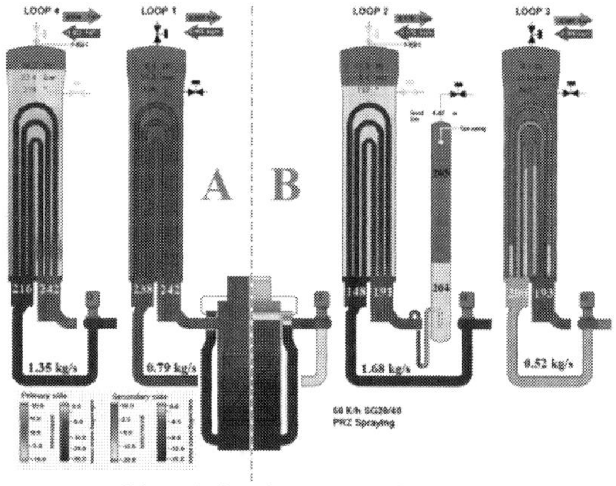

Fig. 9. Conditions during the test; A - after establishing of NC in all loops; B – after lowering the primary side pressure

- In case of lowering the primary side pressure with the pressurizer spray of water from CVCS, evaporation of liquid in isolated SG U-tubes was observed although the subcooling of water at the core outlet was 10 °C (figure 9B)

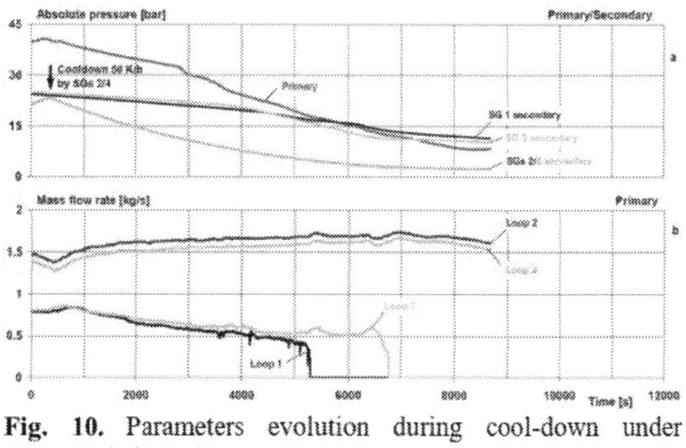

Fig. 10. Parameters evolution during cool-down under asymmetric boundary conditions test

- In the test run without pressurizer spraying much higher subcooling was maintained at the core outlet. This prevented boiling in the SG U-tubes but the flow stagnation in all or some of the U-tubes was still observed in the isolated SGs
- The stepwise cool-down resulted in maintaining the NC also in the isolated SGs. However, the test results revealed that in order to preserve NC, the cooling must be conducted in a certain way, paying attention to the temperature difference between the inlet and outlet plenums. This temperature difference is a gauge for the driving force in the opposite direction and thus, an indicator of maintaining the NC in the loop (figure 11).

 https://doi.org/10.1051/e3sconf/201913701016

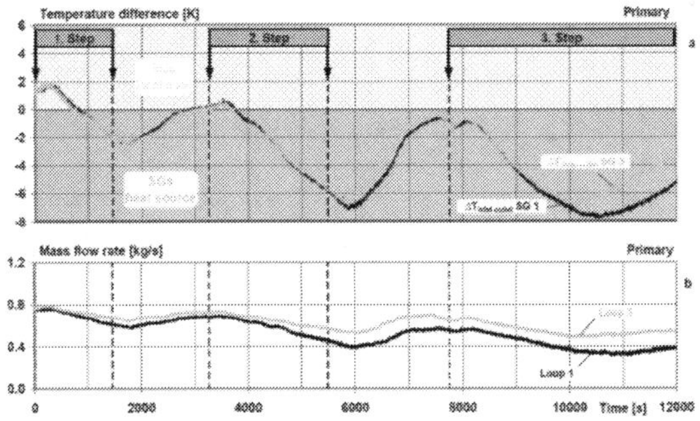

Fig. 11. Mass flows and temperature SG in-out temperature difference during the third run

5.3 Influence of secondary side parameters on heat transfer under reflux condenser condition

Reflux condenser (RC) mode is a cooling mechanism that can occur in SGs of PWRs in defined phases of SB-LOCA only if safety injection systems are operating under reduced availability. In such scenario steam produced in the core flows through the HL to the SG U-tubes and condenses. The condensed water flows due to gravity back to the RPV and supports cooling of the core. During the RC cooling mode counter current flow occurs in the SG tubes and in the HL since steam flows forward into the SG and liquid flaws backward to the RPV.

The performance of heat removal under RC-conditions depends on the condensation mechanism. The film or dropwise condensation are the most effective ones and they contribute to emerging of low temperature differential between the primary and secondary sides, what in turn enables coupling both sides temperature and pressure levels facilitating the cool-down of the primary side.

During an analysis of LOCA scenarios in PWR a conservative assumption of additional failure, reduced availability or late initiation of safety injection system is postulated. Under these assumptions the most relevant parts of cooling procedure are presumed to be performed under RC-conditions. Therefore, the precise replication of the phenomenon is of central demand of thermal-hydraulic codes employed by licensing authorities and expert organizations.

The procedure for dealing with the unlikely event of the LOCA scenario as set out in the operating manuals for PWRs is to reduce RCS pressure and temperature by cooling down the SGs secondary sides.

The test on the RC focuses on investigation of the influence of the secondary side boundary conditions (fill level, cool-down gradients) on heat removal from the primary to the secondary side in conditions resembling course of events of LOCA scenarios with lately initiated cool-down procedures. Therefore, rather than modeling a specific accident scenario, the test was designed as a parametric study to investigate how individual system parameters affect heat removal. Figure 12 shows the initial conditions of the test.

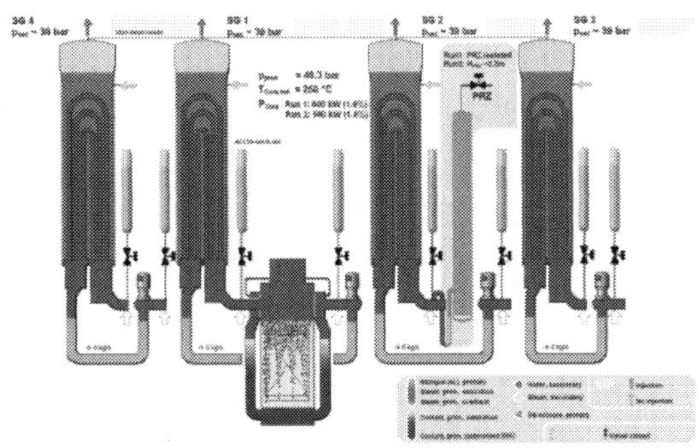

Fig. 12. Initial conditions of the RC conditions test

The variation of the parameters was distributed among two runs:

- Variation of secondary side fill levels
- Different cool-down gradients

The main objective of the test was to provide a data basis for thermal-hydraulic codes validation in the scope the following issues:

- Evolution of heat transfer and temperature difference

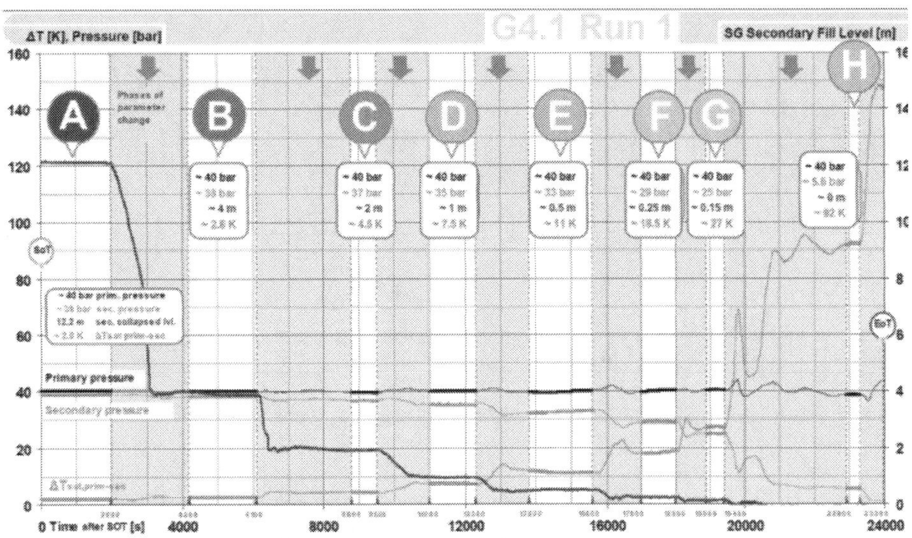

Fig. 13. RC mode test run 1 procedure and parameters evolution

between the primary and secondary sides at different secondary side fill levels and during cool-down procedures under transient boundary conditions and different cool-down gradients

- Temperatures on secondary side during depressurization and injection of cold emergency feedwater in saturated water or saturated steam
- Occurrence and evolution of counter current flow limitation (CCFL) under RC conditions at different cool-down gradients
- Magnitude of displacement of coolant from the RPV into the SGs due to CCFL

In order to achieve the goals of the test scenario two runs of the experiment were conducted. The first run was dedicated to parameter study for investigation of heat removal decrease as a function of the secondary fill level. For this analysis multiple stepwise variations of SGs secondary side liquid water level with in-between periods (A – H) of steady state operation under RC conditions (figure 13).

The second run comprised a sequence of cool-down procedures through the secondary side from 250 °C at the core exit temperature and 40 bars at the primary side. The procedures were conducted with different cool-down gradients, i.e. 100, 250 and 470 K/h (figure 14).

Across both test runs, the following significant test results may be summarized:

- Despite very low water level at the secondary side of the SGs, the heat transfer to the secondary side is preserved provided the SG feed is unbroken
- Reflux condenser operation mode assures a very effective heat removal. It requires only a low temperature difference between the primary and secondary sides to keep the pressures closely coupled
- During the cool-down sequences, substantial coolant displacements (from HLs towards SGs and in the RPV from the downcomer into the core region) occur. They do not however hinder core cooling – even for 470 K/h cool-down gradient
- CCFL was not observed in the SGs U-tubes for the pressure in the primary circuit over 10 bars for cool-down gradient of 100 K/h and coolant inventories of approximately 40 %.

6 PKL tests results benefits and implementation

As the PKL is a worldwide unique test facility, it delivers solid data for a wide spectrum of implementation and it contributes to overall development of the nuclear industry – both in the context of human resources and in the field of general knowledge.

The major goal of all tests conducted at the PKL is supplying robust and comprehensive data for validation of thermal-hydraulic codes like RELAP, ATHLET, CATHARE and TRACE. Due to its large instrumentation density, the PKL has been substantially contributing to the development of the numerical representation of both, singular phenomena and integral systems. The computer codes verified against the PKL measurements are subsequently implemented for extrapolation of the results to the real PWRs. PKL project partners have been using the data base for validating their codes. During regular meetings within the PKL working group they can also provide their feedback and express proposals for the next tests. On the basis of simulations results presentations provided by various organizations from many countries implementing different codes, the conclusions for directions of further development are drawn. The working group gathered around the PKL projects became a unique platform of international debate on nuclear safety and for experiences exchange between organizations working on similar issues.

The program review group of the PKL projects has also grown into an important forum for other test facilities operators. Counterpart testing at other test rigs is a crucial aspect for investigation of various phenomena and scenarios replicated with different scaling principles. This in turn extends the knowledge and understanding in the scope of extrapolating results to full-scale PWRs. Thus, counterpart testing against other integral facilities such as LSTF, BETHSY, LOBI, PMK, PACTEL, ROCOM and JULIETTE was performed so far. Furthermore, counterpart testing against full-scale Upper Plenum Test Facility [15] (UPTF) proven that the results obtained at both facilities are complementary.

Another field, where the PKL results have been playing an important role since the construction of the

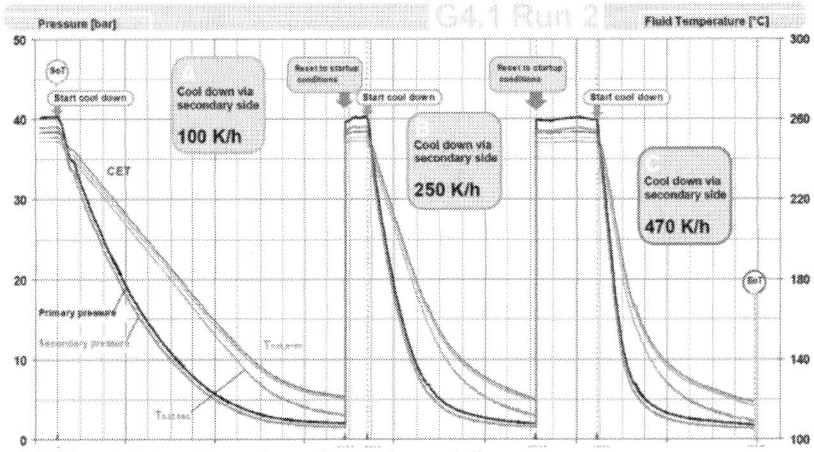

Fig. 14. RC mode test run 2 procedure and parameters evolution

facility is addressing issues raised by licensing authorities. Over the years there have been many instances of contribution of the PKL to clarification of concerns related to emergencies and accidents procedures. Among the others, safety concerns in the scope of the influence of nitrogen on the heat transfer in the SGs raised by an authorized inspection agency were assuaged by means of the dedicated test.

Due to the fact, that the actual boron acid is used and detailed measurements of the local concentration are made in the PKL, the experiments performed to date delivered unique data base on PWR system behavior for boron dilution, boron precipitation and concentration at the core inlet. Further examples of addressing similar problems are described in [15].

PKL tests results are also used for verification and extension of operating manuals and cool-down procedures. Since the procedures are constantly improved and extended to cover all possible scenarios (even those of infinitesimal likelihood), the PKL outcomes largely contribute to inputs to the operating manuals.

The last but not least benefit of the PKL operation is the nuclear power plant operating personnel training. Theoretical trainings are conducted together with practical courses. The tests results provide graphic interpretation of the processes during operational transients and accidents and they are utilized to indicate the performance of the operator actions in a practical way.

7 Conclusions and outlook

PKL is a worldwide unique test facility replicating the entire primary side and the most relevant parts of the secondary side of PWR in the scale of 1:1 in height and 1:145 in volume and power. Since the very first test in 1977, over 200 integral experiments conducted at the PKL have contributed to a deeper understanding of highly complex phenomena occurring during operational transients and different accident scenarios. These tests led also to a better assessment of the countermeasures implemented for emergency control, development and extension of operating manuals and cool-down procedures as well as they provided unique and comprehensive data for codes validation.

Since 2001, the tests at PKL have been conducted in the frames of international cooperation within OECD/PKL projects. Up to now a number of parameter studies and cool-down procedures were examined and established assuming various scenarios, e.g.:

- Loss of RHRS under shut-down conditions
- Cool-down under natural circulation and asymmetric boundary conditions
- Influence of secondary side parameters on heat transfer under RC conditions
- Cold water transients following main steam line break
- Boron dilution events following SB-LOCA
- Boron precipitation following LB-LOCA
- Formation and behavior of upper head void during cool-down under NC conditions
- Effectiveness of secondary side depressurization, performance of the core exit temperature
- Effectiveness of accident management procedures during station black-out.

In the course of the currently conducted tests campaign the upper plenum and upper head of the PKL were significantly modified so that the facility did not only correspond to specific German-type PWR but also to American- and French-types which are nowadays widely constructed around the world. These modifications comprised installation of a new upper core plate, control rod guide assemblies, rod cluster control assemblies, upper support plate and the whole upper head vessel. Since the old configuration can be brought back, it is now possible to replicate any western-type PWR configuration.

The current PKL III i test campaign conducted in the frame of OECD/PKL-4 project will come to an end in the middle of 2020. In order to address still open issues, prove the reliability and investigate the performance of innovative passive decay power removal systems, a new program will be proposed. The final content of the next project will be agreed with the project partners and it will aim at specific needs of involved organizations.

References

1. Mandl R., Brand B., Watzinger H., PKL reflood tests including end-of-blowdown, *Proceedings of the 13th Water Reactor Safety Research Information Meeting*, (1985).
2. Mandl R., Weiss P., PKL tests on energy transfer mechanisms during small-break LOCAs, *Nuclear Safety* Vol. 23 no. 2, pp. 146 – 158, (1982).
3. Umminger K., Kastner W., Mandl R., Weber P., Thermal hydraulic behavior of a PWR under beyond-design-basis accident conditions: conclusions from an experimental program in a 4-loop test facility (PKL), *Proceedings of the 2nd ASME-JSME Nuclear Engineering Joint Conference (ICONE '93)* Vol. 1, pp. 409 – 416, (1993).
4. Umminger K., Mull T., Brand B., Integral effect tests in the PKL facility with international participation, *Nuclear Engineering and Technology*, Vol. 41, no. 6, pp. 765 – 774, (2009).
5. Umminger K., Dennhardt L., Schollenberger S., Integral Test Facility PKL: Experimental PWR Accident Investigation, *Science and Technology of Nuclear Installations*, Vol. 2012 (2012).
6. Mull T., Schoen B., Umminger K., Final Report of the PKL Experimental Programm within the OECD/SETH Project, Framatome ANP GmbH, (2004).
7. http://www.oecd-nea.org/jointproj/seth.html
8. https://www.oecd-nea.org/tools/abstract/detail/csni2001
9. Umminger K., Mull T., Schoen B., Experiments on Boron Dilution in the Integral Test Facility PKL, *Nureth-10: the Tenth International Topical Meeting on Nuclear Reactor Thermal Hydraulics*, (2003).
10. Umminger K., Schoen B., Mull T., PKL Experiments on Loss of Residual Heat Removal

https://doi.org/10.1051/e3sconf/201913701016

under Shutdown Conditions in PWRs, *2006 International Concgress on Advances in Nuclear Power Plants (ICAAP '06)*, (2006).
11. https://www.oecd-nea.org/jointproj/pkl-2.html
12. https://www.oecd-nea.org/jointproj/pkl-3.html
13. https://www.oecd-nea.org/jointproj/pkl-4.html
14. Nuclear Energy Agency Committee on Nuclear Regulatory Activities, Loss of residual heat removal (RHR) while at mid-loop conditions corrective actions, NEA/CNRA/R(2006)4 (2006).
15. Liebert J., Brand B., Schwarz W., Sgarz G., Umminger K., Results of UPTF and PKL research projects for PWR plant operation, *VGB PowerTech*, Vol. 79 no. 8, pp. 20 – 27, (1999).

RESEARCH & DEVELOPMENT IN POWER ENGINEERING, 2019

Evaluation of electricity generation subsystem of Power-to-Gas-to-Power unit using gas expander and heat recovery steam generator

Daria Katla[1,*], *Łukasz* Bartela[2],and *Anna* Skorek-Osikowska[3]

[1]Silesian University of Technology, Akademicka 2A 44-100 Gliwice , daria.katla@polsl.pl, Poland
[2]Silesian University of Technology, Akademicka 2A 44-100 Gliwice , lukasz.bartela@polsl.pl, Poland
[3]Silesian University of Technology, Akademicka 2A 44-100 Gliwice , anna.skorek-osikowska@polsl.pl, Poland

Abstract. In the last years, the European energy policy has required to increase the share of renewable energy sources in the national energy systems. It is important to diversify the energy system not to bring about a global crisis resulting from the fundamental lack of electricity. Unfortunately renewable sources are unstable and generate several problems during integration with the power grid. The solution is to store additional energy produced from renewable sources. In this way, energy can be used when there is a need. The paper discusses the study of the Power-to-Gas-to-Power installation using electrolysis and methanation processes at the energy storage stage and gas expanders during energy discharges. In addition, a part of the Heat Recovery Steam Generation installation has been implemented. The purpose of the work was to determine the impact of a given Heat Recovery Steam Generation installation on the efficiency of the entire installation and flue gas temperature at the outlet from Heat Recovery Steam Generator.

1 Introduction

The demand for energy is constantly growing. Globally the energy sector mainly focuses on the exploitation of fossil fuels for energy production. Fossil resources are exhaustible, and the cost of their exploitation is constantly increasing, as is the difficulty of their extraction. Combustion of fossil fuels also results in the emission of various harmful substances such as nitrogen, sulfur or carbon oxides into the environment, which strongly affect the increase of the greenhouse effect. Energy policy draws attention to improving energy efficiency, increasing the security of fuels and energy supply, diversifying the structure of electricity generation through the introduction of nuclear energy, developing the use of renewable energy sources (RES), developing competitive fuel and energy markets and reducing the impact of energy on the environment.

Turning to alternative energy sources is inevitable. Their unquestionable advantage over non-renewable sources is their unlimited access. Among other things, water, tidal waves, and sea currents, as well as temperature differences in the deep sea, are used. Wind energy put the wind turbine in the motion. Solar energy is used both for the production of electricity in photovoltaic cells and for the production of heat in solar collectors. Another example is geothermal energy used, among others, in heat pumps. Although biomass is considered in many countries in the world as a renewable source, its energetic use is associated with the emission of harmful substances. For this reason, currently, in accordance with the law of the European Union (which is courtially questioned), biomass is not a renewable source. Energy from coal, oil and natural gas still prevails, but in recent years there has been an increase in the share of different energy sources in the global energy mix. There are a number of investments in the nuclear energy sector but also in renewable energy sources. Among others, the member states of the European Union have committed to increasing the share of energy from renewable sources to 30% by 2030 [1-3]. Such legal files as [4,5], are focused not only on increasing the use of renewable energy sources but also pay attention to reducing pollutant emissions and increasing energy security, understood as the diversification of fuels used for energy production. The advantages of renewable energy sources include, universality, free-flow and reduction of pollutant emissions. However, with the use of renewable energy sources installations, several unfavourable aspects are associated, such as production unpredictability, lack of continuity of energy supply or terrain conditions.

The legal regulations established by the EU for the Member States, support the production of energy from RES and impose its priority in grid connection. Sometimes this is problematic because, on days where there is a lot of energy produced from renewable energy, conventional producers are rejected. Which is another problem related to the necessity of discontinuing conventional blocks and generating huge costs. A complication is a fact that the energy produced by the RES is intermittent and dependent on particular conditions. The wind farm capacity rate for European countries is around 20-40%, which depends, among others on area configuration. Also, electricity produced from renewable sources cannot always provide

https://doi.org/10.1051/e3sconf/201913701017

an immediate reaction to demand, because these sources do not generate energy on a regular basis. The increase of this decentralized production means more problems with the stability of grid operation. But investing in renewable energy sources and hence, wind energy will continue to take place [6]. Therefore, a compromise should be found between increasing the share of renewable energy sources in the global energy mix and the possibility of their implementation in the national power systems without major losses. The answer could be storing of surplus renewable energy without rejecting conventional technologies. Considering the energy storage systems, it is possible to match the production potential with the current demand.

2 Power-to-Gas-to-Power process chain description

Power-to-Gas (P-t-G) technology is a technology of energy storage, so it is a solution that could implement energy from renewable sources and released it to the grid during summit hours [7,8]. The storage of electricity requires its conversion into a form that is easier to accumulate. The conversion of electric energy into chemical energy occurs during the process of electrolysis of water, which products are hydrogen and oxygen in the gas form [9]. Hydrogen produced in the electrolysis process can be stored and then used in the production of heat as well as electricity (Fig. 1). Hydrogen, due to its properties, is suitable as fuel. It has the highest Lower Heating Value. As conventional fuel, it is mainly used in rockets. The automotive industry is dominated by engines based on fuel cells [10].

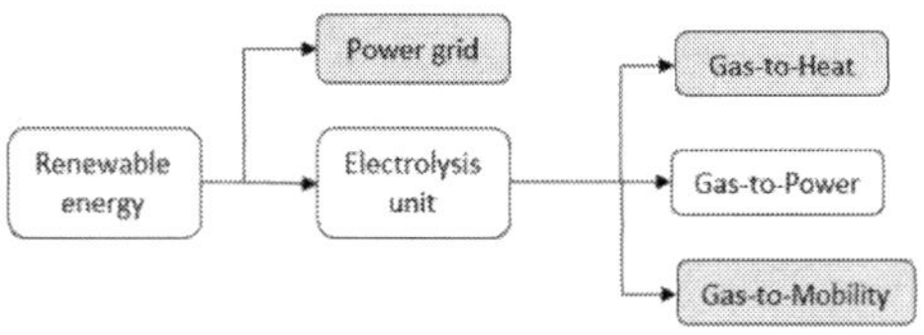

Fig. 1. Simplified block diagram of Power-to-Gas technology

Unfortunately, there are many problems related to the production, buffering, transport and use of pure hydrogen. For example, it requires the use of expensive and still imperfect technology for the installation of fuel cells. The hydrogen can be also synthesized to different types of hydrocarbon fuels. The alternative is to introduce a methanation installation into the process chain of Power-to-Gas installation (Fig. 2) to produce Synthetic Natural Gas (SNG) from hydrogen and carbon monoxide or carbon dioxide in the Sabatier process [11,12]. The great advantage of methane synthesis from hydrogen is that it can be introduced into the gas distribution system without any restrictions and also the storage in pressurized tanks is less problematic [13-15]. Synthetic methane can be actually used in the same infrastructure as commercial natural gas. It means that this technology requires less expensive expenditures of exploitation than hydrogen technology, where conditions of combustion or use in fuel cells cannot be defined as reliable and known technology [16,17]. Generated SNG can be used both to heat as well as to electricity production. The most common method to produce power from Synthetic Natural Gas are gas turbine generators [18].

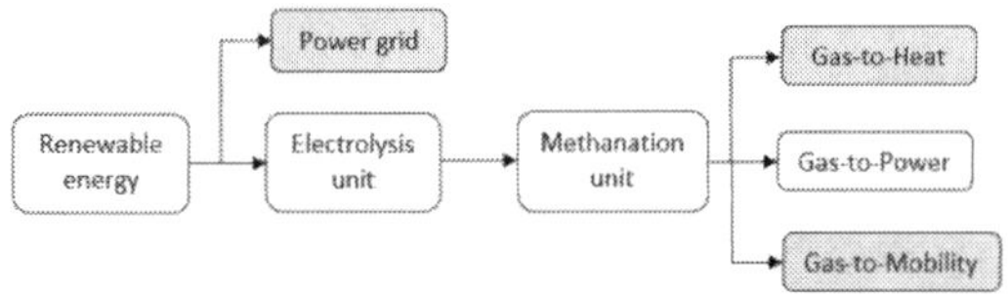

Fig.2. Simplified block diagram of Power-to-Gas technology with methanation unit

3 Description of the mathematical model of a considered energy storage system

Power-to-Gas-to-Power installation with methanation unit was a subject of analysis, assuming the use of standard gas expanders. The aim was to determine the thermodynamic efficiency of these systems.

It was assumed that the surplus renewable energy produced during the night valleys is supplied to the electrolyzer. The hydrogen then supplies the methanation unit in the appropriate cycles. The SNG produced in this process can be stored without restrictions in an existing gas network or stored and used during increased demand for energy production. Renewable energy is provided to the electrolyzer unit and the products of the reaction are hydrogen and oxygen. Generated gases are buffered. Hydrogen is introduced to the methanation unit and the product of this process is SNG. Synthetic methane combustion takes place in the combustion chamber in an atmosphere of oxygen and recirculated carbon dioxide. The combustion is stoichiometric and there is no oxygen behind the combustion chamber. Thanks to the oxy-combustion process, we can separate carbon dioxide through condensation, thanks to which the installation is emission-free, this carbon dioxide is recirculated to the combustion chamber ensuring proper conditions for natural gas combustion. Non-recirculated carbon dioxide can be also stored and further treated as a commercial product, used in broadly understood chemistry. Exhaust gases are directed to the gas expander system, where electricity is generated. Behind the expander, there is a heat exchanger that cools the hot exhaust gases and dissipates the heat. From the gases leaving the heat exchanger, carbon dioxide is separated by condensation of water vapor. The condensed water is stored to be used in the electrolysis process and the remaining part of the separated carbon dioxide is buffered for and used in the methanation unit for the production of SNG.

A big disadvantage in terms of efficiency of this system is a very high heat flux, which is dispersed in the environment by cooling the exhaust gases leaving the expander to separate the carbon dioxide from the water. The exhaust gases leaving the gas expander system have a high temperature, about 500 °C, while the outlet temperature of the heat exchanger is 30 °C. In this way, a very large amount of heat is not used. To increase the efficiency of the system, it was proposed to use high temperature heat for the production of electricity. Therefore, six systems with heat recovery steam generator (HRSG) system were considered. In Figure 3 a conceptual scheme of such a model is presented.

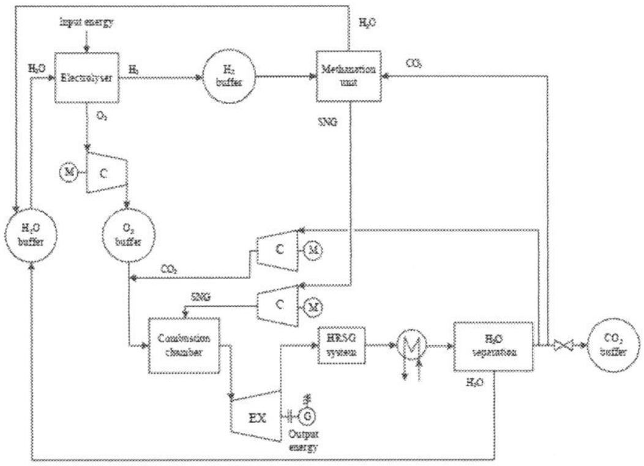

Fig.3. Simplified diagram of the considered Power-to-Gas-to-Power case with HRSG system

3.1. Energy production system

Two possible configurations of the Gas-to-Power system have been taken into account in simulations, without and with HRSG system. The second configuration differs from the first with one element, which is the heat recovery steam generator (HRSG) system after the gas expander. The HRSG is a set of heat exchangers that include economizers, evaporators, and superheaters, which recover heat from the hot exhaust gases. Water is the working fluid and during this process, steam is generated. The steam is used to drive the steam turbine, so in this configuration, there is an additional energy generation. Six cases of HRSG systems were examined in this work. The impact of various constructions of the HRSG was analysed. The results were compared to the results obtained for the case without the HRSG system.

Before the simulation of the energy discharge process, a model of the methanation process in the Aspen Plus software was prepared. The composition of the obtained gas was implemented for further calculations (Tab 1).

Table 1.SNG composition accepted for further calculations

Component	Mole fraction
H_2	0.037076
CO_2	0.009061
CH_4	0.945649
H_2O	0.008212
CO	$1.63 \cdot 10^{-6}$

In Table 2, all assumptions used to model the cases using the Gate Cycle software were summarized. Variables that determine the results of the analysis are the gas temperature at the outlet of the combustion chamber and gas pressure at the outlet of the expander. The rest of the assumptions are constant for all calculations.

Table 2. Main assumptions for the energy production analysis

Description	Value	Unit
Combustion chamber outlet temperature	800, 900, 1000	°C
Combustion chamber O_2 and CO_2 inlet pressure	3	MPa
Combustion chamber SNG inlet pressure	4.5	MPa
Gas expander outlet pressure	100-500	kPa
Heat exchanger water inlet temperature	15	°C
Heat exchanger gas outlet temperature	30	°C
Compressor efficiency	88	%
Gas expander efficiency	90	%
Condenser outlet pressure	10	kPa
Lower Heating Value of SNG	48.66	MJ/kg

Six cases of HRSG systems were considered:

- HRSG1 - 1 pressure HRSG with steam extraction for deaerator,
- HRSG2 - 1 pressure HRSG with integral deaerator,
- HRSG3 - 2 pressure HRSG,
- HRSG4 - 2 pressure HRSG with reheat,
- HRSG5 - 3 pressure HRSG with reheat,
- HRSG6 - 3 pressure HRSG with reheat.

In general, the structure of each stage of the recovery boiler consists of an economizer (EC), an evaporator (EV) and a superheater (SHT), optionally a reheater (RH). Exhaust gases exiting the gas expander system (EX) are directed to the system of three heat exchangers: economizer, evaporator, and superheater. The economizer heats the water, which is then directed to the evaporator, the steam leaving the evaporator is overheated in superheater and directed to a steam turbine with steam bleed to deaerator (except for HRSG2 case where deaerator is fed with steam generated by the low-pressure evaporator of HRSG). Behind the turbine, there is a condenser, and the condensate also fed the deareator. The water pump closes the circuit by supplying the economizer. Each of them is presented in the Figure 4.

 https://doi.org/10.1051/e3sconf/201913701017

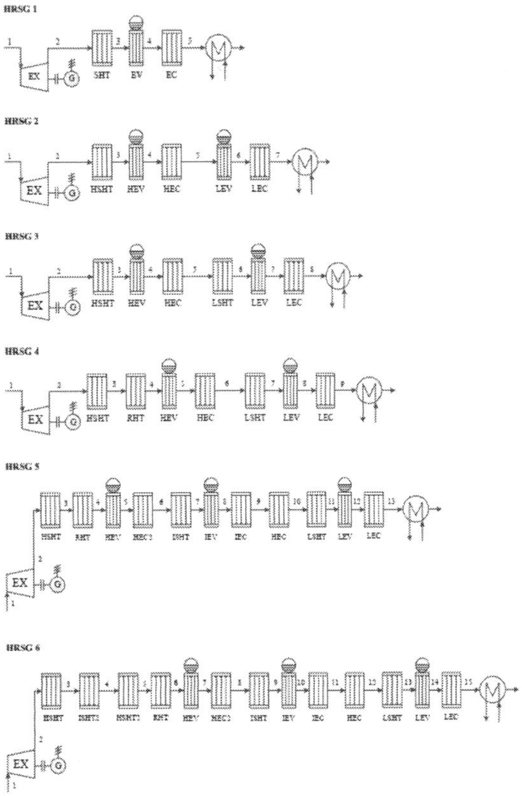

Fig.4. Graphic representation of all considered HRSG cases

Table 3 presents all the assumptions needed to perform the simulation of each system work, required by the Gate Cycle software. For each of the analyzed cases, the same assumptions were adopted for individual elements of the system. The inlet stream to the gas expander is compatible as from the model without the HRSG system.

Table 3. Main assumptions for the energy production analysis for cases with HRSG installation

Description	Value	Unit
Gas expander inlet temperature	1000	°C
Gas expander inlet pressure	3	MPa
Gas expander isentropic efficiency	90	%
High and intermediate-pressure economizer exit subcooling	3	°C
Low-pressure economizer effectiveness	50	%
Evaporator pinch delta temperature	8	°C
Superheater approach temperature	30	°C
Heat exchangers' pressure loss	1	%
Steam turbine isentropic efficiency	90	%
Steam turbine generator efficiency	98	%
Condenser desired pressure	5	kPa
Condenser fixed cooling water temperature rise	10	°C
Pump isentropic efficiency	85	%

4 Results

In order to compare all cases, the definition of energy storage efficiency has been introduced. Electrical energy obtained from the gas expander system is reduced by the work of the CO_2 compressor and the work of a pump (if exists). The input effort is defined as the power of the hydrogen generator. Derived dependencies are true only when the charging time is the same as the discharging time of the unit. The energy storage efficiency is defined by the following equation:

$$\eta_{ES} = \frac{N_{el_{EX}}+N_{el_{ST}}-N_{el_C}-N_{el_P}}{N_{el_{HG}}}, \quad (1)$$

$N_{el_{EX}}$ Gas expander power output,
$N_{el_{ST}}$ Steam turbine power output,
N_{el_C} Compressors power input,
N_{el_P} Pumps power input,
$N_{el_{HG}}$ Input power of hydrogen generators.

Analysis of the system without HRSG in the Gate Cycle program was performed. The calculations were made for three temperatures behind the combustion chamber: 800 °C, 900 °C, and 1000 °C. The expander outlet pressure was changed from 100 to 500 kPa with a step of 10 kPa. The results are presented in Figure 5. The output power of the system is expressed as the difference of the gas expander power output and the CO_2 compressor power input. The highest output power occurs for a combustion temperature of 1000 °C and expander outlet equal to 100 kPa. Data from this configuration have been implemented for further analysis.

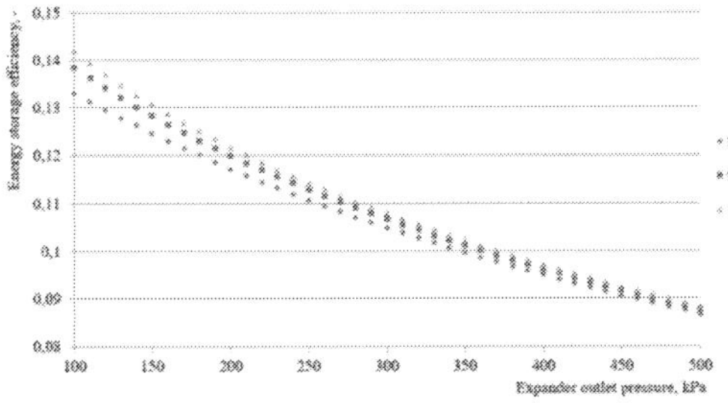

Fig.5. Dependence of gas expander power output as a function of expander outlet pressure for different start-up temperatures of the process

Then, the data was implemented to the models of HRSG systems in order to estimate, in which model the temperature at the outlet of the recovery boiler would be the lowest. The lowest temperature was calculated in the third case of the HRSG system. This is the most beneficial case from the point of view of the heat exchanger located before the H_2O separator because the heat it receives is dissipated.

Table 4 presents data from all models that provide the generated power taking into account the needs of the systems. The highest efficiency is achieved in the fifth case of the models with the HRSG system.

Table 4. The results of performance for all considered cases

Case		T, °C	η_{ES}, %
with HRSG	1	197.8	20.14
	2	148.0	20.51
	3	94.2	21.08
	4	95.9	21.13
	5	96.3	21.39
	6	97.2	21.36
without HRSG		522.9	14.18

5 Summary and conclusions

From the political and ecological point of view, the development of renewable sources will be promoted because it leads to a decrease in consumption of non-renewable primary energy. Nevertheless, the greater is the share of wind energy in the power system, the greater are the losses of conventional power plants. There are complications related to the introduction of RES to the network. Due to the priority of including the RES in the network, it is necessary to discontinue the work of coal-fired steam plants. Also, the price of electricity becomes crucial, for example when the production of electricity is much higher than the consumption there is a necessity of selling it at lower prices. The storage of surplus energy and its use during increased demand promotes a conscious use of resources and sustainable development. Therefore, energy storage technologies are possibly the most beneficial solution to this problem and these technologies should be developed.

Referring to the analyzed systems, the use of a heat recovery system from the exhaust gases behind the expander increases the efficiency of the systems under consideration. Still, the efficiency values for the above cases are quite low. They could not be an alternative to Power-to-Gas-to-Power systems using fuel cells. In practice, fuel cells achieve energy efficiency of 40 – 60% [19] therefore the case defined as a reference assuming the use of fuel cell in combination with electrolyser achives total efficiency of energy storage of 22.8 – 34.2% [20]. The issue of improvement of analyzed cases in order to increase efficiency should be considered. However, the big advantage of the combined cycle over the fuel cells is that the operation of fuel cells requires high financial expenditures, while the proposed systems consist of available elements and constitute less expensive technology. Moreover, exhaust gases generated by analyzed systems consist mostly of H_2O and CO_2 (used in the electrolysis and methanation processes or stored), so it can be assumed that these systems are emission free. In these analyses, many simplifications were used, which probably still inflates the obtained values of efficiency of the systems.

6 Acknowledgement

The scientific work is funded by the National Science Centre within the framework of the research project no. 2017/27/B/ST8/02270.

References

1. European Council. 2030 climate & energy framework, (2014). Available at: <https://ec.europa.eu>
2. An IPCC Special Report on the impacts of global warming of 1.5 °C above pre-industrial levels and related global greenhouse gas emission pathways, in the context of strengthening the global response to the threat of climate change, sustainable development, and efforts to eradicate poverty. 2018 Intergovernmental Panel on Climate Change.
3. Transforming our world: the 2030 Agenda for Sustainable Development. United Nations A/RES/70/1 General Assembly.
4. Directive 2012/27/EU of the European Parliament and of the Council of 25 October 2012 on energy efficiency, amending Directives 2009/125/EC and 2010/30/EU and repealing Directives 2004/8/EC and 2006/32/EC, *Official Journal of the European Union L315/1*, (2012).
5. Communication from the commission to the European Parliament, the council, the European economic and social committee and the committee of the regions. Energy 2020, A strategy for competitive, sustainable and secure energy.
6. The European Wind Initiative Wind Power Research and Development to 2020. Available at: <http://www.ewea.org>
7. Ibrahim H., Ilinca A., Perron J., Energy storage systems: Characteristics and comparisons, *Renewable and Sustainable Energy Reviews*, Vol. 5, pp.1221-1250, (2008).
8. Götz M., Lefebvre J., Mors F., McDaniel Koch A., Graf F., Bajohr S., Reimert R.,. Kolb T, Renewable Power-to-Gas: A technological and economic review, *Renewable Energy*, Vol. 85, pp.1371-1390, (2016).
9. Jentscha M., Trosta T., Sternerb M., Optimal Use of Power-to-Gas Energy Storage Systems in an 85% Renewable Energy Scenario, *Energy Procedia*, Vol. 46, pp. 254–261, (2014).
10. Lehner M., Tichler R., Steinmüller H., Koppe M., Power-to-Gas: Technology and Business Models. *SpringerBriefs in Energy*, (2014).
11. Benjaminsson G., Benjaminsson J., Rudberg R.B., Power to Gas – a Technical Review, *Swedish Gas Technology Centre Rapport*, (2013).
12. Ghaib K., Ben-Fares F.-Z., Power-to-methane: A state-of-the-art review, *Renewable and Sustainable Energy Reviews*, Vol. 81, pp. 433-446, (2018).
13. Ghaib K., Nitz K., Ben-Fares F.-Z., Chemical methanation of CO_2: A review, *ChemBioEng Reviews*, Vol. 3, pp. 266–275, (2016).
14. Iskov H., Rasmussen N.B., Global screening of projects and technologies for Power-to-Gas and Bio-SNG, Hørsholm, Denmark, Danish Gas Technology Centre, 2013, ISBN : 978-87-7795-373-6.
15. Gahleitner G., Hydrogen from renewable electricity: an international review of Power-to-Gas pilot plants

for stationary applications, *International Journal of Hydrogen Energy*, Vol. 38, pp. 2039–2061, (2013).
16. Su X., Xu J., Liang B., Duan H., Hou B., Huang Y., Catalytic carbon dioxide hydrogenation to methane: A review of recent studies, *Journal of Energy Chemistry*, Vol. 25, pp. 553-565, (2016).
17. Kopyscinski J., Schildhauer T. J., Biollaz S. M., Production of synthetic natural gas (SNG) from coal and dry biomass – a technology review from 1950 to 2009, *Fuel*, Vol. 89, pp. 1763-1783, (2010).
18. Rönsch S., Schneider J., Matthischke S., Schlüter M., Götz M., Lefebvre J., Praseeth P., Bajohr S., Review on methanation – From fundamentals to current projects, *Fuel,* Vol. 166, pp. 276–296, (2016).
19. Edwards P.P., Kuznetsov V.L., David W.I.F., Brandon N.P., Hydrogen and fuel cells: Towards a sustainable energy future, *Energy Policy,* Vol. 36, pp. 4356–4362, (2008).
20. Bartela Ł., Katla D., Skorek-Osikowska A., Evaluation of conceptual energy storage systems based on the electrolysis process using gas expanders. Proceedings of the 32nd International Conference on Efficiency, Cost, Optimization, Simulation and Environmental Impact of Energy Systems. ECOS 2019, Wrocław, Poland, 23-28 June 2019, Ed. by Wojciech Stanek, Paweł Gładysz, Sebastian Werle, Wojciech Adamczyk. Gliwice: Institute of Thermal Technology, Silesian University of Technology, 2019, pp. 3523-3534, ISBN: 978-83-61506-51-5.

RESEARCH & DEVELOPMENT IN POWER ENGINEERING, 2019

EXERGY-BASED INVESTIGATION OF A COAL-FIRED ALLAM CYCLE

Jing Luo[1,*], *Ogechi* Emelogu[1], *Tatiana* Morosuk[2], and *George* Tsatsaronis[2]

[1,2] Technical University of Berlin, Institute for Energy Engineering, Marchstr. 18, 10587, Berlin, Germany
E-mail: Jing.luo@campus.tu-berlin.de; emelogu.ogechi@gmail.com; tetyana.morozyuk@tu-berlin.de; tsatsaronis@mailbox.tu-berlin.de

Abstract. Allam cycle is a novel cycle that capitalizes on the unique thermodynamic properties of supercritical CO_2 and the advantages of oxy-combustion for power generation. It is a high-pressure supercritical carbon dioxide cycle designed to combust fossil fuels such as natural gas or syngas (from coal gasification systems) with complete CO_2 separation at a high-efficiency and zero atmospheric emissions. This semi-closed cycle produces sequestration-ready/pipeline quality CO_2 by-product, and thus eliminates the need for additional CO_2-capture system. The Coal-fueled Allam cycle is targeted to deliver between 51-52% net efficiency (lower heating value) for coal gasification. In this study, the expected energetic efficiency is verified by simulating the system in Ebsilon professional software and the result showed that the net efficiency of the simulated coal-fired plant is 30.7%, which is significantly lower than the targeted value. The lower efficiency maybe as a result of the missing heat integration in the system, the high power demand of the oxidant compressor and CO_2 compressors. And an exergy analysis based on published cycle data is employed, to investigate the cycle performance, identify the sources of the cycle's thermodynamic inefficiencies at the component level; a sensitivity analysis is also performed to study the effects of selected thermodynamic parameters on the overall performance of the coal-fired Allam cycle.

1 Introduction

Coal is regarded as one of the largest resources for power generation worldwide. According to International Energy Agency (IEA), coal-fired power generation increased by approximately 3% in 2017 and in 2018, reaching a new peak above 10,000 TWh [1]. Globally, over 30% of the global energy demand and over 40% of the electricity generated is derived from coal, owing to the fact that it is a secure and relatively low-cost source of energy, and the resources are abundant and broadly distributed geographically [2]. However, with the significantly increasing demand in global energy, comes the challenge of increasing CO_2 emissions, of which coal-fired power plants are major contributors.

In a bid to reduce CO_2 emissions from power generating plants, several solutions have been proposed and developed. The carbon capture and storage (CCS) is one of such. Incorporation of this solution however, presents detrimental efficiency penalty and increased price of electricity due to the high cost of additional equipment. Therefore, novel technologies that generate electricity from coal with reduced CO_2 emissions and high efficiency are being developed.

The core Allam cycle was developed and patented by Rodney Allam in 2013 and demonstrated with natural gas by NET Power LLC. It capitalizes on the unique thermodynamic properties of supercritical CO_2 as a working fluid in an oxy-combustion processes for power generation, is one of such promising technologies. This semi-closed cycle employs high-pressure supercritical CO_2 in a highly recuperated cycle to reduce emissions, producing only pipeline grade CO_2, hence eliminating the common necessity of additional capture, clean-up, and compression system [3]. The results of this technological setup include: significantly higher efficiencies, lower capital cost due to the simplicity of the Allam cycle core. The high efficiency stems from the moderate critical point of carbon dioxide at 30.98 ℃ and 73.8 bar, which compared to water can be easily compressed; as a result, the work of compression can be significantly reduced and invariably the plant efficiency enhanced [4].

In 2016, Lu et al. [5] introduced the coal variant of the Allam cycle system; the study focused on the unique considerations, possible hurdles and the advantages of integrating a commercially-available gasifiers with the Allam cycle. The efficiency of the Coal Allam cycle varied between 43.3% - 49.7% (HHV, or 45% -52% on the LHV basis).

2 System Description

In order to utilize coal, the core Allam cycle requires several additional units, namely:

a) a coal gasifier and ancillary systems (to produce clean syngas)

b) Syngas pre-combustion clean up systems and post-combustion clean-up
c) Unique considerations for a syngas combustion turbine.

The power plant in the present study comprises of three (3) main subsystems: the coal gasification island, the supercritical CO_2 power cycle, and the air separation unit (ASU) as shown in Fig. 1. It is based on the flow sheet proposed by Rodney et al. [6] with few variations which are described in subsequent sections.

Due to proprietary and intellectual property, detailed flow sheet, component assumptions and boundary conditions used in the design of Allam cycle are yet to be disclosed in literature. Hence, available process data [6-8] as well as best practice guidelines are used to obtain somewhat comparable results.

The environmental conditions according to [9] as well as the air compositions (default composition from Ebsilon professional software) are presented in Table 1 and 2 respectively. The composition of the coal (Table 3) used for the simulation is according to NETL's Quality Guidelines for Energy System Studies.

Table 1. Environmental conditions used for the simulation [9].

Site Conditions	
Ambient Pressure	1.050 bar
Ambient Temperature	15.0 °C
Relative Humidity	60 %
Cooling water temperature	15.0 °C

Table 2. Air composition[a]

Air Composition	
Nitrogen	0.7544
Oxygen	0.2312
Argon	0.1289
Carbon Dioxide	0.0004995
Water	0.001000

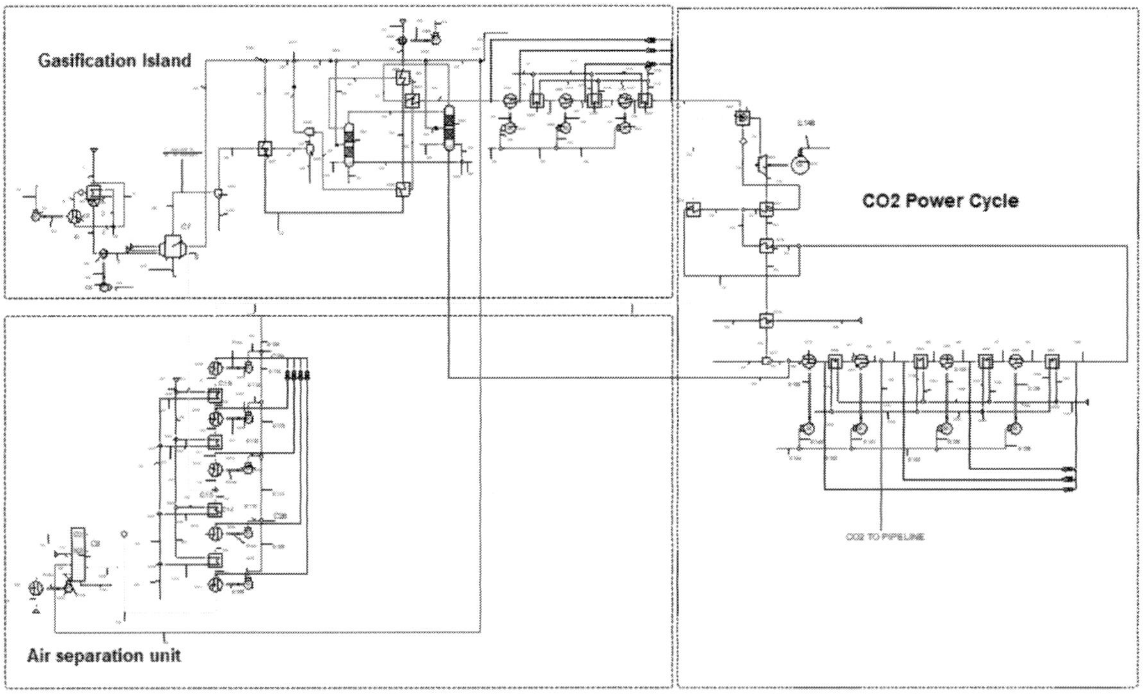

Fig. 1. Flowsheet of the Coal-fired Allam cycle simulated in Ebsilon

[a] According to the default air composition in Ebsilon professional

Table 3. Bituminous Coal Composition [10]

Coal Composition	
Ultimate analysis	**As received, wt%**
Carbon	63.75
Hydrogen	4.50
Nitrogen	1.25
Sulfur	2.51
Chlorine	0.29
Ash	9.70
Moisture	11.12
Oxygen	6.88

The coal is dried by steam to 5% moisture content and is then fed to the gasifier. The steam and the oxygen (from the ASU) needed for the gasification process are also fed into the gasifier. The produced fuel gas passes through a series of heat exchangers, fine particle filters and acid gas removers to absorb CO_2 and H_2S. The cleaned syngas is then compressed and sent to the combustor of the power cycle where it reacts with pure oxygen and recycled CO_2. The turbine inlet temperature of 1150 °C by controlling the mass flow rate of the recycled CO_2.

The air separation unit is modelled as a black box, with specific power demand specified in order to calculate the power that must be supplied by the main cycle. The pure oxidant stream exiting the ASU is split into two streams, one to be fed into the coal gasifier and the other, to the oxidant compression train. In order to limit the net power consumption of the oxidant compressors, a multistage compression with interstage cooling is employed. The residual heat of compression is transferred to a recycle CO_2 stream, lowering the temperature of the pure oxygen fed into the combustor to 750 °C. The supercritical CO_2 (SCO_2) power cycle is modelled according to the SCO_2 power cycle of the pulverized coal power station retrofit example in [6].

After leaving the heat exchangers, the flue gas is further cooled to condense water contained within the stream. A selective splitter is used to separate liquid, and gaseous water, unburnt oxygen and nitrogen, leaving the flue gas as a pure CO_2 stream. This CO_2 stream is then compressed using a multistage inter-cooled compressor, to the needed high pressure, thus closing the power cycle.

3 Methodology

The Ebsilon Professional software was used for the simulation of the coal-fired Allam cycle and the Microsoft Visual Basic Application (VBA) was used to conduct the exergetic analysis of the stream and the components. Ebsilon professional provides information about the thermodynamic efficiency of the system, while the Microsoft VBA tool was used obtain the chemical exergy of the individual streams and to conduct the component-wise exergy analysis, this analysis provides detailed data about the main sources of inefficiencies within the cycle. The simulation parameters for the coal-fired Allam cycle are shown in table 4 below.

Table 4. Simulation parameters for the analysis of the coal-fired Allam cycle

Parameter	Value	Reference
Turbine Inlet Temperature °C	1200	[7]
Turbine Inlet Pressure bar	300	[7]
Turbine Pressure Ratio	10	[7]
Combustor Pressure drop %	1.6	[7]
Oxygen Purity %	99.5	[11]
ASU, Specific Power Demand kWh/kgO$_2$	245	[11]
Generator efficiency	99	[11]
Mechanical Efficiency for all Compressor	98	[11]
Compressor Isentropic Efficiency	85	[9]

3.1 Thermodynamic analysis

In using the Ebsilon Professional modelling software, the mass and energy balances for the simulation model are solved. The electrical efficiency of the power cycle is defined as follows:

$$\eta = \frac{\dot{W}_{net}}{\dot{m}_{fuel} * LHV} \quad (1)$$

The reference temperature and pressure used for the exergetic analysis are the same as listed in Table 1.

3.2 Exergy Analysis

Exergy analysis of a system provides vital information that reveal thermodynamic inefficiencies within a system, which are not obvious when analyzed using conventional thermodynamic analyses. It is defined as the maximum theoretical useful work obtained from a thermal system as it is brought into thermodynamic equilibrium with the environment while interacting with the environment only. Exergy can be destroyed within a system due to chemical reactions, friction, mixing of streams at different temperatures, pressure and composition and heat transfer.

The exergy of each stream is further split into its chemical and physical exergy as follows:

$$\dot{E} = \dot{E}^{CH} + \dot{E}^{PH} \quad (2)$$

Since the process operates at steady-state conditions, the exergy balance of each component k is used to calculate the exergy destruction peculiar to each component k.

$$\sum_j \dot{E}_{q,j,k} + W_k + \sum_i \dot{E}_{i,k} - \sum_e \dot{E}_{e,k} - \dot{E}_{D,k} = 0 \quad (3)$$

To determine the real thermodynamic performance of a system's component k as well as for the overall system, the exergy destruction $\dot{E}_{D,k}$ can further be used. The exergetic efficiencies of the k-th component ε_k is calculated using the following equation:

$$\varepsilon_k = \frac{\dot{E}_{P,k}}{\dot{E}_{F,k}} = 1 - \frac{\dot{E}_{D,k}}{\dot{E}_{F,k}} \quad (4)$$

In addition to the exergy destruction that occurs within the system components, the analysis of possible exergy losses to the environment is also considered with the equation below.

$$\varepsilon_{tot} = \frac{\dot{E}_{P,tot}}{\dot{E}_{F,tot}} = 1 - \frac{\sum_j \dot{E}_{D,k} + \dot{E}_{L,tot}}{\dot{E}_{F,tot}} \quad (5)$$

where $\dot{E}_{P,tot}$ and $\dot{E}_{F,tot}$ are the exergy rates of product and fuel of the overall system. The exergy rates $\dot{E}_{P,k}$ and $\dot{E}_{F,k}$ determine the respective exergy rates of product and fuel of a specific component k. By calculating the total component exergy destruction $\dot{E}_{D,k}$ and the exergy losses $\dot{E}_{L,tot}$ of the overall system, the real thermodynamic efficiencies of the system are determined.

Another important parameter is the exergy destruction ratio $y_{D,k}$, which quantifies the contribution of the exergy destruction within a component k to the reduction in the overall exergetic efficiency ε_{tot} of the system.

$$\varepsilon_{tot} = \frac{\dot{E}_{D,k}}{\dot{E}_{F,tot}} \quad (6)$$

Data obtained from the exergy analysis is used to define the thermodynamic performance of a system and the various components that make up the system.

4 Results and Evaluation

This section presents the findings of the thermodynamic and exergetic analysis of the coal-fired Allam cycle which forms the bedrock for further discussions on the effects of important process parameters on cycle performance.

4.1 Results of simulation

The net efficiency of the coal-fired Allam cycle in this study is 30.7 %; the principal parasitic loads of the plant (these were considered in the efficiency calculation) as shown in Table 5 below include: The Air Separation Unit, the CO_2, syngas, and oxidant compressor.

Table 6 shows the most important components in the exergetic analysis to be considered for improvement. The results of the exergetic efficiency as shown in table 6, indicate that the main inefficiencies can be found with the power cycle. The overall exergetic efficiency is 29%. This means that only 29% of the available energy is converted into useful work. An exergetic efficiency of 29% is considerably low, implying that there are large amounts of inefficiencies within the system. Which are as a result of high destruction and losses. Exergy destruction within a component could be a function of friction and irreversibilities within the components and in the overall system itself, whilst losses are the heat transferred to the environment.

Table 5. Results of the simulation

Parameter	Unit	Value
Mass Flow Rate of Syngas	Kg/s	49.9
Gross energy produced (LHV)	[MW]	644.077
CO2 compression train	[MW]	142.393
Oxidant compression train	[MW]	119.218
Net Energy Produced	[MW]	255.643
Efficiency (LHV)	[%]	30.7

The result in Table 5 shows that the net efficiency of the coal-fired plant is 30.7% which is significantly lower than the NETL reported value. The lower efficiency maybe as a result of the missing heat integration in the system, the high power demand of the oxidant compressor and CO_2 compressors. In addition to this, the parameter uncertainty has a very huge influence on the efficiency.

Table 6. Results of exergy analysis

Component	$\dot{E}_{D,k}$ [MW]	$y_{D,k}$ [%]	ε_k [%]
Combustion Chamber	132.70	27	82.4
Expander	13.74	3	97.98
Flue Gas HE (1)	6.57	1	41.65
Flue Gas HE (2)	4.39	1	98.74
Cold Water HE	21	4	93.26
ASU	16	3	93.05
Overall Cycle	[%]	49.3	29.20

The component with the highest exergy destruction is the combustion chamber, this is due to irreversibilities in the chemical reaction occurring within, and the mixing of streams at different temperatures. Within the combustion chamber, about 27% of the exergy of the fuel is destroyed; this implies that it is a clear candidate for improving the overall exergetic efficiency. Even a slight increase in this components exergetic efficiency will lead to a significant decrease in exergy destruction. This can be achieved by incorporating a cooling technology as described in literatures. However, the Ebsilon professional software limits the modelling of a combustion chamber and expander cooling.

Other candidates for improvement include the gas expander (over 13% exergy destruction), the ASU, the series of two heat exchanger used in cooling the flue gas as shown in Fig. 2 and the heat exchanger which increases the temperature of recycle CO_2 by using the heat from the compressed pure oxygen. These heat exchangers including the gas expander, however, shows high exergetic efficiency compared to the combustion chamber.

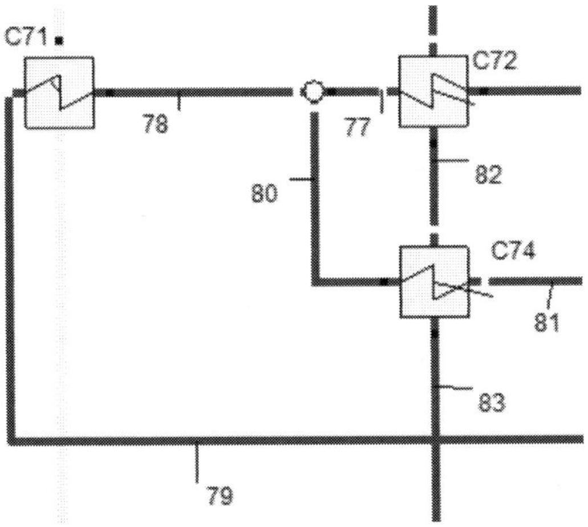

Fig. 2. Heat Exchanger with high exergy destruction rates.

4.2 Sensitivity Analysis

The result of the thermodynamic and exergetic analysis show that most of the important units are the combustion chamber, the expander and the heat exchangers within the cycle. Bases on this information, some important parameters are varied in orders to understand their influence on the cycle performance.

4.2.1 ASU Specific Power Consumption

The specific energy consumption for producing oxygen of 99.5% purity vary from literature to literature. However, in this study, the specific energy consumption of 245 kWh/t was used according to [11]. The effect of the ASU specific energy consumption on the cycle performance is observed in this section over a range of 200–250 kWh/t and the results is shown below in Fig. 3.

The results show that the efficiency decreases with every 10 kWh/t increase in the specific power consumption by approximately 1%. Invariably, the lower the specific power consumption of the air separation unit, the higher the efficiency of the plant. This unit has significant influence on the plant efficiency.

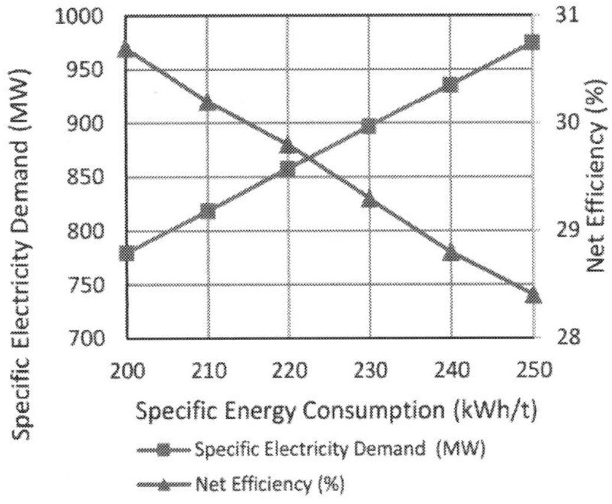

Fig. 3. Effect of Specific Power Consumption of the Air Separation Unit on the cycle performance.

4.2.2 Turbine Inlet and outlet Pressure

According to [11], the net efficiency of the plant increases with increased turbine inlet temperature and a fixed turbine outlet pressure. This section studies the effect of varying the turbine inlet and outlet pressure keeping the pressure ratio of the outlet to inlet constant at 10. It also studies the effect on the CO_2 recompression power demand which inevitable affects the overall cycle performance. The turbine inlet pressure was varied from 240 to 300 bar with 10 bar increase in each step as higher inlet pressures require adjustments in the heat recuperation and design of the heat exchangers. The results as shown in Fig. 4 indicates that, the lower the turbine inlet pressure, the higher the power demand for the CO_2 compression and resultantly, the lower the cycle efficiency.

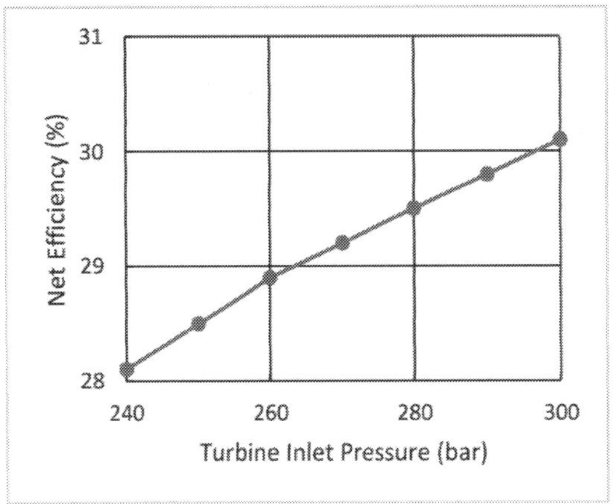

Fig. 4. Effect of varied turbine inlet and outlet pressure at a constant pressure ratio of 10.

5 Conclusion

This study modelled the coal-fired Allam cycle in Ebsilon professional software based on the limited data open to the public and subsequently, analyzes the exergetic performance of the cycle. The results obtained in terms of efficiency is considerably low (30.7%) compared to the targeted value of above 51% LHV reported in [5]. The major missing links, being the heat integration loops, combustion and the turbine cooling as indicated by Lu et al. [5], which have beneficial effect on the cycle performance have not been included in this study. The analyses showed that the efficiency of the simulated Allam cycle can be improved by further cooling the flue gas to near ambient temperatures; this condition will lower the work done by the compressor, the power demand which would increase the efficiency of the plant. The study also shows that the efficiency of the plant is directly related to the inefficiencies of components like the combustion chamber, gas turbine, recuperative heat exchangers and the power consumption of the air separation unit. To improve the efficiency of the plant, particular attention should be given to the above in future research.

References

1. Tracking Clean Energy Power 2019. https://www.iea.org/tcep/power/coal/
2. International Energy Agency, "21st Century Coal: Advanced Technology and Global Energy Solution," OECD/IEA, Paris, (2013). https://www.iea.org/publications/insights/insightpublications/21stCenturyCoal_FINAL_WEB.pdf
3. R. J. Allam, S. Martin, B, Forrest, J. Fetvedt, X. Lu, D. Freed, G. W. Brown Jr., T. Sasaki, M. Itoh, J. Manning, "Demonstration of the Allam Cycle : An update on the development status of a high efficiency supercritical carbon dioxide power process employing full carbon capture", *Energy Procedia*, vol. 114, pp. 5949-5953, 2017.
4. Cheng W-L, Huang W-X, Nian Y-L. Global Parameter Optimization and Criterion Formula of Supercritical Carbon dioxide Brayton Cycle with Recompression. Energy Convers Manage 2017; 150(SupplementC):669–77. http://dx.doi.org/10.1016/j.enconman.2017.08.055
5. X. Lu, B. Forrest, S. Martin, J. Fetvedt, M. McGroddy, D. Freed, "Integration and Optimization of Coal Gasification Systems With A Near-Zero Emissions Supercritical Carbon Dioxide Power Cycle", in *Proceedings of ASME Turbo Expo 2016: Turbomachinery Technical Conference and Exposition*. Paper No. GT2016-58066, 2016.
6. R. J. Allam, G. W. Brown, and M. R. Palmer, "System and method for high efficiency power generation using a carbon dioxide circulating working fluid." U.S. Patent : US 8,959,887 B2, 2015.
7. R. J. Allam, M. R. Palmer, G. W. Brown, J. E. Fetvedt, D. Freed, H. Nomoto, M. Itoh, N. Okita, and C. Jones, "High efficiency and low cost of electricity generation from fossil fuels while eliminating atmospheric emissions, including carbon dioxide," *Energy Procedia*, vol. 37, pp. 1135–1149, 2013.
8. R. J. Allam, J. E. Fetvedt, B. A. Forrest, and D. A. Freed, "The oxy-fuel, supercritical CO2–Allam cycle : New cycle developments to produce even lower-cost electricity from fossil fuels without atmospheric emissions," in *Proceedings of the ASME Turbo Expo 2014 : Power for Land, Sea and Air*, Paper No. GT2014-26952, 2014.
9. M. Penkuhn, G. Tsatsaronis, "Exergy Analysis of the Allam Cycle", in *The 5th International Symposium – Supercritical CO2 Power Cycles*, pp. 6-7, 2016.
10. N. T. Weiland, C. W. White, "Techno-economic analysis of an integrated gasification direct-fired supercritical CO2 power cycle", *Fuel, vol. 212, pp. 613-625,* 2018. https://www.sciencedirect.com/science/article/pii/S0016236117312681
11. Y. Zhao, B. Wang, J. Chi, Y. Xiao, "Parametric study of a direct-fired supercritical carbon dioxide power cycle coupled to coal gasification process", *Energy Conversion and Management Vol. 156, pp. 733-745,* 2018. https://www.sciencedirect.com/science/article/pii/S0196890417310968?via%3Dihub

RESEARCH & DEVELOPMENT IN POWER ENGINEERING, 2019

Improvements perspectives of cryogenics-based energy storage

Jimena Incer[1,*], *Sarah* Hamdy[2], *Tatiana* Morosuk[3], and *Georges* Tsatsaronis[4]

[1] Energy Engineering Department, Campus El Gouna, Technische Universität Berlin, Marchstr.18, 10587 Berlin, jimena.incervalverde@tu-berlin.de, Germany
[2] Energy Engineering Department, Campus El Gouna, Technische Universität Berlin, Marchstr.18, 10587 Berlin, sarah.hamdy@tu-berlin.de, Germany
[3] Chair of Exergy-based methods for refrigeration systems, Technische Universität Berlin, Marchstr.18, 10587 Berlin, tetyana.morozyuk@tu-berlin.de, Germany
[4] Chair of Energy Engineering and Environmental Protection, Technische Universität Berlin, Marchstr.18, 10587 Berlin, georgios.tsatsaronis@tu-berlin.de, Germany

Abstract. Advanced exergy-based analyses provide the information for potential of improvement of energy-conversion systems from exergetic, economic and environmental point of view. These analyses are applied to Cryogenic-based Energy Storage (CES) also known as Liquid Air Energy Storage (LAES). Advantages such as (a) lack of geographical restrictions, (b) low environmental impact and the fact that it is (c) based on mature technology, drive further the research on this energy storage system. An adiabatic LAES system charged with Heylandt liquefaction of air process is analysed. Parameters such as exergy destruction, investment cost, cost associated with the exergy destruction, as well as the environmental impact associated with the thermodynamic irreversibilities are split into avoidable/unavoidable and endogenous/exogenous parts. Aspen Plus® software was used to simulate the LAES system and Engineering Equation Solver was used to conduct the conventional and advanced exergy-based analyses. The dependence of the improvement of each component with the rest of the system was found and all components present higher exogenous exergy destruction than endogenous. The component with the highest potential for improvement is the main heat exchanger in the discharge unit. Focusing on improvement of the components that were found to be the most inefficient ones with the highest exergy destruction, CES is expected to become thermodynamically and economically feasible.

1 Introduction

In the past years, especially since the Paris Agreement at the 21st Conference of the Parties of the UNFCC, 195 countries have signed the agreement to keep the increase in global average temperature below 2 °C according to the pre-industrial levels and also to limit the increase to 1.5 °C.

The rapid increase of renewables technologies such as wind and solar power integrated to the power grid is therefore promoted by many countries. Increasing interest in energy storage systems that do not endanger the power network stability while introducing fluctuating renewable energies to the grid are of great importance and are being rapidly developed. The share of renewable energy worldwide is supposed to reach 12.4 % in 2023 in all sectors combined [1].

This paper discusses a feasible solution: Cryogenic-based Energy Storage, currently at precommercial state and still under development with a Technology Readiness Level of 8 (TRL=9 is the maximum) [5]. Contrary as the two most common and commercial large-scale energy storage technologies; Pumped Hydro Systems (PHS) and Compressed Air Energy Storage systems (CAES), CES does not present any geographical limitations and can be constructed much faster [2,3]. Efficiencies reported until now are promising for LAES, but the criteria where CES outstands the most is the considerably higher energy density than PHS and CAES.

The operation of the system consists in three units: charge, storage and discharge. The charging unit liquefies air when excess electricity is available, for this process different and well-proven air configuration systems can be used such as Kapitza, Linde, Heylandt Claude and variations of these configurations. The air is compressed to pressures around 150 bar and further expanded and liquefied [5]. The liquid air is stored at ambient pressure and low temperature of approximately -196 °C in storage vessels like those for liquid natural gas. Finally, the discharge unit can produce electricity when the demand is higher in a Rankine cycle, where the liquid air is re-gasified, superheated and expanded [4].

2 State of the art

In present, two LAES pilot plants exist in the United Kingdom, one research plant at the University of Birmingham of 350 kW/2.5 MWh and one of the private company Highview Power Storage of 2 MW. Currently, a bigger plant of 200 MW is being developed by the same company [4].

Even though, the CES concept was first introduced by the University of Newcastle more than forty years ago and it was first built more than twenty years ago by Mitsubishi Heavy Industries, it is still a relevant topic since considerably improvements were found that made the technology more attractive to investors and researchers. The first LAES pilot plant of 2.6 MW air liquefaction unit and power recovery system operated independently reported a low roundtrip efficiency and decreased the interest in this technology. Later, Hitachi researchers proposed a LAES plant that could reach 70 % round trip efficiency if the integration of air liquefaction unit and power recovery regenerator is considered. Implementing solid materials and liquids as cold carriers in the heat recovery network are the key to reach this efficiency [4]. These results caught the attention of the private company, Highview Power Storage, that decided to work together with the University of Leeds in 2009 to demonstrate the advantages of this technology by building a 350 kW/2.5 MWh LAES system in Slough, Scotland. Two years later the plant became operational and is now in the research institute for CES in the University of Birmingham. The company completed the construction of a 5 MW/15MWh pre-commercial LAES system last year. The enterprise is also planning a 200 MW/1.2 GWh GigaPlant that would prove the large-scale capacity of this system [6]. The costs are estimated to decrease at a learning rate of 17.5 % [7]. The initial investment costs of a first-of-a-kind daily cycling unit are expected to reduce significantly from 880 – 2.580 €2017/kW to 555 – 1.480 €2017/kW when the technology reaches its maturity.

Recently, great potential in cost reduction and efficiency improvement through CES system integration was found [8].

This paper introduces advanced exergy-based methods split into avoidable/unavoidable and endogenous/exogenous to not only identify the limit to thermodynamic and cost improvements, but also to calculate the independency of the components with each other. The environmental impact of CES systems has also not yet been discussed in any of the reviewed literature. This paper aims to identify the economic potential of adiabatic CES systems through revealing the limit – the unavoidable part – of thermodynamic performance, interaction and dependence of results between each component – the endogenous part- and cost-effectiveness with advanced exergy-based methods.

3 Methodology

3.1 Exergy-based analyses

The exergy balance for the overall system can be written as [9]:

$$\dot{E}_{F,tot} = \dot{E}_{p,tot} + \dot{E}_{D,tot} + \dot{E}_{L,tot} \tag{1}$$

For the exergoeconomic analysis the cost associated with the exergy destruction is calculated according to [10]:

$$\dot{C}_{D,k} = c_{F,k}\dot{E}_{D,k} \tag{2}$$

With help of an economic analysis, the investment cost rate $\dot{Z}_k$ is calculated to proceed with the cost balances and the specific cost per unit of exergy of the streams exiting each component is calculated as [9]:

$$\dot{C}_{P,k} = \dot{C}_{F,k} + \dot{Z}_k \tag{3}$$

An exergoenvironmental analysis was also performed considering both exergy analysis and Life Cycle Assessment [11]. An environmental impact rate $\dot{B}_j$ and an environmental impact b are assigned for each exergy stream. The Eco-indicator 99 is used as an environmental indicator assigning Eco-indicator points (Pts) when the categories ecosystem, human health and natural resources are compromised [13]. With this analysis, exergy destruction is related to the environmental impact for each component with the environmental impact per unit of exergy of fuel, $b_{F,k}$, as showed [10]:

$$\dot{B}_{D,k} = b_{F,k}\dot{E}_{D,k} \tag{4}$$

The environmental impact of electricity for low voltage electricity in Europe is 26 mPt/kWh or 7.22 mPt/MJ [14]. Depending on the country where the energy storage system is located this impact may vary. The air entering the air compressors have an associated environmental impact of 0 mPt/MJ.

3.2 Advanced exergy-based analyses

In the advanced exergy analysis the exergy destruction, cost associated with the exergy destruction, environmental impact and the investment cost are split into avoidable and unavoidable parts [12]. The unavoidable part is the part of the exergy destruction that is associated to technical limitations of a component. The exergy destruction can also be split into endogenous and exogenous. Operating at these conditions the system is referred to as the most efficient adiabatic LAES process. The avoidable part can be calculated as:

$$\dot{E}_{D,k}^{\ AV} = \dot{E}_{D,k} - \dot{E}_{D,k}^{\ UN} \tag{5}$$

Where at a specific designed point, A:

$$\dot{E}_{D,k,A}^{\ UN} = \dot{E}_{P,k,A}\left(\frac{\dot{E}_{D,k}}{\dot{E}_{P,k}}\right)^{UN} \tag{6}$$

Applying the same approach, the unavoidable investment cost and unavoidable cost rates can be calculated [16].

For the calculation of the endogenous and the exogenous parts of the exergy destruction, the theoretical fuel is calculated when the system operates at ideal conditions with no exergy destruction.

$$\dot{E}^{T}_{F,tot} = \dot{E}^{R=T=constant}_{P,tot} + 0 + \dot{E}^{T}_{L,tot} \quad (7)$$

To calculate the endogenous conditions, where all components behave at ideal conditions, except the component under study, hybrid conditions are calculated [12].

$$\dot{E}^{H}_{F,tot} = \dot{E}^{R=T=H}_{P,tot} + \dot{E}^{H}_{D,k} + \dot{E}^{H}_{L,tot} \quad (8)$$

The part from the exergy destruction that is not endogenous is known as exogenous and combination of the avoidable and unavoidable part of the exergy destruction can be obtained through the term $\dot{E}^{UN,EN}_{D,k}$ [15].

$$\dot{E}^{UN,EN}_{D,k} = \dot{E}^{EN}_{P,k} \times (\frac{E_{D,k}}{E_{P,k}})^{UN} \quad (9)$$

4 Simulation

The simulated process consists of a Heylandt process to liquefy the air in the charging process and a Rankine cycle for the discharge process. The detailed simulation results were reported in [8]. Figure 1 shows the flow diagram of the evaluated system. This configuration was simulated using Aspen Plus® software.

Table 1 shows the values for the design parameters of the components at a (a) real, (b) optimal and (c) worst adiabatic liquid air energy storage system simulation process.

Detailed simulation for the advanced analysis was performed on component level for the three conditions mentioned above.

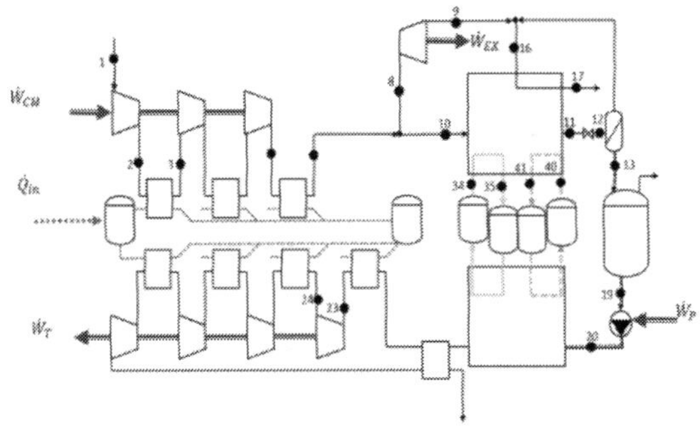

Fig. 1. Flow diagram of the adiabatic CES system

Table 1. Parameters for the real, unavoidable exergy destruction and unavoidable investment cost conditions

Component		Real	$\dot{E}_{D,k,A}^{UN}$	$\dot{Z}_{k,A}^{UN}$
Compressors	η (%)	85	90	70
Expander	η (%)	84	88	60
Turbines	η (%)	90	92	70
Pump	η (%)	75	80	65
Main HE 1	ΔT (K)	3.6	0.5	10
Main HE 2	ΔT (K)	4.8	0.5	10
Re-heaters	ΔT (K)	3	1	15
Intercoolers	ΔT (K)	3	1	15
Flasher	p (bar)	1.1	1.2	1.014

5 Results

Figure 2 shows the results obtained from the advanced exergy analysis. The compressors are the components with the highest share of the exergy destruction within the system. The only component with a higher avoidable than unavoidable exergy destruction is the main heat exchanger in the discharge unit, more than 60 % of the exergy destruction could be avoided if the component operates with a lower minimum temperature difference.

60 % of the exergy destruction within the compressors and 70 % of the exergy destruction within the turbines is unavoidable 95 % of the exergy destruction of the re-heaters in the discharge unit is unavoidable, representing the components with the highest percentage of unavoidable exergy destruction.

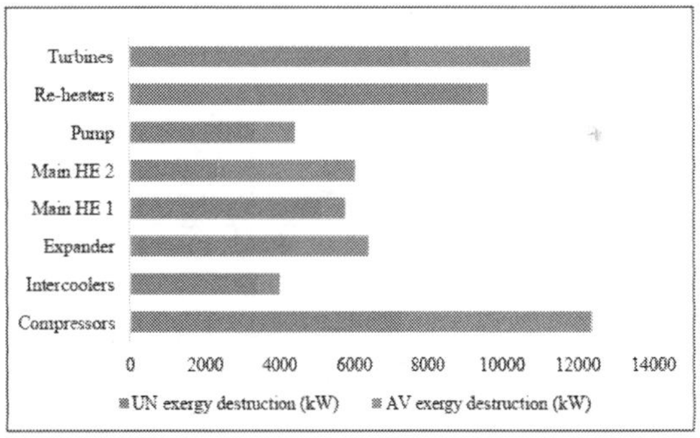

Fig. 2. Exergy destruction of every component divided by avoidable and unavoidable parts

Figure 3 represents the further split into endogenous and exogenous part for the exergy destruction. For all components, the avoidable exogenous exergy destruction is higher than the avoidable endogenous exergy destruction. In order to reduce the exergy destruction within a component the remaining components must be improved. For instance, to reduce the exergy destruction within the main heat exchanger in the discharge unit, the remaining components must be improved as well, since they are highly dependent from each other.

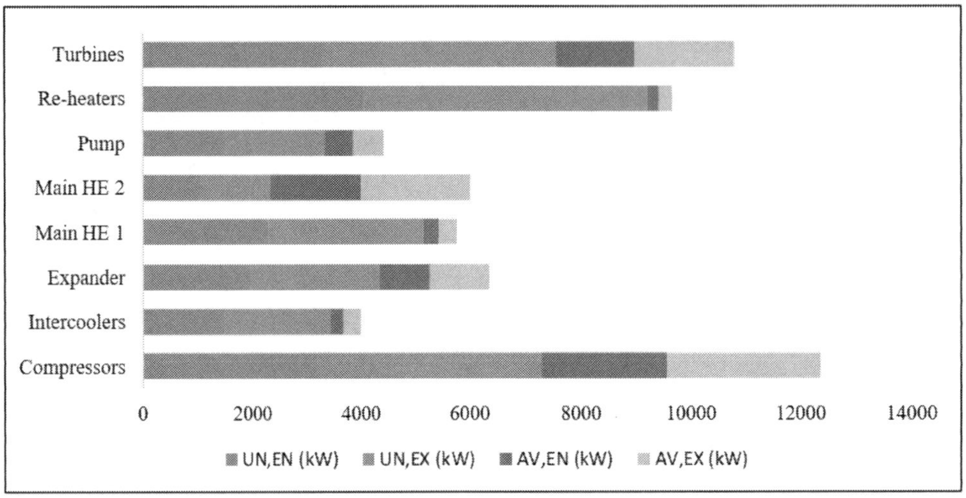

Fig. 1. Split of avoidable/unavoidable and endogenous/exogenous exergy destruction

Table 2 summarizes the values of the unavoidable and avoidable investment costs within the components with the highest avoidable investment cost in the LAES system under study. For the turbomachinery 80 % of unavoidable exergy destruction was consider and for the heat exchangers equations cost (previously published and discussed in detail in [6]) were used to calculate the new investment cost in the conditions of worst performance [8]. The component with the highest possibility for improvement is the re-heater 6, since 40 % of its investment cost could be avoided if the heat exchanger operates with a minimum temperature approach of 15 K and not 3 K.

Table 2. Splitting the investment cost into avoidable and unavoidable for the LAES system

Component	$\dot{Z}_{k,A}^{R}$ (€/h)	$\dot{Z}_{k,A}^{UN}$ (€/h)	$\dot{Z}_{k,A}^{AV}$ (€/h)	$\dot{Z}_{k,A}^{AV}$ (%)
Main heat exchanger 2	2621	1960	661	25
Turbine 1	975	780	195	20
Turbine 2	955	764	191	20
Turbine 3	949	759	190	20
Turbine 4	948	758	190	20
Cryogenic pump	844	675	169	20
Main heat exchanger 1	772	712	61	8
Re-heater 6	233	139	94	40

The turbines represent the highest share of total cost of exergy destruction within the system for its elevated cost, as shown in Figure 4. By increasing the efficiency of the turbines 2 %-point (from 90 % to 92 %), the cost associated with the exergy destruction for these components can be reduced by 30 %. The most relevant component is main heat exchanger 2 in the discharge unit, which has 60 % of avoidable cost of exergy destruction. However, this component operates at cryogenic temperatures and a more efficient component would be significantly expensive. The intercoolers and reheaters operate at quasi optimal conditions, because almost all cost of exergy destruction is unavoidable.

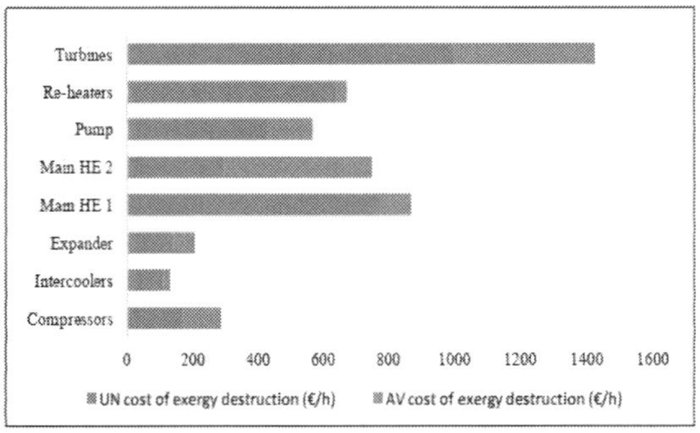

Fig. 2. Distribution of the avoidable and unavoidable cost of exergy destruction

The sum $(\dot{Z}_k + \dot{C}_{D,k})$ and its distribution between avoidable and unavoidable parts is presented in Figure 5. The main heat exchangers and the turbines play an important role because of their high cost.

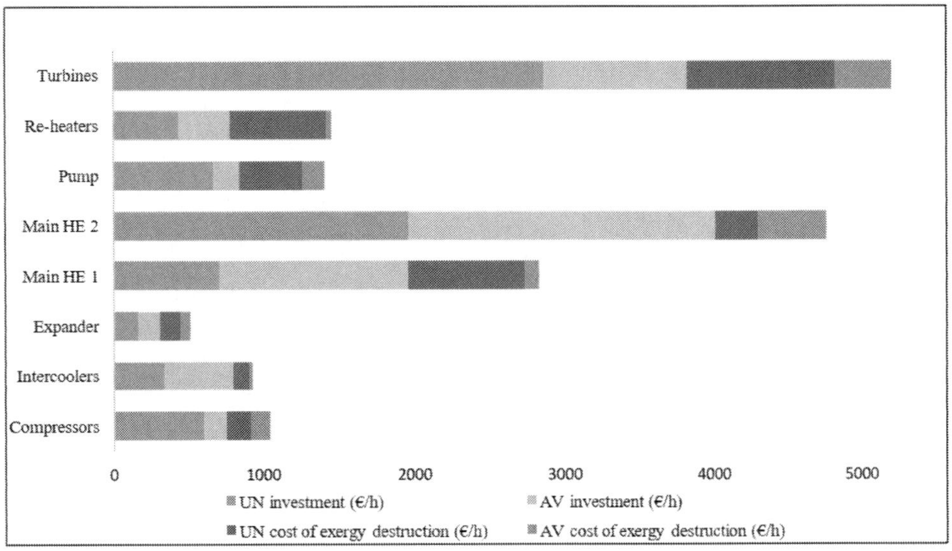

Fig. 3. Division of the sum of investment cost and cost of exergy destruction into avoidable and unavoidable parts ($\dot{Z}_{k,} + \dot{C}_{D,k,}$)

Within the exergoenvironmental analysis, the environmental impact due to the exergy destruction of each component was calculated. Operating at the best performance conditions, the system has an environmental impact of 66 % (calculated using Eco-Indicator 99), that cannot be avoided. Nevertheless, some components such as the compressors have 43 % of avoidable exergoenvironmental impact. The component that has the highest avoidable environmental impact is main heat exchanger in the discharge unit, it has 58 % of avoidable exergoenvironmental impact. Figure 6 represents the distribution between avoidable and unavoidable exergoenvironmental impact due to the exergy destruction in each component. The turbomachinery has a high share in the total environmental impact of the LAES system, as well as the reheaters.

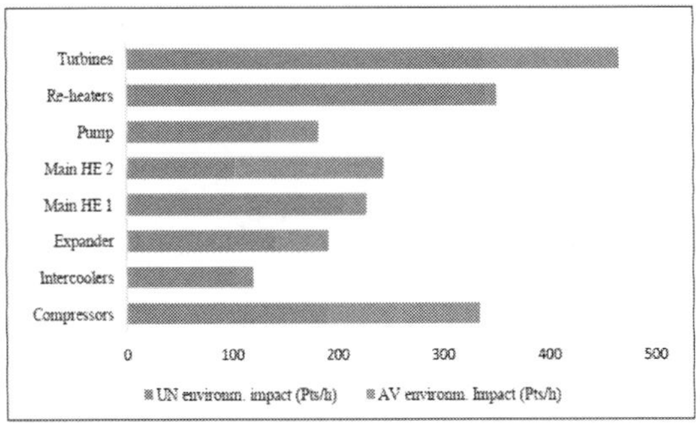

Fig. 4. Avoidable and unavoidable environmental impact associated with exergy destruction

6 Conclusions

1. 59-70% of the exergy destruction in the turbomachinery is unavoidable with exception of the cryogen pump (40 %).
2. The component with the highest possibility for improvement in performance and cost is the main heat exchanger 2 in the discharge unit.
3. The improvement of the isentropic efficiency in the turbomachinery avoids 30-41 % of the cost of exergy destruction in the respective component.
4. The total LAES system could prevent 34 % of its environmental impact due to avoidable exergy destruction. Main contributors are the turbines and compressors.
5. Exogenous exergy destruction is higher than endogenous for all components.
6. The exergy destruction of any of the components can be reduced by improving the remaining components.

Nomenclature

b	specific environmental impact, Pts/kJ
B	environmental impact rate, Pts/h
c	specific cost, $/kW
C	cost rate, $/h
E	exergy, kWh
p	pressure, bar
T	temperature, K
CAES	Compressed Air Energy Storage
CES	Cryogenic-based Energy Storage
HE	heat exchanger
LAES	Liquid Air Energy Storage
Pump	cryogenic pump
PHS	Pumped Hydro Storage
Z	investment cost

Greek symbols

ε	exergetic efficiency, %
η	isentropic efficiency, %
Δ	Difference

Subscripts and superscripts

$\cdot$	time rate
AV	avoidable
D	destruction
EN	endogenous
EX	exogenous
F	fuel
H	hybrid
k	refers to a component
L	losses
P	product
R	real
T	theoretical
tot	total
UN	unavoidable

Acknowledgement

The authors acknowledge the financial support of the Bundesministerium für Bildung und Forschung/ Federal Ministry of Education, the Research (BMBF) under the Transnational Education project (ID 57128418) of the German Academic Exchange Service (DAAD) and Jimena Incer Valverde acknowledge the economical support of the University of Costa Rica during the preparation of this research.

References

1. IEA, International Energy Agency, 5 February 2019. [Online]. Available: https://www.iea.org/renewables2018/ (2018).
2. The Linde Group, Air separation plants. History and technological progress, Linde AG, Pullach, Germany, (2017).
3. Pérez, J., Sarasúa, J., Wilhelmi, J., Contribution of a hydraulic short-circuit pumped-storage power plant to the load–frequency regulation of an isolated power system, *Science Direct ELSEVIER,* pp. 199-211, (2014)
4. Howe, T., Pollman, A., Gannon, A., Operating Range for a Combined, Building-Scale Liquid Air Energy Storage and Expansion System: Energy and Exergy Analysis, *entropy,* (2018).
5. Hamdy, S., Morosuk, T., Tsatsaronis, G., Cryogenics-based energy storage: Evaluation of cold exergy recovery cycles, *Energy,* pp. 1069-1080, (2017).
6. Sciacovellia, A., Smitha, D., Navarroa, H., Lia, Y., Dinga, Y., Liquid air energy storage – Operation and performance of the first pilot plant in the world, *ECOS 29Th International Conference on Efficiency, Cost and Optimization,* (2016).
7. HighviewPowerStorage, Liquid Air Energy Storage, United Kingdom, (2017).
8. Hamdy, S., Morosuk, T., Tsatsaronis, G., Exergetic and economic assessment of integrated cryogenic energy storage systems, *Cryogenics journal,* (2019).
9. Bejan, A., Tsatsaronis, G., Moran, M., Thermal Design and Optimization, New York, United States: Wiley, (1996).
10. Lazzaretto, A., Tsatsaronis, G., A systematic and general methodology for calculating efficiencies and costs in thermal systems, *Energy Int. J.,* pp. Vol. 31, pp.1257–1289., (2006).
11. Meyer, L., Tsatsaronis, G., Buchgeister, J., Schebek, L., Exergoenvironmental analysis for evaluation of the environmental impact of energy conversion systems, *Energy Int. J.,* pp. Vol. 34, pp.75–89, (2009).
12. Tsatsaronis, G., Morosuk, T., Understanding and improving energy conversion systems with the aid of exergy-based methods, *Int. J. Exergy,* (2012).
13. Goedkoop, M., Spriensma, R., The eco-indicator 99: a damage oriented method for Life Cycle Impact assessment, *Methodology Report,* (2000).
14. S. P. a. t. E. Ministry of Housing, " Eco-indicator 99 Manual for Designers: A damage oriented method for Life cycle impact assessment," 6 March 2019. [Online]. Available: : https://www.pre-sustainability.com/download/EI9
15. Kelly, S., Tsatsaronis, G., Morosuk, T., Advanced exergetic analysis: approaches for splitting the exergy destruction into endogenous and exogenous parts, *Energy Int. J.,* pp. Vol. 34, pp.384–391., (2009).
16. Tsatsaronis, G., Moung-Ho, P., On avoidable and unavoidable exergy destructions and investment costs in thermal systems, *Energy Conversion and Management,* pp. 1259-1270, (2002).

RESEARCH & DEVELOPMENT IN POWER ENGINEERING, 2019

Development of operation strategy for recompression supercritical CO_2 cycle with intercooled main compressor under off-design condition

Yuegeng Ma[1,2], *Tatiana* Morozyuk[2], *Jiping* Liu[1*]

[1] MOE Key Laboratory of Thermal Fluid Science and Engineering, Xi'an Jiaotong University, Xianning West Road. 28, Xi'an 710049, PR China
[2] Institute for Energy Engineering, Technische Universität Berlin, Marchstr. 18, Berlin 10587, Germany

Abstract. The supercritical carbon dioxide (S-CO_2) cycle is regarded as a potential option for the next generation power conversion system. Concentrated solar power (CSP) plant is one of the promising scenarios to adopt the S-CO_2 cycle due to the appealing thermal efficiency and the ability to integrate thermal storage and dry cooling. Among various cycle configurations of S-CO_2 cycle, the recompression S-CO_2 cycle with intercooled main compressor is one of the optimal choices that can provide superior efficiency and a large enough temperature differential for thermal input, which together contribute to the minimization of the overall levelized cost of electricity (LCOE) of the whole CSP plant. The off-design performance and the associated control scheme have important effects on the CSP plant. This paper develops an off-design model for the recompression S-CO_2 cycle with intercooled main compressor for the commercialized hundred-megawatt CSP plant. The effects of different off-design conditions on cycle performance are first evaluated. Different operating strategies regarding the control of cycle maximal pressure and preventing abnormal compressor conditions during off-design operation are then presented and compared. This work is expected to provide knowledge for the optimal control of recompression S-CO_2 cycle with intercooled main compressor during the real operation of the CSP plant.

1. Introduction

Concentrated solar power (CSP) is deemed as a promising renewable energy technology that is likely to play the role of middle-load or even base-load power source in the future energy mix [1]. The CSP plant utilizes heating media such as molten salts to transfer solar energy to the power block. Depending on the upper-temperature limits of the adopted heating media, the power block could be operated at various maximal temperature levels ranging from 350 to over 700 °C [2]. In this temperature range, the S-CO_2 cycle has been regarded as a superior option for the power block in the CSP plant due to its better efficiency and simpler configuration among various choices [3,4].

While many published works focusing on the parametric analyses and optimization of the S-CO_2 cycle, further studies on the off-design characteristics and developments of cycle control strategies were not gained enough attention. The S-CO_2 cycle in the CSP plant is usually operated under off-design conditions due to the changes of ambient conditions in the cold end of the cycle and the inlet temperature and mass flowrate of heating media in the cycle hot end. Therefore, the investigation of the off-design performance and development of the associated control strategies are expected to contribute to the improvement of the actual performance during operation. According to the authors' knowledge, the previous off-design performance studies were only done for the recompression S-CO_2 cycle and simple-recuperated cycle with limited consideration on the effects of various control strategies on the cycle performance [5–7]. Many previous studies suggested that the recompression with intercooled main compressor (ICMC) was the most promising cycle configuration due to the superior performance and large temperature differential for the integration of the thermal storage system [8,9].

In this work, an off-design model is proposed for the recompression cycle with ICMC. The operational issues of the main compressor (MC) are highlighted in the off-design model. four control strategies and the associated configurations are proposed and compared to determine the optimal off-design operating solution with superior efficiency and no risks of damaging MC.

2. System description

The schematic of the S-CO_2 recompression cycle with ICMC and its corresponding T-S diagram are presented in Fig.1. The design parameters of the S-CO_2 recompression cycle with ICMC are listed in Table 1. The elaborative introduction of the thermodynamic characteristics of this cycle is presented in our previous work [9]. Compared to the classic recompression cycle, the introduction of ICMC lead to both higher energetic efficiency and larger temperature differential across the primary heat exchanger (PHX), which contributes to lower LCOE of the whole CSP plant [9]. Split shaft configuration is used for the cycle system due to the benefits in terms of efficiency and control perspectives for such large-scale plants [10]. A synchronous generator is configured for this turbine and variable-speed drive motors are used for the recompressor and main compressor. A buffer tank is configured upstream of the MC to stabilize the compressor inlet pressure at the specified value by inventory control [11].

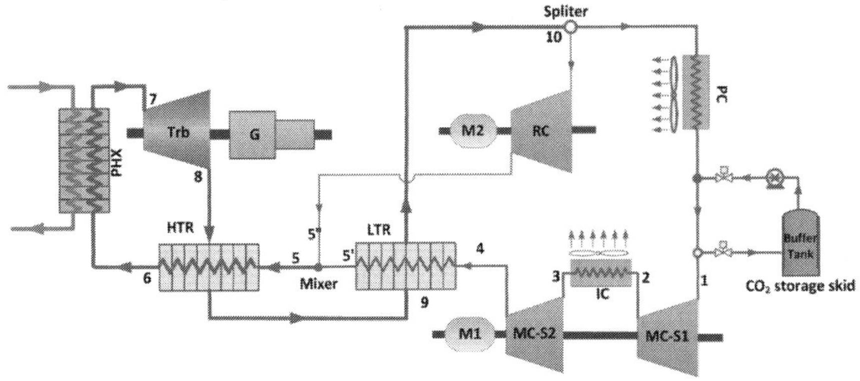

Fig. 1 Configuration of recompression cycle with ICMC

Table 1 Input parameters for the design parameters of the S-CO_2 recompression cycle with ICMC [9]

Input Parameter\ Symbol	Unit	Value
Net output power	MW	100
Temperature of hot storage tank	°C	565
Turbine inlet temperature	°C	550
Main Compressor outlet pressure	MPa	25
Cooler cold-end temperature approach	°C	15
Turbine isentropic efficiency	%	93
Compressor isentropic efficiency	%	89
Pressure drop in PHX	kPa	50
Hot-side Pressure drop of HTR	kPa	60
Cold-side Pressure drop of HTR	kPa	30
Hot-side pressure drop of LTR	kPa	20
Cold-side pressure drop of LTR	kPa	40
Pressure drop in the cooler	kPa	20
Pinch-point temperature approach	°C	5

3. Model implementation

The model of each component and the implementation of cycle simulation under design condition has been detailed in the previous work [9]. This section presented the off-design model of each crucial components in the cycle and the implementation of cycle off-design simulation. These off-design models for components are established in MATLAB 2018a using an objected-oriented approach. REFPROP is used to obtain the necessary thermal properties of CO_2 [12].

3.1 Component model

▪ *Turbine*

Due to the hundred-megawatt level output power, the multi-axial flow turbine is chosen for the cycle in views of higher efficiency and relatively steady flow [13]. Stodola's ellipse method is applied to obtain the turbine inlet pressure under off-design conditions [14,15]. The turbine is assumed to work under the sliding mode with a fixed nozzle area [14]. According to Stodola's ellipse method, the relationship between the mass flow coefficient under design (ϕ_d) and off-design (ϕ_{od}) conditions can be described as follow.

$$\frac{\phi_{od}}{\phi_d} = \frac{\sqrt{1-\left(\frac{p_{out,od}}{p_{in,od}}\right)^2}}{\sqrt{1-\left(\frac{p_{out,d}}{p_{in,d}}\right)^2}} \tag{1}$$

where ϕ is defined as

$$\phi = \dot{m}_{in} \cdot \frac{\sqrt{T_{in}}}{p_{in}} \tag{2}$$

Substituting the above two equations, $p_{in,od}$ can be obtained as

$$p_{in,od} = \sqrt{\dot{m}_{in,od}^2 \cdot T_{in,od} \cdot Y_d + p_{out,od}^2} \text{ with } Y_d = \frac{p_{in,d}^2 - p_{out,d}^2}{p_{in,d}^2 \cdot \phi_d^2} \tag{3}$$

The turbine isentropic efficiency under off-design conditions ($\eta_{T,od}$) can be obtained as follows.

$$\eta_{T,od} = \eta_{T,d} \cdot \sin\left[0.5\pi\left(\frac{\dot{m}_{in,od} \cdot \rho_{in,od}}{\dot{m}_{in,d} \cdot \rho_{in,d}}\right)^{0.1}\right] \tag{4}$$

where $\rho_{in,od}$ and $\rho_{in,d}$ are the density of CO_2 under design and off-design conditions.

▪ *Compressor*

The off-design model of compressor adopted here is based on the nondimensionalized empirical equation regressed out of the experimental data from SNL compressor developed by Dyreby [5]. The flow coefficient of the compressor is defined as

$$\phi_C = \frac{\dot{m}}{\rho_{in} U_{in} D_{in}^2} \tag{5}$$

where U_{in} is the inlet velocity and D the inlet diameter of the compressor.

$$U_{in} = \frac{D_{in}}{2} N \quad (6)$$

The ideal head coefficient (ψ_C) is defined as

$$\psi_C = \frac{(h_{out} - h_{in})}{U_{in}^2} \quad (7)$$

where h_{in} and h_{out} are the enthalpies of CO_2 at the inlet and outlet of the compressor.

To consider the effect of shaft speed (N) on the off-design performance, ϕ_C, ψ_C and the compressor isentropic efficiency (η_C) are modified as Eqs. (8) through (10), and the performance of compressor can be mapped with polynomial regression as Eqs. (11) and (12)

$$\phi_C^* = \frac{\dot{m}}{\rho_{in} U_{in} D_{in}^2}\left(\frac{N}{N_d}\right)^{1/5} \in [0.02, 0.05] \quad (8)$$

$$\psi_C^* = \frac{(h_{out} - h_{in})}{U_{in}^2}\left(\frac{N_d}{N}\right)^{(20\phi_C^*)^3} \quad (9)$$

$$\eta_C^* = \eta_C\left(\frac{\eta_{C,D,0}}{\eta_{C,D}}\right)\left(\frac{N_d}{N}\right)^{(20\phi^*)^5} \quad (10)$$

$$\eta_C^* = -0.7069 + 168.6\phi_C^* - 8089\phi_C^{*2} + 182725\phi_C^{*3} - 1638000\phi_C^{*4} \quad (11)$$

$$\psi_C^* = -0.04049 + 54.6\phi_C^* - 2505\phi_C^{*2} + 53224\phi_C^{*3} - 498626\phi_C^{*4} \quad (12)$$

- *Heat Exchanger*

The model of heat exchanger here is applied to the high-temperature recuperator (HTR), low-temperature recuperator (LTR) and the primary heat exchanger (PHX). The heat sink is assumed to keep constant cold-end temperature difference through the adjustment of the mass flow rate of cooling air through the cooler. The off-design model of heat exchanger consists of two parts, i.e., the pressure sub-model for the pressure loss (Δp) and heat transfer sub-model for the conductance of heat transfer (UA). Depending on the thermodynamic states of CO_2 through the heat exchanger, the model considers the effects of thermal property variation on the Δp and UA differently. The CO_2 thermal properties through the HTR and PHX change uneventfully because the operating condition is far away from the critical region. The conductance of heat transfer (UA_{od}) and pressure loss (Δp_{od}) under off-design conditions is therefore calculated without considering the effects of thermal properties variation as follows.

$$\frac{UA_{od}}{UA_d} = \frac{\dot{m}_{c,d}^{-0.8} + \dot{m}_{h,d}^{-0.8}}{\dot{m}_{c,od}^{-0.8} + \dot{m}_{h,od}^{-0.8}} \quad (13)$$

$$\frac{\Delta p_{od}}{\Delta p_d} = \left(\frac{\dot{m}_d}{\dot{m}_{od}}\right)^{7/4} \quad (14)$$

Unlike the cases for the HTR and PHX, the CO_2 through the LTR is close to its critical point. The thermal properties relating to heat transfer and hydraulic characteristic undergo drastic changes with temperature and pressure should be considered in the UA calculation under off-design conditions. Therefore, the UA_{od} and Δp_{od} are calculated as follows [7].

$$\frac{UA_{od}}{UA_d} = \frac{\alpha_{c,d}^{-1} + \alpha_{h,d}^{-1}}{\alpha_{c,od}^{-1} + \alpha_{h,od}^{-1}} \text{ with } \alpha = \dot{m}^{0.8} c_p^n k^{(1-n)} \mu^{(n-0.8)} \quad (15)$$

$$\frac{\Delta p_{od}}{\Delta p_d} = \left(\frac{\dot{m}_d}{\dot{m}_{od}}\right)^{7/4} \cdot \left(\frac{\mu_d}{\mu_{od}}\right)^{1/4} \cdot \left(\frac{\rho_d}{\rho_{od}}\right)^{-1} \quad (16)$$

where c_p is the specific heat capacity, k is the coefficient of heat conductivity. μ is the dynamic coefficient of viscosity. n=0.4 for heating fluid and n=0.3 for cooling fluid.

4. Results and discussion

The cycle performance under various off-design conditions is investigated in this section. Parametric optimizations are first implemented for three cases with different MC inlet temperature ($T_{MC,in,d}$) to obtain optimal design points. The effects of $T_{salt,in}$, $\dot{m}_{salt}$ and $T_{MC,in}$ on the off-design performance are then studied with sensitivity analysis in their respective ranges as presented in Table. 2. In the following off-design calculation under different conditions, only the results without abnormal conditions of the MC are presented in the figures.

Table 2 Variation ranges of parameters for off-design study

Parameter	Unit	Design point	Off-design range
$T_{salt,in}$	°C	565	535-595
Relative m_{salt}	\	1	0.6–1.2
$T_{MC,in}$	°C	32,41,50	32–50

4.1 Off-design performance under varying heat source conditions

Fig. 2–3 presents the off-design performance with the change of $T_{salt,in}$ under the three conditions of $T_{MC,in,d}$. It is found that the cycle can be operated normally in the whole range of $T_{salt,in}$ with these four control schemes under three conditions of $T_{MC,in,d}$. The η_{cyc} and $\dot{W}_{net}$ increase linearly with the rise of $T_{salt,in}$ for all the cases. The variation of the studied variables show similar variation tendencies with $T_{salt,in}$ for the three $T_{MC,in,d}$ cases with slight differences observed among different control schemes. As illustrated in Fig. 2, the variation of p_{max} in the range of $T_{salt,in}$= 535 °C–595 °C is less than 0.3MPa for the cases of the FP control scheme and the change of relative shaft speed (RN) is less than 0.3%. The comparisons between B_FP and M_FP and between B_SP and M_SP show that the configuration of MC, i.e., basic configuration or modified configuration have no effects in control strategy and hence no effects on the cycle performance because only the control of N is implemented for both basic and modified configurations during off-design condition of $T_{salt,in}$. The comparisons between B_FP and B_SP and between M_FP and M_SP indicate that the $\dot{W}_{net}$ with different pressure control schemes have slight difference in the rate of change with $T_{salt,in}$. As shown in Fig.3, the $\dot{W}_{net}$ increases at higher rate under SP control than under FP control. This can be attributed to the higher decreasing rate of $\dot{m}_{CO2}$ with FP control which partially counteracts the effects of η_{cyc} on $\dot{W}_{net}$.

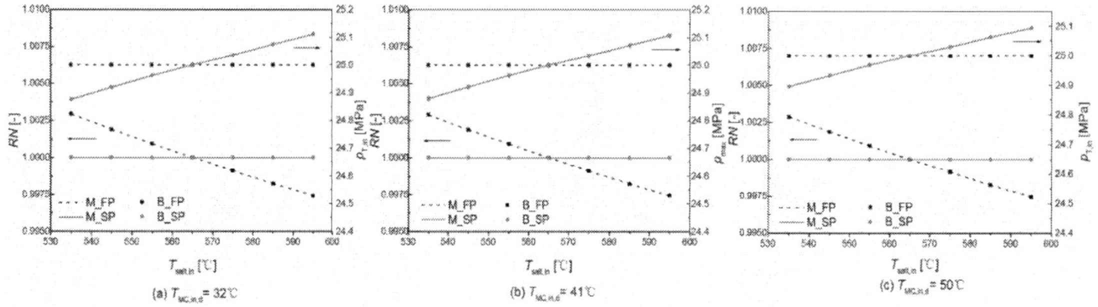

Fig. 2 The variations of RN and $p_{T,in}$ with $T_{salt,in}$

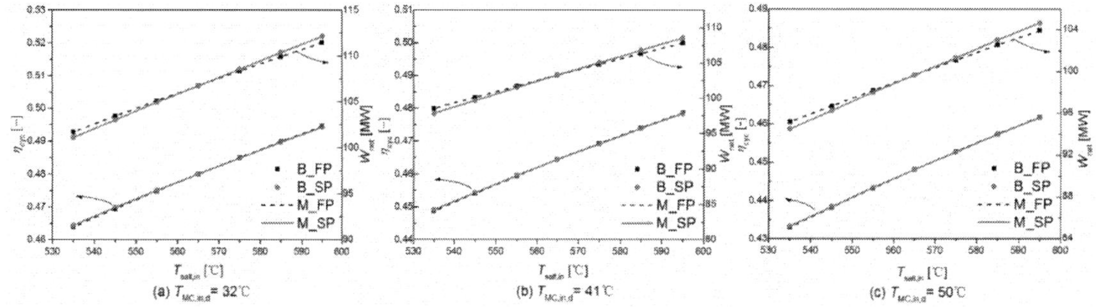

Fig. 3 The variations of $\dot{W}_{net}$ and η_{cyc} with $T_{salt,in}$

Fig. 4–5 presents the effect of $\dot{m}_{salt}$ on the off-design performance of cycle. The cycle can be operated in the whole range of $T_{salt,in}$ with these four control schemes and three $T_{MC,in,d}$ conditions. As shown in Fig. 4, The η_{cyc} and $\dot{W}_{net}$ increase monotonously with $\dot{m}_{salt}$ at a decreasing rate. The comparisons between B_FP and M_FP and between B_SP and M_FP indicate that the configurations have no effect on the off-design performance as only the N is used as control variable during off-design operation. The comparisons between B_FP and B_SP and between M_FP and M_FP indicate that the FP control leads to lower η_{cyc} yet higher $\dot{m}_{CO2}$ than SP control when $\dot{m}_{salt}$ decreases, and these differences are more significant as $\dot{m}_{salt}$ reduces. As the decreased η_{cyc} offsets the effect of increased $\dot{m}_{salt}$, the $\dot{W}_{net}$ of FP control is only slightly higher than that of SP control when $\dot{m}_{salt}$ is lower than the design value and the difference in $\dot{W}_{net}$ become less significant as $\dot{m}_{salt}$ rise. (see Fig. 5)

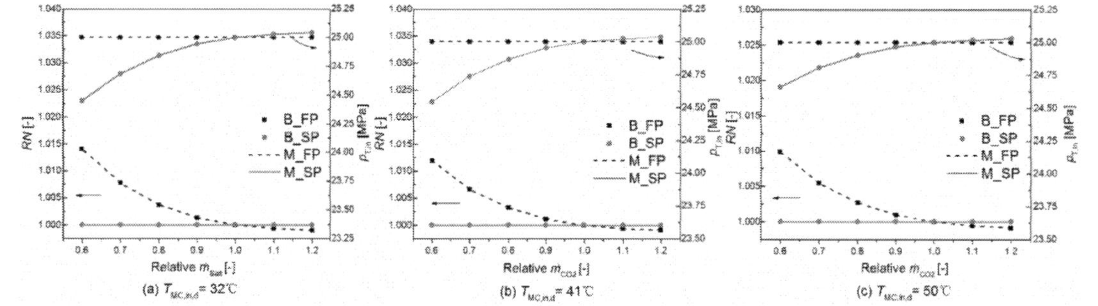

Fig. 4 The variations of RN and $p_{T,in}$ with $\dot{m}_{salt}$

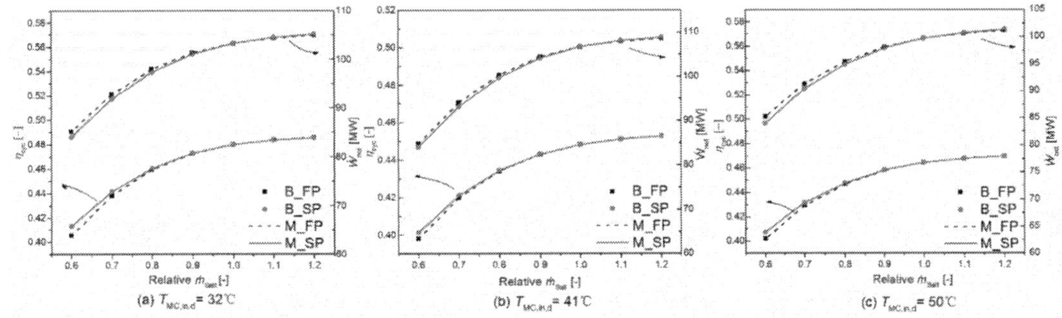

Fig. 5 The variations of $\dot{W}_{net}$ and η_{cyc} with $\dot{m}_{salt}$

4.2 Off-design performance under varying ambient conditions

Fig. 6–7 presents the cycle off-design performance under different MC inlet temperature conditions. Unlike the relative uneventful effects of $\dot{m}_{salt}$ and $T_{salt,in}$, the impact of $T_{MC,in}$ variations on cycle performance is significant. The change of $T_{MC,in}$ results in variations of cycle performance with different fashions from one another for these cases with different control schemes and $T_{MC,in,d}$. The comparisons between B_FP and M_FP and between B_SP and M_SP show that the basic configurations cannot be applied under low $T_{MC,in}$ off-design conditions for the cases of $T_{MC,in,d}$=41 °C and 50 °C, with a larger applicable range for B_FP than B_SP; on the contrary, the modified configurations can cover all the studied range in three cases with recirculation system of MC being activated to prevent surge for low $T_{MC,in}$ off-design conditions when $T_{MC,in,d}$=41 and 50 °C. According to the comparisons of B_FP versus B_SP and M_FP versus M_SP, it is found that the pressure control strategy has a significant effect under the off-design conditions of $T_{MC,in}$. The FP control generally results in better η_{cyc} than that of FP control especially when the $T_{MC,in}$ is significantly deviated from the $T_{MC,in,d}$. Under off-design conditions of $T_{MC,in}$, the $\dot{W}_{net}$ under SP control deviates apparently from the design point value whereas the $\dot{W}_{net}$ under FP control undergoes relatively mild change. This is partially due to the different trends of $\dot{m}_{CO2}$ variation during off-design operation besides the effect of η_{cyc}. The $\dot{m}_{CO2}$ is controlled at almost constant value under FP control. By comparison, the $\dot{m}_{CO2}$ under SP control varies significantly as $T_{MC,in}$ changes under off-design conditions due to the varying $p_{T,in}$ under off-design conditions. The use of SP control also lead to more significant variations in $T_{T,in}$ and $T_{salt,out}$, which may cause adverse effects on the performance of solar components.

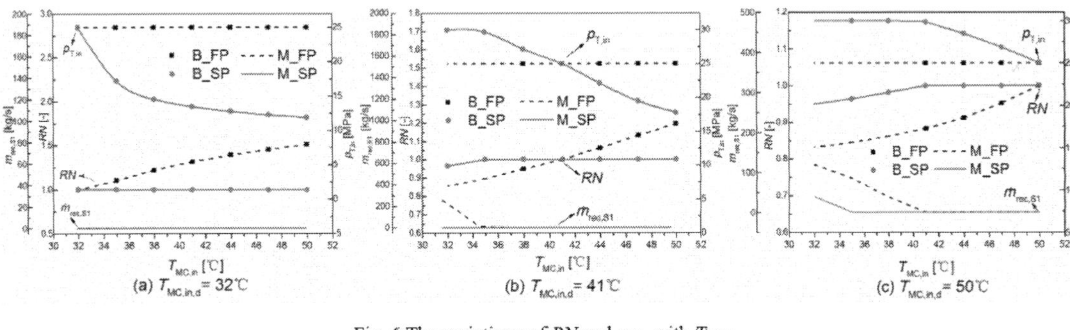

Fig. 6 The variations of RN and $p_{T,in}$ with $T_{MC,in}$

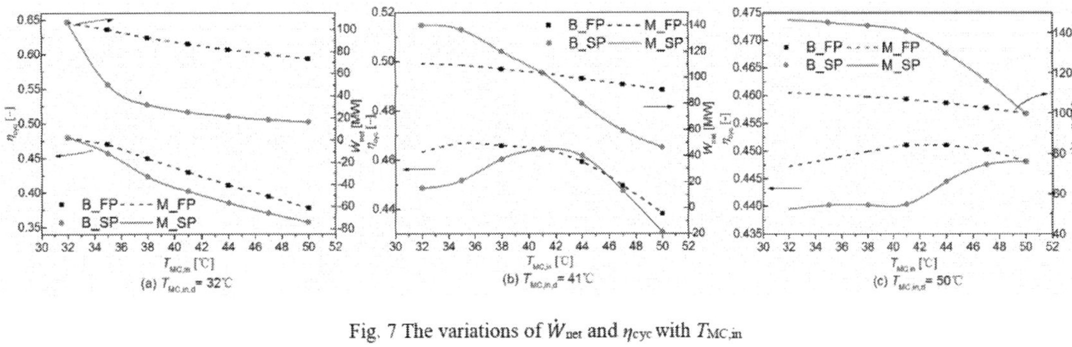

Fig. 7 The variations of $\dot{W}_{net}$ and η_{cyc} with $T_{MC,in}$

5. Conclusion

This study establishes an off-design model for the S-CO_2 cycle with ICMC. The potential issues of the MC abnormal operation were highlighted, four control schemes and the associated configurations for MC are developed and compared. The off-design performance under conditions of varying $T_{salt,in}$, $\dot{m}_{salt}$ and $T_{MC,in}$ are investigated for three cases with different $T_{MC,in,d}$. The following conclusions are drawn after this study:

- The changes of $T_{salt,in}$ and $\dot{m}_{salt}$ under off-design conditions in the cycle hot end lead to uneventful effects in terms of the MC control. No abnormal operation conditions are observed for MC. The MC configuration cycle performance has no effect on cycle performance, while the pressure control strategy has very slight effects.
- The change of $T_{MC,in}$ under off-design conditions significantly affects the MC control. The S-CO_2 cycle with a high $T_{MC,in,d}$ will run into the abnormal condition of MC surge under low $T_{MC,in}$ conditions when integrated with basic configuration of MC. In contrast, the cycle with modified MC configuration can prevent the surge risk by recirculating partial working fluids.
- During all these off-design analyses on $T_{salt,in}$, $\dot{m}_{salt}$ and $T_{MC,in}$, only single mode is required for MC. The parallel compressor is not activated for either S1 or S2 of the modified configuration. The risk of zero pressure head under the conditions of $T_{MC,in} > T_{MC,in,d}$ can be prevented by N control without the use of parallel compressor, and the N can be over 1.5 times of the N_d.
- The off-design analyses show that the M_FP as the control scheme can results in superior efficiency, steady $\dot{W}_{net}$ and no risk of abnormal MC condition under off-design conditions than the other three schemes. However, the cycle performance may be further improved when the real-time parametric optimization is applied for the cycle during off-design conditions.

References

[1] Mehos M, Turchi C, Vidal J, Wagner M, Ma Z, Ho C, et al. Concentrating Solar Power Gen3 Demonstration Roadmap. *NREL Tech Rep*, pp.1–140, (2017).

[2] Philibert, C. Technology roadmap: concentrating solar power. *OECD/IEA*, (2010).

[3] Turchi CS, Ma Z, Neises T, Wagner M. Thermodynamic Study of Advanced Supercritical Carbon Dioxide Power Cycles for High Performance Concentrating Solar Power Systems. *Proc. ASME 2012 6th Int. Conf. Energy Sustain.*, San Diego, CA, USA, (2012).

[4] Crespi F, Gavagnin G, Sánchez D, Martínez GS. Supercritical carbon dioxide cycles for power generation: A review. *Appl Energy*, Vol. 195, pp.152–83. (2017);

[5] Dyreby JJ. Modeling the Supercritical Carbon Dioxide Brayton Cycle with Recompression. *The University of Wisconsin-Madison*, (2014).

[6] Wright SA, Davidson CS, Husa C. Off-design performance modeling results for a supercritical CO_2 waste heat recovery power system. *6th Int. Supercrit. CO_2 Power Cycles Symp.*, pp.1–10. Pennsylvania, USA (2018)

[7] de la Calle A, Bayon A, Soo Too YC. Impact of ambient temperature on supercritical CO_2 recompression Brayton cycle in arid locations: Finding the optimal design conditions. *Energy*, Vol.153, pp.1016–27. (2018)

[8] Wang, Kun, Ming-Jia Li, Jia-Qi Guo, Peiwen Li, and Zhan-Bin Liu. A systematic comparison of different S-CO2 Brayton cycle layouts based on multi-objective optimization for applications in solar power tower plants. *Appl Energy*, Vol.212, pp.109–21, (2018).

[9] Ma Y, Morozyuk T, Liu M, Yan J, Liu J. Optimal integration of recompression supercritical CO_2 Brayton cycle with main compression intercooling in solar power tower system based on exergoeconomic approach. *Appl Energy*, Vol.242, pp.1134–54, (2019)

[10] Weiland N, Thimsen D. A Practical Look at Assumptions and Constraints for Steady State Modeling of sCO_2 Brayton Power Cycles. *No. NETL-PUB-20271. NETL*, (2016).

[11] Held, Timothy J. Initial test results of a megawatt-class supercritical CO_2 heat engine. *4th International Symposium on Supercritical CO_2 Power Cycles*, (2014).

[12] Lemmon EW, Huber ML, Mclinden MO. NIST Reference Fluid Thermodynamic and Transport Properties. *REFPROP database* 23: v7, (2002)

[13] Sienicki JJ, Moisseytsev A, Fuller RL, Wright SA, Pickard PS. Scale Dependencies of Supercritical Carbon Dioxide Brayton Cycle Technologies and the Optimal Size for a Next-Step Supercritical CO_2 Cycle Demonstration. *3th International Symposium on Supercritical CO_2 Power Cycles*, (2011).

[14] Li H, Yang Y, Cheng Z, Sang Y, Dai Y. Study on off-design performance of transcritical CO_2 power cycle for the utilization of geothermal energy. *Geothermics*, Vol. 71, pp.369–79, (2018);

[15] D. COOLKE. Modeling of off-design multistage turbine pressures by Stodola's ellipse. *Energy Incorporated PEPSE User's Group Meeting*. Richmond, VA, Nov, pp. 2-3. (1983).

RESEARCH & DEVELOPMENT IN POWER ENGINEERING, 2019

Formation of ammonia bisulfate in coal-fired power plant equipped with SCR reactors and the effect of reduced load operation

Maciej Żyrkowski[1,2,*] , Monika Motak[1]

[1] AGH Akademia Górniczo – Hutnicza, al. Mickiewicza 30, 30-059 Kraków, mzyrkow@agh.edu.pl, Poland
[2] PGE Energia Ciepła S.A., ul. Złota 59, 00-120 Warszawa, Poland

Abstract. Coal fired power plant are still responsible for vast amount of electricity and heat production in Poland in 2019. Increasing number of units is equipped with SCR reactors in order to mitigate NOx emission. Older units have been designed to work in a base load operation, however increasing number of intermittent energy sources on the market, forces power plant owners to operate such units in reduced load for significant number of hours in a year. Most of parameters in coal-fired boiler would change in reduced load operation, particularly flue gas parameters. That would affect the formation of ammonia bisulfate (NH_4HSO_4) as well as operation of SCR reactors in the boiler. The NH_4HSO_4 formation is highly undesired as it can plug the catalyst or heating surfaces in the boiler. To investigate this phenomenon, the NH_4HSO_4 formation was comprehensively studied. Moreover, reduced load operation of boiler in a medium size CHP plant was simulated, using computer modelling tools. Obtained parameters were used to analyse, how different flue gas parameters affect NH_4HSO_4 formation and what consequences does it have for boiler operation. In particular, effect related with operation of SCR reactors equipped with V_2O_5-WO_3/TiO_2 catalyst was deeply studied.

1. Introduction

Ammonia bisulfates formation is one of the major problems related to DeNOx installations, such as SCR (Selective Catalytic Reduction) or SNCR (Selective Non-Catalytic Reduction), in solid fuels power plants. Ammonia (NH_3) that can escapes from these installations reacts with sulfur trioxide, forming compounds such as ammonium bisulfates NH_4HSO_4 and ammonium sulfates $(NH_4)_2SO_4$. The problem is related with the first one in particular - salt which at a temperature below its condensation point, forms sticky substance that covers porous structure of the SCR catalysts, as well as it can contaminate surfaces of the rotary air heater and the electrostatic precipitator [1] [2] [3] [4]. Because it condensates at a certain temperature, region that can be affected by plugging depends on the boiler load. The NH_4HSO_4 also passes in solid form to the fly ash, which results in increased ammonia content in the combustion by products.

2. SO3 formation

The SO_3 in flue gas exists because of SO_2 to SO_3 conversion. Due to the fact that SO_3 concentration is crucial for ammonia bisulfate formation, it is vital to discuss this mechanism in details. Sulfur in the combustion chamber is delivered together with the fuel. Modern solid fuels boilers use a division of the combustion process into a reduction zone (with oxygen deficiency) and a post-combustion zone (with excess oxygen). During combustion, sulfur contained in coal is oxidized according to the following reaction:

$$S + O_2 \rightarrow SO_2 \quad (1)$$

The largest amount of SO_2 is formed in the first stage of combustion. Then, some part of SO_2 is converted to SO_3. The main parameters that support this process is temperature, the residence time of the particle, oxygen concentration, as well as the composition and size of the mineral fraction in fuel. The effect of temperature on the degree of conversion begins to be visible above 700°C. However, the influence of temperature on the conversion process is small compared with catalytic influence, discussed in below [5]. Overall conversion in combustion zone is up to 2% [6].

Above the area of the boiler where the combustion process has been completed, sulfur is mainly present in the form of SO_2. The SO_3 concentration depends on the sulfur content in fuel. Also, elements such as iron, nickel, wolfram, titanium and vanadium show catalytic properties for the conversion of SO_2 to SO_3. Thus, along with the increase of these elements in the fuel, the potential SO_3 emission will also increase. On the other hand, the presence of alkaline elements may prevent this process by binding both SO_2 and SO_3 in the form of sulfates (e.g. K_2SO_4, Na_2SO_4) at temperature below 1000°C. The highest intensity of catalytic reactions is recorded at a temperature of about 700°C and their impact on the overall level of SO_2 to SO_3 conversion is much greater than other factors. Table 1 shows the influence of temperature and the above-mentioned catalysts on the SO_2 conversion process. It can be seen, that at temperature of 700°C as much as 95% of the total conversion occurs due to the action of catalysts [7].

Table 1. Effect of temperature and catalysis on SO_2 to SO_3 conversion rate, based on *[7]*

Temperature	Conversion rate [%]		Catalytic effect on conversion [%]
	Temp. effect	Total conversion	
400°C	0.041	0.042	1
500°C	0.062	0.077	19
700°C	0.098	1.784	95
900°C	0.361	1.308	72
1000°C	0.773	1.791	57

As noted above, vanadium oxide (V_2O_5) being part of the SCR catalyst has catalytic properties when converting SO_2 to SO_3. In modern catalysts, the conversion rate on SCR installations is roughly 1-2%. According to Lu et al. [6], conversion increases in the presence of ammonia in flue gas as well as in the presence of V_2O_5, WO_3 and TiO_2. These compounds constitutes a catalyst, that is most often applied in SCR reactors. Also, if the ratio NH_3/NO is higher than 0.6, the conversion raises. The effect of oxygen concentration, NO concentration as well as Al_2O_3 is relatively small on the other hand. Also the conversion rate drops when the amount of SiO_2 and BaO increases [6]. Taking this into account, it is important to ensure that SCR installation operates at its optimal conditions. Operational parameters (such as NOx concentration, temperature and flue gas flow before installation) should be observed. The essence of the exhaust gas temperature before the SCR installation is particularly important - when it is too low, there is an increased slip of ammonia, when it is too high - increased amount of sulfur trioxide can be formed [8].

In the zone of the rotary air heater, the sulfur trioxide SO_3 reacts with the steam to form sulfuric acid (VI) according to the equation:

$$SO_3\,(g) + H_2O \rightarrow H_2SO_4\,(g) \qquad (2)$$

This reaction takes place in temperature range between 204°C and 426°C. At temperatures below 204°C, sulfuric acid is also in the form of an aerosol [8]. Sulfuric (VI) acid is dangerous at temperatures below its condensation point, bringing risk of corrosion occurrence. The temperature of this point increases proportionally to the partial pressure of water vapor and SO_3 in exhaust gases. A typical temperature range in which H_2SO_4 condensates is in the range of 95-160°C. The increased SO_3 content can therefore lead to sulfuric acid condensation at the cold end of the air heater. The relationship between the sulfuric acid condensation temperature and SO_3 concentration in the flue gas after the rotating air preheater, is presented in the Table 2 [9] [10].

Table 2. The relationship between the sulfuric acid condensation temperature and SO3 concentration

Acid dew point	SO_3 concentration
155°C	60 ppm
146°C	30 ppm
124°C	5 ppm (desired value)

Although there is a high probability that the flue gas will cool below the acid dew point temperature, the first signs of corrosion caused by sulfuric acid condensation should be expected only at temperatures below 70-80°C [10]. Above this temperature, condensing sulfuric acid will rather bind sticking ash particles together which, however, does not lead to the corrosion of particular surfaces. Another problem for rotating air preheater is further discussed occurrence of liquid ammonium bisulfate [5].

3. SO3 concentration in flue gas

Considering the pulverized coal boiler, it can be estimated that the level of sulfur dioxide emissions is, for the nominal load operation, in the range from about 1200 mg / $_Nm^3$ (458 ppm) to about 2500 mg / $_Nm^3$ (955 ppm), depending on the sulfur content in the fuel (from 0.7% to 1.1%). Assuming the SO_2 to SO_3 conversion (before SCR) equals to 2%, the corresponding range of SO_3 concentration in the furnace chamber is from about 9.2 ppm to 19.1 ppm. Part of SO_3 reacts with alkaline elements in fly ash to form compounds that settle on the boiler heating surfaces (e.g. K_2SO_4, Na_2SO_4). The associated SO_3 reduction will therefore be directly proportional to the alkali content in fly ash. Part of SO_3/H_2SO_4 is also adsorbed on the fly ash particles. After the flue gas passes through successive layers of the SCR installation, the SO_3 concentration increases accordingly: about 5 ppm in the first catalyst layer, about 6 ppm in the second catalyst layer and about 6.5 ppm in the third catalyst layer [8]. The above values may increase when the SCR installation is turned off and the exhaust gas is not bypassed (increase by approx. 30%). The reason for this is the main component of SCR catalysts - vanadium oxide - which also acts as a catalyst for the conversion of SO_2 to SO_3. Besides that, the increased flue gas temperature and reduced flue gas flow can also influence conversion [8].

Occurrence of ammonia bisulfate in the boiler is important from power plant operation point of view, because it can easily deposit on heating surfaces as well as on SCR catalysts. Moreover, it aggregates fly ash particles carried by flue gas. Plugged heating surfaces would affect heat transfer while plugged catalyst surface would have lower activity. Temperature of ABS condensation has been subjected to numerous studies. There is no consistency for the results though. According to Shi et al. [11] temperature of ABS condensation is in the range of 190°C - 240°C, but other authors indicate, that risk of ABS formation may occur at any temperature. Ammonia bisulfate can be removed when flue gas temperature is higher than 400°C, but this can bring about issues related with catalyst sintering on the other hand. Temperature of ABS condensation is also related to the pores structure of the catalyst. For smaller pores, temperature may increase due to the capillary forces [12] [13]. It can be concluded, that the lower the temperature of SCR operation is, the higher the risk of ammonia bisulfate condensation will be. It is also important to notice, that SCR catalysts producers indicate that proper temperature for the installation should be between 300°C and 400°C.

4. Modelling of coal-fired boiler operation

Modelling of the coal-fired boiler was made using Ebsilon Professional, the software for thermodynamic simulation of power units [14]. By utilizing this tool, it is possible to design and simulate single devices as well as technological systems for power or heat production. In coal-fired boiler, temperature of flue gas is not measured in every single region of the boiler. Besides that, in the first pass of the boiler, temperatures are too high to install traditional measurement sensors, which leaves room for acoustic temperature measurement systems only. The latter are usually used in combination with SNCR technologies, in order to seek for desired temperature window in real time. However, if the boiler is not equipped with proper flue gas temperature measurements, the only way to find out desired value of temperature, is to simulate the boiler operation. Because, from the SCR operation point of view, it is crucial to understand the temperature distribution, an pulverized coal-fired units with nominal capacity of 430 t/h was simulated. The schematic view of the boiler is shown on Fig. 1 while the corresponding thermodynamic model layout is shown on Fig. 2. What can be notice from Fig. 1, is that simulated boiler has only one stage of steam superheating, and the SCR installation is installed downstream the water heater. What can also be noticed from Fig. 1, is that SCR reactor has two catalytic layers and the third one is left empty. That is often the case for new installations, where the third layer is added after the overall activity of the installation drops under the required limit.

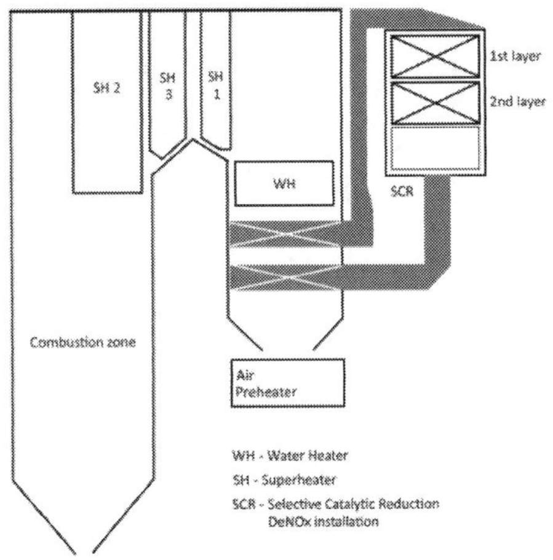

Fig. 1. Schematic view on the pulverized coal boiler with 430 t/h nominal capacity, equipped with DeNOx SCR installation.

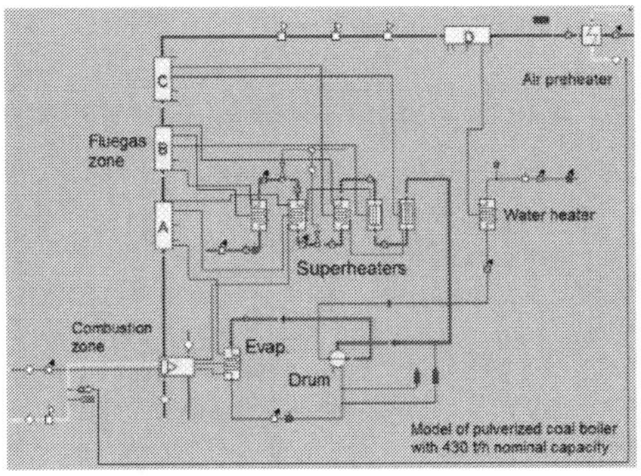

Fig. 2. Layout of the coal-fired boiler model in Ebsilon Professional software.

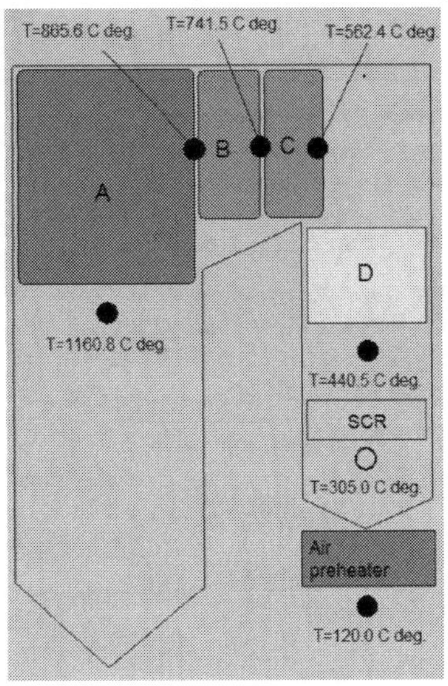

Fig. 3. Flue gas temperature distribution in 100% boiler load.

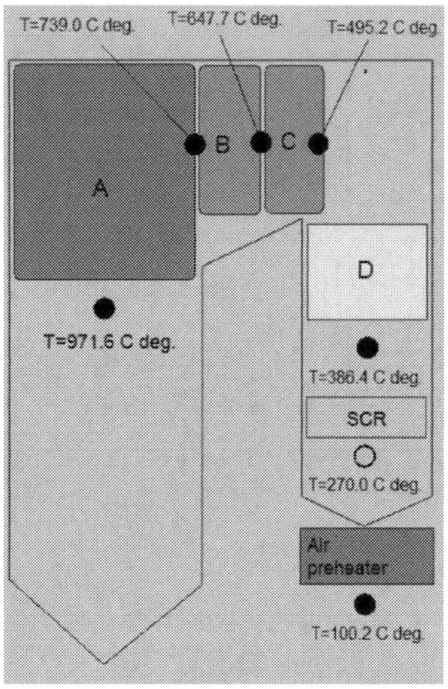

Fig. 4. Flue gas temperature distribution in 60% boiler load.

5. Model outcomes and discussion

Based on the model outcomes, flue gas temperature distribution along the boiler was determined for nominal (100%) and minimal (60%) load. Results are shown on Fig. 3 and Fig. 4. Rectangles marked with capital letter (A-D), represent areas of the boiler, from where heat is conveyed (through the radiation and convection mechanism) to superheaters and other heat exchangers installed in the boiler. Temperatures indicated by black dots, are calculated as an average value for the cross section between regions A-D. In order to simulate exact temperature in every single place of the boiler, more advance simulation techniques, like Computational Fluid Dynamics (CFD), must be applied. Because the region D represent heat absorbed the water heater, that means temperature calculated downstream this region is also the temperature at the inlet of SCR installation. The temperature downstream the SCR installation shown on Fig. 3 and Fig. 4 (transparent dot), is adopted from power plant measurement system, as the model does not include calculations for temperature drop at the SCR installation.

6. Conclusion

In this paper, mechanism of ammonia bisulfate formation was comprehensively described. Also, the mechanism of the SO_2 to SO_3 conversion in coal-fired boiler was discussed. In order to investigated temperature of SCR installation operation, the thermodynamic model of an 430 t/h capacity coal-fired boiler was simulated in nominal and minimal load operation. The ammonia bisulfate is dangerous for boiler operation, because it contaminates active surface of the SCR catalyst as well as the cold end of rotating air preheaters. To avoid, or at least reduce issues related with ammonia bisulfate, it is crucial to maintain proper flue gas temperature as well as prevent excessive ammonia slip and SO_3 formation.

Acknowledgements

M. Żyrkowski acknowledges the financial grant of the Ministry of Science and Higher Education no. 0039/DW/2018/02

References

[1] G. Qi i R. T. Yang, „Performance and kinetics study for low-temperature SCR of NO with NH3 over MnOx-CeO2 catalyst," *Journal of Catalysis, 217,* pp. 434-441, 2003.

[2] P. S. Hammershoi, P. N. Vennestrom, H. Falsig, A. D. Jansen i T. V. Janssens, „Importance of the Cu oxidation state for the SO2 poisoning of a Cu-SAPO-34 catalyst in the NH3-SCR reaction," *Applied Catalysis B: Environmental, 236,* pp. 377-383, 2018.

[3] L. Song, J. Chao, Y. Fang, H. He, J. Li, W. Qiu i G. Zhang, „Promotion of ceria for decomposition of ammonia bisulfate over V2O5-MoO3/TiO2 catalyst for selective catalytic reduction," *Chamical Engineering Journal, 303,* pp. 275-281, 2016.

[4] S. Youn, I. Song, H. Lee, S. J. Cho i D. H. Kim, „Effect of pore structure of TiO2 on the SO2 poisoning over V2O5/TiO2 catalysts for selective catalytic reduction of NOx with NH3," *Catalysis Today, 303,* pp. 19-24, 2018.

[5] R. Sporl, J. Maier i G. Scheffknecht, „Sulphur Oxide Emissions from Dust-Fired Oxy-Fuel Combustion of Coal," *Energy Procedia, 37,* pp. 1435-1447, 2013.

[6] J. Lu, Z. Zhou, H. Zhang i Z. Yang, „Influenced factors study and evaluation for SO2/SO3 conversion rate in SCR process," *Fuel, 245,* pp. 528-533, 2019.

[7] P. Marier i H. P. Dibbs, „The catalytic conversion of SO2 to SO3 by fly ash and capture of SO2 and So3 by CaO and MgO," *Thermochem. Acta., 8,* pp. 155-165, 1974.

[8] Breen Energy Solutions, „Comprehensive acid gas management," 2012. [Online]. Available: http://breenes.com/wp-content/uploads/2017/04/Comprehensive_Condensable_Management_Program-RevA.pdf. [Data uzyskania dostępu: July 2019].

[9] W. Wei, F. Sun, Y. Shi i L. Ma, „Theoretical prediction of acid dew point and safe operating

temperature of heat exchangers for coal-fired power plants," *Applied Thermal Engineering, 123,* pp. 782-790, 2017.

[10] Z. Li, F. Sun, Y. Shi, F. Li i L. Ma, „Experimental study and mechanism analysis on low temperature corrosion of coal fired boiler heating surface," *Applied Thermal Engineering, 80,* pp. 355-361, 2015.

[11] Y. Shi, H. Shu, Y. Zhang, H. Fan, Y. Zhang i L. Yang, „Formation and decomposition of NH4HSO4 during selective catalytic reduction of NO with NH3 over V2O5-WO3/TiO2 catalysts," *Fuel Processing Technology, 150,* pp. 141-147, 2016.

[12] C. Zheng, Y. Wang, Y. Liu, Z. Yang, R. Qu, D. Ye, C. Liang, S. Liu i X. Gao, „Formation, transformation, measurement, and control of SO3 in coal-fired power plants," *Fuel, 241,* pp. 327-346, 2019.

[13] Charles A. Lockert, Breen Energy Solution., „Dynamic Control of SCR Minimum Operating Temperature," w *CoalGen*, 2009.

[14] „EBSILON®Professional for the design of power plants," STEAG Energy Services, [Online]. Available: https://www.steag-systemtechnologies.com/en/products/ebsilon-professional/. [Data uzyskania dostępu: 2019].

[15] L. Muzio, S. Bogseth, R. Himes, Y. Chien i D. Dunn-Rankin, „Ammonium bisulfate formation and reduced load SCR operation," *Fuel, 206,* pp. 180-189, 2017.

[16] J. Menasha, D. Dunn-Rankin, L. Muzio i J. Stallings, „Ammonium bisulfate formation temperature in a bench-scale single-channel air preheater," *Fuel, 90,* pp. 2445-2453, 2011.

[17] S. Shwan, W. Partridge, J. Choi i L. Olsson, „Kinetic modeling of NOx storage and reduction using spatially resolved MS measurements," *Applied Catalysis B: Environmental, 147,* pp. 1028-1041, 2014.

[18] G. Dolanc, S. Strmcnik i J. Petrovcic, „NOx selective catalytic reduction control based on simple models," *Journal of Process Control, 11,* pp. 35-51, 2001.

 https://doi.org/10.1051/e3sconf/201913701022

RESEARCH & DEVELOPMENT IN POWER ENGINEERING, 2019

Hazards associated with syngas storage

Katarzyna Stolecka[1*], *Andrzej* Rusin[2,*]

[1,2] Silesian University of Technology, Institute of Power Engineering and Turbomachinery, Konarskiego 18, 44-100 Gliwice, Poland, katarzyna.stolecka@polsl.pl, andrzej.rusin@polsl.pl

Abstract Energy needs of many countries are largely covered by energy obtained from fossil fuels. This in turn involves environmental pollution and greenhouse gas emissions. The growing environmental awareness and the need to prevent climate changes mean that clean energy and alternative energy sources are still a significant research issue. One of the most important technologies for efficient and low-carbon energy generation is the gasification process and synthesis gas production. Worldwide, there are now more than 270 such installations. More installations are under construction. Syngas is a mixture of hydrogen and carbon monoxide. Depending on the feedstock, it can also contain smaller amounts of carbon dioxide, methane and nitrogen. The gasification process consists of four stages: syngas production, storage, transport and utilization, e.g. as fuel. Because syngas is mainly composed of flammable and toxic gases, in the event of an uncontrolled release into the atmosphere these processes may pose a potential hazard to humans and the environment. The paper presents the results of analyses related to hazards resulting from an uncontrolled release of gas at the stage of the gas storage, before it is transported or finally used. Hazard scenarios are presented and the probability of their occurrence as well as the consequences for humans and the environment are determined.

1. Introduction

Energy needs of many countries are covered primarily by energy obtained by firing fossil fuels, which involves environmental pollution and greenhouse gas emissions.

The effect of the growing awareness of the hazards, the stricter requirements concerning the reduction of emissions and the need to protect the environment is that clean energy and alternative energy sources are still a significant research issue. The shortage of conventional energy sources and the advancement in fuel conversion have stimulated interest in research on fuel conversion and gasification technologies as clean, reliable and energy-efficient processes [1,2].

It is assumed that gasification will be one of the main energy sources in future – it will become the tool for the transition from coal-based energy to energy based on hydrogen [1,2].

Syngas is the end product of the gasification process. The term relates to gases being the effect of gasification of different feedstocks, such as coal, biomass, waste, liquid hydrocarbons, etc. The process main gasifying medium is oxygen, steam or air. Synthesis gas is a mixture of carbon monoxide and hydrogen in the first place. It can also contain other gases, e.g. methane, nitrogen or carbon dioxide. The composition depends on the feedstock or the gasifying medium [3,4]. An example composition of syngas is presented in Table 1 [2,5].

Table 1. Syngas example composition

	biomass	brown coal
gasification process	**fluidized bed**	**fluidized bed**
CO	20-30	30-50
CO_2	25-40	13-25
H_2	20-30	35-46
CH_4	5-10	1-3
N_2	0-1	-

Syngas can be used as fuel to produce electricity. It can also be used as a semi-finished product in the chemical industry to produce synthetic natural gas, synthetic petroleum, ammonia and methanol. Currently, about 25% of the world© ammonia and 30% of the world© methanol are produced using the gasification process [6]. Syngas is also used to make biofuels and biomaterials.

The research on the scope of syngas application is therefore centred on the integrated gasification combined cycle (IGCC) technology, alternative internal combustion engines (ICE) and fuel-chemical production [1,4,7]. Such studies are presented e.g. in [2,7]. An analysis of syngas application in a microturbine is conducted in [8]. Syngas with a high content of hydrogen can also be used in high-temperature solid oxide fuel cells (SOFC) or in molten carbonate fuel cells (MCFC) [9].

Regardless of the final method of syngas application and of the technology advancement, each stage of syngas production, storage, transport and use requires additional attention focused on safety issues. Such analyses are presented for example in [4,10]. Safety is an important factor to consider because the presence of flammable and toxic gases in syngas creates potential hazards to humans and the environment in the event of an installation failure or an uncontrolled gas release. The paper presents an analysis of hazards related to the process of syngas storage in tanks.

2. Synthesis gas production

Many processes are now used to produce synthesis gas. As already mentioned, syngas can be produced from coal, biomass, wood, waste or from natural gas. Gasification plants can be found on every continent, and most of them are located in China [6].

The selection of the syngas production process depends on many factors, such as the availability of raw materials, their cost and, first of all, the final product target composition. The basic parameter that defines syngas composition is the H_2/CO ratio [11].

The gasification dominant feedstock at present, as well as in predictable future, is coal. Biomass and feed waste are also gaining importance and the number of plants gasifying them is on the rise. Last but not least, there are also syngas production plants based on petroleum residues gasification and steam reforming [6].

In the process of steam-methane reforming, the heated mixture of steam and methane flows through a nickel catalyst. In such conditions methane undergoes a strong endothermic process referred to as steam reforming, which makes it possible to obtain hydrogen-rich synthesis gas with the typical H_2/CO ratio of 3:1 to 5:1 [6].

The next method – coal gasification – is one of the most developed technologies of the gasification process. Considering the type of the feedstock bed, gasification technologies can be divided into processes taking place in a fixed bed, fluidized bed or entrained-flow bed reactor. In the fixed bed technology the coal grain size is included in the range of 5-80 mm. Coal is on the fuel bed and the gasifying medium, i.e. air and/or steam, is fed from below (updraft gasification). In fluidized bed gasification the coal grain size is up to 10 mm. The processes occurring in this case are similar to combustion processes typical of a fluidized bed boiler, and gasification proceeds under atmospheric pressure. In entrained-flow reactors, pulverized fuel (<0.1 mm) is fed into the reaction zone in a jet of steam and oxygen. The fluidized layer is created as coal dust is lifted by gasifying mediums [1,3,12].

Biomass is usually gasified using air in fixed bed (con- and countercurrent) reactors or fluidized bed reactors. In concurrent reactors, the feedstock and the gasifying medium are transported in the same direction, whereas in countercurrent gasifiers – in opposite ones. Such systems are characterized by a number of requirements concerning the degree of biomass comminution, and their main disadvantages are low efficiency and the tar content. Two bed types are used in the fixed bed technology of biomass gasification: the circulating fluidized bed and the bubbling fluidized bed. Good-quality gas is obtained.

Many different kinds of reactors can be used for waste gasification. They differ in size and the feedstock type. Some are intended for solid municipal waste gasification, others – for gasification of construction and demolition debris. Many of them often require the feedstock pretreatment, e.g. breaking up or drying, or removal of non-organic materials that cannot be gasified [2,8,9].

3. Synthesis gas storage

Syngas storage is not currently a common practice. This is due to the fact that synthesis gas is usually fed directly for use. However, there is an extra potential in syngas storage compared to its immediate use. Storage enables wider applications, additional supplies and - ultimately - economic advantages to both producers and end consumers. For example, stored synthesis gas can be used to produce electricity in peak-demand periods. It can be a method to improve productivity, reliability and availability of IGCC power plants by increasing syngas availability during scheduled and unscheduled downtimes [7,13].

However, it has to be remembered that syngas storage involves technical difficulties because it contains hydrogen enhancing metal embrittlement. It is also necessary to consider specific safety issues in case of an uncontrolled release or corrosion. Therefore, the technical feasibility and economic attractiveness of syngas storage lie first of all in the gas properties, such as energy density or composition. Low energy density of synthesis gas, which varies from about a sixth to a third of that of natural gas, means that more syngas has to be produced to generate the same amount of electricity. This in turn makes it necessary to design large storage tanks with a high working pressure. Syngas can be stored in low- and high-pressure ground tanks, in existing pipelines or in underground sites [13].

The most essential large-scale stationary syngas storage system is compressed gas storage. This is a simple way to store syngas which generally needs only a pressure tank and a compressor. The costs are therefore lower compared to condensing, for example [13].

4. Hazards related to synthesis gas use

Syngas composition and the physicochemical properties arising therefrom have a substantial impact on the safety of the gas utilization. The analysis of the risk and hazards created by the process of syngas production, transport and storage should take account of the properties of the two basic components: hydrogen and carbon monoxide. Due to the content of the two gases, released synthesis gas can pose a fire hazard or explosion risk. The presence of a toxic gas can also present a toxic hazard.

The consequences of an uncontrolled failure-related release of syngas depend on the course of the event, i.e. whether the release is prompted by complete or partial damage to the pipeline (rupture or puncture) and whether ignition of the gas occurs. The consequences of the failure will also depend on the installation type and operating parameters, such as the tank pressure for example.

https://doi.org/10.1051/e3sconf/201913701022

If there is a failure of a synthesis gas installation and immediate or delayed ignition of the mixture occurs, a number of dangerous events may follow, such as:
• jet fire - caused due to a release and ignition of gas flowing through the hole (puncture) under high pressure; it is characterized by a long and stable flame;
• flash fire - the cloud of released gas moves and ignites suddenly, sometimes far from the failure site;
• BLEVE - a violent phenomenon related to a release of liquid vapours to the environment with a temperature exceeding the boiling point; it is most often caused by flames washing the storage tank, which results in an increase in the tank inside temperature, the tank rupture and a violent release of the tank contents; if the failure involves a volatile flammable substance, the BLEVE phenomenon is usually accompanied by a fireball;
• explosion - i.e. a violent oxidation or decomposition reaction causing a rise in pressure and/or temperature [3,4,10,11].

The negative effects of the scenarios presented above are the fire-generated heat flux affecting humans and the environment (cf. Table 2) and the explosion-generated pressure wave (cf. Table 3) [14]. The effect of a release of syngas without ignition is the toxic hazard related to the toxicity of carbon monoxide contained in the mixture.

Table 2. Effects of the heat flux on humans and facilities

heat flux [kW/m^2]	effects
35-37.5	100% death rate within 1 min; destruction of buildings
25-32	deformation of steel
23	100% death rate within 1 min; serious injuries within 10 s
12.5	1% death rate within 1 min; first-degree burns within 10 s
4.7	pain if exposure time exceeds 20 s
4	glass cracking after 30 min of exposure
2.5	threshold value causing pain if exposure time exceeds 1 min

Table 3. Effects of explosion-related overpressure on humans and facilities

overpressure [kPa]	effects
500 – 800	100% death rate
350 – 500	50% death rate
199.8	99% death rate due to lung damage
34.4	lung damage
20.7	minor damage to heavy machinery and equipment
17.2	demolition of 50% of brick buildings
4.8	damage to the structure of buildings
0.21	cracking of large window panes

	(ordinary glass)

5. Consequences of a syngas tank failure

The potential hazards related to failures of syngas storage installations were analysed using the PHAST software [15].

Depending on the production process, synthesis gas composition may vary. This in turn has an impact on different properties creating fire-, explosion- or toxicity-related hazards. The following composition of the gas mixture is adopted for the purposes of the analysis:
• CO - 19%, CO_2 - 12%, H_2 - 19%, CH_4 - 2%, N_2 - 48% (mixture I)
• CO - 23%, CO_2 - 29%, H_2 - 38%, CH_4 - 9.5%, N_2 - 0.5% (mixture II).

The mixtures are obtained by gasifying biomass and coal, respectively. Fig. 1 and Fig. 2 present hazard zones within which humans feel pain if they happen to be there (heat flux exceeding 4.7 KW/m^2). The analysis assumes a catastrophic complete rupture of a 10 m^3 tank and syngas pressure of 25 bar. Due to the gas release and ignition, a fireball is created (BLEVE). The wind speed is 1.5 m/s.

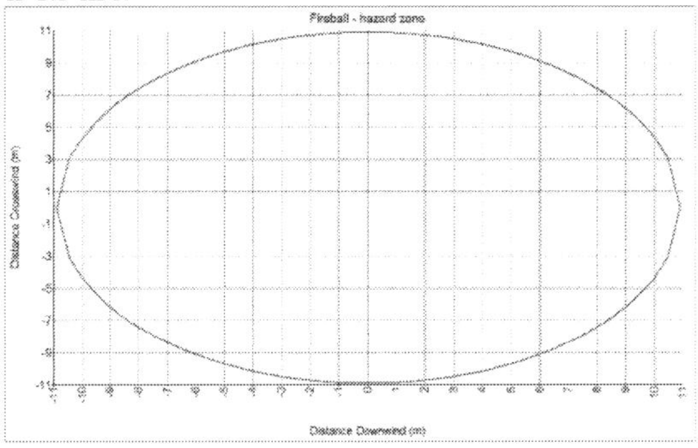

Fig. 1. Fireball hazard zone (mixture I)

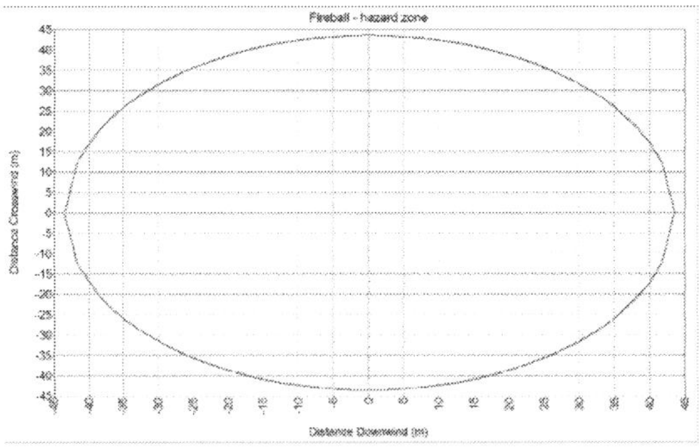

Fig. 2. Fireball hazard zone (mixture II)

The charts presented above indicate that a change in the content of flammable gases in the syngas mixture has a substantial impact on the level of the hazard related to a tank failure. If the hydrogen content is doubled, the hazard zone gets longer by about 35 metres.

Another important parameter that affects the range of hazard zones related to a fire and the generated heat flux is syngas pressure in the storage tank. Fig. 3 presents heat flux values depending on the distance from the site of the tank failure, assuming different values of syngas pressure.

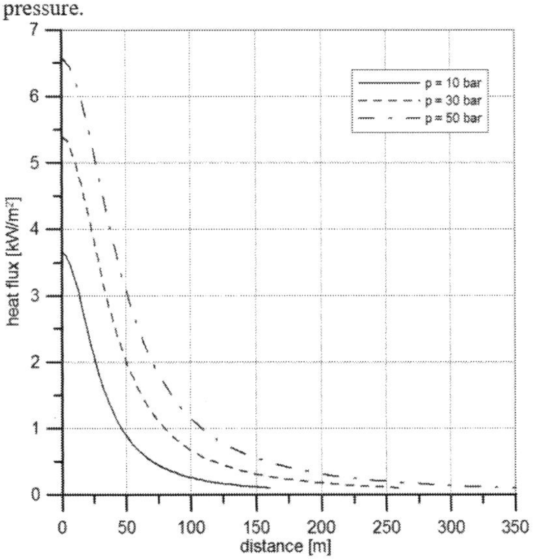

Fig. 3. Heat flux depending on the distance from the tank failure site (mixture I)

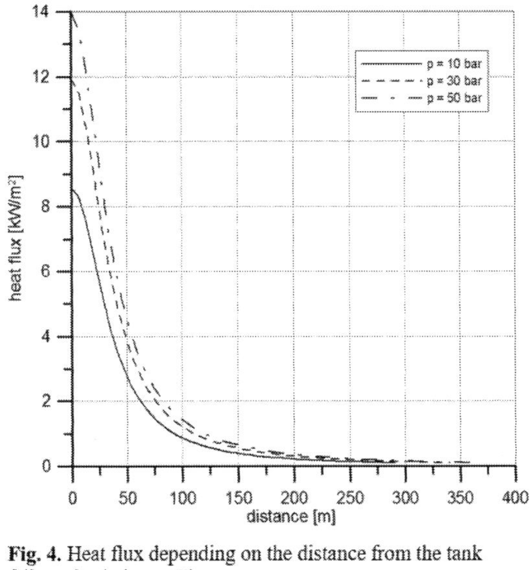

Fig. 4. Heat flux depending on the distance from the tank failure site (mixture II)

The tank immediate and complete rupture, in the case of delayed ignition, may also pose an explosion hazard. The next two figures present hazard zones related to an explosion of released synthesis gas. The parameters of the gas and of the storage tank assumed for the analysis are the same as in the analysis of the fireball hazard. The hazard zone corresponds to the explosion-generated pressure wave at the level of 34.4 kPa and higher, which causes damage to human lungs.

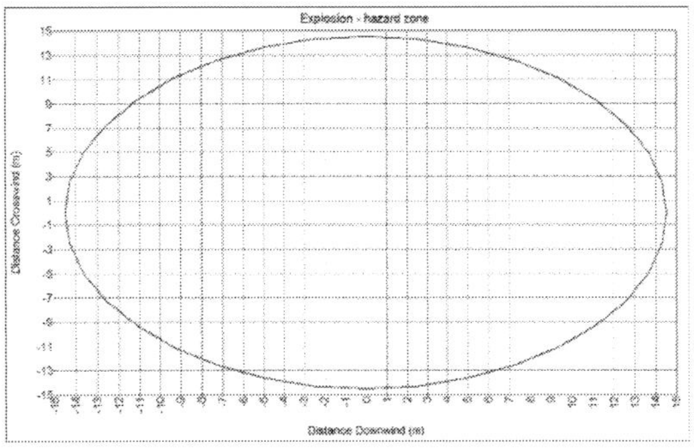

Fig. 5. Hazard zone for syngas explosion (mixture I)

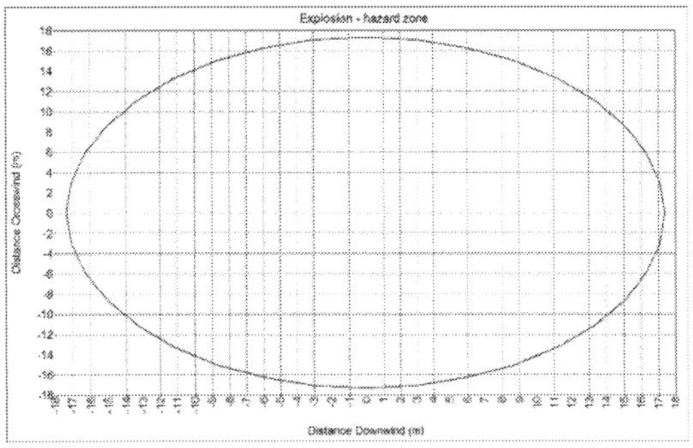

Fig. 6. Hazard zone for syngas explosion (mixture II)

As mentioned above, if the tank is partially damaged (tank puncture), a jet fire may occur. The jet of pressurized gas is ignited immediately, creating a long and stable flame with strong thermal radiation. The figures below illustrate hazard zones related to this failure type for a tank containing 10 m^3 of synthesis gas under the pressure of 25 bar. The assumed diameter of the puncture is 5 cm. The hazard zones correspond to the heat flux of 4.7 and 23 kW/m^2, causing pain and 100% death rate, respectively. If syngas contains a bigger content of hydrogen, an additional zone occurs with a heat flux value causing human death, i.e. the value exceeding 37.5 kW/m^2.

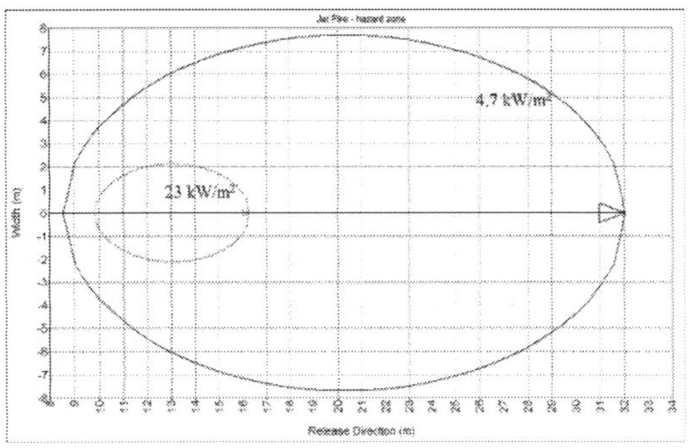

Fig. 7. Jet fire hazard zone (mixture I)

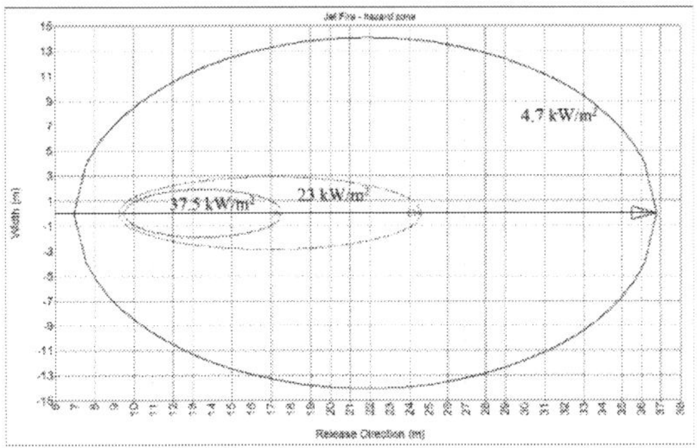

Fig. 8. Jet fire hazard zone (mixture II)

Analysing the figures presented above, it can be noticed that syngas fire hazard zones get bigger if the gas mixture contains bigger contents of flammable constituents.

In the case of a jet fire, the level of the hazard presented to humans and the environment will also be affected by the size of the puncture through which the gas is released. Fig. 9 presents changes in the heat flux generated by the fire of syngas with a higher content of hydrogen depending on the distance from the failure site for three different diameters of the puncture.

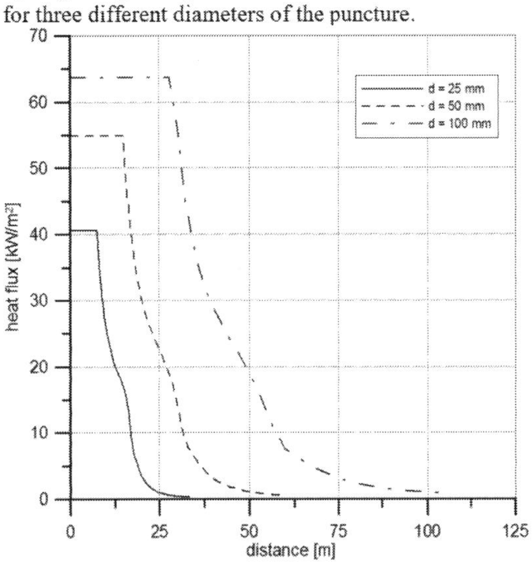

Fig. 9. Heat flux depending on the distance from the tank failure site for three different diameters of the puncture (mixture II)

Apart from the operating parameters of the installation and the size of the damage, the factors that have an essential impact on the consequences of a potential failure involving synthesis gas are the gas composition and the nature of the properties related thereto. For this reason, any consideration of issues related to the safety of syngas use, transport and storage should take account of the parameter defining the mixture composition, i.e. the H_2/CO ratio. Fig. 10 presents the change in the heat flux generated after a failure resulting in a fireball (BLEVE) and depending on the distance from the failure site for a storage tank with the H_2/CO ratio of 1:1, 2:1 and 3:1.

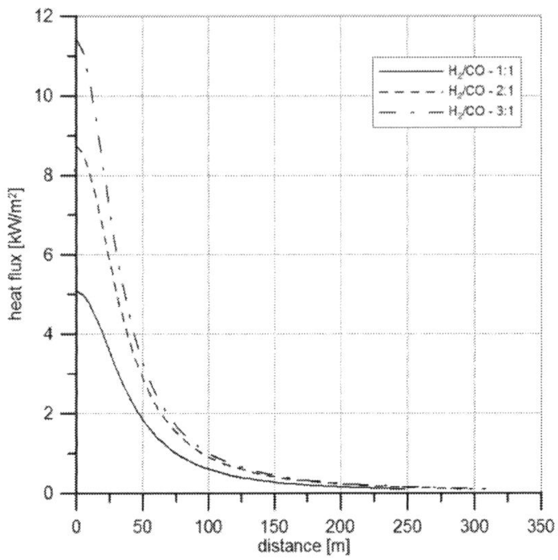

Fig. 10. Fireball-generated heat flux depending on the distance from the failure site for three different H_2/CO ratios

Analysing the chart presented above, it can be seen that a change in the H_2/CO ratio in the syngas mixture causes about twice as high a rise in the generated heat flux in the immediate vicinity of the fire.

6. Summary and conclusions

Undoubtedly, fuel gasification and conversion processes are very important in technologies aiming at a reduction in greenhouse gas emissions and clean energy production. They also play an essential role in the development of the chemical industry.

Synthesis gas can be used as an independent fuel, or it can be processed and utilized as an energy carrier. Any processes related to syngas processing, transport and storage should take account of its low energy density, which may create technical problems. The fact that it contains toxic and flammable gases should also be taken into consideration, as it may pose serious hazards.

The paper focuses on the hazards related to syngas storage. Storage processes may for example be a part of electricity production in IGCC power plants, where stored synthesis gas can be used to improve the power unit productivity or reliability.

A release of syngas may cause a situation with no ignition at all, but it is also possible that immediate or delayed ignition will occur [4].

If the released gas does not ignite, it will disperse in air creating no fire hazard or explosion risk. In the case of immediate ignition, a jet fire will be a potential hazard. If ignition is delayed, an explosion may occur. The level of the hazard presented to humans and the environment will depend on a number of factors, such as the tank geometry, the tank operating parameters and syngas composition. The heat flux generated by a syngas jet fire for the tank damage in the form of a hole with the diameter of 2.5, 5 and 10 cm will vary from about 40 kW/m^2 to 65 kW/m^2 in the immediate vicinity of the tank. Such values pose an essential hazard to human life and structural strength. The hazard zones for humans arising due to the pressure wave generated during a syngas explosion (and causing lung damage) will reach the range of about 15 metres. In the case of syngas storage installations, the hazard level will also depend on the gas composition and the H_2/CO ratio in the mixture. If the ratio is raised from 1:1 to 3:1, the hazard related to the impact of the heat flux generated by the fire will be about twice as high.

The presented work was supported by the Silesian University of Technology within statutory research funds.

References

1. Mishra A., Gautam S., Sharma T., Effects of operating parameters on coal gasification, *International Journal of Coal Science & Technology*, 5(2), pp. 113–125, (2018)
2. Rauch R., Hrbek J., Hofbauer H., Biomass gasification for synthesis gas production and applications of the syngas, *Energy and Environment*, 3(4), pp. 343-362, (2014)
3. Stolecka K., Rusin A., Analysis of hazards related to syngas production and transport, *Renewable Energy*, 146, 2535-2555, (2020)
4. Pierorazio A.J., Baqer Q., A., Hazards for Syngas Fires and Explosions, *Process Safety Progress*, 29(4), pp. 288-292, (2010)
5. Kordylewski W., Spalanie i paliwa [Combustion and fuels], Oficyna Wydawnicza Politechniki Wrocławskiej, Wrocław (2008)
6. Global Syngas Technologies Council, World Gasification Database
7. Cocco D., Sierra F., Tola V., Assessment of energy and economic benefits arising from syngas storage in IGCC power plants, *Energy*, 58, pp. 635-643, (2013)
8. Corrêa P.S.P., Yhang J. Lora E.E.S., Andrade R.V., Pinto R.L., Ratner A., Experimental study on applying biomass-derived syngas in a microturbine, Applied Thermal Engineering, 146(5), pp. 328-227, (2019)
9. Pańczyk M., Borowiecki T., Otrzymywanie i zastosowanie gazu syntezowego [Syngas production and application], *Adsorbenty i katalizatory: wybrane technologie a środowisko[Adsorbents and catalysts: selected technologies vs. the environment]*. Edited by J. Ryczkowski, the University of Rzeszów, pp. 275-287, (2012)
10. Molino A., Braccio G., Fiorenza G., Marraffa F.A., Lamonaca S., Giordano G., Rotondo G., Stecchi U., La Scala M., Classification procedure of explosion risk areas in presence of hydrogen-rich syngas: Biomass gasifier and molten carbonate fuel cell integrated plant, *Fuel*, 99, pp. 245-253, (2012)
11. Nandan A., Siddiqui N.A., Mondal P., Chaudhar K., Pandey R., Hazards associated to synthesis gas and its mitigation measures, Research Journal Engineering and Tech., 5(3), pp. 144-146, (2014)
12. Chmielniak T., Stelmach S., Współczesne technologie zgazowania węgla [Modern technologies of coal gasification] in: *Problemy ekologii*, 13(2), pp. 69-76, (2009)
13. An Engineering-Economic Analysis of Syngas Storage, DOE/NETL-2008/1331 Draft Final Report, (2008)
14. Rusin A., Stolecka K., Reducing the risk level for pipelines transporting carbon dioxide and hydrogen by means of optimal safety valves spacing, *Journal of Loss Prevention in the Process Industries*, 33, pp.77-87, (2015)
15. Phast v6.7, DNV Software

RESEARCH & DEVELOPMENT IN POWER ENGINEERING, 2019

Thermodynamic analysis of the Compressed Air Energy Storage system coupled with the Underground Thermal Energy Storage

Rafał Hyrzyński[1], *Paweł* Ziółkowski[2,*], *Sylwia* Gotzman[1], *Bartosz* Kraszewski[1] and *Janusz* Badur[1]

[1] Institute of Fluid-Flow Machinery, Polish Academy of Sciences, Fiszera 14, 80-231 Gdańsk, rafal.hyrzynski@imp.gda.pl, Bartosz.kraszewski@imp.gda.pl, jb@imp.gda.pl, Poland

[2] Gdańsk University of Technology, Faculty of Mechanical Engineering, Narutowicza 11/12, 80-233 Gdańsk, e-mail: pawel.ziolkowski1@pg.edu.pl, Poland

Abstract. Improvement of flexibility is one of the key challenges for the transformation of the Polish Power System aiming at a high share of renewable energy in electricity generation. Flexible and dispatchable power plants will contribute to this ongoing transformation process as they compensate for fluctuations in electricity generation from renewable energy sources such as wind and photovoltaics. In this context, CAES storage tanks are currently the only alternative to storage facilities using pumped-storage hydroelectricity due to the possibility of obtaining the appropriate energy capacity of the storage facility. However, a relative disadvantage of these plants is the heat loss caused by the cooling of the air after compression. The basic elements of the CAES warehouse are: an air compression station, a compressed air reservoir that is also a storage facility (in the existing solutions, these are underground caverns), an expansion station with combustion chambers and gas turbines, and a generator. A key aspect of CAES is the optimal configuration of the thermodynamic cycle. In this paper, the situation of cooperation between the current conventional power plants and wind farms is first analysed, and then, based on thermodynamic models, the process of storing thermal and electrical energy in the CAES system coupled with heat recovery after the gas turbine is analysed. A solution with a ground heat exchanger was also proposed, as the soil, due to its properties, may serve as a thermal energy storage. The paper also analyzes the discharge of the heat storage based on CFD approaches. The ground can be charged during the cooling down of the compressed air. On the other hand, thermal energy was recovered when water flowing to the heat customers was heated. On the basis of non-stationary calculations, the heat stream received from the underground thermal energy storage was estimated.

1 Introduction

Currently, due to the development of renewable energy, the variability of operation of conventional units will be one of the most noticeable consequences of the transformation of the Polish Power System (PSE, Polskie Sieci Elektroenergetyczne SA) [1], and this involves one of the most important aspects of PSE, namely its stability. Therefore, it is worthwhile to integrate energy storage into the intermittent energy sources in order to increase their flexibility, *inter alia,* by building CAES (Compressed Air Energy Storage) systems.

Additionally, it should be remembered that the activities of the UN (through the Conference of the Parties, COP) significantly affect the global energy policy of most countries or even regions of the world, and in particular the policy of the European Union. It is in the framework of climate and energy policy that the EU has defined three key objectives until 2030:

1. a reduction of at least 40% in greenhouse gas emissions (compared to 1990 levels),

2. ensure at least a 27% share of energy from renewable sources in total energy consumption,

3. increase energy efficiency by at least 27%.

Precisely these targets were adopted by the European Council of 23-24 October 2014 [2]. Therefore, more or less successful synergy between solar and wind energy will inevitably occur, characterised by significant unpredictability and intermittence of work with traditional systems that are able to store either electricity or mechanical energy. It partly forces also the situation, when conventional sources connected to the power system have to compensate the fluctuating electricity generation from the weather-dependent energy sources such as wind and photovoltaics [3]. In order to illustrate the volatile character of electricity generation from wind sources, the authors selected two significant profiles of electricity generation from wind farms, starting from 1 January 2019 until 21 May 2019. The full range of profiles can be found in the work by Badur et al. [4]. For Figures 1 and 2, the left-hand axis refers to the "Load of Polish Power System" and "Load covered without WFs (Wind Farms)", while the right-hand axis refers to "Total WFs Generation". Figure 1 shows the Polish Power System operation from 1 January 2019 to 21 May 2019, which shows a characteristic variability depending on the work schedule of customers, in particular industrial ones, and on weather conditions, seasons, cultural, social events, etc. [5].

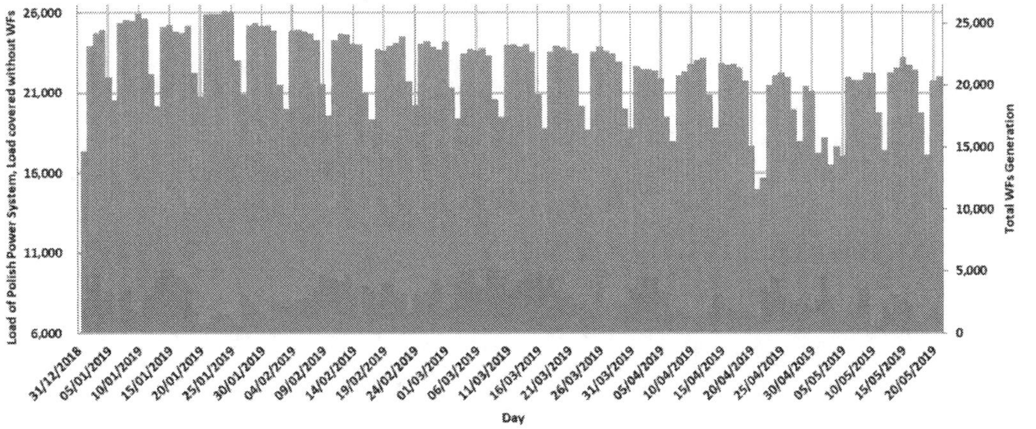

Fig. 1. Electricity production in Poland in the period from 1 January 2019 to 21 May 2019 (MW). The data may differ from the data presented by the TSO due to the fact that the data available on the TSO platform are published with 1 h resolution (not 15 min).

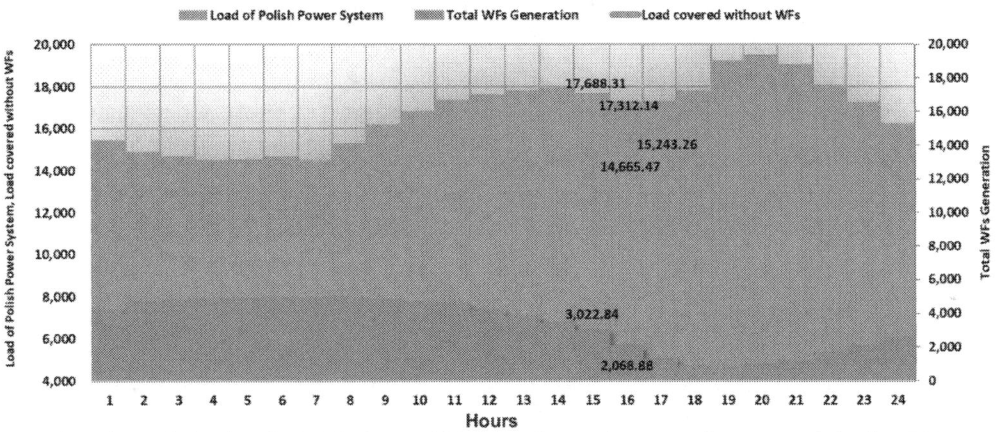

Fig. 2. Electricity production in Poland on the day on which the maximum 2-hour generation power variation from WFs occurred (954 MW; 10 March 2019 between 3 p.m. and 4 p.m.) [MW]. The data may differ from the data presented by the TSO due to the fact that the data available on the TSO platform are published with 1 h resolution (not 15 min).

However, the wind energy generation throughout the Sunday of March 10, 2019 (Figure 2), deserves a more detailed analysis. This is because on that day the maximum hourly variability of power generation from wind farms occurred. This situation took place between 3 p.m. and 4 p.m. During the hour the change of wind power output amounted to -954 MW. At that time, the change in the load of the Polish Power System alone amounted to -376 MW. This means that this sudden reduction in the power generation from the wind farms forced other sources included in the power system to increase their generation by 578 MW over the course of 60 minutes. This value is close to the total installed capacity of the three units of Ostrołęka power plant (690 MW) and, what is more, required on short notice. This single case of wind farm capacity reduction is obviously not too much of a challenge for the Transmission System Operator (TSO) (in the absence of emergency situations, e.g. sudden failure of a large unit in the system or failure of several units). Such a power loss (frequency drop in the system) can be quickly replenished by launching several hydropower units in pumped-storage power plants for generation operation mode, using the possibility of increasing capacity on cross-border connections in the direction of import, as well as by forcing the opening of regulatory valves in individual conventional generating units, which results in adjusting generation to the load and restoring balance [7].

On the other hand, Fig. 3 presents electricity generation and consumption on the day when there was a minimal difference between the generation from WFs and the load of the Polish Power System (8258.24 MW; 1

January 2019 at 9 a.m.). The following types of generating units should are presented on the graph: industrial combined heat and power plants (ICHP), power sector combined heat and power plants (PCHP), centrally dispatched power plants (PP), and pumped-storage hydroelectricity (PSE). During the whole year, the energy generated in RES is dispatched first, and according to the data from [6] it amounted to: 4,735 MW. Next in a series of energy conversions, especially in winter, are both industrial and power sector CHP plants, which at that time reached the capacity of 289 MW and 1,162 MW respectively. On the other hand, centrally dispatched power plants generated electricity both for pumped storage power plants (1,423 MW) and for average consumers (4,635 MW). This figure shows that the only valuable energy storage at this point in time is pumped storage and there are no alternatives.

An increase in the installed capacity of sources which are difficult to forecast will require more sophisticated countermeasures from the TSOs (in case of sudden reduction or increase in the capacity generated in these sources). On the other hand, an increase in the capacity of variable RES will force operators of conventional units to modernize their generating units in order to increase their flexibility understood as [8]: the ability to safe unit operation in planned transient conditions, while taking actions to control the generating unit. This notion applies both to changes in the load of the unit, as well as to its provisioning or restarting. This issue is so complex and still open, and one of the alternatives is CAES-type energy storage.

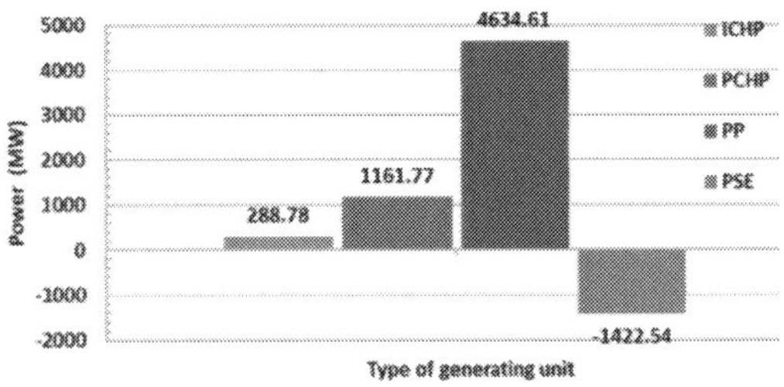

Fig. 3. Electricity production on the day when there was a minimal difference between the generation from WFs and the load of the Polish Power System (8258.24 MW; 1 January 2019 at 9 a.m.). (MW).

CAES storage tanks are currently the only alternative to storage facilities using pumped-storage hydroelectricity due to the possibility of obtaining the appropriate energy capacity of the storage facility. However, a relative disadvantage of these plants is the heat loss caused by the cooling of the air after compression. There are currently two commercial systems in operation worldwide on a large scale. First commissioned in Huntorf in 1978 in Germany with a capacity of 290 MW and 580 MWh, and McIntosch with a capacity of 110 MW and 2860 MWh in the US.

In general, three systems can be distinguished [9]:

a) standard system (without using the exhaust heat from the gas turbine exhaust - the system used in the Huntorf Power Plant),

b) with recuperation (using the exhaust heat from the gas turbine exhaust - the system applied in the McIntosh Power Plant),

(c) adiabatic (non-fuel).

However, new CAES-based energy storage technologies have developed in recent years, namely Underwater Compressed Air Energy Storage also called Ocean Compressed Air Energy Storage [10], Second Generation Compressed Air Energy Storage [11], Isothermal Compressed Air Energy Storage [12], Low Temperature Compressed Air Energy Storage [13], Advanced Adiabatic Compressed Air Energy Storage [14], Liquid Air Energy Storage [15]. However, simple CAES systems with and without heat regeneration are of the greatest practical importance [16].

The CAES system without heat recuperation is associated with large losses due to the high temperature of the flue gases behind the gas turbine, which are directed directly to the chimney. Another source of losses is the compressor, from which the thermal energy released during the compression of the air is taken up in the compressed air tank (cavern). Therefore, the efficiency of energy storage in such a system is low, and the only benefits result from improving the efficiency of the gas turbine cycle operating in a simple system. In order to obtain a relatively high efficiency index of the energy storage cycle, i.e. about 0.50, heat recovery systems from exhaust gases are used [9.16].

An adiabatic (fuel-free) system remains in the sphere of research and as such has not yet been implemented, but its advantage is theoretically possible to achieve a storage cycle efficiency of approximately 0.7, without the need to use fuel. The main technological limitation is the necessity of effective accumulation of thermal energy from the air compression process [16].

Taking into account the above arguments, it was decided to analyze the system with heat regeneration. The aim of this paper is to analyse thermodynamic CAES and preliminary thermal-FSI (Fluid-Solid Interaction) analysis of the heat storage system, which in this case is the ground. The next section presents a diagram and model for CFM (Computational Flow Mechanics) analyses, while the third section is an introduction to thermal-FSI analyses and contains the first step, which in this case is the analysis of heat exchange in the ground.

2 System description

The basic elements of CAES system are: 1) air compression station, 2) compressed air reservoir being at the same time a mass storage (in the existing solutions they are underground caverns), 3) expansion station with combustion chambers and gas turbine and generator. The basic elements are a low-pressure compressor (C_1), which, together with high-pressure compressors (C_2-C_4) and intersection cooling exchangers (IC_1-IC_3), increases the pressure of the storage medium up to 60 bar. A cavern (CAV) is the storage place for mass and mechanical energy, which is emptied at the moment of electricity demand. For that purpose, the air is first heated in a regenerative heat exchanger (HE) and then in the first combustion chamber (CC_1) to increase its enthalpy and thus prepare it for partial expansion in the high-pressure turbine (GT_1). Finally, the second combustion chamber (CC_2) is used to increase enthalpy of exhaust gases, which expands into a low-pressure turbine (GT_2). The underground thermal energy storage (UTES) is a novelty in this case and it is analysed as 2D geometry in Computational Fluid Dynamics (CFD) codes.

The working pressure in the CAES system determines the required volume of the storage cavern depending on the amount of accumulated energy - for the air pressure in the cavern of about 8 MPa the density of stored energy is 7.5 kWh/m^3, while for the pressure of 2MPa the density of energy stored decreases to only 2 kWh/m^3.

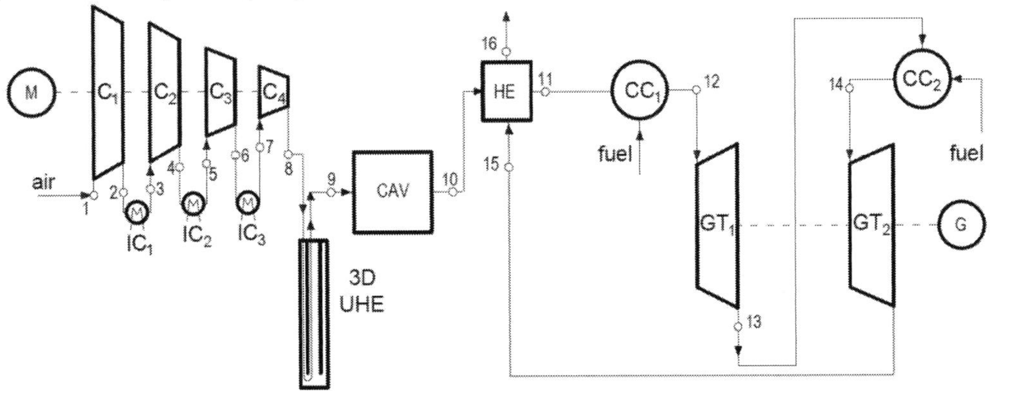

Fig. 4. Scheme of the CAES system, where: C_1 - low pressure compressor, C_2-C_4 - high pressure compressors, IC_1-IC_3 - intersection cooling exchangers, CC_1-CC_2- combustion chambers, GT_1 - high pressure turbine, GT_2 - low pressure turbine, HE - regenerative heat exchanger, UHE - underground heat exchanger, G - electric generator, M - motor.

The thermodynamic model has the following assumptions: regenerative heat exchanger efficiency η_{HE} =0.8, isentropic efficiency of compressors η_{iC} =0.81, mechanical efficiency η_{mC} =0.99, isentropic efficiency of expanders η_{iGT} ==0.85, generator efficiency η_g =0.99. An additional assumption may be the power output by a gas turbine and in such case the mass streams are the resultant parameters. For compressor calculations such parameters as internal efficiency η_{iC} and mechanical efficiency η_{mC}, compression ratio π and mass flow rate of air $\dot{m}_a$ are given as input data. Knowing the gas parameters at the compressor inlet, e.g. (T_1, p_1) we easily calculate the pressure p_2 after the compression change:

$$p_2 = \pi\, p_1 \tag{1}$$

The next step is to determine the temperature at the end of compression, where for the ideal process we use the equation of isentropes, *s = idem*:

$$T_{2s} = T_1 \left(\frac{p_2}{p_1}\right)^{\frac{\kappa-1}{\kappa}} \tag{2}$$

where the κ of the isentropic exponent is determined on the basis of thermodynamic property tables of a given working medium. In order to determine the actual temperature of the end of compression T_2, one should take into account the isentropic efficiency of the compressor's stages η_{iC}, which is expressed by the formula:

$$\eta_{iC} = \frac{l_{t1-2s}}{l_{t1-2}} = \frac{h_1 - h_{2s}}{h_1 - h_2} \tag{3}$$

where: l_{t1-2s} is the isentropic specific compression work, l_{t1-2} means the actual specific compression work, h_1, h_2, h_{2s} are the specific enthalpy of the working substance in characteristic points (1), (2), (2s). The expander parameters are defined in the same way. The power consumed by the 4 compressors' stages , N_C, is determined by the following formula:

$$N_C = \frac{\dot{m}_a}{\eta_{mC}} [(h_2 - h_1) + (h_4 - h_3) + (h_6 - h_5) + (h_8 - h_7)] \tag{4}$$

where: $\dot{m}_a$ - air mass flow rate, η_{mC} - mechanical efficiency of the compressor, and $h_1 \div h_8$ are the specific enthalpy of the working substance in characteristic points $1 \div 8$ (Fig. 4). For energy storage, the amount of stored energy becomes relevant, which can be determined by taking into account the time τ using the following relationship:

$$E_C = N_C \tau_C \tag{5}$$

where: τ_C cycle time of compressors to fill a cavern with air at a pressure of 6 MPa and may be different from the cycle time of a turbine τ_{GT}. Equations (1-5) can be used in a similar way for expanders description, but it is worth repeating the most important parameter, which is the energy generated by the turbine during one cycle in cooperation with the cavern:

$$E_T = N_{GT} \tau_{GT} \tag{6}$$

where: N_{GT} is the total power of the turbines. The mass and energy balance equations of combustion chambers were used to determine the parameters at the inlet to individual turbines. It is worth recalling the energy-flux balance equation of the combustion chambers (CC_1):

$$\eta_{CC}(\dot{Q}_{chem1} + h_{11}\dot{m}_a + h_f\dot{m}_{f1}) = h_{12}\dot{m}_{12} \quad (7)$$

where: η_{CC} - combustion chamber efficiency, h_{11} - oxidising agent enthalpy before combustion chamber (CC_1), h_f - fuel enthalpy, $\dot{m}_{f1}$ - fuel mass flow rate, h_{12} - flue gas enthalpy, $\dot{m}_{12}$ - exhaust gases mass flow rate and $\dot{Q}_{chem1}$ - chemical energy transfer rate defined as:

$$\dot{Q}_{chem1} = \dot{m}_{f1}LHV \quad (8)$$

The lower heating value (LHV) is the amount of heat obtained from the total and complete combustion of a unit quantity of fuel without condensation of water vapour and was assumed to be 50,03 MJ/kg. Equations (7-8) can be used in a similar way for second combustion chamber (CC_2). Of course, total chemical energy transfer rate is equaled $\dot{Q}_{chem} = \dot{Q}_{chem1} + \dot{Q}_{chem2}$. Duration of combustion in combustion chambers (CC_1 and CC_2) relates to expansion process in both expanders (GT_1 and GT_2). Therefore, the time of transferring the heat during combustion is equal to the time τ_{GT}, which means that the energy supplied to the combustion chambers is expressed as:

$$E_F = \dot{Q}_{chem}\tau_{GT} \quad (9)$$

Bearing in mind the above data, the efficiency of electricity storage in the CAES system can be defined as follows [17,18]:

$$\eta_{CAES} = \frac{E_T}{E_C + E_F} \quad (10)$$

Taking into account Fig. 4 and the presented model, it was assumed that the temperature of t_{12}=600°C will be reached in the first combustion chamber (CC_1) and t_{14}=1100°C in the second combustion chamber (CC_2). Therefore, assuming thermodynamic parameters as for the existing installations, both for the gas turbine and for the compressor, the calculations were carried out with the assumption of a generator capacity of 30 MWe. The efficiency of this system according to equation (10) amounted to η_{CAES} =0.52, which is a satisfactory value given the volatility of renewable energy sources such as wind and photovoltaics.

3 Preliminary thermal-FSI analysis

The preliminary thermal-FSI analysis was carried out for a uniform soil with the following parameters: density 1900 kg/m^3, heat conduction coefficient 2.2 W/(mK), specific heat 980 J/(kgK). A natural temperature gradient of 3°C/100m was assumed. The computational domain of the diameter of 1 km reached into the ground at the depth of 5 km. The Field's ground exchanger, on the other hand, according to its geometry, can draw or deliver a heat stream at a depth of 4 km, and it consists of an inner pipe of 76 mm diameter and an outer pipe of 318 mm diameter. The pressure at the inlet has been set at the outer pipe as 3,1 MPa and the temperature as 70°C with the water mass flow rate of 5,3 kg/s. The outlet pressure at the inner pipe, taking into account the calculated pressure drop, was 2 MPa. The system of equations solved to determine the transient liquid temperature profiles is worth quoting here [19,20]:

$$\frac{\partial}{\partial t}\begin{Bmatrix}\rho\\ \rho\mathbf{v}\\ \rho e\\ \rho k\\ \rho\omega\end{Bmatrix} + \mathrm{div}\begin{Bmatrix}\rho\mathbf{v}\\ (\rho\mathbf{v}\otimes\mathbf{v}) + p\mathbf{I}\\ (\rho e + p)\mathbf{v}\\ \rho\mathbf{v}k\\ \rho\mathbf{v}\omega\end{Bmatrix} = \mathrm{div}\begin{Bmatrix}0\\ \mathbf{t}^c\\ \mathbf{t}^c\mathbf{v} + \mathbf{q}^c\\ \mathbf{J}_k\\ \mathbf{J}_\omega\end{Bmatrix} + \begin{Bmatrix}0\\ \rho\mathbf{b}\\ \rho S_e\\ \rho S_k\\ \rho S_\omega\end{Bmatrix} \quad (11)$$

where: ρ - density, $\mathbf{v}$ - velocity, p - pressure, $\mathbf{I}$ - unit tensor, $\mathbf{t}^c$ – viscous stress tensor, $\mathbf{b}$ – the force of earth's gravitation, $e = u + \frac{1}{2}\mathbf{v}^2$ – the sum of internal and kinetic energy, $\mathbf{q}^c$- total heat flux, $\mathbf{J}_k$, $\mathbf{J}_\omega$ – diffusion flux of turbulent kinetic energy k and diffusion flux of kinetic energy dissipation ω . In equation (11) is source of turbulent kinetic energy S_k and source of turbulent kinetic energy dissipation S_ω. This equation is complemented by heat conduction in the solid - the soil. The analyzed domains, on the example of the end of the exchanger, have been presented in Fig. 5.

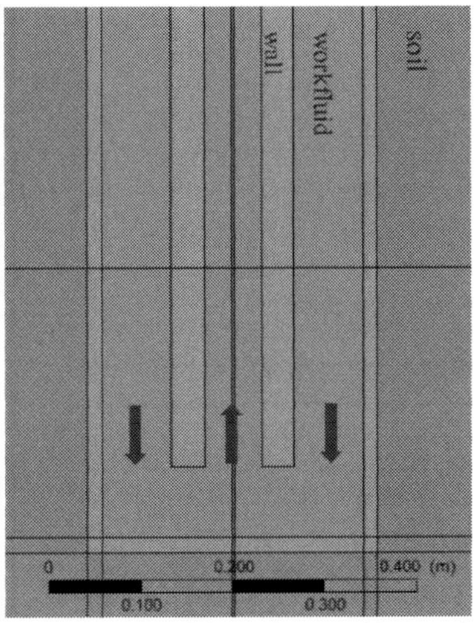

Fig. 5. The lower end of the underground heat exchanger.

The analysis was made in 2D axially-symmetrical form - the domains' axis coincides with the axis of the exchanger. Considering mixing and recurrence of the medium at the lower end of the exchanger, the k-omega SST turbulence model has been chosen. The 2D mesh (0.5 x 5 km) rotated around the pipe axis consists of 825 956 elements. The sample of the discretization of the ground, pipes and working fluid is shown in Fig. 6.

To minimalize simulation errors the heat exchanger, the ground and the working fluid under consideration has been divided into some blocks that have been discretized by means of a structured numerical grid, steeply refined in the normal wall direction. Initial tests allowed to use the numerical grid to ensure that further refinement did not influence the computational results. It has been assumed that the surface structure of the pipes can be treated as a homogeneous one. Wall function has been implemented at the level 50. The standard SIMPLE (semi-implicit method for pressure-linked equations) method has been employed for pressure-velocity coupling. The second order upwind schemes have been chosen for the solution of the convection term in governing equations. The diffusion terms have been central-differenced with the second order accuracy as well. The detailed methodology of numerical integration regarding the set of governing equations can be found in work [20].

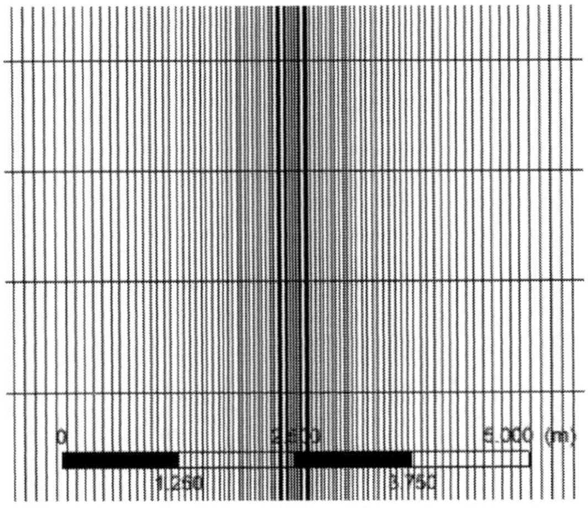

Fig. 6. A sample of discretization of the analyzed domains.

In the preliminary calculations, which are performed at the present level of detail, water is used as the working fluid. It was also assumed that the inner pipe is an ideal insulator and that there is no heat transfer between the fluids. The average temperature at the outlet of the ground heat exchanger over time is presented in Fig. 7. The calculations assume that the pipes are initially filled with the working substance and for about 1 hour there is an outflow of the medium accumulated earlier, and only then the average outlet temperature is formed as a result of heating of water from the temperature of 70°C.

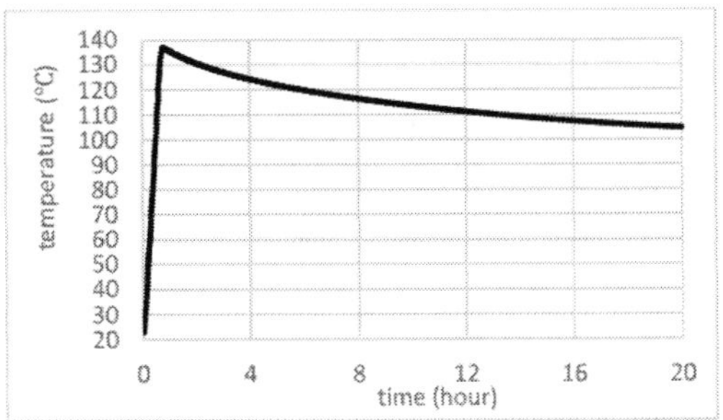

Fig. 7. Average outlet temperature for non-stationary heat transfer.

Fig. 8 shows the balance of the heat transfer rate obtained from the ground at the outlet from the exchanger. It indicates that at the beginning, the working medium transfers the heat to the soil. After some time the exchanger starts to play the proper role and draws the heat from the ground. The heat transfer rate from the maximum value of 1.4 MW starts to decrease, but does not show any sudden changes, which should be considered positive given the fact that the underground heat exchanger (UHE) will operate on a cyclical basis.

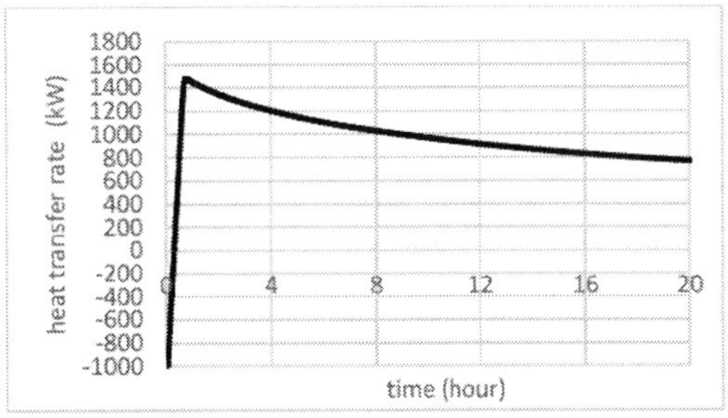

Fig. 8. Balance of the heat flow obtained from the ground at the outlet of the underground heat exchanger.

The influence of the heat exchanger on the temperature distribution in the ground is presented in Fig. 9. The resulting temperature funnel does not exceed 3 m in diameter, which suggests the use of smaller computational domains and higher heat exchanger capacities in future analyzes. However, in the literature [21-23] we can find examples of numerical calculations of heat transfer between exchanger and the environment in which the soil domain area is large - similarly scaled as in this paper.

During the charging process, the ground temperature will return to its initial state and in the next operating cycle it will be possible to generate a similar heat flux. In the next stage of work, the dimensions of the underground heat exchanger, loading and unloading time and the number of exchangers will be optimized.

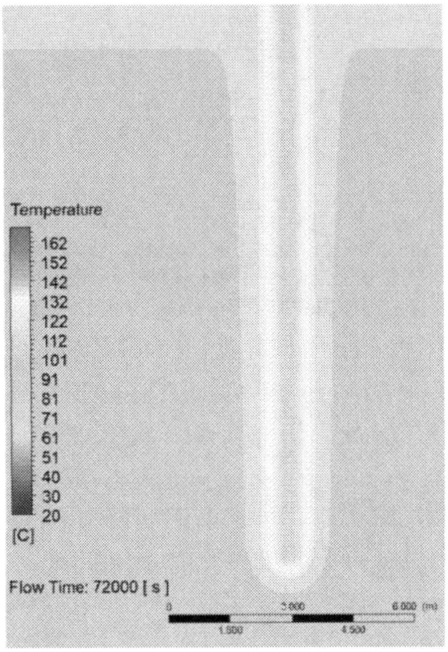

Fig. 9. The temperature distribution at the lower end of the underground heat exchanger after 20 h of the simulation

4 Conclusions

In this paper, the information concerning the cooperation of variable RES sources with conventional steam power plants were presented. Two situations (Fig.1 and Fig.2) are shown which indicate that there is a need to develop CAES systems. Next, a system is proposed which may improve the operation of the power system to some extent by storing energy in CAES systems. The efficiency of this process of charging and discharging the cavern has been estimated according to the formula (10) at η_{CAES}= 0.52, which is a satisfactory value, taking into account the irregularity of the work of hardly predictable renewable energy sources such as wind and photovoltaic, whose irregularity will be compensated for. The underground heat exchanger (UHE) may be an element reducing heat losses to the environment from the CAES system and thus increasing the efficiency of energy conversion, but further work is needed on its design and integration into the system.

ACKNOWLEDGEMENTS Paper was created as part of the statutory work of Institute Fluid-Flow Machinery Polish Academy of Sciences. However, other part of this work was partially funded by research subvention from Faculty of Mechanical Engineering (No. 033740/DS), Gdańsk University of Technology.

References

1. Hyrzyński R., Badur J., Jaroszewska M., Ziółkowski P., Gotzman S., Froissart M., Impact of wind turbines on climate, *Energetyka*,Vol. 776, No 2, pp. 77-83, (2019), [in Polish].
2. The European Council. European Council, 23-24.10.2014. Climate and Energy Policy Framework Up to 2030, (2014). https://www.consilium.europa.eu/ pl/meetings/ european-council/2014/10/23-24/.
3. Klonowicz P., Witanowski Ł., Jędrzejewski Ł., Suchocki T., Surwiło J., Stepniak D., Wstepna analiza potencjału zasobników energii typu UWCAES w Zatoce Gdańskiej, *Rynek Energii*, Vol. 124, No 3, pp. 100-107, (2016).
4. Badur J., Hyrzyński R., Kraszewski B.. Ziółkowski P., Dudda W., Analysis of electricity generation variability in the first five months of 2019, with particular emphasis on wind energy generation, *Nowa Energia* Vol. 68, No 3, pp. 40-45, (2019), [in Polish].
5. Wasiak I., *Elektroenergetyka w zarysie. Przesył i rozdział energii elektrycznej*, Politechnika Łódzka, Łódź, (2010).
6. Polskie Sieci Elektroenergetyczne SA. System data, Pr KSE 2019. https://www.pse.pl/dane-systemowe/praca-kse/informacje-ogolne/opis-systemu, [in Polish].
7. Zajczyk R., *Regulacja częstotliwości i mocy w systemie elektroenergetycznym*, Politechnika Gdańska, Gdańsk, (2002).
8. Polish Wind Energy Association; Lower Silesian Institute of Energy Studies. Cooperation of conventional coal and large scale RES sources, (2019), [in Polish].
9. Li B., DeCarolis JF., A techno-economic assessment of offshore wind coupled to offshore compressed air energy storage, *Appl Energy*, Vol. 155, pp. 315–322, (2015).
10. Cheung BC., Carriveau R., Ting DSK., Multi-objective optimization of an underwater compressed air energy storage system using genetic algorithm, *Energy*, Vol. 74, pp. 396–404, (2014).
11. Salvini C., Techno-Economic Analysis of Small Size Second Generation CAES System, *Energy Procedia*, Vol. 82, pp. 782–788, (2015).
12. Iglesias A., Favrat D., Innovative isothermal oil-free co-rotating scroll compressor–expander for energy storage with first expander tests, *Energy Convers Manag*, Vol. 85:565–572, (2014).
13. Wolf D., Budt M., LTA-CAES – A low-temperature approach to Adiabatic Compressed Air Energy Storage, *Appl Energy*, Vol. 125, pp.158–164, (2014).
14. Hartmann N., Vöhringer O., Kruck C., Eltrop L., Simulation and analysis of different adiabatic CAES plant, *Appl Energy*, Vol. 93, pp. 541–548, (2012).
15. Krawczyk P, Szablowski L, Karellas S, Kakaras E, Badyda K., Comparative energy and exergy analysis of compressed air and liquid air energy storage systems, *Proceedings of ECOS 2016 - the 29th International Conference on Efficiency, Cost, Optimization, Simulation and Environmental Impact of Energy Systems*, June 19-23, 2016, Portorož, Slovenia.
16. Liu W., Liu L., Zhou L., Huang J., Zhang Y., Xu G., Yang Y., Analysis and Optimization of a Compressed Air Energy Storage—Combined Cycle System, *Entropy*, Vol. 16, pp. 3103-3120, (2014).
17. Badyda K., Milewski J., Thermodynamic analysis of compressed air energy storage working conditions, *Arch Energ*, Vol. 42, pp. 53-68, (2012).
18. Szablowski L, Krawczyk P, Badyda K, Karellas S, Kakaras E, Bujalski W., Energy and exergy analysis of adiabatic compressed air energy storage system. *Energy*, Vol. 138, pp. 12-18, (2017).
19. Badur J., Ziółkowski P., Sławiński D., Kornet S., An approach for estimation of water wall degradation within pulverized-coal boilers, *Energy*, Vol. 92, pp. 142-152, (2015).
20. Badur J., Numerical modelling of sustainable combustion in gas turbines, *IFFM Publishers*, Gdańsk, 2003.
21. Lous M., Larroque F., Dupuy A., Moignard A., Thermal performance of a deep borehole heat exchanger: Insights from a synthetic coupled heat and flow model, *Geothermics* Vol, 57, pp. 157–172, (2015).
22. Hanuszkiewicz-Drapała M., Składzień J., Heating system with vapour compressor heat pump and vertical U-tube ground heat exchanger, *Archives of thermodynamics*, Vol. 31, No. 4, pp. 93–110, (2010).
23. Naldi C., Zanchini E., Full-Time-Scale Fluid-to-Ground Thermal Response of a Borefield with Uniform Fluid Temperature, *Energies*, Vol. 12, 3750; doi:10.3390/en12193750, (2019).

RESEARCH & DEVELOPMENT IN POWER ENGINEERING, 2019

Preheating procedure for fast start-up of a steam turbine from a cold state

Wojciech Kosman[1,*], *Andrzej* Rusin[2]

[1] Silesian University of Technology, ul. Konarskiego 18, Gliwice, wojciech.kosman@polsl.pl, Poland
[2] Silesian University of Technology, ul. Konarskiego 18, Gliwice, andrzej.rusin@polsl.pl, Poland

Abstract. The paper describes a procedure that allows to start up a steam turbine in a significantly shorter period. The procedure is developed for start-ups that begin from a cold state, when the temperature of the parts of the turbine is close to the ambient temperature. The pre-heating rises the parts temperature before the actual start-up begins. It changes the thermal state of the turbine and causes smaller maximal stress during the initial period of the start-up. The procedure involves a pre-heating of the turbine with a hot air generated in an electric heater. The paper describes the requirements of the process. It presents the possible configurations of the flow in the turbine and the analysis of the thermal and the strength state of the turbine during the pre-heating.

1. Introduction

During the changes in the structure of the power generating industry in the recent years steam cycles survived as the core systems [1-3]. However their operation has changed considerably [4]. It is now affected by the operation of the renewable energy sources including solar and wind power [5]. The power generation based on these sources is not constant. They depend largely on the ambient conditions. The core systems, including the steam cycles, compensate the changes in the load of the renewable sources [6]. The operational conditions of steam turbines change much more frequently.

The conditions that undergo frequent changes result in an intensive fatigue process in steam turbines [7,8]. The number of start-ups is larger than ten or twenty years ago. In many cases it exceeds 50 start-ups per year. Also the load ramps happen very often.

The negative effects of the fatigue process may be reduced if the change of the stresses in the machines occurs at slower rate and with a lower amplitude. In steam turbines this may be achieved if the temperature of the working fluid changes in an appropriate way.

The larges amplitudes of the stress occur in steam turbines during start-ups from a cold state, when the initial temperature of the machines is equal to the ambient temperature. The start-up from these conditions is also the longest one.

The aim of the research presented here is to shorten the start-up period of a steam cycle while maintaining a safe level of stresses. The suggested procedure is a rearranged version of the original sequence designed by the manufacturer. The turbine and the boiler begin their start-up procedures at the same time as opposed to the sequential order. The process is presented for the 13K215 steam turbines type.

2. Modification of the start-up procedure

A typical start-up procedure of a steam cycle includes three basic steps. They are shown schematically in the fig. 1a and listed below.

1. Boiler start-up.
2. Heating of the steam pipes.
3. Turbine start-up.

This sequence depends on the availability of the steam. At the end of the first step the boiler already generates steam, although its temperature and pressure are much lower than the design, full load values.

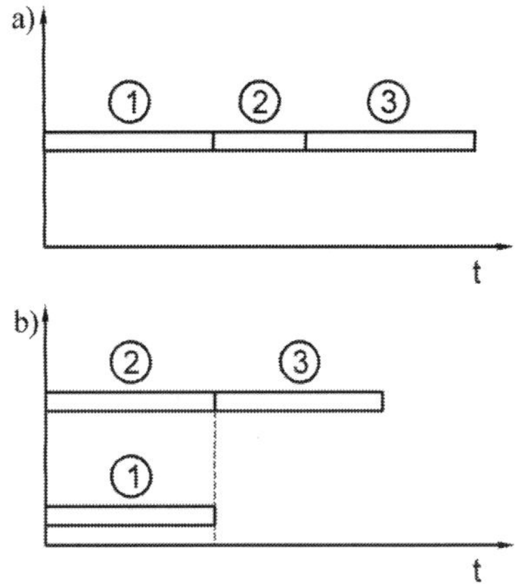

Fig. 1. Turbine start-up sequence: original (a), with a pre-heating (b)

The steam is then used to heat the pipes between the boiler and the turbine. In some turbines a small fraction of the steam is delivered into the space between the inner and the outer casing in the high pressure and the intermediate pressure sections of the turbine in order to heat up the casings and the rotor. Even though the rotor is not directly exposed to the steam at this point, it is running to avoid its deformation. The larger fraction of the steam bypasses the turbine through the pressure reduction valve and the cooling system.

In the start-up sequence described above the heating of the pipes and the turbine may not begin before the boiler starts to deliver the steam in the required amount.

The approach towards the optimization of the start-up begins with the observation that the heat source for the heating of the pipes and the turbine does not have to be the steam generated in the boiler. If the heat may be delivered from another source then the start-up sequence may be rearranged.

The suggested sequence of the start-up is shown in fig. 1b. The heating of the turbine may start at the same time, when the boiler begins its operation. This step is now called a pre-heating to distinguish it from the heating with the steam from the boiler. The whole period of the start-up becomes much shorter.

In the optimized procedure the heating of the turbine from an external source is independent of the available steam temperature from the boiler. It means that the heating process itself may be optimized. The key idea is to raise the temperature of the turbine during the pre-heating to such a level that allows the admission of the steam from the boiler at high temperature. If this is achieved then the cold start-up becomes similar to a start-up from a warm state, when the period between a preceding shut-down and the start-up is less than 50 hours. The whole start-up becomes even shorter. This is explained in fig. 2. The figure shows the change of the live steam temperature during a start-up. The dashed line is a manufacturer's curve and the solid line corresponds to the optimized procedure.

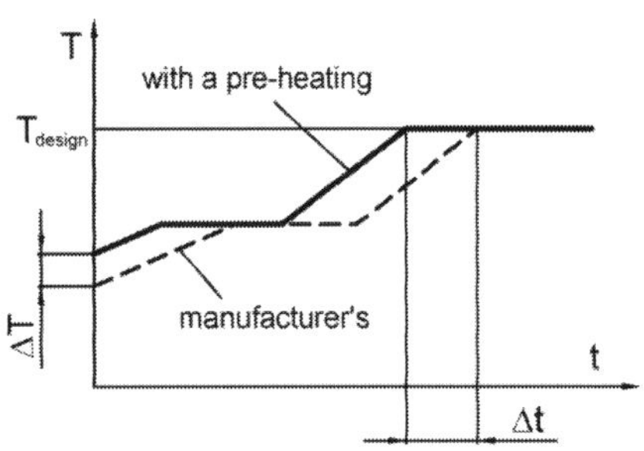

Fig. 2. Start-up temperature curves

If the turbine is pre-heated then the live steam temperature may be higher in the moment when the cut-off valves open and the live steam starts to flow into the turbine. Hence the temperature difference ΔT in the fig. 2. The result is the time difference Δt, which means that the design value of the live steam temperature is reached earlier than in the manufacturer's start-up. This effect occurs even without any other optimization. The rate of the increase of the temperature is the same for the manufacturer's curve and for the optimized one - the slopes of the lines in the fig. 2 are exactly the same.

3. Heat source options

A very important issue is the choice of the external heat source for the pre-heating. There are several possible options described below.

a) Steam from an external source

The turbine parts may be exposed to the steam from a different turbine in the same power plant. The pressure and the temperature of the steam may be adjusted to the required levels through the throttling and the water injection.

Most power plants have several steam turbines and at least one is always running. It means that there is always steam available for the pre-heating.

The usage of the steam from an external source is related to several issues that make the pre-heating process problematic. First of all the pre-heating system involves two power generating units. One delivers the heat and the other receives it. This means that the two cycles must be adapted to both: the delivery and the reception of the heat. The application would involve the modifications of the cycles in two separate points each.

Secondly the heat delivery system would have to include at least one steam extraction. The temperature of steam for the heating must be relatively high (250 - 400°C). This excludes a common pipe that is usually fed from each turbine in a power plant because of the low temperature.

There are also operational issues. The steam parameters in an extraction strongly depend on the load in the turbine. In order to provide stable steam parameters for the pre-heating, the turbine that delivers the heat would have to maintain stable load for the period of the pre-heating.

Further the problem of the steam condensation must be taken into the account. The heating steam would be delivered in small streams to a cold turbine. This would result in some amount of the water gathering in the lower areas of the turbine and the water would have to be removed through the drainage system. The condensation influences the temperature of the parts exposed to the steam and makes the process more difficult to control.

b) Electric heating mats

A relatively fast temperature increase may be achieved if turbine parts are covered with electric mats. The pre-heating process may be conducted without any hot fluid.

Unfortunately the choice of the locations in turbines where the mats may be installed is very restricted. The only place inside a turbine is the external surface of the inner casing. The rotor would then be heated indirectly through the inner casing. Also the mats located in the area between the inner and the outer casings would be exposed to the leakage steam that creates erosion conditions.

c) Hot air

The use of the hot air eliminates the problems caused by the two previously described methods. The air flow may be arranged just like the steam flow. At the same time the problem of the steam condensation and the necessity of the water removal is significantly reduced. The air may be used to the direct heating of the casings and the rotor.

The hot air pre-heating is considered in the further analysis. The pre-heating system consists of an electric air heater and the compressor that provides a pressure level required to overcome the pressure losses in the pipes and the turbine. The process may be easily controlled. The two main parameters that are adjusted is the amount of the air and its temperature. The heating conditions may be matched to the requirements that depend on the thermal state of the turbine parts.

4. Existing thermal systems

The 13K215 turbines have two systems that affect the thermal state of the machine parts. The first one is used during a forced cooling of the turbine after the shut-down of the power generating unit. The system allows to decrease the temperature of the metal of the turbine parts and begin the maintenance works in a much shorter period when compared to natural cooling. A simplified diagram of the system is shown in fig. 3.

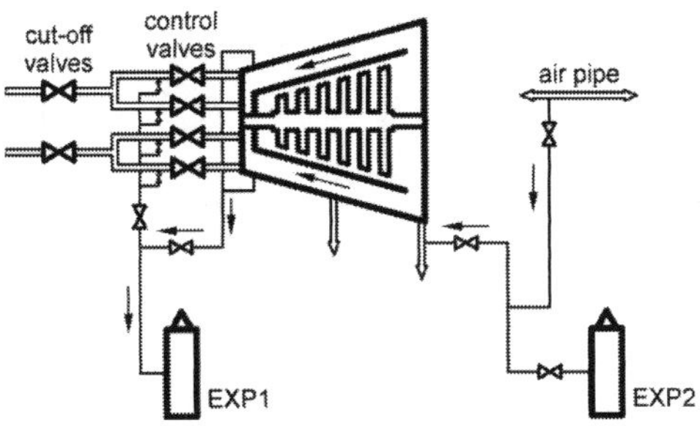

Fig. 3. Forced cooling system

After the control valves are closed the compressed clean air is delivered from the air pipe. The air flows through the expander EXP2, which under normal operation is a part of the drainage system. The cooling air flows from the outlet of the turbine to the inlet - the flow is opposed to the steam flow. The air leaves the turbine through the pipes connected to the expander EXP1 that is also a part of the drainage system. The system does not include a separate compressor because the air pressure in the delivery pipe is high enough to force the flow.

Due to the closed position of the control valves the air does not flow through the blades path but into the area between the inner and the outer casing. The rotor is cooled indirectly and the heat from this part is rejected through the inner casing. This arrangement of the flow is necessary to prevent the damage. In this particular turbine the rotor is of a disc type and the discs act as fins that intensify the heat transfer. Therefore if the rotor was exposed directly to the cooling fluid during the forced cooling it would shrink faster than the casing. In the result the rotating blades on the rotor could come into the contact with the stationary nozzle blades and cause damage.

The second system that affects the thermal state of the turbine is the heating system for the casings. As mentioned before the exposure of the rotor to a fast flowing fluid results in a rapid increase of the metal temperature. During a start-up when the rotor is exposed to hot steam the relative elongations between the rotor and the casings may become very large. This may occur especially during a start-up from a cold-state.

In order to prevent large relative elongations a small fraction of the steam is delivered to the area between the inner and the outer casings. This intensifies the heating of the casings and allows to keep the relative elongations between the rotor and the casings in a safe range. This small, additional flow is shown in the fig. 4.

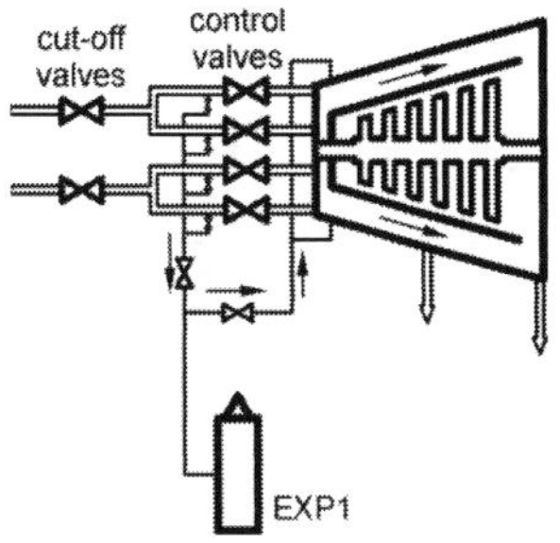

Fig. 4. Casings heating system

The steam is taken from the main live steam pipe from small extractions located in front of the control valves. The extracted steam flows through the pipes that are a part of the drainage system and is delivered into the front area of the turbine. The casings heating system uses the same pipes as the forced cooling system.

The casing heating system may operate during any phase of a start-up. The only restriction is the steam temperature. If it exceeds 86 percent of the design value of the live steam temperature the system must be cut off. Usually the operation of the system is determined on-line on the basis of the measurements of the casings temperature and the relative elongations between the casings and the rotor.

The pre-heating system may use the parts of both systems described above. Since the systems are able to handle flows of the temperature up to about 86% of the design live steam temperature there is no danger of overheating the pipes. Also no modifications to the existing turbines are required, so the suggested system may be applied to older machines as well.

5. Air flow

The rate of the heat transfer depends on the following three issues related to the flow of the hot air:

- temperature,
- mass flow rate,
- flow arrangement.

 https://doi.org/10.1051/e3sconf/201913701024

In the proposed design the temperature of the air and its flow rate are to be controlled during the process. The flow rate is adjusted either by the valves or by matching operation of the compressor to the required amount of the air. The temperature may be controlled through the electric power for the coils in the air heater.

The base values for the flow rate and the temperature are determined in the following way. The mass flow rate may be chosen equal to the flow during the forced cooling after the power unit shut-down. This ensures that the amount of the air is correct for the particular size of the turbine. The maximal temperature of the air is restricted by the value given in the description of the casings heating system in the previous section. For the sake of safety the maximal air temperature should be set at an even lower level. In the calculations presented in this paper the maximal air temperature was either 350 or 400°C. It should be noted here that the pre-heating is not meant to substitute the heating with the steam from the boiler. Its task is simply to shorten the period of turbine start-up.

The air pressure should be high enough to force the flow through the turbine.

The key feature is the arrangement of the air flow in the turbine. Figure 5 shows the basic designs.

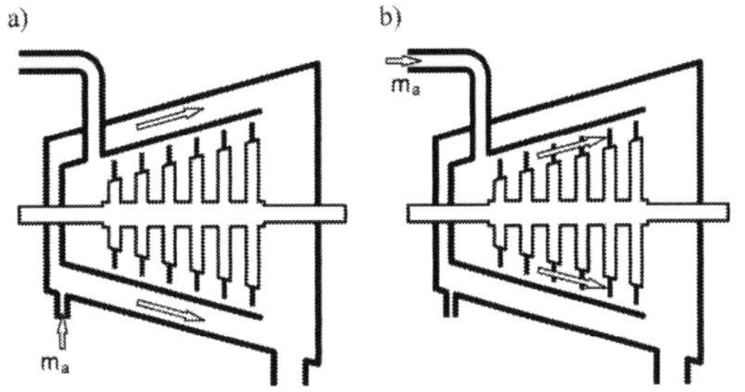

Fig. 5. Basic air flow in the turbine

In the design presented in the fig. 5a the air flows in the same direction as the steam in the casing heating system (see the previous section). The steam enters the area between the inner and the outer casings. It flows towards the rear end of the turbine. The temperature of the inner and the outer casings is increased first and then the temperature of the rotor.

The design from the fig. 5b assumes the same direction of the flow as during the operation of the turbine. The air enters the nozzle box and then flows through the blades path from the inlet to the outlet. Due to the low pressure of the flow the expansion process is negligible. The air must overcome the pressure losses related to the flow through ducts.

In this arrangement of the flow the rotor is heated directly while the outer casing is heated indirectly. According to the notes given in the preceding sections this related to the most intensive heat transfer process on the surface of the rotor. The relative elongations between the rotor and the casings may change very fast. However in contrast to the cooling process where turbine parts shrink the thermal expansion here does not lead to a direct contact between the stationary and the rotating parts of the turbine. This is due to the clearances that are larger between a rotor blade and the following nozzle blade than between the rotor and the preceding nozzle. Also the air heating may easily be controlled - easier than heating with steam.

Nevertheless the design from the fig 5b is treated here only as a reference to determine the limiting parameters of the air.

Figures 6 and 7 present additional cases of the air flow through the turbine. The case from the figure 6a is used in some turbines during the forced cooling after the shut-down of the steam cycle. The air enters the area between the casings and flows towards the rear end of the turbine. The valves at the outlet pipes are closed and the air cannot leave the turbine there. The air flows towards the front of the turbine through the blades path and exits through the nozzle box and live steam pipes. This arrangement exposes all three of the main parts of the turbine to the air: the rotor and both casings.

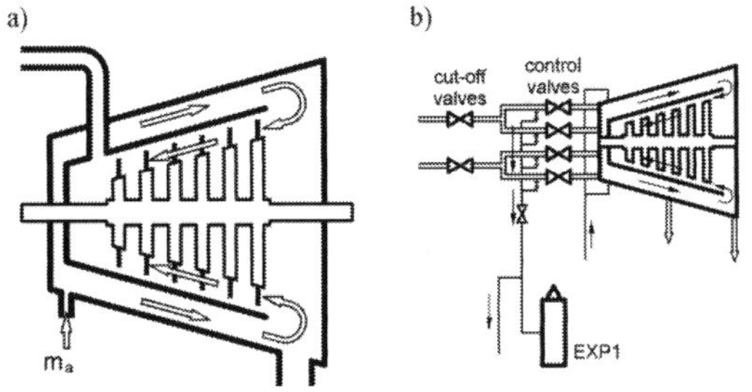

Fig. 6. Pre-heating with a reversed air flow

The application of the case from the fig. 6a requires small modifications in the drainage system of the turbine. The entry and exit locations of the air must be modified according to the fig. 6b.

The flow arrangement from the fig. 6 has one major disadvantage: the heating of the rotor and the casings is always conducted proportionally. If the heat transfer rate is increased for the casings then the proportional increase occurs for the rotor. That is why yet another designed is considered. It is shown in the fig. 7.

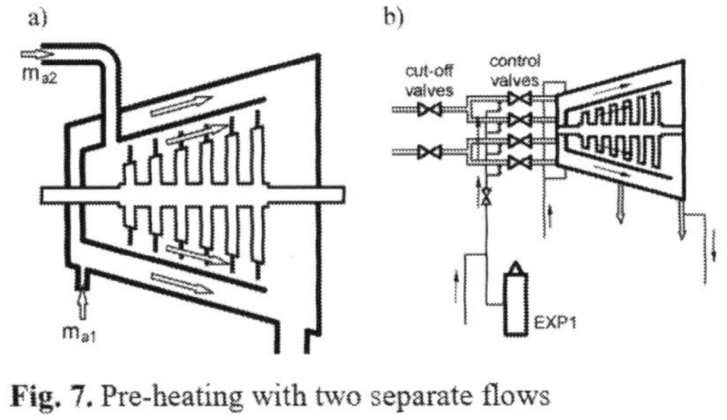

Fig. 7. Pre-heating with two separate flows

There are two flows in the turbine, one through the blades path and the other through the area between the casings. The flows may be adjusted separately. Even if the hot air is generated in an electric heater there is always a possibility to adjust each flow rate. The disadvantage here is that there are two entry points of the

 https://doi.org/10.1051/e3sconf/201913701024

air into the cycle. They are shown in the fig. 7b. The exit point is just one for both streams as they mix in the rear end of the turbine.

In every configuration presented in this section the rotor must be running during the whole period of the pre-heating. It must be driven be an electric engine in order to prevent damage.

6. Numerical modelling

The pre-heating configurations were analysed against different conditions. The research involved numerical simulations of the thermal and the strength states of the main turbine parts. The rotor and the casings models were axisymmetric. Their shapes are shown in the fig. 8. The axisymmetric models are simplified especially for the casings. However the experience indicates that the stress in the rotor is always higher than in the casings. It is the rotor that determines the limits for the transient thermal states of the whole turbine. Therefore simplified models of the casings do not influence the limiting values of the stresses. The axisymmetric models of the casings provide practically the same results of the thermal elongations as the three dimensional versions.

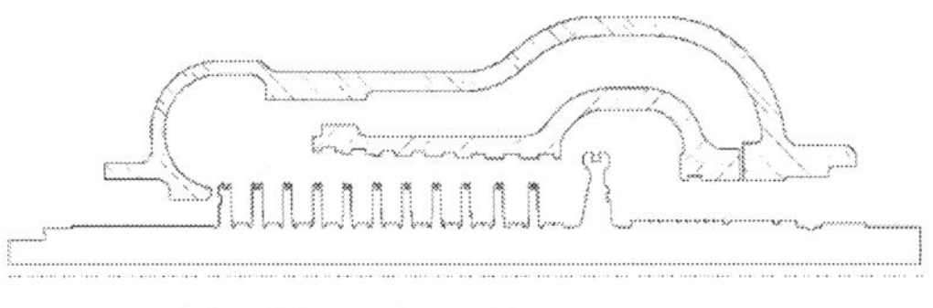

Fig. 8. Models of the main turbine parts

The numerical calculations were conducted with the finite elements method that allowed to determine the distribution of the temperature, stress and displacements of the turbine parts. The most important displacement here are the elongations along the main axis of the turbine.

The surfaces of the parts were divided into sections that corresponded to similar boundary conditions. The convective heat transfer coefficients were calculated for each section from semi-empirical formulas [9-11].

7. Analysis results

The graphs in the following figures are a fraction of the whole research and are presented here to explain the investigated conditions of the pre-heating. Four cases (variants) were analysed and their symbols are given below.

- W1 - air delivered between the casings (fig. 5a).
- W2 - air delivered to the blades path (fig. 5b).
- W3 - air delivered between the casings and reversed in to the blades path (fig. 6).
- W4 - two streams of the air (fig. 7).

The preliminary simulations allowed to narrow the values of the analysis parameters to a reasonable ranges. The primary criteria for the selection of the conditions were the stress level and the magnitude of the relative elongations between the casings and the rotor. The time of the pre-heating was between 180 and 240 minutes. The maximal air temperature was within the range of 350-400°C. The air flow rate was kept constant.

There are several points of the stress concentration. The most important ones are shown in the fig. 9 together with their symbols.

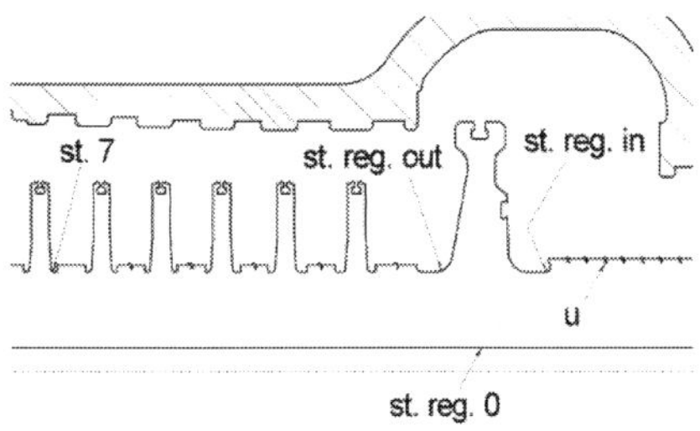

Fig. 9. Points of the stress concentration

Other symbols in the figures that show the numerical results include:

- inlet - front side of the turbine (steam inlet)
- KWWP - high pressure inner casing,
- KZWP - high pressure outer casing,
- outlet - rear side of the turbine (steam outlet)
- t - pre-heating time
- T - temperature in the blade path,
- T_{mk} - temperature in the area between the inner and the outer casings.
- WWP - high pressure rotor.

The detailed graphs shown in the following five figure (figs. 10 - 14) correspond to the case number 2 with the period of the pre-heating lasting 210 minutes. The figures show the change of the chosen values over time.

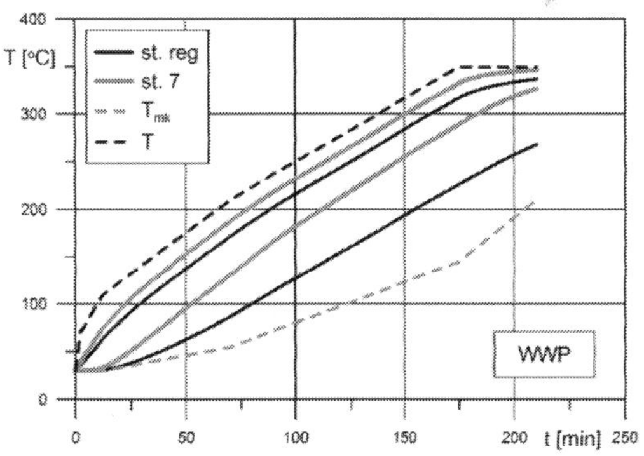

Fig. 10. Rotor temperature, case W2, t = 210 min

The graphs that show the temperature of the turbine parts include two lines of the same colour and type for the locations in the specific part. They refer to both sides of a part. The line that is located higher in the graph is related to the temperature on the surface that is exposed to the hot air. For example in the case number 1 the outer surface of the inner casing is exposed to the hot air. In the case number two it is the inner surface of the inner casing that is exposed to the hot flow. The line that is located lower is the temperature on the opposing side of the part not exposed to the flowing air.

The rotor is always exposed to the hot flow at its external surface. The second line in the graphs refers to the point located in the central void of the rotor in the same cross section as the point on the external surface.

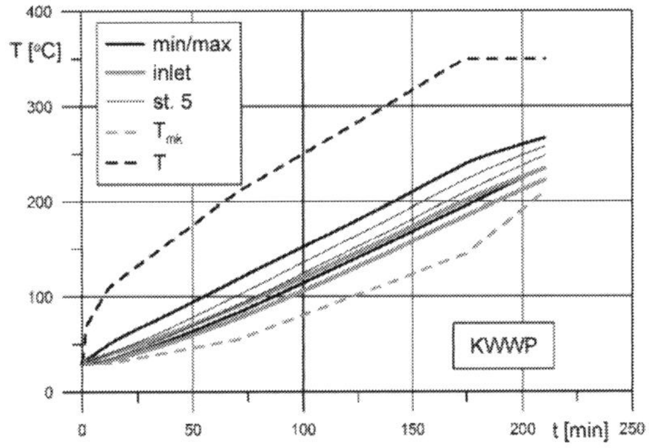

Fig. 11. Inner casing temperature, case W2, t = 210 min

The calculations of the relative elongations assume that the inner casing is attached to the outer casing in the cross-section where the live steam inlet is located. In the figures there are two symbols for two distinct relative elongations:

- WWP-KWWP - relative elongation between the rotor and the inner casing calculated in the outlet cross-section of the casing,
- WWP-KZWP - relative elongation between the rotor and the outer casing calculated in the rear end of the casing.

The second elongation from the above list is usually measured in most steam turbines.

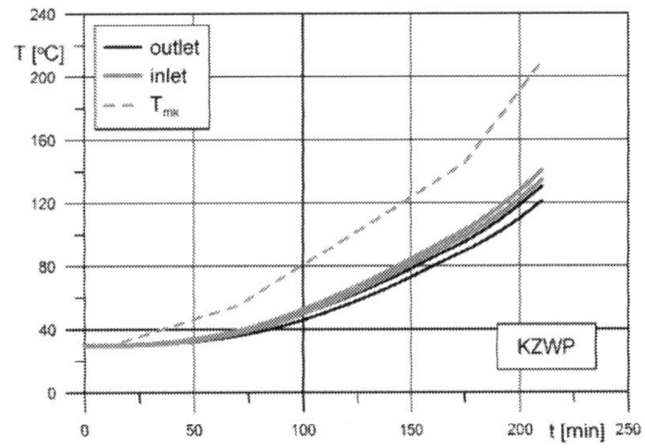

Fig. 12. Outer casing temperature, case W2, t = 210 min

In the cases 1 and 2 there is only one area filled with a flowing air. It is either the area between the casings or the blades path. The other one is filled with air, which does not flow. The heat transfer conditions there are similar to the natural convection. The exception is the surface of the is running rotor. Even if it is not exposed to the flowing air the convection is still intensive due to the rotational velocity.

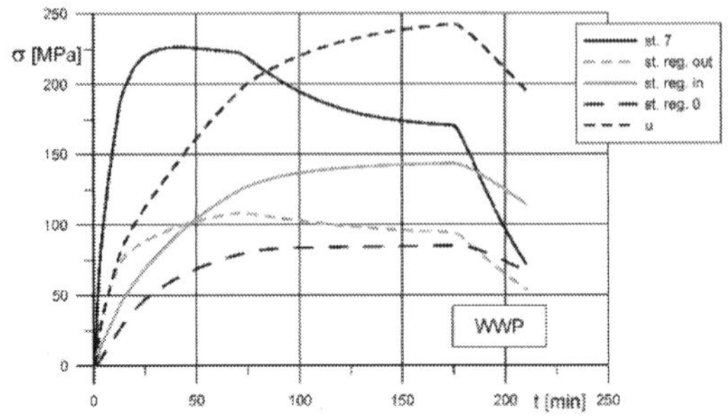

Fig. 13. Stress in the rotor, case W2, t = 210 min

It may be observed in the fig. 13 that the local maximal values of the stress occur in different moments in different sections of the rotor. This is the reason that five different characteristic locations are shown in order to fully present the change of the stress.

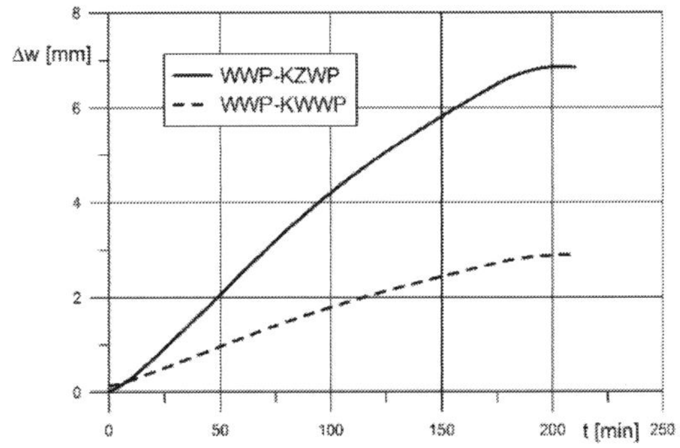

Fig. 14. Relative elongations, case W2, t = 210 min

The values of the elongations like the ones shown in fig. 14 must be confronted against the clearances set in the turbine in the axial direction.

8. Comparison of the cases

The result show that the most favourable thermal state in the turbine is achieved for the case number 4 of the air flow configuration.

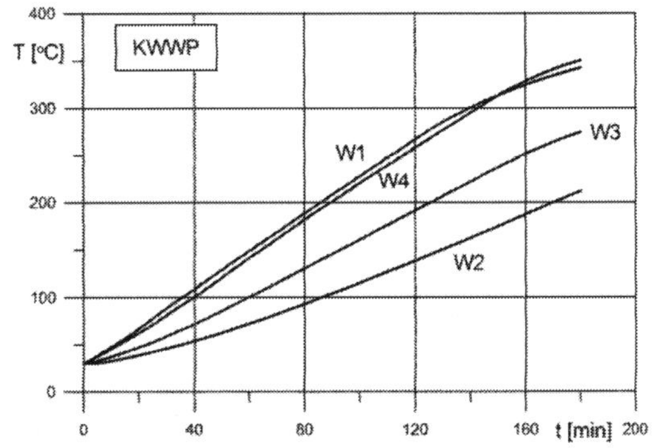

Fig. 15. Temperature of the inner casing

https://doi.org/10.1051/e3sconf/201913701024

If two streams of the hot air are delivered to the turbine then the temperature of the casings and the rotor increase simultaneously. At the end of the simulated period the inner casing and the rotor have the same temperature. This is clear when the lines for W4 (case 4) are compared between the figures 15 and 16. The relative elongations are the smallest of all of the cases, which is visible in figs. 17 and 18. This is because the temperature increase is the same in the casings and in the rotor.

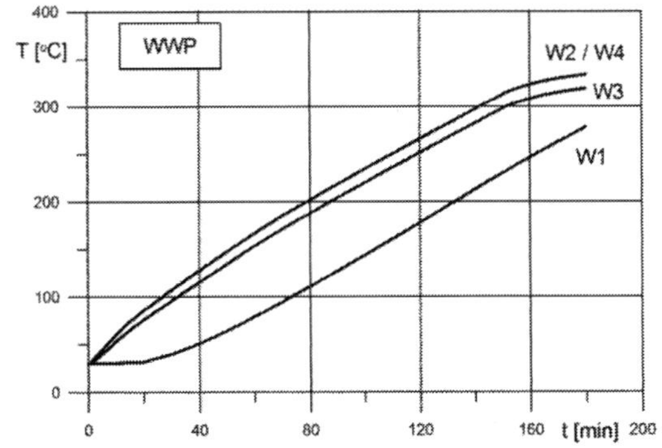

Fig. 16. Temperature of the rotor

Unfortunately the forth case causes some difficulties in the practical application. The heat source must deliver the air at two different temperature levels. In the simulations the maximal temperature between the casings was 50 degrees lower than the temperature in the blades path for the whole time of the pre-heating. It means that the electric air heater would require an air extraction.

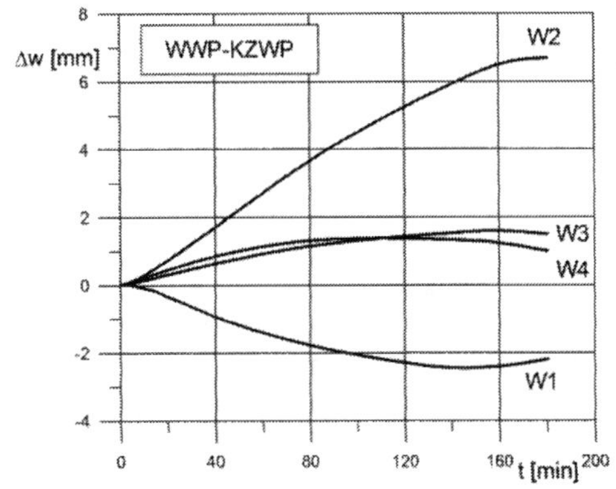

Fig. 17. Relative elongations rotor - outer casing

A solution that is very close to the optimal case number 4 is the case number 3, where the air flows between the casings and then in the blade path. The final temperature of the rotor at the end of the pre-heating is very close to the temperature of the rotor in the case 4, see fig. 16. The temperature of the inner casing is lower (fig. 15). This however is not a problem since the highest values of the stress occur in the rotor and not the casing.

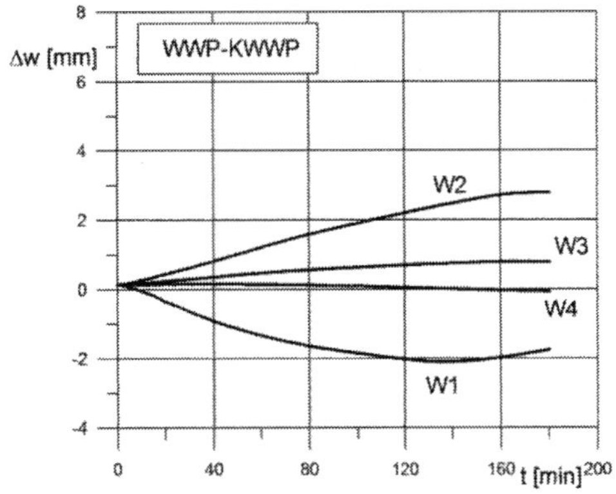

Fig. 18. Relative elongations rotor - inner casing

The case number 2 should be treated as an equivalent of the heating with a steam during a typical start-up. There are two differences. The first is that the pre-heating may begin earlier and that shortens the whole start-up. The second difference is the control of the process. It is much easier to control the flow and the temperature of the air than the steam that is generated in the boiler. Due to the size of the boiler its response is slower. The steam temperature may be decreased quite rapidly with the water injection but it cannot be raised so fast. The electric air heater allows much better flexibility in that regard.

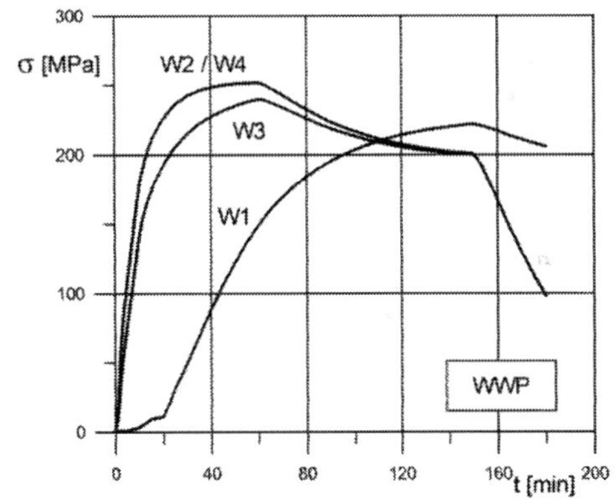

Fig. 19. Stress in the rotor for different flow cases

The pre-heating conducted according to the case number 1 results in negative relative elongations. Even though the absolute values are small they pose a threat of damage. The case number 1 is regarded here as an option for the control of the process. If the control valves are opened and closed accordingly the flow configuration changes between the cases number 1 and 3. This allows to handle the thermal stress in a better way.

The comparison of the stress shown in the fig. 19 prove that the maximal level of the stress is quite similar for different cases. The graph in the figure was obtained for the groove near the disc of the seventh stage. The largest stress occurs for the air flow configurations when the rotor is directly exposed to the hot fluid. In the case number 3 the air that reaches the surface of the rotor has lower temperature than in other cases because it cools down in the area between the casings before entering the blade path.

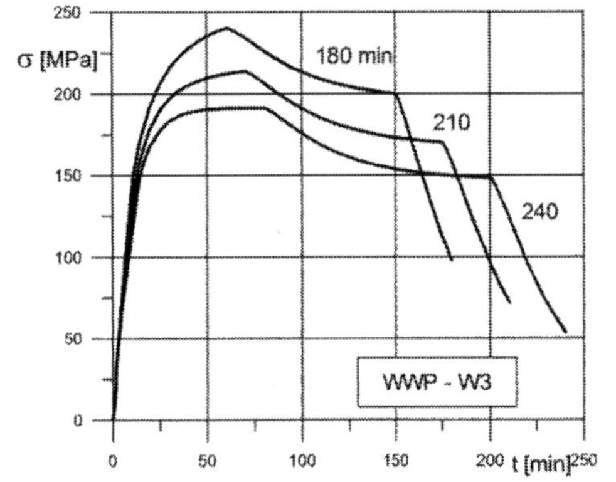

Fig. 20. Stress in the rotor for different length of the pre-heating period

In the case number 1 the maximal value of the stress is in the right side of the graph in the fig 19. This is a moment late in the process, when the air in the blades path finally reaches the temperature that results in the stress concentration in the rotor.

The basic time of the pre-heating process investigated in the simulations was 180 minutes. The stress in the rotor may be lowered if the period of the process is longer. The comparison of the stress in one point of the rotor for different length of the pre-heating is shown in fig. 20.

The maximal stress for the basic case is over 200 MPa. It should be noted that this is a local value that corresponds to a very small area of the rotor.

9. Conclusions

The analysis presented here indicates that the optimized procedure allows to conduct a start-up from a cold state much faster than the manufacturer's curves. The possible decrease of time derives from two modifications. Firstly an external heat source is used to increase the temperature of the parts of the turbine. This procedure may begin at the same time the boiler is started. Secondly the pre-heating raises the turbine temperature above the standard level. When the valves are open and the live steam is delivered to the turbine for the first time the turbine temperature is already quite high. The rate of the temperature increase may be higher.

The costs of the preheating system includes the electric air heater and the compressor. The latter may be omitted if a power plant has got an air collector pipe with the pressure at an appropriate level. This is usually the case if the force cooling is used after the shut-downs. The estimated cost of the electric heater is later covered by the reduced costs of the start-up.

The pre-heating procedure may be applied to the existing steam cycle. The key advantage of the presented method is the low number of the required modifications. The turbine itself is not modified at all. The system uses the existing pipes with one or two new inlets and outlets depending on the chosen air flow configuration.

References

1. Tumanovskii A.G., Shvarts A.L., Somova E.V., Verbovetskii E.Kh., Avrutskii G.D., Ermakova S.V., Kalugin R.N., Lazarev M.V.: Review of the Coal-Fired, Over-Supercritical and Ultra-Supercritical Steam Power Plants. *Thermal Engineering*, Vol. 64, No. 2, 2017, pp. 83-96.
2. Abe F.: Research and Development of Heat-Resistant Materials for Advanced USC Power Plants with Steam Temperatures of 700C and Above. *Engineering*, 1(2), 2015, pp. 211-224.
3. Crespi F., Gavagnin G., Sanchez D., Martinez G.S.: Supercritical carbon dioxide cycles for power generation: A review. *Applied Energy*, Vol. 195, 2017, pp. 152-183.
4. Topel M., Guedez R., Laumert B.: Impact of increasing steam turbine flexibility on the annual performance of a direct steam generation tower power plant. „Energy Procedia" Vol. 69, 2015, pp. 1171-1180.
5. Chmielniak T.: Technologie energetyczne. *Wydawnictwa Naukowo-Techniczne*, Warszawa 2015.
6. Arkadyev B.A.: Features of Steam Turbine Cooling by the Example of an SKR-100 Turbine for Supercritical Steam Parameters. *Thermal Engineering*, Vol. 62, No. 10, 2015, pp. 728-734.
7. Zhao N., Wang W., Zhang J., Liu Y.: Numerical investigation on life improvement of low-cycle fatigue for an ultra-supercritical steam turbine rotor. „Journal of Mechanical Science and Technology", Vol. 30 (4), 2016, pp. 1747-1754.
8. Rusin A., Łukowicz H., Lipka M.: Thermal states and stresses in selected structures of the supercritical and ultra-super critical turbines rotors. *Journal of Power Technologies* Vol. 96, Iss. 6, 2016, pp. 420-431.
9. Bergman L. B., Lavine A. S., Incropera F. P., Dewitt D. P.: Fundamentals of Heat and Mass Transfer. Wiley and Sons, 2011.
10. Dunn M.G.: Convective Heat Transfer and Aerodynamics in Axial Flow Turbines. *Journal of Turbomachinery*, Vol. 123, pp. 637-686.
11. Lienhard IV J.H., Lienhard V J.H.: A Heat Transfer Textbook. Third Edition. Phlogiston Press, Cambrige, Massachusetts 2008.

E3S Web of Conferences **137**, 01025 (2019) https://doi.org/10.1051/e3sconf/201913701025
RDPE 2019

RESEARCH & DEVELOPMENT IN POWER ENGINEERING, 2019

Selecting optimal conditions for the turbine warm and hot start-up

Andrzej Rusin[1*], *Grzegorz* Nowak[2], *Henryk* Łukowicz[3], *Wojciech* Kosman[4], *Tadeusz* Chmielniak[5],

[1,2,3,4,5] Silesian University of Technology, Institute of Power Engineering and Turbomachinery, Konarskiego 18, 44-100 Gliwice, Poland, andrzej.rusin@polsl.pl, grzegorz.nowak@polsl.pl, henryk.lukowicz@polsl.pl, wojciech.kosman@polsl.pl, tadeusz.chmielniak@polsl.pl

Abstract. The changing role of coal-fired power units which are now used more and more often as sources balancing the demand for electricity in the power system calls for a change in the character of their operation. They have to be able to respond appropriately to rapid and substantial variations in loads and they must have the capacity for shortening the start-up time, especially after a few or several hours of downtime. The paper presents an analysis of changes in thermal states in the turbine main elements after short downtime periods and shows the possibilities and consequences of the turbine fast start-up from such states. Optimal warm and hot start-up characteristics are developed. Detailed deliberations are devoted to elements of 200 MW turbines, most of which are and will be used as flexible sources of electricity generation in the power system.

1 Introduction

The changing structure of power systems and the new role of coal-fired power units which more and more often now act as sources balancing the demand for electricity call for a change in the character of their operation [1]. This is due to the increasing share of renewable energy sources, wind power in particular, in the current energy mix. The present share of different kinds of power plants in the total production of electricity in Poland is presented in Fig. 1.

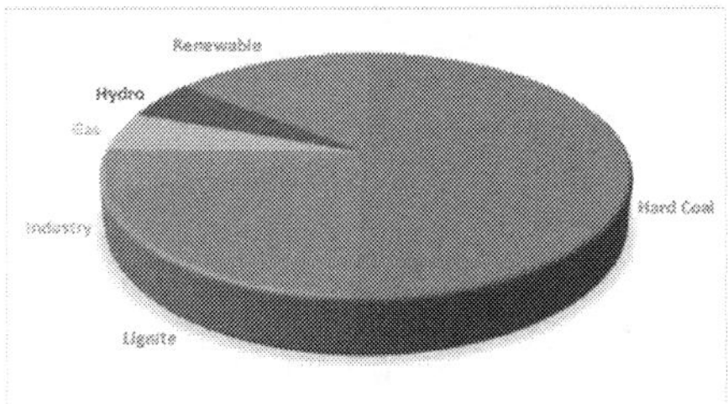

Fig. 1. Percentage share of individual groups of power generation in electricity production in Poland – July 2019

As it can be seen, the lion©share in energy generation falls to coal-fired power plants, which at the same time have to increase electricity production if no wind energy is available. This means that coal-fired power units have to be able to respond appropriately to rapid and substantial variations in loads and they must have the capacity for shortening the start-up time, especially after a few or several hours of downtime. Considering that these requirements increase the thermal load of individual elements of power units and taking account of the relatively long service life, new and optimal operating conditions have to be determined to ensure adequate safety of the power plant future operation in the new regime [2]. The new conditions should make it possible to improve the operation flexibility and, in particular, enable faster start-ups. On the other hand, they should ensure operating safety, i.e. thermal stresses in the main elements of the turbine and of the boiler must be kept within allowable values during the start-up entire procedure. The issues related to optimization of the power unit operation have been the subject of many studies and scientific publications.

For example, the results of studies on optimization of the heating of the power boiler thick-walled elements are included in [3, 4]. Optimal characteristics of steam temperature and pressure that enable fast start-ups of the boiler are presented in [5].

The results of studies on optimization of the heating of the turbine elements are included in [6÷11]. Optimal on-line control schemes of the steam inlet temperature taking account of low-cycle fatigue are presented in [6]. In [7], a genetic particle swarm optimization algorithm is used to find the best rates of raising steam temperature. Paper [8] focuses on integrating a steam accumulator into the power plant process to increase the power plant flexibility. Paper [9] adopts input and output data of a finite-element analysis under cold start-up conditions to establish an on-line damage-monitoring model of a steam turbine rotor. Optimization of the start-up of a high-pressure rotor for a large-power steam turbine is discussed in [10]. Different cold start-up optimization schedules for a 320 MW subcritical condensing steam turbine are made in [11].
The issues of turbine operation optimization described in literature are focused primarily on cold start-up optimization. The further part of this paper presents a discussion of problems related to the optimization of warm and hot start-ups, as start-ups from these thermal states will become dominant in the current operating conditions of power plants. An analysis is conducted of changes in thermal states in the turbine main elements

https://doi.org/10.1051/e3sconf/201913701025

after short downtime periods, and the possibilities and consequences of the turbine fast start-up from such states are demonstrated. The analyses make use of computer simulations of thermal and strength states in turbine elements performed by means of the finite element method. Numerical models of the turbine elements, as well as details of thermal and flow analyses and the selection of boundary conditions, are described e.g. in [12, 13].

2 Optimization of the turbine warm start-up

The turbine warm start-up optimization is more complex compared to start-ups from the cold state because such a process includes start-ups carried out after several or a few dozen hours of downtime. The thermal state of elements after such cooling times can be different. For this reason, the optimization procedure in the first step must include an analysis of the cooling of the turbine elements. Considering that even in a single power unit turbines may differ in a number of characteristics, such as design solutions, organization of the steam flow to the sealing areas or the quality of insulation, the cooling process can proceed in a different way. The analysis should provide curves illustrating temperature changes of the turbine individual elements as they cool. Determination of the temperature change is particularly important in inlet regions of the elements, rotors included, because when the turbine is re-started, these places will come into contact with steam entering the flow system. The temperature difference between steam and the element metal (ΔT_{s-m}) will have an impact on thermal stresses generated in this area. Therefore, temperature difference ΔT_{s-m} will be the first quantity to optimize. Since the manufacturer's start-up characteristics are broken lines it is assumed to maintain their shape and to change only the slopes. Another reason is that linear change in parameters (mainly temperature) which means a constant temperature change ratio results in minimum stress within the heated element.

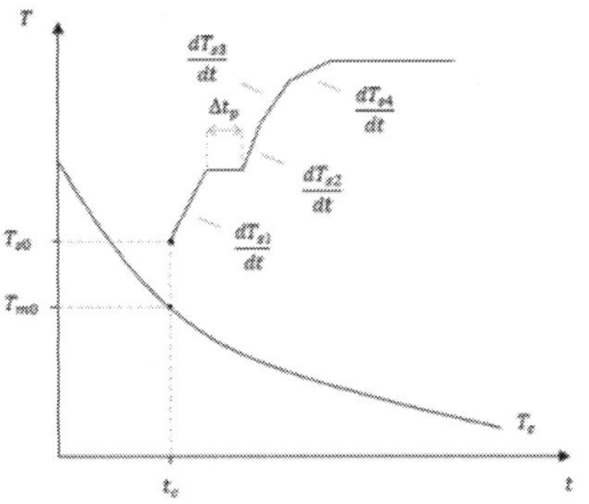

Fig. 2. Optimized quantities of the characteristic of the rotor heating from the warm state

So the other quantities optimized as the rotor is heated during start-ups from the warm state are derivatives (slopes) of the steam temperature change dT_{s1}/dt, dT_{s2}/dt and dT_{s3}/dt, as well as the time of the steam temperature stabilization Δt_p (cf. Fig. 2). In the next step, curves illustrating changes in pressure, power and the rotor rotational speed are selected for the optimized temperature curve.

2.1 Optimization of the turbine heating from the warm state after 36 hours of downtime

The rotor heating process in a 200 MW turbine after 36 hours of downtime is discussed as an example of the selection of the optimal parameters of running start-ups from the warm state. The first stage consisted in computer simulations of the HP rotor cooling process. Example results of the analysis are presented in Fig. 3. In its top part, Fig. 3 shows the temperature distribution on the surface of the rotor central bore as cooling proceeds. The bottom part illustrates the distribution of temperature in the rotor after 36 hours of cooling.

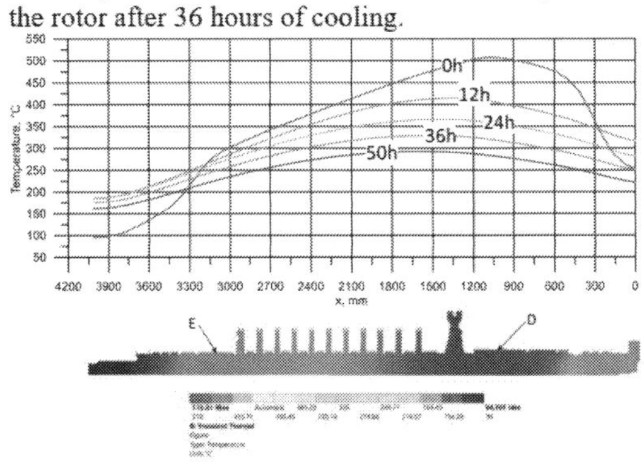

Fig. 3. Temperature field after 36 hours of the rotor cooling (bottom) and the temperature distribution on the central bore surface during the cooling process (top)

It can be seen that the heat is transferred from the hottest inlet part outbound in order to equalize the rotor temperature. However, the process is not very fast and one can observe that even after 50 hours of cooling there is still about 150K temperature difference within the element. What is more, the back part of the rotor is heated up within the first 12 hours by about 80K. It can result with enhanced stress at this area if too cold steam is introduced.

In the second stage, a heating curve was selected for this thermal state of the rotor that enables the rotor fastest heating without exceeding allowable values of maximum stresses. The allowable stress values, included in the range of 250-300 MPa depending on temperature, was established based on the results of small punch testing (SPT) of micro specimens [14] of the rotor material . The curve illustrating changes in live steam temperatures optimized for such conditions and the curves representing changes in pressure and in the rotor rotational speed are shown in Fig. 4. Time-dependent history of effective stresses in the sealing area, where the stresses reach maximum values, is shown in Fig. 5.

https://doi.org/10.1051/e3sconf/201913701025

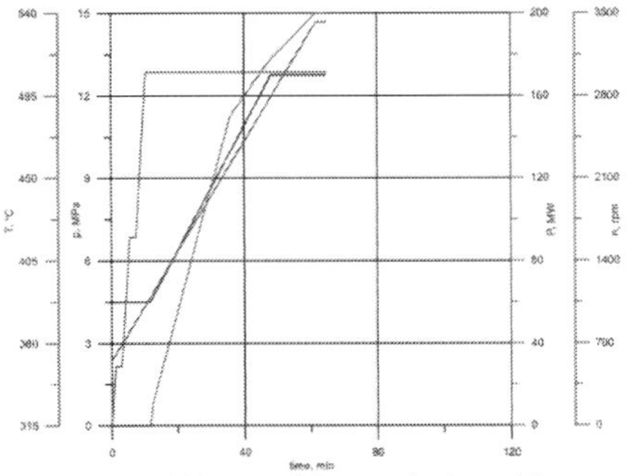

Fig. 4. Optimized history of changes in the turbine start-up parameters after 36 hours of cooling

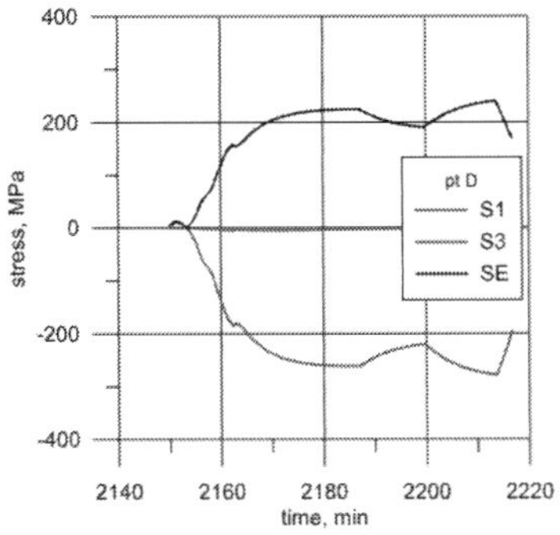

Fig. 5. Histories of stress components and equivalent (von Mises) stresses in the rotor sealing area

The optimized start-up procedure results with maximum stress close to the limit value. One can see, that the element heating at the initial stage of start-up results in quite a rapid growth in stresses up to about 220 MPa (von Mises) and then they stay at more less the constant level. This is the effect of a radial temperature gradient within the rotor. The higher the steam-metal temperature difference the greater the resulting stress level. After the turbine reaches the nominal operation condition, the gradient changes direction to axial and the stress observed decrease (only the beginning of the process visible).

3 Optimization of the turbine hot start-up

The turbine hot start-up optimization process includes start-ups carried out after several or a few dozen hours of downtime. The thermal state of elements after such cooling times can be different. For this reason, the optimization procedure in the first step must include an analysis of the cooling of the turbine elements. The analysis should provide curves illustrating changes in the temperature of the turbine individual elements as they cool.

A detailed analysis of 200 MW turbine start-ups after a few hours of downtime indicates that the most common method is composed of two stages. Steam first flows through and heats the turbine IP part, and then it passes through the HP part. This fact is taken into consideration in the developed optimal start-up. The start-up model is presented in Fig. 6. Like in the case of warm start-ups, the process of cooling HP and IP rotors is modelled first. The change in the HP rotor maximum temperature is illustrated by curve T_{cHP}. Curve T_{cIP} relates to the IP rotor. After time t_{cIP}, the start-up through the IP part begins, which is run according to the curve illustrating changes in the temperature of reheated steam and presented in Fig. 6.

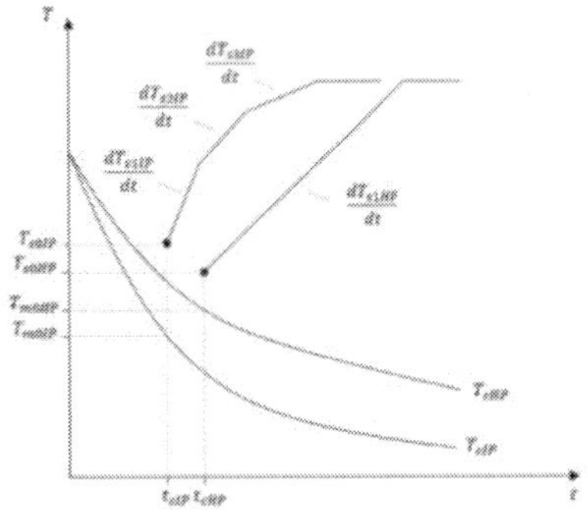

Fig. 6. Optimized quantities of the hot start-up characteristic

The steam initial temperature T_{sOIP} and the temperature gradients $\left(\frac{dT_{s1IP}}{dt}, \frac{T_{s2IP}}{dt}, \dots, \frac{dT_{snIP}}{dt}\right)$ are the optimized variables. After time t_{cHP}, steam begins to flow through the HP part. The steam initial temperature T_{sOHP} and the temperature rise gradient dT_{s1HP}/dt are the optimized quantities. After the optimal histories of changes in the temperature of steam flowing into the IP and the HP part are developed, the other curves are selected of changes in the steam pressure, the rotor revolutions and the turbine power.

3.1 The turbine optimal hot start-up after 8 hours of downtime

According to the optimization model discussed above, the turbine optimal start-up curves were developed for the thermal state after 8 hours of downtime. Using this procedure, the process of the cooling of the turbine elements was simulated. The time-dependent changes in the temperature of the HP and the IP rotor are presented in Fig. 7 and Fig. 8, respectively.

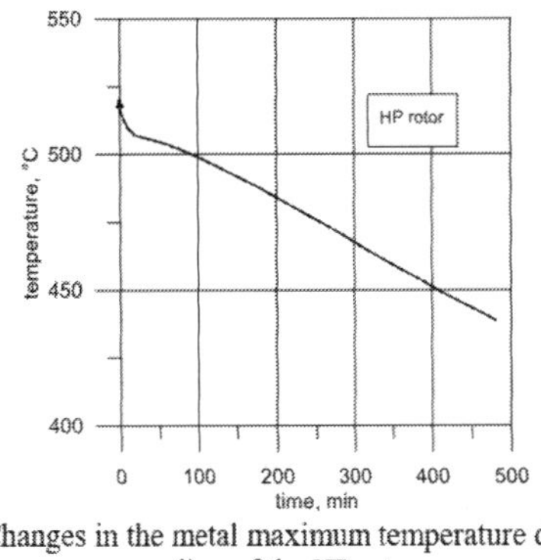

Fig. 7. Changes in the metal maximum temperature during the cooling of the HP rotor

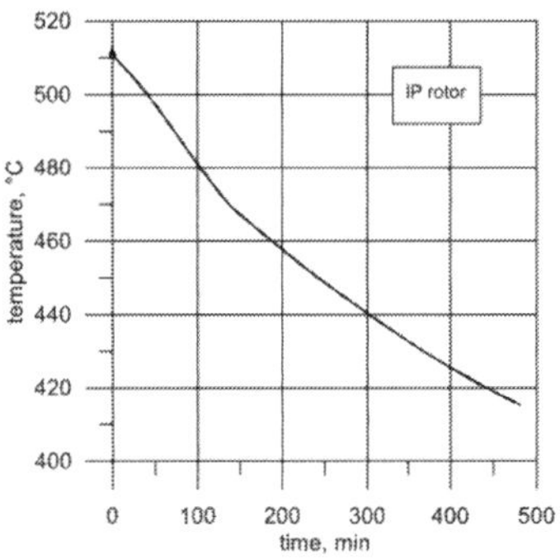

Fig. 8. Changes in the metal maximum temperature during the cooling of the IP rotor

The temperature distributions in the two rotors at the end of the 8-hour cooling process are shown in Figs. 9 & 10. In the next step of the optimization procedure, the optimal curves illustrating the heating of the IP and the HP rotor were developed for these initial thermal states of the turbine main elements (curve Trs and curve Tls in Fig. 11).

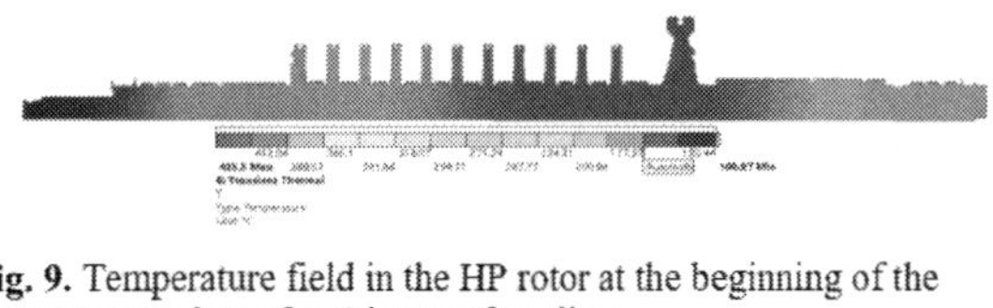

Fig. 9. Temperature field in the HP rotor at the beginning of the start-up procedure after 8 hours of cooling

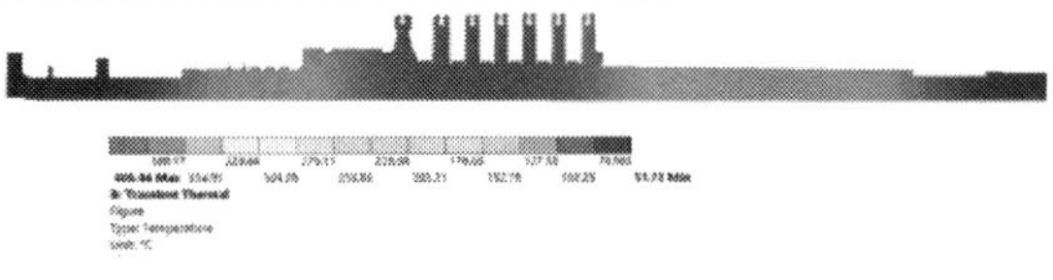

Fig. 10. Temperature field in the IP rotor at the beginning of the start-up procedure after 8 hours of cooling

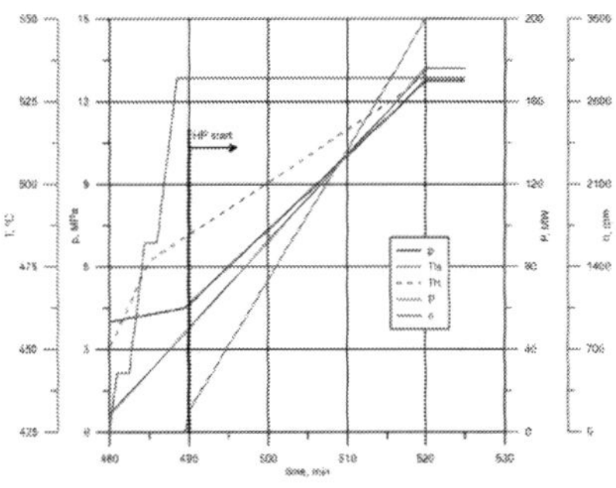

Fig. 11. Optimal changes in parameters during start-up

The HP start-up begins 10 minutes after the beginning of the start-up of the IP part and proceeds with a constant gradient. The reheated steam temperature changes along the 4-segment broken line, where the slopes differ by factor 5. The entire optimized start-up characteristic is shown in Fig. 11. The stress states in the most loaded area of the IP rotor are presented in Fig. 12. The maximum values do not exceed 250 MPa. The histories of the values of stresses in the HP rotor are presented in Fig. 13 and Fig. 14. Fig. 13 shows the changes in stresses in the front sealing area whereas Fig. 14 – in the rear sealing area.

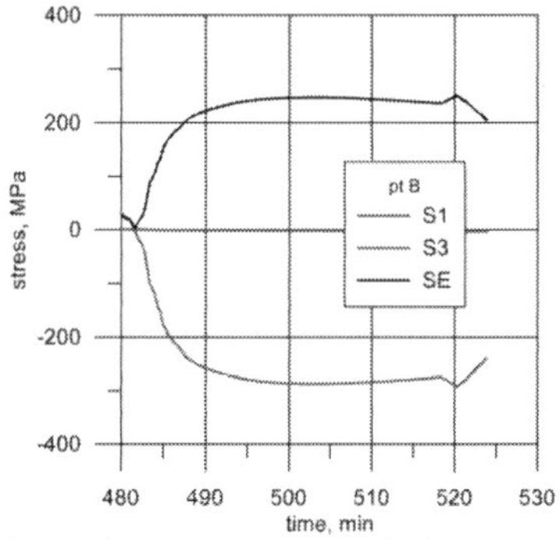

Fig. 12. Changes in maximum stresses in the IP rotor

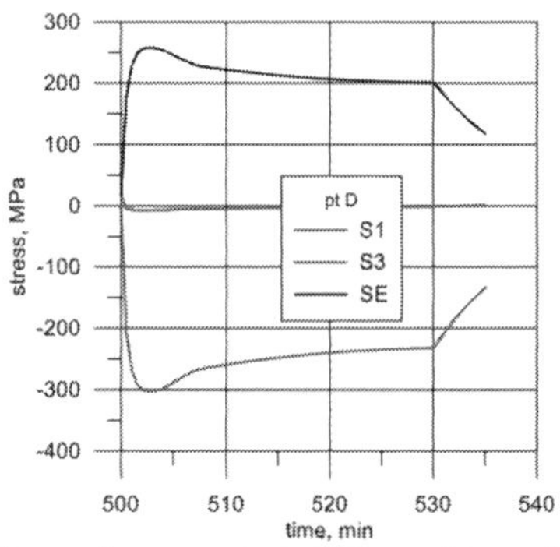

Fig. 13. Changes in stresses in the HP rotor front sealing area

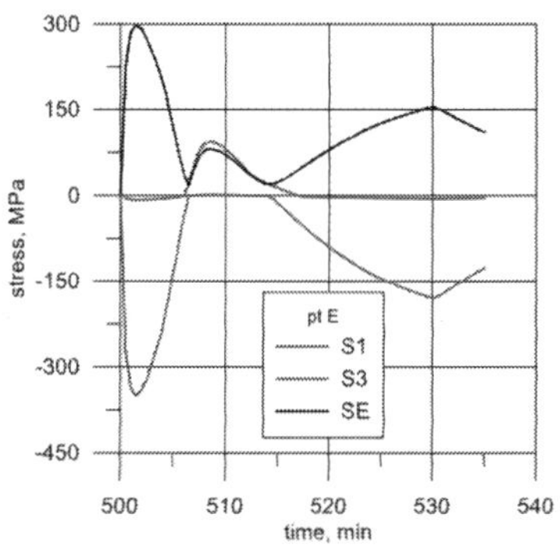

Fig. 14. Changes in stresses in the HP rotor rear sealing area

Similarly as in the case of warm start-up the most crucial stress magnitudes (objective variables) for both rotors growth rapidly to the assumed limit, and stay at more-less the same level till the nominal operating conditions are reached. Then stress tends to decrease to the steady-state magnitude. A bigger stress rise can be observed for a short time at the rear sealing, where hot steam (very small pressure drop in the turbine flow system at the beginning of the start-up) comes into contact with relatively cold rotor. Soon after, the turbine starts to produce power and steam expansion along the flow system becomes vital. The temperature at the rear sealing goes down, so as the stress level.

4 Conclusions

The new conditions of the power system operation where priority is given to renewable energy sources impose the need to improve flexibility of the operation of the other sources of energy generation, including coal-fired power units. It is therefore necessary to develop new methods of operation of both boilers and turbines in such units. The model of the turbine start-up optimization presented above indicates that the process of heating the turbine main elements can be accelerated safely, which means that it is possible to shorten the start-up time. The presented results concerning warm and hot start-ups suggest that it is possible to shorten the start-up time from 35% to as much as 45% compared to start-ups run according to the manufacturer© characteristics. Both the manufacturer© start-up characteristics and the optimized characteristics discussed above should be treated as a reference standard that sets out the optimal method of the start-up process realization (i.e. the method of changing steam temperature and pressure and the rotor rotational speed) for a specific thermal state of the turbine elements and for the expected time of the start-up duration. As shown above, start-up characteristics which are correctly developed and suited to the turbine specific thermal state make it possible to shorten the time of the turbine start-up substantially, keeping the stress level below allowable values determined for each individual element.

References

1. Rusin, A., Wojaczek, A., 2015, Trends of changes in the power generation system structure and their impact on the system reliability, Energy, 92, pp. 128-134.
2. Łukowicz, H., Rusin, A., 2018, The impact of the control method of cyclic operation on the power unit efficiency and life, Energy, 150, pp. 565-574.
3. Taler, J., Węglowski, B., Taler, D., Sobota, T., Dzierwa, P., Trojan, M., Madejski, P., Pilarczyk, M., 2015, Determination of start-up curves for a boiler with natural circulation based on the analysis of stress distribution on critical pressure components, Energy, 92, pp. 153-159.
4. Taler, J., Węglowski, B., Zima, W., Grądziel, S., Zborowski, M., 1999, Analysis of Thermal Stresses in a Boiler Drum During Start-up, Transactions of the ASME, Journal of Pressure Vessel Technology, 121, pp. 84-93.
5. Taler, J., Dzierwa, P., Taler, D., Harchut, P., 2015, Optimization of the boiler start-up taking into account thermal stresses, Energy, 92 (1), pp. 160-170.
6. Zhang, H., Xie, D., Yu, Y., Yu, L., 2016, Online optimal control schemes of inlet steam temperature during start-up of steam turbines considering low-cycle fatigue, Energy, 117, pp. 105-115.
7. Ji, D., Sun, J., Sun, Q., Guo, H-C., Ren, J., 2018, Optimization of start-up scheduling and life assessment for a steam turbine, Energy, 160, pp. 19-32.
8. Richter, M., Oeljeklaus, G., Gorner, K., Improving the load flexibility of coal-fired power plants by the integration of a thermal energy storage, Applied Energy 236, 2019, pp. 607-621.
9. Sun, Y-J., Liu, X-Q., Hu, L-S., Tang, X-Y.: Online life estimation for steam turbine rotor, Journal of Loss Prevention in the Process Industries 26, 2013, pp. 272-279.
10. Du, Q-W., Zheng, Z-L., Xie, Y-H., 2016, Optimization on start-up process of high-pressure rotor for large power steam turbine, Thermal Science, 20, pp. 815-822.
11. Ji ,D-M., Sun, J-Q., Dui, Y., Ren, J-X., 2017, The optimization of the start-up scheduling for a 320 MW steam turbine, Energy, 125, pp. 345-355.
12. Rusin, A., Łukowicz, H., Kosman, W., 2015, Transient Temperature and Thermal Stresses in Turbine Components, a chapter in: R. Hetnarski (ed.), Encyclopaedia of Thermal Stresses, Springer Science, Netherlands.
13. Chmielniak, T., Nowak, G., Łukowicz, H., Rusin, A., 2017, Analysis of Thermal and Stress States in Transient Operation of a Turbine Co-operating with Twinboiler, Heat Transfer Engineering, 39, (13-14), pp. 1251-1259.
14. Hähner, P., Soyarslan, C., Çakan, B., Bargmann, S., 2019, Determining Tensile Yield Stresses from Small Punch Tests: A numerical-based scheme, Materials & Design, Volume 182, pp. 1-11

RESEARCH & DEVELOPMENT IN POWER ENGINEERING, 2019

Analysis and management of operating risk created by turbine operation under flexible regimes

Andrzej Rusin[1,*], *Adam* Wojaczek[2], and *Martyna* Tomala[3]

[1,2,3] Silesian University of Technology - Institute of Power Engineering and Turbomachinery, Konarskiego 18, 44-100 Gliwice, Andrzej.Rusin@polsl.pl, Adam.Wojaczek@polsl.pl, Martyna.Tomala@polsl.pl,

Abstract. The new conditions in which coal-fired power plants, especially 200 MW units, have to operate require a considerable increase in the dynamics of their operation. The power unit start-up frequency increases and so does the frequency of changes in loads. This intensifies some wear processes, such as low-cycle fatigue and crack propagation in particular. Therefore, further operation of power units which have already been in service for a long time has to be supplemented with results of analyses and tests taking account of the intensification of wear processes. The paper presents a proposal for an extension of standard diagnostic testing of turbines by adding small punch tests (SPT) of the rotor material micro specimens. The SPT method enables a fast quasi non-destructive assessment of changes in mechanical properties, especially rotor steel embrittlement due to the turbine previous operation. The other element of the proposed testing is the analysis of the propagation rate of potential cracks in the rotor and assessment of the rotor failure probability for different scenarios of the power unit further operation.

1 Introduction

The new conditions in which coal-fired power plants, especially 200 MW units, have to operate require a considerable increase in the dynamics of their operation. Considering the priority of energy generation from renewable sources, more and more often now conventional power plants have to play the role of sources balancing the demand for electricity in the power system. This in turn necessitates a change in the operation of individual power units from constant-load operation to cyclical operation characterized by frequent shutdowns and start-ups, as well as variable loads [1, 2]. Another operating issue is that a large proportion of the power units have already been in service for longer than their design service life and, despite many upgrades, their further operation requires special supervision. The age structure of Polish power plants presented in Fig. 1 indicates that even though state-of-the-art low-emission supercritical power units have been gradually replacing the 200 MW and 400 MW power units with a long design service life, the "old" power plants will remain dominant in the power system and they will be operated for another several years.

On the other hand, ensuring appropriate reliability of energy supplies requires high availability of energy sources, including coal-fired power units. And this requires actions preventing a serious failure. This is where diagnostic testing has a crucial role to play. Its basic aim is to detect symptoms pointing to the degree of wear of the element under consideration. The wide variety of wear processes involving degradation of the elements of power machinery and equipment results in a similar variety of wear symptoms. A common symptom of the material increasing degradation is the appearance of micro and then macro cracks. In the case of turbine rotors, the cracks further propagation and reaching critical dimensions will potentially create the most dangerous scenario of a failure. Diagnostic testing must therefore enable identification of the location and size of potential cracks on the one hand, and assessment of ageing-related changes in mechanical properties of turbine steel on the other.

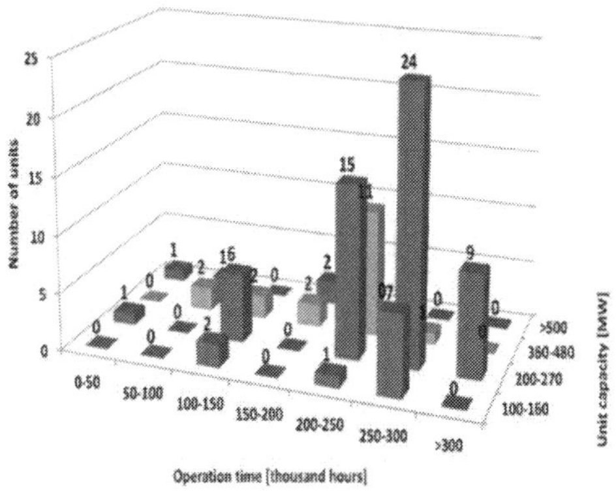

Fig. 1. Age structure of coal-fired power units in the power system in Poland

The paper presents an analysis of the risk created by propagation of potential cracks in turbine rotors due to variable operating conditions. Small punch testing (SPT) of the rotor material micro specimens is proposed as the method enabling a fast quasi non-destructive assessment of changes in mechanical properties, including first and foremost rotor steel embrittlement due to the turbine previous operation. Methods are discussed of estimation of the propagation rate of potential cracks in the rotor and calculation of the rotor failure probability for different scenarios of the power unit further operation.

2 Assessment of the risk related to long-term operation of turbine rotors

2.1 Crack propagation in the rotor

As already mentioned, one of the most dangerous scenarios of a turbine failure is an uncontrolled increase in defects being the consequence of technological processes or cracks arising due to material ageing. For this reason, the analysis of the development of potential discontinuities in the rotor has become an important element of life assessment and prediction of the time of further safe operation. How the cracks behave in the material depends on a number of factors, including for example the crack location and the material stress state [3]. Crack propagation under variable loads will be related to fatigue processes, whereas under constant loads – it is the creep process that matters. For a cracked element subjected to fatigue and creep processes, the crack propagation rate can be correlated to the stress intensity coefficient K_I defined as:

$$K_I = M\sigma\sqrt{a} \tag{1}$$

where:
M – non-dimensional function of geometrical parameters of the crack and the element,
σ – tensile stress acting on the crack,
a – crack dimension.

Crack propagation under variable loads is described by the following Paris-Erdogon equation [4]:

$$\frac{da}{dN} = C(\Delta K)^m \tag{2}$$

where:
C and m – material constants
$\Delta K = K_{Imax} - K_{Imin}$
N – number of fatigue cycles.

Crack propagation in a material under creep conditions can be described using the following general relation [4]:

$$\frac{da}{dt} = AK^n \tag{3}$$

where:
A and n – material constants,
t – steady-state operation time.

In real operating conditions turbine elements are subjected to both variable loads during transient operation and constant loads during operation with a constant power output. For this reason, both the above-mentioned relations should be used in the assessment of potential crack propagation.

The crack grows steadily until stress intensity coefficient K reaches the limit value of fracture toughness K_{IC}, which means that the crack has reached the critical dimension at a given level of tensile stresses acting on the crack surfaces. Therefore, an important element in the assessment of potential threats related to the occurrence of cracks in rotors is the assessment of the stress level in the area where defects and cracks may develop. Another necessary element of the safety analysis is the assessment of changes in rotor steel properties due to previous operation, including in particular the assessment of the change in fracture toughness.

2.2 Using small punch testing (SPT) of the rotor material micro specimens for the assessment of changes in mechanical properties

The material actual mechanical properties, including increased embrittlement of a rotor operated for a long time, can be identified and assessed by performing small punch tests of micro specimens of rotor steel (the SPT method).

The SPT method is the material punch test reduced to a micro scale. It consists in punching a small specimen using a hemispherical indenter. The SPT method enables determination of the material basic mechanical properties, otherwise specified typically by means of static tensile tests and impact strength testing, such as the yield point, tensile strength, brittle-to-ductile transition temperature and the material fracture toughness [5-11]. The SPT specimen is sampled from the rotor using a special device shown in Fig. 2. A specimen sampled from a 200 MW turbine rotor and the micro specimens in the form of discs with the diameter of 8 mm and thickness of 0.5 mm are shown in Fig. 3.

Fig. 2. Specimen sampling from a 200 MW turbine rotor disc

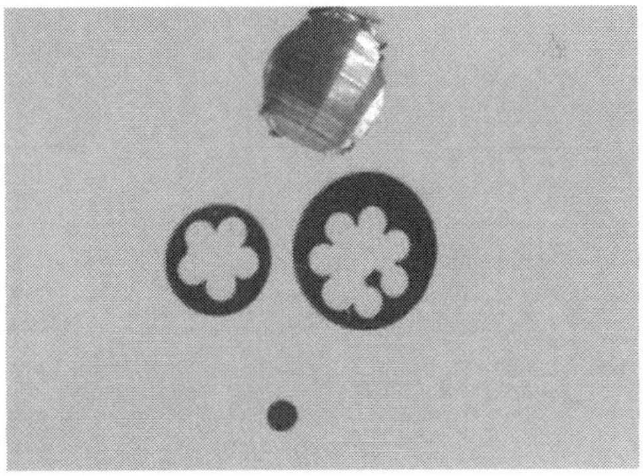

Fig. 3. Rotor material specimen and the micro specimens made of it

The specimens sampled from the rotor inlet and outlet part are then tested by means of the SPT method in ambient temperature and in reduced temperatures. The results of the testing of the properties of the rotor outlet part, which is not loaded thermally, are considered as close to rotor steel initial properties. Example curves illustrating SPT results obtained in ambient temperature for micro specimens sampled from the HP rotor inlet and outlet are shown in Fig. 4.

https://doi.org/10.1051/e3sconf/201913701026

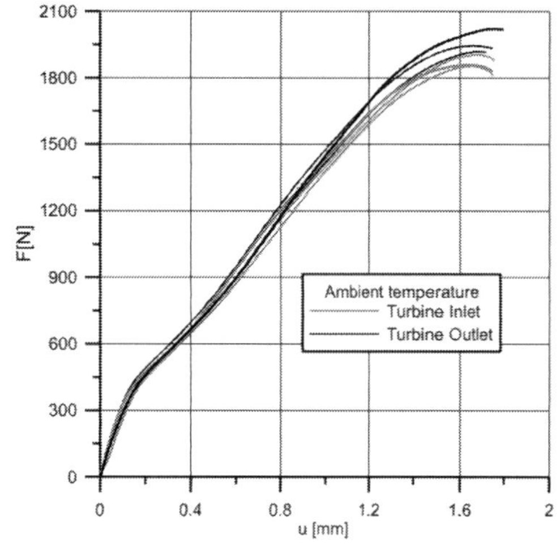

Fig. 4. Testing in ambient temperature – inlet and outlet part specimens

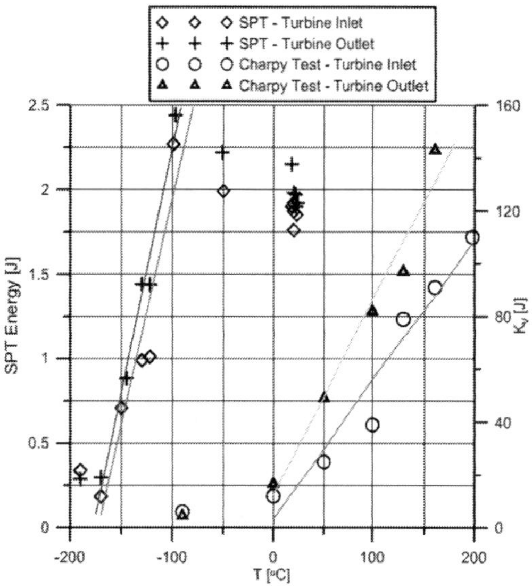

Fig. 5. Comparison of the work of breaking and punching of specimens sampled from the rotor cold and hot end

The testing of micro specimens from the rotor inlet and outlet part makes it also possible to calculate the work of the specimen punching in different temperatures. Using correlations that relate the micro specimen punching work determined in this way to the work of the specimen breaking in a typical impact strength test, it is possible to evaluate temperatures of the FATT brittle-to-ductile transition. By conducting such tests of specimens sampled from the rotor inlet and outlet part, it is possible to assess the decrease in rotor steel fracture toughness due to previous operating conditions. First and foremost, such information decides whether the rotor qualifies for further operation. It also enables prediction of how much longer the element can be operated safely.

3 Analysis of the risk of the turbine rotor failure

3.1 Stress states in the rotor

According to the remarks made above, the development of potential defects and cracks in the rotor depends, among others, on the stress state in the crack vicinity. The crack propagation rate will depend, among others, on the value of the stress tensile components. For this reason, the stress-state analysis in all phases of operation becomes an indispensable element of the assessment of the safety of the rotor operation. The maximum stresses arising in transient states of operation, i.e. during start-ups and shutdowns, will determine the amplitude of stresses occurring in a working cycle. And this very amplitude has an impact on the size of the crack increment during the cycle. An example distribution of the circumferential component of stresses in the HP rotor during the heating process is shown in Fig. 6.

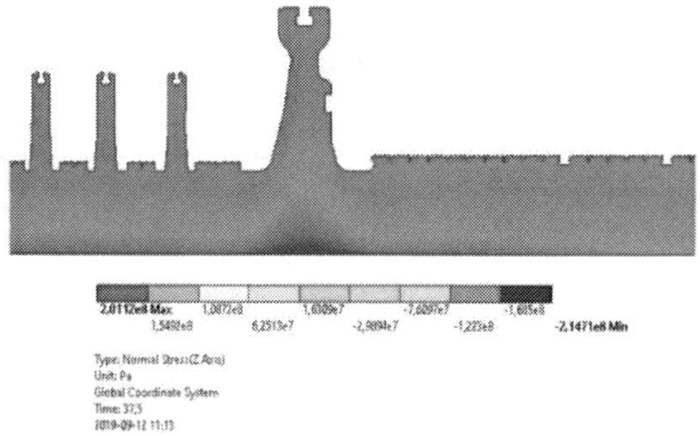

Fig. 6. Distribution of circumferential component of stress in the heating rotor

Fig. 7 presents the distribution of equivalent (von Mises) stresses in the rotor during steady-state operation, when creep-related stress relaxation occurs. So-calculated stress values are used in further analyses of crack propagation in the rotor.

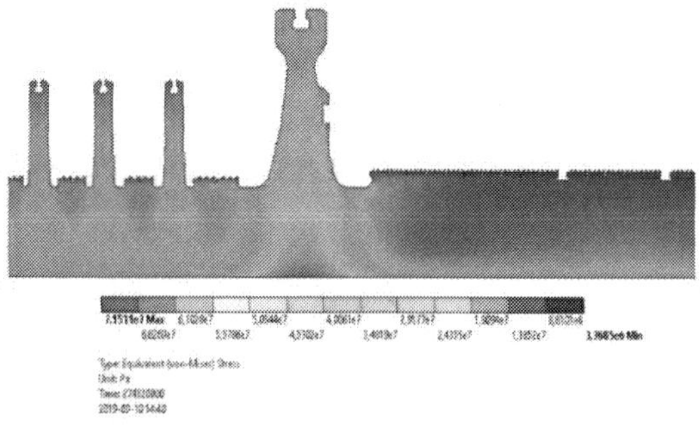

Fig. 7. Equivalent (von Mises) stress distribution in the rotor during creep relaxation process

3.2 Probability of the rotor failure

Among others, the risk analysis procedure includes:
- definition of potential scenarios causing the element failure,
- estimation of the probability of the scenarios,
- estimation of the consequences of the element failure,

 https://doi.org/10.1051/e3sconf/201913701026

- assessment of the risk understood as the product of the probability of the element failure and the failure consequences.

Many different scenarios which can cause a failure are taken into account in the analysis of risk related to the power unit operation. In the case of turbines, considering the new, flexible conditions of their operation, the scenarios must take account of processes related to fatigue wear and creep wear, as well as to intensification of other factors, such as erosion and corrosion. When it comes to turbine rotors, the highest risk is related to the scenario of the rotor failure due to crack propagation and the crack reaching critical dimensions.
For this particular scenario, the rotor failure criterion can be written as:

$$g = a_{cr} - a \tag{4}$$

where $\boldsymbol{a_{cr}}$ is assumed as the crack critical dimension, which results in the rotor immediate destruction. It is calculated from the following relation:

$$a_{cr} = (K_{IC}/M\sigma)^2 \tag{5}$$

Due to the random character of the quantities in the above formula, i.e. fracture toughness K_{IC} and the maximum tensile stresses acting on the crack surface σ, the crack critical dimension should also be considered as random. The nature of the crack current dimension a in formula (4) calculated using relations (1) and (2) is also random, and the quantity depends e.g. on the number and type of working cycles, i.e. on the number of start-ups from different thermal states and the level of stresses occurring in subsequent working cycles.

Using these assumptions, the formula defining the probability of the rotor failure can be written as:

$$P_f = P(g \leq 0) \tag{6}$$

Determination of the probability requires calculations of all the quantities used in formula (4) and formula (5). Example results of calculations of the propagation of a crack located in the HP rotor central hole are shown in Fig. 8 and Fig. 9. The calculations are performed assuming the rotor service life of 20 years, with an average of 6 thousand hours of operation per year. Each year the turbine is started 200 times on average, from three different thermal states. Each of those three types of start-ups involves the occurrence of stresses in the rotor central hole with average stress values of 200 MPa, 250 MPa and 300 MPa, respectively. The results shown in the figure below relate to the service life of 20 years assuming operating scenario #1, according to which each year there are 50 start-ups producing stresses of 200 MPa, 100 start-ups causing stresses of 250 MPa and 50 start-ups resulting in 300 MPa stresses. Fig. 9 illustrates the crack propagation calculation results obtained assuming operating scenario #2, according to which there are 200 start-ups a year producing average stresses of 200 MPa. Depending on the size of the initial defect or crack, the crack propagation rate and the crack dimensions can take different values. Therefore, in each specific case the crack initial value should be taken as the value obtained from measurements or the value corresponding to the sensitivity of the instruments used in the detection of defects and cracks.
In the next stage of the analysis, the rotor failure probability is calculated according to relation (6). The mean values and standard deviations of input data treated as random quantities and used in the calculations are listed in Table 1.

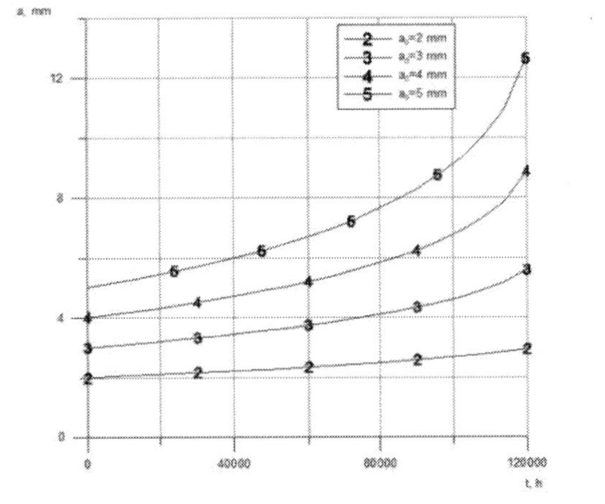

Fig. 8. Crack propagation in the rotor central hole for the crack different initial dimensions – operating scenario #1

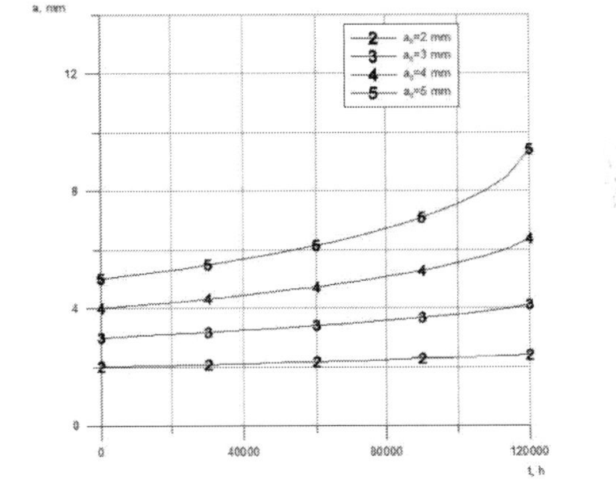

Fig. 9. Crack propagation in the rotor central hole for the crack different initial dimensions – operating scenario #2

Table 1. Mean values and standard deviation of input data

Input data	Mean value	Standard deviation
$\Delta\sigma_1$	200 MPa	10 MPa
$\Delta\sigma_2$	250 MPa	12,5 MPa
$\Delta\sigma_3$	300 MPa	15 MPa
C	2e-12	1e-13
m	3,4537	0,173
σ_{creep}	65 MPa	3,25 MPa
A	3e-14	1,5e-15
n	5,6572	0,283
M	1,98	0,086
a_0	2-5 mm	0,5 mm
K_{Ic0}	100 MPa$\sqrt{m}$	5 MPa$\sqrt{m}$

The change in the mean value and standard deviation of the crack dimensions during operation is calculated using the Monte Carlo method [3]. The probability expressed by means of relation (6) is calculated using the FOSM methods [3]. The results of the rotor failure probability calculations for scenario #1 and scenario #2 are presented in Fig. 10 and 11.

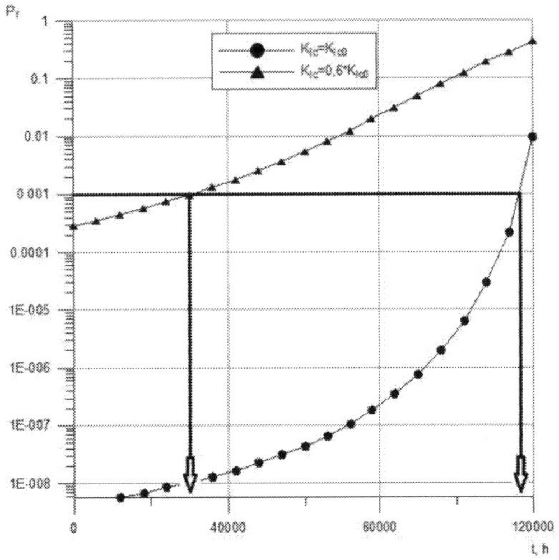

Fig. 10. Probability of the turbine rotor failure for a_0=4mm (scenario #1)

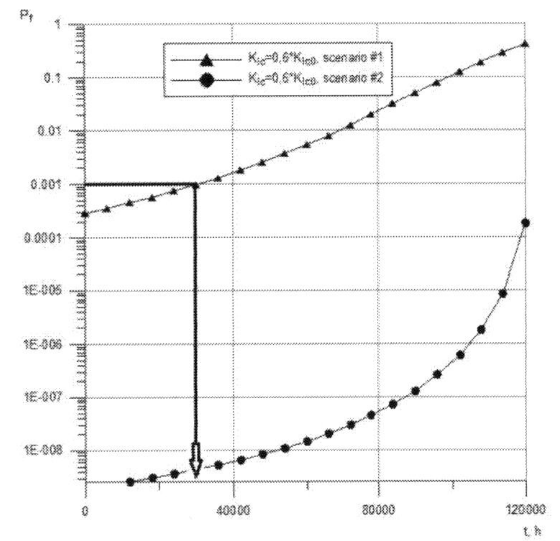

Fig. 11. Probability of the turbine rotor failure for a_0=4mm (scenario #1 and scenario #2) for reduced crack resistance $K_{Ic} = 0{,}6 \cdot K_{Ic0}$

Assuming a value of 0.001 as the permissible probability of rotor failure, we can estimate based on the graphs shown in Figure 10 that for a material with unchanged fracture toughness, such a probability will be achieved after approx. 117 000 h of operation. For rotor steel with reduced fracture toughness, such damage probability will occur after approx. 30 000 h of operation. Analysis of the graphs shown in Figure 11 shows that when operating the turbine according to scenario 2, this limit probability will not be achieved at all.

4 Summary and conclusion

The presented method of analysing the risk related to further operation of turbines which have already been in service for a long time makes use of both materials testing and computer simulations. The result is an estimation of the turbine rotor failure probability for different operating scenarios.

The probability calculated in this way will be a quantity rising over time. As soon as it reaches values assumed as critical, actions have to be taken to reduce the probability and, thereby, lessen the risk related to further operation [12]. In order to protect the rotor from failure, preventive diagnostic testing should be carried out to confirm the crack real dimensions and enable a rational decision whether to continue the turbine operation or replace some of the turbine elements.

References

1. Rusin, A., Wojaczek, A., Trends of changes in the power generation system structure and their impact on the system reliability, *Energy*, Vol. 92, pp. 128-134, (2015).
2. Łukowicz, H., Rusin, A., The impact of the control method of cyclic operation on the power unit efficiency and life, *Energy*, Vol. 150, pp. 565-574, (2018).
3. Rusin A., Awaryjność, niezawodność i ryzyko techniczne w energetyce cieplnej, Wydawnictwo Politechniki Śląskiej,Gliwice, (2008).
4. British Electricity International, Turbines Generators and Associated Plant, Pergamon, (1991).
5. Altstadt E., Houska M., Simonovski I., Bruchhausen M., Holmström S., Lacelle R.: On the estimation of ultimate tensile stress from small punch testing, *International Journal of Mechanical Science*, Vol. 136, pp. 85÷93, (2018).
6. Altstadt E., Ge H.E., Kuksenko V., Serrano M., Houska M., Lasan M., Bruchhausen M., Lapetite J.M., Dai Y.: Critical evaluation of the small punch test as a screening procedure for mechanical properties, *Journal of Nuclear Materials*, Vol. 472, pp. 186÷195, (2016).
7. Ortiz-Mariscal A., Saucedo-Munoz M.L., Naveena, Komazaki S.: Application of small punch creep testing for evaluation of creep properties of as-received and artificially aged 5Cr-0,5Mo steel, *Materials Science and Engineering Vol. A 709*, pp. 322÷329, (2018).
8. Kumar K., Madhusoodanan K., Singh R.N.: Miniature test techniques for life management of operating equipment, *Nuclear Engineering and Design*, Vol. 323, pp. 345÷358, (2017).
9. Moreno M.F.: Application of small punch testing on the mechanical and microstructural characterizations of P91 steel at room temperature, *International Journal of Pressure Vessels and Piping*, Vol. 142-143, pp. 1÷9, (2016).
10. Rasche S., Kuna M., Improved small punch testing and parameter identification of ductile to brittle materials, *International Journal of Pressure Vessels and Piping*, Vol. 125, pp. 23÷34, (2015).
11. Rodríguez C., Fernández M., Cabezas J.G., Garciá T.E., Belzunce F.J.: The use of the small punch test to solve practical engineering problems, *Theoretical and Applied Fracture Mechanics*, Vol. 86, pp. 109÷116, (2016).
12. Rusin A., Bieniek M.: Maintenance planning of power plant elements based on avoided risk value, *Energy*, Vol. 134, pp. 672÷680, (2017).

RESEARCH & DEVELOPMENT IN POWER ENGINEERING, 2019

Influence of three surface condensers connection setup on power plant unit performance

Ewa Dobkiewicz-Wieczorek[1,*]

[1] Silesian University of Technology / Valmet Automation, edobkiewicz@wp.pl, Poland

Abstract. This paper presents a comparison of three surface condenser connection setups on the cooling water side. Serial, mixed and parallel connections were considered. The thermodynamic justification for the use of more complex configurations was verified. The analysis was conducted based on the calculated heat balances of verified power units for nominal and not nominal parameters for tested connections. The exhaust steam pressure was calculated using the technical data of the surface condenser and cooling water parameters. Three methods of calculating the heat transfer coefficient based on characteristic numbers, HEI method, and the ASME standard, were used. The most advantageous model was indicated and used in heat balance calculations. The assumptions and simplifications for the calculations are discussed. Examples of the calculation results are presented.

The direction of development and transformation in the Polish power industry are a response to the new European legal regulations, the aim of which is the environmental protection. As a result, diversification of electricity production was provoked, additionally coal-fired units were forced to apply new solutions. New units are getting greater, technically and technologically more advanced. Introduced improvement that increases unit efficiency by as much as 0.1 percent are important considering the actual requirements and unit efficiency of 45% net.

In the case of great unites, when few surface condensers are used, it is essential to verify, which connection setups on the cooling water side would give the highest unit efficiency.

When research, it is necessary to consider the entire steam-condensate cycle to analyze exhaust steam pressure changes effect on steam flow and thermodynamic parameters fluctuations. The correct calculation of the exhausted steam pressure value is one of key steps to ensure the calculation's correctness. This paper presents analysis of three exhausted steam pressure calculation algorithms, comparison of their complexity and the set of the data needed for these calculations. The model of the surface condenser was based on three methods of calculating the heat transfer coefficient: dimensionless equation with characteristic numbers, the HEI method and the ASME standard. The most advantageous model was indicated after verification with the data from the site.

Study of three surface condenser connection setups on the cooling water side aims to help to find the answer on the rationale behind using more complex configurations, to analyze their advantages and disadvantages, and to give advice on which system is the best from the thermodynamics perspective.

1 Exhaust steam pressure calculation

Condensation turbine exhaust steam pressure was calculated using the heat transfer equations.

Model assumed isobaric heat exchange, no condensate subcooling. Calculations were made for steady state. Thermodynamic calculations in accordance with IAPWS IF-97 [1].

1.1 Heat transfer equations:

Heat transfer equations are:

$$\dot{Q} = \dot{m}_s(i_s - i_c) \quad (1)$$

$$\dot{Q} = \dot{m}_g Cp_g(T_2 - T_1)\frac{1}{\eta_c} \quad (2)$$

Nomenclature: $\dot{Q}$ – heat transfer rate, $\dot{m}_s$– exhausted steam flow rate, i_s– exhausted steam enthalpy, i_c– condensate enthalpy, $\dot{m}_g$– cooling water flow rate, Cp_g – water specific heat, T_1 – inlet cooling water temperature, T_2 – outlet cooling water temperature, η_c– condenser efficiency

1.2 Condenser heat load: [2],[4],[5].

Condenser heat load equation is:

$$\dot{Q} = UA_sLMTD \quad (3)$$

$$LMTD = \frac{T_2-T_1}{ln\frac{T_c-T_1}{T_s-T_2}} \quad (4)$$

Nomenclature: $\dot{Q}$ – condenser heat load, U – heat transfer coefficient, A_s– surface tube area, $LMTD$ – logarithmic mean temperature difference, T_c– condensate temperature, T_s– steam temperature, T_1– inlet cooling water temperature, T_2– outlet cooling water temperature.

1.3 Heat transfer coefficient - Characteristic numbers [2].

Heat transfer coefficient, when characteristic number method used, equation is:

$$U = \frac{1}{R_m + R_t \frac{D_{out}}{D_i} + R_s + R_f} 10^{-3} \quad (5)$$

Tube-Wall resistance was computed as follows:

$$R_m = D_{out}\, ln \frac{D_{out}}{D_i} \frac{1}{2K_m} \quad (5.1)$$

Shellside resistance was computed based on dimensionless equation for heat transfer when the steam condenses on the outside horizontal pipe's surface.

$$R_s = (\frac{Nu\, K_s}{D_{out}})^{-1} \quad (5.2)$$

$$Nu = 0.725\, C_v^{\;0.25} \quad (5.3)$$

$$C_v = \frac{D_{out}^{\;3} \delta_c^{\;2} g\, di_{vap}^{\;2}}{K_t \mu_c dT_{ct}} \quad (5.4)$$

dT_{ct} - The difference in condensate and wall temperatures depends on the thickness of the condensate layer, therefore on the heat transfer coefficient. It is indicating that the most appropriate calculation method is the iterative method, but with satisfactory accuracy, the value can be calculated as $dT_{ct} = \frac{LMTD}{2}$ [3].

Physical properties of condensate: δ_c, K_t, μ_c are determined for surface and saturation average temperature $T_f = \frac{T_s + (T_s + dT_{ct})}{2}$

Tubeside resistance was computed based on dimensionless equation for forced convection for turbulent flow inside a circular pipe:

$$R_t = (\frac{Nu\, K_t}{D_i})^{-1} \quad (5.5)$$

$$Nu = 0.021\, Re^{0.8} Pr_g^{\;0.43} \left(\frac{Pr_g}{Pr_t}\right)^{0.25} \quad (5.6)$$

$$Re = \frac{V_g D_i \delta_g}{\mu_g} \quad (5.7)$$

$$V_g = \frac{W_g}{\delta_g} \frac{4}{N \pi D_i^{\;2}} \quad (5.8)$$

$$Pr = \frac{\mu\, Cp}{K} \quad (5.9)$$

Pr_g / Pr_t – calculated for water temperature T_g and wall temperature $T_t = T_s - dT_{ct}$

Nomenclature: U– heat transfer coefficient, D_i– tube inside diameter, D_{out}– tube outside diameter, K_m- tubewall resistance, K_t– tubeside thermal conductivity, K_s– shellside thermal conductivity, R_m– tubewall resistance, R_t– tubeside resistance, R_s– shellside resistance, R_f– fouling resistance, Nu– Nuselt number, Pr– Prandl number, Re– Reynolds number, T_g– average cooling water temperature, V_g– cooling water velocity, Cp_g– water specific heat, di_{vap}– enthalpy of exhausted steam vaporization, dT_{ct}– The difference in condensate and wall temperatures, μ_c – condensate viscosity, μ_g – cooling water viscosity, δ_g– cooling water density, δ_c– condensate density

1.4 Heat transfer coefficient - HEI standard [4]

In this case, the calculation of heat transfer coefficient is based on design guidelines of Heat Exchange Institute (HEI). The proposed function uses the data from experimental research. The heat transfer coefficient was computed as follows

$$U = U_1 F_w F_m F_c \quad (6)$$

Nomenclature: U_1 – uncorrected heat transfer coefficients, as a function of tube diameter and cooling water velocity, F_w– inlet water temperature correction factor, F_m– tube material and gauge correction factors, F_c – cleanliness factor.

U_1, F_w, F_m are read from HEI table. U_1values are based on clean, 1.245 mm tube wall gauge, Admiralty metal tubes with 21.1°C cooling water temperature. Uncorrected heat transfer coefficient is describing as a function of tube diameter and water velocity. F_w introduces a water temperature correction and F_m introduces a tube material and gauge correction.

1.5 Heat transfer coefficient - ASME PTC 12.2 codes: [5]

Heat transfer coefficient, when ASME codes used, equation is:

$$U = \frac{1}{R_m + R_t \frac{D_{out}}{D_i} + R_s + R_f} 10^{-3} \quad (7)$$

Tube-Wall resistance was computed as follows:

$$R_m = D_{out}\, ln \frac{D_{out}}{D_i} \frac{1}{2K_m} \quad (7.1)$$

Tubeside resistance was computed as follows:

$$R_t = (\frac{Nu\, K_t}{D_i})^{-1} \quad (7.2)$$

$$Nu = 0.0158\, Re^{0.835} Pr^{0.426} \quad (7.3)$$

$$Re = \frac{V_g D_i \delta_g}{\mu_g} \quad (7.4)$$

$$V_g = \frac{W_g}{\delta_g} \frac{4}{N \pi D_i^{\;2}} \quad (7.5)$$

$$Pr = \frac{\mu\, CP}{K} \quad (7.6)$$

Shellside resistance for the first iteration was computed as follows:

$$R_s = \frac{1}{U\, 10^3} - R_m - R_t \frac{D_{out}}{D_i} - R_f \quad (7.7)$$

Shellside resistance for the next iteration was computed as follows:

$$R_s = R_{s0} (\frac{\dot{Q}_0}{\dot{Q}})^{\frac{1}{3}} (\frac{\mu_0}{\mu})^{\frac{1}{3}} \frac{K_{s0}}{K_s} (\frac{\delta_0}{\delta})^{\frac{2}{3}} \quad (7.8)$$

Nomenclature determined as in point 1.3.

Index 0 means the value from the previous iteration. Physical properties of condensate:δ, K, μ are determined for condensate film $T_f = T_s - 0.2LMTD$.

Condenser technical data: Steel 1.4401 - $K_m = 15\ \frac{W}{m\,K}$ was assumed to be used, tube dimension Ø24x0.7mm, condenser efficiency $\eta_c = 0.99$, cleanliness factor k $F_c = 0.95$; Calculation was done for two pass surface condenser with surface tube area $A_s = 19177\ m^2$, quantity of tubes $N = 31920$.

1.6 Turbine's isentropic efficiency

In calculation turbine's isentropic efficiency was used

$$\eta_1 = \frac{i_{s\_LP} - i_s}{i_{s\_LP} - i_{s0}} \quad (8)$$

Nomenclature: i_{s_LP}– inlet LP turbine steam enthalpy, i_s– exhaust steam enthalpy, i_{s0}– exhaust steam enthalpy when isentropic flow, p_s– exhaust steam pressure,

 https://doi.org/10.1051/e3sconf/201913701027

η_1 was set as a constant because in external test, small impact of vary this value as a function of load, for main calculation, was verified.

1.7 Calculation procedure and example calculation results

Variables calculation was based on the iterative algorithm. Calculation procedure is the same for dimensionless equation with characteristic number (CN) and HEI methods but different for ASME standard.
Input data: A_s– surface tube area, N – quantity of tubes, F_c– cleanliness factor, D_i– tube inside diameter, D_{out}– tube outside diameter, K_m– tubewall thermal conductivity, $\dot{m}_g$– cooling water flow rate, p_g– cooling water pressure, T_1– inlet cooling water temperature, p_{s_LP}– inlet LP turbine steam pressure, T_{s_LP}– inlet LP turbine steam temperature, $\dot{m}_s$– exhausted steam flow rate, dT_c– condensate subcooling, η_1– turbine efficiency, η_c– condenser efficiency, condenser pass number

Calculation procedure

CN, HEI method
- Initialization parameters $\dot{Q}, U$

Calculation
- T_2 based on (2)
- $LMTD$ based on (3)
- T_s based on (4)
- $p_s = f(p\_sat(T_s))$, $i_{s\_LP} = f(p_{s\_LP}, T_{s\_LP})$
- i_s based on (8)
- $x_s = f(p_s, i_s)$
- $\dot{Q}$ based on (1)
- U based on (5) / (6)

Next iteration

ASME method
- Initialization parameters p_s

Calculation
- $T_s = f(T\_sat(p_s))$, $i_{s\_LP} = f(p_{s\_LP}, T_{s\_LP})$
- i_s based on (8)
- $x_s = f(p_s, i_s)$
- $\dot{Q}$ based on (1)
- T_2 based on (2)
- $LMTD$ based on (4)
- U based on (3)/ (7)
- $LMTD$ based on (3)
- T_s based on (4)
- $p_s = f(p\_sat(T_s))$

Next iteration

In table 1 input data were presented. In table 2 example calculation results compare with real exhausted steam pressure were shown.

1.8 Discussion of the results and exhaust steam pressure calculation method selection.

Verifying calculations have been made for data from real units: 65MW and 460MW. This paper presents results for

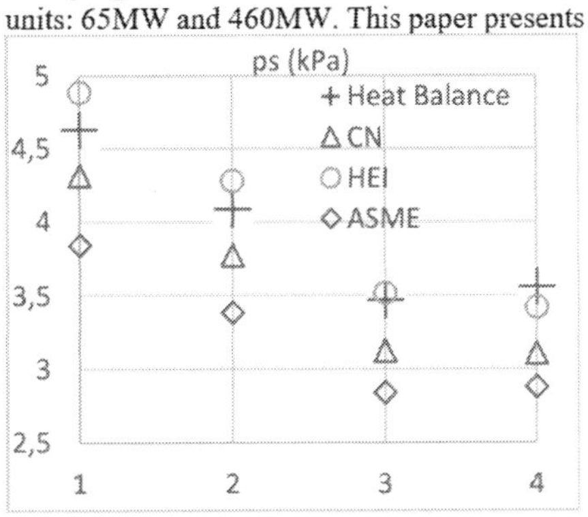

Fig. 1. Exhausted steam pressure

Table 1. Input data for calculations

series		1	2	3	4
load		100%	90%	75%	60%
$\dot{m}_g$	$\frac{t}{h}$	48060	48060	48060	48060
T_1	°C	18.3	16.8	15.2	16.7
$\dot{m}_s$	$\frac{kg}{h}$	753240	697910	587350	491770
T_{s_LP}	°C	278.7	273.6	280	272.6
p_{s_LP}	kPa	579	526	441	358
η_1		0.82	0.82	0.82	0.82

Nomenclature: $\dot{m}_g$– cooling water flow rate; T_1 – inlet cooling water temperature; $\dot{m}_s$ – exhausted steam flow rate; T_{s_LP}– inlet LP turbine steam temperature; p_{s_LP} – inlet LP turbine steam pressure, η_1 – LP turbine's isentropic efficiency

Table 2. Example calculation results

series		1	2	3	4
load		100%	90%	75%	60%
p_{s_REF}	kPa	4.63	4.09	3.47	3.56
CN					
p_s	kPa	4.31	3.77	3.12	3.10
$\dot{Q}$	kW	466641	432688	368136	310295
U	$\frac{kW}{m^2K}$	3.45	3.41	3.38	3.45
T_2	°C	26.6	24.5	21.7	22.2
HEI					
p_s	kPa	4.88	4.28	3.51	3.41
$\dot{Q}$	kW	467298	433316	368667	310664
U	$\frac{kW}{m^2K}$	2.59	2.52	2.45	2.52
T_2	°C	26.6	24.5	21.7	22.2
ASME					
p_s	kPa	3.84	3.38	2.83	2.87
$\dot{Q}$	kW	466023	432138	367714	310000
U	$\frac{kW}{m^2K}$	5.19	5.11	5.02	5.06
T_2	°C	26.6	24.5	21.7	22.2

Nomenclature: p_{s_REF}- reference exhausted steam, p_s- calculated exhausted steam, $\dot{Q}$ – condenser heat load, U- heat transfer coefficient, T_2 – outlet cooling water temperature

the bigger one. Figure 1 presents calculated exhausted steam pressure compared with reference value. The most similar results to the expected value gave HEI method although results of characteristic numbers method are also correct. The square root error of exhaust steam pressure was used to assess the series of results. For calculation based on HEI method it is 0.176kPa, for characteristic numbers method 0.361kPa, for ASME method 0.704kPa. For second reference unite the best results gave characteristic numbers and HEI method. Least accurate results give ASME method. Although this method largely based on similar equations as characteristic numbers method, significant results difference follows on from shellside resistance calculation. Reviewing the actual reference data, it is concluded that the results with the best accuracy were obtained using the HEI method. It is also the simplest method when considering the complexity of the calculations.

2 Comparison of three surface condenser connection setups on the cooling water side.

Four condenser connection configurations were tested: I- parallel (Fig.2), II-serial (Fig.3), III- parallel-to-serial (Fig.4) and IV- serial-to-parallel (Fig.5). For proposed thermal cycle (Fig.6) nominal load heat balance was calculated. Next heat balance for 70% and 40% of nominal load was computed. Considering steam flow change and thermodynamic parameters fluctuations, the influence of the tested connections on improving the unit efficiency was verified.

2.1 Calculation procedure [6]

The unit shown as a Figure 6 was described by energy and mass balances equations. The coefficients of the system of equations were appointed by the enthalpy value at the determined points. Enthalpy was calculated from the thermodynamics dependence in accordance with IAPWS IF-97. Exhausted steam pressure was calculating based on algorithm with HEI heat transfer coefficient. By iterating these three calculation steps, the pressure, temperature, enthalpy and mass flow were computed for the determined points at nominal load.

Calculations input data were: p_0– live steam pressure, t_0– live steam temperature, t_{20}– reheated steam temperature, N_{el}– electric power and value needed to exhausted steam pressure calculation presented in first part of this thesis.

Using Stodola-Flügel turbine passage equation calculation for 70% and 40% of nominal load were done. Calculations input data were: t_0 , t_{20}, N_{el} and value needed to exhausted steam pressure calculation.

To compare the results operation following indicators were calculated: gross unit heat rate and unit efficiency.

$$q = 3600\frac{\dot{Q}_d}{N_{el}}\left[\frac{kJ}{kWh}\right] \tag{9}$$

$$\eta = \frac{N_{el}}{\dot{Q}_d} \tag{10}$$

Where $\dot{Q}_d = m_0(i_0 - i_{100}) + m_{19}(i_{20} - i_{19})$

Comparing the proposed configurations, the following assumptions were made: equal steam distribution to the

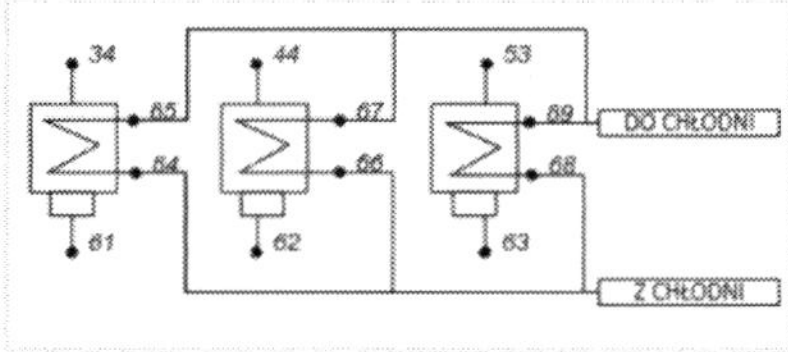

Fig. 2. Parallel configuration

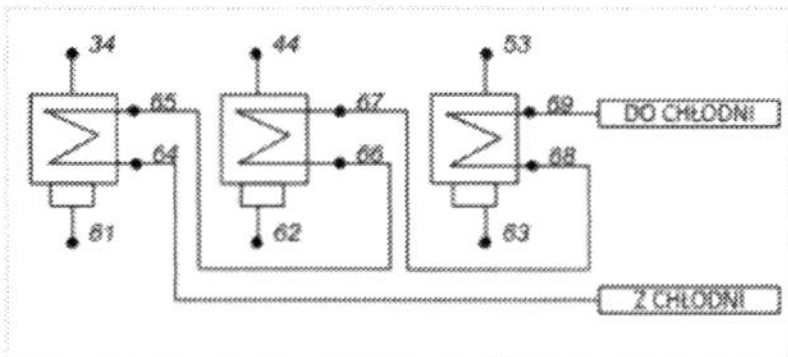

Fig. 3. Serial configuration

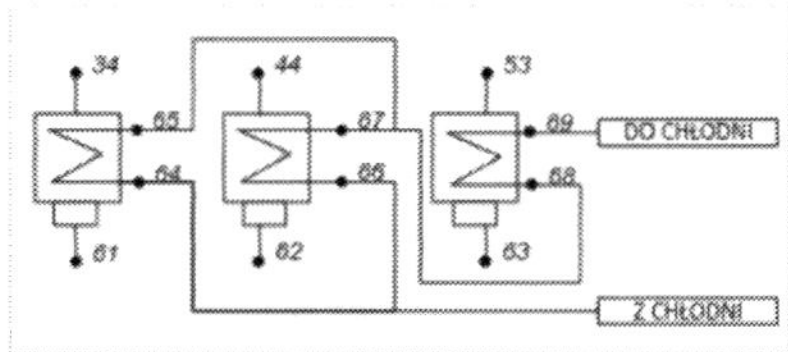

Fig. 4. Parallel-to-serial configuration

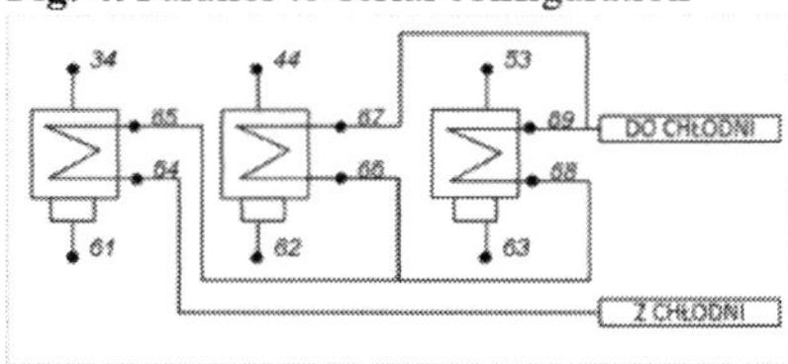

Fig.5. Serial-to-parallel configuration

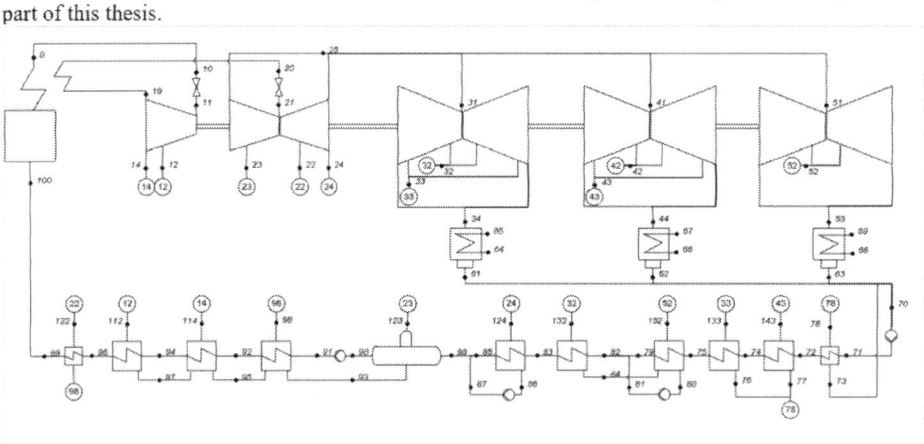

Fig.6. Tested thermal cycle scheme

 https://doi.org/10.1051/e3sconf/201913701027

LP turbine part, the total surface tube in each configuration is similar, the amount of cooling water is the same and the number of tubes has been chosen so that the cooling water velocity does not exceed 2.6 $\frac{m}{s}$.

It was assumed: one pass surface condenser except the parallel configuration where two pass surface condenser was selected; use of steel 1.4401 - $K_m = 15\ \frac{W}{m\,K}$., tube dimension Ø 22 x 0.5 mm, condenser efficiency $\eta_c = 0.99$, cleanliness factor $F_c = 0.95$;

In table 3.1 and 3.2 input data for the calculations were presented. Following, proposed configurations were compared after changing parameters: temperature or flow of cooling water, heat exchange surface, temperature of live and reheated steam, cleanliness factor

Tabela 3.1 Input data for the calculations (the same for all configuration)

p_0	MPa	28.5	A_s	m^2	~48700
t_0	°C	600	$\dot{m}_g$	$\frac{t}{h}$	81000
t_{20}	°C	610/600*	η_1	-	0.89
T_1	°C	16.0			

Tabela 3.2. Input data for the calculations (different for each configuration)

configuration		I**	II**	III**	IV**
A_{s1}	m^2	16252	16242	13935	20903
A_{s2}	m^2	16252	16242	13935	13935
A_{s3}	m^2	16252	16242	20903	13935
N_1		19640	25000	16840	25260
N_2		19640	25000	16840	16840
N_3		19640	25000	25260	16840

**) Temperature for 40% load*
***) Proposed configuration: I-parallel, II-serial ,III- parallel-to-serial and IV- serial-to-parallel*

Nomenclature: p_0– live steam pressure, t_0 –live steam temperature, t_{20}–reheated steam temperature, $\dot{m}_g$– cooling water flow rate, T_1– inlet cooling water temperature, A_s– surface tube area N – quantity of tubes η_1 – LP turbine's isentropic efficiency

2.2 Result discusion

The results of calculations for 100% and 40% loads are presented in tables 4.1-4.2. Figures 7.1-7.2 show the gross unit heat rate for the nominal load when the value of cooling water flow (7.1) and surface tube area (7.2) was changed. The results were compared with the results from the initial calculations (marked by x).

The best results in thermodynamic terms were obtained for a serial connection. In this case efficiency was improved by 0.15% compared to the parallel connection for nominal load and 0.1% for minimum load. Series-parallel connection was also somewhat more favorable, while other configurations are least beneficial.

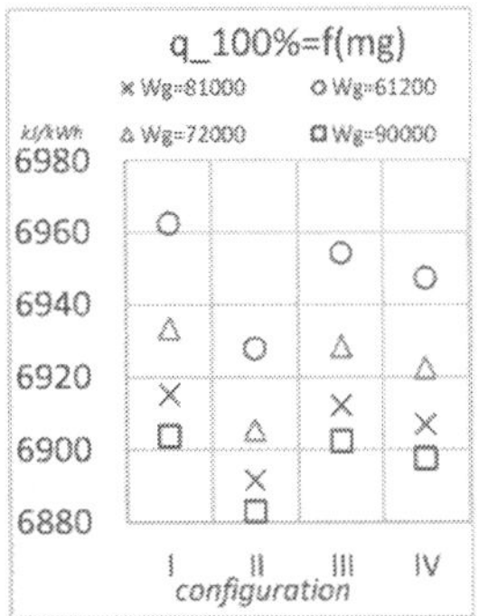

Fig. 7.1. Gross unit heat rate tested configuration when $\dot{m}_g$ change for nominal load

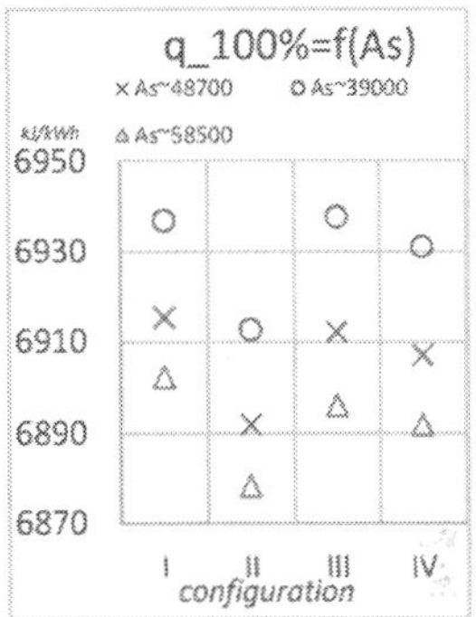

Fig. 7.2. Gross unit heat rate tested configuration when A_s $\dot{m}_g$ change for nominal load

Figures 7.1-7.2 show the impact of changing the relevant parameters on the gross unit heat rate q. On the one hand, the results show that regardless of the tested parameter, the serial system is the most advantageous. On the other hand, the charts show the savings this configuration gives. For example, a similar indicator q for 72000$\frac{t}{h}$ cooling water flow for serial configuration was obtained for 90000$\frac{t}{h}$ using a parallel configuration (Fig.7.1). This gives a 20% reduction in the amount of cooling water. Figure 7.2 shows that the same indicator q as for the base data series in parallel configuration can be obtained by reducing the surface tube area by 20% for the serial configuration - this can be interpreted as a decreasing surface tube area during operation. Similar conclusions were reached when analyzing subsequent results for the previously described parameters changes.

Table 4.1. Calculations results for nominal load

configuration	I-parallel		II-serial		III-parallel-to-serial		IV-serial-to-parallel	
	m	p	m	p	m	p	m	p
i	$\frac{kg}{h}$	MPa	$\frac{kg}{h}$	MPa	$\frac{kg}{h}$	MPa	$\frac{kg}{h}$	MPa
0	2416392	28.50	2393172	28.50	2413044	28.50	2406960	28.50
34	496260	0.00384	492192	0.00291	495648	0.00368	494604	0.00270
44	482724	0.00376	478836	0.00341	482148	0.00362	481176	0.00424
53	432900	0.00350	429552	0.00386	432432	0.00366	431568	0.00398
100	2416392	32.41	2393172	32.41	2413044	32.41	2406960	32.41
Operation indicators								
q	6915		6892		6912		6907	
η	0.521		0.522		0.521		0.521	

Table 4.2. Calculations results for 40% of nominal load

configuration	I-parallel		II-serial		III-parallel-to-serial		IV-serial-to-parallel	
	m	p	m	p	m	p	m	p
i	$\frac{kg}{h}$	MPa	$\frac{kg}{h}$	MPa	$\frac{kg}{h}$	MPa	$\frac{kg}{h}$	MPa
0	910116	10.82	903996	10.85	909288	10.82	907560	10.83
34	222300	0.00260	221796	0.00228	222264	0.00255	222120	0.00220
44	205992	0.00254	204624	0.00244	205848	0.00249	205452	0.00271
53	184248	0.00245	183024	0.00259	184068	0.00252	183744	0.00262
100	910116	12.31	903996	12.34	909288	12.31	907560	12.32
Operation indicators								
q	7433		7416		7431		7427	
η	0.484		0.485		0.484		0.485	

3 Conclusions

This paper presents calculation and verification of which connection setups of surface condensers on the cooling water side is the best from thermodynamics perspective. The study was not easy, because the phenomena occurring in the last stage of the turbine and in the condenser are complex and difficult to describe using mathematical formulas. Therefore, in the first part of work, the focus was on describing and choosing the best method for calculating the exhausted steam pressure of condensing turbine. Three calculation methods were compared, results were verified with data from the factual site. Considering the correctness of the results and the complexity of the calculations, the method based on HEI standard has been identified as the most advantageous method for calculating the turbine exhaust steam pressure. It needs to be highlighted that using this method to calculate heat transfer coefficient requires only the cooling water and condenser technical parameters. There is no need to enter the parameters of exhausted steam what simplifies the calculation.

In the next step, four condenser connection configurations were tested. In each case, the serial configuration was the most thermodynamically favourable. For nominal parameters, obtained improvement of unit efficiency was around 0.15%. The use of this configuration can improve unit efficiency or reduce design or operating costs by reducing surface tube area, cooling water quantity, superheated steam temperature.

However, for a serial connection, the problem of unequal operation of the LP turbine part attention should be paid to. The design of each parts of the turbine is the same, but when serial configuration is used, the exhausted steam pressure of each part is different, so they do not work at their optimal point. This is a significant problem when assuming the work of the unit mainly with nominal parameters. When serial connection is used, there is also a large dependence of the steam parameters of next LP turbine parts, which is not present for a parallel system. Incorrect assumptions or design calculations may have a greater impact to the operation then in parallel configuration. When choosing a series system, attention should also be paid to the useful power of the condensate pump, which will certainly be higher due to the greater drop in water pressure due to the flow in the tubes.

Summarizing the researches, it has been proven that the most advantageous configuration for thermodynamic reasons is the serial configuration. Although this setup has several important disadvantages that can have a significant impact on the final result.

References

1. The International Association for the Properties of Water and Steam, *Revised Release on the IAPWS Industrial Formulation 1997 for the Thermodynamic Properties of Water and Steam*
2. Kostowski E.: *Przepływ Ciepła*, (2006)
3. Szlachtin P.N.: *Turbiny Parowe*, (1953)
4. *HEI Standards for Steam Surface*, (2012)
5. *ASME PTC 12.2-2010 Steam Surface Condensers Performance Test Codes*, (2007)
6. Łukowicz H.: *Podstawy siłowni cieplnych*, Wykłady

RESEARCH & DEVELOPMENT IN POWER ENGINEERING, 2019

Balancing axial force in centrifugal pumps with pump out vanes

Marek Szlaga[1]

[1] Institute of Heat Engineering, Warsaw University of Technology. 25 Nowowiejska Street, Warsaw 00-665, szlagamarek@gmail.com, Poland

Abstract. Application of pump out vanes is one of the solutions to reduce hydraulic axial force generated during centrifugal pump operation. This article presents the cause of the occurrence of hydraulic axial force and method of calculating pressure distribution at the rear of the impeller used to design pump out vanes properly. It illustrates results of pump out vanes CFD calculations and its validation by measurements. The article reviews the methods of reducing pressure on the rear wall of the centrifugal pump's rotor using pump out vanes. It presents empirical formulae allowing calculation of pressure depending on the geometrical parameters of the blades. The article presents various design solutions of pump out vanes.

NOMENCLATURE

c_{wl}	$[m/s]$	absolute velocity
$c_{f,w}$, $c_{f,R}$	[-]	friction coefficient of casing and impeller shroud
k	[-]	rotation factor
Q_{th}	$[m^3/s]$	theoretical pump capacity
ω	$[1/s]$	angular velocity
ρ	$[kg/m^3]$	density

1 Introduction

The rotational movement of the pump creates a complex pressure field around the impeller. As a result, during the pump operation, the impeller is subjected to radial and axial forces. The ability to calculate these forces is necessary for the correct design of the shaft and sizing pump bearing system.

Many different methods and solutions are used to balance the hydraulic axial force. How to deal with axial thrust depends on the type of pump, its design, size, and its rotational speed. In pumping devices of small size, operating at low speeds, all axial thrust can be absorbed by the capacity of thrust bearings since the magnitude of this force is small. Larger pumps operating at high speeds may already require the use of specially designed balancing systems. In multi-stage pumps balance disk or balance piston is used. For pumps with a double-entry impeller, the symmetry of impeller reduces axial thrust. In single-stage pumps, two methods of reducing the axial force are most often used: using pump out vanes located on the rear shroud or using an annular sealing ring with balance holes on the rear shroud.

In the impeller sidewall gaps, at the walls of the rotating impeller, there are radially outward fluid flows caused by inertial forces acting on the liquid by the rotor. For reasons of continuity fluid flows radially inwards at the walls of the stationary casing (Fig. 1). The occurrence of leaks (e.g. at the annular seal on the front shroud) resulting from the pressure difference on both sides of the seal also has a very large impact on the velocity field in these areas. Depending on the pump design, leaks can also occur in the space behind the impeller (including relief holes, shaft seal, multi-stage pumps). The effect of these phenomena is the presence of complex, three-dimensional flow phenomena in the impeller sidewall gaps.

As a consequence, there are different pressure distributions on front and rear impeller shroud (Fig. 2), which is the main reason for the hydraulic axial force in the pumps.

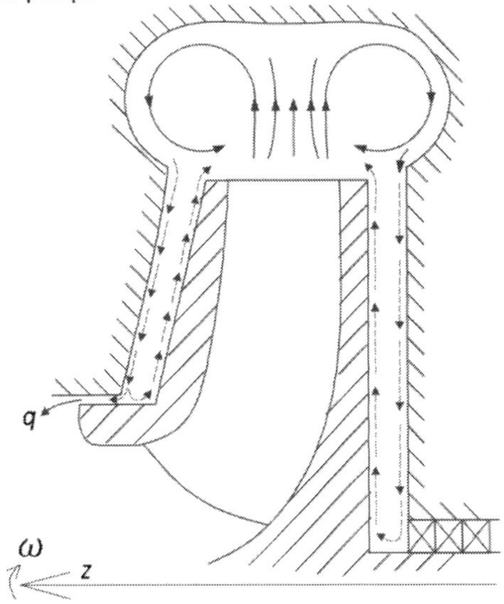

Fig. 1. Velocity distribution in the impeller sidewall gaps [5]

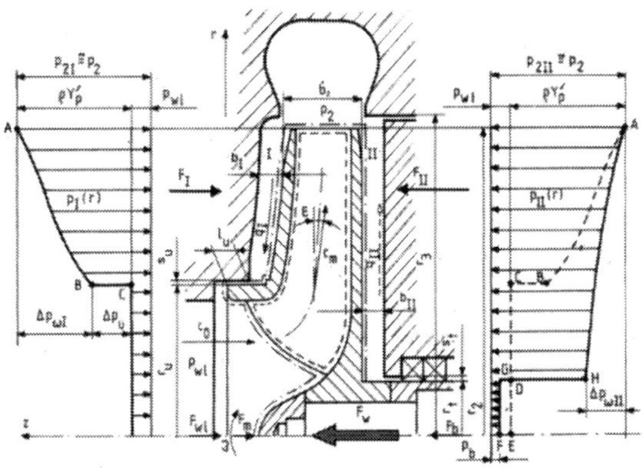

Fig. 2. Pressure distribution and axial forces on an impeller of a single-stage pump [4]

2 Axial thrust calculation

Assuming that the pressure distributions are axially symmetrical, which based on many research carried out has been proven for operation at the nominal pump capacity, the forces F_I and F_{II} acting on the surfaces of the front and rear impeller shrouds can be determined based on the following equation

$$F_I = 2\pi \int_{r_u}^{r_2} p_I(r) r dr \tag{1}$$

$$F_{II} = 2\pi \int_{r_t}^{r_2} p_{II}(r) r dr \tag{2}$$

Where r_2 is the impeller outlet radius, r_u is the annular seal radius on the front shroud, r_t is the impeller hub radius.

The other forces acting on the impeller are:

• force resulting from the pressure at the impeller inlet p_{wl}

$$F_{wl} = \frac{\pi}{4} d_u^2 p_{wl} \tag{3}$$

• force from atmospheric pressure p_b acting on the pump shaft

$$F_b = \frac{\pi}{4} d_{t\_shaft}^2 p_b \tag{4}$$

• reaction force resulting from changing the flow direction in impeller by 90^0 from axial to radial

$$F_Q = \rho \frac{Q_{th}^2}{A_{wl}} = \rho Q_{th} c_{wl} \tag{5}$$

Net hydraulic axial force acting on the rotor results from the sum of the listed forces, including their directions

$$F_w = F_{II} - F_I - F_{wl} - F_Q + F_b \tag{6}$$

3 Fluid rotation factor

Based on many research, it can be concluded that across areas I and II, the pressure practically does not change, that means

$$\frac{\partial p}{\partial z} \cong 0 \tag{7}$$

Momentum equation in the radial direction takes the form

$$\frac{\partial p}{\partial r} = \rho \frac{v_\varphi^2}{r} \tag{8}$$

To determine the pressure distribution in the in the impeller sidewall gaps the velocity v_φ is needed with which the fluid is rotating. Assuming that pressure distributions $p_I(r)$ and $p_{II}(r)$ depend only on the local fluid rotation

$$k_{I,II}(r) = \left(\frac{v_\varphi}{\omega r}\right) = \left(\frac{\omega_{cI,II}}{\omega}\right) \tag{9}$$

For impeller sidewall gaps following equation can be written based on (8) and (9)

$$\frac{1}{\rho}\frac{\partial p_{I,II}}{\partial r} = \frac{v_\varphi^2}{r} = \omega_c^2 r = k_{I,II}^2(r)\omega^2 r \tag{10}$$

To calculate the net hydraulic axial force and obtain pressure distributions in the I and II areas of the impeller, a rotation factor is needed as $k = k(r)$.

Many experimental and theoretical studies show that the circulation coefficient is a complex function of the geometrical dimensions of the flow area, roughness and the pump operating parameters.

In the vast majority of cases, the flow in the sidewall areas is turbulent. Rotation factor can be determined by numerically solving the Reynolds equations. For the purposes of preliminary calculations of the hydraulic axial force in the pump design process, these relationships are too complicated. However, the assumption that $k_I = k_{II} = 0{,}5$ used e.g. in books [1,2] is too simplified, which generates errors in calculating the forces acting on the rotor reaching several tens of per cent [3]. This approach does not take into account the complexity of the flow phenomenon in the impeller sidewall areas.

Jedral [4] suggests to use the following values of rotation factor depending on the net flow through the impeller sidewall gap:

• if the leakage is directed radially outwards $k = 0{,}3 \div 0{,}4$,
• if the leakage is directed radially inwards $k = 0{,}7 \div 0{,}9$,
• for lack of the leakage ($q = 0$) $k = 0{,}45 \div 0{,}5$.

For the purpose of rotation factor calculation Gulich [5] suggests using the following equation when there is no net through-flow

$$k = \frac{1}{1+\left(\frac{r_3}{r_2}\right)^2 \sqrt{\left(\frac{r_3}{r_2}+5\frac{t_{ax}}{r_2}\right)\frac{c_{f,w}}{c_{f,R}}}} \tag{11}$$

Where t_{ax} is the cylindrical part of the casing with r_3 radius in the area of sidewall gap.

In [5] a separate procedure can also be found to calculate the circulation rate in case of net flow through the impeller sidewall gap.

For averaged k_I and k_{II} values, after integrating the equation (10), pressure distributions can be obtained in the areas in front and behind the rotor

$$p_I(r) = p_{I2} - \rho\omega^2 \int_{r_u}^{r_2} k_I^2(r) r dr \tag{12}$$

$$p_{II}(r) = p_{II2} - \rho\omega^2 \int_{r_t}^{r_2} k_{II}^2(r) r dr \qquad (13)$$

After inserting formulae for pressure distribution into formulae (1) and (2), the force F_I i F_{II} can be calculated.

4 Balancing axial force with expeller vanes

Among the forces acting on the impeller, the largest component is the force F_{II} acting on the rear shroud. To reduce the hydraulic axial force, F_{II} must be reduced by reducing the pressure in the area of rear shroud. In single-stage pumps for this purpose expeller vanes on the rear impeller shroud can be used. Pump out vanes increase the rotation factor of the liquid and, as a result, reduce the pressure acting on the impeller in this area (area ABC as shown in Fig. 3).

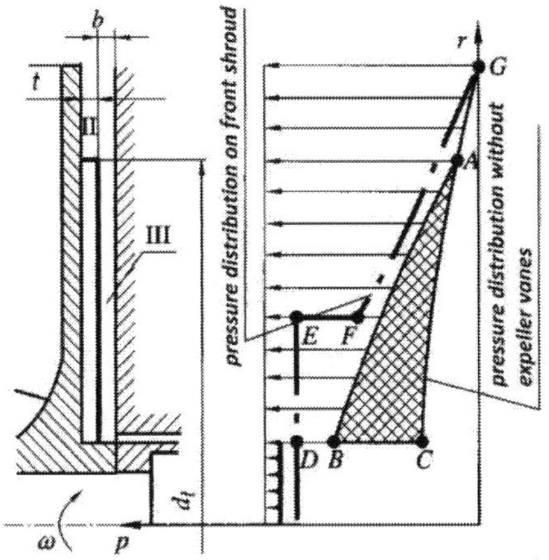

Fig. 3. Pressure distribution for the impeller with expeller vanes[4]

The advantages of reducing axial force with pump out vanes are:

• reducing the amount of pump parts (reduced production costs and subsequent repairs) - expeller vanes are the least costly design,
• easier standardization of rear cover since no seal,
• stuffing box protection when pumping solids contaminated fluids,
• the possibility of completely balancing the axial force over a wide range of performance and inlet pressures.

Expeller vanes are sometimes used also on the front impeller shroud. In this solution, their purpose is not to reduce the net axial force, but to protect the annular seal on the front shroud against solids entering the impeller sidewall gap.

These advantages make the expeller vanes widely used for single-stage pumps for pumping water, mixtures of liquids and solids, aggressive fluids and as chemical pumps.

The use of expeller vanes increases the disk friction losses and, as a result, reduces the efficiency of the pumps. Oversizing of pump out vanes in extreme cases can cause a change in the direction of the axial force and its increase in relation to the operation without them. Considering these aspects, the use of expeller vanes require accurate calculation formulae.

Various equations and approaches can be found in the literature for determining the value of the rotation factor when using expeller vanes. They differ in the scope of application and the number of simplifying assumptions.

Lazarkiewicz and Troskolański [1] suggest using 4 to 6 vanes with a height of 4 to 7 mm and very small gaps between blades and the casing. Authors assume that the liquid between expeller vanes rotates at a constant angular velocity equal to the impeller speed. If expeller vanes are trimmed ($d_t < d_2$), authors assume that the fluid rotation factor in the plain outer part of the rear shroud decreases immediately up to $k = 0.5$.

Stepanoff [6] assumed that the fluid rotation coefficient between expeller vanes and the fixed wall of the housing does not change along the radius and expressed it by the formula

$$k = 0{,}5\left(1 + \frac{t}{t+b}\right) \qquad (14)$$

Where t and b are shown in Fig. 3. In the plain outer part of the rear shroud, the rotation coefficient changes rapidly to a value of $k = 0.5$.

Gulich [5] proposes a formula determining the value of the circulation coefficient for a disc bladed through the following equation

$$k_t = \frac{1}{1 + 0{,}13\frac{t}{t+b}\sqrt{\frac{r_2}{ti}}} \qquad (15)$$

For trimmed expeller vanes ($d_t < d_2$) following equation is used to determine the average value of the rotation coefficient

$$k_{sr} = \sqrt{\left(\frac{d_t}{d_2}\right)^{\left(2 - 0{,}9\frac{d_t}{d_2}\right)} [k_t^2 - k^2] + k^2} \qquad (16)$$

In [5] following recommendations for the design of vanes can be found:

• number of expeller vanes equal to number of impeller blades,
• shape of expeller vanes same as of impeller blades (mostly back swept),
• outer diameter: $d_t = 0{,}85 d_2$,
• vane height: $t/r_2 = 0{,}04$ to $0{,}05$,
• axial clearance $b/r_2 = 0{,}01$,
• width of expeller vanes equal to height h or equal to nominal blade thickness (whatever is greater).

As part of the research work carried out at Institute of Heat Engineering, measurements of pressure distribution on the bladed rotating disc were carried out (rotating disc is marked with number 8 in Fig. 4). During tests, pressures on several disc diameters were measured, and then local average circulation coefficient values were determined based on the following equation:

$$k_i = \frac{30}{\pi n}\sqrt{\frac{2(p_{i+1} - p_i)}{\rho(r_{i+1}^2 - r_i^2)}} \qquad (17)$$

https://doi.org/10.1051/e3sconf/201913701028

The radius of the rotating disc on the stand was $r_2 = 160mm$, and the radius of the cylindrical casing $r_3 = 172{,}5\text{mm}$.
Following blade radii were examined $r_t = 160;\ 140;\ 120;\ 100;\ 80\ mm$, blade height $t = 4;\ 6{,}3;\ 8;\ 10\ mm$. Axial clearance $b = 1;\ 2;\ 4;\ 8\ mm$. Discs with different number of blades were tested; $i = 4;\ 8;\ 12;\ 16$ at rotational speeds $n = 2250;\ 1500;\ 750\ \frac{1}{min}$ and the net flow through the rotating domain $Q_s = 0;\ 0{,}3;\ 0{,}6\ \frac{m^3}{h}$.

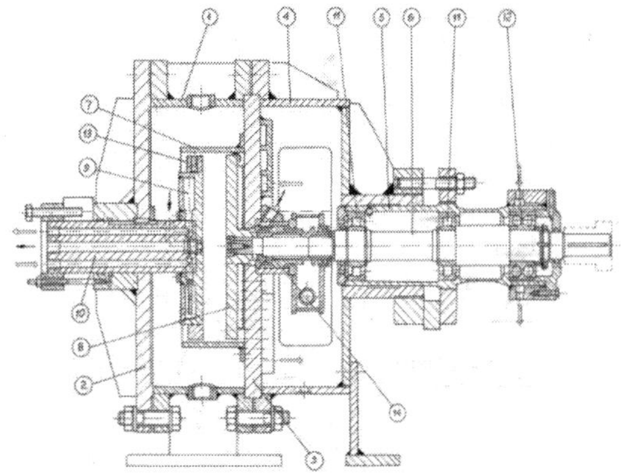

Fig. 4. Sectional drawing of the test stand

Selected measurement results were compared with results obtained by the author of numerical simulations performed in the Ansys Fluent software. The chart below shows the rotation coefficient obtained on the basis of measurements with values obtained on the basis of numerical calculations and calculated from formulae found in the literature.

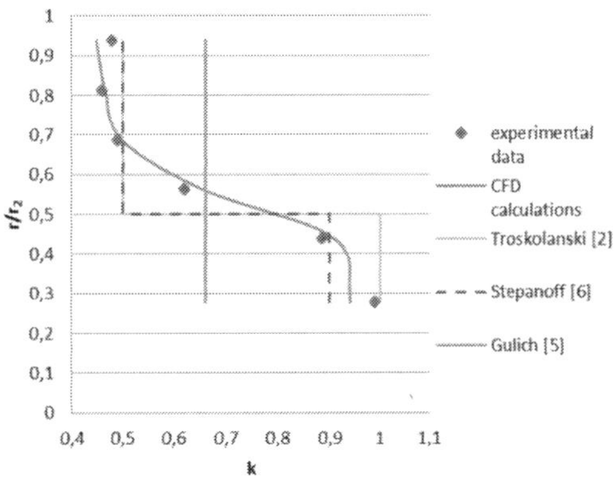

Fig. 5. Rotation factor for $r_t = 80\ mm; t = 4\ mm; b = 1\ mm;\ i = 12;\ n = 2250\ \frac{obr}{min};\ Q_s = 0$

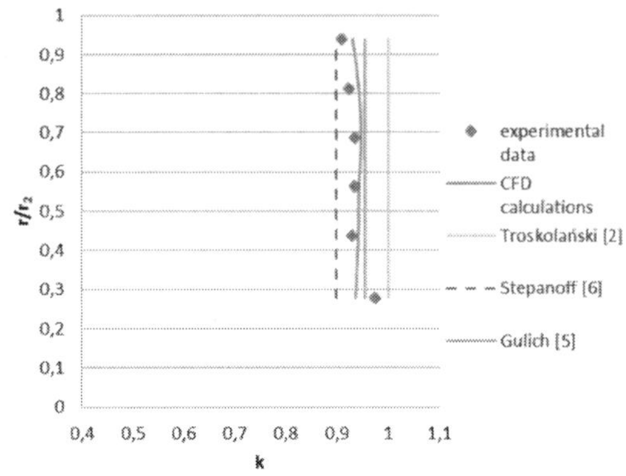

Fig. 6. Rotation factor for $r_t = 160\ mm;\ t = 4\ mm;\ b = 1\ mm;\ i = 12;\ n = 2250\ obr/min;\ Q_s = 0$

Comparing measurements results and CFD calculations with the formulae found in the literature, there is a very large discrepancy in the case of trimmed vanes. The model of the step change of the rotation factor at the transition from the area with vanes to the flat disc turns out to be completely inadequate. The rotation factor changes gradually and begins to decrease already in the area of occurrence of the vanes. In no case was $k = 1$ reached.
In the case of a fully bladed disc, test results are close to calculated values. CFD calculations allow to determine the pressure distribution with much greater accuracy than empirical formulae.
In [5] information can be found that in the case of expeller vanes, a solution generating lower energy losses in relation to straight radial blades is the use of curved blades. Such solutions can be found in manufactured pumps (Fig. 7).

Fig. 7. Different solutions in design of expeller vanes

There is a lack of literature confirming this thesis and presenting the impact of other types of unloading blades (other shapes, positioning, angles) on the rotation factor and disk friction losses. It also marks a certain area of solutions and research that should be addressed in scientific works. The goal of further research should be to create guidelines and mathematical equations that allow designing optimal unloading blades in terms of effectiveness in reducing hydraulic axial force and minimizing energy losses.

5 Conclusions

Expeller vanes are the simplest and the least expensive solution in case the reduction of axial thrust is required. For expeller vanes design, accurate formulae shall be used to diminish the increase of disk friction losses.
This paper presents several formulae used to determine the rotation factor in the impeller sidewall gaps in case pump out vanes are used. Selected formulae for rotation factor calculation were compared with experimental data and CFD calculations. Significant discrepancies were found especially for trimmed expeller vanes.
Further research should be carried out to determine the impact of expeller vanes shape and positioning on rotation factor and disk friction losses.

References

1. Pfleiderer C.: *Die Kreiselpumpen fur Flüssigkeiten und Gase*, 5. Aufl, Springer, Berlin, 1961.
2. Troskolański A.T., Łazarkiewicz S., *Pompy wirowe*, WNT, Warsaw (1973).
3. Jedral W., *Metoda obliczania sił wzdłużnych w pompach wirowych*, Przegląd Mechaniczny, Vol. 13, pp. 11÷14 (1992).
4. Jedral W., *Rotodynamic Pumps*, PWN, Warsaw (2001).
5. Gülich J.F., *Centrifugal Pumps*, Springer-Verlag, Berlin Heidelberg (2014).
6. Stepanoff A.J., *Centrifugal and axial flow pumps. Theory design and application*, J. Wiley Sons Inc., New York (1957).

RESEARCH & DEVELOPMENT IN POWER ENGINEERING, 2019

Investigation of off-design characteristics of solid oxide electrolyser (SOE) operated in endothermic conditions

Konrad Motylinski[1,2,*], *Michał* Wierzbicki[1,2], *Stanisław* Jagielski[1,2], *Jakub* Kupecki[1,3]

[1] Department of High Temperature Electrochemical Processes (HiTEP), Institute of Power Engineering, Augustowka 36, 02-981 Warsaw, Poland
[2] Institute of Heat Engineering, Warsaw University of Technology, Nowowiejska 21/25, 00-665 Warsaw, Poland
[3] National Fuel Cell Research Center (NFCRC), University of California, Irvine, Engineering Laboratory Facility, Irvine, California 92697-3550, United States
[*] corresponding author: konrad.motylinski@ien.com.pl

Abstract. One of the key issues in the energy production sector worldwide is the efficient way to storage energy. Currently, more and more attention is focused on Power-to-Gas (P2G) installations, where excess electric power from the grid or various renewable energy sources is used to produce different kind of fuels, such as hydrogen. In such cases, generated fuels are treated as energy carriers which, in contrast to electricity, can be easy stored and transported. Currently, high temperature electrolysers, based solid oxide cells (SOC), are treated as an interesting alternative for P2G systems. Solid oxide electrolysers (SOE) are characterized as highly efficient (~90%) and long-term stable technologies, which can be coupled with stationary power plants. In the current work, the solid oxide cell stack was operated in electrolysis mode in the endothermic conditions. Based on the gathered experimental data, the numerical model of the SOC stack was created and validated. The prepared and calibrated model was used for generation of stack performance maps for different operating conditions. The results allowed to determine optimal working conditions for the tested stack in the electrolysis mode, thus reducing potential costs of expensive experimental analysis and test campaigns.

1 Introduction

Currently more attention in power and energy sector is focused on minimizing the wide usage of fossil fuels, mainly due to their negative impact on the environment. At the same time, various solutions concerning power generation and efficient ways to storage energy are correlated with renewable energy sources (RES). The most commonly used technologies world-wide are solar and wind based power systems. The main issue with these solutions is dependence on local weather conditions, thus such energy sources are intermittent, which in many cases might by highly undesirable [1-3]. In response to that, many scientific and R&D entities are analysing different counter-solutions. Many specialists treats hydrogen as the gas of the future, which can be used as one of the main substrates in creating different and more calorific fuels, or directly convert it as the main power source for production of energy. This technologies are highly suggested for application in systems aiming to minimize pollution emissions to the atmosphere.

At present, steam reforming is the most commonly used method to generate hydrogen [4,5], which is not a preferable solution in long-term perspective. As a result, in recent years new methods for efficient and clean hydrogen production are developed [6,7]. Steam electrolysis is currently considered as an preferable technology for application in mid- and large- scale hydrogen production installations, due to its high efficiency [8-10]. Among the existing electrolysers, commercially available and most advanced are alkaline and proton exchange membranes (PEM), which are used for low temperature electrolysis processes [11]. In recent years however, high temperature steam electrolysis based on solid oxide cells (SOC) proves to be a much better solution, due to lack of expensive catalysts and high hydrogen generation efficiency (~90%) [12,13]. The comparison of performance between electrolysis technologies is presented in Fig. 1, showing polarization areas for specific device.

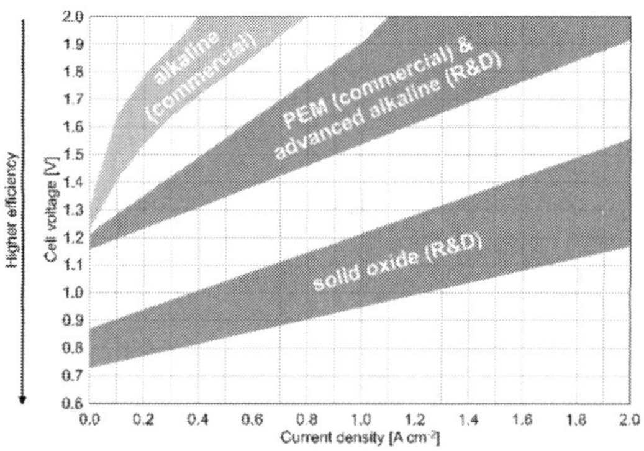

Fig. 1. State-of-the-art steam electrolysers (based on [13]).

The performance of the electrolysers can be evaluated by comparing the need of electric power for the electrolysis reaction, generating the same amount of hydrogen. The operating areas presented in Fig. 1 show that high temperature electrolysers, based on solid oxide cells, require less power than low temperature technologies, thus their hydrogen production efficiency is higher. The difference is caused mainly by higher internal resistances between low- and high-temperature cells, hence the usage of the second technology might be more economically profitable in long-term perspective.

Solid oxide cells can operate as fuel cell which generates heat electric power or as an electrolyser for production of clean hydrogen and oxygen, when the specific conditions are met. Additionally, one of the main advantage of SOC-based technologies is the possibility to simply integrate them with existing stationary systems, either dedicated for production of electric power and heat or hydrogen [14].

Efficient way for storage of energy is currently an important and urgent topic in power sector world-wide. For that reason, more attention is dedicated for solutions that allows to both generate and distribute energy with minimization of potential losses. In such cases, electrolysers proves to be a preferable option, where generated hydrogen, treated as energy carrier, can be stored and transported for long distances. Systems with such topology are known as Power-to-Gas (P2G) installations where, besides hydrogen, methane and other, more calorific gases might be generated [15]. This concept can be further expanded by implementing proper chemical reactors, thus allowing to produce various liquid fuels (Power-to-Liquid (P2L)).

2 High temperatures electrolysis

Water electrolysis reaction was discovered at the end of XIX century [16], but is still used in moderns technologies, such as in high temperature solid oxide electrolysers (SOE). The operating temperature of SOE cells is from 600°C to 1000°C, which is the result of used materials, such as ZrO_2, for the electrolyte layer [14]. The steam electrolysis reaction is presented in Eq. 1:

$$H_2O_{(g)} \rightarrow H_2 + 0.5O_2$$
$$\Delta H = 247.9\ kJ \cdot mol^{-1}\ (at\ 750°C) \tag{1}$$

where, ΔH is the reaction enthalpy. Operation of solid oxide electrolysers is shown in Fig 2.

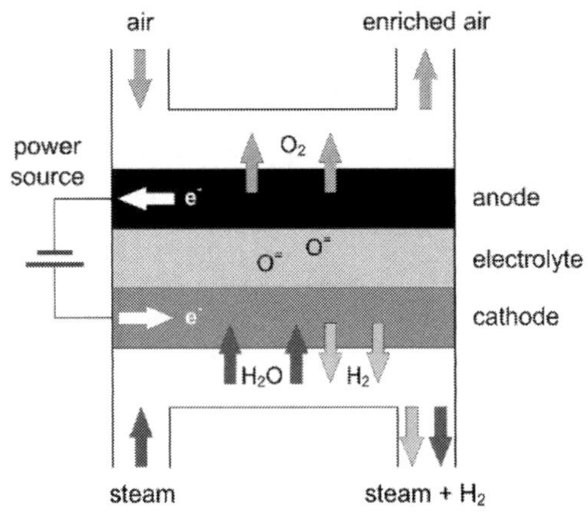

Fig. 2. SOE cell working principle.

Solid oxide cell consists of three different layers: cathode, anode and electrolyte. For the electrolysis operation, steam and electric power are delivered to the porous, fuel electrode. This results in reduction of water molecules for oxygen ions (O^{2-}) and hydrogen in the cathode. Oxidant electrode is usually supplied with air but different gases might also be delivered or vacuum conditions might be applied, depending on the wanted SOE operation. In every case, oxidant molecules are formed on the anode side. The electrolyte is a thin layer which allows to pass only oxygen ions. Its structure has to be properly prepared in order to block diffusion of hydrogen and other, potential gases [17].

The SOE stack can operate in three different operating modes: thermo-neutral, endothermic and exothermic, which are strictly correlated with temperature and applied current [18,19]. During the electrolyser operation, part of electric power is used in the steam electrolysis reaction, while the rest is lost in from of the heat. As a result, temperature distribution in the stack depends on the compromise between the energy consumed by the reaction and the generated heat. When the amount of heat caused by the electric power exceed the required heat for the electrolysis reaction, the stack operates in the exothermic conditions, thus in-stack temperature rises. For the endothermic conditions, electrochemical reactions uses more heat than is delivered to the stack via electric current, thus lowering the stack operating temperature. In such case, additional heat needs to be delivered to the stack [20]. The switching point between exothermic and endothermic operation mode is known as the thermo-neutral voltage. Its value is dependent on the stack operating temperature and type of the electrolysis reaction, such as steam electrolysis, carbon dioxide electrolysis or mixed. Thermo-neutral voltage is calculated via Eq. 2, which is based on the Faraday's Law:

$$E_{th} = \frac{\Delta H}{n \cdot F} \tag{2}$$

where, ΔH is the reaction enthalpy, F is Faraday's constant and n is the number of electrons needed in the reaction. For the steam electrolysis at 800°C, the thermo-neutral voltage is 1.28V.

Depending on the power system configuration with the installed SOE electrolyser, different operating modes are more preferable [21]. The main disadvantage of endothermal operation is need to deliver both heat and electricity for the electrolysis reaction. However, this can be compensated in case, where the SOE stack is connected to the external heat source. For the exothermal state, more electric current must be applied to the stack, which might result in damaging the cells in long-term operation. In the previous studies it was observed that for high currents in the electrolysis mode, the delamination of the oxidant electrode from the electrolyte layer might occur. In this process, interface of anode-electrolyte suffers build up of oxygen partial pressure, resulting in irreversible damage on the cell [22-24]. At the same time, high current affects the nickel-based cathode, which might suffer more rapid degradation [25-27]. As a result, stable and long-term operation of SOE-based systems, which is the aim for potential commercialization of the technology, is more difficult to achieve in exothermal state. For that reason, working in endothermal or near thermo-neutral conditions are more preferable.

In the current study, the numerical model of solid oxide electrolysis stack was designed and compared with gathered experimental data during endothermal operation of physical device. Next, the tuned model was used to generate stack performance maps for different operating conditions. The results allowed to determine optimal working conditions for the tested stack in the electrolysis mode, thus reducing potential costs of expensive test campaigns.

3 Methodology

The solid oxide electrolysis technology is under research and development for more than 40 years, but there are not many publications related to modelling of their operation both in stationary and transient cases. Most of the SOE numerical tools are based SOFC mathematical methodologies, due to comparable mechanisms, such as gas diffusion and similar electrochemical behaviour. The proper SOFC modelling approach can be modified in order to simulate SOE operation [28-30].

In the chosen methodology for the present study, solid oxide cells electrical operation is calculated as equivalent electric circuit. This solution was already used to predict performance of SOFC-based power units both in steady state [31] and in transient operations [32]. In both cases, proposed methodology was validated and calibrated based on the experimental data, resulting in prediction error below 5%. This concept was currently adopted for simulation of solid oxide electrolyser. The electric circuit equivalent of SOE is shown in Fig. 3.

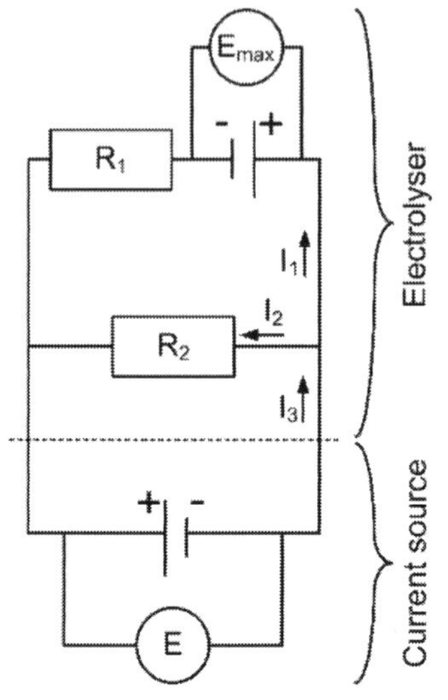

Fig. 3. Electric circuit equivalent of SOE.

In the proposed diagram, representing electrochemical cell operation, current delivered from the power source to the SOE is shown as I_3. Next, it is divided into flow of electrons used in the electrochemical reaction (I_1) and I_2 which represents the total electronic conductivity.

In this approach, the voltage of the single cell is calculated by solving the electric circuit based on Ohm's and Kirchhoff's Laws :

$$E_{cell} = \frac{E_{max} + i_{max} r_1 \eta_s}{\frac{r_1}{r_2}(1 - \eta_s) + 1} \tag{3}$$

where, E_{max} is the highest possible voltage obtained during isothermal process; η_s is the steam utilization factor, which is defined by the current working conditions of the cell; i_{max} is the maximum current density correlated with amount of steam delivered to the cells; r_1 is the area specific internal ionic resistance representing oxygen ions permeability through the electrolyte and r_2 is the area specific internal electronic resistance which corresponds to solid electrolyte electric conductance. The full description of the used methodology is presented elsewhere [33].

In the current analysis, selected SOE module consists of two separated stacks, 30 cells each, as shown in Fig 4.

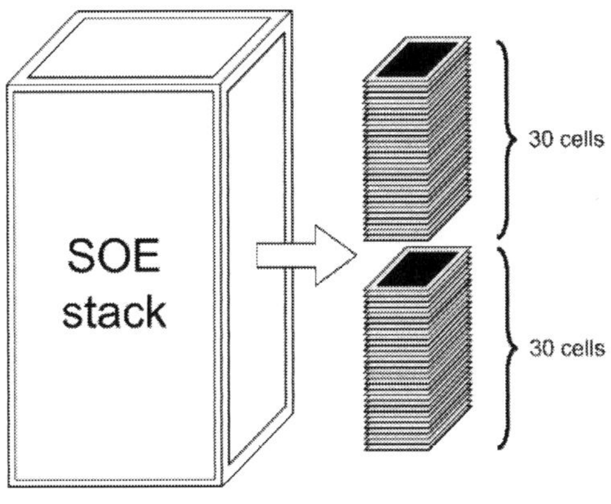

Fig. 4. SOE module concept.

The single stack contains 30 cells, where the cathode is porous NiO/GDC, the electrolyte is dense 3YSZ and the anode is double layered 8YSZ/LSM–LSM. The active area of a single cell is 128 cm^2. The chosen modeling approach includes technical specification of the stack, such as physical properties of the used materials for cells and interconnectors, thus minimizing the usage of empirical and semi-empirical constants and equations.

The model of the SOE stack was designed and implemented in Aspen Hysys modeling software. In order to properly simulate operating in the electrolysis mode, the proposed numerical tool was enhanced with additional in-house codes, which were determined based on the performed experiments with single cells and fully-dimensional stack.

4 Results

The numerical model of the stack has undergone a validation process, based on the data gathered during measurements on the SOE module with two 30-cell stacks connected electrically in series. The experiments were performed in the electrolysis mode, operating mainly in the endothermic conditions.

The polarization curve of the SOE module was generated in constant 750°C. In order to maintain stable temperature during measurements, proper power of inlet air heaters was set for each saved point. The validation of the proposed model is shown in Fig. 5.

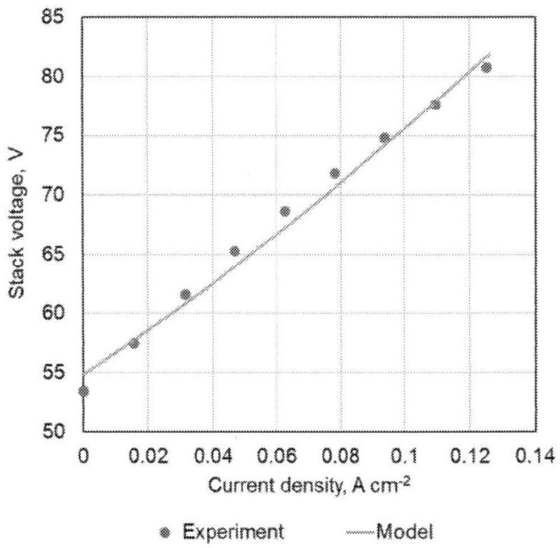

Fig. 5. Model validation

The average prediction error was less than 1.41%, where the relative prediction error did not exceed 3%. While operating in the electrolysis mode, inlet gas for the cathode was composed of steam and additionally hydrogen and nitrogen, in order to maintain reductive atmosphere on the Nickel-based part of the cell. For the anode side, air was used as the oxidant.

Based on the known parameters and technical limitation of the stack, the validated model was used to generate performance maps of the SOE module. Such approach allows to determine optimal operating conditions for the stack, without having to realize expensive and time-consuming experimental campaigns. The main limiting factor included in the simulations is the maximum steam utilisation factor (SU), which shows the molar amount of steam used in water electrolysis reaction. There is no strict limiting value concerning SU reported in the available literature, but it is commonly assumed that 80% is the highest value [37].

In the realized numerical activities, two performance maps were generated for different input parameters: stack power and stack voltage. The amount of air and mixture of hydrogen and nitrogen were constant for every case. Due to the fact, that experimental analysis of the stack were performed only at 750°C, the same temperature was used in the simulations. The only variables in the realized simulations were current density and flow of delivered steam. The obtained stack voltages and power are shown in Fig. 6 and 7, respectively.

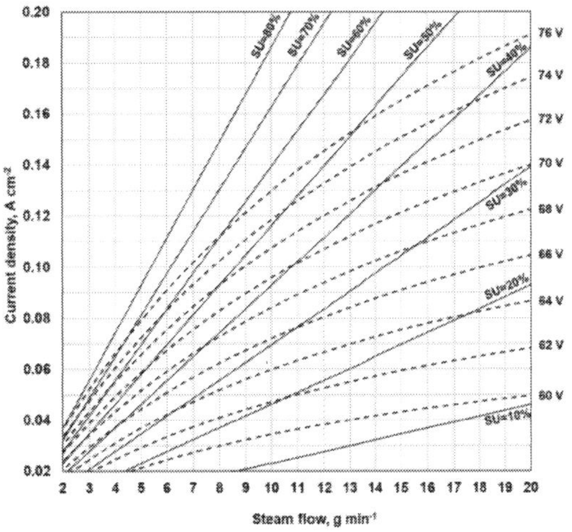

Fig. 6. SOE module performance map: voltage.

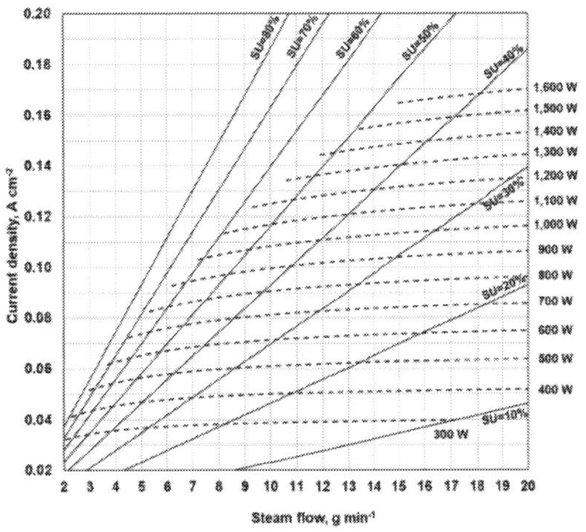

Fig. 7. SOE module performance map: power.

In order to operate in the endothermic conditions, maximal stack voltage was set to 76V, which corresponds to ~1.27V per cell. In Fig. 7, the dashed lines, representing different values of generated power, were presented only in the area which did not exceed 76V. Operating above this conditions, SOE module would work in exothermic mode, generating more heat than required for the electrolysis reaction, but resulting in potentially more rapid degradation of the cells.

5 Conclusions

In the current work, SOE module consisting of two stack with 30 cells each, was modelled and simulated in endothermic conditions. Operation below thermo-neutral point is more preferable solution compared to exothermic states, due to extended long-term durability of the stack and lower risks for degradation and potential damages in the cells. Based on the generated performance maps of the SOE module, it is possible to determine preferable operating conditions, depending on the wanted results. This allowed to perform simulations for different steam flows and current densities, without the need to realize expensive and long experimental campaigns. For the current study, SOE module was tested only in 750°C. Further work in this topic will be dedicated in performing additional tests with the presented stacks in different operating temperatures. Obtained data will be used to further tune the proposed numerical model and generate additional performance maps, thus allowing to determine SOE module behaviour in wider operational range.

Acknowledgment

Authors acknowledge the financial support from the Ministry of Science and Higher Education of the Republic of Poland for Statutory Grant CPE/040/STAT/19 in the Institute of Power Engineering.

References

1. Ding Z., Hou H., Yu G., Hu E., Duan L., Zhao ., Performance analysis of a wind-solar hybrid power generation system, *Energy Conversion and Management,* Vol. 181, pp. 223-234, (2019).
2. Ma Z., Wang S., Li S., Shi Y., Long-term coordination for hydro-thermal-wind-solar hybrid energy system of provincial power grid, *Energy Procedia,* Vol. 158, pp. 6231-6235 (2019).
3. Salomone F., Giglio E., Ferrero D., Santarelli M., Pirone R., Bensaid S., Techno-economic modelling of a Power-to-Gas system based on SOE electrolysis and CO2 methanation in a RES-based electric grid, *Chemical Engineering Journal*, (2019), in press.
4. Balat M., Balat M., Political, economic and environmental impacts of biomass-based hydrogen, *International Journal of Hydrogen Energy*, Vol. 34, Issue 9, pp. 3589-3603, (2009).
5. Konieczny A., Mondal K., Wiltowski T., Dydo P., Catalyst development for thermocatalytic decomposition of methane to hydrogen, *International Journal of Hydrogen Energy*, Vol. 33, Issue 1, pp. 264-272, (2008).
6. Talawar M.B., Sivabalan R., Mukundan T., Muthurajan H., Sikder A.K., Gandhe B.R., Subhananda Rao A., Environmentally compatible next generation green energetic materials (GEMs), *Journal of Hazardous Materials*, Vol. 161, Issues 2-3, pp. 589-607, (2009).
7. Yan Y., Fang Q., Blum L., Lehnert W., Performance and degradation of an SOE stack with different cell components, *Electrochimica Acta*, Vol. 258, pp. 1254-1261, (2017).
8. Bo Y., Wenqiang Z., Jingming X., Jing C., Status and research of highly efficient hydrogen production through high temperature steam electrolysis at INET, *International Journal of Hydrogen Energy*, Vol. 35, Issue 7, pp. 2829-2835, (2010).
9. Mingyi L., Bo Y., Jingming X., Jing C., Thermodynamic analysis of the efficiency of high-temperature steam electrolysis system for hydrogen production, *Journal of Power Sources*, Vol. 177, Issue 2, pp. 493-499, (2008).

10. Li Q., Zheng Y., Guan W., Jin L., Xu C., Wang W.G., Achieving high-efficiency hydrogen production using planar solid-oxide electrolysis stacks, *International Journal of Hydrogen Energy*, Vol. 39, Issue 21, pp. 10833-10842, (2014).
11. Mocoteguy P., Brisse A., A review and comprehensive analysis of degradation mechanisms of solid oxide electrolysis cells, *International Journal of Hydrogen Energy*, Vol. 38, Issue 36, pp. 15887-15902, (2013).
12. Fang Q., Blum L., Menzler N.H., Performance and Degradation of Solid Oxide Electrolysis Cells in Stack, *Journal of The Electrochemical Society*, Vol. 162, Issue 8, pp. F907-F912, (2015).
13. Graves C., Ebbesen S.D., Mogensen M., Lackner K.S., Sustainable hydrocarbon fuels by recycling CO2 and H2O with renewable or nuclear energy, *Renewable and Sustainable Energy Reviews*, Vol. 15, Issue 1, pp. 1-23, (2011).
14. Gruber M., Wienbrecht P., Biffar L., Harth S., Trimis D., Brabandt J., Posdziech O., Blumentritt R., Power-to-Gas through thermal integration of high-temperature steam electrolysis and carbon dioxide methanation - Experimental results, *Fuel Processing Technology*, Vol. 181, pp. 61-74, (2018).
15. Nielsen S., Skov I.R., Investment screening model for spatial deployment of power-to-gas plants on a national scale – A Danish case, *International Journal of Hydrogen Energy*, Vol. 44, Issue 19, pp. 9544-9557, (2018).
16. Yan X.L., Hino R., Nuclear hydrogen production handbook. Green Chemistry and Chemical Engineering, *CRC Press*, 1st edition, (2016).
17. Ni M., Leung M.K.H., Leung D.Y.C., Technological development of hydrogen production by solid oxide electrolyser cell (SOE), *International Journal of Hydrogen Energy*, Vol. 33, Issue 9, pp. 2337-2354, (2008).
18. Navasa M., Yuan J., Sunden B., Computational fluid dynamics approach for performance evaluation of a solid oxide electrolysis cell for hydrogen production, *Applied Energy*, Vol. 137, pp. 867-876, (2015).
19. Udagawa J., Aguiar P., Brandon N.P., Hydrogen production through steam electrolysis: Model-based dynamic behaviour of a cathode-supported intermediate temperature solid oxide electrolysis cell, *Journal of Power Sources*, Vol. 180, Issue 1, pp. 46-55, (2008).
20. Petipas F., Brisse A., Bouallou C., Model-based behaviour of a high temperature electrolyser system operated at various loads, *Journal of Power Sources*, Vol. 239, pp. 584-595, (2013).
21. Santhanam S., Heddrich M.P., Riedel M., Friedrich K.A., Theoretical and experimental study of Reversible Solid Oxide Cell (r-SOC) systems for energy storage, *Energy*, Vol. 141, pp. 202-214, (2017).
22. Virkar A.V., Mechanism of oxygen electrode delamination in solid oxide electrolyser cells. International, *Journal of Hydrogen Energy*, Vol. 35, Issue 18, pp. 9527-9563, (2010).
23. Sun X., Chen M., Liu Y.L., Hendriksen P.V., Life Time Performance Characterization of Solid Oxide Electrolysis Cells for Hydrogen Production, *ECS Transactions*, Vol. 68, Issue 1, pp. 3359-3368, (2015).
24. Sun X., Chen M., Liu Y.L., Hjalmarsson P., Ebbesen S.D., Jensen S.H., Mogensen M.B., Hendriksen P.V., Durability of Solid Oxide Electrolysis Cells for Syngas Production, *Journal of the Electrochemical Society*, Vol. 160, Issue 9, pp. F1074-F1080, (2013).
25. Tietz F., Sebold D., Brisse A., Schefold J., Degradation phenomena in a solid oxide electrolysis cell after 9000 h of operation, *Journal of Power Sources*, Vol. 223, pp. 129-135, (2013).
26. Chen M., Sun X., Chatzichristodoulou C., Koch S., Hendriksen P.V., Thermoneutral Operation of Solid Oxide Electrolysis Cells in Potentiostatic Mode, *ECS Transactions*, Vol. 78, Issue 1, pp. 3077-3088, (2017).
27. Hoerlein M.P., Schiller G., Tietz F., Friedrich K.A., Systematic Parameter Study on the Influence of Humidification and Current Density on SOEC Degradation, *ECS Transactions*, Vol. 68, Issue 1, pp. 3553-3561, (2015).
28. Njodzefon J.C., Klotz D., Kromp A., Weber A., Ivers-Tiffee E., Electrochemical Modeling of the Current-Voltage Characteristics of an SOFC in Fuel Cell and Electrolyser Operation Modes, *Journal of The Electrochemical Society*, Vol. 160, Issue 4, pp. F313-F323, (2013).
29. Petipas F., Grisse A., Bouallou C., Model-based behaviour of a high temperature electrolyser system operated at various loads, *Journal of Power Sources*, Vol. 239, pp. 584-595, (2013).
30. Iora P., Chiesa P., High efficiency process for the production of pure oxygen based on solid oxide fuel cell–solid oxide electrolyser technology, *Journal of Power Sources*, Vol. 109, Issue 2, pp. 408-419, (2009).
31. Kupecki J., Off-design analysis of a micro-CHP unit with solid oxide fuel cells fed by DME, *International Journal of Hydrogen Energy*, Vol. 40, Issue 35, pp. 12009-12022, (2015).
32. Kupecki J., et al., Dynamic numerical analysis of cross-, co-, and counter-current flow configurations of a 1 kW-class solid oxide fuel cell stack, *International Journal of Hydrogen Energy*, Vol. 40, Issue 45, pp. 15834-15844, (2015).
33. Milewski J., Świrski K., Santarelli M., Leone P., Advanced methods of solid oxide fuel cell modeling, Green Energy Technology, *Springer*, (2014).
34. Iora P., Taher M.A.A., Chiesa P., Brandon N.P., A novel system for the production of pure hydrogen from natural gas based on solid oxide fuel cell–solid oxide electrolyser, *International Journal of Hydrogen Energy*, Vol. 35, Issue 22, pp. 12680-12687, (2010).
35. Cai Q., Luna-Ortiz E., Adjiman C.S., Brandon N.P., The Effects of Operating Conditions on the Performance of a Solid Oxide Steam Electrolyser: A

Model-Based Study, *Fuel Cells From Fundamentals to Systems*, Vol. 10, Issue 6, pp. 1114-1128, (2010).

36. Kim J.S., Boardman R.D., Bragg-Sitton S.M., Dynamic performance analysis of a high-temperature steam electrolysis plant integrated within nuclear-renewable hybrid energy systems, *Applied Energy*, Vol. 228, pp. 2090-2110, (2018).
37. Wang Z., Mori M., Araki T., Steam electrolysis performance of intermediate-temperature solid oxide electrolysis cell and efficiency of hydrogen production system at 300 Nm3 h−1, *International Journal of Hydrogen Energy*, Vol. 35, Issue 10, pp. 4451-4458, (2010).

RESEARCH & DEVELOPMENT IN POWER ENGINEERING, 2019

Thorium application in the medium-sized sodium-cooled fast reactor

Eeshu Raaj Saasthaa Arumuga Kumar[1], *Piotr* Darnowski[1*], *Mihir* Kiritbhai Pancholi[1] and *Aleksandra* Dzido[1]

[1] Warsaw University of Technology, Institute of Heat Engineering, Nowowiejska 21/25,00-665, Warsaw, Poland
*Corresponding author: piotr.darnowski@pw.edu.pl

Abstract. The report presents an analysis of the medium-sized Sodium-Cooled Fast Reactor (SFR) core with Thorium-based Mixed-Oxide fuel. The introduction of Transuranics (TRU) to the fuel was to allow long-lived nuclear waste incineration. The studied core is based on the modified Advanced Burner Reactor (ABR) 1000MWth core design, which was analysed in the OECD/NEA "Benchmark for Neutronic Analysis of Sodium-Cooled Fast Reactor Cores with Various Fuel Types and Core Sizes". The full-core simulations with SERPENT 2.1.31 Monte Carlo computer code and ENDF library were performed, including static criticality and fuel burnup calculations for five fuel cycles. The core inventories at the Beginning of Cycle (BOC) and End of Cycle (EOC) were studied, and the impact of thorium fuel was assessed. The proposed core design is a burner reactor which uses thorium fuel. The excess core reactivity stays positive for long time despite large net consumption of transuranic elements as new fissile Uranium 233 is constantly breed from Thorium 232. Breeding of uranium allows longer fuel cycles.

Introduction

In 2002, the Generation IV International Forum (GIF) was founded to propel the future of the Nuclear Power industry through international co-operation using knowledge exchange and research. The activities of which are continued till date, the primary goal of the GIF was to design the next generation of reactors that will enable us to achieve the sustainable and efficient usage of uranium resources. The GIF identified six potential reactor designs that meet its basic goal which will also provide the highest level of nuclear safety, resistance to nuclear proliferation, minimization of radioactive waste production and other applications such as industrial heat, hydrogen production and so on [1],[2].

One of the most promising and matured Gen-IV design is the Sodium Cooled Fast Reactor (SFR). Many countries, such as Russia, France, China and India, have shown their interest in SFR technology building more than half a century of experience. As of 2019, there are six reactors under operation: the Russian (BN-600, BN-800, BOR-60), the Chinese (CEFR) and the Indian (PFBR, FBTR) SFRs [2]. The most advanced operating Generation-IV reactor is the Russian BN-800 which started its operation in 2016. The European project ASTRID based in France was unfortunately frozen in later stage of its development in 2019 [3]. The USA recently raised interest in the SFR technology, as in March 2019 the Department of Energy (DOE) announced development of sodium Versatile Test Reactor (VTR) based on the GE-Hitachi PRISM technology [4].

Thorium was identified as a prospective nuclear fuel and an alternative to uranium way back in the 1950s. Thorium naturally occurs as Th-232 and is about three times more abundant than uranium [5]. Currently, there is no commercial-scale thorium fuelled nuclear reactors under operation, meaning all that resource is left practically unused for energy production. One of the main reasons is that thorium has no fissile isotope; therefore it needs either uranium or plutonium as drivers for reactor cores to initiate breeding of fissile U233. It can be breed in both fast and thermal spectrum which makes it different than uranium-plutonium which demands fast neutron spectrum for breeding [2]. In the early days of nuclear power, the development was directed into usage of uranium. Partly because early reactors were built to produce military plutonium and thorium produces weapons-grade material, which is difficult to handle, which means it is a profit when you look from the proliferation resistance point of view. Thus, uranium technology was matured and preferred to be used as it was more problematic to start development of new thorium technology with the resources back then. Renewal interest in new reactor designs and trend of decarbonisation makes the time more apt for thorium than ever before.

Thorium's ability to breed fissile material and produce relatively small traces of plutonium and minor actinides makes it considerable fuel for the Gen-IV requirements creating a larger interest to use thorium-based nuclear fuel [2],[5]. Most of the reactor designs suit thorium as a fuel, especially Molten Salt Reactors and High-Temperature Reactors, but it can also be utilized in Sodium Cooled Fast Reactors. Although, it is believed that the MSR will be most efficient, but the technology is still in early development phase, and it will take many years of research before it can make it to the commercial stage [6].

https://doi.org/10.1051/e3sconf/201913701030

There are arguments that the application of thorium in fast reactors is not very competitive when comparing its application in thermal designs. Those arguments are mainly due to the possibility to use the large depleted uranium stockpiles, which are the heritage of the Cold War and by-product of enrichment process. In the future, in the event of uranium shortage thorium can be a serious alternative and its application in fast reactors should not be discarded. In their paper [6], Gyorgy and Czifrus studied different Gen-IV designs where they showed that fast spectrum reactors (SFR and LFR) have neutronic advantages in comparison to other designs using thorium. Moreover, due to its lower MA production thorium can be used in nuclear waste burner reactors which leads to the scope and motivation of this paper.

The inspiration of the current work was the paper by Gyorgy and Czifrus [7], who studied the application of thorium in large 3600MWth core based on OECD design with oxide fuel which was proposed in "Benchmark for Neutronic Analysis of Sodium-Cooled Fast Reactor Cores with Various Fuel Types and Core Sizes" [8]. On the contrary, in this paper, the OECD 1000MWth medium-size oxide core design was used as a reference. This design is based on the Advanced Burner Reactor (ABR) [9],[10], a core dedicated for nuclear waste incineration and electricity production studied for several years by Argonne National Laboratory [9].

The main objective of this project was to propose a core design that can use mixed oxide fuel with thorium, plutonium and minor actinides [11] and that can be used for long-lived nuclear waste incineration. Then, the objective was to study the developed core from the neutronic point of view. In this paper an emphasis was put on the isotopic inventory changes in time.

2 Methodology

2.1 Tools

The three-dimensional continuous energy Monte Carlo reactor physics code, the Serpent 2 was applied [12]–[14]. It solves neutron-transport equation allowing to study static-criticality properties of a core, and it also allows to simulate fuel depletion using the solution of Bateman equations. All calculations were performed using the ENDF-VII nuclear data library.

2.1 Studied reactor cores

The reference core was the 1000MWth oxide core with (U,TRU)O2 fuel described in the SFR OECD Benchmark [15]. It was three-zone design with 30 inner, 90 middle and 60 outer fuel assemblies. The reference core was solved using the Serpent 2.1.28 code in a previous study [16], which served as the base for our current expedition. The thorium core studied in this work has the same number of assemblies and dimensions but the core pattern, fuel materials and enrichment zones are different – see Figure 1. The design of assemblies for reflectors, control and structure are the same as the previous study [16].

The radial core layout consisted of two rings: inner and outer core rings with slightly higher fuel enrichment in the outer core to distinguish them. The number of inner and outer assemblies were 108 and 72 respectively. In total 180 fuel assemblies were equipped in the core design. Such a design was chosen firstly, to have enough inner irradiated fuel for the next cycle. Secondly, the inner core should contain more thorium Th232 due to the fact that neutron population is higher towards the centre of the core leading to higher transmutation to Pa233 which indeed would lead to higher production of fissile U233 [7].

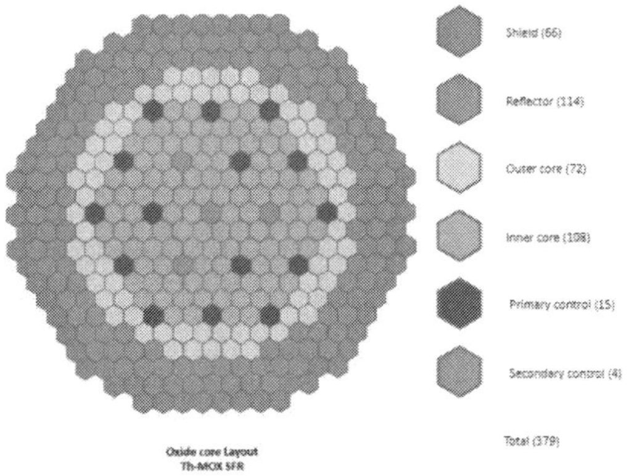

Fig. 1. 1000MWth Th-MOX SFR core map.

2.2 Thorium fuel

The fuel type applied was Thorium Mixed Oxide fuel. It contains reprocessed spent nuclear fuel that contains Plutonium but also Minor Actinides mixed with thorium feed. It can be abbreviated as $(Th,TRU)O_2$ or $(Th,Pu+MA)O_2$. The TRU fuel isotopic vector is the same as oxide charge fuel used by Kim et al. in their ABR study [9] (see Table 2). Kim's mixture was used in OECD study as a fresh fuel in fuel cycle calculations to prepare the equilibrium core used in the benchmark [8].

Studies have shown that thorium mixed fuel can be used in nuclear reactors effectively, and in this work, we extrapolate this result to TRU based mixed oxide [6],[11]. Thorium charge composition is based on the natural thorium mixture.

The fuel pin design and dimensions are the same as in the reference core. It was designed with a smeared density of 85%, to allow the possibility of fuel swelling during normal operation. For simplicity, Oxide-to-Metal ratio was assumed to be 2.0. The thorium fuel density was estimated using the thorium thermophysical correlations implemented in the X-Core Matlab Toolbox [17]. Thorium in the inner core had an atomic fraction of 73% and 72% in the outer core. The remaining (27% and 28%) consisted transuranic elements (TRU), both plutonium and minor actinides. It is to be noted that Pu239 and Pu241 are the main fissile isotopes in the vector shown in Table 2. These two are the driver fuel for the core, what is more, is that the fissile isotopes

 https://doi.org/10.1051/e3sconf/201913701030

fraction is less than 50% of the total TRU fraction. Which means the enrichment for the fuel chosen is less than 20% thus abiding the non-proliferation standards.

Table 2. Isotope vectors used in Th-MOX/TRU fuel charge and at BOL. TRU fuel vector is based on [9].

Isotope	Atomic Fraction [-]
Thorium	
Th230	0.0002
Th232	0.9998
Oxygen	
O16	0.99757
O17	0.00243
TRU	
Np237	0.015
Pu238	0.031
Pu239	0.411
Pu240	0.313
Pu241	0.057
Pu242	0.081
Am241	0.036
Am242m	0.002
Am243	0.026
Cm243	0.0
Cm244	0.018
Cm245	0.005
Cm246	0.003

Enrichment was selected based on initial criticality and fuel burnup studies. The goal was to obtain BOL core eigenvalue comparable to the multiplication factor (k-eff) of the reference core and to keep the reactivity margin to critical state being at least of 0.001 SD guaranteeing that the reactor is operational, and the effect of Monte Carlo statistical error impact is minimized.

2.3 Fuel cycle simulations

The core was assumed to run a cycle with 328.5 effective full-power days, corresponding to one year with 90% capacity factor. Eight burnup steps were applied corresponding to: 0, 1, 5, 25, 75, 115, 185, 250 and 328.5 full power days with a power density 0.06521 kW/gHM throughout the cycle. After each cycle a 90 day decay period was assumed to let the decay of protactinium into uranium and perform proper core reloading (Figure 2). In total five cycles were computed to estimate reaching the equilibrium core state. Static criticality calculations were performed to check whether the excess reactivity stays large enough to allow operation.

The fresh fuel was introduced in the first cycle, and after the end of cycle (EOC) inner fuel was reshuffled to the outer core upon completion of the decay period (see Figure 2). The same reshuffling pattern was used throughout the remaining cycles. In other words, the fuel resides two cycles in the core, one cycle in the inner core; then it is moved to the outer core and finally to reprocessing or spent fuel storage. The one cycle irradiated fuel composition was decayed using the decay function of SERPENT. The decay steps chosen were 1,5,15,25,35,45,65,75 and 90 days. The fission products and the HM isotopes tracked during the decay process are essential for calculating the irradiated fuel composition which will be the fuel input for the outer assemblies in the following cycle. The decay process entirely focuses on the decay of Pa233 to U233; the FP tracked were same as that from our previous study [16].

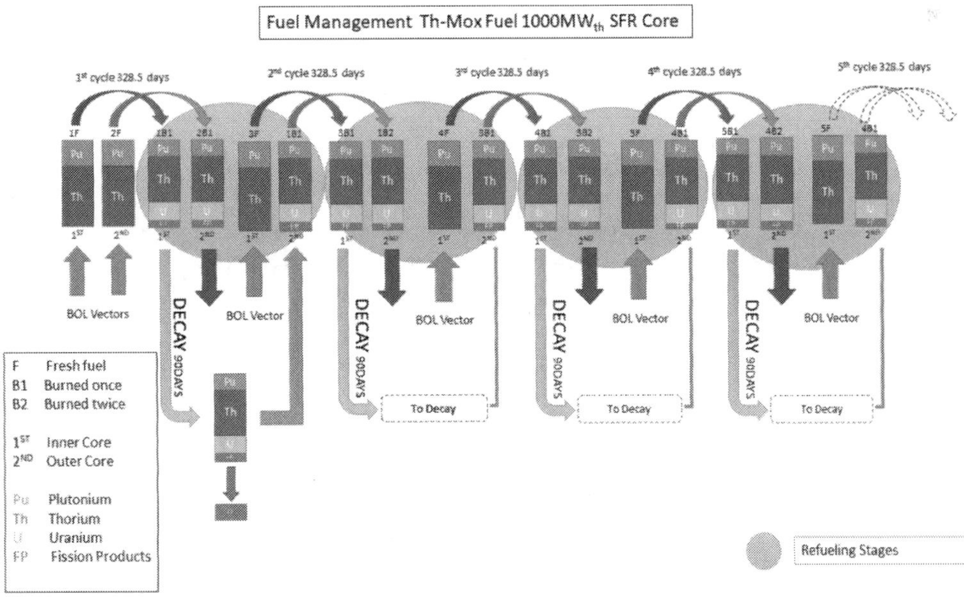

Fig. 2. Fuel management overview. Flowchart inspired from [18].

Figure 2 shows the fuel management strategy implemented in this work and such a strategy was chosen comes down to two primary objectives: the possibility to use the plutonium (and minor actinides) stockpiles appropriately and to reduce the usage of fresh fuel thereby reducing fuel production effort. Bearing in mind the scope of the paper, the applied approach was simplified to study only the fuel usage in the reactor, not the whole fuel cycle with spent fuel and reprocessing strategy. In future research it is possible to address various applicable cycles in more detail. Moreover, the studied core can operate much longer without reloading adding flexibility to the fuel management cycle, and fuel can be reloaded more than once as achieved burnup is relatively low.

For fuel burnup calculations, core was divided into two radial zones and five axial zones, hence, totally ten different burnup zones. Depletion computations for five fuel cycles were performed with relatively low neutron population, as this work is a preliminary study. Therefore, a neutron population of 30000 in 250 cycles and 100 inactive cycles was introduced to establish a balance between the computational time and power at hand while making sure that the maximum Relative Statistical Error (RSE) did not exceed 0.00057.

3 RESULTS AND DISCUSSION

The effective multiplication factor (k-eff or eigenvalue) along with the change in the fuel inventory were observed closely. The time evolution for five cycles is shown in Figure 3, and core inventory at BOC and EOC is presented in Table 1.

The core is approaching the equilibrium state during the 2nd cycle; hence, both k-eff and core inventory above second cycle are similar. Results are compared with reference equilibrium core - light blue in Figure 3 [16]. The BOC eigenvalue is close to the reference core, as it was one of the design constraints during selection of the core configuration. One can observe that the k-eff for EOC state is also very close to the reference core as burnup configuration was similar.

The substantial initial drop in eigenvalue for each cycle (Figure 3) is due to the production of Pa233, neutron poison from Th232. Pa233 decays to U233, until it reaches saturation concentration after which the poisoning effect stays constant. In every cycle, Pa233 reaches an equilibrium after ~100 days and then breeding of fissile U233 continues steadily introducing additional reactivity to the core.

The conversion ratio (CR) for each cycle was ~0.96 (Table 1). This parameter is the ratio of fissile materials breeding rate per fissile materials consumption rate, and it met the criteria for sustainable design (CR~1). Generally, for a core to be a breeder the CR should be higher than one, for break-even (ideal converter) core it should be close to one and for converter core CR is lower than one.

The reference ABR core conversion ratio was slightly lower CR~0.9 and CR_{TRU} ~0.75. The behaviour of the reference core is different as eigenvalue is steadily decreasing during the cycle (Figure 3) as fissile materials are consumed. For the studied thorium core after the decay period, the eigenvalue stays constant, and it is caused by the breeding of fissile U233 which compensates burnup.

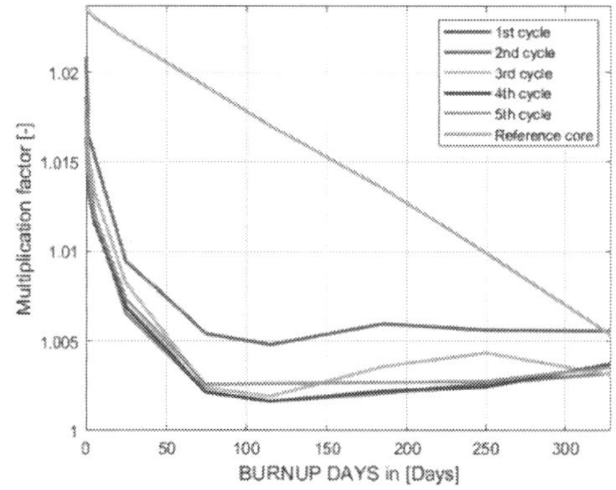

Fig. 3. Effective multiplication factor for five cycles of thorium core compared with reference uranium-plutonium core.

The overall mass balance for basic elements and groups of elements, between the BOC and EOC (at shutdown) is shown in Table 1. We can observe that

Table 1. Cumulative masses of isotope groups in [kg]. TRU consumption is in [kg/TWh]. CR is dimensionless [-].

Cycle	1		2		3		4		5	
Stage	**BOC**	**EOC**	**BOC**	**EOC**	**BOC**	**EOC**	**BOC**	**EOC**	**BOC**	**EOC**
U233+Pu239+Pu241	1939	1876	1906	1848	1902	1843	1906	1865	1901	1841
Fissile	2507	2416	2456	2375	2387	2367	2456	2397	2387	2366
U	0	219	116	316	175	362	116	268	167	354
Th	10682	10395	10666	10395	10626	10361	10666	10463	10634	10369
Pu	3712	3401	3535	3251	3470	3198	3535	3320	3482	3208
MA	441	423	427	409	423	405	427	414	423	405
TRU	4153	3824	3962	3660	3894	3603	3962	3735	3905	3614
TRU consumption	41.8		38.4		36.8		28.9		36.9	
CR	0.9648		0.9669		0.9616		0.9681		0.9641	

https://doi.org/10.1051/e3sconf/201913701030

about ~250 kg of fertile Th232 is transmuted during each cycle, and it is transformed at least partially to fissile U233. In the considered design there is relatively no traces of U235 and U238 compared to U233, and the production of higher actinides is reduced in comparison to cores based on uranium-plutonium cycle. We can find that mass of fissile materials stays almost constant and it agrees with observed conversion ratio. The net balance of plutonium indicates consumption, and it is similar for minor actinides (elements with atomic number >92), as they are effectively consumed. As a consequence, the burner side of the core comes into play with the burning TRU (Pu & MA) isotopes mentioned. TRU consumption rate measures the amount of transuranic elements consumed per terawatt hour of power produced, during each cycle it was ~28-42 kg/TWh (Table 1) and it higher than for the reference design [9] which was ~14 kg/TWh.

The calculated burnup increase at the end of each cycle was only ~22 GWd/tHM. Hence, after two cycles, fuel is far from reaching high burnups (~150 GWd/tHM) possible for fast reactors. Otherwise, the reference core design is not able to reach higher burnups as it will lose reactivity – see Figure 3. As an expected outcome of U233 production, there is no substantial reduction in k-eff but also the thorium core can hypothetically operate without refuelling for more extended time period than reference design. Longer operation was tested for the first cycle, and reactivity margin is high enough that the core can still operate for at least 500days and possibly even more – see Figure 4.

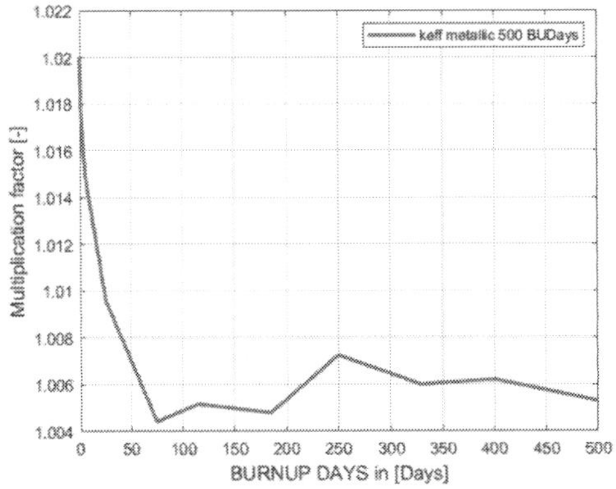

Fig. 4. First Fuel Cycle Simulation Effective Multiplication Factor for 500 EFPDs.

4 CONCLUSIONS AND SUMMARY

A medium-sized 1000MWth Sodium Cooled Fast Reactor operating on Thorium Mixed Oxide (addition of Plutonium with Minor Actinides) fuel was studied. The emphasis was put into perspective the fuel cycles, actinides incineration and core inventory for the proposed core design.

The uniqueness of the studied thorium core is the fact that despite of high TRU consumption, we are not losing fissile material, and reactivity removal is substantially slower than in reference ABR uranium-plutonium core.

The irradiated thorium is transmuted to U233, which is not a TRU element but is fissile, and it induces additional positive reactivity. In traditional uranium-plutonium cycle to produce reactivity it would be necessary to produce plutonium and other TRU from uranium. What is more, irradiation of U-233 produces less higher actinides than U-235 and U-238 in fast spectra as it demands more neutron captures.

A point worth mentioning is: the TRU consumption is a goal for burner reactors as TRU elements are responsible for long term radiotoxicity of nuclear waste.

The presented design can hypothetically be used to burn the existing TRU stockpiles (plutonium and other higher actinides) and breed uranium from thorium. The criticality calculations showed that the reactor is critical throughout all five cycles with fissile material enrichment less than 20%. The core can be operated well beyond its planned outage after 327 days, and it is expected that it could operate for at least 500 days with proper excess reactivity margin.

The study of the core safety parameters is ongoing, and it was not reported in this paper. In principle, it seems that the presence of thorium slightly degrades safety parameters. The presented results are preliminary and simulations were performed for relatively low neutron populations. In the next stage of the work simulations will be performed for much larger neutron populations to obtain results with substantially reduced Monte Carlo statistical errors. Further research will aim to assess safety aspects, study alternative core configurations, extended core operation and develop more optimized fuel cycles. The complete study of the whole fuel cycle with reprocessing would be a challenging endeavour.

References

1. OECD/nea, "Technology Roadmap Update for Generation IV Nuclear Energy Systems: Preparing Today for Tomorrow's Energy Needs" (2014).
2. Alan e. waltar, D. R. Todd, and P. V. Tsvetkov, *Fast Spectrum Reactors*, Springer (2013).
3. IAEA, "Liquid Metal Cooled Reactors: Experience in Design and Operation," *Atomic Energy*,(IAEA-TECDOC-1569) (2007); https://doi.org/IAEA-TECDOC-1569.
4. World nuclear news, "US DoE's Versatile Test Reactor Project"; http://www.world-nuclear-news.org/Articles/US-launches-test-reactor-project.
5. World nuclear association, "Thorium;" https://www.world-nuclear.org/information-library/current-and-future-generation/thorium.aspx.
6. H. György and S. Czifrus, "The utilization of thorium in Generation IV reactors," *Progress in Nuclear Energy*, **93**, pp. 306 (2016); https://doi.org/10.1016/j.pnucene.2016.09.007.
7. H. György and S. Czifrus, "Investigation on the potential use of thorium as fuel for the Sodium-cooled Fast Reactor," *Annals of Nuclear Energy*, **103**, pp. 238 (2017); https://doi.org/10.1016/j.anucene.2017.01.030.
8. N. E. Stauff et al., "Benchmark for Neutronic

Analysis of Sodium-cooled Fast Reactor Cores with Various Fuel Types and Core Sizes," pp. 87 (2016).

9. T. K. Kim et al., "Annals of Nuclear Energy Core design studies for a 1000 MW th Advanced Burner Reactor," *Annals of Nuclear Energy*, **36** (3), pp. 331, Elsevier Ltd (2009); https://doi.org/10.1016/j.anucene.2008.12.021.
10. R. Grandy, c., seidensticker, "Advanced Burner Reactor 1000MWth Reference Concept" (2007).
11. K. Insulander björk, *Thorium Fuels for Light Water Reactors Steps towards commercialization* (2015).
12. J. Leppänen, "PSG2 / Serpent – a Continuous-energy Monte Carlo Reactor Physics Burnup Calculation Code," pp. 157 (2011).
13. J. Leppänen, *Development of a new Monte Carlo reactor physics code*, in VTT Publications(640) (2007).
14. J. Leppänen et al., "The Serpent Monte Carlo code: Status, development and applications in 2013," *Annals of Nuclear Energy*, **82**, pp. 142 (2015); https://doi.org/10.1016/j.anucene.2014.08.024.
15. C. Paper et al., "Evaluation of Large 3600MWth Sodium-Cooled Fast Reactor Neutronic OECD Benchmarks" (2014).
16. P. Darnowski and A. Mikołajczak, "Monte Carlo simulations of the 1000 MWth SFR OECD/NEA benchmark with SERPENT code," *AIP Conference Proceedings*, **2116** (July), pp. 450067 (2019); https://doi.org/10.1063/1.5114534.
17. P. Darnowski, "X-Core Nuclear Engineering Matlab Toolbox," GitLab Repostory; https://gitlab.com/darczu/x-core.
18. P. Stanisz, J. Cetnar, and M. Oettingen, "Radionuclide neutron source trajectories in the closed nuclear fuel cycle," *Nukleonika*, **64** (1), pp. 3 (2019); https://doi.org/10.2478/nuka-2019-0001.

https://doi.org/10.1051/e3sconf/201913701031

RESEARCH & DEVELOPMENT IN POWER ENGINEERING, 2019

Characteristics of agro and wood biomass combustion in the stream of inert material

Konrad Kaczyński[1,*], Katarzyna Kaczyńska[1], and Piotr Pełka[1]

[1] Czestochowa University of Technology, Faculty of Mechanical Engineering and Computer Science, Institute of Thermal Machinery, al. Armii Krajowej 21, 42-201 Czestochowa, kkaczynski@imc.pcz.pl, Poland

Abstract. Agricultural residual biomass presents a high potential for energy use around the world, often not utilized to a large extent due to its significant differences with respect to other biomass types, such as the one of wood origin. These differences are mainly related to the characteristics of its ashes (quantity and composition) which increase certain problematic phenomena during combustion, among them ash sintering. The main purpose of this article is the experimental study of these issues for various agro pellets and wood pellets, analyzed in various operating conditions in a laboratory reactor with a circulating bed.

1 Introduction

In view of the world's population continues to grow at a high rate, such that today's population is twice that of 1960, and is projected to increase further to 9 billion by 2050 [1] increased energy demand is inevitable. In the year 2018, the world's primary energy consumption increased almost twice as fast as the average rate recorded since 2010 and is approx. to 14.3 billion tons of oil equivalent which is about 599 EJ (exajoules), along with an average growth rate of 2.3% per year [2]. The production of energy by burning solid fuels as the main item in energy mixes is prevalent across many countries in the world, including the largest energy producers, such as the USA, China, India and most of the European Union (EU) countries. It seems that coal fuels will remain an important element of their economies for at least several dozen years [3]. However, a key policy solution to mitigate climate change is to increase renewable energy production [4]. The renewable energy market is expected to grow dynamically in the near future, and play a significant role in the world® future energy mix. The Paris agreement-2015 is the first agreement in the world to bind the treaty to reduce climate change [5]. The sustainable development goals of the United Nations also underline the importance of energy. 'Affordable and clean energy' in Goal 7 and 'Climate action' in Goal 13, highlight the significance of new and renewable energy to transform the world [6]. In 2018, renewable energy accounted for around 26% of global electricity production. Hydropower remains the largest renewable source, satisfying 16% of global electricity demand, followed by wind (5%), solar farm (2%) and bioenergy (3%) [2]. Due to the wide availability of biomass around the world, it is a renewable energy source with high growth potential [7]. Biomass is referred to as a carbon neutral fuel because there is no net addition of carbon dioxide in the atmosphere unlike fossil fuels [8]. Biomass can be transformed into solid, liquid and gaseous form using modern technologies, and thus becomes an efficient and clean energy supplier for all sectors such as heat, power and transport fuel [9]. There are two routes for biomass conversion, namely the biochemical route and the thermochemical route. The biochemical conversion process uses enzymes, bacteria or other microorganisms to convert lignocellulosic biomass into liquid biofuels. In the thermochemical conversion process, biomass converts to any form of energy in the presence of heat and with a controlled supply of oxygen. In comparison to the biochemical route, the thermochemical route has certain advantages, such as the utilization of entire biomass, faster kinetics, and flexibility in the feedstock [10]. There are three main thermochemical pathways, such as direct combustion [11,12], gasification [13], and pyrolysis [14]. Until now, mainly biomass from forest origin (residue after logging and wood processing up to 27% of wood mass) was used for energy purposes [4,15]. However, along with the growing demand, the non-wood biomass market developed - originating, among others, from agricultural crops, energy plantations. In the energy sector, processed biomass is most commonly used, in the form of woodchips, sawdust, pellets and briquettes. The advantage of using biomass as fuel is the possibility of rapid technological adaptation of coal installations. It is very beneficial to use fluidized bed combustion technology due to the possibility of reducing emissions of environmental pollutants arising during combustion and the possibility of combusting inferior fuels, e.g. biomass. In the case of wood biomass, comprehensive knowledge on thermal energy conversion has long been accumulated and a number of studies have been published [16-22], however, in the case of agricultural biomass as well as energy plantations, more research and analysis is still needed due to the fact that thermal the conversion of these fuels is significantly different from the conversion of wood biomass. A lot of studies concerned the combustion of pellets in grate boilers. At work [23] NO_X emission behavior of three pelletized energy crops was assessed.

The experiment found that the ash content and chemical composition of fuels had a significant impact on the need for combustion air. During the experiment [24] various fuels obtained from new energy crops - herbaceous plant, poplar and manioc - were burned. Particular emphasis was placed on explaining the role of the main elements forming ash, sediments and emissions of solid particles (fractions of fine and coarse ash particles) based on experimental data. The main purpose of the work [25] was to assess the suitability of multi-fuel combustion technology of various pelletized energy plants in terms of the required adaptations of four combustion parameters, i.e. the ratio of charge, lambda, primary / secondary air distribution and residence time on the grate, leading to the best combustion efficiency in the system. The publication [26] quantifies the release of K, Cl, S and P for two species of energy plants and their impact on the ash flow phenomenon. However, the issue of agro biomass combustion in a small-scale fluid bed was discussed in a review article [27], where the most important problems concerning biomass combustion were presented, paying special attention to the presence of alkali metal compounds. The mechanisms leading to slagging and contamination of the heating surfaces of boilers have been explained. Experimental research on the biomass of various origins was presented in [28], where the process of combusting spherical "briquettes" (obtained in the process of grinding and pressing) of agro biomass (wheat straw and energetic willow) and wood biomass (Scots pine) in the circulating fluid layer was analyzed at 850 ° C. In order to determine the fuel combustion kinetics, a single briquette was introduced into the combustion chamber, which was placed in a specially constructed measuring system. Surface and pellet center temperatures were measured with thermocouples. Simultaneously with the measurement of the sample temperature, ignition time, volatile component burning time and total pellet burning time were measured. However, for the analysis of pollutant emissions, briquettes were burned directly in a fluidized layer at 850 ° C. Emissions of pollutants (NO, N_2O, NO_2, SO_2, CO, HCl, HF and HCN) were measured and oxygen concentration measurements were made. However, to characterize such a versatile fuel as biomass, further experimental research is necessary. There are few analyzes in the literature regarding the combustion of a single biomass particle. In contrast to research [28], the experiment described in this paper was conducted using processed biomass fuels specially dedicated to combustion in the power industry.

2 Biomass as fuel

The components of biomass include cellulose, hemicelluloses, lignin, lipids, proteins, simple sugars, starches, water, hydrocarbon, ash, and other compounds. The concentration of each class of compound varies depending on species, type of plant tissue, stage of growth, and growing conditions. [29]. The bulk composition of biomass in terms of carbon, hydrogen, and oxygen (CHO) does not differ much among different biomass sources. Typical (dry) weight percentages for C, H, and O are 30 to 60%, 5 to 6%, and 30 to 45% respectively. [30].

2.1 Effects of certain elements of fuel

Certain elements in biomass deserve special attention. These include chlorine, potassium, sulfur, nitrogen, and silicon. Perhaps the most important element, with regard to its behavior in different combustion related problems is chlorine. Chlorine, which is found in high quantities in certain biomass types, such as straw, may affect operation by corrosion. The high chlorine and alkali content of some biomass fuels can cause severe damage to the combustion units. Chlorine concentration often dictates the amount of alkali (potassium) vaporized during combustion as strongly as it dictates the transport of alkali from the fuel to surfaces, where the alkali often forms sulfates. In the absence of chlorine, alkali hydroxides are the major stable gas phase species in moist, oxidizing environments i.e., combustion gases [29]. High percentages of silicon together with potassium and chlorine also play their role in causing severe ash deposition problems at high or moderate combustion temperatures. Potassium is the dominant alkali source in most biomass fuels [29, 31, 32]. Biomass is usually low in sulfur. The high percentages of these elements (Cl, K, Si) also pose grave concerns for bed agglomeration problems in fluidized bed boilers.

2.2 Ash

The major inherent ash forming elements in biomass include Si, Al, Ti, Fe, Ca, Mg, Na, K, S, and P. Table 1 shows the ash analysis of some selected biofuels and bituminous coal [27]. High alkali (potassium) and Si content typically gives low ash melting temperatures while Mg and Ca increase ash melting temperature [27,31]. Although a major fraction of low melting point alkali is released into the gas phase, the part left in the ash may deliver detrimental effects (bed agglomeration) during fluidized bed combustion. Ash sintering, softening, and melting temperatures can differ significantly among biofuels, and this characteristic is essential in determining temperature control to avoid sintering or slagging (resulting in deposits on, e.g., boiler tubes).

3 Motivations

The aim of the experimental research was:

- analysis of the combustion process of selected biomass fuels dedicated to combustion in technology with CFB,
- comparison of the agro and wood biomass combusting process,
- comparison of the combustion mechanism of selected pellets in clean air and the conditions of the circulating fluidized bed,
- detailed analysis of the mass loss of pellets combusted at variable fluidized bed temperatures.

Table 1. Ash analysis of some biofuels and bituminous coal (mass basis % ash) [27].

Fuel	SiO_2	Al_2O_3	Fe_2O_3	Mn	MgO	CaO	Na_2O	K_2O	TiO_2	P_2O_5	SO_3
Wood pellets	4.3	1.3	1.5	5.9	8.5	55.9	0.6	16.8	0.1	3.9	1.3
Sunflower pellets	2.9	0.6	0.8	0.1	21.6	21.6	0.24	22.8	0.1	15.2	14.0
Straw	53.1	3.6	1.2	---	3.0	17.7	4.5	30.0	---	4.1	---
Greenhouse residue	28.4	3.9	18.4	0.3	5.7	25.8	0.8	9.7	0.8	3.8	---
Bituminous Coal	59.7	20.3	7.0	<0.01	1.9	1.8	1.0	2.3	0.9	0.1	1.3

4 Experiment

During experimental research was analyzed the mass loss of biomass pellets. In the research two types of fuels were used: wood and agro-biomass. The research was carried out in three different temperatures of the combustion chamber: 850°C, 750°C and 650°C. The research was performed without inert material and in a stream of inert material: Gs=2,5kg/m2s and Gs=5kg/m2s. The values of the stream of inert material are characteristic for the upper and middle areas of the real zone of the fluidized bed boilers. The temperature of 850 ° corresponds to the combustion temperature of fuels in professional fluidized bed boilers. The conducted experiments may also prove to be helpful due to the possibility of using biomass in other processes, among others rapid pyrolysis and gasification [13,14].

4.1 Experimental stand

The research was carried out on a specially constructed test stand. The experimental stand made it possible to model the conditions of the circulating fluidized bed. the experimental stand is presented in Fig. 1. [33].

4.2 Research material

Four types of biomass pellets, each in the shape of a cylinder, were used for the study. Oak sawdust pellets, a mixture of beech and oak sawdust and straw had a diameter of 6mm, while the diameter of pellets from sunflower husk was 8mm. The mass of the samples was 0.73 ± 0.03g. Table 2 contains the results of the technical analysis of the tested fuels.

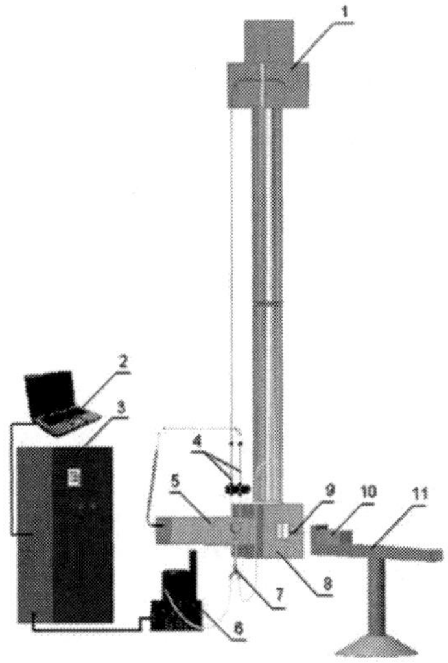

Fig. 1. The layout of the test apparatus. (1) vessel of inert material, (2) computer (3) control panel, (4) rotametres, (5) gas heater, (6) ventilator, (7) T-connector, (8) combustion chamber, (9) fuel particle, (10) tensometric branch scale, (11) support.

Table 2. Results of technical analysis of biomass fuels.

Pellets	Volatile [%]	Moisture [%]	Ash [%]	Fixed coal [%]	Combustion heat [MJ/kg]
Oak sawdust	79.6	8.7	1.2	10.5	18.2
30%beech and 70%oak sawdust	77.7	8.8	1.3	10.7	17.9
Sunflower husk	73.8	8.4	5.5	12.3	19.8
Straw	71.8	8.7	12.2	7.3	16.2

5 Results of the experiment

Fig. 2. presents a comparison of pellets combustion time combusted at different conditions in a circulating fluidized bed. Fig. 3. presents percentage comparison of pellets combustion time during reduction of the combustion chamber temperature compared to the reference temperature (850°C). As expected, when combusting pellets without inert material, lowering the temperature in the combustion chamber results in longer combusting time for all types of biomass fuels. The results of mass loss of wood biomass pellets (pellets from oak sawdust and pellets from a mixture of beech and oak sawdust) combusted at different temperatures and at Gs = 2.5 kg/m^{2}s indicate that, as in the case of combustion without inert material, lowering the temperature in the combustion chamber increases the pellets combusting time. It is different in the case of agro biomass combustion, where the combusting time at 850 ° C is the longest. At this temperature the phenomenon of ash softening is manifested by the sticking of sunflower husk pellets through quartz sand. The sand forms a durable surface surrounding the incinerated pellet, consequently preventing the oxidizer from entering the combusted

https://doi.org/10.1051/e3sconf/201913701031

pellet. The residue after combusting sunflower husk pellet at 850°C, tests were carried out in a stream of inert material at Gs = 2.5 kg/m^{2}s was shown in Fig. 4. A similar effect was observed when combusting straw pellets at 850°C. The straw pellets combusted at 850°C were sintered. The residue after combusting straw pellet at 850 ° C, tests were carried out in a stream of inert material at Gs = 2,5 kg/m^{2}s was shown in Fig. 5. Analyzing the combusting of pellets at an increased value of the inert material stream Gs = 5 kg/m^{2}s, in the case of wood biomass it can again be seen that lowering the temperature in the combustion chamber results in longer combusting time. In the case of agro biomass combustion, as in the case of combustion in Gs = 2.5 kg/m^{2}s, during the combusting of pellets from sunflower husk, the formation of a coating was observed, which was partially broken down as a result of the mechanical impact of inert material - Fig. 6. On the other hand, the sinter was formed during combustion of straw was completely broken down. The longest combusting time of agro biomass was recorded at 750 ° C, and the shortest at 650 °C. Which is due to the increased erosive effects of the inert material on the pellet surface at a lower temperature at which ash softening does not occur.

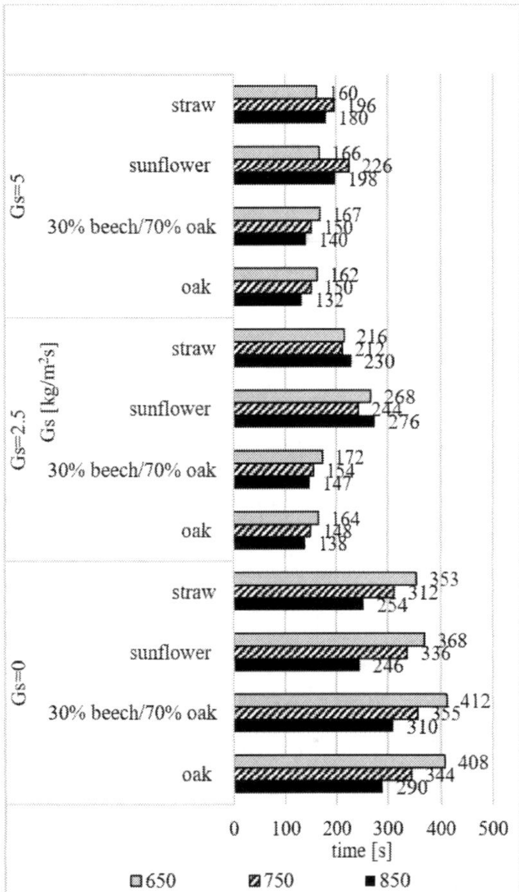

Fig. 2. Comparison of pellets combustion time combusted at different conditions in a circulating fluidized bed.

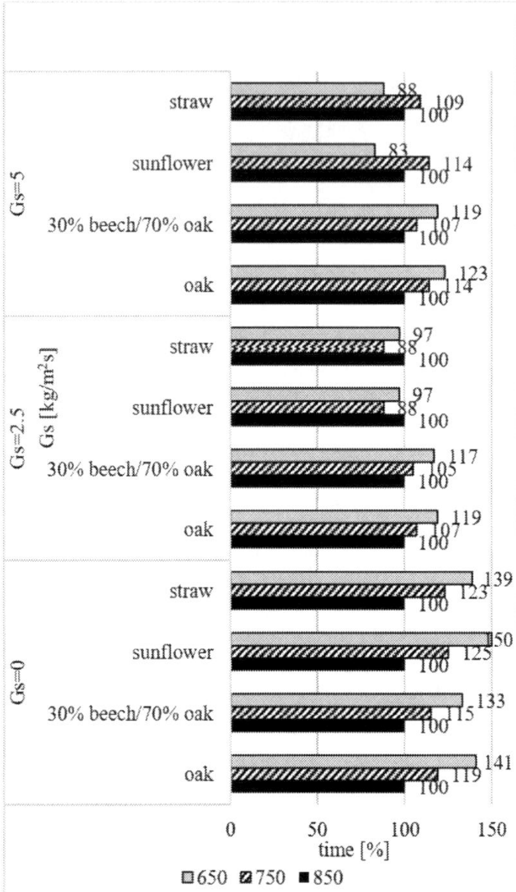

Fig. 3. Percentage comparison of pellets combustion time during reduction of the combustion chamber temperature compared to the reference temperature (850°C).

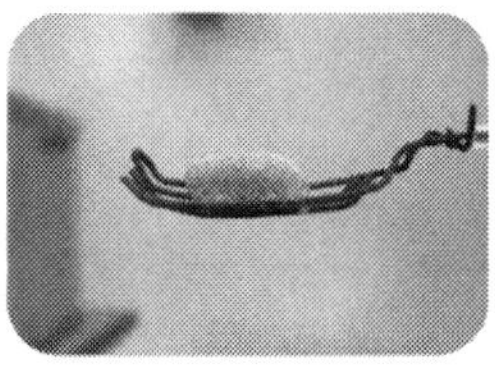

Fig. 4. The residue after combusting sunflower husk pellet at 850°C, tests were carried out in a stream of inert material at Gs = 2,5kg/m^{2}s.

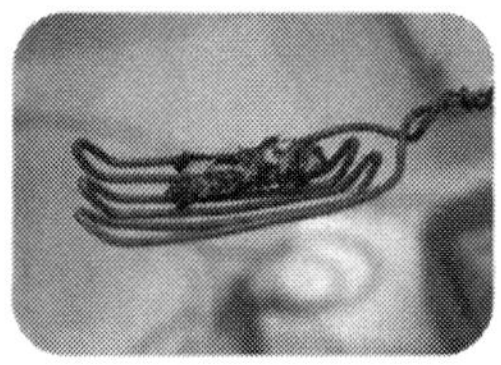

Fig 5. The residue after combusting straw pellet at 850 ° C, tests were carried out in a stream of inert material at Gs = 2,5 kg/m^{2}s.

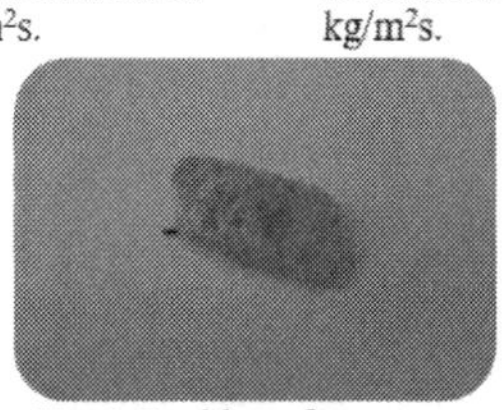

Fig 6. Residue after combusting sunflower husk pellets at 850°C and Gs = 5 kg/m^{2}s.

Figure 7. shows a comparison of pellet combustion times depending on the stream of inert material combusted at different temperatures (Gs = 0 reference level). The presence of inert material accelerates the mass loss of pellets from oak sawdust. In comparison with combustion without inert material, the combustion time in the stream of material Gs = 2,5 kg/m^{2}s at any temperature caused a reduction of the combustion time by about 40-50%. On the other hand, a further increase of the inert material stream to Gs = 5 kg/m^{2}s to a negligible extent accelerated the combustion process. In the case of combusting sunflower seed pellets compared to combustion without inert material, the combustion time in the Gs = 2,5 kg/m^{2}s stream at 650°C and 750°C has been reduced by approx. 30%, and for 850°C combustion time it increased by 12%. Increasing the value of the stream of inert material to Gs = 5 kg/m^{2}s resulted in shortening the combustion time at all temperatures. In the case of straw pellets compared to combustion without inert material, the combustion time in the stream of inert material Gs = 2,5 kg/m^{2}s at temperatures of 650°C and 750°C has been reduced by approximately 30-20%, and for temperatures of 850°C the combustion time increased by 9%. Increasing the value of the stream of inert material resulted in shortening the combustion time at all temperatures.

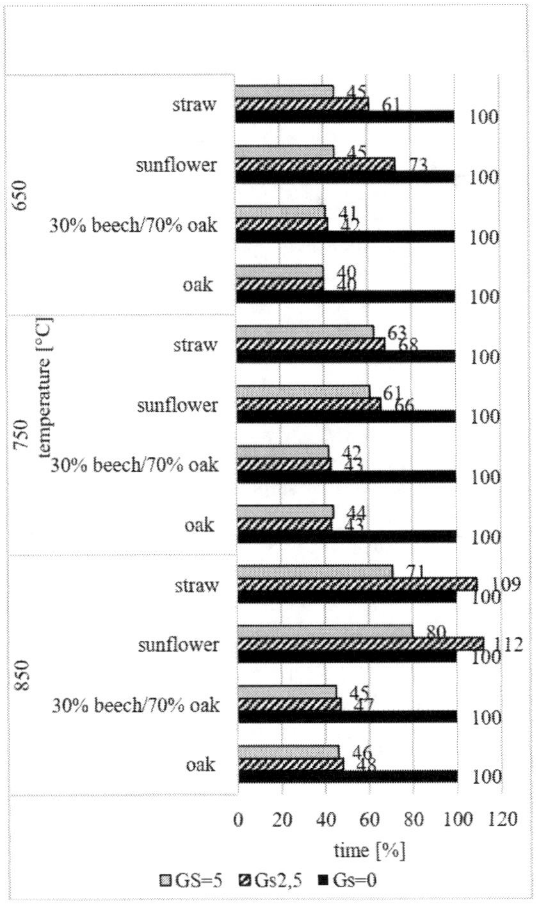

Fig. 7. Comparison of pellet combustion times depending on the value of the Gs material stream combusted at different temperatures (Gs = 0 reference level).

6 Conclusions

The analysis carried out as part of the work facilitates the formulation of the following conclusions:

1. Combustion of all types of pellets without inert material and at different temperatures proceeded in a similar and predictable manner described in subject-related literature. As the temperature dropped, the combustion time was increased.
2. The appearance of inert material caused very significant differences in terms of time and the combustion mechanism. These differences were mainly observed at 850 °C for wood and agro-biomass.
3. Combustion of oak sawdust pellets in the stream of inert material (2.5kg/ m^{2}s and 5kg/m^{2}s) in each analysed chamber temperature intensifies the process of mass loss of the combusted particles - the amount of air reaching directly to the particle increases as a result of removing ash from its surface of gaseous combustion products through a stream of quartz sand.
4. In the case of sunflower husk and straw pellets, combustion in a stream of inert material in Gs = 2.5 kg/m^{2}s lasts the longest at the highest temperature of the furnace (850 °C). In both cases, the ash softening temperature is exceeded and the quartz sand covers the pellets. Quartz sand coating hinders the flow of the oxidizer to the fuel particle. The combustion time is extended.
5. When the inert material stream was increased in the case of sunflower husk pellets, the coating partially disintegrated, and in the case of straw pellets, the sinter was completely disintegrated.
6. During the combustion of pellets from oak sawdust, it was noticed that the stronger parameter determining the intensity of the combustion process in comparison with increasing the inert material stream intensity with Gs = 2.5 kg/m^{2}s per Gs = 5 kg/m^{2}s is increasing the temperature. Increasing the temperature prevailing in the furnace chamber accelerates the mass loss of the pellets combusted (almost 20%) more than the increase of the inert material stream (only 1-2%).
7. Experimental research indicates that pellets obtained from various types of biomass (agro and wood) are characterized by a different nature of combustion in fluidic conditions. The ash of different types of biomass contains compounds of alkaline metals whose content affects its softening and melting point.
8. Therefore, it is necessary to operate biomass boilers that are adapted to the properties of biomass combusted. Incorrect boiler temperatures and Gs intensity to the type of fuel supplied may cause an increase in the loss of incomplete combustion and build-up of sediments on the heating surfaces of boilers, as well as disturb the work of circulatory fluidized bed boilers due to the change in the particle distribution of the layer material.

References

1. Desai, B.G. CO2 emissions—Drivers across time and countries. *Curr. Sci.*, 115, 386–387, (2018).

2. The International Energy Agency, Global Energy & CO2 Status Report 2018, The latest trends in energy and emissions in 2018, (2018).
3. Tchapda A.H., Pisupati S.V., A Review of Thermal Co-Conversion of Coal and Biomass/Waste, *Energies*, 7, 1098–1148, (2014).
4. Greinert A., Mrówczyńska M., Szefner W., The Use of Waste Biomass from the Wood Industry and Municipal Sources for Energy Production, *Sustainability*, 11, 3083, (2019).
5. UNFCCC, United Nations Framework Convention on Climate Change, (2019)
6. UNSDG, United Nations Sustainable Development Goal report, (2019).
7. Li Y., Rezgui Y., Zhu H., District heating and cooling optimization and enhancement—towards integration of renewables, storage and smart grid. *Renew. Sustain. Energy Rev.*, 72, 281–294, (2017).
8. Pradhana P., Mahajanib S.M., A. Aroraa, Production and utilization of fuel pellets from biomass: A review, *Fuel Processing Technology*, 181, 215–232, (2018).
9. Chen W.H., Peng J., Bi X.T., A state-of-the-art review of biomass torrefaction, densification and applications, *Renew. Sust. Energ. Rev.*, 44, 847–866, (2015).
10. Singh R., Krishna B.B., Mishra G., Kumar J., Bhaskar T., Strategies for selection of thermo-chemical processes for the valorisation of biomass, *Renew. Energy*, 98, 226–237, (2016).
11. Geng A., Yang H., Chen J., Hong Y., Review of carbon storage function of harvested wood products and the potential of wood substitution in greenhouse gas mitigation, *Forest Policy Econ.*, 85, 192–200, (2017).
12. Zhang J., Liu J., Evrendilek F., Xie W., Kuo J., Zhang X., Buyukada M., Kinetics, thermodynamics, gas evolution and empirical optimization of cattle manure combustion in air and oxy-fuel atmospheres, *Appl. Therm. Eng.*, 149, 119–131, (2019).
13. Weiland P., Biogas production: Current state and perspectives. *Appl. Microbiol. Biotechnol.* 85, 849–860, 2010.
14. Johansson A.-C., Wiinikka H., Sandström L., Marklund M., Öhrman O.G.W., Narvesjö J., Characterization of pyrolysis products produced from different Nordic biomass types in a cyclone pilot plant, *Fuel Process. Technol.*, 146, 9–19, (2016).
15. Mboumboue E., Njomo D., Biomass resources assessment and bioenergy generation for a clean and sustainable development in Cameroon. *Biomass Bioenergy*,118, 16–23, (2018).
16. Johansson LS., Leckner B., Gustavsson L., Cooper D., Tullin C., Potter A., Emission characteristics of modern and old-type residential boilers fired with wood logs and wood pellets. *Atmos Environ*, 38(25):4183–95, (2004).
17. Obernberger I., Biedermann F., Widmann W., Riedl R., Concentrations of inorganic elements in biomass fuels and recovery in the different ash fractions. *Biomass Bioenergy*, 12(3):211–24. (1997).
18. Wierzbicka A., Lillieblad L., Pagels J., Strand M., Gudmundsson A, Gharibi A, et al. Particle emissions from district heating units operating on three commonly used biofuels. *Atmos Environ*, 39(1):139–50 (2005).
19. Wiinikka H., Gebart R., Boman C., Bostrom D., Ohman M., Influence of fuel ash composition on high temperature aerosol formation in fixed bed combustion of woody biomass pellets. *Fuel*, 86(1–2):181–93 (2007).
20. Wiinikka H., Gebart R., Experimental investigations of the influence from different operating conditions on the particle emissions from a small-scale pellets combustor. *Biomass Bioenergy*, 27(6):645–52, (2004).
21. Brunner T., Obernberger I., Scharler R., Primary measures for low-emission residential wood combustion-Comparison of old with optimised modern systems. *Proceedings of 17th European Biomass Conference and Exhibition*, Hamburg, Germany, p. 1319–28 (2009).
22. Obernberger I., Brunner T., Barnthaler G., Chemical properties of solid biofuels–significance and impact. *Biomass Bioenergy*, 30(11):973–82, (2006).
23. Diaz-Ramirez M., Sebastian F., Royo J., Rezeau A., Influencing factors on NOX emission level during grate conversion of three pelletized energy crops. *Appl Energy*, 115:360–73, (2014).
24. Diaz-Ramirez M., Boman C., Sebastian F., Royo J., Xiong S, Bostrom D., Ash characterization and transformation behavior of the fixed-bed combustion of novel crops: poplar, brassica, and cassava fuels. *Energy Fuels*, 26(6):3218–29 (2012)
25. Diaz-Ramirez M., Sebastian F., Royo J., Rezeau A., Combustion requirements for conversion of ash-rich novel energy crops in a 250 kWth multifuel grate fired system. *Energy*,46(1):636–43, (2012).
26. Diaz-Ramirez M., Frandsen F.J., Glarborg P., Sebastian F., Royo J., Partitioning of K, Cl, S and P during combustion of poplar and brassica energy crops. *Fuel*, 134:209–19 (2014).
27. Khana A.A.,. de Jong W, Jansens P.J., Spliethoff H., Biomass combustion in fluidized bed boilers: Potential problems and remedies, *Fuel Processing Technology*, Volume 90, Issue 1, 21-50, (2009).
28. Kosowska-Golachowska M., Wolski K., Gajewski W., Kijo-Kleczkowska A., Musiał T., Środa K., Spalanie biomasy agro i leśnej w cyrkulacyjnej warstwie fluidalnej, *Rynek Energii*, 3(124), (2016).
29. Jenkins B.M., Baxter L.L., Miles T.R. Jr., Miles T.R., Combustion properties of biomass, *Fuel Processing Technology*, 54 (1–3) 17–46, (1998).
30. Faaij A.P.C., Biomass combustion, *Encyclopedia of Energy* 1 175–191, (2004).
31. Demirbas A., Combustion characteristics of different biomass fuels, *Progress in Energy and Combustion Science*, 30 (2) 219–230, (2004).
32. Demirbas A., Fuel and combustion properties of bio-wastes, *Energy Sources*, 27 (5) 451–462, (2005).
33. Pełka P., Analysis of a coal particle mass loss burning in flow of inert material. *Combust. Flame*, 156, 1604, (2009).

RESEARCH & DEVELOPMENT IN POWER ENGINEERING, 2019

Oxy-combustion of agro and wood biomass in a fluidized bed

Katarzyna Kaczyńska[1,*], *Konrad* Kaczyński[1], and *Piotr* Pełka[1]

[1] Czestochowa University of Technology, Faculty of Mechanical Engineering and Computer Science, Institute of Thermal Machinery, al. Armii Krajowej 21, 42-201 Czestochowa, kgajewska@imc.pcz.pl, Poland

Abstract. Present trends in the global energy sector effectively drive the development of technologies using renewable energy sources and the improvement of combustion technology towards increasing its efficiency and removing environmentally harmful flue gases. One solution is to replace fossil energy sources with renewable sources, including biomass. The degree of biomass utilization depends on the size of resources and processing technology. There are more and more different types of biomass on the market. There are many technologies for its energetic use. In recent years, fluidized combustion technology has been considered as one of the main directions of development of professional energy in the world. Many units in the world in fluidized combustion technology use as primary or supplementary fuel: sludge, wood waste and biomass, as well as municipal waste, etc. Fluidized bed boilers also allow the implementation of the dynamically developing oxy-combustion technology, which perfectly fits the prevailing trends due to the key advantages of increased energy conversion, as well as the possibility of direct sequestration of carbon dioxide. During experimental research, the mass loss of biomass pellets of various origins was analyzed in an oxidizing and inert atmosphere in a two-phase flow with the use of inert material. The obtained results allow to state that the large variety of biomass offered by suppliers requires a thorough knowledge of its properties and determination of its impact on the process and installations when used in the power industry.

1 Introduction

Growing concerns on the greenhouse gas emissions and their potential impact on climate change demand not only the application of CO_2 capture and storage (CCS) technologies to large point anthropogenic CO_2 emission sources such as coal and natural gas fired power plants but also the implementation of CO_2 negative combustion technologies such as Bio-energy with Carbon Capture and Storage (BECCS) with in the next decades. [1,2]. Biomass is considered as a renewable fuel, a carbon-neutral energy source [3], and hence its combustion integrated with CCS can lead to negative CO_2 emissions. Biomass has already captured worldwide attention in the context of greenhouse gas control even though fossil fuels are expected to retain their dominant role in the world energy supply in the coming decades [4]. Oxy-combustion is one of the most developed CCS technologies and considered as technically feasible and economically competitive for future commercial applications [5, 6, 7]. Oxy-combustion refers to fuel being burned in a mixture of oxygen and recycled flue gas (RFG). Unlike conventional air combustion plants that use air as the oxidant, an oxy-fired plant employs an Air Separation Unit (ASU) to produce an almost pure oxygen stream. The oxygen stream is then combined with RFG to produce an oxygen enriched gas as the oxidant. The flue gas recycle is necessary to moderate the otherwise excessively high flame temperature that would result from fuel combustion in pure oxygen. After the removal of water and other impurities from the flue gas exhaust stream, high-purity CO2 (up to 95%) is produced and almost ready for sequestration [1].

Among all the available combustion technologies, fluidized bed combustion (FBC) is often considered as the best choice for the combustion and/or co-combustion of biomass, waste and other low-quality solid fuels due to its fuel flexibility, long residence times, and uniform combustion temperatures. The characteristics of FBC also offers several advantages for its application in oxy-fuel systems [8]. Firstly, the difficulty of flue gas recirculation for temperature control in pulverized fuel (PF) applications could be reduced in circulating fluidized bed (CFB) by means of the bed material recirculation, since the specific heat of the solids is much higher than that of the recycled flue gas. Secondly, lower NOx emissions and better sulphur removal are possible. Finally, it is easier to retrofit a fluidized bed boiler from air to oxy-fuel combustion as there will be no need for a new burner. The effects of oxy-fuel atmosphere and O_2 concentration in fluidized bed systems firing different kinds of coal have been thoroughly investigated during the past years by a number of researchers [8-18] in which the advantages of oxy-combustion have been confirmed, namely: increased energy conversion and the possibility of carbon dioxide sequestration. In general, the experimental results showed that NOx emissions in oxy-fuel combustion with low O2 concentrations are lower than those obtained under air-firing atmosphere Furthermore, NOx emissions were found to increase with the increasing O_2 concentrations in the oxy-fuel oxidant, which is as a result of a) the increase of the temperature in the furnace which elevates the concentrations of O and OH radicals and enhances NO formation and b) the lower gas velocity in the riser and longer residence time of fuel particles in the combustor, which may promote the fuel-N conversion into NOx precursors [13-16]. In some studies, an opposite trend was found, i.e. a decrease of NOx with the increasing O_2 concentration in the oxy-fuel oxidant, as in the work of de las Obras-Loscertales et al. [17], The authors explained this trend by means of the different operational procedure used, comparing with other investigations: in their work [15], an increase in the oxygen concentration was compensated with an increase of the coal flow rate fed to the reactor, keeping constant the total gas flow rate and excess oxygen coefficient in all tests. As a result, more unconverted char was present in the bed, favouring the NO reduction on the char surface.

So far, few researchers have analyzed oxy-fuel combustion in fluidized bed reactors fueled with 100% biomass fuels, so further research is needed.

Biomass properties differ from those of coal in many important ways which results in different combustion behaviours [19]. For example, biomass generally has less carbon, more oxygen, higher hydrogen content and lower heating value. There are huge differences in volatile matter contents between biomass and coal: biomass can lose up to 90% of their masses (as volatiles) in its first stage of combustion [20].

Duan et al. [21] conducted a series of experiments firing three kinds of Chinese biomass fuels, i.e. rice husk, wood chips and dry wood flour, under air and oxy-fuel atmosphere in a 10 kWth CFB combustor. The main objective of their study was to investigate the pollutant emissions of the co-firing of biomass with coal under oxy-fuel combustion conditions although experiments firing only the biomass fuels were also carried out for comparison purposes. They observed lower NO emissions in the oxy-fuel atmosphere compared with those with air combustion. This behaviour was explained as the result of the reduced yield of NOx precursors like NH_3 during the devolatilization process and the enhanced NO reductions via char/NO/CO reactions under the oxy-fuel combustion conditions. The objective of the work [1] was oxy-fuel combustion of biomass fuels firing three kind of biomass fuels, two non-woody (miscanthus and straw) and one woody (wood), in a 20 kWth bubbling fluidized bed (BFB) combustor, studying the effects of the combustion atmosphere (air and oxy-fuel) and the oxygen concentration in the oxy-fuel oxidant on the gas emissions and temperature profiles. In contrast, work [21], unlike the abovementioned, was devoted to the analysis of the combustion of a single biomass particle in a fluidized bed in different oxidizing atmospheres. In this research, spherical 8-mm particles of wooden biomass (willow, Salix viminalis) were used. Biomass was combusted in a fluidized bed at 850 ° C and various atmospheres: air atmosphere: 21% O_2 + 79% N_2 and oxidizing atmospheres: 21% O_2 + 79% CO_2, 30% O_2 + 70% CO_2 and 40% O_2 + 60% CO_2. The results of the experiment indicate that the composition of the oxidizing atmosphere strongly affects the combustion process of biomass fuels. The replacement of N_2 in the combustion environment by CO_2 resulted in a slight ignition delay and lower maximum mass loss during the combustion of raw biomass. Increasing the amount of oxidant resulted in shorter burning time. Unlike studies [21], the experiment described in this work was conducted using applications on the market of processed biomass fuels specifically dedicated to combustion in the energy sector.

2 Motivations

The aim of the experimental research was:

1) analysis of the oxy-combustion process of selected biomass fuels dedicated for combustion in technology with CFB,
2) comparison of the agro and wood biomass combusting process,
3) comparison of the combustion mechanism of selected pellets without inert material and in the conditions of the circulating fluidized bed,

3 Experiment

During the experimental research, the impact of inert material and oxidizing atmosphere on the mass loss of pellets of two types of biomass was analyzed: wood biomass - oak sawdust pellets and pellets from a mixture of sawdust 30% beech / 70% oak) and agro-pellets from sunflower husk and pellets from straw. The temperature of 850 ° was corresponds to the combustion temperature of fuels in professional fluidized bed boilers. The research was performed in air atmosphere and gas compound: 21% O_2 and 79% CO_2, 25% O_2 and 75% CO_2, and 30% O_2 and 70% CO_2. The research was carried out in a stream of inert material Gs=0 kg/m²s, 2,5kg/m²s and 5kg/m²s. The values of the stream of inert material are characteristic for the upper and middle area of the real zone of the fluidized bed boilers.

3.1 Experimental stand

The research was carried out on a specially constructed test stand. The experimental stand made it possible to model the conditions of the circulating fluidized bed.
the experimental stand is presented in Fig. 1. [11].

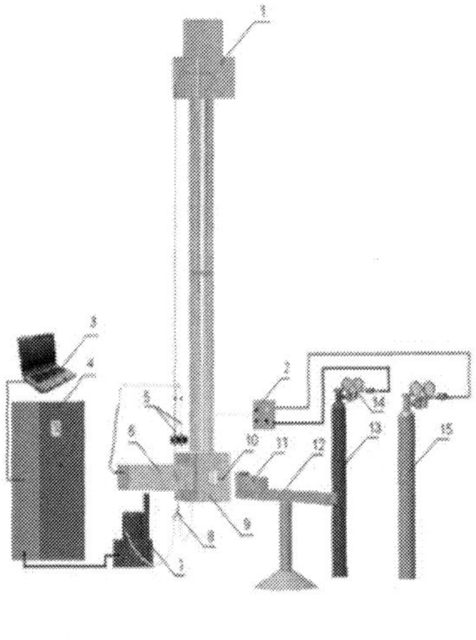

Fig. 1. The scheme of test apparatus. (1) vessel of inert material, (2) gas mixer, (3) PC-computer, (4) control panel, (5) rotametres, (6) gas heater, (7) ventilator, (8) T-connector, (9) combustion chamber, (10) coal particle, (11) tensometric branch scale, (12) support, (13,15) technical gases, (14)-reducer [1].

3.2 Research material

Four types of biomass pellets, each in the shape of a cylinder, were used for the study. Oak sawdust pellets, a mixture of beech and oak sawdust and straw had a diameter of 6mm, while the diameter of pellets from sunflower husk was 8mm. The mass of the samples was 0.73 ± 0.03g. Table 1 contains the results of the technical analysis of the tested fuels.

Table 1. Results of technical analysis of biomass fuels.

Pellets	Volatile [%]	Moisture [%]	Ash [%]	Fixed coal [%]	Combustion heat [MJ/kg]
Oak sawdust	79.6	8.7	1.2	10.5	18.2
30%beech and 70%oak sawdust	77.7	8.8	1.3	10.7	17.9
Sunflower husk	73.8	8.4	5.5	12.3	19.8
Straw	71.8	8.7	12.2	7.3	16.2

4 Results of the experiment

Fig. 2. presents a comparison of pellets combustion time combusted at different conditions in a circulating fluidized bed. Fig. 3. presents percentage comparison of pellets combustion time during reduction of the combustion chamber temperature compared to the reference reference atmosphere (21% oxygen). In all cases, according to the theory, a change in the combustion atmosphere consisting of an increase in the concentration of the oxidizer results in the reduction of the total loss of mass of the biomass pellets combusted. Supplying more oxidizer to the combustion area causes acceleration of the combustion process, therefore the loss of mass is accelerated due to the increased concentration of oxidizer and temperature, which is caused by the occurring chemical reactions. During combusting pellets without inert material, the wood biomass pellets were combusted the longest, and straw pellets the shortest. In the case of pellets combusting in the stream of inert material Gs = 2.5 kg/m^{2}s, wood biomass pellets were the shortest combusted, the longest combusted pellets from sunflower husk. At this temperature the phenomenon of ash softening is manifested by the sticking of sunflower husk pellets through quartz sand. The sand forms a durable surface surrounding the incinerated pellet, consequently preventing the oxidizer from entering the combusted pellet. The residue after combusting sunflower husk pellet at 850°C, tests were carried out in a stream of inert material at Gs = 2.5 kg/m^{2}s was shown in Fig. 4. A similar effect was observed when combusting straw pellets at 850°C. The straw pellets combusted at 850 °C were sintered. The residue after combusting straw pellet at 850 °C, tests were carried out in a stream of inert material at Gs = 2,5 kg/m^{2}s was shown in Fig. 5. With an increased value of the inert material stream to Gs = 5 kg/m^{2}s, similar to Gs = 2.5 kg/m^{2}s of inert, the longest combusted pellets from sunflower husk, and the shortest pellets from forest biomass. In the case of agro biomass combustion, as in the case of combustion in Gs = 2.5 kg/m^{2}s, during the combusting of pellets from sunflower husk, the formation of a coating was observed, which was partially broken down as a result of the mechanical impact of inert material - Fig. 6. On the other hand, the sinter was formed during combustion of straw was completely broken down.

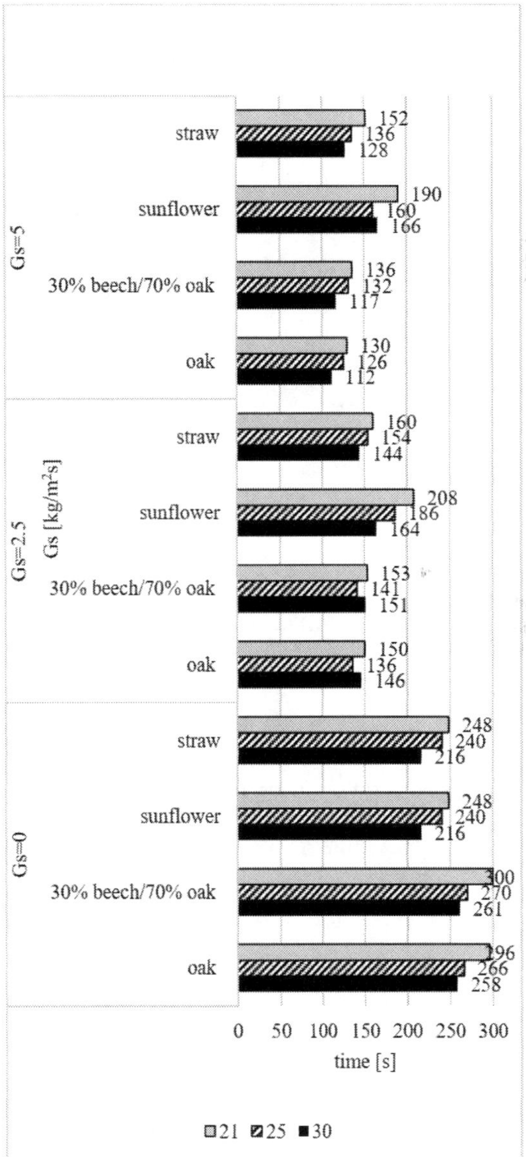

Fig. 2. Comparison of pellets combustion time combusted at different conditions in a circulating fluidized bed.

https://doi.org/10.1051/e3sconf/201913701032

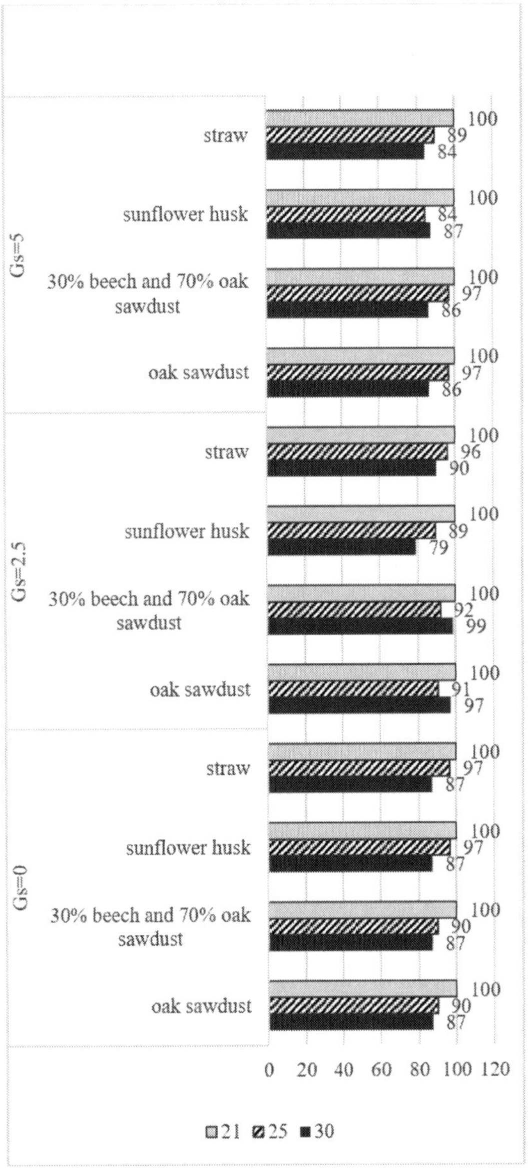

Fig. 3. Combustion times during the change of the oxidizing atmosphere compared to the reference atmosphere (21% oxygen) of all types of pellets.

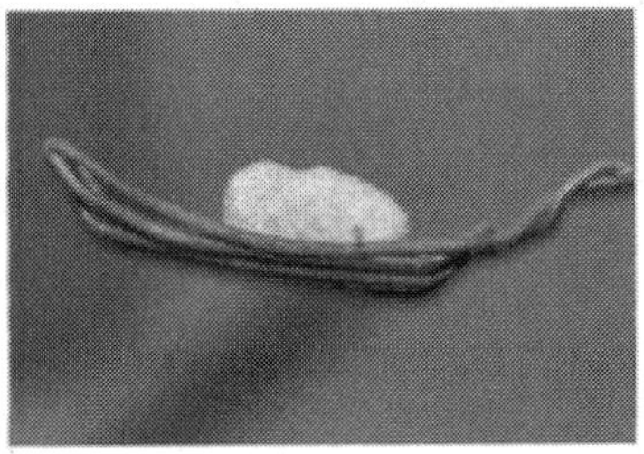

Fig. 4. The residue after combusting sunflower husk pellet at 850°C, tests were carried out in a stream of inert material at Gs = 2,5kg/m^{2}s.

Fig 5. The residue after combusting straw pellet at 850 ° C, tests were carried out in a stream of inert material at Gs = 2,5 kg/m^{2}s.

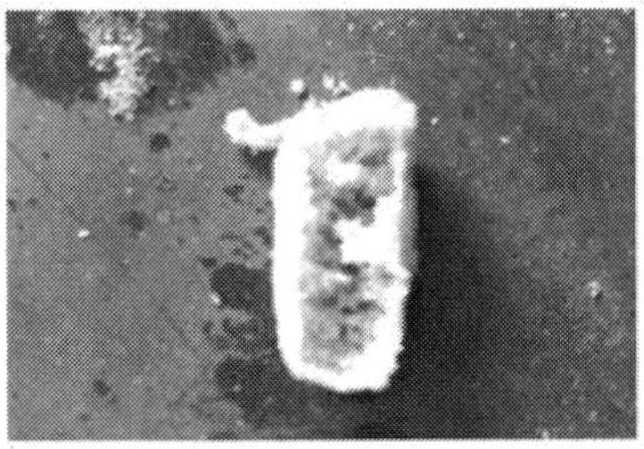

Fig 6. Residue after combusting sunflower husk pellets at 850°C and Gs = 5 kg/m^{2}s.

Fig. 7. shows a comparison of pellet combustion times depending on the stream of inert material combusted at different at different atmosphere (Gs = 0 reference level). The presence of inert material to varying degrees accelerates the mass loss of all types of pellets in all atmospheres. In the case of wood biomass, the combustion in the stream of material Gs = 2.5 kg/m^{2}s in each atmosphere is shorter by about 50% compared to incineration without inert material. However, further increase of the inert material stream to Gs = 5 kg/m^{2}s accelerated the combustion process by a further 10%. In the case of combusting sunflower husk pellets compared to incineration without inert material, the combustion time in the stream Gs = 2.5 kg/m^{2}s is reduced by about 20-30%. Increasing the value of inert material stream to Gs = 5 kg/m^{2}s resulted in shortening the combustion time in the case of two oxidizing atmospheres of 21% and 25% oxygen by about 10%, while in the case of 30% oxygen concentration, the combustion time was extended by 1%. In the case of straw pellets compared to incineration without inert material, the combustion time in the stream of inert material Gs = 2.5 kg/m^{2}s is reduced by about 35%. Increasing the value of the stream of inert material caused a reduction of the combustion time by about 10%.

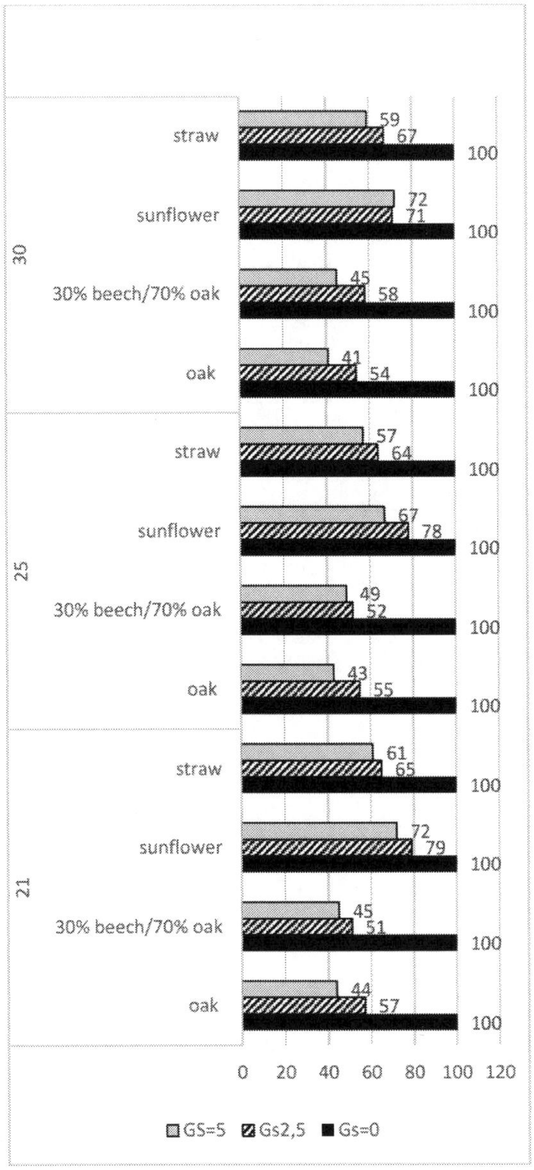

Fig. 7. Comparison of pellet combustion times depending on the value of the Gs material stream combusted at different atmosphere (Gs = 0 reference level).

5 Conclusions

Biomass is a source of energy characterized by dispersion and different fuel properties compared to fossil fuels, among which there should be mention of lower calorific value and high moisture content. Nevertheless, biomass has the highest energy potential as it accumulates a large amount of energy that can be obtained from renewable sources. One of the best methods for direct biomass combustion is that of fluidized bed technology, characterized by high efficiency and stable working conditions as well as the possibility of easily adapting the installation for oxygen combustion of fuels.

The analysis carried out as part of the work facilitates the formulation of the following conclusions:

a) Increasing the concentration of oxidant in the atmosphere of combustion causes the reduction of the time of total mass loss of combusted particles of all tested biomass fuels. According to the theory of providing more oxidizer to the combustion area, it accelerates the combustion process, therefore the mass loss is accelerated.

b) The introduction of inert material into the combustion zone of the particle resulted in the reduction of the total combustion time for all tested fuels in comparison with the combustion time without its participation in each analysed oxidizing atmosphere.

c) In case of combustion of oak sawdust pellets, the inert material at Gs = 2.5 kg / m^2s, the total combustion time lasts about 45% shorter than without the inert material in each analysed oxidizing atmosphere. A further increase in the value of the inert material stream results in the reduction of the combustion time by approximately 10%.

d) In the case of the combustion of the sunflower husk pellets in an inert material stream, the ash softening temperature was exceeded and the sample sand was glued through the quartz sand in each analysed oxidizing atmosphere. In the case of combusting sunflower husk pellets in all atmospheres compared to incineration without inert material, the combustion time in the stream Gs = 2.5 kg / m^2s is reduced by about 20-30%. Increasing the value of inert material stream to Gs = 5 kg/m^2s resulted in shortening the combustion time by another 10%.

e) During the combustion of straw pellets in the flow of inert material, in each analysed oxidizing atmosphere the formation of sinters was observed, which at the lower value of the stream of inert material at the level of Gs = 2.5kg / m^2s did not break down. During the intensification of the flow of inert material, the sinters were fragmented due to the interaction of inert material. In the case of straw pellets compared to incineration without an inert material, the combustion time in the stream of inert material Gs = 2.5 kg / m^2s is reduced by about 35%. Increasing the value of the stream of inert material caused a reduction of the combustion time by about 10%.

Based on the conducted experimental research, it was found that forest and agricultural biomass can be successfully used during oxy-combustion in boilers with a circulating fluidized bed under several conditions. Due to the chemical composition agricultural biomass (high content of chlorine and elements alkaline) and low temperatures softening and melting ash, it should be combusted under appropriate controlled conditions. Due to significant differences in biomass quality first of all, quality should be monitored fuel supplied to the combustion chamber.

References

1. Sher F., Pans M.A., Sun Ch., Snape C., Liu H., Oxy-fuel combustion study of biomass fuels in a 20 kWth

fluidized bed Combustor, *Fuel*, 215, 778–786, (2018).

2. Greinert A., Mrówczyńska M., Szefner W., The Use of Waste Biomass from the Wood Industry and Municipal Sources for Energy Production, *Sustainability*, 11, 3083, (2019).
3. Pradhana P., Mahajanib S. M., Aroraa A., Production and utilization of fuel pellets from biomass: A review, *Fuel Processing Technology*, 181, 215–232, (2018).
4. The International Energy Agency, Global Energy & CO_2 Status Report 2018, The latest trends in energy and emissions in 2018, (2018).
5. Toftegaard M.B., Brix J., Jensen P.A., Glarborg P., Jensen A.D., Oxy-fuel combustion of solid fuels. *Prog Energy Combust Sci*, 36(5):581–625 (2010).
6. Scheffknecht G., Al-Makhadmeh L., Schnell U., Maier J., Oxy-fuel coal combustion – anreview of the current state-of-the-art. *Int J Greenhouse Gas Control*, 5:S16–35, (2011).
7. Wall T., Liu Y., Spero C., Elliott L., Khare S., Rathnam R., et al. An overview on oxyfuel coal combustion—State of the art research and technology development. *Chem Eng Res Des*; 87(8):1003–16 (2009).
8. Mathekga H.I., Oboirien B.O., North B.C., A review of oxy-fuel combustion in fluidized bed reactors. *Int J Energy Res*, 40(7):878–902 (2016).
9. Bis Z., Czakiert T., Nowak W., Spalanie w atmosferach modyfikowanych tlenem kierunkiem rozwoju dla kotłów CWF; *Energetyka i Ekologia*; str. 713-718 (2008).
10. Bisio G., Bosio A., Rubatto G., Thermodynamics applied to oxygen enrichment of combustion air, *Energy Conversion and Management*, Vol. 43, pp. 2589-2600 (2002).
11. Pełka P., Analysis of a coal particle mass loss burning in flow of inert material. *Combust. Flame*, 156, 1604, (2009).
12. Kaczyński K., Kaczyńska K., Pełka P., Analiza ubytku masy ziarna węgla w atmosferze utleniającej oraz obojętnej w zmiennej temperaturze panującej w komorze paleniskowej w dwufazowym przepływie z udziałem materiału inertnego, *Rynek Energii*, Nr 1(140), 87-97, (2019).
13. Duan L., Zhao .C., Zhou W., Qu C., Chen X.. Effects of operation parameters on NO emission in an oxy-fired CFB combustor. *Fuel Process Technol*, 92(3):379–84, (2011).
14. Singh R.I., Kumar R., Current status and experimental investigation of oxy-fired fluidized bed. *Renewable Sustainable Energy Rev*, 61:398–420 (2016).
15. Roy B., Chen L.G., Bhattacharya S., Nitrogen oxides, sulfur trioxide, and mercuryemissions during oxyfuel fluidized bed combustion of victorian brown coal, *Environ Sci Technol*, 48(24):14844–50, (2014).
16. Hofbauer G., Beisheim T., Dieter H., Scheffknecht G., Experiences from oxy-fuel combustion of bituminous coal in a 150 kWth circulating fluidized bed pilot facility. In: Rokke NA, Svendsen H, editors. 7th Trondheim Conference on CO_2 capture, transport and storage. Amsterdam: *Elsevier Science Bv*, p. 24–30, (2014).
17. de las Obras-Loscertales M., Mendiara T., Rufas A., de Diego L.F., García-Labiano F., Gayán P., et al. NO and N_2O emissions in oxy-fuel combustion of coal in a bubbling fluidized bed combustor, *Fuel*, 150:146–53 (2015).
18. Jia L., Tan Y., Anthony E.J., Emissions of SO_2 and NOx during oxy-fuel CFB combustion tests in a mini-circulating fluidized bed combustion reactor, *Energy Fuels*, 24:910–5 (2010).
19. Khana A.A., de Jong W., Jansens P.J., Spliethoff H., Biomass combustion in fluidized bed boilers: Potential problems and remedies, *Fuel Processing Technology*, Volume 90, Issue 1, 21-50, (2009).
20. Niu Y.Q., Tan H.Z., Hui S.E., Ash-related issues during biomass combustion: alkaliinduced slagging, silicate melt-induced slagging (ash fusion), agglomeration, corrosion, ash utilization, and related countermeasures. *Prog Energy Combust*, Sci,52:1–61, (2016).
21. Kosowska-Golachowska M., Kijo-Kleczkowska A., Luckos A., Wolski K., Musiał T.: Oxy-combustion of biomass in a circulating fluidized bed. *Archives of Thermodynamics*, vol. 37, no. 1, p.17-30, (2016).

RESEARCH & DEVELOPMENT IN POWER ENGINEERING, 2019

Investigation of Applicability of Polyimide Membranes for Air Separation in Oxy-MILD Zero-Emission Power Plants

Leszek Remiorz[1], *Grzegorz* Wiciak[2], and *Krzysztof* Grzywnowicz[3,*]

[1] Silesian University of Technology, Institute of Power Engineering and Turbomachinery, 44-100 Gliwice, ul. Konarskiego 18; leszek.remiorz@polsl.pl; Poland

[2] Silesian University of Technology, Institute of Power Engineering and Turbomachinery, 44-100 Gliwice, ul. Konarskiego 18; grzegorz.wiciak@polsl.pl; Poland

[3] Silesian University of Technology, Institute of Power Engineering and Turbomachinery, 44-100 Gliwice, ul. Konarskiego 18; krzysztof.grzywnowicz@polsl.pl; Poland

Abstract. Primary element of an oxy-combustion plants is ambient air separation unit. This paper presents the results of experimental research concerning the parameters of the separation of N_2/O_2 from ambient air, using capillary polymer membranes, potentially applicable in oxy-combustion technology, under variable operational conditions. Collected data were utilized to approximate continuous functions describing the variability of essential parameters of the air separation based on such membranes. The functions were introduced to develop a complete mathematical model of the separation unit, intended to be applied in oxy-Moderate or Intense Low Oxygen Dilution (oxy-MILD) zero-emission plants. Computational analyses were performed for three variants of the unit's configuration: serial connection of membrane modules, unit with retentate recirculation and unit with permeate recirculation. The results of the research, in the form of sets of characteristic curves, depicting parameters of the separation process as a function of the variable operational conditions, show that crucial differences to the subsequent separation parameters (permeate purity, real selectivity coefficient, recovery coefficient) and with regard to the power consumed, were obtained. The highest parameters of the module were gained for serial connection, whereas the lowest – for permeate recirculation. The lowest energy consumption was acquired for the retentate recirculation variant.

1 Introduction

Technological processes of energy generation are mainly based on the combustion of fossil fuels, such as hard coal, lignite, natural gas, or crude oil. The by-products of those processes include a series of volatile chemical compounds. Among them are CO_2, N_2, H_2O, SOx, and NOx, as well as volatile organic compounds. In addition, particulate matter, formed due to the imperfect oxidation of coal, which is commonly lifted with the above-mentioned gases, should be listed from the outset [1,2,3,4]. Due to their toxic influence on humans and the natural environment, a number of such compounds should be separated from the stream of exhaust gases before its emission to the atmosphere. Similarly, compounds affecting global warming should be separated as well [1,5,6,7].All technologies, devoted to reduce emissions of potentially dangerous compounds from the solid fossil fuels, are commonly denoted as the clean coal technologies (CCT) [1,2].

Following the concept of the common utilization of clean coal technologies in commercial power generation units, zero-emission power plants units are designed. These units are characterized by the almost neutral environmental impact of their operation [6,8]. Considering the reduction of CO_2 emissions, the above-mentioned units utilize technology such as carbon capture and storage (CCS) installations. In order to introduce such installations, three technological concepts are used: pre-combustion – capturing CO_2 before the combustion of the fuel, post-combustion – capturing CO_2 after the process of combustion, and oxy-combustion – combustion of the fuel under high oxygen concentration conditions [8,9].

The main aim of the oxy-combustion process is an increase in the percentage share of carbon dioxide in the total stream of exhaust gases [9]. The oxygen is produced in an air separation unit, where ambient air is separated into streams of oxygen and nitrogen with remaining residual gases. Rise in the efficiency of the unit with simultaneous decrease in emissions of selected pollutants, as NOx and CO, is characteristic for the innovative Moderate or Intense Low Oxygen Dilution (MILD) technology [10]. In the oxy-MILD combustion technology, comprising both abovementioned technologies, the oxygen concentration in the oxidizer cannot exceed 40%, since excess oxygen content results in a drop of content of heat absorbing gases, such as nitrogen and carbon dioxide. The main effect of their shortage is a crucial increase in the temperature of the exhaust gases [10].The separation of oxygen from ambient air using membrane methods uses semipermeable films (membranes) made from polymeric and other materials [11,12]. Their vital advantage, comparing to commonly implemented, commercially mature technology of pressure swing adsorption (PSA), is lack of permanent adsorbents and requirement for alternating depressurization of the adsorber vessel.

The membrane separation module (membrane separator) is formed by a set of single membrane modules, connected in parallel, in series, or with any combination of these simple connections. Such constructions are required to have a compact size, which is a crucial feature in the case of widespread introduction of carbon capture installations to existing power plants and other industrial units [13,14]. The selection and type of membrane modules, dedicated to be used during the construction of a given membrane separator, depend mainly on the available expenditure and the nominal operational parameters of the unit [3,6,15]. Thus, the optimal configuration of a membrane separation unit with regard to its internal construction remains an essential issue.

The main objective introduced with the paper is experimentally-based analysis of operational parameters of selected variants of the separators' configuration. As a novel approach, semi-empirical model, constructed on the basis of the experiment, was utilized to predict complex parameters of the separator.

2 Materials and Methods

2.1 Methodology of the experimental research

The scheme of the experimental setup, utilized during the experimental part of the research, is presented in Fig. 1. The methodology of experimental research consisted in supplying the module (fig. 1, point 1) with a compressed gaseous mixture at initial temperature of 15°C, stored in external tanks (fig. 1, point 2) of 10m^3 total volume under a pressure of 200 bar. In order to reduce and stabilize the pressure in the experimental unit, multiple two-stage pressure reducers were applied (fig. 1, point 3). The setup was equipped with the feed heater (fig. 1, point 9) to increase and stabilize temperature of the inflowing mixture at stable level of 23°C. The synthetic mixture, utilized during the investigations, simulated dry air and consisted exclusively of nitrogen (79%) and oxygen (21%). At the outflow duct of the tank, a set of instruments (fig. 1, point 4), including a gauge manometer and a thermocouple, were located. At the same duct, a throttling valve (fig. 1, point 5), an additional flow stabilizer, was located. The gaseous mixture was supplied to the membrane module, which consisted of a NM-B02A capillary membrane produced by UBE Industries Ltd, Japan. The choice was based on its widespread commercial availability and relatively low price compared to similar membranes produced by other manufacturers, crucial for the investment costs in the case of large-scale separation units [11,14,16].

The gaseous products of the separation process were transferred to the measurement and control units, containing pressure sensors (fig. 1, point 6), thermocouples, and rotameters (fig. 1, point 7). The control function of the unit was fulfilled by additional throttling valves (fig. 1, point 8), responsible for the regulation of the flow of the gases transported to certified gas composition analyzers.

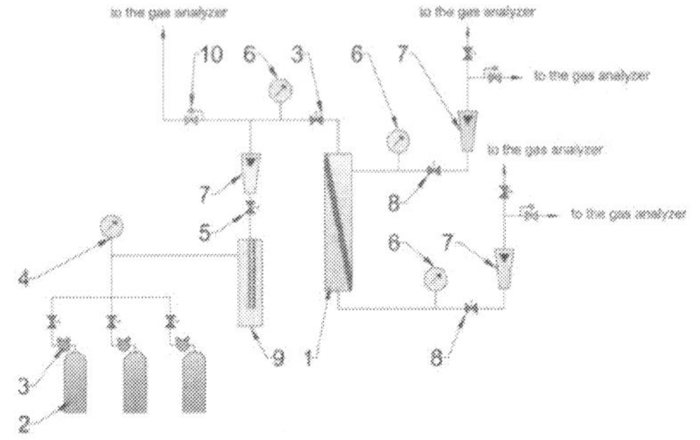

Fig. 1. Scheme of the experimental test stand: 1 - membrane module, 2 - external tanks of gaseous mixture, 3 - pressure reducers, 4 - unit of measuring instruments, 5 - throttling valve, 6 - gauge manometer, 7 - rotameter, 8 – control/throttling valves, 9 – the feed heater and humidity control unit, 10 – safety valve.

During the experimental phase of the research, the values of selected separation process parameters were identified, including: concentration of O_2 (YO_2) or N_2 (YN_2) in the separated flow, recovery coefficient of O_2 or N_2 (R), and the real separation factor (α). According to the literature [10,11,17,18], the parameters mentioned were described as:

a) purity of the permeate, the concentration of oxygen in the permeate stream, YO_2

b) recovery coefficient R, the ratio of elementary oxygen flows in the permeate and feed streams, defined by equation (1):

$$R = \frac{n_P Y_{O_2}}{n_F X_{O_2 F}} \quad (1)$$

c) real separation factor α, defined as the ratio of the concentrations of the ingredients of the mixture in the permeate stream relative to the ratio of the concentrations of the ingredients in the retentate stream, defined for oxygen with respect to nitrogen, as given by equation (2):

$$\alpha = \left(\frac{Y_{O_2}}{Y_{N_2}}\right) \bigg/ \left(\frac{X_{O_2 R}}{X_{N_2 R}}\right) \quad (2)$$

These values were obtained on the basis of direct measurements of the temperatures and capacities of the feed, permeate, and retentate streams, performed by measurement and control units located in subsequent ducts, as well as the oxygen and nitrogen concentrations identified by the gas analyzers.

2.2 Description of the model

On the basis of the experimental research, crucial empirical functions, describing the dependence of vital parameters of the separation process quality on the environmental conditions, were identified. Regression functions, obtained for a single membrane module, stated average regression functions characterized by the highest value of the Pearson product-moment correlation coefficient and were acquired using the respective empirical data analysis. The acquired regression equations were further used as fundamental functions of the mathematical model of the separation unit. As discussed in [19], to obtain results of satisfactory accuracy, a computational model of a separation unit was extended to include description of the Joule-Thomson effect, appearing in the case of real units.

The calculations were realized by executing series of computational steps each iteration. The steps mentioned included:

a) identification of thermophysical properties of the gaseous starting mixture used for the separation (including its density, viscosity, and thermal conductivity) on the basis of the real-gas model,

b) determination of nominal maintenance parameters of the membrane module, basing on the properties of the separated gas and empirical regression equations,

c) estimation of the active area of the membrane, basing on the pressure drop condition,

d) estimation of the influence of the Joule-Thomson expansion effect on the behavior of the gas flowing through the separation module,

e) recalculation of parameters basing on results of the Joule-Thomson expansion effect estimation,

f) estimation of power demand of the unit under set operational conditions.

The scheme of algorithm of the computations performed by mathematical model of the unit is presented in Fig. 2.

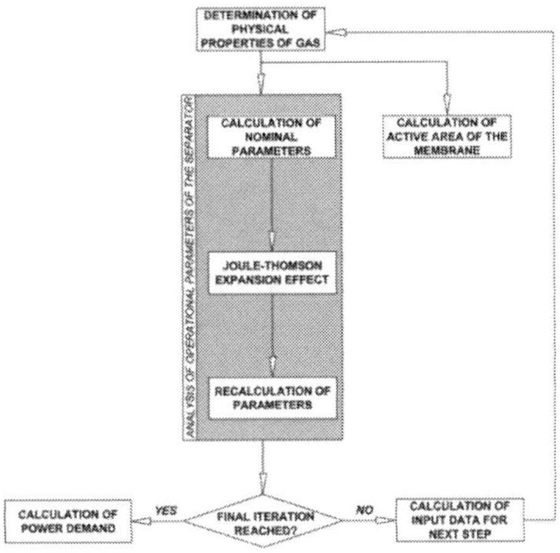

Fig. 2. Algorithm of calculations within the computational research

3 Results

3.1 Results of the experimental research

In fig. 3a to fig. 3b, the characteristic curves of recovery coefficient and permeate purity, performed for a constant retentate stream flow (Q_R) equal to 500l/h, at varying feed pressure are indicated. Results of the permeate purity (YO_2) and the oxygen recovery coefficient (R) of the permeate as a function of the retentate stream flow (Q_R), at a constant pressure equal to 5 or 6 bar, are presented in fig. 4a and fig. 4b, respectively. In fig. 4c to fig. 4d results from the NM-B02A membrane module, concerning the dependence of variable permeate flow and retentate pressure on the real separation factor (α), are indicated.

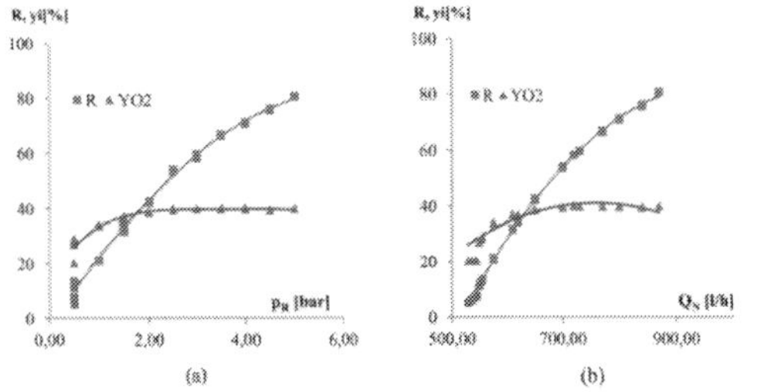

Fig. 3. Characteristic curves for: (a) permeate YO_2 and R with respect to retenate pressure P_R, (b) for permeate YO_2 and R with respect to feed flow Q_N

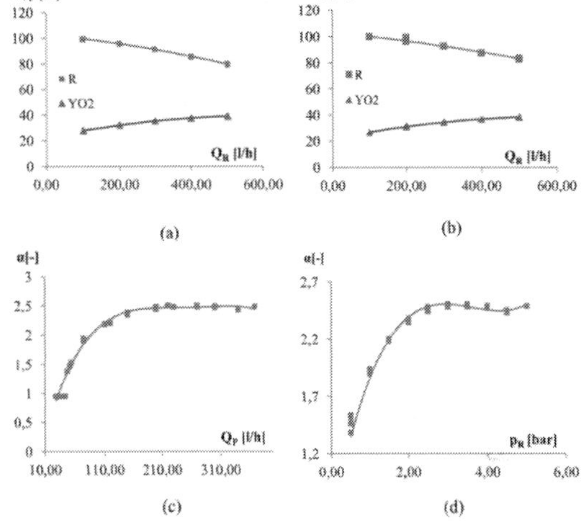

Fig. 4. Experimental characteristic curves: (a) for the permeate YO_2 and R as a function of retenate flow Q_R for a pressure of 5 bar, (b) for the permeate Y_{O2} and R as a function of retenate flow Q_R for a pressure of 6 bar, (c) for α with respect to permeate flow Q_P, (d) for the real separation factor α with respect to retenate pressure p_R.

3.2 Results of the computational analysis

The first of the investigated configuration variants involved the serial connection of successive modules by the permeate duct (stream flow of the permeate from one module became the feed for the next one). The configuration is schematically presented in fig. 5.

In fig. 6a and fig. 6b, the results of the model-based analysis concerning the serial connection of modules are presented. The graph shown in fig. 6a depicts an estimation of the separation factor on number of modules, depending on feed flow Q_N. The results of further investigation, shown in Fig. 6b, indicate the influence of the number of modules connected in series on the permeate purity Y_{O2}. measurements of the temperatures and capacities of the feed, permeate, and

 https://doi.org/10.1051/e3sconf/201913701033

retentate streams, performed by measurement and control units, as well as the oxygen and nitrogen concentrations identified by the gas analyzers.

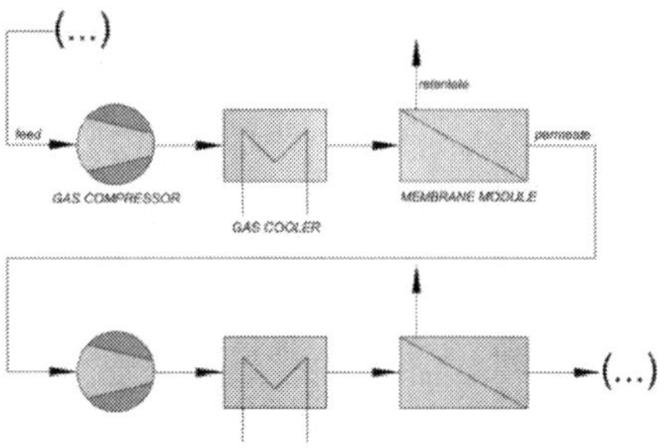

Fig. 5. Scheme of the membrane separator with serial connection of modules (variant I)

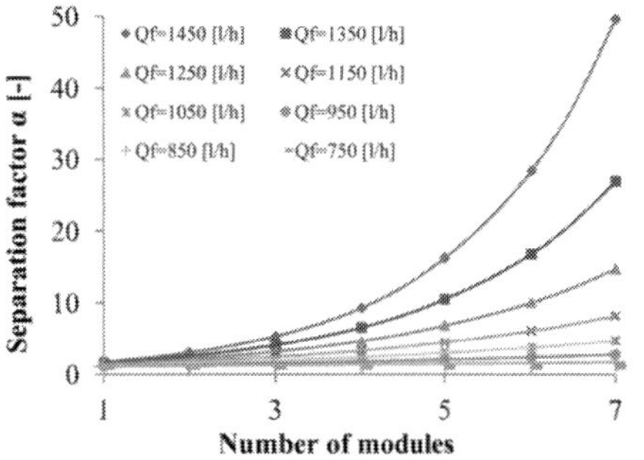

(a)

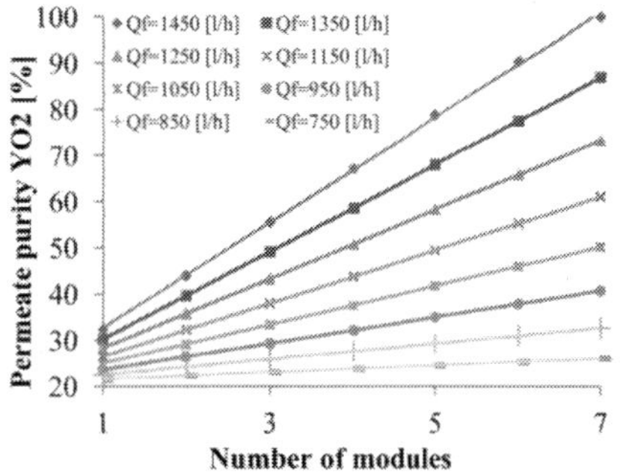

(b)

Fig. 6. Characteristic curves of the variant with serial connection of modules: (a) dependence of separation factor on number of modules in series for varying feeds, (b) dependence of purity of the permeate on number of modules in series for varying feeds

The results of further analysis, concerning the second of the investigated separator configurations, based on a single-module separation unit with multiple retentate recirculation (fig. 7), are shown in fig. 8a to fig. 8b.

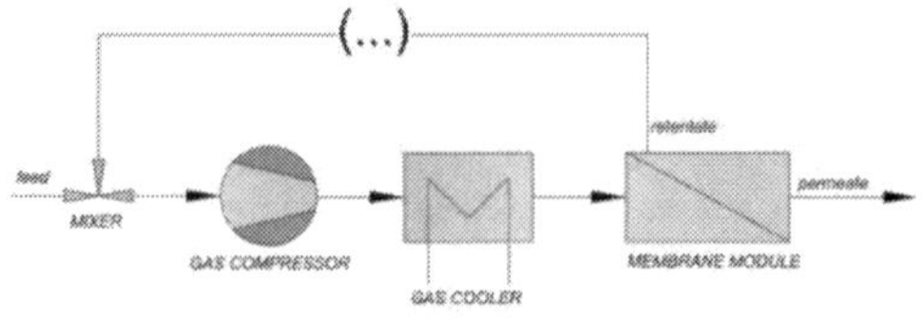

Fig. 7. Scheme of the separator with multiple recirculation of the retentate (variant II)

Figure 8a depicts the dependence of oxygen concentration X_{O2R} in the retentate as a function of the number of retentate recirculation stages n_C and primary feed flow values Q_N. The dependence of permeate purity on the number of stages of retentate recirculation for varying feed flows Q_N is indicated in Fig. 8b.

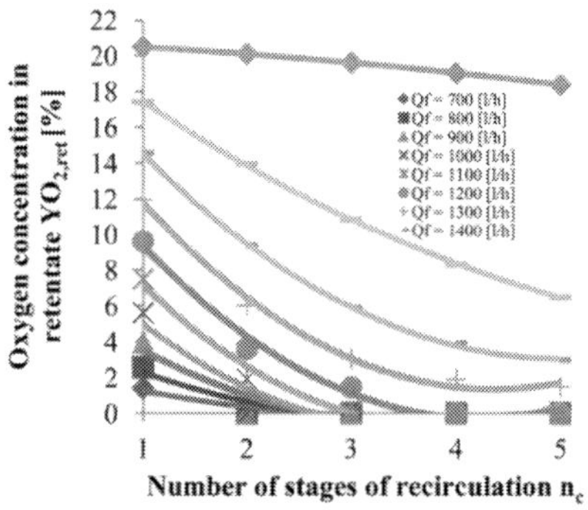

(a)

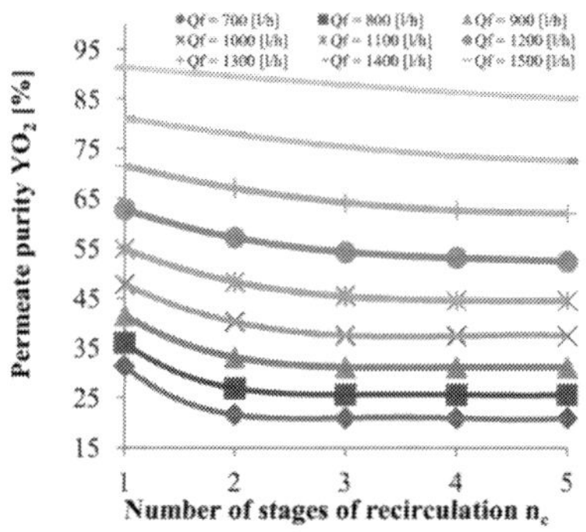

(b)

Fig. 8. Characteristic curves of the variant with multiple retentate recirculation: (a) oxygen concentration in the stream flow as function of retentate recirculation stages for variable feed flows, (b) permeate purity as function of retentate recirculation stages for variable feed flows

The last of the analyzed variants of the membrane separation unit configurations was the single-module unit with multiple permeate recirculation n_{CP}, presented schematically in fig. 9. Results of the computation of minimal and maximal power demand of the configurations: serial connection of the modules (variant I), retentate recirculation (variant II) and the permeate recirculation (variant III) are presented in Table 1.

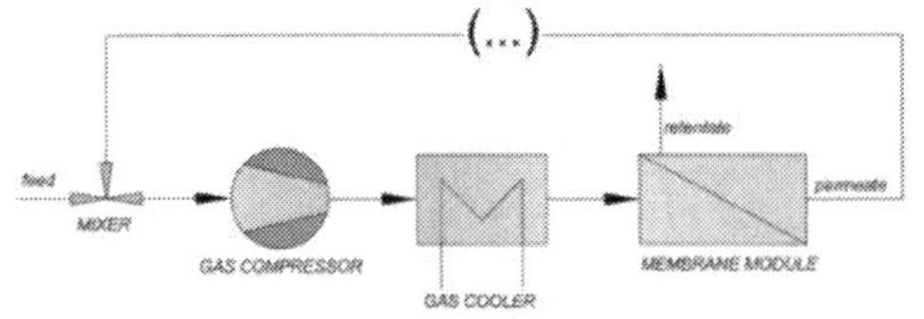

Fig. 9. Scheme of the separator with multiple recirculation of the permeate (variant III)

The graph, presented in fig. 10 indicate the dependence of the permeate purity YO_2 on the number of recirculations for this case of the separator configuration.

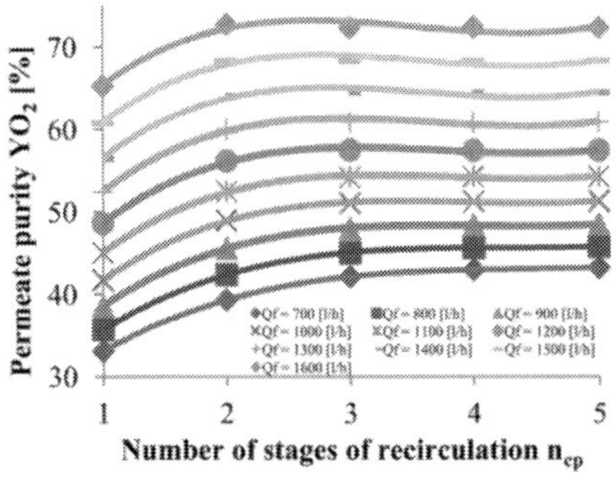

Fig. 10. Dependence of permeate purity on number of recirculation stages for variable feed flows

Table 1. Estimation of minimal and maximal power demand of investigated variants

	Min. power demand, W	Max. power demand, W
Serial connection of the modules	12	90
Retentate recirculation	7	58
Permeate recirculation	20	201

4 Discussion

At a constant retentate stream flow (Q_R) equaling 500l/h, with an increasing pressure of the retentate, the rise of the recovery coefficient (R) and the permeate purity (Y_{O2}) was observed, as seen in fig. 3a. Figure 4c, indicating the characteristic curve for the separation factor (α) with respect to the permeate stream flow (Q_P), shows that at constant retentate stream flow value (Q_R) equal to 500l/h with increasing permeate flow (Q_P), a rapid increase in separation factor (α) appeared.

The rise in the number of modules in serial connection, forming the variant I of the membrane separator (fig. 5), resulted in an exponential increase in the separation factor of the system, as indicated in fig. 6a. However, visible changes in the separation factor depended on the feed flow supplying the system: for feed flows below 1000 l/h, the value of the factor for a seven-module unit equaled roughly twice the value of the separation factor for a single-module system. significant fluctuations of the dependence were observed for any of the investigated variants. The almost linear rise in the purity of the permeate was observed for this configuration of the separator, as presented in fig. 6b. Nevertheless, slope of the discussed trend vitally depended on the feed flow – the highest values of oxygen concentration in the permeate were observed for the largest feed flow, exceeding 1000 l/h.

As indicated in Fig. 8a, multiple retentate recirculation (fig. 7) might be used to reduce the oxygen concentration in the retentate. However, the crucial dependence of the drop in oxygen concentration both on the number of recirculation stages and the feed flow should be emphasized – for flows below 1000 l/h, triple retentate recirculation allowed to obtain retentate with almost no oxygen content. Nevertheless, an increase in the feed flow resulted in a significant rise in the number of recirculation stages required to obtain such an effect. For flows exceeding 1300 l/h, the oxygen fraction in the retentate was significant, even in the case of its recirculation for five times. Nevertheless, rise in the number of retentate recirculation stages resulted in drop of the permeate purity, especially visible for low feed flows. As indicated in Fig. 8b, for feed stream flow of 700 l/h, introduction of double retentate recirculation lead to drop permeate purity exceeding 10 percentage points.

Considering the variant III of the investigated separator (fig. 9), the characteristic curve, indicated in fig. 10, showed a rise in purity with an increase in the number of stages for lower feed flows, but significant fluctuations of purity were seen for higher flows (exceeding 1300 l/h), with the peak of the purity observed for double permeate recirculation. The stable rise in the permeate purity with rise in permeate recirculation stages was observed only for feed flows below 900 l/h.

Results of estimation of the power demand, presented in Table 1, indicate, that the separator with permeate recirculation (variant III) consumes up to 2.23 times more energy than the serial connection-based unit. The retentate-recirculation based separator might be deemed as the most energy efficient construction, considering both for the minimal and maximal power demands

5 Conclusions

Research concerning the variant I configuration of the membrane separation unit, consisting of a number of modules connected in series, indicated a linear rise in the power demand of the unit with an increase in the number of module membranes present. A similar influence was indicated on increasing the feed flow. A rise in the number of modules also resulted in an exponential increase in the separation coefficient of the system, clearly depending on the feed flow. This observation led to a vital conclusion: that the serial connection of modules in a compound separation unit might lead to significant operational benefits in the case of large volumetric flows of the feed. Considering all results of the research discussed, utilization of a serial connection of single modules is the best in case of large exhaust gas flows, since this particular configuration leads to satisfactory separation process parameters with acceptable power demand. Nevertheless, considering low feed flows, which might occur for instance during the start-up of the unit, the operational parameters of such units are insufficient for oxy-MILD technology applications.

Considering the results of the computational analysis of the system based on multiple retentate recirculation, the vital result of a non-linear fall in the oxygen concentration in the permeate following a rise in the number of recirculation stages was proven by the results acquired. The rate and value of the reduction in oxygen concentration depended strongly on the feed flow. Thus, the above-mentioned observation suggests that the introduction of retentate recirculation leads to a decrease in the operational parameters of the separation unit with regard to the purity of the permeate. This is caused by the significantly greater pressure losses between the feed and the permeate outlet than between the feed and retentate outlet in the membrane modules being considered. Therefore, the real applicability of such a configuration in oxy-MILD zero-emission power units would be marginal.

Analysis of the results of the third variant of the membrane separation unit with the multiple permeate recirculation showed the positive influence of an increase in the number of permeate recirculation stages on the permeate purity, especially considering low feed flows. Nevertheless, for high feed flows, use of the unit with further gas recirculation resulted in unstable performance and fluctuations of the permeate purity, damping with a rise in the number of stages. This observation might be explained by locally exceeding the maximum oxygen transport capacity of the investigated membrane. However, besides its potentially attractive characteristics of operational parameters, applicability of such construction of the membrane separator is marginal due to essentially higher power demand, than other constructions compared.

Considering general application of investigated units in oxy-MILD zero-emission power plants, results of the research indicate, that investigated membrane separation technology is only partially able to compete with mature cryogenic methods, basically due to low purity of the permeate. Furthermore, in the case of membrane separators, utilization of several stages of cascade seems to be required, leading to high number of compressors and vital rise in overall power demand of the unit.

References

[1] Wiciak G., Kotowicz J.: *Experimental stand for CO2 membrane separation*, Journal of Power Technologies, Vol. 91, pp. 171-178 (2011)

[2] Chmielniak T., Łukowicz H.: *Condensing power plant cycle - assessing possibilities of improving its efficiency*, Archives of Thermodynamics, Vol. 31, pp. 105-113 (2010)

[3] Janusz-Szymańska K., Dryjańska A.: *The analysis of economic efficiency of oxy-type power plant on supercritical parameters with a capacity of 600 MW*, Archivum Combustionis, Vol. 33, pp. 109-123 (2013)

[4] Lin L. et al. (2019): *Combustion characteristics of lignite char in a fluidized bed under O2/N2, O2/CO2 and O2/H2O atmospheres*, Fuel Processing Technology, Vol. 186, pp. 8-17

[5] Fujimori T., Yamada T.: *Realization of oxyfuel combustion for near zero emission power generation*, Proceedings of the Combustion Institute, Vol. 34, pp. 2111-2130 (2013)

[6] Meyer J. et al.: *Techno-economical study of the Zero Emission Gas power concept*, Energy Procedia, Vol. 4, pp. 1949-1956 (2011)

[7] Saeidi S. et al. (2016): *Energy and exergy analyses of a novel near zero emission plant: Combination of MATIANT cycle with gasification unit*, Applied Thermal Engineering, Vol. 108, pp. 893-904

[8] Nami H. et al.: *Thermodynamic assessment of zero-emission power, hydrogen and methanol production using captured CO2 from S-Graz oxy-fuel cycle and renewable hydrogen*, Energy Conversion and Management, Vol. 161, pp.53-65 (2018)

[9] Chen W. et al.: *Membrane-integrated oxy-fuel combustion of coal: Process design and simulation*, Journal of Membrane Science, Vol. 492, pp. 461-470 (2015)

[10] Perrone D. et al.: *Numerical simulations on Oxy-MILD combustion of pulverized coal in an industrial boiler*, Fuel Processing Technology, Vol. 181, pp. 361-374 (2018)

[11] Sunghoon L. et al.: *Development of novel sub-ambient membrane systems for energy-efficient post-combustion CO2 capture*, Applied Energy, Vol. 238, pp. 1060-1073 (2019)

[12] Rezakazemi M. et al.: *Thermally stable polymers for advanced high-performance gas separation membranes*, Progress in Energy and Combustion Science, Vol. 66, pp. 1-41 (2018)

[13] Banaszkiewicz T. et al.: *Comparative analysis of oxygen production for oxy-combustion application*, Energy Procedia, Vol. 51, pp. 127-134 (2014)

[14] Toftegaard M.B. et al.: *Oxy-fuel combustion of solid fuels*, Progress in Energy and Combustion Science, Vol. 36, pp. 581-625 (2010)

[15] Zhang D. et al. (2017): *Modeling of purge-gas recovery using membrane separation*, Chemical Engineering Research and Design, Vol. 125, pp. 361-366

[16] Chen W. et al.: *Oxygen-selective membranes integrated with oxy-fuel combustion*, Journal of Membrane Science, Vol. 463, pp. 166-172 (2014)

[17] Remiorz L. et al.: *Numerical modeling of CO2 separation process*, Archives of Thermodynamics, Vol. 34, pp. 41-53 (2013)

[18] Yamauchi Y., Akiyama K.: *Innovative Zero-emission Coal Gasification Power Generation Project*, Energy Procedia, Vol. 37, pp. 6579-6586 (2013)

[19] Bounaceur R. et al.: *Rigorous variable permeability modeling and process simulation for the design of polymeric membrane gas separation units: MEMSIC simulation tool*, Journal of Membrane Science, Vol. 523, pp. 77-91 (2017)

RESEARCH & DEVELOPMENT IN POWER ENGINEERING, 2019

Modeling of influence of vibration on intensification of heat transfer within the absorber of the vacuum solar collector

Krzysztof Grzywnowicz[1,*], *Łukasz* Bartela[2], *Leszek* Remiorz[3] and *Bartosz* Stanek[4]

[1] Silesian University of Technology, Institute of Power Engineering and Turbomachinery, 44-100 Gliwice, ul. Konarskiego 18; krzysztof.grzywnowicz@polsl.pl; Poland

[2] Silesian University of Technology, Institute of Power Engineering and Turbomachinery, 44-100 Gliwice, ul. Konarskiego 18; lukasz.bartela@polsl.pl; Poland

[3] Silesian University of Technology, Institute of Power Engineering and Turbomachinery, 44-100 Gliwice, ul. Konarskiego 18; leszek.remiorz@polsl.pl; Poland

[4] Silesian University of Technology, Institute of Power Engineering and Turbomachinery, 44-100 Gliwice, ul. Konarskiego 18; bartosz.stanek@polsl.pl; Poland

Abstract. Due to escalating role of mitigation of climate change in power and energy sector, power units based on renewable energy sources (RES) became vital part of global power and heat market, including distributed heat generation as well. Significant number of such installations belong to individual users, commonly using solar collectors to prepare domestic hot water directly at their sites. However, the vitally variable solar irradiance makes the solar energy difficult in efficient harvesting considering long time period. Thus, maximization of power gain from single solar absorber, when the solar radiation flux is temporarily high, might lead to further rise in overall, year-averaged efficiency of such units. The paper concerns modelling of intensification of heat transfer, taking place within the absorber of a vacuum solar collector, due to insertion of vibrating element inside the thermal oil canal, compared to static turbulization method. Different geometries of vibrating elements and amplitudes of oscillatory motion, as well as heat carriers, are investigated using commercially available CFD software. Results indicate rise in solar power acquired within the absorber using vibrating elements, with essential difference between circulating media, and suggest vital benefits from utilization of static turbulizing devices.

1 Introduction

Although fossil fuels remain fundamental for worldwide energy sector, their continuous depletion lead to the widespread introduction to the global power and heat sector of power technologies based on renewable energy sources (RES) in last decades [1]. Due to their high seasonal and day-by-day instability of operation, the RES-based technologies were widely introduced to the distributed energy sector [2]. Among all primary energy sources, the solar energy shares an essential part within the distributed RES-based units [1]. Solar irradiance might be obviously utilized both to generate current within the photovoltaic (PV) cells, as well as to produce useful heat within the thermal collectors; nevertheless, hybrid units, producing both forms of the useful energy, are applied as well [1]. Apart from global rise of PV modules manufactured and installed, observed recently, the thermal technology continues to play significant role in utilization of solar energy, mainly due to its typical higher efficiency [1]. Nevertheless, due to the short-time, seasonal and year-by-year variability of radiation, followed by essential fluctuations caused by weather conditions and location-based sunrays accessibility, number of technologies, devoted to enhance energy output from PV or solar thermal units, has been investigated. One of these technologies focuses on concentration of direct solar radiation on strictly identified part of the energy receiver, introducing limitation to radiation dissipation and other energy losses [3].

The solar irradiance concentrators, due to their high applicability both within the PV, thermoelectric and thermal units [4], state matter of recent profound investigation. Research, discussed in the literature, focuses on broad variety of topics, beginning with geometry and materials, through modelling of the concentrators' optical properties, up to design and optimization of complex systems.

Strong investigational emphasis is put on optical analysis of geometry of parabolic trough collectors (PTC), primarily due to its potentially low manufacturing costs and easy implementation [5,6,7]. Current investigation focuses on application of advanced measuring devices [5] to determine accurate optical properties of reflective surfaces and introduce fundamentals of their optimization [6]. Experimental research on other traditional geometry of the parabolic satellite dish (PSD) has been performed recently as well [8], concerning mainly comparison of such concentration devices with other available geometries. Profound study has been given to the conical solar concentrators [9] as well, including design strategy of complete installations. Nevertheless, due to no strict limitations of length of the absorbing body, the aplanat geometry concentrators [10] seem to be comparably more beneficial. Apart from classical geometry studies, vital focus has been recently given to rapidly developing two-stage solar concentrators [4] and spectrum-splitting units [11,12], which might introduce significant benefits to the solar thermal power devices in the future. Alternatively, optimization of the renowned solutions [6,7,13] might introduce the vital benefits for state-of-the-art technology and bridge the gap between current market-availability and laboratory-scale research.

Nevertheless, in spite of numerous advantages, utilization of solar concentrators leads to essential challenges, including high risk of photovoltaic cells overheating or local phase transition or degradation of working fluid within the solar thermal collectors [14,15]. Primary drawback connected to appearance of gaseous phase or local degradation of properties of the working fluid, even considering it as present only within a thin film located directly at the irradiated walls, might be vital drop in convective heat transfer between the internal wall of solar thermal collector's absorber and circulating medium. This is followed by danger of potential damage within the device, as well as sudden drop of thermal properties of the heat carrying fluid [3,14]. Therefore, intensification of heat transfer within the absorbers of solar collectors is vital especially for installations concentrating solar irradiance.

The literature derives description of several investigations, focused on the discussed issue, concerning utilization of nanoparticles of carbon structures [16], copper [17] or copper oxide [18] to enhance the conduction through the interior of the circulating heat carrier. As the initial results indicate, utilization of thermal oil-based nanofluids within the solar collectors might introduce a breakthrough; nevertheless, the long-term investigation of influence of such fluids on total lifetime of installation has not been reported yet. Therefore, research on geometry of twisted or coil-like turbulizing elements [19] along with strongly aligned ribs [20] and wedge–shaped vertices producing elements [21] remains essential.

However, the investigations has not considered utilization of vibrating elements to enhance the heat transfer due to turbulization of the flow of external circulating medium. Although considerably positive effect of vibrations on heat dissipation has been investigated and discussed [22,23], its applicability within the solar thermal collectors has not be investigated yet.

The main aim of the paper was to perform computational analysis of the effect of utilization of vibrating elements, as the source of turbulences enhancing the convective heat transfer, on operational parameters of the medium-temperature parabolic trough collector (PTC). The research was performed as the introductory investigation, preceding further experimental analysis, performed on currently designed and constructed, dedicated test-stand. In order to compare the proposed method of enhancement of convection, which might introduce certain complexity to the traditional installations, two reference cases of static turbulizing elements, were modelled as well. In order to increase accuracy of the simulation, a 3D model was prepared and numerically analyzed using market-available computational fluid dynamics (CFD) software.

2 Computational Model

The prepared numerical model considered the geometry of a concentrated solar irradiance receiver in the form of cylindrical absorber, made of structural steel. The assumed geometry of the absorber is presented in Fig. 1.

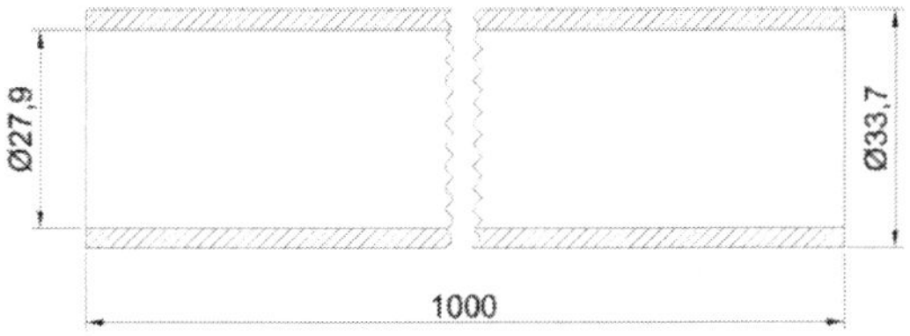

Fig. 1. Scheme of the modelled absorber geometry.

In order to simplify the geometry, the parabolic trough, assumed as the concentrating element of the parabolic trough collector (PTC) unit, was not considered in CFD modelling. Nevertheless, in order to estimate the boundary conditions on the absorber's external surface, computational ray tracing and determination of irradiance was performed using the APEX™ optical analysis software. The computations, performed in the APEX™, included preparation of three-dimensional geometry of the parabolic concentrator, assumed to be used in a standard PTC installation, selection of the coating, basing on available library of properties of coatings, and final calculations of direct sunlight reflection, including advanced ray tracing using Monte Carlo method and the ray energy averaging. The optical model of the parabolic trough with part of the cylindrical absorber, presenting the ray tracing results, is shown in Fig. 2.

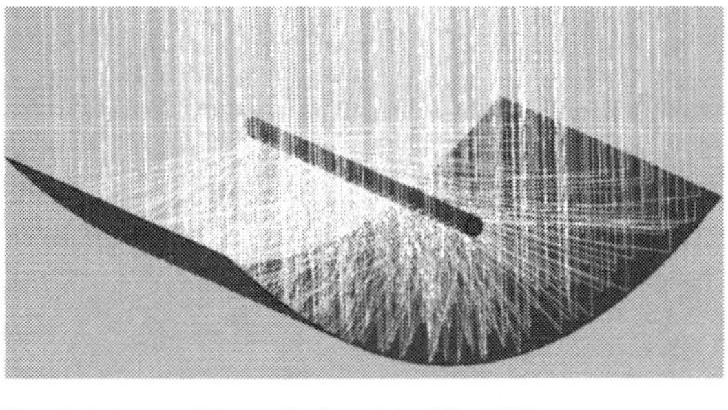

Fig. 2. Scheme of the optical model of the PTC.

However, the optical analysis resulted in identification of essential boundary conditions, required to be inserted within the CFD investigation. These included the average angle, which under the concentrated irradiance is transferred to the absorber's

 https://doi.org/10.1051/e3sconf/201913701034

surface, and the mean radiative heat flux, enforcing the appearance of thermal phenomena within the absorber.

Further boundary conditions of the model, including primarily inlet and outlet parameters, were identified on the basis of holistic analysis of the general dataset for parabolic trough solar collectors [3]. The properties of the circulating fluid, vitally influencing both heat transfer conditions, scale of a viscous boundary layer and pressure losses, were based on the literature [24]. In order to investigate influence of circulating fluid properties, two different media were assumed – as a reference case, the hydrogenated terphenyl-based thermal oil, and pure water as an additional case. Furthermore, in accordance with data derived by numerical optical analysis of the unit and initial experimental tests of similar installation, two different concentrated radiation angles were identified. This resulted in preparation of two absorber geometries, differing in slicing planes angle, which was used to separate areas radiated by concentrated and direct non-concentrated light.

The geometrical and environmental boundary conditions, set during the simulation, are concluded in Table 1.

Table 1. The most relevant boundary conditions

	Value
Angle of concentrated irradiance lighting, degrees	166.0
Total area of absorber, irradiated by concentrated light, mm^2	48.5× 10^3
Emissivity of the absorber walls, -	3.5 × 10^{-1}
Irradiance transferred to the directly irradiated surface, W/m^2	1000.0
Radiative heat flux on the absorber surface, W	708.6
Initial temperature of the absorber' walls, Celsius degrees	314.6
Mean pressure of the circulating fluid, atm	1.1
Initial temperature of the circulating fluid, Celsius degrees	30.0
Mass flow rate of circulating fluid at the absorber's inlet, kg/s	9.9 × 10^{-3}

Apart from the boundary conditions, set at the external surfaces of the control volume of the numerical model, vital consideration has been given to the model equations. The main issue focused on selection and implementation of highly-reliable, stable and accurate models of turbulences, expected to arise within the boundary layer both by implementing the static turbulizing devices and vibrating elements. Essential challenge, considering such selection, is the partially unknown and variable nature of turbulent vertices, occurring within the fluid domain due to presence of static or vibrating elements. Thus, either two separate models, covering separate cases of turbulence sources might be introduced, or one complete model, covering possibly wide range of turbulence nature might be implemented. Following the last concept, the Reynolds stress model, indicated by Eq. (1), in the SSG variant [25], was assumed. In order to define pressure and velocity distribution at the entrance and outlet of the absorber, the zero gradient model [26], given by Eq. (2) has been applied to these boundaries.

$$\frac{\partial}{\partial t}\left(\rho\overline{u'_iu'_j}\right)+\frac{\partial}{\partial x_k}\left(\rho u_k\overline{u'_iu'_j}\right)=-\frac{\partial}{\partial x_k}\left[\rho\overline{u'_iu'_ju'_k}+\overline{p'\left(\delta_{kj}u'_i+\delta_{ik}u'_j\right)}\right]+\frac{\partial}{\partial x_k}\left[\mu\frac{\partial}{\partial x_k}\left(\overline{u'_iu'_j}\right)\right]-\rho\left(\overline{u'_iu'_k}\frac{\partial u_j}{\partial x_k}+\overline{u'_ju'_k}\frac{\partial u_i}{\partial x_k}\right)+\overline{p'\left(\frac{\partial u'_i}{\partial x_j}+\frac{\partial u'_j}{\partial x_i}\right)}-2\mu\overline{\frac{\partial u'_i}{\partial x_k}\frac{\partial u'_j}{\partial x_k}}-2\rho\Omega_k\left(\overline{u'_ju'_m}\varepsilon_{ikm}+\overline{u'_iu'_m}\varepsilon_{jkm}\right) \quad (1)$$

$$\begin{cases}-\frac{\partial P}{\partial x}+\frac{d}{dy}\left(\mu\frac{dU}{dy}-\rho\overline{u'v'}\right)=0\\ -\frac{\partial P}{\partial y}+\frac{d}{dy}\left(-\rho\overline{u'v'}\right)=0\end{cases} \quad (2)$$

where: t – time, $\overline{u'_iu'_ju'_k}$ – individual Reynolds stresses, ρ – density of the fluid, x_k - boundary layer coordinate, p – pressure, μ – dynamic viscosity of fluid, ε_{ikm} – dissipation term, ν – kinematic viscosity of fluid.

Further consideration has been given to the selection of radiation model – since the radiation is the only way of energy propagation outside the external absorber' surface and might vitally contribute to the heat transfer at the absorber-fluid internal interface, high accuracy of modelling was strongly required. Thus, the Monte Carlo and Discrete Transfer models, available within the ANSYS™ CFX™ software, were considered. However, since utilization of the Monte Carlo model, in spite of its accuracy, especially considering complex systems [27], leads to high computational cost, the more time- and energy-efficient model of Discrete Transfer, given by Eq. (3), was assumed [27]. Furthermore, in order to increase the accuracy of results and cover fundamental radiation physics within the domain of circulating heat carrier, the spectral model of gray body was assumed as well.

$$\frac{dI}{dS}+aI=\frac{a\sigma T^4}{\pi} \quad (3)$$

where: a – fluid absorption coefficient, I – radiation intensity, σ – Stefan-Boltzmann constant, T – local temperature.

Finally, the convective heat transfer between the circulating fluid and the absorber was modelled as well. Determination of convective heat transfer coefficient, fundamental for accurate inclusion of thermal effect in total energy balance, was assumed to be performed using the correlation formula. Due to highly unstable conditions within the absorber, as well as development of viscous boundary layer due to sudden enforcement of thermal fluid motion, the correlation for simultaneously thermally and hydrodynamically developing flow, given by literature [28] was selected. Prepared model was also tested under variable meshing parameters and different timesteps to validate by reaching stable outputs.

https://doi.org/10.1051/e3sconf/201913701034

3 Results and Discussion

The results of the numerical simulations are presented in the form of temperature distributions at the absorber's outlet, indicated on Fig.3 to Fig.6.

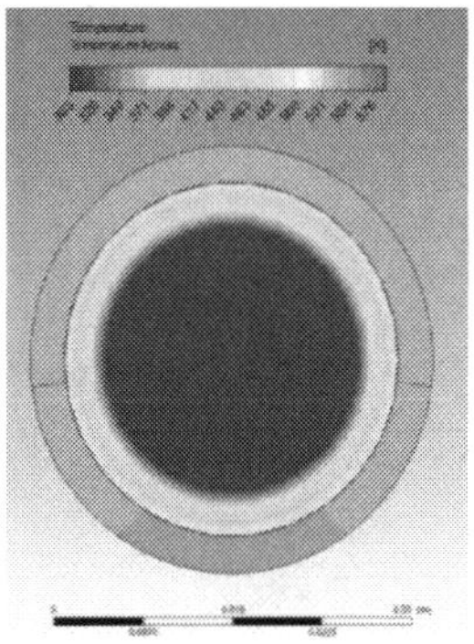

Fig. 3. Temperature distribution at the absorber's outlet for thermal oil used as a working medium and vibration amplitude of δ=0.1mm

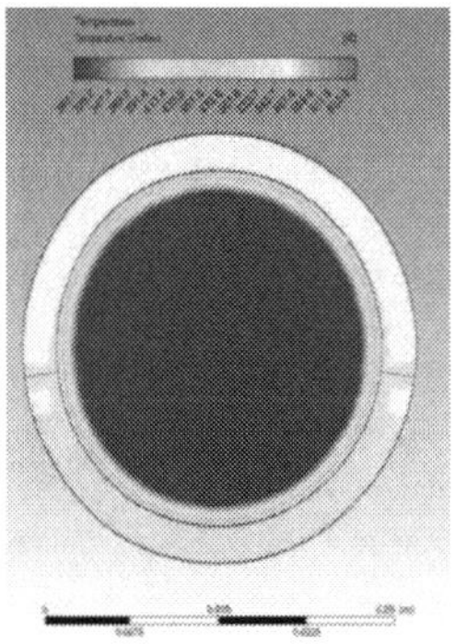

Fig. 4. Temperature distribution at the absorber's outlet for thermal oil used as a working medium and vibration amplitude of δ=0.9mm

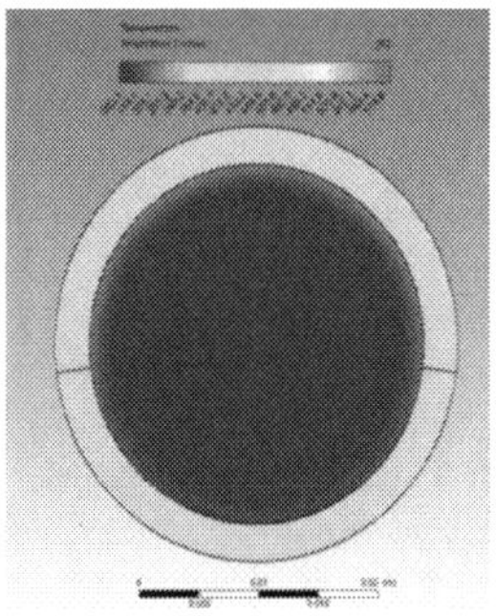

Fig. 5. Temperature distribution at the absorber's outlet for water used as a working medium and vibration amplitude of δ=0.1mm

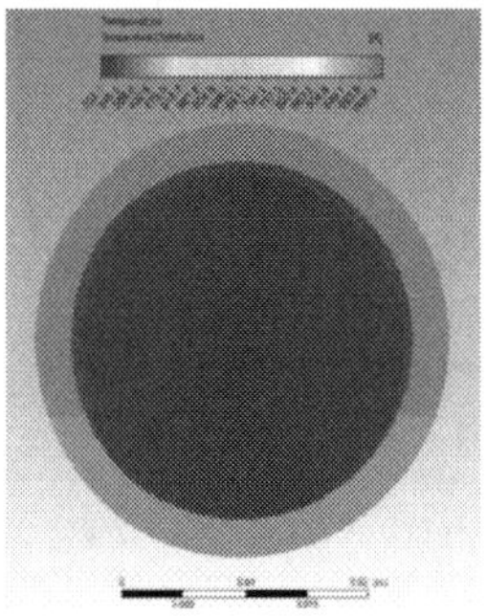

Fig. 6. Temperature distribution at the absorber's outlet for water used as a working medium and vibration amplitude of δ=0.9mm

Results of the analysis, concerning influence of the increase in the frequency of vibrations of the absorber on the operational parameters of the absorber, are indicated in Fig. 7 and Fig. 8. Results of the simulation of the operation of static turbulizing devices, as well as the detailed analysis of the turbulence caused, depicted as distributions of the turbulence kinetic energy, are depicted on Fig. 9 to Fig. 12. Table 2 presents mean values of the circulating medium temperatures at the outlet of the absorber for all simulated cases.

Considering the results acquired for the thermal oil, used as a working medium (Fig. 3 and Fig. 4), no significant difference within temperature distribution, especially concerning the viscous boundary layer within interior if the absorber, is visible. This observation suggests, that the investigated amplitude of vibration does not derive energy required to induce the boundary-layer turbulences, affecting the significant change in temperature distribution around boundary.

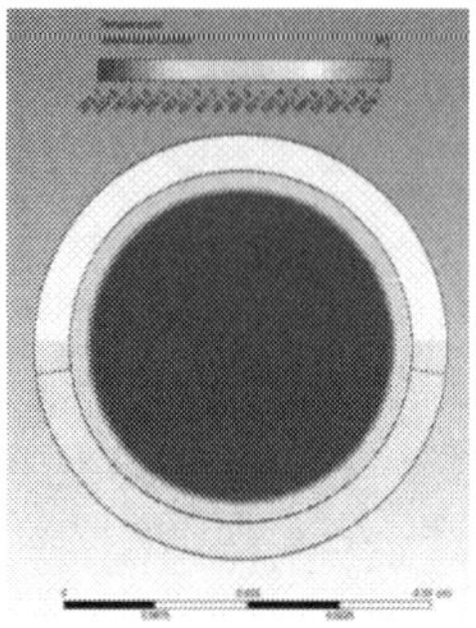

Fig. 7. Temperature distribution at the absorber's outlet for thermal oil used as a working medium and vibration amplitude of δ=0.1mm and frequency f=75Hz

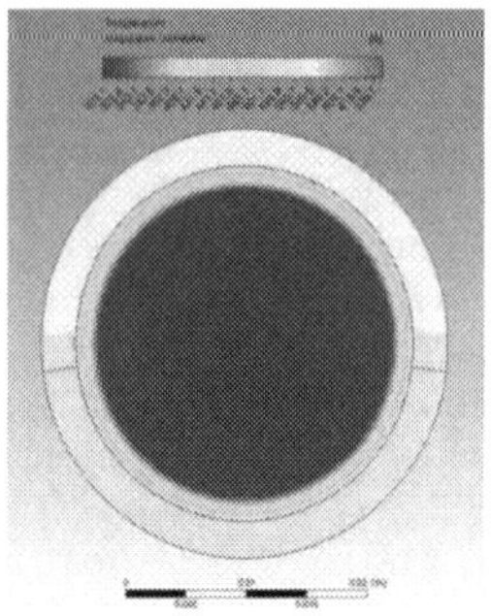

Fig. 8. Temperature distribution at the absorber's outlet for thermal oil used as a working medium and vibration amplitude of δ=0.9mm and frequency f=75Hz

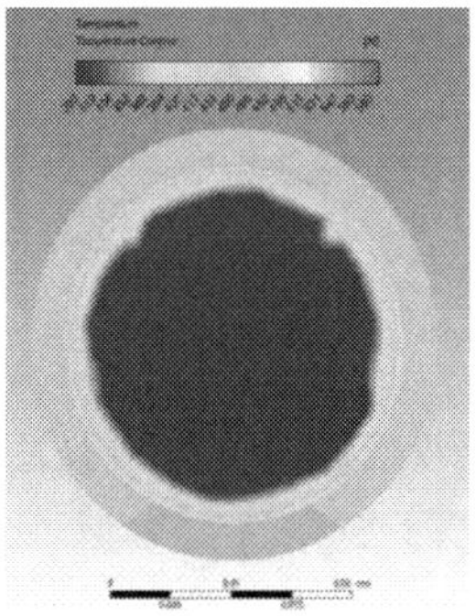

Fig. 9. Temperature distribution at the absorber's outlet for thermal oil used as a working medium and static turbulizing device with 7 revolutions/meter

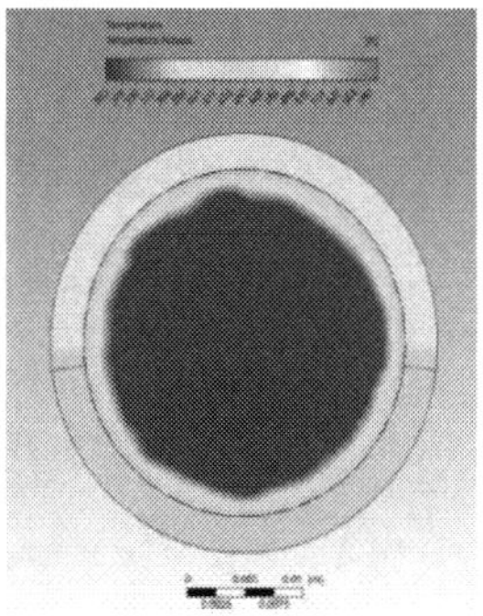

Fig. 10. Temperature distribution at the absorber's outlet for thermal oil used as a working medium and static turbulizing device with 12 revolutions/meter

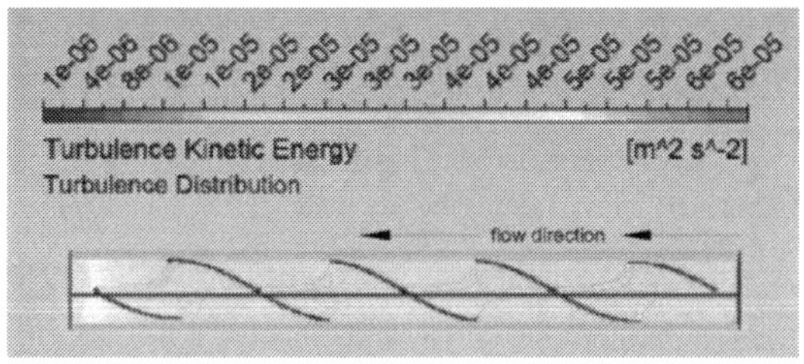

Fig. 11. Turbulence kinetic energy distribution on vertical co-axial plane for static turbulizing device with 7 revs/meter

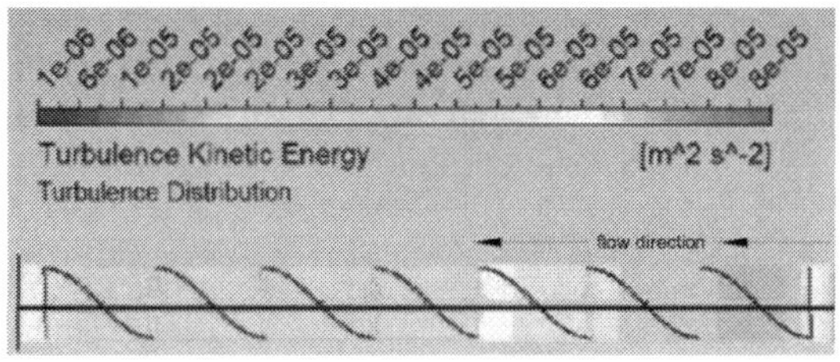

Fig. 12. Turbulence kinetic energy distribution on vertical co-axial plane for static turbulizing device with 12 revs/meter

Table 2. Mean temperature of the working fluid at the outlet of the absorber – comparison of simulated cases.

Case description	Mean temperature, K
Thermal oil, no turbulization [29]	305.0
Thermal oil, vibration of f=50Hz and δ=0.1mm	304.5
Thermal oil, vibration of f=50Hz and δ=0.9mm	304.6
Water, no turbulization [29]	307.9
Water, vibration of f=50Hz and δ=0.1mm	307.9
Water, vibration of f=50Hz and δ=0.9mm	307.8
Thermal oil, vibration of f=75Hz and δ=0.1mm	305.0
Thermal oil, vibration of f=75Hz and δ=0.9mm	305.0
Thermal oil, static turbulizing device with 7 revs/m	309.9
Thermal oil, static turbulizing device with 12 revs/m	310.8

The proposed summary of the initial results is supported by data collected for analogical simulation, with water used as a circulating fluid (Fig. 5 and Fig. 6), on which the distortion of the boundary-layer temperature distribution is visible. The difference in temperature of the absorbers between the water and oil fluids is caused by vital difference in transient heat transfer conditions, taking place at the initial phase of the simulation. However, since the heat transfer analysis for water-cooled absorber did not state objective of the research, no detailed investigation has been performed.

Although the vibration influenced the boundary layer in the discussed case, energy derived still was not sufficient to vitally turbulize the flow – despite induction of oscillations, the laminar flow was maintained. Therefore, blending the colder and warmer layers of the interior and the boundary-layer of the circulating fluid, leading to vital intensification of heat transfer within the absorber, require higher parameters of the oscillatory motion. This observation is in accordance with the results acquired for simulations performed with the increased frequency of the oscillations (Fig. 7 and Fig. 8). Independently on the amplitude of the vibration, the rise of the frequency of 50% comparing to the initial case, which results are presented in Fig. 3 and Fig. 4, no significant change within the structure of the boundary layer and the outlet temperature distribution was observed. This is followed by insignificant change in the mean temperature of the working medium between discussed cases, as indicated in Table 2. Vibrations characterized by low acoustic frequencies cannot introduce significant effect on intensification of heat transfer within the solar absorber, shall be stated. This is influenced primarily by high viscosity of the circulating fluid (approximately 120 times greater than the viscosity of water at the ambient temperature [24]), resulting in creation of large viscous boundary layer at the absorber-fluid interface, as well as low range of the frequencies assumed for the simulation [23].

Considering results collected for the static turbulizing devices, analysis of data depicted on Fig. 9 and Fig. 10 proves significant destabilization of laminar flow at the absorber-wall boundary. Considering both cases, essential distortion of the uniform temperature field is indicated/ The distortion directly leads to rise in the mean thermal oil temperature, as data in Table 2 indicates. Nevertheless, regarding results of the turbulence intensity analysis (Fig. 11 and Fig. 12), the element characterized with 12 revolutions per meter introduces slightly higher turbulence kinetic energy, compared to the other investigated element. Furthermore, for this geometry, the highest values of the turbulence kinetic energy are observed at the inlet to the absorber and close to the inlet, whereas for the second of discussed geometries, the highest energy was observed for the absorber's outlet plane. Nevertheless, for both cases, significant flow turbulization at the absorber's outlet is visible.

4 Conclusions

Results of the performed simulations indicate, that utilization of vibrations of low acoustic frequencies in enhancement of heat transfer within the absorber of vacuum solar collector is strongly limited. The main reason is high viscosity of the terphenyl-based thermal oil, used as the working fluid for high temperature installations, resulting in large thickness of the viscous boundary layer. The simulated vibrations, characterized by frequency not exceeding f=75Hz and amplitude below δ=1mm do not introduce sufficient kinetic energy to create the boundary-layer vertices, resulting in significant turbulization of the fluid. Nevertheless, further investigation on effect of induction of vibrations of frequency close to the resonant frequency of the system, inspired by the promising results of the analysis discussed in [22], states subject of the future research. Furthermore, due to the introductory nature of the presented computational research, discussed results are aimed to be verified during future experimental investigation.

Data collected for the static turbulizing devices proves possibility of its successful application in absorbers of the solar collectors. Both the temperature distribution at the absorber's outlet, as well as mean temperature of the circulating fluid, prove positive effect of introduction of both investigated geometries of the turbulizing elements. However, detailed analysis of the turbulence kinetic energy suggests, that increase in the number of revolutions per unitary length leads to rise in the turbulence kinetic energy of the created fluid vertices. Furthermore, for the twelve-revolutions geometry, the majority of turbulent vertices were produced at the region of higher average heat flux, comparing to the case of seven-revolution geometry. The optimization of the constructional features of the turbulizing elements states matter of further investigation.

Acknowledgments

The scientific work is funded by the National Science Centre within the framework of the research project No. 2018/29/B/ST8/02406.

References

[1] E. Klugmann, E. Klugmann-Radziemska: Ogniwa i moduły fotowoltaiczne oraz inne niekonwencjonalne źródła energii, Ekonomia i Środowisko, Białystok (2005)

[2] G. Lobaccaro, S. Croce, C. Lindkvist, M.C. Munari Probst, A. Scognamiglio, J. Dahlberg, M. Lundgren, M. Wall: A cross-country perspective on solar energy in urban planning: Lessons learned from international case studies, *Renewable and Sustainable Energy Reviews*, Vol. 108, pp. 209-237 (2019)

[3] W. Gogół, O. Skonieczny, L. Zakrzewski: Niektóre zagadnienia wymiany ciepła w kolektorach energii promieniowania słonecznego, Politechnika Warszawska, Warszawa 1979

[4] N. Bushra, T. Hartmann: A review of state-of-the-art reflective two-stage solar concentrators: Technology categorization and research trends, *Renewable and Sustainable Energy Reviews*, Vol. 114, 109307 (2019)

[5] M.E. Ydrissi, H. Ghennioui, E.G. Bennouna, A. Farid: Geometric, optical and thermal analysis for solar parabolic trough concentrator efficiency improvement using the Photogrammetry technique under semi-arid climate, *Energy Procedia*, vol. 157, p. 1050-1060 (2019)

[6] L. Evangelisti, R. De Lieto Vollaro, F. Asdrubali: Latest advances on solar thermal collectors: A comprehensive review, *Renewable and Sustainable Energy Reviews*, Vol. 114, 109318 (2019)

[7] N. B. Desai, S. B. Kedare, S. Bandyopadhyay: Optimization of design radiation for concentrating solar thermal power plants without storage, *Solar Energy*, Vol. 107, pp. 98-112 (2014)

[8] J. Ruelas, D. Sauceda, J. Vargas, R. García: Thermal and concentration performance for a wide range of available offset dish solar concentrators, *Applied Thermal Engineering*, Vol. 144, pp. 13-20 (2018)

[9] G. H. Lee: Construction of conical solar concentrator with performance evaluation, *Energy Procedia*, Vol. 153, pp. 137-142 (2018)

[10] E.T.A. Gomes, N. Fraidenraich, O.C. Vilela, C.A.A. Oliveira, J. M. Gordon: Aplanats and analytic modeling of their optical properties for linear solar concentrators with tubular receivers, *Solar Energy*, Vol. 191, pp. 697-706 (2019)

[11] C. Michel, P. Blain, L. Clermont, F. Languy, C. Lenaerts, K. Fleury-Frenette, M. Décultot, S. Habraken, D. Vandormael, R. Cloots, G. K. V.V. Thalluri, C. Henrist, P. Colson, J. Loicq: Waveguide solar concentrator design with spectrally separated light, *Solar Energy*, Vol. 157, pp. 1005-1016 (2017)

[12] G. Wang, Y. Yao, Z. Chen, P. Hu: Thermodynamic and optical analyses of a hybrid solar CPV/T system with high solar concentrating uniformity based on spectral beam splitting technology, *Energy*, Vol. 166, pp. 256-266 (2019)

[13] J. Chen, L. Yang, Z. Zhang, J. Wei, J. Yang: Optimization of a uniform solar concentrator with absorbers of different shapes, *Solar Energy*, Vol. 158, pp. 396-406 (2017)

[14] P. Selvakumar, P. Somasundaram, P. Thangavel: Performance study on evacuated tube solar collector using Therminol D-12 as heat transfer fluid coupled with parabolic trough, *Energy Conversion and Management*, Vol. 85, pp. 505-510 (2014)

[15] J. Qin, E. Hu, G. J. Nathan, L. Chen: Concentrating or non-concentrating solar collectors for solar Aided Power Generation?, *Energy Conversion and Management*, Vol. 152, pp. 281-290 (2017)

[16] M. Sabiha: An experimental study on Evacuated tube solar collector using nanofluids, *International Conference on Advances in Science, Engineering, Technology and Natural Resources (ICASETNR-15) Sabah, Malaysia*, Vol. 2, pp. 42-49

[17] M.A. Sharafeldin, G. Gróf, E. Abu-Nada, O. Mahian: Evacuated tube solar collector performance using copper nanofluid: Energy and environmental analysis, *Applied Thermal Engineering*, Vol. 162, 114205 (2019)

[18] H.Fathabadi: Novel solar collector: Evaluating the impact of nanoparticles added to the collector's working fluid, heat transfer fluid temperature and flow rate, *Renewable Energy*, DOI: https://doi.org/10.1016/j.renene.2019.10.008 (2019)

[19] A. García, R. Herrero-Martin, J.P. Solano, J. Pérez-García: The role of insert devices on enhancing heat transfer in a flat-plate solar water collector, *Applied Thermal Engineering*, Vol. 132, pp. 479-489 (2018)

[20] D. Jin, S. Quan, J. Zuo, S. Xu: Numerical investigation of heat transfer enhancement in a solar air heater roughened by multiple V-shaped ribs, *Renewable Energy*, Vol. 134, pp. 78-88 (2019)

[21] F. A.S. da Silva, D. J. Dezan, A. V. Pantaleão, L. O. Salviano: Longitudinal vortex generator applied to heat transfer enhancement of a flat plate solar water heater, *Applied Thermal Engineering*, Vol. 158, 113790 (2019)

[22] W. Liu, Z. Yang, B. Zhang, P. Lv: Experimental study on the effects of mechanical vibration on the heat transfer characteristics of tubular laminar flow, *International Journal of Heat and Mass Transfer*, Vol. 115, Part A, pp. 169-179 (2017)

[23] M. Setareh, M. Saffar-Avval, A. Abdullah: Experimental and numerical study on heat transfer enhancement using ultrasonic vibration in a double-pipe heat exchanger, *Applied Thermal Engineering*, Vol. 159, 113867 (2019)

[24] *Therminol® 66 High Performance Highly Stable Heat Transfer Fluid*, Solutia Europe S.A./N.V., Louvain-la-Neuve (2016)

[25] C.G. Speziale, S. Sarkar, T.B. Gatski: Modeling the Pressure-Strain Correlation of Turbulence: an Invariant Dynamical Systems Approach, *Journal of Fluid Mechanics*, Vol. 227, pp. 245-272 (1991)

[26] Bredberg J.: *On the Wall Boundary Condition for Turbulence Models*, Chalmers University of Technology, Göteborg (2000)

[27] *ANSYS CFX-Solver Theory Guide. Release 15.0*, ANSYS Inc., Canonsburg (2013)

[28] G.F. Nellis., S.A. Klein: *Heat Transfer*, Cambridge University Press, New York (2009)

[29] Bartela Ł., Remiorz L., Stanek B., Grzywnowicz K.: Kierunki intensyfikacji przepływu ciepła w absorberze parabolicznego koncentratora promieniowania słonecznego, *Konferencja Zarządzanie Energią i Teleinformatyką, Rynek Gazu, Rynek Ciepła*, Nałęczów 17-19.06.2019

RESEARCH & DEVELOPMENT IN POWER ENGINEERING, 2019

Experimental investigation of LWR passive safety systems performance at the INKA test facility

Rafał Bryk[1,2*], *Thomas* Mull[2] and *Holger* Schmidt[2]

[1] Institute of Heat Engineering, Warsaw University of Technology, Nowowiejska 21/25, 00-665 Warsaw, Poland
[2] Framatome GmbH, Paul-Gossen-Straße 100, 91052 Erlangen, rafal.bryk@framatome.com, Germany

Abstract. INKA is a test facility designed by Framatome and built in the technical center in Karlstein. The original objective for establishing this test rig was the investigation of the performance of the passive safety systems developed in a new Framatome Boiling Water Reactor (BWR) design – KERENA. INKA was constructed in the scale of 1:1 in heights while the total volume of the containment was replicated in 1:24. Since the geometries of particular safety systems are faithfully reflected, their actual performance in the original plant can be investigated at the full scale. Due to the unquestionable interest of the nuclear community in the inherent safety, not only new BWR and PWR designs are equipped with the passive systems, but also particular passive solutions are considered to be applied into the already existing Light Water Reactors (LWR). In this context and due to the fact that both, single component tests and integral tests can be conducted at INKA, the facility can be employed for a demonstration/qualification of a large range of passive safety systems foreseen for quite different types of LWRs. Hence, the goal of the EASY project was the experimental confirmation of the passive systems performance and the analysis of their interactions between each other in the integral tests. Besides, the overarching target of all tests performed at INKA is provision of data for codes validation. This paper presents major outcomes and conclusions drawn on the basis of EASY project results.

1 Introduction

Passive systems increase inherent safety of nuclear power plants by utilizing basic laws of physics such as gravity and natural circulation for removal of residual heat from the reactor pressure vessel (RPV) in case of an accident. Properly designed passive solutions enable bringing the plant to safe and stable condition without the need for active systems and corresponding engines, electric instrumentation and control equipment etc. This in turn decreases the total investment cost as well as lowers the maintenance effort. The concept of the passive safety eliminates also the human factor, which is considered to be the reason of the most relevant accidents in the nuclear industry.

Following the trend of the inherent safety increase and setting standards in this field at the same time, Framatome developed a new boiling water reactor (BWR) design – KERENA. KERENA is a medium size reactor equipped with both, passive and active systems. The development of the reactor was based on combining the state of the art knowledge from experiences collected during elaborating solutions for other products as well as on the implementation of new ideas [1].

In order to examine the performance of the passive systems of KERENA, a dedicated test facility was built at the Components Testing Department of Framatome in Karlstein, Germany. The Integral Test Facility Karlstein (INKA) represents the KERENA containment with a volume scaling of 1:24. Components heights and levels over the ground are kept in the full scale. The test rig setup allows to investigate both, each component individually and the whole system during integral tests. For the purpose of testing the integral system, INKA is equipped with several large vessels corresponding to the individual pools and vessels of KERENA. The RPV vessel has a storage capacity of 1/6 of the KERENA RPV. In order to simulate the decay heat of the core, the vessel is fed with steam by the Benson boiler with maximum power output of 22 MW. The vessel is designed for operation pressure up to 160 bars. Compartments of the KERENA RPV are represented by separate vessels: flooding pool vessel, pressure suppression pool vessel and drywell vessel. An additional tank is used as a representation of the shielding/storage pool vessel [2]. Figure 1 illustrates replication of KERENA by the INKA test facility.

Since several passive systems are installed at the INKA facility and some solutions are utilized also in other light water reactors (LWR), the test facility itself can be perceived as a test rig for investigation of passive systems in general and not only as a one replicating a single reactor. This is particularly valid in the light of the possibility of investigation of individual components as well as interactions between them. Thus, for instance the passive pressure pulse transmitter (PPPT) which was analyzed at INKA is also considered as an actuator for the Safety Condenser (SACO) system for the pressurized water reactor (PWR) steam generators (SG).

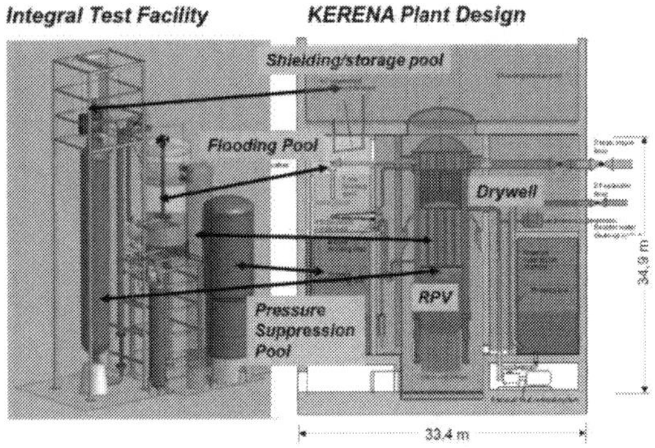

Fig. 1. KERENA replication by the INKA facility

Furthermore, besides the new BWR and PWR designs particular passive solutions are considered to be applied into already existing LWRs. In this context INKA can be employed for demonstration/qualification of a large range of passive safety systems foreseen for quite different types of LWRs. With this in mind, the Integral Experimental and Analytical Safety Cases for Design-Basis Accidents in Nuclear Power Plants Featuring Passive Safety Systems (EASY) project was initiated in March 2015. In the frames of the project four integral tests were conducted with the main objective of data base generation for behavior of and interaction between the different passive systems and components under realistic accident conditions. Besides, the overarching target of the project was provision of data for codes validation.

All tests performed in the frames of EASY project were conducted as integral tests, so not only the performance of individual components was investigated but also interactions between them and their influence on the entire system. In each test different accident scenario was simulated and the response of the facility was observed. The following events were covered by the tests campaign:

- Loss of Coolant Accident (LOCA) by a break of the Main Steam Line (MSL) with simulation of the decay heat.
- Simulation of Feedwater Line Break (FWLB) with taking into account the decay heat
- Simulation of RPV bottom leak with taking into account the decay heat
- Simulation of Station Blackout (SBO) with taking into account the decay heat

All the above tests showed a demonstrable reliability of the passive systems. They also provided a comprehensive data base for better understanding of the involved phenomena. The test data constitute a robust validation cases for thermal-hydraulic systems codes.

2 Passive systems of KERENA and INKA

KERENA reactor is equipped with several passive safety systems which are replicated in the full scale at the INKA facility. However, if the original plant comprises more than one set of the same devices, only one was built at the test facility. For example, although the KERENA reactor is equipped with four Emergency Condensers (EC) and four Containment Cooling Condensers (CCC) only one EC and CCC were installed at the INKA. Figure 2 illustrates all passive systems of KERENA which were replicated at INKA.

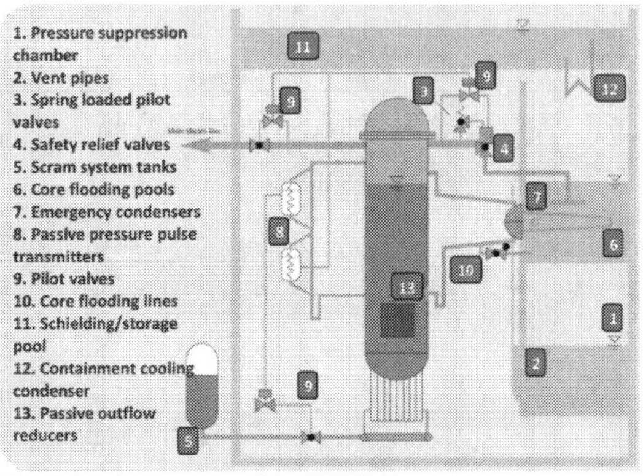

Fig. 2. Passive systems of the KERENA reactor

In case of an accident the reactor scram in the KERENA design is triggered by Passive Pressure Pulse Transmitters (PPPT). The PPPTs are small heat exchangers that operate according to the principle of communicating vessels (figure 3).

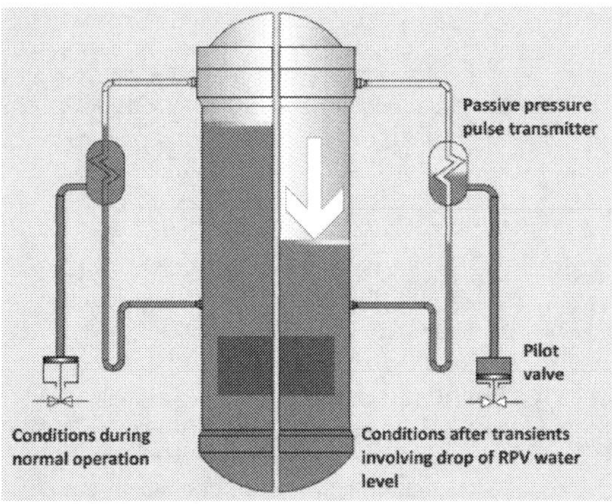

Fig. 3. PPPT operation

Upon a drop of the reactor water level, steam enters the top connective line and when it reaches the PPPT, pressure builds up on its secondary side due to the liquid expansion or evaporation caused by the temperature increase which in turn is a result of the heat transfer from the steam at the primary side. This pressure is then used to initiate safety-related switching operations without any need for electric power or manual or I&C signals. PPPTs are installed in the KERENA reactor for safety-classified initiation operations. Depending on the water level in the RPV, the PPPTs actuate different procedures (figure 2). Apart from reactor scram, they trigger automatic depressurization of the RPV by opening of the safety relief valves and they activate the isolation of the containment at the main steam lines. Their integration into the plant's systems engineering is (configuration:

two 2-out-of-2) such that spurious actuation of a single PPPT does not lead to initiation of actions, but also such that loss of one PPPT cannot prevent initiation. PPPTs can be also applied to any kind of other systems triggered passively in other types of nuclear power plant. For instance PPPT is considered to serve as an actuator of the Safety Condenser (SACO) – passive system for heat removal from the PWR steam generators.

The Emergency Condenser (EC) system depressurizes and removes the decay heat from the RPV after the scram. The condenser is composed of a large number of slightly inclined tubes submerged in the core flooding pool and filled with cold water during normal operation of the plant (figure 4).

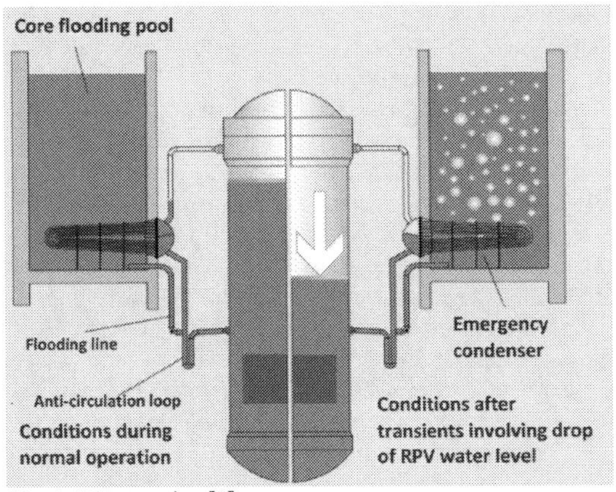

Fig. 4. EC operation [2]

Should the water level in the RPV decrease, the water drains from the tubes and the steam from the RPV enters the EC tubes. Due to the contact of external tube walls with the cold water in the core flooding pool, the steam at the primary side condenses leading to removal of substantial amount of heat. The heat transfer capacity of each EC is about 60 MW. Since the tubes of the EC are slightly inclined with respect to the horizontal orientation, the condensate flows back to the RPV due to gravity. The EC is connected with the RPV without any isolating elements and it is activated automatically without any need for electric power and according to the natural law of communicating vessels [2]. Therefore, since the EC system responses immediately to the liquid water drop in the RPV, the high pressure injection system is no longer necessary and it was eliminated from the KERENA design.

Due to its large capacity the Flooding Pool Vessel (FPV) acts not only as a sink for the EC but also for the Safety Relief Valve (SRV) system (figure 5).

The SRVs are of a diverse design: four of them operate according to the pressurization principle, four according to the pressure relief principle, thus excluding a loss of all safety relief valves due to a common mode failure.

The SRV system is located inside the containment and consists of the safety relief valves and relief lines with steam quenchers, which are installed in the core flooding pools. This system follows proven concepts used in the existing Framatome's BWR plants.

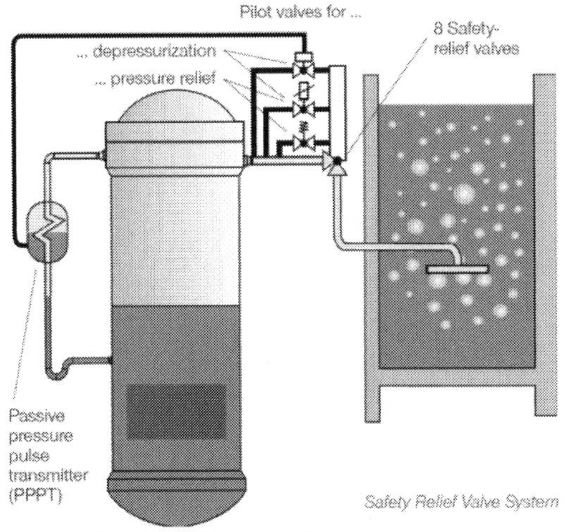

Fig. 5. Safety Relief Valve operation

The FPV acts also as a water reservoir for passive core flooding. After the sufficient reduction of pressure in the RPV a check valve opens and it enables the water to flow gravitationally to the RPV (figure 6A). Despite the extremely low likelihood of a severe accident (8.0 * 10^{-8} core damage frequency), the design of the KERENA reactor mitigates the consequences of such an event by employing the FPV also as a containment flooding pool. Therefore, even if it comes to the extremely improbable event of core meltdown, the core itself would remain in the RPV, which would be cooled-down from the external side (figure 6B).

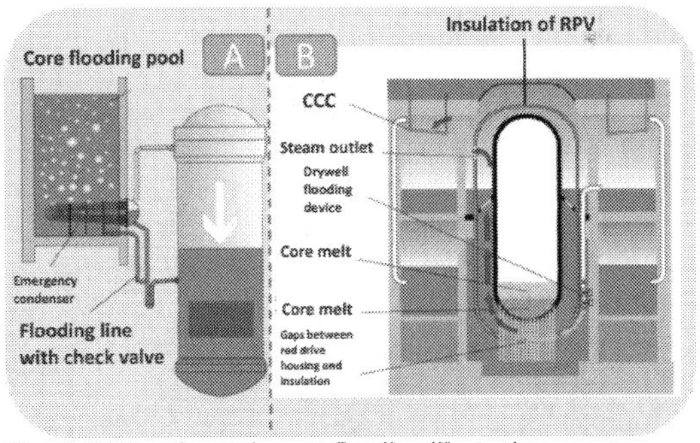

Fig. 6. RPV and containment flooding illustration

The Containment Cooling Condenser (CCC) is a system responsible for long-term (approximately 3 days) heat removal from the containment of the reactor. The whole CCC system consists of four heat exchangers hanged at the top of the containment vessel (Figure 2 and 6B). During normal operation of the reactor the condenser tubes are filled with cold water from the Shielding/Storage Pool Vessel (SSPV) and the CCC systems is not in operation. In case of a LOCA the steam could be released from the reactor loop increasing temperature and pressure inside the containment. Due to the temperature difference between the medium (steam-air mixture) in the containment and the water inside the CCC tubes, the heat transfer from the containment to the water in the tubes would be initiated (Figure 7).

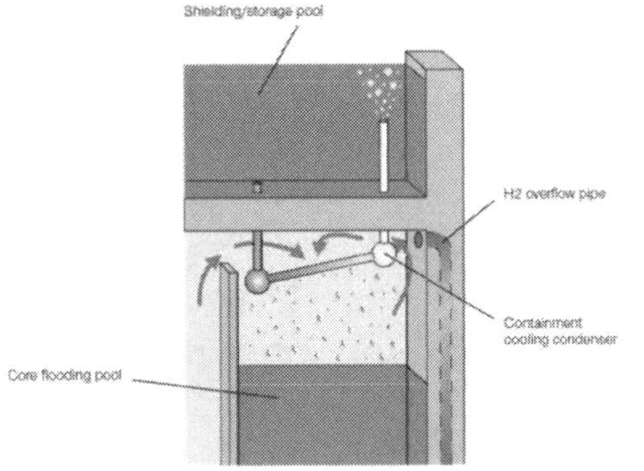

Fig. 7. CCC operation

Due to the density decrease resulting from the temperature increase in the CCC tubes the buoyancy force establishes natural circulation in the tubes. The fluid flow facilitates the heat removal to the long-term heat sink. At the primary side i.e. in the containment the steam released from the reactor loop condenses at the CCC tubes. Since condensation is a very efficient mechanism of heat transfer, substantial amount of heat can be removed from the containment. Furthermore, condensation at the CCC tubes supports the depressurization of the containment. Due to the placement of the CCCs at the top of the containment above the Flooding Pool Vessels, the condensate dribbles gravitationally from the CCC tubes to the FPV.

Depressurization and heat removal from the containment is also supported by the Pressure Suppression Pool (PSP). The PSP is a vessel filled with water up to a certain level and it is connected with the drywell with the vent pipe. Should the pressure in the containment increase, the water column in the vent pipes would be "pushed" downwards so that the steam released during a LOCA could enter the PSP and condense in the cold water in the pool. The PSP serves also as a water reservoir for the active residual heat removal system.

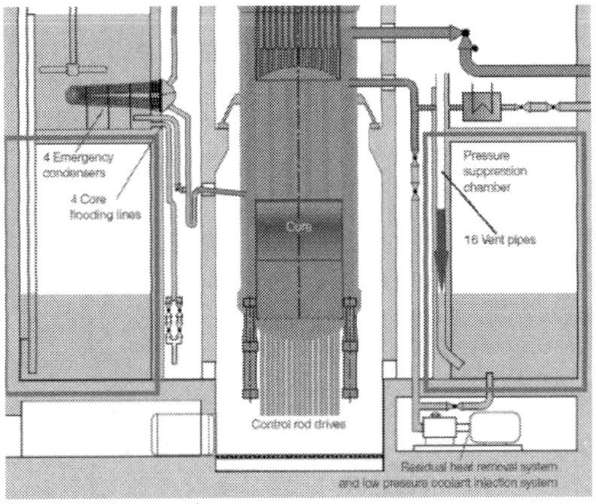

Fig. 8. PSPV operation

3 INKA setup

3.1 Components arrangement

The INKA test facility was designed to perform steady-state, full-scale tests of the passive systems (EC, CCC, passive core flooding, PPPT) and for simulating transient LOCA scenarios in a reduced scale [3, 4]. Thus, the individual passive systems geometry at the INKA facility as well as the geodetical heights faithfully correspond to those at the KERENA reactor whereas the volumes of the containment are modeled to a scale of 1:24. The piping between the components was accurately designed for correct replication of those in the KERENA reactor. For a detailed determination of the heat transfer capacities as well as thermodynamic conditions in the individual components a dense network of instrumentation was provided.

The INKA facility is integrated into the large valve test facility Grossarmaturen-Prüfstand (GAP) that has been in operation in Karlstein for many years. The GAP test facility consists of a steam accumulator (height: 21.7 m, volume: 125m^3) fed by a 22-MW boiler. This accumulator supplies the steam needed for the experiments and replicates the RPV in the integral tests. An additional vessel with a water volume of about 50m^3 is placed inside the GAP support frame at the top to simulate the shielding/storage pool. This vessel supplies therefore water for the secondary side of the CCC.

The entire containment of the KERENA plant is simulated by three vessels. The Flooding Pool Vessel (FPV) represents the core flooding pools that contains the EC and the CCC. The Passive Core Flooding System connects the FPV with the return line of the EC. The PPPT is connected to the downcomer line. The Drywell Vessel (DWV) simulates the residual gas volume of the containment. The Pressure Suppression Pool (PSP) is simulated by a third vessel [3].

Figure 9 shows a simplified Piping and Instrumentation Diagram (P&ID) of the INKA test facility with the containment. Two pipes connect the FPV with the DWV and represent the connections between the gas spaces of these compartments. The FPV is connected to the PSP vessel via the overflow pipe limiting the FPV water level. A second connection is the hydrogen overflow pipe used for pressure limitation during severe accident mitigation at KERENA. The DWV and the PSP vessel are connected via a full-scale vent pipe. Additionally, the function of the safety and relief valve is included in the design. The system is connected to the downcomer line and it goes into the FPV. For LOCA test scenarios, two different lines enter the Drywell Vessel, one for the simulation of breaks of a water line and the other for breaks of a steam line [3].

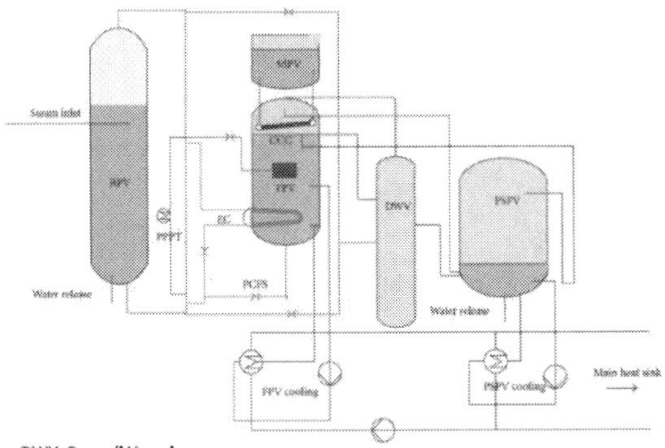

Fig. 9. Simplified P&ID of the INKA facility [3]

3.2 Instrumentation of the INKA facility

The INKA facility is comprehensively instrumented with over 300 measurement sensors. Most of the instrumentation comprises widely used temperature, pressure, differential pressure and mass flow sensors. For two-phase flow measurement thermo-needle probes and gamma densitometer are installed at the INKA facility. The gas mixture in the vessels is measured via a probe sampling system analyzed by a mass spectrometer.

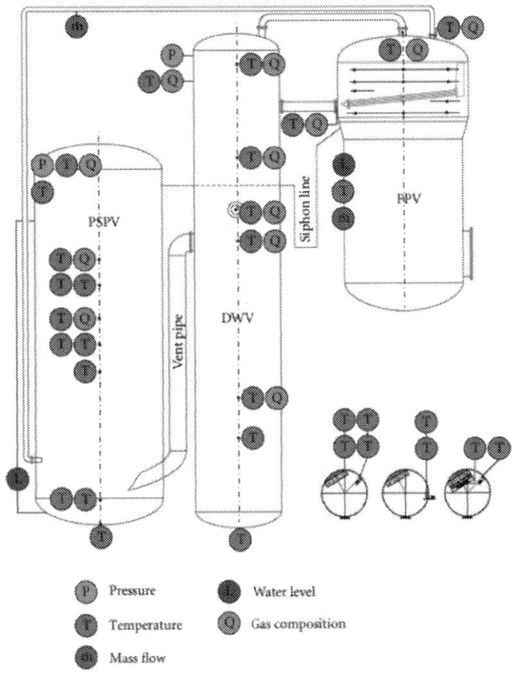

Fig. 10. INKA vessels instrumentation [3]

An overview of the INKA vessels (PSP, DWV and FPV) instrumentation is illustrated in the figure 10. For the obtainment of the complete information about the conditions inside the PSP, the vessel is instrumented with the pressure, water level as well as the gas composition and temperature sensors in several positions. Similarly, the DWV instrumentation provides detailed information about the conditions inside the vessel. Since the FPV houses the EC and CCC passive systems and monitoring the conditions inside the vessel is essential for drawing the correct conclusions regarding operation of the two systems, the FPV is equipped with the largest number of measurement sensors (figure 11).

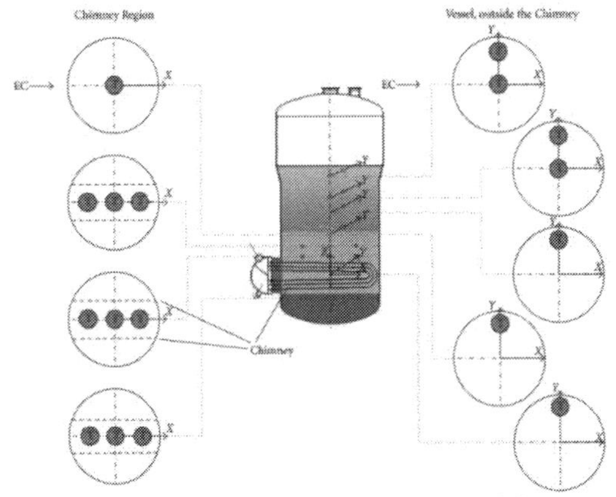

Fig. 11. Water volume instrumentation of the FPV [3]

The temperature of the water pool is measured at nine elevations. At each level up to three sensors are installed to provide an overview of the cross-sectional temperature distribution. Apart from temperature, the water level and pressure inside the vessel are measured. However, due to the overflow pipe connecting the FPV with the PSP, the water inventory cannot exceed a certain level. For investigation of the conditions around the CCC, the space above the water in the FPV is also intensively instrumented. This region is particularly important in the light of the gas composition influence on the heat transfer capacity of the CCC.

The interface between the containment vessels and the RPV is the Emergency Condenser. The tube bundle of the EC is submerged in the water of the FPV and its inlet and outlet pipes are connected to the standpipe representing the downcomer of the KERENA RPV. The standpipe and the RPV are extensively instrumented with temperature, absolute pressure and mass flow sensors. Furthermore, a number of differential pressure measurements is installed both, in the vessels and in the EC piping for provision of accurate data regarding pressure drops and water levels. Two outer EC tubes are also densely instrumented with temperature sensors and thermo-needle probes. Most of the thermocouples are located at the axis of the pipes. Additionally, at several positions four supplementary sensors are located at the inner and outer sides of the tubes. Such a combination of the local void fraction with the temperature in five points per location enables a detailed description of the system.

Similarly to the EC, the CCC is densely instrumented for obtainment of the detailed description of the component performance. Thus, apart from the absolute pressure, the differential pressure measurements are installed at the inlet and outlet lines as well as at the heat exchange section. Thermocouples are located at several positions over different cross sections. The two-phase flow can be directly identified via the gamma densitometer installed in the return line.

4 Test results

The test program named Integral Experimental and Analytical Safety Cases for Design-Basis Accidents in Nuclear Power Plants Featuring Passive Safety Systems (EASY) was initiated in March 2015. In the frames of the project four tests were performed simulating LOCA and non-LOCA scenarios. In particular, the main steam line break, feedwater line break, station blackout and RPV bottom leak tests were performed. In this paper the main steam line break will be discussed and the performance of each passive system will be analyzed in details.

4.1 Test boundary conditions

During the conditioning phase of the test the INKA facility was pressurized up to 75 bars and heated up to corresponding saturation temperature. These conditions are consistent with the normal BWR operation.

In order to assure the faithful simulation of conditions inside the facility, the test comprised several activities. The break initiation was performed 350 s. after the Start of the Test (SOT). Simultaneously, cooling of the Shielding/Storage Pool Vessel (SSPV) was initiated. This activity was necessary because the SSPV is relatively small in comparison to other components of the facility. Therefore, in order to provide SSPV heat capacity appropriately scaled with reference to the original plant, the vessel was additionally cooled.

Simultaneously, the decay heat simulation was initiated by enabling the steam flow from Benson boiler to the RPV/GAP vessel. At the same time corresponding amount of liquid was discharged from the vessel so that the latent heat of water was supplied to the system with possibly low change of the water inventory.

4.2 Primary side

The primary side of the INKA test facility consists of the RPV-simulator, the Standpipe, the PPPT and the EC including inlet and outlet lines.

Initiation of the break caused a rapid decrease of the pressure at the primary side. Correspondingly, the pressure at the secondary side, i.e. in the containment vessels increased. The pressures were equalized approximately 2330 s after the SOT. Figure 12 shows the pressures both at the primary and secondary side.

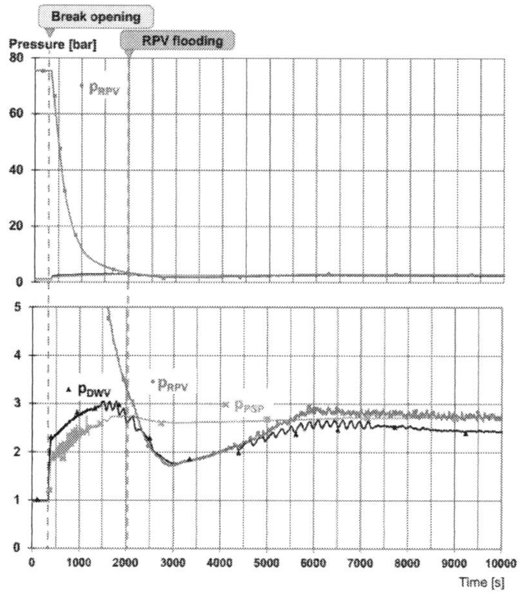

Fig. 12. Pressures in the RPV, DWV and PSP

Since during the normal operation the RPV is filled with medium under saturation conditions, rapid depressurization of the vessel leads to intensive flash boiling at the primary side. This in turn results in a loss of liquid water inventory and the fill level drop. Figure 13 depicts the water levels in the RPV and standpipe together with the dashed line illustrating the uppermost, middle and lowermost levels of the EC. The uppermost level of the EC heat exchange section is 14.3 m. This means that once the water level decreased to this value, the EC was triggered.

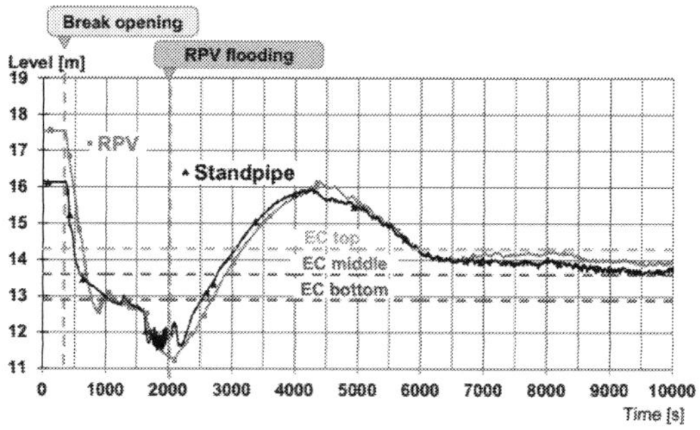

Fig. 13. Water levels in the RPV and Standpipe

In the wake of the depressurization of the system, the temperature decreased too. Shortly after the SOT, the steam and liquid temperature declining was following the saturation temperature curve. As the depressurization proceeded, the steam temperature deviated from the saturation curve leading to a significant difference after several hundred seconds. This disparity was caused by the heat transfer between hot vessel structure and steam (figure 14).

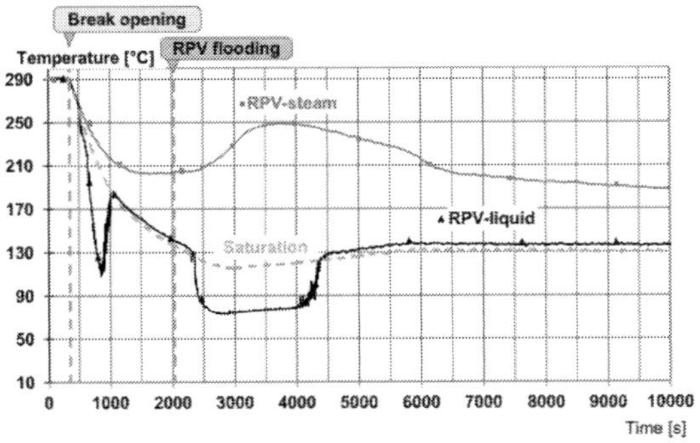

Fig. 14. RPV temperatures

In contrast, the liquid temperature followed the saturation curve most of the time. Two quick and significant drops of the liquid temperature correspond to injections of cold water. The first injection came from the liquid water occupying the EC piping during normal operation. Once this cold inventory reached the RPV, it caused a meaningful temperature decrease. Since the decay heat was being still provided to the RPV, the temperature drop was equalized within approximately 500 s. The second decrease of the liquid water temperature was caused by triggering of the core flooding at 2062 s after the SOT. As the flooding flow stagnated between 4000 and 4500 s, the liquid inventory was heated up again to saturation temperature due to the decay heat.

Figure 13 indicates that the EC was activated at t=426 s, so about 75 s after the break initiation. Due to the RPV inventory loss, the water from the EC system was drained into the RPV. Consequently, the steam from the RPV entered the EC system. Figure 15 shows the measured mass flow together with the calculated power of the EC.

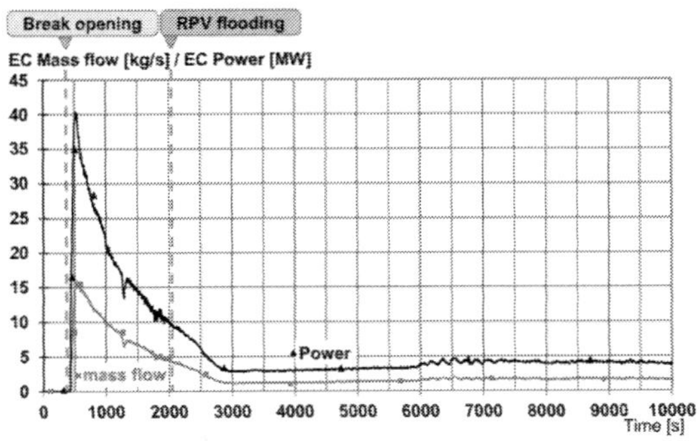

Fig. 15. EC mass flow and power

The EC power was calculated on the basis of the mass flow and the inlet and outlet enthalpies. The maximum power of 41 MW was reached shortly after the break initiation. As the test was proceeding the power was steadily decreasing to 4 MW and this value was kept in the long term.

Both, figure 13 and 14 are consistent with respect to the RPV flooding. Triggering of the flooding at 2062 s resulted in immediate increase of the fill level and slightly delayed decrease of temperature. In fact the inventory level was increased above the uppermost level of the EC piping. This means that the EC piping was completely covered with liquid water, condensation did not occur anymore and the flow through the EC should be decreased to zero. However, the condensation in the inlet pipe (between the Standpipe and the EC heat exchange section) was maintained and a low mass flow could still be observed it the EC system.

The temperature distribution in the EC piping strictly corresponded to the conditions in the other parts of the primary side. Figure 16 illustrates the temperatures at the inlet and outlet of the EC system.

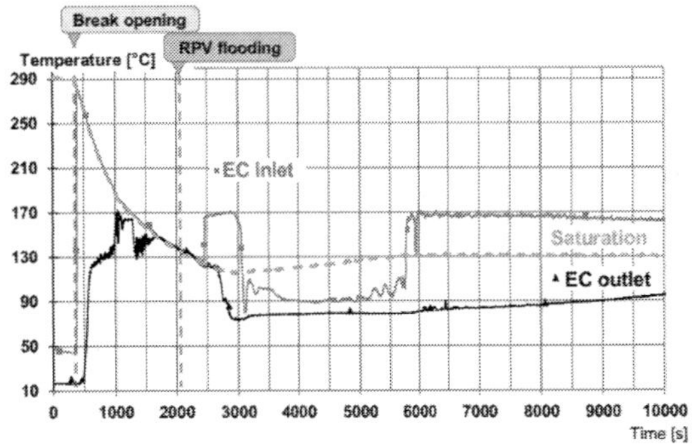

Fig. 16. EC temperatures

The start of the EC operation was indicated by the sharp increase of temperature in the top pipe row of the EC piping system from 44 °C to the saturation temperature. At this early stage the amount of condensate was not enough to cause a significant mass flow from the EC to the RPV. As the RPV inventory level was decreasing more steam entered the EC piping leading to enlarging of the EC piping uncover and the heat transfer area. This in turn led to more intensive condensation and larger mass flow through the EC.

Consistently with the RPV steam region, the temperature in the EC inlet line was also higher than the saturation temperature. Figure 16 shows also that after initiation of the RPV flooding the temperatures drop to the subcooled region. This comes from the fact that due to the equalization of the liquid levels the steam from the RPV does not enter the EC heat exchange section anymore. Therefore, EC acts in this case as a water cooler which leads to establishing of the inlet/outlet temperature difference of approximately 15 K. After the end of the flooding and corresponding later decrease of the RPV liquid level (due to the evaporation cause by the decay heat), steam entered again the EC piping.

4.3 Secondary side

The secondary side of the INKA facility simulates the reactor containment and it consists of the FPV, DWV, PSPV, the vent pipe, the hydrogen overflow pipe and the siphon.

The bottom part of the figure 12 illustrates pressures at the secondary side. After the break opening an instant growth of secondary side pressure was observed. The pressure in the DWV and in the FPV increased to 3.1 bar whereas in the PSPV up to 2.7 bar. The difference of around 0.4 bar corresponds to the submerging of the vent pipe in the PSP water inventory.

Once the pressure at the secondary side reached its peak of 3.1 bar at t=1600 s, it decreased to 1.7 bar within approximately 1400 s. Figure 12 shows that this decrease was accompanied by significant oscillations at the initial state. Both, the decrease and the oscillations are caused by the operation of the CCC. Additionally, the RPV flooding deepened the pressure decrease since it limited boiling in the RPV. Approximately 2160 s after the SOT, the pressures at the primary and secondary side equalized. For a certain amount of time the pressure in the RPV was even lower than in the containment vessels. This inversed difference was partly caused by injection of the cold water from the flooding system that hindered flash boiling and lowered the swell level which contributed to the steeper pressure decrease. The pressure decrease in the RPV was also supported by the condensation in the EC.

In the long term, the primary/secondary side pressure difference of around 0.3 bar was established until the end of the test. A rough estimation showed that the pressure loss of the pipe break was in this range.

Together with the pressure the temperature in the containment vessels rose sharply after the break opening. Figure 17 shows the temperatures in the INKA vessels.

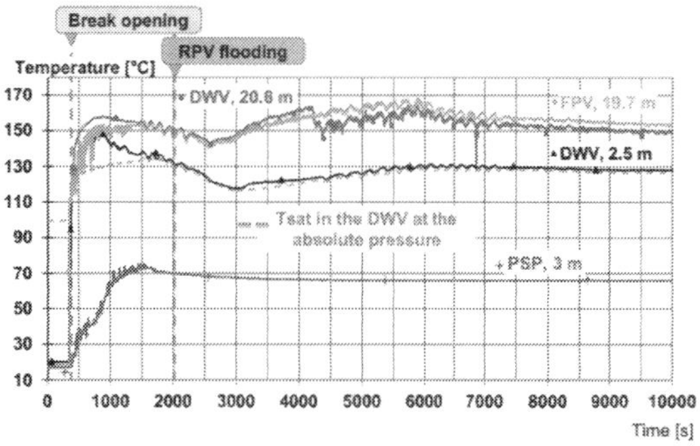

Fig. 17. DWV, FPV and PSP temperatures

Figure 17 shows that the PSP water temperature (PSP, 3 m) grew steadily to approximately 65 °C and this temperature was kept until the end of the test. Due to the non-condensable gas present in the DWV and FPV more complex phenomena were observed in the gas regions of these vessels.

At the very beginning of the transients similar peaks of temperature were measured in both vessels at all elevations. As the experiment proceeded, the temperature at the bottom of the DWV (2.5 m) decreased to the saturation point calculated for the total pressure and it remained at this level. At the upper level of the DWV the temperature was higher and its general trend followed the conditions in the steam region in the RPV, i.e. similarly to the RPV steam (Figure 14), the temperature in the DWV started to increase around 2500 s. On the other hand, mixing with cold inert gas and pressure oscillations caused by CCC operation also had a significant influence on the final shape of the DWR temperature curve.

The temperature in the FPV shows a similar characteristics. It should be emphasized that although the medium inside the vessel was constantly and intensively mixing, the curves have rather local character. It especially refers to the upper region of the FPV, where the steam condenses at the CCC tubes. Steam condensation decreases the partial pressure of vapor and hence the total pressure too. Furthermore, the steam condensation rate depends not only on the conditions in the FPV but also in the CCC piping. Upper plot of the figure 18 shows the mass flow in the CCC system.

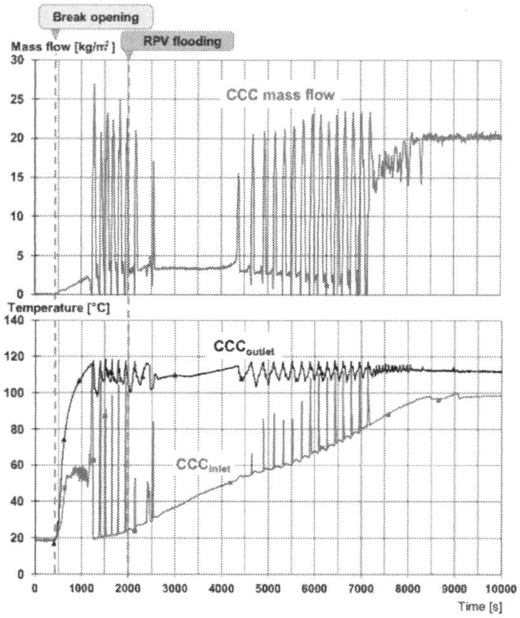

Fig. 18. CCC mass flow (top) and temperatures (bottom)

Figure 18 shows shortly after the break opening (until approximately 1200 s) the mass flow was steadily increasing. During this time the cold water in the CCC was being heated up and it was naturally circulating through the CCC piping. As more and more heat was being transferred from the FPV, the water temperature increased. As the temperature reached the saturation temperature, bubbles started to form and migrate along the CCC piping. The CCC piping is only slightly inclined with respect to horizontal orientation and at the end of the heat exchange section there is a collector which enables the vertical flow. Once the bubbles migrated to this region strong buoyancy force contributed to increase of flow and displacement of the bubble to the SSPV vessel, where it condensed again. At the same time cold water was sucked by the inlet tube decreasing the temperature in the CCC piping. Due to the insignificant inclination of the CCC tubes, the reverse direction of the mechanism is also possible. Since the mass flow sensors measure the flow only in one direction, the reverse flow could not be measured, but it can be unquestionably confirmed by the inlet temperature measurement (figure 18 bottom).

The inlet and outlet tubes are connected to the collectors which then distributes or collect water to/from the CCC piping. Therefore, the temperatures in the figure 18 show only the overall and predominant effects (normal or reverse flow). Flow conditions in single tubes can be different and they can depend on conditions (e.g. gas composition) at the secondary side. This shows the complexity of the whole CCC system.

5 INKA test results interpretation and benefits

The postulated LOCA due to the main steam line break caused rapid depressurization of the RPV and a pressure increase in the vessels simulating the containment of a BWR plant. The passive safety systems investigated during the experiment reacted automatically and immediately. The first system employed for mitigation of the LOCA results was the Emergency Condenser. The most important outcomes from the EC performance are:

- The system reacts immediately to the RPV inventory change and it can be utilized as a safety injection in the whole range of pressure
- It significantly contributes to the depressurization of the primary side. This in turn lowers the pressure difference between the primary and the secondary side in case of the LOCA and it indirectly limits the pressure growth at the secondary side
- The power of the EC is large enough to remove the decay heat with substantial safety margins

The increase of pressure in the containment vessels was also limited by the PSPV. After reaching the pressure difference between the DWV and PSPV of approximately 0.4 bar, the medium from the DWV was discharged to the PSP limiting further increase of the Drywell pressure.

Simultaneously, the passive CCC system was activated contributing to the further decrease of pressure in the containment. The superposition of these systems allowed maintaining the pressure in the containment far below the design limit. Apart from pressure limitation, the CCC was proven as a reliable system for passive, long-term heat removal.

The overall test results prove that the investigated passive systems fulfill their design goals. Each of the systems is characterized with safety margins significantly exceeding the sufficient level. Thus, the whole system can be considered as redundant.

Apart from investigation the passive systems, the experimental data obtained during the test campaign were utilized for system codes validation [5, 6].

6 Conclusions and outlook

INKA is a large-scale test facility replicating an innovative BWR KERENA design of Framatome. It comprises the all passive systems of the original plant and it may be utilized for demonstration of the reference plant performance as well as for investigation of passive systems which can be applied to other LWRs.

For this reason EASY project was conducted at the INKA facility. In the frames of the project simulation of the main steam line break was performed in order to investigate the integral response of the system.

The results of the experiment confirmed the functionality of each component as well as demonstrated their redundancy. They also enabled a detailed analysis of the interactions between the systems and they contributed to better understanding of the functioning of each component integrated into the large system. Furthermore, the experimental database was utilized for thermal-hydraulic codes validation.

In total four tests were performed in the framework of the EASY_1 project. Further investigations are planned to be conducted as the EASY_2 continuation project.

References

1. Stosic Z., Brettschuh W., Stoll U, Boiling water reactor with innovative safety concept: The Generation III+ SWR-1000, *Nuclear Engineering and Design* Vol. 238, pp. 1863 – 1901, (2008).
2. Bryk R., Schmidt H., Mull T., Wagner T., Ganzmann I., Herbst O., Modeling of KERENA emergency condenser, *Archives of Thermodynamics* Vol. 38, no. 4, pp. 29 – 51, (2017).
3. Leyer S., Wich M., The Integral Test Facility Karlstein, *Science and Technology of Nuclear Installations*, Vol. 2012, (2012).
4. Leyer S., Wich M., Schafer H., SWR 1000 integral and full scale tests of the passive safety systems, *Proceedings of the International Conference on Advances in Nuclear Power Plants (ICAPP '08)*, pp. 87 – 92, (2008).
5. Drescher R., Wagner T., Prasser H. M, Leyer S., Passive Integral LOCA Testing at Karlstein Test Facility INKA, *Proceedings of the ICAPP 2014*, (2014).
6. Bryk R., Schmidt H., Mull T., Modeling of emergency condenser system response to loss of coolant accident in a BWR III+ generation, *Eksploatacja i Niezawodność – Maintenance and Reliability*, Vol. 21, no. 3, pp. 468 – 475 (2019).

E3S Web of Conferences **137**, 01036 (2019) https://doi.org/10.1051/e3sconf/201913701036
RDPE 2019

RESEARCH & DEVELOPMENT IN POWER ENGINEERING, 2019

Transient behavior of a plate-fin-and-tube heat exchanger taking into account different heat transfer coefficients on the individual tube rows

Dawid Taler[1], *Jan* Taler[2*], and *Katarzyna* Wrona[3]

[1] Faculty of Environmental and Energy Engineering, Cracow University of Technology, ul. Warszawska 24, 31-155 Cracow, dtaler@pk.edu.pl, Poland

[2*] Faculty of Environmental and Energy Engineering, Cracow University of Technology, Al. Jana Pawła II 37, 31-864 Cracow, jtaler@mech.pk.edu.pl, Poland

[3] Faculty of Environmental and Energy Engineering, Cracow University of Technology, Al. Jana Pawła II 37, 31-864 Cracow, katarzynawrona@interia.com, Poland

Abstract. Plate-fin and tube heat exchangers (PFTHE) are made of round, elliptical, oval or flat tubes to which continuous fins (lamellas) are attached. Liquid flows inside the tubes and gas flows outside the tubes perpendicularly to their axes and parallel to the surface of continuous fins. Experimental studies of multi-row plate-fin and tube heat exchangers show that the highest average heat transfer coefficient on the air side occurs in the first row of tubes when the air velocity in front of the exchanger is less than approximately 3.5 m/s when a Reynolds number based on an equivalent hydraulic diameter equal to the distance between tube rows in the direction of air flow is less than 10,000. In the subsequent rows of tubes up to about the fourth row the heat transfer coefficient decreases. In the fifth and further rows, it can, that the heat transfer coefficient is equal in each tube row. It is necessary to find the relationships for the air-side Nusselt number on each tube row to design a PFTHE with the appropriate number of tube rows. The air-side Nusselt number correlations can be determined experimentally or by CFD modeling (Computational and Fluid Dynamics). The paper presents a new mathematical model of the transient operation of PFTHE, considering that the Nusselt numbers on the air side of individual tube rows are different. The heat transfer coefficient on an analyzed tube row was determined from the equality condition of mass- average air temperature differences on a given tube row determined using the analytical formula and CFD modeling. The results of numerical modeling were compared with the results of the experiments.

1 Introduction

Plate-fin and tube heat exchangers (PFTHEs) are widely used in air conditioning, heating, and many industries [1-4]. Inside the tubes, a liquid usually flows, e.g., water, glycol solution, or oil, and outside perpendicular to the tube axis air or exhaust gas or another gas. PFTHEs are very widely used in practice. For this reason, the number of books and articles in journals concerning their design, mathematical modeling, and experimental studies is very large. So far, both in the calculation of heat exchangers and in experimental studies, it has been assumed that the heat transfer coefficient (correlation for Nusselt number) on the gas side of each row of tubes is the same [1-4]. The fundamental experimental study showing the influence of the number of rows of tubes on the Colburn parameter for the entire exchanger as well as for the individual rows of tubes was performed by Rich [5]. The research carried out by Saboya and Sparrow [6] using the technique of naphthalene sublimation of for one-, two- and three-row system with staggered tube configurations confirmed Rich© observations. If the number of tube rows is less than four, the correlation for the Nusselt number on the gas side is usually determined separately for a one-, two-, three- or four-row exchanger. If the number of tube rows in the heat exchanger is greater than four, it is assumed that the addition of the fifth and further tube rows does not affect the average heat transfer coefficient of the whole heat exchanger [2-3,7]. Usually, the correlation $j_{N=4}$ for the Colburn factor for a four-row heat exchanger is given, assuming that it is also valid for heat exchangers with more rows of tubes. The $j_N / j_{N=4}$ correction for a heat exchanger with a smaller number of tube rows, i.e., a single, double or triple row heat exchanger, is a function of the number of tube rows and the Reynolds number. Correlations for the calculation of Colburn© factors $j_{N=4}$ and j_N were proposed for staggered tube alignment by McQuiston [2-3] and Gray and Webb [3] based, in part, on Rich© experimental results [5].

The literature review shows that for air velocities before a PFTHE lower than 3.5 m/s, the first four rows of tubes exchange higher heat flow rates compared to the subsequent rows. The empirical heat transfer correlations available in the literature permit the calculation of average heat transfer coefficients for the entire heat exchanger with different tube rows. However, there are no

 https://doi.org/10.1051/e3sconf/201913701036

correlations for the calculation of the local air side Nusselt numbers for the individual tube rows, especially for the first six rows. The air-side heat transfer correlations can be developed based on CFD modeling [8-12]. This paper proposes a calculation method of PFTHEs based on the air-side heat transfer correlations, obtained using CFD simulations [12]. The method of determining CFD based correlations should be compatible with the procedure of determining empirical heat transfer correlations, to ensure that the results of PFTHEs calculations are consistent with the results of measurements. Semi-empirical relationships developed by Taler [13-16] were used to calculate the heat transfer coefficients on the internal surfaces of tubes in the transition and turbulent flow regime. The formulas proposed by Gnielinski [17] were used to calculate the heat transfer coefficient in the laminar flow regime. By using reliable CFD correlations to calculate air-side heat transfer coefficients and semi-empirical relationships to calculate heat transfer coefficients on the internal surfaces of tubes, the experimental studies required can be significantly reduced, especially for PFTHEs with new design, e.g. made of tubes with different cross-section shapes or with different designs of continuous fins on the air-side. The PFTHEs numerical models can also take into account different heat transfer coefficients on the air side of each tube row. The method of modeling PFTHEs in steady-state and transient states proposed in the paper together with the method of determining the air-side heat transfer correlations proposed in [12] will significantly reduce the cost and shorten the time of implementation of PFTHEs with new construction and flow system.

2 Mathematical formulation

The basis for the development of the mathematical model of the PFTHE is the set of energy conservation equations for water flowing through tubes, tube walls with fins, and for air flowing perpendicularly to the tube axes (Fig. 1) [18-19]. Due to the low thermal capacity of the air, the time derivative from the air temperature was omitted, i.e., the temperature distribution in the direction of its flow (in the y-axis direction) was determined from the solution of the steady-state energy conservation equation. The set of partial differential equations describing the space and time variations of water T_w, tube wall T_t, and air T_a temperatures are

$$\frac{1}{NTU_w}\frac{\partial T_w}{\partial x^+}+\tau_w\frac{\partial T_w}{\partial t}=-\left(T_w-T_t\right) \tag{1}$$

$$\tau_t\frac{\partial T_t}{\partial t}+\tau_f\frac{\partial \bar{T}_a}{\partial t}+T_t=\frac{h_w U_{in}}{h_w U_{in}+h_o U_{out}}T_w+\frac{h_o U_{out}}{h_w U_{in}+h_o U_{out}}\bar{T}_a \tag{2}$$

$$\frac{1}{NTU_a}\frac{\partial T_a}{\partial y^+}=T_t-T_a \tag{3}$$

where $\bar{T}_a$ denotes the mean air temperature over the row thickness, defined as

$$\bar{T}_a=\int_0^1 T_a\left(x^+,y^+,t\right)\mathrm{d}y^+ \quad . \tag{4}$$

The numbers of heat transfer units NTU_w and NTU_a are defined as follows

$$NTU_w=\frac{h_w A_{in}}{\dot{m}_w \bar{c}_w} \qquad NTU_a=\frac{h_o A_{out}}{\dot{m}_a \bar{c}_{pa}} \tag{5}$$

The time constants τ_w, τ_t, and τ_f are

$$\tau_w=\frac{m_w c_{pw}}{h_w A_{in}}\ , \quad \tau_t=\frac{\left(m_t+\eta_f m_f\right)c_t}{h_w A_{in}+h_o A_{out}}\ ,$$

$$\tau_f=\frac{\left(1-\eta_f\right)m_f c_t}{h_w A_{in}+h_o A_{out}}\ , A_{in}=N_t P_{in} L_t\ ,\ A_{out}=N_t P_{out} L_t\ ,$$

$$A_t=\pi a b\ , \quad A_{oval}=\pi\left(a+\delta_t\right)\left(b+\delta_t\right)\ , \tag{6}$$

$$m_f=N_t N_f\left(p_1 p_2-A_{oval}\right)\delta_f \rho_f\ , \quad m_t=N_t P_m \delta_t L_t \rho_t\ ,$$

$$m_w=N_t A_t L_t \rho_w\ , \qquad P_m=\left(P_{in}+P_{out}\right)/2$$

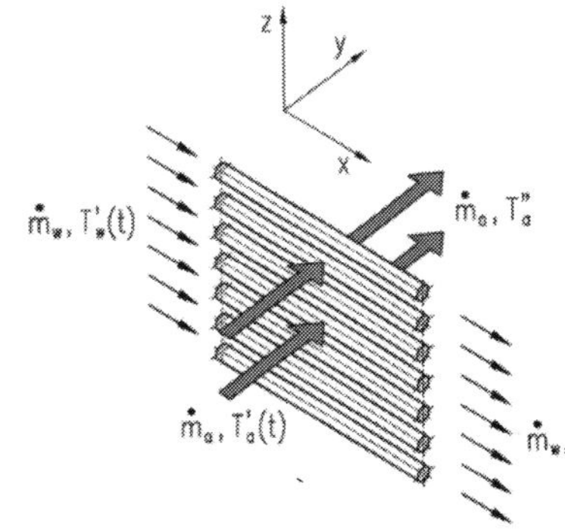

Fig. 1. Cross-flow tube heat exchanger with one row of tubes

The effective weighted heat transfer coefficient h_o is given by

$$h_o=h_a\left[\frac{A_{bf}}{A_{out}}+\frac{A_f}{A_{out}}\eta_f\left(h_a\right)\right] \tag{7}$$

where: a, b - minimum and maximum inner radius of the oval tube; A_f - fin surface area; A_{bf} - area of the tube outer surface between fins; A_{oval}, A_t - outside and inside cross-section area of the oval tube; $A_{in}=N_t P_{in} L_t$, $A_{out}=N_t P_{out} L_t$ - inside and outside surface area of the bare tube; c_p - specific heat at constant pressure; c_t, c_f - specific heat of the tube and fin material; h_w and h_a - water and air side heat transfer coefficients, respectively; h_o-effective weighted heat transfer coefficient from the air side related to outer surface area of the bare tube; L_t - tube length in the heat exchanger; m_w, m_f , and m_t - mass of the water, fins, and tube walls in the heat exchanger, $\dot{m}_w$ - mass flow rate of liquid flowing inside the tubes; $\dot{m}_a$ - air mass flow rate; N_f - number of fins on the tube length; N_t - number of tubes in the heat exchanger; NTU_w and NTU_a - number of heat transfer units for water and air, respectively; p_1 - pitch of tubes in plane perpendicular to flow (height of fin); p_2 - pitch of tubes in direction of flow (fin width); s - fin pitch; t - time; T_w, T_t, and T_a - water, tube wall and air temperature; $\bar{T}_a$ - mean air temperature on the thickness of one tube row; P_{in} and P_{out} - inner and outer tube perimeter of the bare tube, η_f - fin efficiency; $x^+ = x / L_t$, $y^+ = y / p_2$ - dimensionless Cartesian coordinates; δ_f and δ_t - fin and tube thickness, respectively; ρ - density; τ_w, τ_t, and τ_f - time constant of water, tube, and fin, respectively.

The initial temperature distribution of the water $T_{w,0}(x^+)$, air $T_{a,0}(x^+,y^+)$, and the tube walls $T_{t,0}(x^+)$ are known from measurements or the steady-state calculations of the PFTHE. The initial conditions are

$$T_w\left(x^+,t\right)\big|_{t=0} = T_{w,0}\left(x^+\right) \tag{8}$$

$$T_t\left(x^+,t\right)\big|_{t=0} = T_{t,0}\left(x^+\right) \tag{9}$$

$$T_a\left(x^+,y^+,t\right)\big|_{t=0} = T_{a,0}\left(x^+,y^+\right) \tag{10}$$

The equation system (1)-(3) is subject to the following boundary conditions

$$T_w\left(x^+,t\right)\big|_{x^+=0} = T'_w(t) \tag{11}$$

$$T_a\left(x^+,y^+,t\right)\big|_{y^+=0} = T'_a(t) \tag{12}$$

where $T'_w(t)$ and $T'_a(t)$ are functions representing the variations of the inlet temperatures of water and air in time. The initial-boundary problem formulated by Eqs (1–12) applies to the heat exchangers made of smooth tubes as well as to PFTHEs. The transient water, air, and wall temperature distributions in the one-row heat exchanger (Fig. 1) are then determined using the explicit finite difference method.

3 Finite difference method of the PFTHE

The temperature distribution $T_w\left(x^+,t\right)$, $T_t\left(x^+,t\right)$ and $T_a\left(x^+,y^+,t\right)$ were determined by the explicit finite difference method. The time derivatives in Eqs (1)-(2) were approximated by forward differences, while the spatial derivatives were approximated by backward differences. Figure 2 shows a single difference cell.

The nodes where the water temperature is calculated are at the beginning and end of the cell, and the nodes where the air and tube wall temperatures are calculated at the middle of the cell® length. The finite-difference grid is defined as follows

$$x_i = (i-1)\Delta x \;,\quad i = 1,\ldots,N+1, \qquad t_n = n\Delta t \;,\quad n = 0,1,\ldots \tag{13}$$

Eq. (3) was solved analytically, assuming that the tube wall temperature is constant within a single cell and equals

$$\bar{T}_{t,i} = \frac{T_{t,i} + T_{t,i+1}}{2} \tag{14}$$

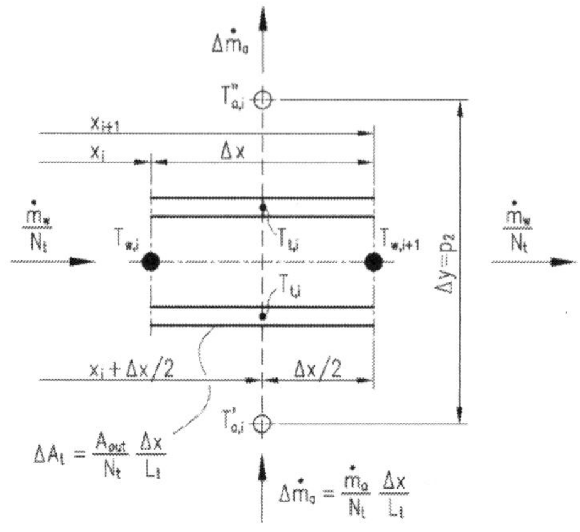

Fig. 2. A single cell (finite volume) in a finite difference grid

The solution of Eq.(3) for the boundary condition $T_{a,i}\big|_{y^+=0} = T'_{a,i}$ (Fig. 2) has the following form

$$T_{a,i} = \bar{T}_{t,i} - \left(\bar{T}_{t,i} - T'_{a,i}\right)\exp\left(-\frac{U_o\,\Delta A_t}{\Delta\dot{m}_a\,c_{pa}}y^+\right) = \bar{T}_{t,i} - \left(\bar{T}_{t,i} - T'_{a,i}\right)\exp\left(-NTU_a\,y^+\right),$$
$$x_i^+ \le x^+ \le x_{i+1}^+,\quad i = 1,\ldots,N,\quad 0 \le y^+ \le 1 \tag{15}$$

where: $\Delta A_t = \frac{A_{out}}{N_t}\frac{1}{N} = P_{out}\,\Delta x$ and $\Delta\dot{m}_a = \frac{\dot{m}_a}{N_t}\frac{1}{N}$.

Substituting $y^+ = 1$ into Eq. (15) yields the air temperature $T''_{a,i}$ after the tube row

$$T''_{a,i} = \bar{T}_{t,i} - \left(\bar{T}_{t,i} - T'_{a,i}\right)\exp\left(-NTU_a\right),\quad x_i^+ \le x^+ \le x_{i+1}^+,\; i = 1,\ldots,N \tag{16}$$

The integral mean temperature of the air over the thickness $\Delta y^+ = 1$ of one tube row can be determined using the definition (4)

$$\bar{T}_{a,i} = \int_0^1 T_{a,i}\left(y^+,t\right)dy^+ = \bar{T}_{t,i} - \frac{1}{NTU_a}\left(\bar{T}_{t,i} - T'_{a,i}\right)\left[\exp\left(-NTU_a\right) - 1\right] \tag{17}$$

Solving of finite difference equations for $T^{n+1}_{w,i+1}$ and $T^{n+1}_{t,i}$ gives

$$T^{n+1}_{w,i+1} = T^n_{w,i+1} - \frac{\Delta t}{NTU^n_w\,\tau^n_w}\frac{T^n_{w,i+1} - T^n_{w,i}}{\Delta x^+} - \frac{\Delta t}{\tau^n_w}\left(\frac{T^n_{w,i} + T^n_{w,i+1}}{2} - T^n_{t,i}\right),\quad i = 1,\ldots,N,\quad n = 0,1,\ldots \tag{18}$$

$$T_{t,i}^{n+1}=T_{t,i}^{n}+\frac{\Delta t}{\tau_t^n}\left\{\frac{h_w^n A_{in}}{h_w^n A_{in}+h_o^n A_{out}}\frac{T_{w,i}^n+T_{w,i+1}^n}{2}+\right.$$

$$\left.\frac{h_o^n A_{out}}{h_w^n A_{in}+h_o^n A_{out}}\bar{T}_{a,i}^n-T_{t,i}^n-\frac{\tau_f^n}{\Delta t}\left[\bar{T}_{a,i}^{n+1}-\bar{T}_{a,i}^n\right]\right\}, \quad (19)$$

$i=1,...,N, \quad n=0,1,...$

The average air temperature over the thickness of one tube row occurring in equation (19) is given by the following expression

$$\bar{T}_{a,i}^n=\bar{T}_{t,i}^n-\frac{1}{NTU_a^n}\left[\bar{T}_{t,i}^n-\left(T'_{a,i}\right)^n\right]\times$$
$$\left[\exp\left(-NTU_a^n\right)-1\right], \quad i=1,...,N, \quad n=0,1,... \quad (20)$$

The air temperature behind the tube row can be calculated using Eq. (16)

$$\left(T''_{a,i}\right)^{n+1}=\left(\bar{T}_{t,i}\right)^{n+1}-\left[\left(\bar{T}_{t,i}\right)^{n+1}-\left(T'_{a,i}\right)^{n+1}\right]$$
$$\times\exp\left(-NTU_a^{n+1}\right), \quad i=1,...,N, \quad n=0,1,... \quad (21)$$

The initial conditions (8–10) and the boundary conditions (11- 12) assume the form:

- initial conditions

$$T_{w,i}^0=T_w\left(x_i^+\right)\big|_{t=0} \quad n=0, \quad i=1,...,N+1 \quad (22)$$

$$T_{t,i}^0=T_t\left(x_i^+\right)\big|_{t=0}, \quad n=0, \quad i=1,...,N \quad (23)$$

$$T_{a,i}^0=T_a\left(x_i^+,y^+\right)\big|_{t=0}, \quad n=0, \quad i=1,...,N \quad (24)$$

- boundary conditions

$$T_{w,0}^n=T'_w\left(t_n\right) \; i=0, \quad n=0,1,... \quad (25)$$

$$T_a^n\left(x_i^+,y^+=0\right)=T'_a\left(x_i^+,t_n\right) \; i=1,...,N, n=0,1,... \quad (26)$$

To assure the calculation stability by the explicit finite difference method, the Courant condition must be fulfilled [19]

$$\frac{\Delta t}{N_w^n \tau_w^n \Delta x^+}\le 1 \quad (27)$$

The water temperature $T_{w,i+1}^{n+1}$, the tube wall temperature $T_{t,i}^{n+1}$ and the air temperature $T_{a,i}^{n+1}$ at the new time step (n+1) were calculated using the expressions (18)-(19) and (21) taking into account the initial conditions (22)-(24) and boundary conditions (25)-(26), and starting at $n=0$. The Δt time step resulting from Courant®condition (27) is very small (in the range of several hundredths of a second). In the case of the car radiator analyzed in this paper, the time step Δt was 0.02 seconds. The time step of data acquisition (sampling step) is usually much longer, e.g., 1 second. The measured water and air temperatures at the inlet to the heat exchanger were interpolated with natural cubic splines to calculate the temperature of the water $T'_w(t)$ and the air temperature $T'_a(t)$ at the inlet to the heat exchanger at a given time step t_n.

4 Two-pass model of the PFTHE with two tube rows

The finite difference model of the car radiator, which is a two-pass, two-row PFTHE, was developed. The mathematical model of the car cooler used in internal combustion engines with a cylinder capacity of 1600 cm^3 was based on the single-row heat exchanger model presented in subsection 2. The flow arrangement of the analyzed car cooler is shown in Fig. 3. The heat exchanger is made of oval aluminum tubes with a minimum and maximum external diameter of respectively $d_{min}=b+\delta_t=6.35$ mm and $d_{max}=b+\delta_t=11.82$ mm. The wall thickness δ_t of the tube is 0.4 mm and the fin thickness δ_f is 0.08 mm. The thermal conductivity of aluminum is 207 W/(m·K) at 40 °C. The transverse pitch of tube spacing is $p_1=18.5$ mm, and longitudinal pitch is $p_2=17$ mm. Plain continuous fins are mounted on tubes with pitch $s=1$ mm. The active length of tubes in the car radiator is $L_t=520$ mm. There are 38 tubes in the first and second pass. In the first pass there are 20 tubes, 10 in each row and the second pass there are 18 tubes, 9 in each row.

4.1 Air-side heat transfer correlation

The CFD based heat transfer correlations proposed in the paper [12] for the first and second row of tubes and the double row heat exchanger were used to calculate the Nusselt numbers on the air side. The following Nusselt numbers derived in [12] using the results of CFD simulation were adopted for the calculation of the air side heat transfer coefficients in the mathematical model of the car radiator

$$Nu_a^I=10.0205\,Re_a^{-0.0623}\,Pr_a^{1/3}, 150\le Re_a\le 330 \quad (28)$$

$$Nu_a^{II}=0.1648\,Re_a^{0.5678}\,Pr_a^{1/3}, \; 150\le Re_a\le 330 \quad (29)$$

$$Nu_a=1.0843\,Re_a^{0.2836}\,Pr_a^{1/3}, \; 150\le Re_a\le 330 \quad (30)$$

The classic formula modified by Kays and London [1] was used to calculate the hydraulic diameter d_a on the air side. The air side Reynolds number was calculated using the relationship $Re_a=w_{max}\,d_a/\nu_a$, where the symbol w_{max} denotes the air velocity in the minimum free flow area. The maximum air velocity w_{max} which, in the case of an inline tube arrangement, occurs between two adjacent tubes in a given tube row is given by

$$w_{max}=\frac{s\,p_1}{\left(s-\delta_f\right)\left(p_1-d_{o,min}\right)}\frac{\bar{T}_a+273}{T'_{am}+273}\,w_0 \quad (31)$$

where w_0 is the air velocity in front of the PFTHE.

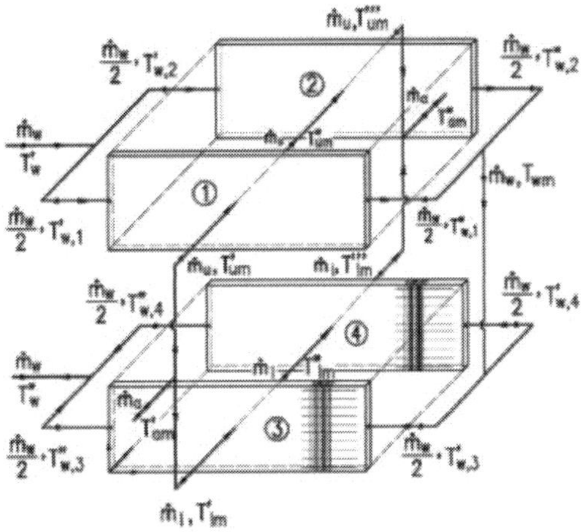

Fig. 3. Flow system of the analyzed two-pass PFTHE (car cooler) with two rows of tubes; 1 - first row of tubes in the first pass, 2 - second row of tubes in the first pass, 3 - first row of tubes in the second pass,4 - second row of tubes in the second pass

All physical properties of the air in the relationships (28) - (30) were evaluated at the average air temperature $\bar{T}_a = (T'_{am} + T''_{am})/2$, where the symbols T'_{am} and T'_{am} mean the average air temperatures before and after the radiator.

4.2 Water-side heat transfer correlation

During transient heat exchanger operation, the flow regime can be laminar, transient, or turbulent.The mean Nusselt number $Nu_{m,q}$ for the laminar flow which is hydraulically and thermally developing in a tube with the constant wall heat flux was calculated using the formula recommended by Gnielinski in VDI Heat Atlas [17]

$$Nu_{m,q} = [Nu^3_{m,q,1} + 0.6^3 + (Nu_{m,q,2} - 0.6)^3 + Nu^3_{m,q,3}]^{1/3}, \quad Re_w \le 2300 \tag{32}$$

The Nusselt numbers $Nu_{m,q,1}$, $Nu_{m,q,2}$, and $Nu_{m,q,3}$ are as follows

$$Nu_{m,q,1} = \frac{48}{11} = 4.364 \ , \ Re_w \le 2300 \tag{33}$$

$$Nu_{m,q,2} = 3^{1/3}\Gamma(2/3)\left(Re_w Pr_w \frac{d_w}{L_t}\right)^{1/3} = 1.9530\left(Re_w Pr_w \frac{d_w}{L_t}\right)^{1/3}, Re_w \le 2300 \tag{34}$$

$$Nu_{m,q,3} = 0.924 Pr_w^{1/3}\left(Re_w \frac{d_w}{L_t}\right)^{1/2}, \ Re_w \le 2300 \tag{35}$$

The Nusselt number in the transition and turbulent flow was calculated using the Taler relationship [13]

$$Nu_w = Nu_{m,q}\big|_{Re_w=2300} + \frac{\frac{\xi_w}{8}(Re_w - 2300)Pr_w^{1.008}}{1.084 + 12.4\sqrt{\frac{\xi_w}{8}}\left(Pr_w^{2/3} - 1\right)} \times \left[1 + \left(\frac{d_w}{L_t}\right)^{2/3}\right], 2300 \le Re_w \le 10^6, 0.1 \le Pr_w \le 1000, \frac{d_w}{L_t} \le 1 \tag{36}$$

The Darcy-Weisbach friction factor in the turbulent flow range, when $3000 \le Re_w$ was calculated using the Taler formula [14]

$$\xi_w = (1.2776 \log Re_w - 0.406)^{-2.246} \quad 3000 \le Re_w \tag{37}$$

The friction factor for the transition flow was determined using the relationship of Taler proposed in [15-16]

$$\xi_w = 0.02783 + 2.2457 \cdot 10^{-5}(Re_w - 2\,300) \quad 2300 \le Re_w \le 3000 \tag{38}$$

The formula (38) represents a linear interpolation of the friction factor between $Re_w = 2300$ and $Re_w = 3000$. The Reynolds number on water side $Re_w = w_w d_w / \nu_w$ is based on the hydraulic diameter d_w . The hydraulic diameter was $d_w = 7.06$ mm. The water physical properties were determined at the average temperature $\bar{T}_w = (T'_w + T''_w)/2$. The relationship (36) was validated experimentally for PFTHE made of oval tubes [18] and for PFTHE made of round tubes [19].

5 Comparison of simulation results with experimental data

Transient behavior of the car radiator after a rapid rise in water volume flow rate $\dot{V}_w$ and at a sudden drop in air velocity w_0 at the inlet to the heat exchanger (Fig.3) was studied experimentally. The description of the test facility is presented in detail in the paper [20]. At first, the simulation of the transient response of the PFTHE was conducted for different heat transfer coefficients on each tube row. The heat transfer coefficients on the air side of the first and second row of tubes were calculated using CFD based correlations (28) and (29). Then, the numerical simulation of the car heat radiator was also carried out for a uniform heat transfer coefficient on the first and second row of tubes. In this case, the value of the heat transfer coefficient on both rows of tubes was calculated by applying the CFD correlation (30). The numerical simulation of the PFTHE was conducted with a time step $\Delta t = 0.02$ s. The temperature of the air T'_{am} and water T'_w at the inlet to the exchanger was constant. The calculation results shown in Fig. 4 were obtained for a uniform heat transfer coefficient in the entire heat exchanger, which was calculated from the CFD based formula (30). Calculated water and air outlet temperatures of PFTHE was compared with the results of the measurements. The conformity of the calculation and measurement results is very good. The following formulas

https://doi.org/10.1051/e3sconf/201913701036

were used to calculate heat flow rates in individual tube rows

$$Q_{1,jHTC} = \left(\dot{m}_w / 2\right)\bar{c}_w \left(T'_w - T''_{w,1}\right) \quad j = 1,2 \tag{39}$$

$$Q_{2,jHTC} = \left(\dot{m}_w / 2\right)\bar{c}_w \left(T'_w - T''_{w,2}\right) \quad j = 1,2 \tag{40}$$

$$Q_{3,jHTC} = \left(\dot{m}_w / 2\right)\bar{c}_w \left(T_{wm} - T''_{w,3}\right) \quad j = 1,2 \tag{41}$$

$$Q_{4,jHTC} = \left(\dot{m}_w / 2\right)\bar{c}_w \left(T_{wm} - T''_{w,4}\right) \quad j = 1,2 \tag{42}$$

The total heat transfer rate from the air to the water was calculated using the expression

$$Q_{t,jHTC} = \dot{m}_w \, \bar{c}_w \left(T'_w - T''_w\right) = Q_{1,jHTC} + Q_{2,jHTC} + Q_{3,jHTC} + Q_{4,jHTC} \quad j = 1,2 \tag{43}$$

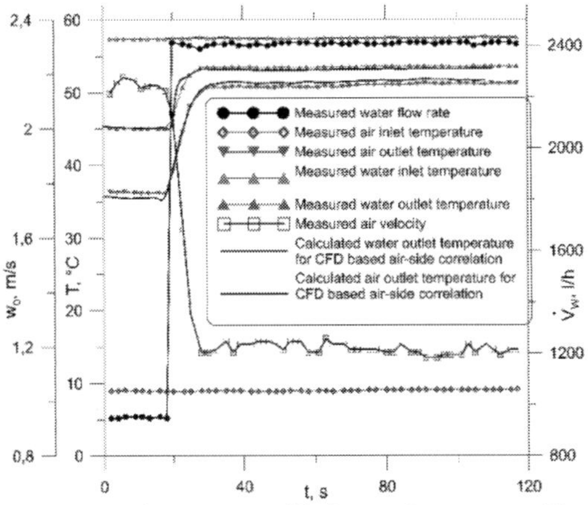

Fig. 4. Transient response of the heat exchanger to a sudden increase in water volume flow rate $\dot{V}_w$ and a decrease in air velocity w_0 at the inlet to the heat exchanger

The symbols $Q_{i,1HTC}$ $i = 1,\ldots,4$ denote the heat flow rate exchanged in the individual rows (Fig. 3) on the assumption that the heat transfer coefficient calculated using formula (30) is uniform throughout the heat exchanger. The symbols $Q_{i,2HTC}$ $i = 1,\ldots,4$ denote the heat flow rate transferred in the individual rows, with the heat transfer coefficient in the first row of tubes calculated using the correlation (28) and in the second row using the correlation (29).

If the heat transfer coefficient on the air side is uniform in the whole heat exchanger then in Eqs (39)-(43) $j = 1$, and if the heat transfer coefficients on the first and second row of tubes are different then $j = 2$. The relative differences between the heat flow rate with different heat transfer coefficients on the first and second row of tubes and the heat flow rate with uniform heat transfer coefficient in the whole heat exchanger were also calculated

$$e_i = \frac{Q_{i,2HTC} - Q_{i,1HTC}}{Q_{i,2HTC}} \cdot 100 \quad i = 1,\ldots,4 \tag{44}$$

The relative e_t difference for the whole heat exchanger was calculated with a similar formula

$$e_t = \frac{Q_{t,2HTC} - Q_{t,1HTC}}{Q_{t,2HTC}} \cdot 100 \tag{45}$$

The obtained results show that the first row of tubes in the first and second pass takes up much more heat than the second row if different heat transfer coefficients in the first and second row of tubes are taken into account. Relative e_1 and e_3 differences in the heat flow rates exchanged in the first row of the first pass and the first row of the second pass are between 7 and 18 %, respectively. Second-row tubes are much less efficient if different heat transfer coefficients are taken into account on both rows of heat exchanger tubes. The tubes in the second row take up between 12.5% and 50.9% less heat compared to the heat flow rates exchanged in the second row, assuming a uniform heat transfer coefficient. Relative differences e_2 and e_4 are between (-50.9%) and (-12.5%). The results of numerical simulations show that if the calculation assumes a uniform heat transfer coefficient on the air side for the whole heat exchanger, then an understated value of the heat flow exchanged in the first row, and an overstated value of the heat flow exchanged in the second row is obtained.

The total heat transfer rate Q_t in the analyzed heat exchanger is similar for a uniform heat transfer rate on the air side, as well as for different heat transfer rates on the first and second row of tubes. This is clear when taking into account the method of determining the uniform heat transfer coefficient for both tube rows. The heat transfer coefficient is selected from the condition of equal increase of air temperature obtained from the CFD simulation and numerical model of the heat exchanger. Heat transfer coefficients on the first and second row of tubes are determined from the condition of equality of air temperature increase on the first and second row of tubes, respectively, determined by CFD simulation and numerical model. Since the sum of air temperature increments on the first and second row of tubes is equal to the total temperature difference on both rows of tubes, therefore the heat flow rates $Q_{t,2HTC}$ and $Q_{t,1HTC}$ must be similar. Relative differences between $Q_{t,2HTC}$ and $Q_{t,1HTC}$ are between (-0.3%) and 1.7%. These small differences result from the procedure for determining correlations (28)-(30) [12], which assumes that the water temperature in the tubes and the heat transfer coefficient on the internal surface of the tube are constant [12]. Under real conditions, the water temperature changes along the length of the tube. Also, the heat transfer coefficient on the water side is lower in the first pass than in the second pass. It should be stressed, however, that slightly different conditions of operation of tubes in the real heat exchanger than those adopted in the method of determining the correlations (28)-(30) [12] only slightly affect the value of the heat flow rate Q_t exchanged in the heat exchanger.

The numerical simulations of the car radiator in steady-state were also carried out for a uniform and non-uniform heat transfer coefficient on the first and second row of tubes. First, calculations were carried out for the analyzed automotive radiator with one CFD correlation given by the formula (30) for the entire radiator. Results of these calculations were compared with the results of experimental tests (Fig. 5).Then, the calculations were carried out for different heat transfer coefficients on the first and second row of tubes. The calculations are intended to confirm the usefulness of CFD correlations for the possible replacement of heat transfer correlations determined based on expensive experimental studies. The comparison of the heat flow rate in the PFTHE determined based of the measured water mass flow rate $\dot{V}_w$ at the inlet of the PFTHE and water temperature at the inlet and outlet of the PFTHE with the results of heat flow rate calculations using the PFTHE mathematical model is shown in Fig. 5. The heat flow rates $Q_{w,\exp}$ and $Q_{w,calc}$ are given by

$$Q_{w,\exp} = \dot{V}_w \rho_w \left(T'_{w,\exp}\right) \bar{c}_w \left(T'_{w,\exp} - T''_{w,\exp}\right) \tag{46}$$

$$Q_{w,calc} = \dot{V}_w \rho_w \left(T'_{w,\exp}\right) \bar{c}_w \left(T'_{w,\exp} - T''_{w,calc}\right) \tag{47}$$

The symbol $\bar{c}_w$ denotes the mean specific heat of the water in the range between the outlet and inlet water temperature. The relative difference e between the heat flow rate $Q_{w,\exp}$ determined experimentally, and the heat flow rate $Q_{w,calc}$ calculated using the mathematical model of the PFTHE was calculated using the formula $e = 100\left(Q_{w,\exp} - Q_{w,calc}\right)/Q_{w,\exp}$.

Fig. 5 also shows a heat flow rate, marked with a solid line, which was calculated for average volumetric flow rate $\bar{\dot{V}}_w = 326.06$ l/h, air inlet temperature $\bar{T}'_{am} = 13.62$ °C, and water inlet temperature $\bar{T}'_w = 59.61$°C . The average values $\bar{\dot{V}}_w$, $\bar{T}'_{am}$ and $\bar{T}'_w$ were calculated based on seven data sets. The flow of water in the tubes was laminar. The Reynolds number $\mathrm{Re}_{w,l}$ in the lower pass varied from 1324 to 1445.

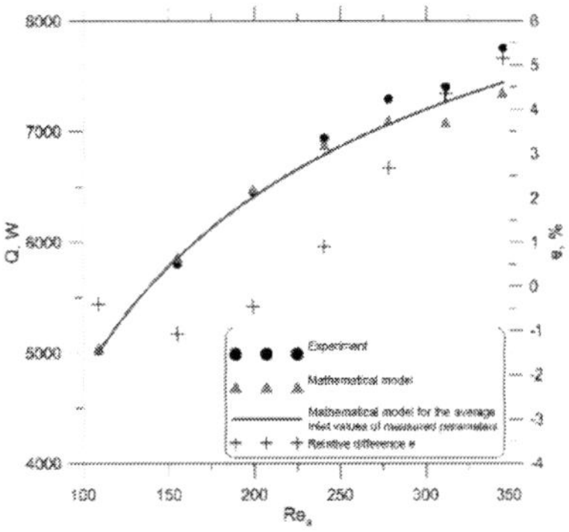

Fig. 5. Heat flow rate transferred from the hot water to the cold air and percentage relative difference $e = 100\left(Q_{w,\exp} - Q_{w,calc}\right)/Q_{w,\exp}$ in %.

Fig. 5 illustrates that the developed numerical model of the heat exchanger using CFD heat transfer correlation to calculate the air-side heat transfer coefficient gives results very close to the experimental data. The largest relative difference e between the heat flow rate determined based on the measurement data and the calculated heat flow rate is 5.14%. Heat flow rates transferred from water to air in the first and second row of tubes in the first and second row of tubes are shown in Figure 6a. The calculations were carried out for a uniform heat transfer coefficient on the first and second row of tubes and different coefficients on each row of tubes. An inspection of the results in Figure 6a indicates that much higher heat flow rate is exchanged in the first row of tubes compared to the second row if the heat transfer coefficients on the first and second tube row are different. This applies to both the first and second heat exchanger passes. The heat flow exchanged in the first row of tubes is about 10% higher than in the second row for low air velocities in front of the radiator in the range of 0.7 m/s when air side Reynolds Re_a number is about 110.

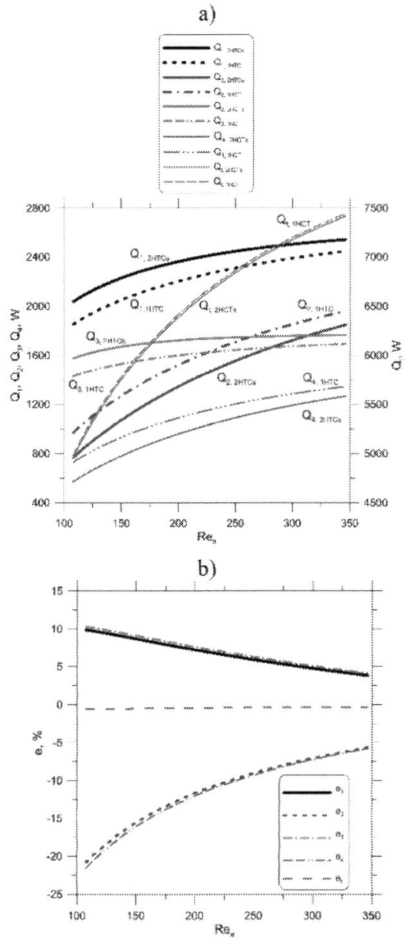

Fig. 6. The heat flow rates exchanged in the first and second row of tubes in the first and second pass for uniform and different heat transfer coefficient on the first and second row of tubes and the relative differences between heat flow rates for different and uniform heat transfer coefficient.

The relative difference between the heat flow absorbed by the air in the first and second row of tubes decreases with increasing air velocity. At an air velocity w_0 of about 2.25 m/s, i.e., for Re_a of around 350, the relative

differences e_1 and e_2 decrease to approximately 3.8% (Fig. 6b). This means that the effect of taking into account different heat transfer coefficients on the individual rows on the transferred heat flow rate decreases as the air velocity increases.
A similar phenomenon of decreasing the relative difference between the exchanged heat flow rate for different and uniform heat transfer coefficients on each tube row is observed for the second tube row. The absolute value of the relative differences e_3 and e_4 decreases from approximately 22.5% for $\mathrm{Re}_a = 110$ to approximately 6% for $\mathrm{Re}_a = 350$.

6 Conclusions

The numerical method of PFTHE modeling under transient conditions is presented. It uses heat transfer correlations for heat transfer coefficients on the air side obtained by CFD simulation. The heat exchanger calculations were carried out for a uniform heat transfer coefficient in the whole heat exchanger or different heat transfer coefficients on the first and second row of tubes. In both cases, a very good agreement was achieved between the heat flow rate transferred from hot water to air in the heat exchanger. The results of the calculation of the heat transfer rate in the heat exchanger match the results of the experimental tests very well. The advantage of calculating the heat exchanger with different correlations for the air side Nusselt number on the individual tube rows is that it is possible to determine the heat flow rate from water to air in each tube row.
This information can be used in the design of a heat exchanger because, for example, a more efficient single-row heat exchanger can be built instead of a double-row heat exchanger by increasing the area of the heat exchange surface in the first row.

References

1. Kays W. M., London A. L., *Compact heat exchangers*, 3rd edition, Krieger, Malabar, (1998).
2. McQuiston F. C., Parker J. D., Spitler J. D., *Heating, ventilating, and air conditioning analysis and design*, Sixth edition, Wiley, Hoboken NJ USA (2005).
3. Webb R. L., Kim N.H., *Principles of enhanced heat transfer*, 2nd. ed., Boca Raton, CRC Press, (2005).
4. Taler D., *Numerical modelling and experimental testing of heat exchangers*, Berlin-Heidelberg: Springer, (2019).
5. Rich D. G., The effect of the number of tube rows on heat transfer performance of smooth plate fin-and-tube heat exchangers, Paper no. 2345, *ASHRAE Transactions*, Part 1, Vol. 81, pp. 307-317, (1975).
6. Saboya F. E. M., Sparrow E. M., Experiments on a Three-Row Fin and Tube Heat Exchanger, *ASME Journal of Heat Transfer*, Vol. 98, pp. 520-522, (1976).
7. Wang C.-C., Lee C.-J., C.-T. Chang, S.-P. Lin, Heat transfer and friction correlation for compact louvered fin-and-tube heat exchangers, *International Journal of Heat and Mass Transfer*, Vol. 42, pp. 1945-1956, (1999).
8. Taler D, Cebula A. Modeling of air flow and heat transfer in compact heat exchangers, *Chemical and Process Engineering*, Vol. 25, pp. 2331-2342, (2004).
9. Taler D., Experimental and numerical predictions of the heat transfer correlations in the cross-flow plate fin and tube heat exchangers, *Archives of Thermodynamics*, Vol. 28 (2007), No. 2, 3-18.
10. Taler D., Effect of thermal contact resistance on the heat transfer in plate finned tube heat exchangers, *Refereed Proceedings of the Conference Heat Exchanger Fouling and Cleaning* VII, 1-6 July 2007, Tomar, Portugal, ECI, New York, pp. 362-371, (2007), http://dc.engconfintl.org/heatexchanger2007/47/.
11. Taler D., Cebula A., A new method for determination of thermal contact resistance of a fin-to-tube attachment in plate fin-and-tube heat exchangers, *Chemical and Process Engineering*, Vol. 31, pp. 839-855, (2010).
12. Taler D., Taler J., Trojan M., Thermal calculations of plate-fin-and-tube heat exchangers with different heat transfer coefficients on each tube row, *14thSDEWES Conference, Dubrovnik, Croatia*, October 1-6, 2019 (paper accepted for presentation).
13. Taler D., A new heat transfer correlation for transition and turbulent fluid flow in tubes, *InternationalJournal of Thermal Science* , Vol. 108, pp. 108-122, (2016).
14. Taler D., Determining velocity and friction factor for turbulent flow in smooth tubes, *International Journal of Thermal Science*, Vol. 105, pp. 109-122, (2016).
15. Taler D., Mathematical modeling and experimental study of heat transfer in a low-duty air-cooled heat exchanger, *Energy Conversion and Management* ,Vol. 159, pp. 232-243, (2018).
16. Taler D., Taler J., Prediction of heat transfer correlations in a low-loaded plate- fin-and-tube heat exchanger based on flow-thermal tests, *Applied Thermal Engineering*, Vol. 148, pp. 641-649, (2019).
17. Gnielinski V., Heat transfer in pipe flow, , in *VDI Heat Atlas*, Second edition, Springer, Heidelberg, Chapter G1, pp. 693-699, (2010).
18. Taler D., Dynamic response of a cross-flow tube heat exchanger, *Chemical and Process Engineering*, Vol. 27, pp. 1053-1071, (2006).
19. Taler D., Direct and inverse heat transfer problems in dynamics of plate fin and tube heat exchangers, in *Heat Transfer – Mathematical Modelling, Numerical Methods and Information Technology*, ed. Aziz Belmiloudi, InTech, Rijeka, pp. 77-100 (2011).
20. Taler D., Experimental determination of correlations for average heat transfer coefficients in heat exchangers on both fluid sides, *Heat Mass Transfer*, Vol. 49, pp. 1125-39, (2013).

RESEARCH & DEVELOPMENT IN POWER ENGINEERING, 2019

Critical review and economic feasibility analysis of electric energy storage technologies suited for grid scale applications

Guido Francesco Frate[1,*], *Lorenzo* Ferrari[2], and *Umberto* Desideri[3]

[1] University of Pisa, Via Largo Lucio Lazzarino 1, 56122 – Pisa, guidofrancesco.frate@ing.unipi.it, Italy
[2] University of Pisa, Via Largo Lucio Lazzarino 1, 56122 – Pisa, lorenzo.ferrari@unipi.it, Italy
[3] University of Pisa, Via Largo Lucio Lazzarino 1, 56122 – Pisa, umberto.desideri@unipi.it, Italy

Abstract. Radical changes in production paradigms are affecting electric systems. This is due to Renewable Energy Sources (RES) introduction, which changed production from being centralized and dispatchable, to be more decentralized and unpredictable. Further transformations are expected as the RES share into electric grids will increase. RES further development will likely rely mostly on additional solar PV capacity, then relevant energy quantities will be needed to be shifted from day hours to evening and night hours. Such "Load Shifting" could be performed with Electric Energy Storage (EES) technologies. Few technologies suited for this are already commercially available, whereas others have been proposed. In this paper, the EES technologies suited for load shifting are reviewed with a focus on economic costs. After that, current and future EES economic feasibility are assessed by using Italian hourly energy prices from 2018. Since EES resulted to be currently uneconomic, the minimum price modification required to make EES feasible is calculated.

1 Introduction

Nowadays, electric systems are changing due to non-dispatchable RES introduction, mostly wind energy and solar PV. Since non-dispatchable RESs are rigid energy sources, their impact is making the grids less flexible. Furthermore, RESs are prioritized on energy markets, hence they are superseding traditional power plants. These are forced to provide the additional flexibility required by non-dispatchable RES by operating in part load and by doing frequent start-ups [1]. Storage might partly solve these issues, because it may smooth RES fluctuations, thus reducing traditional power plant need for part loads and start-ups. Unfortunately, suited sites for Pumped Hydro Energy Storage (PHES) are already exploited [2] and no new storage technologies have been deployed to date. In this situation, as non-dispatchable RESs grow, electric systems might become more and more rigid, up to the point of lacking the flexibility required to be safely operated.

Since most of RES additional capacity will likely come from solar PV [3], traditional power plants might be forced to fast ramping during evening hours to replace solar production. Traditional power plants could be supported in this task by EESs, which could shift energy from daytime to evening. To do that, storages that are able to charge at full power for 4/6 hours to absorb energy during solar production peaks might be required [4]. Such EESs would be characterized by a power/capacity ratio equal to 1/6 – 1/4 h^{-1}. However, not all the EESs may achieve this configuration *efficiently* from the economic point of view, since their capacity-related costs are higher than their power-related ones. Current EESs suited for this are: PHES, Compressed Air Energy Storage (CAES), Liquified Air Energy Storage (LAES), Pumped Thermal Energy Storage (PTES), molten salt (NaS) batteries and Flow batteries. These technologies are here briefly reviewed. By focusing on the economic aspects, current and future economic feasibility of such EESs is assessed. In conclusion, the main contributions of the present analysis are:

- an EES technology review is performed, by including also some often-overlooked technologies like PTES and PHES-based systems like Sea Water PHES (SWPHES) and Underground PHES (UPHES);
- costs are reported and used for assessing economic feasibility of each EES technology in a realistic environment represented by an Italian case study. Based on this estimation, a recommendation on which technologies are the most the suited ones from the economic point of view is provided.

2 EES technologies overview

- Pumped hydro energy storage (PHES) plants are made up of an upper and a lower reservoir, which are connected by one or more penstocks. The storage is charged by pumping water and it is discharged by turbining water. It could seem that the recent interest for storage stems from the fact that easily exploitable sites suited for PHES are already depleted [2]. Actually, PHES has some major drawbacks like the large use of land area and the large impact on river course and flow rates. The use of sea water reservoirs and cliffs (SWPHES), has been proposed to overcome the lack of easily exploitable sites [5], whereas by using abandoned caves and mines as lower reservoirs [6] (UPHES), the environmental impact could be largely reduced. To date, only one SWPHES pilot plant has been built (30 *MW* facility in Okinawa, Japan [7]), but some other facilities are planned [5]. In UPHES, the higher reservoir may be built on top of the facility, thus using an already exploited terrains with no value. Furthermore, by using the water in closed loop, the environmental impact of the facility could be furtherly reduced.

- Compressed Air Energy Storage (CAES) stores electric energy by compressing environment air and by storing it under pressure. The air is then discharged and used to feed a gas turbine combustion chamber. Two CAES are currently in operation: Huntdorf (Germany), commissioned in 1969 and McIntosh in Alabama (USA), commissioned in 1991 [8]. Both facilities use natural gas to power up the turbine during the discharge phase. To cut the costs, salt domes, rocky caves and aquifers are proposed as pressurized air storages, but this makes CAES site dependent [8]. To solve this issue, Under Water CAES (UWCAES), which uses sea/lake water hydraulic head to pressurize the storage [9], has been proposed. To date, only one UWCAES pilot plant has been commissioned in Lake Ontario (Toronto, Canada) [8]. To avoid using fossil fuel, Adiabatic CAES (ACAES) has been proposed. ACAES stores the excess thermal energy from the compression and use it to preheat air during discharge. The first pilot scale ACAES implementation (500 *kW*, 4 *h*) achieved a promising efficiency around 64% [4], whereas the first grid scale facility (90 *MW*, 360 *MWh*) is planned for commissioning [4]. Heat recovery in ACAES can be also done by using water sprays and liquid piston compression/expansion to operate nearly isothermal transformations (Isothermal CAES: ICAES) [10].
- Liquified air energy storage (LAES) stores electric energy in form of cryogenic energy by liquefying air, which is stored and then pumped, heated and expanded in a turbine during discharge phase [7]. Basic LAES are based on Linde cycle, whereas most recent systems are based on Claude or Kapitza cycles. To date, only one LAES pilot plant (300 *kW*, 2.5 *MWh*) has been successfully commissioned [11]. Improvements to basic LAES comprise: internal regenerations and turbine outlet waste heat recovery with ORCs, Brayton cycles or Absorption cooling systems [12]. To cut energy consumption and costs, alternative fluids like CO_2 have been proposed [13]. In fact, Liquified Carbon-dioxide Energy Storage (LCES) might be cheaper as it uses much more compact equipment. LAES has some advantages over CAES, like higher energy density and site independency. However, LAES generally achieves lower efficiency than CAES and PHES. This problem may be overcome by using non-standard configurations, which, however, might increase costs and start-up times, due to added thermal inertia.
- Sodium sulfur (NaS) battery energy storage use molten sodium (Na) and molten sulfur (S) as negative and positive electrodes. The two electrodes are divided by a solid electrolyte of β-alumina which allows for the ionic conduction. NaS batteries are commercially available and they count several *MW* scale applications [14]. NaS batteries are cheap and their building materials are largely recyclable [4]. NaS batteries feature high energy density, fast response, long cycle life (> 2500 cycles at 80 – 100% DOD) and good efficiency [4]. NaS batteries operate between 290 – 350 °C, to maintain electrodes liquid. During discharge phase, losses induced by internal resistance provide the necessary heat, whereas during idle or charge phase, the heat must be provided from the external and this may lead to parasitic losses up to 20% per day [15]. Cold start-up may take up to 15 *h* to allow electrodes to molten [15]. NaS batteries are very promising even though their use may be discouraged by electrodes corrosive behavior, by costs, which are higher than PHES and CAES ones, and by operating temperatures, which should be lower to operate more safely and with lower parasitic losses.
- Flow batteries use liquid electrolytes with dissolved active species, which are pumped through an electrochemical cell that converts chemical energy into electric energy. Electrolytes are stored in tanks, whose size determines battery capacity rating. Similarly, battery power rating is determined by the pumping capacity of the system. Therefore, capacity and power are decoupled, and this is a major advantages over traditional batteries [4,14]. Flow batteries have high efficiency, very quick response, very long cycle life and maximum discharge depth of 100% [4,7]. On the other hand, flow batteries have high cost, low density and greater complexity if compared to other batteries. Flow batteries are commercially available (Vanadium batteries) and they count several *MW* scale applications [7]. One of the largest (25 *MW* and 100 *MWh*) is in Astana (Kazakhstan) [4].
- Pumped Thermal Electricity Storage (PTES) stores electric energy as thermal energy, which is later converted back into electricity. The charge is performed with heat pumps, whereas discharge phase is performed with heat engines. Thermal energy is stored as sensible or latent heat. Both hot and cold reservoirs are used, but sometimes one is replaced by the environment, or by alternative heat sources/sinks. Compared to PHES and CAES, PTES usually has lower efficiency, but it is site independent. PTES technology may use both Brayton cycles (direct and inverse) and Rankine cycles (direct and inverse) [16]. Br-PTES most often uses packed bed sensible heat storage, usually made up of Al_2O_3, which has good conductivity/heat capacity ratio. Br-PTES is characterized by high energy density and very low estimated prices. Ra-PTES could be a valid alternative as it generally achieves higher densities and it stores energy at much lower temperature (~200 °C vs. ~1000 °C), which may allow for the use of phase change materials and the integration of additional low-grade/waste thermal sources to improve efficiency [16].

A summary of the technical and economical features of the reviewed EES technology is reported in Table 1.

3 EES economic feasibility analysis

Based on Table 1 data, the economic feasibility of EES technology can be assessed. Calculations use the 2018 Italian hourly energy prices, taken from [17]. Storage optimal operation is calculated by means of a Mixed

Table 1. EES technical and economic features. η_{rt}: roundtrip efficiency; τ_r: response time; ρ_{en}: energy density; L: operating life; Γ_p: power cost; Γ_c: capacity cost.

Technology	Γ_p [€/kW]	Γ_c [€/kWh]	ρ_{en} [kWh/m^3]	τ_r [min]	η_{rt} [%]	L [y]	References
PHES	400/600 – 1000/2000	1/5 – 100	0.5 – 1.5	1 – 3	65 – 85	30 – 40+	[4,7,15,16,18]
UPHES	400/600 – 1000/2000[a]	85	0.5 – 1.5	1 – 3	70 – 85	30 – 40	[6,15,16,18]
SWPHES	720 – 2200	25 – 30	0.3 – 0.4	1 – 3	70 – 85	15 +	[5,7]
CAES	400/500 – 800/1000	1/2 – 50/100/200	3 – 12	10 – 15	40 – 60	20 – 40	[8,15,16,18]
ACAES	700 – 1000	40 – 80	0.5 – 20	10 – 15	60 – 70	20 – 40	[4,15]
UWCAES	750 – 2000	40 – 200	0.3 – 5	10 – 15	60 – 70	20	[15,19]
ICAES	500 – 1000	10 – 100	1 – 25	1 – 5	60 – 80	20 – 40	[8,15]
LAES	900/1000 –2000	260 – 530	50	5 – 10	50 – 60	20 – 40	[4,15,16]
Br-PTES	600 – 800	20/90 – 60/180	110 – 200	1 – 3	50 – 70	20 – 40	[15,16,20]
Ra-PTES	225[b]/390 – 450	45 – 95/120	140 – 170	1 – 3	40 – 60	20 – 40	[15,21–23]
NaS Batt.	150/200 – 300/900	100/200/300 – 500/600	150 – 250	< 0.005	70 – 90	10 – 15	[4,15,18]
Flow Batt.	300/600 – 500/1500	150/400 – 750/1000	16 – 60	< 0.001	60 – 85	5 – 15	[4,15,18,24]

Integer Linear Programming (MILP) optimization problem. This allows to find the optimal sequence of storage charges and discharges to be performed each day in order to maximize the economic profit. Problem variables are discharge and charge power $P_{d/c}$ in each i-th hour of the day. Problem objective function is the net daily revenue R_{day} (Eq.1):

$$R_{day} = \sum_{i=1}^{24} P_{d,i} \cdot \pi_i \cdot \tau - \sum_{i=1}^{24} P_{c,i} \cdot \pi_i \cdot \tau \qquad (1)$$

where τ is equal to 1 h and π is the hourly energy price in $€/MWh$. Problem constraints are (Eq. 2):

$$\begin{gathered} 0 \leq P_{d,i} \leq P_{nom} \\ 0 \leq P_{c,i} \leq P_{nom} \\ 0 \leq \tau \cdot \eta_c \sum_{n=1}^{i} P_{c,i,n} - \frac{\tau}{\eta_d} \sum_{n=1}^{i} P_{c,i,n} + C_0 \leq C_{nom} \\ \tau \cdot \eta_c \sum_{n=1}^{24} P_{c,i,n} - \frac{\tau}{\eta_d} \sum_{n=1}^{24} P_{c,i,n} = 0 \end{gathered} \qquad (2)$$

where charge and discharge efficiency $\eta_c = \eta_d = \sqrt{\eta_{rt}}$ and C_0 is equal to 0 and P_{nom} = 10 MW and C_{nom} = 40 MWh are the nominal dimensions of the analysed EES. The listed constraints ensure that the storage cannot charge/discharge more than what is allowed by its nominal power and capacity ratings and that the storage is emptied at the end of each day. Not all the storages have $\eta_c = \eta_d$ or minimum charge level equal to 0. For the sake of simplicity, all the EES technologies are treated equally and these hypotheses are enough for first order estimation purposes. By summing the daily revenue over the year, the annual revenue R is found. R may be compared with annualized cost of storage A, which is calculated as in Eq. 3:

$$A = \frac{r(1+r)^L}{(1+r)^L - 1} \left(P_{nom} \cdot \Gamma_p + C_{nom} \cdot \Gamma_c\right) \cdot 1000 \qquad (3)$$

where r = 0.07 is the interest rate and L is the storage operating life. Therefore, different EESs have different revenue, as they have different efficiencies, and different annualized costs, as they have different specific costs and operating lives. In Table 1 a range is reported for η_{rt}, L, Γ_p and Γ_c, then a range for A and R may be defined as well. The Extrema of these ranges represent the best and the worst cases for these quantities. In addition to these, an average value, calculated with the average of each range is investigated. The results regarding the comparison between A and R are reported in Figure 1, where three cost scenarios are reported: base case (Eq 3), 25% discount and 50 % discount. This allows to consider the impact of price reductions or the effect of subsidies. Several conclusions may be drawn from Figure 1:

- None of the listed technologies achieved parity between A and R, not even in the case of 50 % cost reduction. Therefore, none of the listed technologies is currently economically viable, in the selected case study;
- All the technologies have A that is around one order of magnitude higher than R. Therefore, the listed technologies are far from economic feasibility;
- The higher the efficiency, the higher the revenue, therefore batteries should be advantaged. However, since the incomes are well under the required ones, Flow and NaS batteries are actually among the worst performing due to their high costs. This demonstrates that cost and operating life may be more impactful than efficiency, in the context of grid scale load shifting. In other words, the NaS and Flow batteries cost is disproportioned, if compared to benefits brought by their efficiency;
- by calculating to A_{ave} and R_{ave} difference, the most promising technologies can be determined; they are Ra-PTES, ICAES, ACAES and Br-PTES and this justifies the research interest towards these alternative EES. On the other hand, the least promising technologies, i.e. those that have the larger difference between A_{ave} and R_{ave}, are Flow batteries, NaS batteries and LAES.

Based on Figure 1 results, it may be interesting to calculate how much the energy prices should change to make the EES technologies able to repay themselves. To do this, it can be observed that daily revenue R_{day} may be modelled as in Eq. 4:

$$R_{day} \approx R_{model} = \max[0; \beta \cdot (\eta_{rt} \cdot \pi_d - \pi_c)] \qquad (4)$$

where β is an empirical constant to be fitted from data, and $\pi_{c,d}$ are the average energy prices at which the energy is charged and discharged. The data for β estimation come from the solution of Eq. 1 and 2optimization problem, whereas π_c and π_d are calculated by averaging the lowest and highest four π_i values for each day. This may be justified by considering that the storage has a power/capacity ratio equal to ¼ h^{-1}, therefore it charges and discharges for approximately four hours per day.

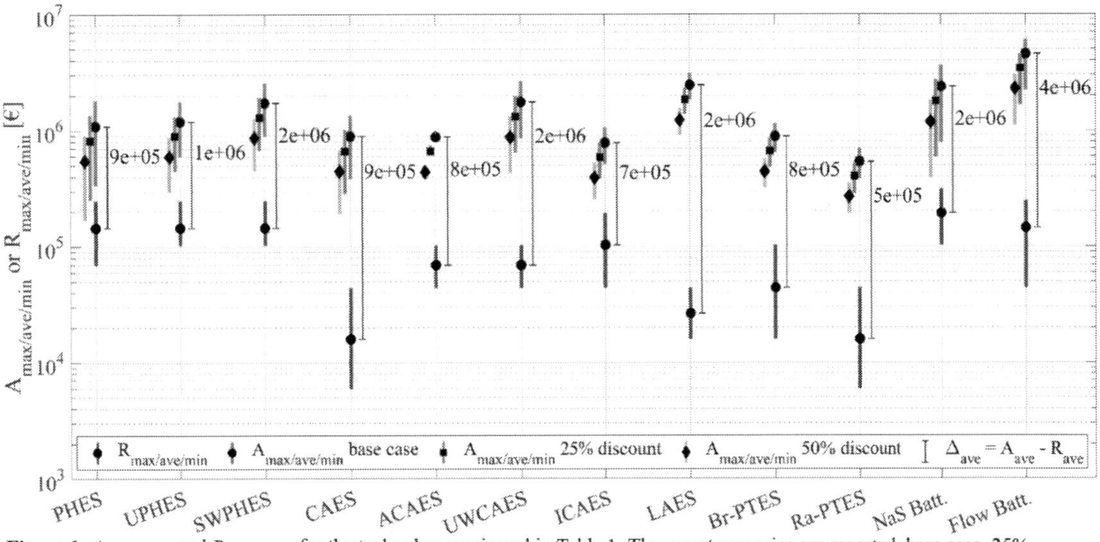

Figure 1. $A_{max/ave/min}$ and $R_{max/ave/min}$ for the technology reviewed in Table 1. Three cost scenarios are reported: base case, 25% discount and 50% discount.

The results related to the fit of the Eq. 4 model are reported in Figure 2, where it is shown how the model can explain reasonably well the daily revenue values.

By using R_{model}, the π_d and π_c values that yield the current average daily revenue may be calculated from Eq. 5:

$$R_{mean} = \frac{\sum_{j=1}^{365} R_{day,j}}{365} = \beta \cdot (\eta_{rt} \cdot \pi_d - \pi_c) \quad (5)$$

Similarly, the π_d and π_c values that yield a daily average revenue, such that the annual revenue is equal to the annuity A_{ave}, may be calculated from Eq. 6:

$$A_{ave} = \frac{\beta \cdot (\eta_{rt} \cdot \pi_d - \pi_c)}{365} \quad (6)$$

If π_c is set, two values of π_d that satisfies respectively Eq. 5 and 6 may be found. These values are function of η_{rt}, then three cases for each may be defined according to maximum, average and minimum η_{rt} values from Table 1. By comparing the two resulting π_d values from Eq. 5 and 6, the price changes required to make the reviewed EES technology viable may be estimated. From Figure 1, it is already known which technology are the most promising ones from the economic point of view. In this regard, by calculating π_d a different estimation may be done. In fact, π_d values from Eq. 5 give an idea of what are the current price situation on the Italian market. On the other hand, π_d values from Eq. 6 provide an estimation of what should be the peak prices on the average for the market to being able to sustain the EES. Of course, in market situations with a strong shortage of dispatchable production, very high prices may arise. Nonetheless, these extreme scenarios may be avoided, or discouraged, by system operators by means of economic incentives, like Capacity Markets, or others. Therefore, it is unclear what π_d values may be realistically achieved. Furthermore, very high π_d values will not suddenly appear on the markets, as the RES development is a slow and steady process. This might be a problem for storage development, as the required economic driver might be absent in all the intermediated steps that divide the current situation from a hypothetical high-RES penetration scenario. A vicious circle might arise, in which storage might be feasible only if a large share of productive capacity is represented by non-dispatchable RESs, but RESs cannot develop up to that point because they would need storage to foster their integration. From this point of view, the fact that none of the current technologies might be economically feasible is very concerning, since it may hinder the RES development and, with this, the power production sector decarbonization.

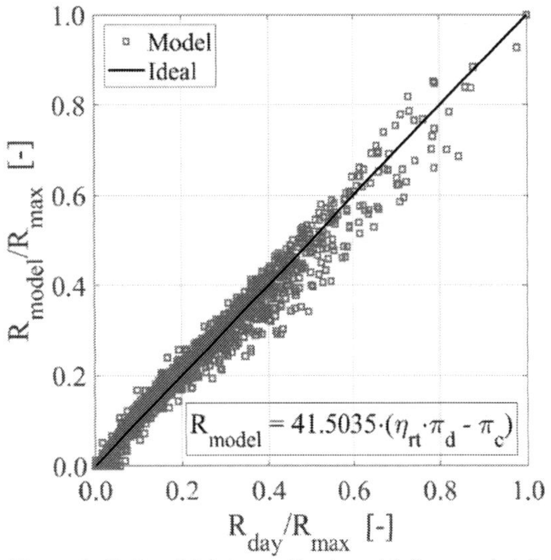

Figure 2. Parity plot between Eq. 4 model R_{model} and daily revenue R_{day} calculated by solving the optimization problem stated in Eq. 1 and 2.

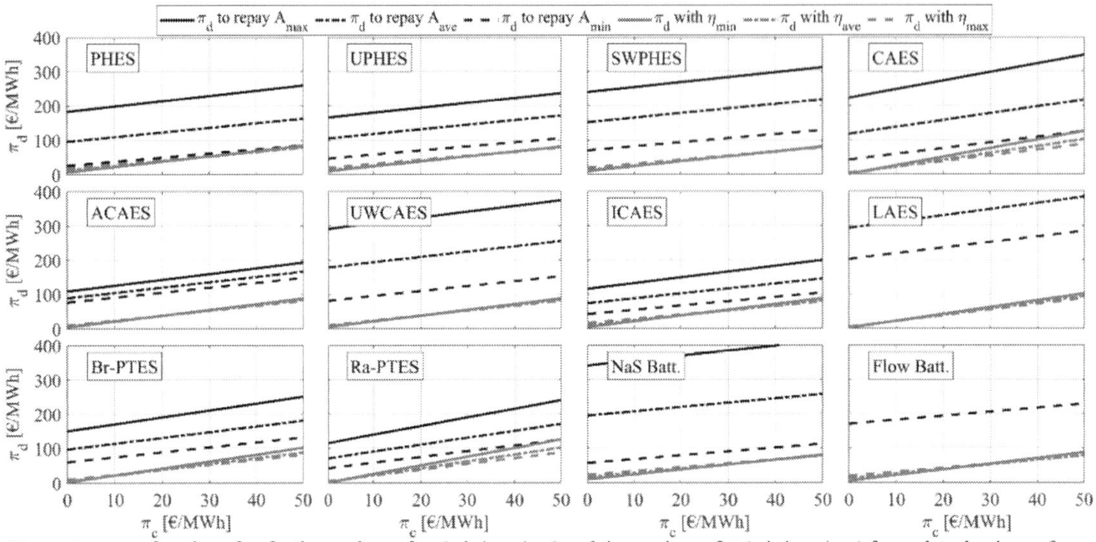

Figure 3. π_d as a function of π_c for three values of η_{rt} (min/max/ave) and three values of A (min/max/ave) for each technology of Table 1. π_d values that satisfy Eq. 5 are in red, whereas π_d values that satisfy Eq. 6 are in black.

The results related to the calculation of π_d through Eq. 5 and 6 are reported in Figure 3. A very interesting case is that with $\pi_c = 0$. As a matter of fact, in a high RES-penetration scenario, charge would occur in the moment of maximum RES production. Due to the large overcapacity required to power up a nation-wide electric system almost with RESs, the demand should not be able to accommodate all the produced energy during production peaks. Therefore, in these moments, energy prices will be very low and even negative, if allowed to be so. Since negative prices are common only in few energy markets, this possibility is disregarded here, and the minimum considered charging price is equal to 0.

With $\pi_c = 0$ and A_{min}, all the reviewed technology, except Flow batteries and LAES, can repay their own costs with an average $\pi_d < 100$ *€/MWh*. Even in the case of $\pi_c = 0$ and A_{ave}, some technologies are still able to repay themselves with an average π_d slightly lower than 100 *€/MWh*. These technologies are: PHES, UPHES, ACAES, ICAES, Br-PTES and Ra-PTES. In this case, NaS battery would require an average π_d equal to 200 *€/MWh*. Finally, for the third case, i.e. $\pi_c = 0$ and A_{max}, few technologies should be able to sustain they self with a minimum π_d lower than 150 *€/MWh*. These technologies are: ACAES, ICAES, Br-PTES and Ra-PTES, thus confirming the strong potentiality of these technologies, based on different operating principles than batteries.

It is worth noting that energy prices around 100 – 150 *€/MWh* should not be considered unrealistically high, as they are currently occurring in several energy market, the Italian one included. However, such prices are not common nowadays, therefore storages cannot gain enough to self-sustain in current market scenarios.

The resulting π_d values may be considered as compatible with realistic future market and energy mix scenarios. Therefore, some EES technologies like PTES, some forms of CAES and some of PHES may realistically become viable in the future. Other technologies like NaS and Flow batteries might encounter much more difficulties in achieving the economic feasibility, even in very positive scenarios.

Finally, as Figure 3 reports, π_d is linear with π_c, therefore all the discussed price threshold should be adjusted, if other values of π_c are considered. For example, if $\pi_c = 10$ *€/MWh* the only technologies that could achieve the economic parity in the worst-case scenario (A_{max}), without surpassing the threshold of $\pi_d = 150$ *€/MWh* are: ACAES, ICAES and Ra-PTES.

3 Conclusive remarks

In this paper a review of EES technologies suited for load shifting was performed. As a result of the review, a range of values for power and capacity costs, roundtrip efficiencies and operating lifetimes were defined for each technology. These data were used to assess the economic feasibility of the reviewed technologies in a realistic European case study (Italian 2018 hourly energy prices). As it resulted, none of the reviewed technologies is currently feasible from an economic point of view and often the annualized costs are of one order of magnitude higher than the revenues.

Based on these results, a simplified model that links daily revenues with the roundtrip efficiency and the charging and discharging energy prices is defined and tuned. By means of this model, the average energy prices required for covering the annualized costs were calculated in function of the charging energy prices.

For low and very low charging energy prices, the most promising technologies resulted to be ACAES, ICAES and Ra-PTES, followed by Br-PTES, PHES and UPHES. All these technologies might become viable with average maximum daily energy prices lower than 150 *€/MWh*. Such values are higher than the current energy prices, but they might be realistically achieved for future energy mixes, as similar prices already occur in few days per year even today.

References

[1] J. Devlin, K. Li, P. Higgins, A. Foley, Gas generation and wind power: A review of unlikely allies in the United Kingdom and Ireland, Renew. Sustain. Energy Rev. 70 (2017) 757–768. doi:10.1016/j.rser.2016.11.256.

[2] M. Gimeno-Gutiérrez, R. Lacal-Arántegui, Assessment of the European potential for pumped hydropower energy storage based on two existing reservoirs, Renew. Energy. 75 (2015) 856–868. doi:10.1016/j.renene.2014.10.068.

[3] IEA, World Energy Outlook 2018, Paris, 2018.

[4] M.C. Argyrou, P. Christodoulides, S.A. Kalogirou, Energy storage for electricity generation and related processes: Technologies appraisal and grid scale applications, Renew. Sustain. Energy Rev. 94 (2018) 804–821. doi:10.1016/j.rser.2018.06.044.

[5] D.A. Katsaprakakis, D.G. Christakis, I. Stefanakis, P. Spanos, N. Stefanakis, Technical details regarding the design, the construction and the operation of seawater pumped storage systems, Energy. 55 (2013) 619–630. doi:10.1016/j.energy.2013.01.031.

[6] W.F. Pickard, The history, present state, and future prospects of underground pumped hydro for massive energy storage, Proc. IEEE. 100 (2012) 473–483. doi:10.1109/JPROC.2011.2126030.

[7] X. Luo, J. Wang, M. Dooner, J. Clarke, Overview of current development in electrical energy storage technologies and the application potential in power system operation, Appl. Energy. 137 (2015) 511–536. doi:10.1016/j.apenergy.2014.09.081.

[8] M. Budt, D. Wolf, R. Span, J. Yan, A review on compressed air energy storage: Basic principles, past milestones and recent developments, Appl. Energy. 170 (2016) 250–268. doi:10.1016/j.apenergy.2016.02.108.

[9] A.J. Pimm, S.D. Garvey, M. de Jong, Design and testing of Energy Bags for underwater compressed air energy storage, Energy. 66 (2014) 496–508. doi:10.1016/j.energy.2013.12.010.

[10] W. He, J. Wang, Optimal selection of air expansion machine in Compressed Air Energy Storage: A review, Renew. Sustain. Energy Rev. 87 (2018) 77–95. doi:10.1016/j.rser.2018.01.013.

[11] R. Morgan, S. Nelmes, E. Gibson, G. Brett, Liquid air energy storage - Analysis and first results from a pilot scale demonstration plant, Appl. Energy. 137 (2015) 845–853. doi:10.1016/j.apenergy.2014.07.109.

[12] A. Tafone, E. Borri, G. Comodi, M. van den Broek, A. Romagnoli, Liquid Air Energy Storage performance enhancement by means of Organic Rankine Cycle and Absorption Chiller, Appl. Energy. 228 (2018) 1810–1821. doi:10.1016/j.apenergy.2018.06.133.

[13] M. Wang, P. Zhao, Y. Yang, Y. Dai, Performance analysis of energy storage system based on liquid carbon dioxide with different configurations, Energy. 93 (2015) 1931–1942. doi:10.1016/j.energy.2015.10.075.

[14] B. Dunn, H. Kamath, J.-M. Tarascon, Electrical Energy Storage for the Grid: A Battery of Choices, Science (80-.). 334 (2011) 928–935. doi:10.1126/science.1212741.

[15] A.B. Gallo, J.R. Simões-Moreira, H.K.M. Costa, M.M. Santos, E. Moutinho dos Santos, Energy storage in the energy transition context: A technology review, Renew. Sustain. Energy Rev. 65 (2016) 800–822. doi:10.1016/j.rser.2016.07.028.

[16] A. Benato, A. Stoppato, Pumped Thermal Electricity Storage: A technology overview, Therm. Sci. Eng. Prog. 6 (2018) 301–315. doi:10.1016/j.tsep.2018.01.017.

[17] GME (Italian energy market operator), Italian Day-ahead energy market prices, (2019). https://www.mercatoelettrico.org/it/download/DatiStorici.aspx (accessed September 1, 2019).

[18] B. Zakeri, S. Syri, Electrical energy storage systems: A comparative life cycle cost analysis, Renew. Sustain. Energy Rev. 42 (2015) 569–596. doi:10.1016/j.rser.2014.10.011.

[19] A.J. Pimm, S.D. Garvey, M. de Jong, Design and testing of Energy Bags for underwater compressed air energy storage, Energy. 66 (2014) 496–508. doi:10.1016/j.energy.2013.12.010.

[20] A. Smallbone, V. Jülch, R. Wardle, A.P. Roskilly, Levelised Cost of Storage for Pumped Heat Energy Storage in comparison with other energy storage technologies, Energy Convers. Manag. 152 (2017) 221–228. doi:https://doi.org/10.1016/j.enconman.2017.09.047.

[21] S. Staub, P. Bazan, K. Braimakis, D. Müller, C. Regensburger, D. Scharrer, B. Schmitt, D. Steger, R. German, S. Karellas, M. Pruckner, E. Schlücker, S. Will, J. Karl, Reversible heat pump-organic rankine cycle systems for the storage of renewable electricity, Energies. 11 (2018). doi:10.3390/en11061352.

[22] M. Morandin, M. Mercangöz, J. Hemrle, F. Maréchal, D. Favrat, Thermoeconomic design optimization of a thermo-electric energy storage system based on transcritical CO2 cycles, Energy. 58 (2013) 571–587. doi:10.1016/j.energy.2013.05.038.

[23] S. Henchoz, F. Buchter, D. Favrat, M. Morandin, M. Mercangöz, Thermoeconomic analysis of a solar enhanced energy storage concept based on thermodynamic cycles, Energy. 45 (2012) 358–365. doi:10.1016/j.energy.2012.02.010.

[24] D.H. Doughty, P.C. Butler, A.A. Akhil, N.H. Clark, J.D. Boyes, Stationary Electrical Energy Storage, Electrochem. Soc. Interface. 19 (2010) 49–53.

RESEARCH & DEVELOPMENT IN POWER ENGINEERING, 2019

Various inlet spiral geometries in 1MW steam turbine

Arkadiusz Koprowski[1], *Dominik* Gotowski[1], *Romuald* Rządkowski[1], and *Ryszard* Szczepanik[2]

[1] Institute of Fluid-Flow Machinery Polish Academy of Sciences, Fiszera 14, 80-952 Gdansk, akoprowski@imp.gda.pl, Poland
[2] Air Force Institute of Technology, Księcia Bolesława 6, 01-494 Warsaw, ryszard.szczepanik@itwl.pl, Poland

Abstract. In small turbines, inlets can be designed using an inlet spiral. This paper analyses the efficiency of eight turbine variants, including seven with three stator and rotor blades and inlet spirals of various geometries, and one variant with a spiral, three stages and an outlet. This involves a 3D steady viscous flow using ANSYS CFX. The analysis shows that the spiral has a considerable influence on turbine efficiency.

1 Introduction

Steam turbine inlets usually comprise inlet boxes for partial and full admission. Inlet design in small turbines is normally simplified to economise on production costs.

Kryłłowicz et al. [1] described in a condensing steam turbine (50 kW) and a back pressure turbine (165 kW). The general parameters are: inlet temperature 340 C, inlet pressure12 bar, mass flow 270 kg/h and rotation speed 6276 rpm. The turbine inlets are of the standard type used in turbines.

Škach and Uher [2] showed design process of stator blades for spiral inlets.

Hecker et al. [3] presented the design of spiral inlets. The structure and aerodynamics analyses were shown.

This paper for the first time studies different inlet spiral geometries for a 1MW, three-stage steam turbine with a 30t/h mass flow.

2 Numerical studies

ANSYS CFX was used to calculate steady viscous flow in four inlet spirals and three stages, each with 32 stator blades and 99 rotor blades. Design Modeller was used to create the geometries of rotor and stator blades as well as turbine inlets and outlets. TurboGrid was used to prepare the blade, inlet and outlet meshes.

The purpose of the analysis was to determine the efficiency of the eight turbine variants.

2.1 Variant one

Here, the circular pipe entered the rectangular channel of the spiral at a low angle to allow for a smoother flow. The outlet spiral leading the steam to the stator blades was in the form tapered diffuser (Fig. 1, 2). The inlet spiral caused a reversal of flow towards in the stator blades. There were 32 stator blades with an S-9012A profile [4].

A stall flow appeared in the inlet pipe and spiral as well as the stator blades when there was a flow velocity perpendicular to stator the pressure side in the first stage (Fig. 3, 4).

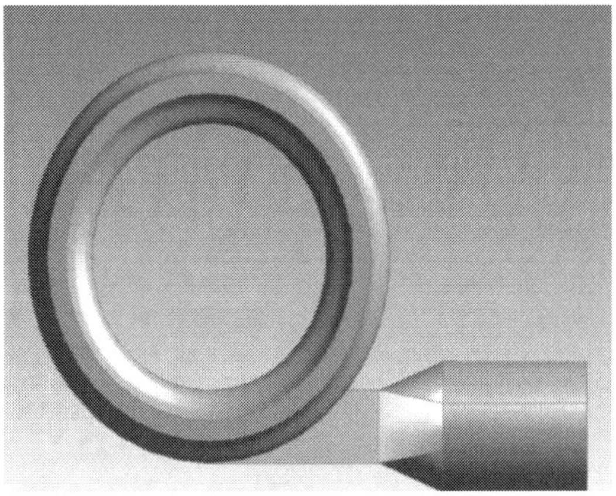

Fig. 1. Side view of Variant one spiral.

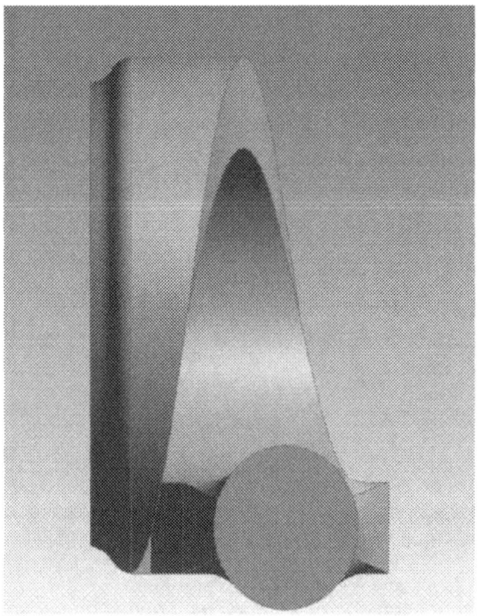

Fig. 2. Front view of variant one spiral.

According to the CFD calculations, the turbine generated 1.368 MW with a mass flow of 9.78 kg/s (35.21 t/h) and an inlet pressure drop of 5.64kPa and was 86.17% efficient.

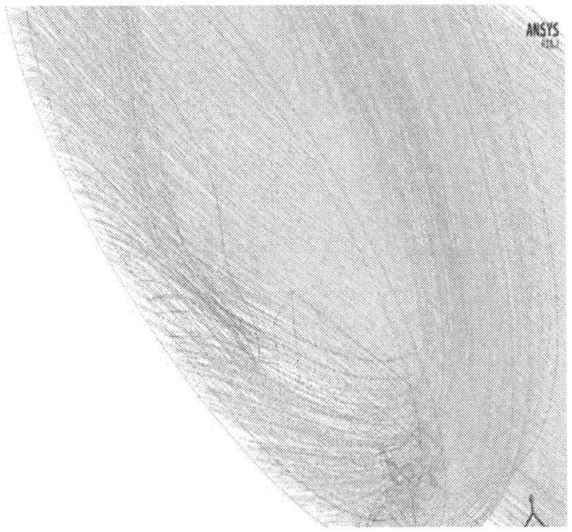

Fig. 3. Spiral inlet flow, variant one.

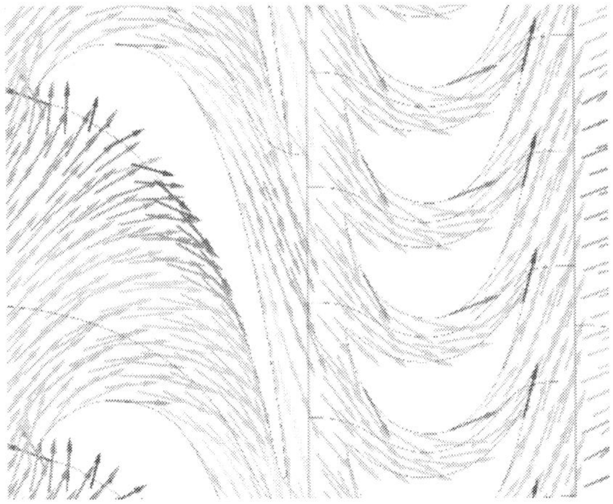

Fig. 4. Velocity field at 0.5 of stator blade, variant one.

2.2 Variant two

The geometry of the spiral in variant two was the same as in variant one. The differences concerned the profile of the first stator (S-5515A) [4] and the number of stator blades was increased to 39. The altered stator blade profile prevented stall in the stator area and the velocity vector was no longer vertical to the pressure side (Fig. 5).

According to the CFD calculations, the turbine generated 1.516 MW with a mass flow of 10.73 kg/s (38.64 t/h) and an inlet pressure drop of 6.55kPa, resulting in 86.9% efficiency. The altered first stator blade profile prevented steam flow stall around the stator blades and caused efficiency to rise from 86.17% to 86.9%.

2.3 Variant three

The circular pipe entered the rectangular spiral channel at a zero angle. The first stator profile was S-9012A [4] and there were 32 stator blades. In contrast to variants one and two, where the steam flowed through the centre of the spiral, in variant three, the steam flowed down the inner side of the spiral. A diffusor was not applied in this case (Fig. 6, 7, 8).

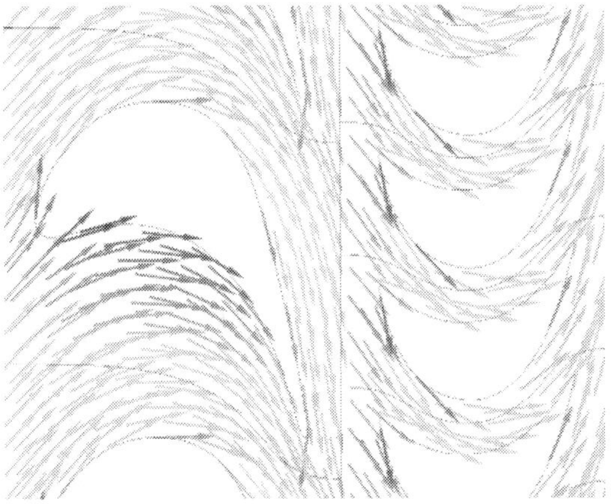

Fig. 5. Velocity field at 0.5 of the stator blade length, variant two.

Fig. 6. Variant three spiral.

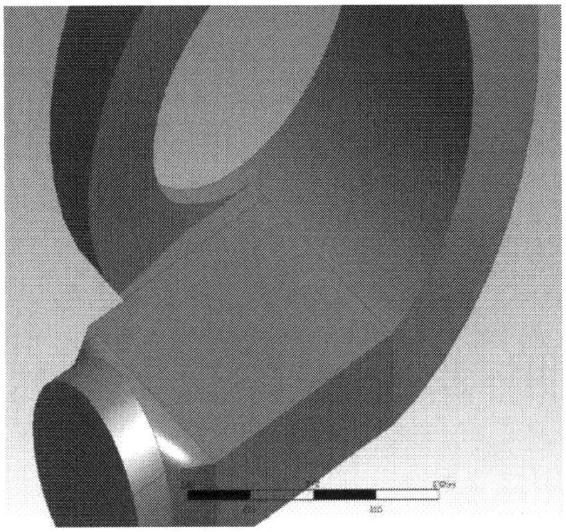

Fig. 7. Variant three spiral, enlarged.

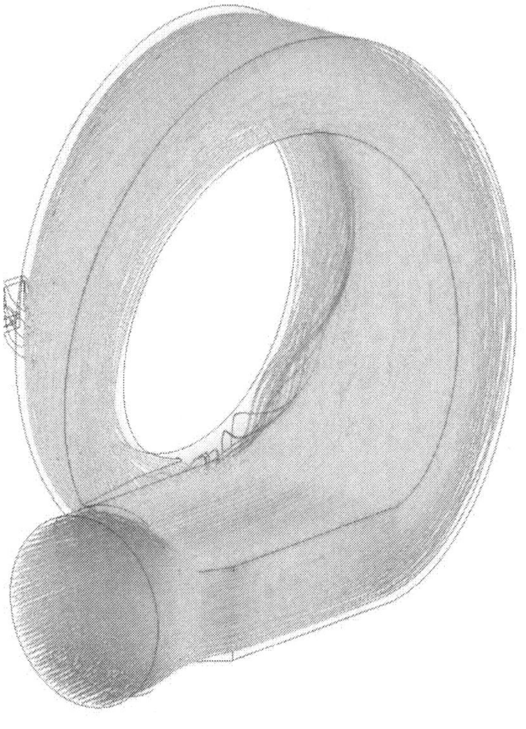

Fig. 8. Streamlines in inlet, variant three

Fig. 9. Variant three velocity field at 0.5 stator blade length.

Due to the altered geometry, a stall appeared in the stator region (Fig. 9), causing efficiency to drop to 83.7%.

According to the CFD calculations, the turbine generated 1.293 MW with a mass flow of 9.50 kg/s (34.21 t/h) and inlet pressure drop of 21.15kPa. Reducing the inlet pipe angle to zero caused a perpendicular flow on the pressure side, as in Variant one, and the reduced efficiency to 83.7%.

2.4 Variant four

Variants four and three had the same geometries, but the profile of the first stator blade in the former was altered to (S-5515A) [4]. As in variant two, there were 39 stator blades.

The altered stator blade profile prevent stall at 0.5 length of the blade but a stall did occur in the root and at the tip (Fig. 10, 11).

The turbine generated 1.458 MW with a mass flow of 10.49 kg/s (37.75 t/h) and an inlet pressure drop of 24.49kPa. The variant four stator blade profile improved turbine efficiency from 83.7% to 85.69%.

Fig. 10. Variant four velocity field at 0.5 of stator blade length.

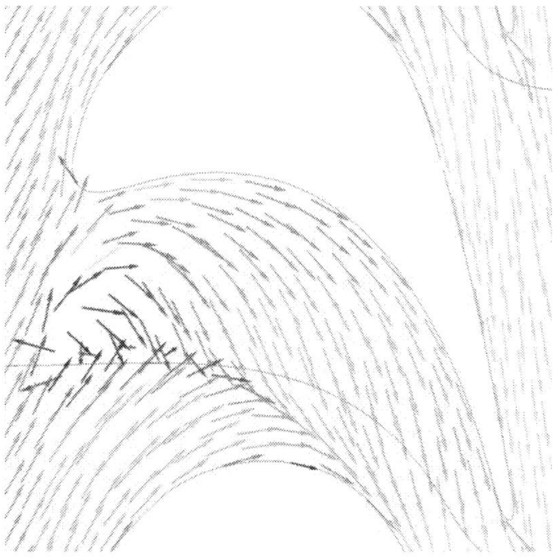

Fig. 11. Variant four velocity field at 0.05 of stator blade length.

2.5 Variant five

Variant 5 inlet spiral was altered to prevent flow reversal (Fig. 12, 13). The length if the first stator (S-9012A) [4] was reduced from 62.5 mm to 56 mm and there were 32 stator blades. In other stages, the number of stator blades

 https://doi.org/10.1051/e3sconf/201913701038

was the same as in variants one to four. Fig.8 presents the flow velocity, which did not stall.

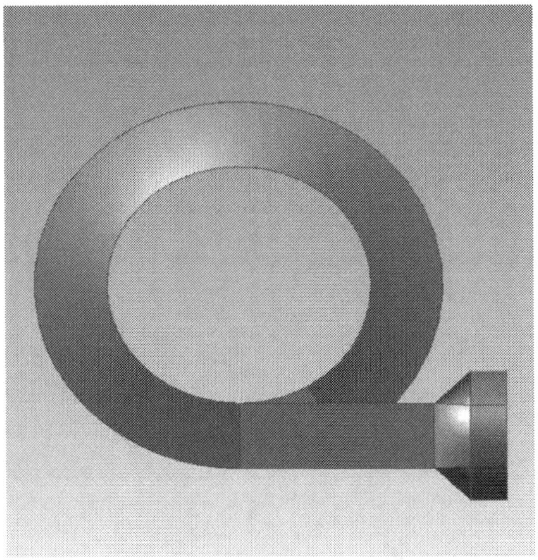

Fig. 12. Side view of variant five spiral.

Fig. 13. Variant five inlet spiral.

The turbine generated 1.373 MW, with a mass flow of 9.68 kg/s (34.86 t/h) and a 20.99kPa pressure drop in the inlet. The variant five alterations improved turbine efficiency from 85.7% to 87.3%. Fig. 14 presents the flow velocity, which did not stall.

2.6 Variant six

Here, the first stator blade profile (S-9012A) [4] was reduced from 62.5 mm to 54 mm, blade thickness was reduced on the suction side and this variant included 32 stator blades.

Fig. 14. Variant five velocity field at 0.5 of stator blade length.

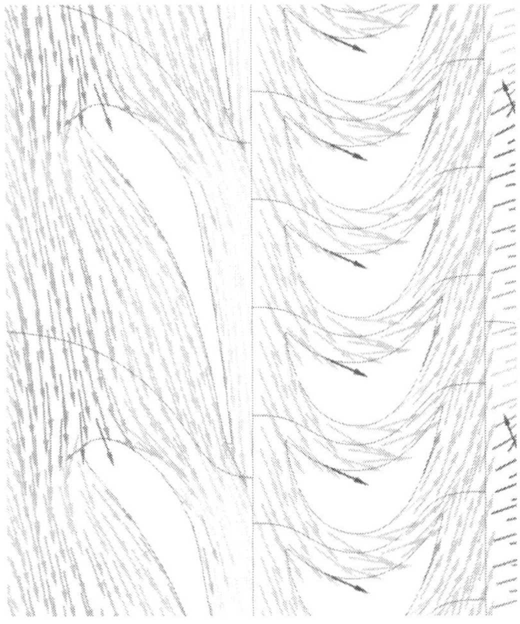

Fig. 15. Variant six velocity field at 0.5 of stator blade length

The turbine generated 1.428 MW with a mass flow of 10.08 kg/s (36.31 t/h) and a 23.02kPa pressure drop at the inlet. The altered first stator blade profile in variant six reduced turbine efficiency from 87.3% to 87.15%. Fig. 15 presents the flow velocity, which did not stall.

2.7 Variant seven

Two inlet pipes were applied in variant seven instead of one (Fig. 16, 17, 18).

The turbine generated 1.178 MW with a mass flow 8.21 kg/s (29.31 t/h) and a 1.24kPa pressure drop at

 https://doi.org/10.1051/e3sconf/201913701038

the inlet. The application of two pipe in variant seven improved turbine efficiency from 87.3% to 88.87%. Fig. 19 presents the flow velocity, which did not stall.

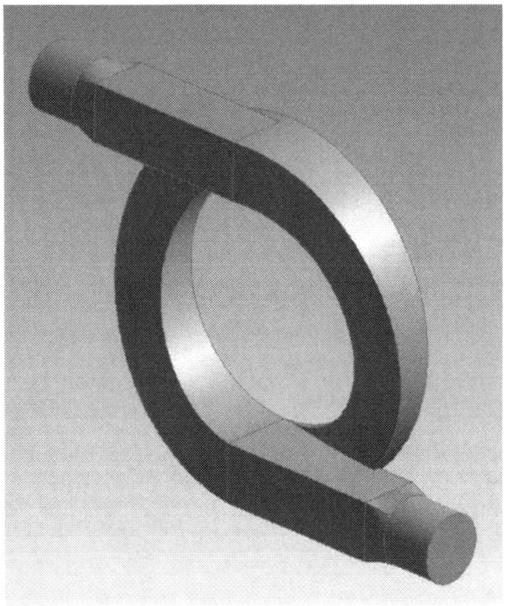

Fig. 16. Variant seven spiral.

Fig. 17. Streamlines in inlet, variant seven

Fig. 18. Streamlines in inlet, variant seven

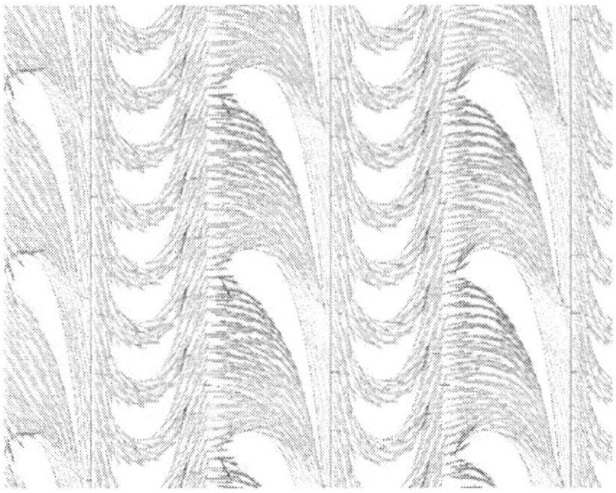

Fig. 19. Variant seven velocity field at 0.5 of stator blade length.

2.8. Variant eight

Variant eight had the same spiral and three stages as in variant seven, and the only difference was the addition of an outlet (Fig. 20, 21).

The turbine generated 1.100 MW with 7.99 kg/s (28.76 t/h) mass flow and 118kPa inlet pressure drop. The inclusion of the outlet reduced turbine efficiency from 88.87% (Variant seven) to 85.31%. This was due to the fact that the outlet was causing the flow to whirl in many place (Fig. 23). Fig. 22 shows the velocity of the flow, which did not stall in the stator and rotor regions.

Fig. 20. Variant eight inlet with 3 stages and an outlet.

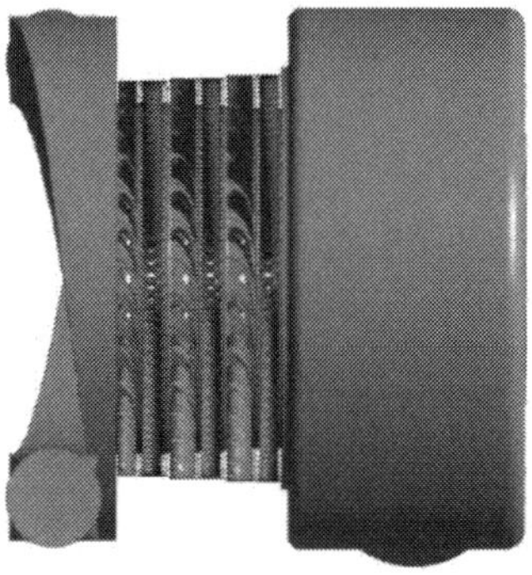

Fig. 21. Side view of variant eight.

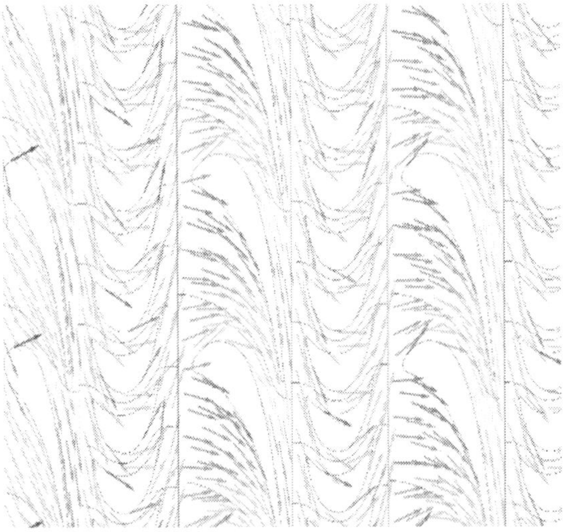

Fig. 22. Variant eight velocity field at 0.5 of stator blade length.

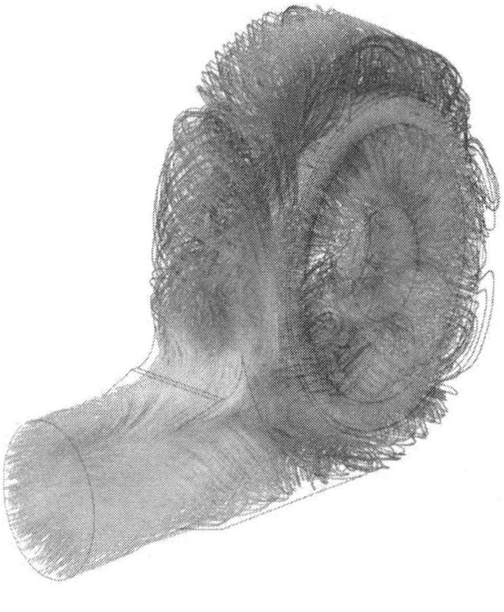

Fig. 23. Streamlines in outlet, variant 8

3 Conclusions

This paper for the first time analyses the flow in various inlet spirals in a 1 MW steam turbine inlets. In addition to altering the spiral geometries, the stator blade profiles were also changed. Adding a second pipe to the inlet spiral improved efficiency from 87.3% to 88.87%.

Efficiency was also increased changing the first stator blade profile from S-5515A to S-9012A.

The changed profile increased the mass flow, which could next be reduced by shortening of the length of the stator blades. In the last variant, an outlet was added to calculate the efficiency of the entire flow system. This reduced efficiency from 88.87% to 85.87%.

Acknowledgement

The authors wish to acknowledge NCBiR for the financial support of this work (POIR.04.01.04-00-0116/17).

References

1. Kryłłowicz W., Liskiewicz G., Szwaja S., Small Steam Turbines for Distributed Power Generation-Technical and Economic Condition, *Energetyka*, November, 719-722, (2015).
2. Škach R., Uher J., Spiral inlets for steam turbines, *36th Meeting of Departments of Fluid Mechanics and Thermodynamics AIP Conference Proceedings 1889*, 020038-1-020038-7, (2017).
3. Hecker S., Rohe A., Stoff H., Steam turbine inlet geometry from a structural and fluid dynamics point of view, *Procedings of ASME Turbo Expo 2012*
4. Dejcz M.E., Filipov G.A., *Profile Collections of Axial Turbine Cascades*, Maszinostrojenie (1965).

RESEARCH & DEVELOPMENT IN POWER ENGINEERING, 2019

Optimization of Curtis stage in 1 MW steam turbine

Arkadiusz Koprowski, and Romuald Rządkowski

Institute of Fluid-Flow Machinery Polish Academy of Sciences, Fiszera 14, 80-952 Gdansk, akoprowski@imp.gda.pl, Poland

Abstract. When operating at 3000 rpm, small turbines do not require a gear box and the generator does not require complex electronic software. This paper analyses the various geometries of the Curtis stage, comprising two rotor and stator blades with and without an outlet, from the efficiency point of view. Presented are 3D steady viscous flows. The results were compared with the performance of an axial turbine.

1 Introduction

Small steam turbine flow paths can have an axial design with several stages, or can comprise one Curtis or more stages.

Described in [1] is the design and CFD analysis of a Curtis Turbine Stage.

[2] showed a Curtis steam Turbine

[3] presented ways of designing the Curtis stage, nozzle/rotor aerodynamic interaction and the effects on Curtis stage performance

This paper for the first time presents eight variants with full and partial admission for the optimization of a Curtis stage in a 1 MW steam turbine.

2 Boundary conditions

ANSYS CFX was used to calculate steady viscous flow through stator and rotor blades. The number of rotor and stator blades is presented in Tab 1. Design Modeller was used to create the geometries of rotor and stator blades as well as the turbine outlet. TurboGrid was used to prepare the blade and outlet meshes.

In the first model, only one blade in each channel, stator1, rotor1, stator2, rotor2, was analysed. The flow in the interfaces between sator1, rotor1, stator2 and rotor 2 was averaged by Stage (Mixing-Plane). In the next model, all the rotor and stator blades were considered. In these cases, the Frozen rotor interface was used. In this interface, only one relative position of the stator and rotor blade was taken into account. An adiabatic wall condition was assumed.

Table 1. Stator and rotor blades in a Curtis stage

	profile	number of blades
First stator	S9012-A	55
First rotor	R2314-A	141
Second stator	S4525-A	85
Second rotor	R5535-A	197

2.1 Variant one

The calculation of the Curtis stage was done for the full admission in order to see the periodicity of the flow.
The blade length was 20 mm. The profiles of the rotor and stator blades was taken from [4], see Table 1.

Fig. 1 presents the flow through the stage. The supersonic and stall regions are visible.

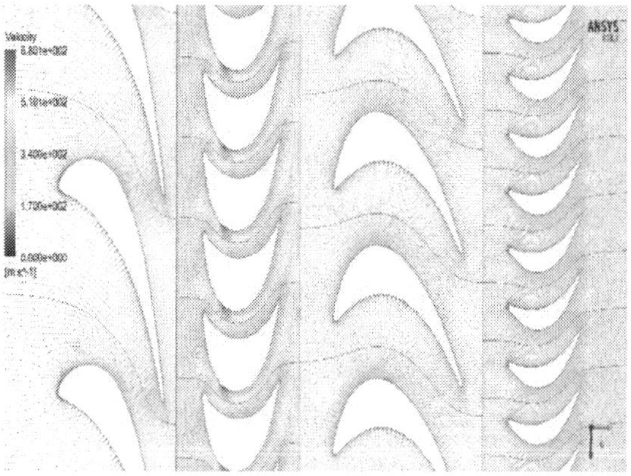

Fig. 1. Velocity field at 0.5 of the stator1 blade length, variant one

According to the CFD calculations, the turbine generated 1.688 MW with a mass flow of 13.98 kg/s (50.33 t/h) and was 74.827% efficient.

2.2 Variant two

In variant two, 1/3 circumference partial admission was analysed.
The number of first stator blades is 17 instead of 55, but the remaining numbers of blades is the same as those presented in Tab 1.

Figs. 2, 3, 4, 5 and 6 the flow through the stage. The supersonic and stall regions are visible.

https://doi.org/10.1051/e3sconf/201913701039

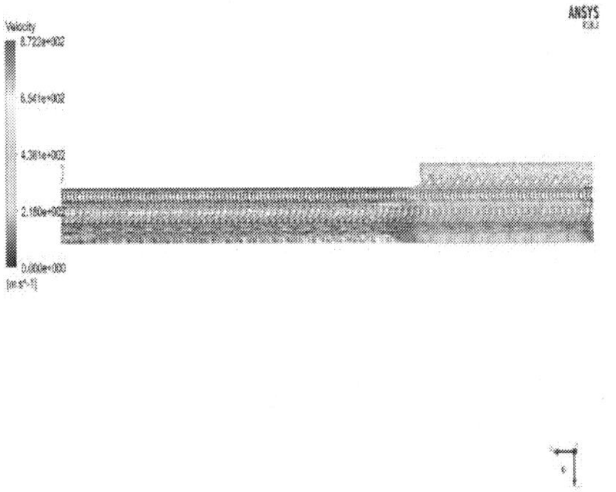

Fig. 2. Velocity field at 0.5 of the stator1 blade length, variant two

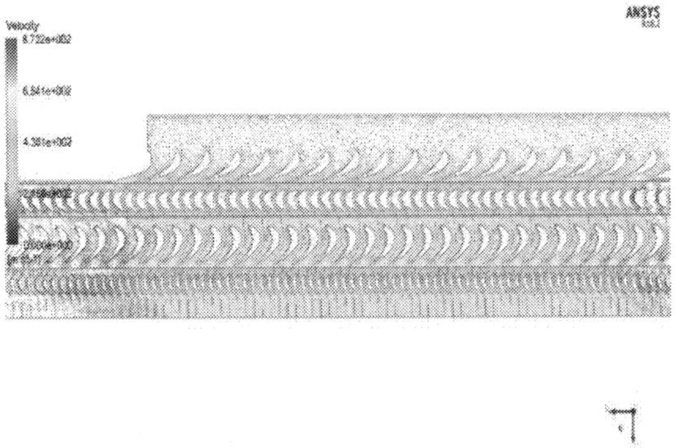

Fig. 3. . Velocity field at 0.5 of the stator1 blade length, variant two

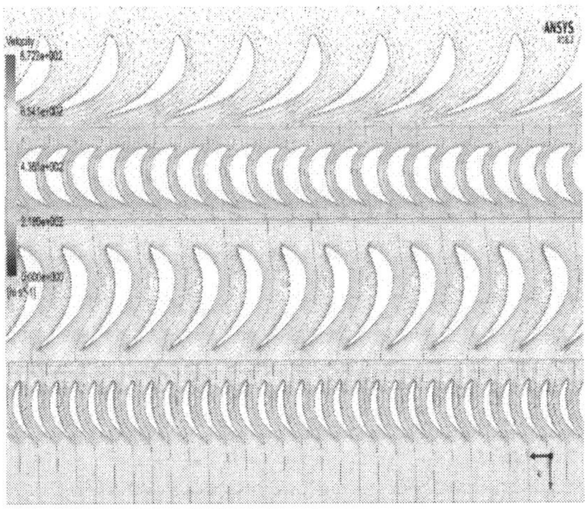

Fig. 4. Pole wektorów prędkości w połowie wysokości łopatek

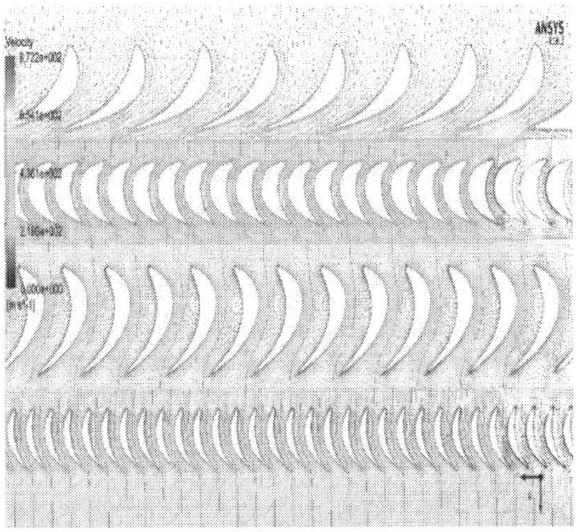

Fig. 5. Velocity field at 0.5 of the stator1 blade length, variant two

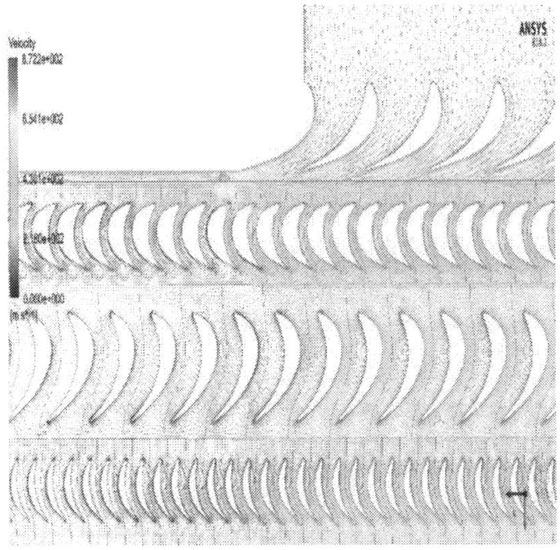

Fig. 6. Velocity field at 0.5 of the stator1 blade length, variant two

In this case, the turbine generated 1,063 MW, with a mass flow of 9.09 kg/s (32.72 t/h) and 72.48 % efficiency. The reduced power, mass flow efficiency were due to the partial admission.

2.3 Variant three

Variant three differed from the second variant with regard to the profile of the first rotor blade. The numbers of rotor and stator blades are shown in Table 2.

Table 2. Stator and rotor blades in a Curtis stage

	profile	number of blades
First stator	S9012-A	17
First rotor	R2117-Bk	141
Second stator	S4525-A	85
Second rotor	R5535-A	197

https://doi.org/10.1051/e3sconf/201913701039

The modification prevented transonic and supersonic flow through rotor 1, but caused stalling in the stator 2 stage (Fig. 7).

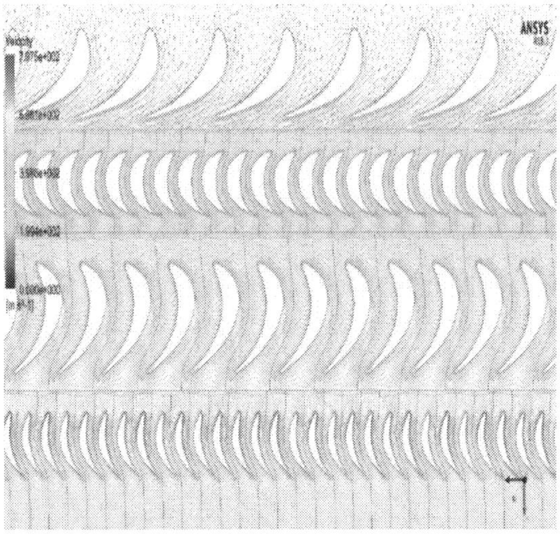

Fig. 7. Velocity field at 0.5 of the stator1 blade length, variant three

In this case, power fell to 0,964 MW, mass flow fell to 8.42 kg/s (30.31 t/h) and efficiency to 70.94 %.

2.4 Variant four

Here, the profile of the stator 2 blades was changed in order to avoid the stalling that occurred in variant three.

Table 3. Stator and rotor blades in a Curtis stage

	profile	number of blades
First stator	S9012-A	17
First rotor	R2117-Bk	141
Second stator	R4629-A	153
Second rotor	R5535-A	197

This measure proved ineffective, and stalling still occurred, as is seen in Fig. 8.

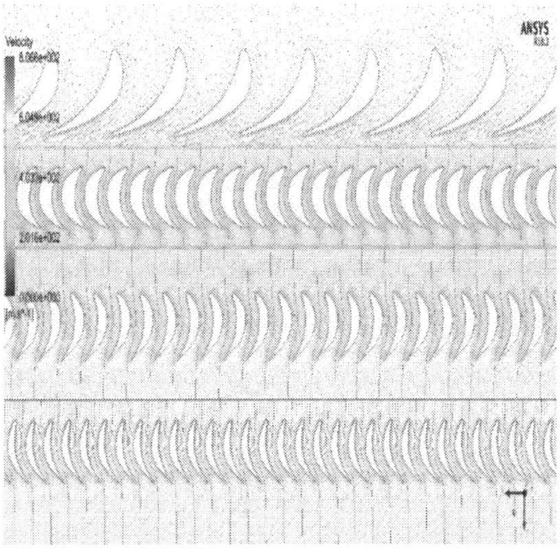

Fig. 8. Velocity field at 0.5 of the stator1 blade length, variant four

Power generation again fell, to 0.958 MW, mass flow increased to 8.50 kg/s (30.6 t/h), but efficiency fell to 68.71 %.

2.5 Variant five

In variant five, the stator 2 blade profile was further altered (Tab. 4), and the inlet of steam now came from two, opposing tubes, and not one, as in the cases of the preceding variants.

Table 4. Stator and rotor blades in a Curtis stage

	profile	number of blades
First stator	S9012-A	16
First rotor	R2117-Bk	141
Second stator	R3525-A	155
Second rotor	R5535-A	197

Figs. 9 and 10 present vector fields. The applied profiles did not prevent stalling, and the application of two opposing inlet tubes increased losses in stator 1.

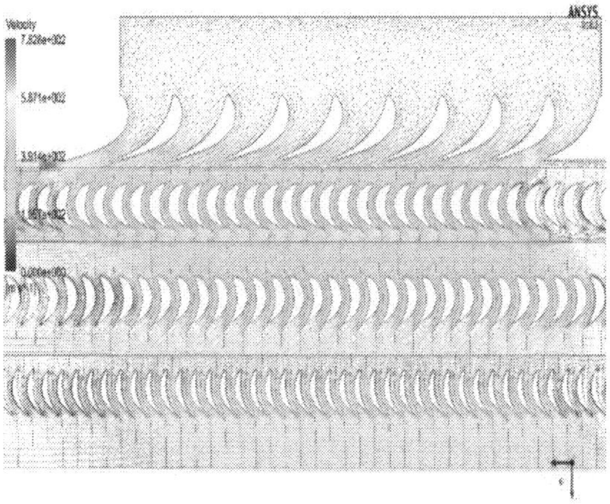

Fig. 9. Velocity field at 0.5 of the stator1 blade length, variant five

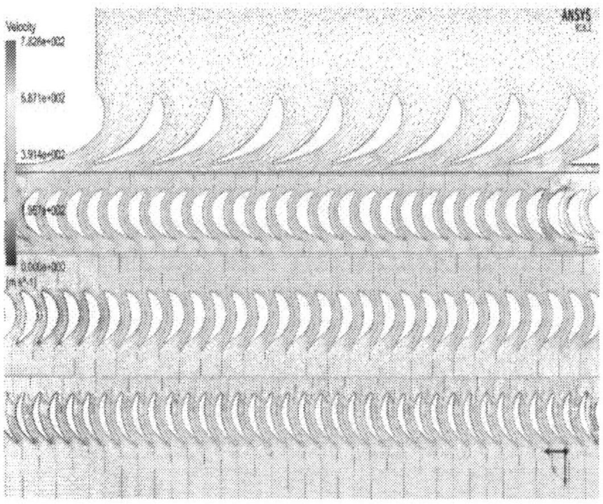

Fig. 10. Velocity field at 0.5 of the stator1 blade length, variant five

In variant five, power generation remained the same as in variant four, 0.958 MW, mass flow increased to 8.66 kg/s (31.18 t/h), but efficiency fell to 68.52 %.

2.6 Variant six

Here, the profiles of all the blades were altered with the exception of those in stator 1 (Tab. 5). These changes were carried out using the polynomial method. The steam entered the turbine the same way as in variants 2, 3 and 4, i.e. without the use of two inlet tubes.

Table 5. Stator and rotor blades in a Curtis stage

	profile	number of blades
First stator	S9012-A	17
First rotor	1818	149
Second stator	3633	234
Second rotor	8731	125

Fig. 11 presents vector fields, where transonic and supersonic flow can be noticed in rotor 1.

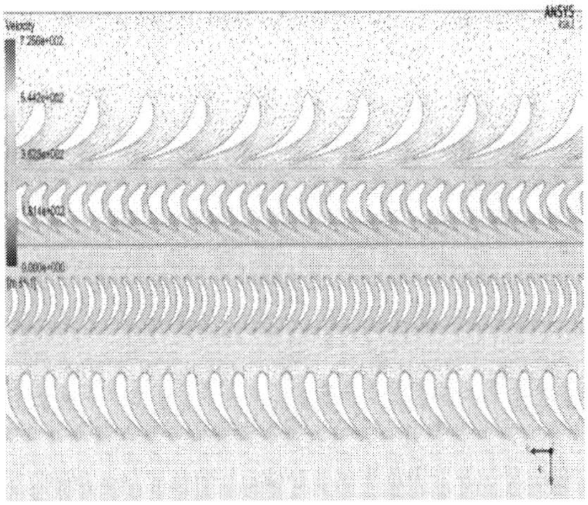

Fig. 11. Velocity field at 0.5 of the stator1 blade length, variant six

In variant six, power generation fell 0.945 MW, mass flow increased to 8.77 kg/s (31.57 t/h) and efficiency fell to 66.74%.

2.7 Variant seven

Here, the profile of the rotor 1 blades was altered, because the shape of the interblade channel in the previous variant resemble the de Laval nozzle (Tab. 6)

Table 6. Stator and rotor blades in a Curtis stage

	profile	number of blades
First stator	S9012-A	17
First rotor	R2117-Bk	141
Second stator	3633	234
Second rotor	8731	125

The altered rotor blade profile prevented transonic and supersonic flow through rotor 1. However, stalling did appear in stator 2. The leading edge of rotor 2 blades could also be improved.

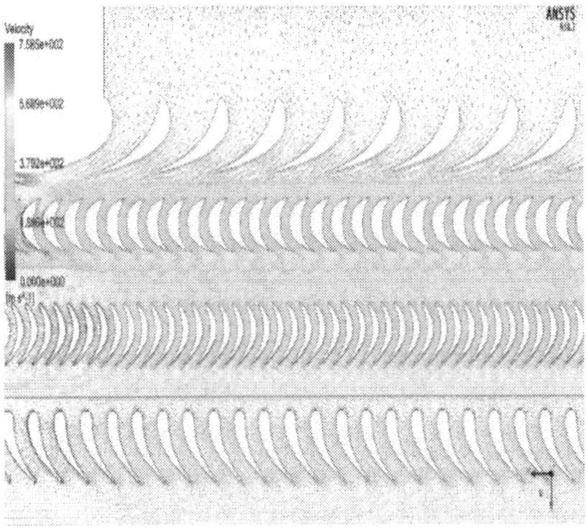

Fig. 12. Velocity field at 0.5 of the stator1 blade length, variant seven

Here, power generation fell to 0.943 MW, mass flow increased to 8.51 kg/s (30.64 t/h), but efficiency increased to 68.75%

2.8 Variant eight

Here, the profile of the rotor 2 blades was modified (Tab. 7).

Table 7. Stator and rotor blades in a Curtis stage

	profile	number of blades
First stator	S9012-A	17
First rotor	R2117-Bk	141
Second stator	3633	234
Second rotor	R5535-A	197

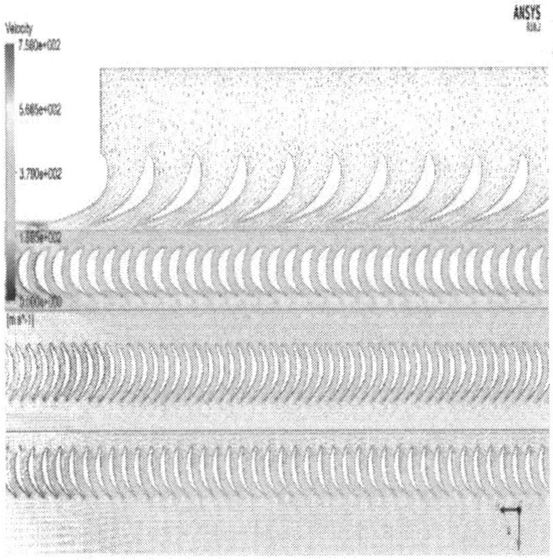

Fig. 13. Velocity field at 0.5 of the stator1 blade length, variant eight

Here, power generation increased to w 0.953 MW, mass flow increased to 8.52 kg/s (30.67 t/h) and efficiency increased to 69.34%.

3 Conclusions

In this paper, for the first time, the geometries of stator and rotor blades in a 1 MW steam turbine Curtis stage were optimized. The most efficient variants for partial admission were variant two, with 1/3 circumference partial admission from one inlet tube and the blade numbers: s1 17, r1 141, s2 85 and r2 197. This gave an efficiency of 72.48 %. This was despite the fact that a standard s2 profile was applied and stalling occurred in r1.

The application of two, opposing inlet tubes deceased efficiency. However, only one such variant was tested and further research would be required. Generally, the efficiency of the Curtis stage is lower than that of typical axial stages.

Optimized flow without transonic and supersonic regions only occurred in variant eight, where half of the blade profiles were modified.

Table 8 presents all the power, mass flow. efficiency and partial admission rate data for all the variants.

Table 8. Power, efficiency, mass flow, partial admission rate of variants one to eight

	Power [MW]	Efficiency [%]	Mass flow [kg/s]	Mass flow [t/h]	Admission
1	1,688	74,82	13,98	50,33	1
2	1,063	72,48	9,09	32,72	1/3
3	0,964	70,94	8,42	30,31	1/3
4	0,958	68,71	8,50	30,60	1/3
5	0,958	68,52	8,66	31,18	2 * 1/6
6	0,945	66,74	8,77	31,57	1/3
7	0,943	68,75	8,51	30,64	1/3
8	0,953	69,34	8,52	30,67	1/3

References

1. Achille M., Cardarelli S., Pantano F., Zito M., Design and CFD Analysis of a Curtis Turbine Stage, *ECOS 2016: Proceedings of the 29th International Conference on Efficiency, Cost, Optimisation, Simulation and Environmental Impact of Energy Systems, June 19. - 23. 2016, Portorož, Slovenia,* (2016)
2. Emmet W. L. R., The Curtis Steam Turbine, *Proceedings of the American Philosophical Society,* Vol. 42, No. 172 (Jan. - Apr.,1903), pp. 68-84, (1903)
3. Rashid S., Tremmel M., Waggott J., Moll R., Curtis Stage Nozzle/Rotor Aerodynamic Interaction and the Effect on Stage Performance, *Journal of Turbomachinery 129 (3),* (2007)
4. Dejcz M.E., Filipov G.A., *Profile Collections of Axial Turbine Cascades*, Maszinostrojenie (1965).

RESEARCH & DEVELOPMENT IN POWER ENGINEERING, 2019

New blade tip-timing system for measuring rotor blade vibration of steam and gas turbines

Piotr Kowaleczko[2] ,Romuald Rządkowski[1,2*], Leszek Kubitz[1], Paweł Troka[1], Paweł Kowaleczko[2], Ryszard Szczepanik[2]

[1] Institute of Fluid-Flow Machinery Polish Academy of Sciences, ul. Józefa Fiszera 14, 80-231 Gdańsk, Poland, z3@imp.gda.pl
[2] Air Force Institute of Technology, ul. Księcia Bolesława 6, 01-494 Warsaw, Poland, piotr.kowaleczko@itwl.pl

Abstract. One of the crucial issues regarding turbine maintenance is registering blade vibrations. These vibrations can cause serious damage to the engine. Turbine blade vibrations were measured during nominal speed as well as during run up and run down. A new, low cost Blade Tip Timing (BTT) is presented in this paper. It composes of two main modules: the FPGA unit and PC unit. The system is based on the TerasIC DE0-CV development board controlled by the Cyclone V 5CEBA4F23C7 chip. Units communicate via an Ethernet interface. The system measures a signal for every revolution as well as up to three signals coming from independent rotor blade sensors. The PC unit records these data in .csv files. The system can be adapted to process the signals of additional sensors. The measurements of the 1st stage compressor blade vibrations in an SO-3 engine prove that the system works correctly, with no data loss during transmission between system units, and compares well with other measurement systems as well as numerical results.

1 Introduction

Blade Tip Timing (BTT) is one of the methods of measuring blade vibrations.

In the industry, each company has its own tip-timing system [1] [2] [3] [4] [5]. These system consist of electronic hardware to measure the time of blade arrival and the signal at every rotor revolution. The LabVIEW was used in [6]. Electronic systems are generally very expensive.

In this paper, for the first time, a very cheap system will be presented. The measurements of this system for rotor blade amplitudes in the 1st stage of an SO-3 engine compare well with the measurements of other systems and numerical results.

2 Blade tip-timing measurement system

The current configuration of the proposed BTT system is the third version of the initial concept, which was originally designed for low-speed steam engine turbines only. In the first version of the system, the FPGA and PC unit communicated via an RS-232 serial connection. As the basic non-functional requirement for the system was its low cost, the FPGA platform based on a TerasIC DE-1 board [7] with a Cyclone II device [8] were selected. System operation was stable and the error rate was at acceptable level. Notwithstanding this, due to the low RS-232 connection speed, data acquisition was only possible for low turbine rotation frequencies (up to 50 Hz) and only one blade sensor. Hence, a second version of the system was developed. The acquired data from four sensors were stored on SD cards. Only after PC processing could this data be read. It was also impossible to save all the data (even from a single turbine sensor), because of the limitations of the SPI mode in SD cards. As the speed test (monitored with an oscilloscope) have shown, data integrity was highly dependent on the SD cards. Multiple data packets were lost because of the prolonged time it took for SD cards to record the data. The error rate was not acceptable even for turbine rotation frequencies lower than 50 Hz. Hence, it was impossible for the second version of the system to meet jet engine blade vibration requirements, such as: the maximum turbine rotation frequency – 250 Hz, the number of turbine blades – 75 and the number of blades' sensors – 3.

The 3rd version of the system was designed to meet the above requirements and also allow for these parameters to be altered. The limitations of the 1st and 2nd versions did not result from FPGA architecture, but the way in which the data were transmitted. Therefore, the 3rd version uses the Ethernet standard for communication between the FPGA and PC unit, supporting data transmission speeds up to 100 Gb/s. The 3rd version of the system composes of two main parts of communicating via UDP protocol: a FPGA-based unit and a receiver desktop application called the PC unit. Both modules have been developed for the first time using commercial off-the-shelf (COTS) hardware components.

2.1 FPGA unit

The measurement component of the system is based on the TerasIC DE0-CV development board [9]. The core of the hardware is the Cyclone V 5CEBA4F23C7 FPGA device [10]. The platform is fed by a 50 MHz clock signal, which is downscaled in the system to 10 MHz – the

 https://doi.org/10.1051/e3sconf/201913701040

resolution used during measurements. The software has been developed using Quartus Prime 18.1.0, with both IP Core functionalities provided by the IDE [11] and with modules manually programmed in Verilog and VHDL. The Quartus project consists of 6 types of software modules, described below.

2.1.1 Time of Arrival measurement modules

The system allows for the measurement of signals from one 'once per revolution' sensor and 3 independent blade sensors. In other words, four Time of Arrival (ToA) measurement modules have been developed in the system. Each of them measures rising edge impulse ToA with 10 MHz precision. The system's internal main clock has 64-bit architecture. This solution allows for measurements to be unaffected by overflow errors (time to first error > 58k years). The minimal turbine revolution frequency in the current version should be greater than 0.03 Hz in order to guarantee error-free ToA acquisition. This applies only if impulses coming from blade sensors are of a clear quadratic shape. This limitation results from system's architecture and it can be easily changed if necessary. The upper frequency limit depends on the number of rotor blades in the turbine and on 10Base-T standard implementation in the Ethernet module. The system was originally designed for monitoring turbines with 75 blades and rotation speeds of between 10 Hz to 250 Hz. Laboratory tests have shown that the system easily delivers error-free ToA values for 75 blades with a rotation frequency of even 500 Hz. The theoretical maximum frequency allowed by the system for 75-blade turbine results from 10Base-T standard's limitations and is equal to ca. 1100 Hz.

2.1.2 Data assembling modules

Quartus Prime project includes two modules responsible for the preparation of acquired data to be sent. ToA values for each of the 4 sensors are uniquely marked (in order to be distinguished at a later stage) and stored in a FIFO buffer (IP Core implementation).

2.1.3 Data sending modules

The system is compatible with the 10Base-T Ethernet standard. Once per revolution, ToA values acquired for four sensors are sent to the PC unit together with an unique revolution number. The UDP protocol has been implemented for the Ethernet connection.

2.1.4 System clock module

In order to guarantee 10 MHz resolution of measurements, a PLL module has been implemented in the project. An IP Core component was used. The base clock signal is provided by a 50 MHz oscillator installed on DE0-CV development board.

2.1.5 User I/O modules

Interaction between the user and the FPGA unit is processed by two modules. These modules monitor start/stop buttons and control board LEDs.

2.1.6 Internal simulator of sensor signals

For the purpose of testing the system after minor or major changes in the architecture (customised for user needs), an internal simulator module has been implemented. The module provides fake sensor impulses of any frequency up to the capabilities of PLL component. One 'once per revolution' and three blade sensor signals can be simulated.

The FPGA unit requires four 3.3V input signals provided to appropriate DE0-CV pins. Ethernet transmission is realized by two dedicated output pins. As the DE0-CV platform provides two 40-pin GPIO ports, the system can be easily adapted to monitor additional blade sensors. The FPGA unit is presented in Fig. 1.

Fig. 1. FPGA unit – TerasIC DE0-CV development board.

2.2 PC unit

Apart from the FPGA data acquisition unit, the system includes a module run on a standard PC. The PC unit is a desktop application developed in C# (.NET) in MS Visual Studio 2017 IDE. It requires a 64-bit MS Windows operating system and a computer with Ethernet RJ-45 interface. The application collects measured ToA values on 1024 port, de-serializes them, checks for transmission errors and saves the values in .csv files. An error is registered when two (or more) consecutive packets are not identified with appropriate rotation numbers (also included in data transmission). Application interface allows the user to start/stop data acquisition and to monitor the number of transmission errors, total measurement time and acquired data size. In order to avoid accidental data loss, it is saved in several .csv files, every 30 seconds of measurement. To improve the readability of data, each sensor has been assigned its own set of .csv files. The PC application window is presented in Fig. 2

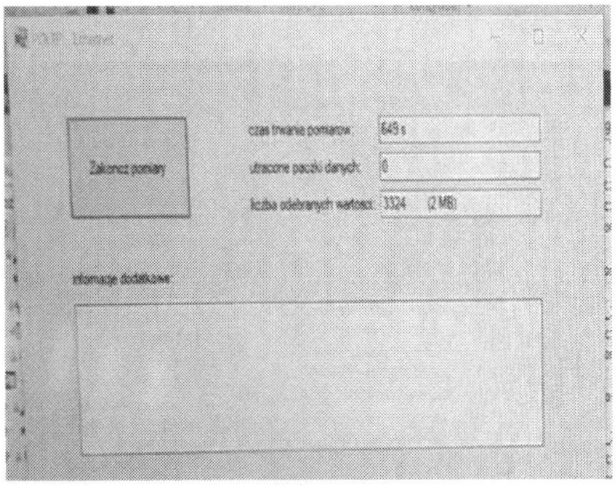

Fig. 2. PC unit – .NET desktop application for receiver PC.

2.3 Measurement system set

The system presented in this paper has been designed to perform measurements for various types of gas and steam turbine. Depending on the configuration of the sensors in turbine, the set of additional equipment for the system can vary. Tests made in a 380 MW steam turbine in Bełchatów required to use optical-to-TTL converters in order to provide 3.3V impulses for system's inputs [6]. During the measurements conducted for an SO-3 jet engine turbine, input impulses were generated by an external converter module. Due to SO-3 test environment specifications (noise, safety), a fibre optic Ethernet connection with dedicated converters between FPGA and PC units was used. The distance between both system units in this case was up to 50 m. Fig. 3 presents the basic system (FPGA and PC unit, PC USB-to-Ethernet adapter, a cable for the FPGA-PC connection and a cable connecting the FPGA-unit with 3.3V input signals coming from sensors).

Fig. 3. Basic system: FPGA and PC units together with cables and adapters.

2.4 Advantages of the system and future development

Requirements for the system have been presented at the beginning of chapter 2. Nevertheless, values for number of turbine blades, number of sensors and maximum frequency should not be treated as unchangeable. The biggest advantage of the system, apart from its now cost, is its flexibility. FPGA-architecture chips have been designed to perform high-speed data processing. Every effort has been undertaken to make the implementation of the system as efficient as possible, for which reason only 3% of the device's ALMs, 1% of block memory bits and 62% of the total number of I/O pins have been used. The summary of the compilation for the system in Quartus Prime is presented in Fig. 4.

Flow Summary	
Flow Status	Successful - Mon Aug 26 14:59:29 2019
Quartus Prime Version	18.1.0 Build 625 09/12/2018 SJ Lite Edition
Revision Name	Pomiary_czasow
Top-level Entity Name	Pomiary_czasow_schemat
Family	Cyclone V
Device	5CEBA4F23C7
Timing Models	Final
Logic utilization (in ALMs)	504 / 18,480 (3 %)
Total registers	761
Total pins	138 / 224 (62 %)
Total virtual pins	0
Total block memory bits	32,768 / 3,153,920 (1 %)
Total DSP Blocks	0 / 66 (0 %)
Total HSSI RX PCSs	0
Total HSSI PMA RX Deserializers	0
Total HSSI TX PCSs	0
Total HSSI PMA TX Serializers	0
Total PLLs	1 / 4 (25 %)
Total DLLs	0 / 4 (0 %)

Fig. 4. Compilation data for system implemented in Cyclone V 5CEBA4F23C7 FPGA device.

The number of 62% I/O pins used is in fact inflated, as only 3% (8 pins in total) are needed for measurement and data transmission tasks. The rest of the pins were used to visualize the state of the system by controlling LEDs, and should be treated as non-essential to the system. The fact that the system uses so little of the resources in the device indicates great potential for future development. What is more, the design of the system was prepared by programming and connecting block modules, making the solution highly scalable. Only minor changes in the system are necessary in order to add processing of signals from additional turbine sensors. All that is required to monitor the ToA of more than 75 turbine blades is an increase in the size of FIFO buffer and changes in the constant values of less than 10 lines of source code of one module block.

Apart from system's flexibility, another (and maybe the most important for the user) feature of the system is its total cost. TerasIC DE0-CV, as COTS hardware is an easily accessible development board and costs about \$150. If the end user uses their own PC, they will only need a few additional items, such as cables, Ethernet adapter (if necessary) and a turbine-specific converter to the TTL 3.3V signal voltage level. Hence, the total cost of the system should be lower than \$200, which is about 50 times cheaper than the price of other comparable systems from vendors like ex. National Instruments. What is more, the system can be implemented without buying any

https://doi.org/10.1051/e3sconf/201913701040

commercial licences for the IDE/software environment like LabVIEW or Matlab.

Future development of the system includes its implementation in the System on Chip (SoC) FPGA device. TerasIC DE1-SoC platform [12] has already been chosen. The core chip of the board comprises two modules: FPGA (Cyclone V) and an ARM-based (Cortex-A9) microcontroller. In addition to the standard functionality provided by FPGA devices, it gives the user the possibility of implementing more complex functions, as the ARM-based module can be controlled by a source code developed in higher-level (than VHDL) programming languages, such as C or C++. The idea is to implement in the FPGA all the modules responsible for ToA measurements, while all components for data transmission are included in the ARM microcontroller. TCP protocol is planned to be implemented. The user will also gain the possibility of remotely controlling the system via the Internet.

3 Measurements results

3.1 Testing the system with simulated impulses

Before testing the system on real engine turbines, three phases of tests were conducted. A brief description of each test is presented below.

3.1.1 Tests with internal signal simulator

As was described in 2.1.6, the design of the system includes an internal simulator module. Fake impulses simulate sensor signals. The data generated during the tests proved that system works correctly. In the internal simulator module blade vibrations do not exist.

3.1.2 Tests with external signal generators

Two external signal generators were used in the experiment. Four signals were generated in total: one 'once per rotation' signal and three independent signals simulating blade sensor impulses. The workbench for the measurements is presented in Fig. 5.

Fig. 5. Testing the system with external signal generators.

The signals were monitored on an oscilloscope. The measurement data was correct, although, as in 3.1.1, also in this case, no ToA deviations could be noted as blade vibrations do not exist.

3.1.3 Tests with turbine simulator

The third stage of system test was based on a turbine simulator developed at the Air Force Institute of Technology (AFIT) in Warsaw. It simulates signals from three sensors: one 'once per revolution' and two blade sensors (Fig. 7). No vibrations were registered, because the blades were simulated as a rectangular metal inserted into a plastic cylinder. The measurements workbench is presented in Fig. 6.

Fig. 6. Testing the system with turbine simulator at AFIT.

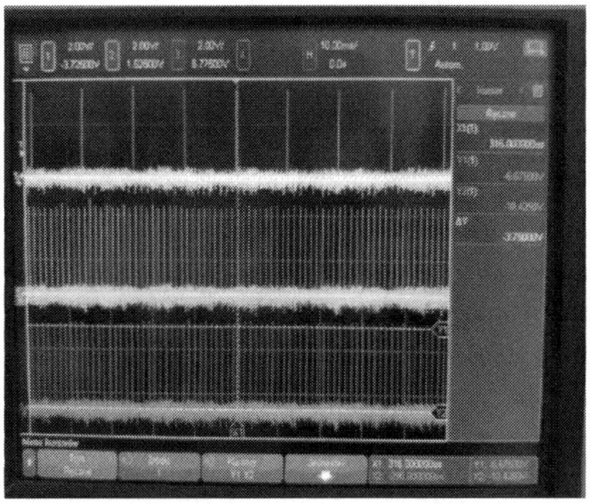

Fig. 7. Signals generated by turbine simulator: 'once per revolution' (yellow) and two blade sensor impulses (green and blue).

The simulator allows the user to change cylinder rotation speed and this feature was used when testing the system. The analysis of ToA data gathered during the test proved that system meets the design requirements.

3.2 SO-3 jet engine

Vibration measurements for an SO-3 jet engine turbine were made after getting the positive results described in 3.1.1, 3.1.2 and 3.1.3. Impulses were generated by 4 sensors in total (one 'once per revolution' and three blade sensors). A dedicated analogue-to-TTL 3.3V converter was used for system inputs. The SO-3 engine measurements workbench is presented in Fig. 9.

https://doi.org/10.1051/e3sconf/201913701040

Vibrations were noted for each sensor. A fragment of a .csv file opened in MS Excel is presented in Fig. 9. The first column is a set of rotation identification numbers, the second (TW) is a set of ToA values for the 'once per rotation' sensor and the remaining columns present the ToA for subsequent turbine blades. It should be noted that the ToA for each successive blade increases.

	A	B	C	D	E	F	G	H
1	nr_obr	TW	TB1_1	TB1_2	TB1_3	TB1_4	TB1_5	TB1_6
2	67720	39992	1203	2627	4047	5478	6901	8333
3	67721	39988	1203	2627	4047	5479	6901	8334
4	67722	39991	1203	2625	4046	5478	6901	8333
5	67723	39985	1203	2626	4046	5478	6901	8332
6	67724	39989	1204	2627	4047	5478	6901	8333
7	67725	39987	1203	2627	4047	5479	6901	8333
8	67726	39991	1204	2627	4047	5479	6901	8333
9	67727	39988	1203	2626	4046	5479	6900	8333

Fig. 8. ToA measurement results in .csv file.

Fig. 9. SO-3 jet engine on stand.

Fig. 10 and Fig. 11 present all rotor blades vibrations. The maximal amplitude of 2.5 mm is in 42 s for rotating stall. The region of stall is from 0 – 42 s (Fig. 10). The second region of maximal amplitude, from (105 – 120 s) with maximum 1.5 mm at 115 s, is connected with 2EO excitation [5].

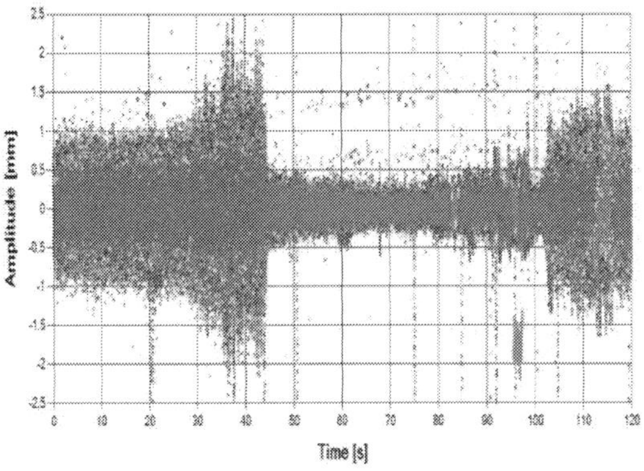

Fig. 10. All rotor blade amplitude of the first stage of SO-3 engine

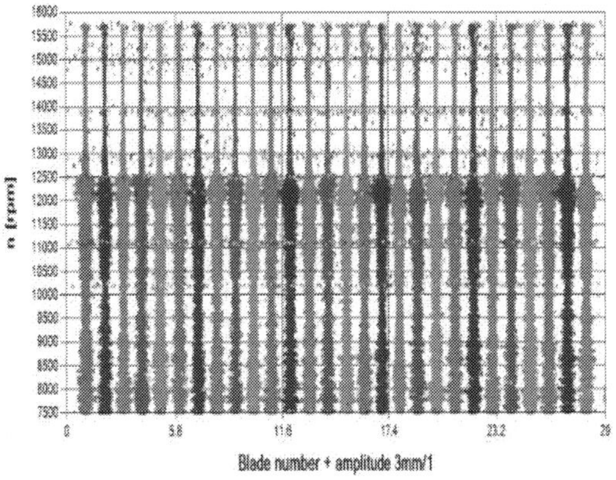

Fig. 11. Separate rotor blade amplitudes of the first stage of SO-3 engine

5 Conclusions

The system described in this paper is effective at 15000 rpm. Its main advantage, a cost of less than $200, sets it apart from other comparable solutions. The system was thoroughly tested before making blade vibration measurements of an SO-3 jet engine compressor. The analysis of the results proved its usability as the vibrations were accurately detected and measured. The architecture of the system is highly scalable and suitable for further developments, such as, using additional sensors, changing rotation speeds, and allowing the user to control the system remotely. No packets were lost while transmitting data from the FPGA module to the PC unit during the tests and final measurements. Measurements of rotor blade amplitude in the first stage of the SO-3 engine using the new system were similar to those of other systems and numerical results.

References

[1] H. S. Dhadwal i A. P. Kurkov, „Simultaneous Optical Measurements of Axial and Tangential Steady-State Blade Deflections," *NASA TM-1999-209051,* (1999).

[2] Y. Zablotskiy, A. Korostelev i L. Sviblow, „Device for Non-Contact Analysis of Vibration Movement of Turbomachinery Blades," *Biuletyn, Otkrytia, izobretenija, promyszlennyje obrazcy, tovarnyje znaki,* Avt. Svid. No 160886, no 5, (1964).

[3] M. Zielinski, „MTU Perspective on Tip-Timing, Panel Session, Tip-Timing: State-of-the-Art and Future Developments," *ASME TURBO-EXPO, Denmark, Copenhagen,* GT2012-70184, June 11-15, (2012).

[4] P. Russhard, „Perspective on Tip-Timing, Panel Session, Tip-Timing: State-of-the-Art and Future Developments," *ASME TURBO-EXPO, Denmark, Copenhagen,* June 11-15, (2012).

[5] R. Szczepanik, Rzadkowski R.Dynamic Experimental Analysis of Dynamic Properties of Aircraft Engine Rotor Blades in Various

Operating Conditions," *PBI Radom*, (2012) in Polish.

[6] R. Przysowa, „Blade Vibration Monitoring in a Low-Pressure Steam Turbine," *ASME TURBO-EXPO: Power for Land, Sea and Air,* Volume 6: Ceramics; Controls, Diagnostics and Instrumentation; Education; Manufacturing Materials and Metallurgy ():V006T05A025, pp. 1-11, (2018).

[7] Altera, DE1 Development and Education Board User Manual, Altera Corporation, (2006).

[8] Altera, Cyclone II Device Handbook, Volume I, Altera Corporation, (2008).

[9] TerasIC, DEo-CV User Manual, Terasic Inc., (2015).

[10] Altera, Cyclone V Device Handbook,, Altera Corporation, (2019).

[11] Intel, „Intel FPGA IP Portfolio," Intel, (2019). [Online]. Available: https://www.intel.com/content/www/us/en/products/programmable/intellectual-property.html. (2019).

[12] TerasIC, DE1-SoC User Manual, TerasIC, (2015).

RESEARCH & DEVELOPMENT IN POWER ENGINEERING, 2019

Analysis of long-term operation of the steam boiler live-steam outlet header and assessment of the feasibility of the header operation during accelerated start-ups

Bohdan Węglowski[1,*], *Marcin* Pilarczyk[1]

[1] Institute of Thermal Power Engineering, Cracow University of Technology, Cracow, Poland

Abstract The energy mix in most European countries is now characterized by a growing share of renewable energy sources (RES). The random nature of RES entails the need to balance the differences between the current amount of power generated in the system and the actual power demand. For this reason, power units incorporating steam boilers are most often operated under sub-peak rather than base load regimes. This creates many challenges to boiler operators due to the advanced age of the power units in service. In Poland, for instance, 77% of power units are more than 20 years old and 45% have been working for more than 30 years, which means that most of them are approaching their design life time. Further operation of these units requires a detailed analysis of the degree of wear of critical pressure components, including a thermal and mechanical stress-state analysis. The paper aims to present a methodology of wear degree calculations incorporated in the power unit control system. The performed calculations are based on real measurement data obtained from 8 months of operation of a boiler unit with the steam output of $650 \cdot 10^3$ kg/h. Wear degree calculations and a thermal and mechanical FEM-based analysis are performed to assess the possibility of shortening start-ups to address the dynamic requirements that power units operating in the Polish power generation system will have to meet.

1. Introduction

If much electricity is to be generated from renewable sources, wind power in the first place, due to the random character of RES-based production, the power system has to be provided with stable energy sources [1, 2, 3]. The effect of the obligation to adopt electricity generation from renewable sources is that a conventional power plant reserve capacity is held as a hot or a cold reserve, which means that boilers are kept in a standby mode, ready for a quick start-up, or they are shut down completely. In Poland, this role is played by centrally dispatched generating units (CDGU©). These are typically small-capacity (200 MW) power units. However, there are more and more plants which, due to their large capacity, have been operating in base-load regimes and now are used for balancing the demand for electricity in the national power system. This is the effect of new highly efficient large-capacity supercritical power plants operating under the base load all year round. As their older power units operate with lower efficiency, the plants find it profitable to use them to satisfy the peak and sub-peak demand because it is possible to sell generated energy at higher intervention prices on the power market [4]. The operation of generating units in the power system involves the need to run start-ups as fast as possible. The other reasons for shortening the boiler start-up time are economic and ecological aspects. Considering that the boiler start-up process is carried out using heavy fuel oil (mazout), efforts are made to quickly achieve parameters that will make it possible to turn off the oil burners and use pulverised coal burners only. Consequently, the consumption of expensive mazout is reduced and start-up boiler's losses are minimized [5]. Another important issue is limiting the emissions of pollutants arising when heavy fuel oil is fired. The pollutants do not pass through the flue gas purification system and get into the atmosphere directly.

Frequent shutdowns and start-ups, as well as sub-peak and peak load operation of boilers involve degradation of their pressure elements, which shortens their life significantly. Issues related to the wear degree assessment are essential in terms of the safety of the boiler current operation and the possibility of extending the boiler safe operation beyond the design life.

2. Calculation of the wear degree of a pressure element

Each power boiler contains thick-walled pressure elements which determine their heating/cooling rates and the rate of load changes. In the case of subcritical steam boilers, the main critical elements are the drum and the live steam outlet header. The element selected for the analyses is the outlet header of a 5th stage superheater of an OP-650 boiler, which is a very popular power unit in Poland. As the header operates in high temperatures, the element is affected by creep wear and, additionally, by low-cycle fatigue. The live steam outlet header is made of steel 10CrMo9-10 grade and is presented in Fig. 1.

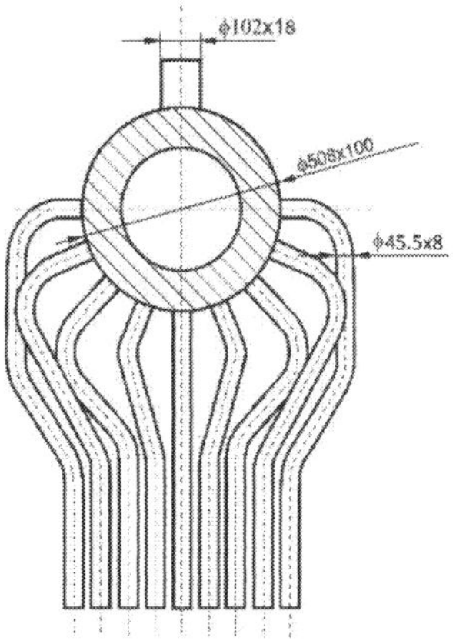

Fig. 1. Cross-section of the outlet header of the final stage (5th stage) superheater of the OP-650 boiler

The header is a horizontal element with the following dimensions: length – 10,400 mm, diameter – 508 mm, wall thickness – 100 mm (cf. Fig. 1).

To demonstrate the method of calculating the wear degree caused by low-cycle fatigue, a long time of the boiler operation of 350,000 minutes (8 months) was selected for the analysis. The operating data obtained from the control system include temperatures measured on the outer surface of the outlet header and values of steam pressure. The data are presented in Fig. 2.

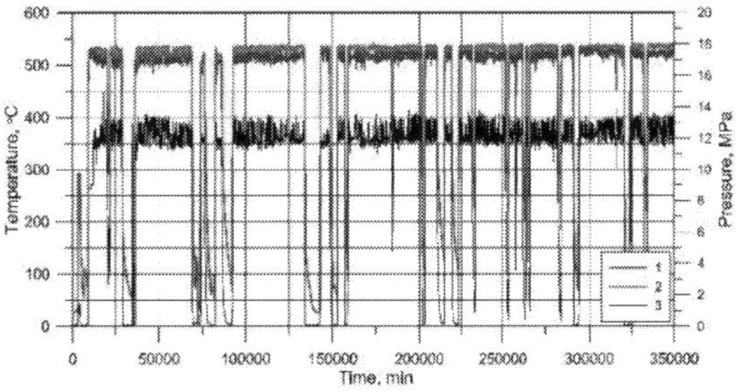

Fig. 2. Changes in the temperature of the live steam outlet header: header top (1); header bottom (2); changes in the header pressure (3)

During the period under analysis the boiler was started up 23 times (11 cold start-ups, 3 warm start-ups and 9 hot start-ups). Temperature was measured in seven points on the outer surface of half of the header perimeter. For better visibility only selected temperature measurements are presented in Fig. 2., i.e. on the header top (1) and bottom (2). To reconstruct the spatial-time temperature field in the cross-section of a pressure element, especially on its internal surface, a inverse heat conduction solution is applied, using measured temperatures on an easily accessible, insulated external surface. Then, it allows to calculate the distribution of thermal stresses and, after taking into account the change in factor pressure, to calculate the total thermal and pressure stresses. [6,7,8]

The data shown in Fig. 2 are used to calculate stresses in the header using an in-house program written in the Fortran environment. Fig. 3 and Fig. 4 illustrate axial and hoop (circumferential) stresses, respectively, on the header top and bottom inner surface for a cross-section not weakened by holes.

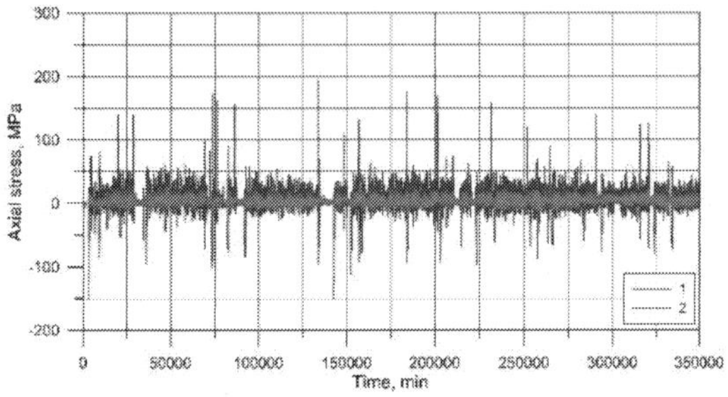

Fig. 3. Changes in axial stresses on the inner surface of the live steam outlet header: header top (1), header bottom (2)

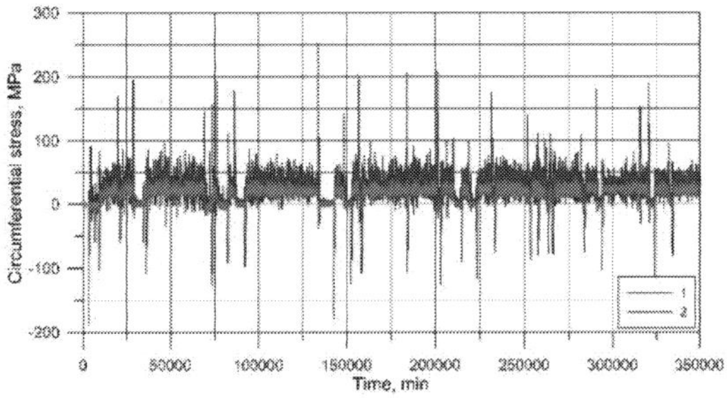

Fig. 4. Changes in hoop stresses on the inner surface of the live steam outlet header: header top (1), header bottom (2)

Fig. 3 and Fig. 4 indicate that axial stresses vary in the range of -/+ 50 MPa, with the highest stress values reaching -150 to 200 MPa. The typical values of hoop stresses are included in the range of -50 to 100 MPa, and the maximum values reach -180 to 250 MPa.

2.1. Wear degree due to low-cycle fatigue

The wear degree related to low-cycle fatigue is calculated using hoop stresses because they are higher compared to axial ones. The outlet header load cycles calculated using the "rainflow method" [9] are shown in Fig. 5 and Fig. 6. Fig. 5 presents time-dependent values of the absolute sum of load cycle amplitudes (2σ) for measured operating data.

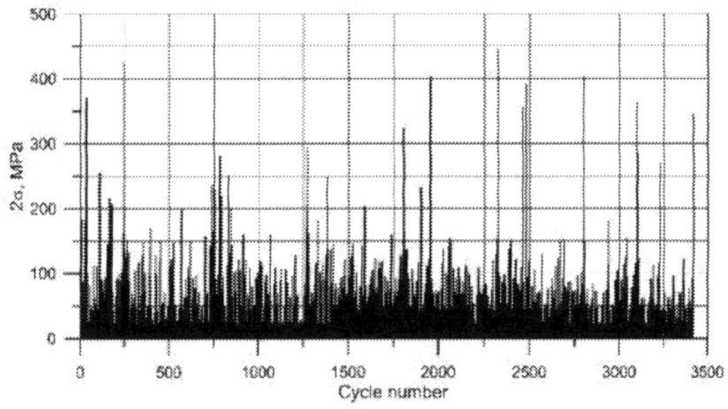

Fig. 5. Load cycles of the live steam superheater outlet header

To enable a fuller picture of the analysed element loading, load count calculations are carried out for cycles of 20 MPa and higher. In practice, cycles with the highest values are taken so that, taking account of the 20% error in low-cycle curves determination, the number of hours preceding the specimen rupture can be read from the chart. If the significant cycles were those with the value of 220 MPa and higher, for the case under analysis there would be 24 such cycles and the calculated wear due to low-cycle fatigue would total 0.002. It can be seen more clearly in Fig. 6, where the cycles are ordered from the biggest to the smallest.

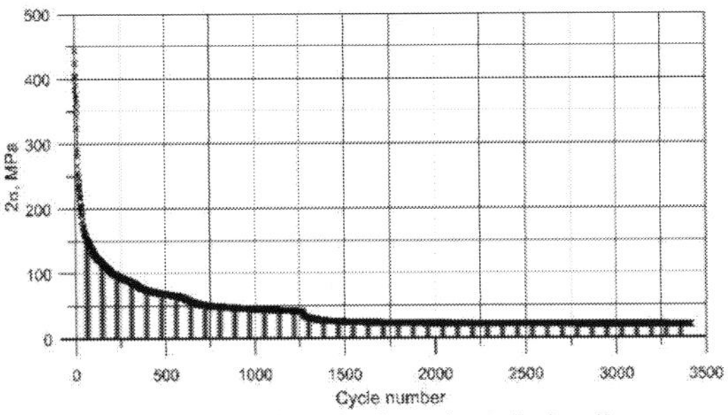

Fig. 6. Biggest-to-smallest load cycles of the live steam superheater outer header

Linear summation hypothesis is used to take account of the variable amplitude: for the amplitude of 2σ the number of cycles N_{dj} is determined at which the element destruction occurs. If the number of cycles with the doubled amplitude 2σ (cf. Fig. 7) is n_j, the element wear degree due to low-cycle fatigue totals:

$$D_z = \sum_{j=1}^{N_z} \frac{n_j}{N_{dj}} \tag{1}$$

Fig. 7 presents characteristics of low-cycle fatigue strength of ferritic steel for the room temperature of 20°C and for the temperatures of 100 to 600°C every 100°C.

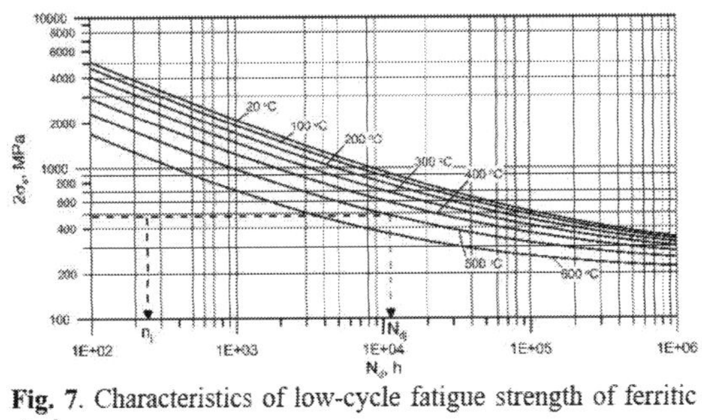

Fig. 7. Characteristics of low-cycle fatigue strength of ferritic steel

Fig. 7 presents an example of determining the number of cycles to destruction N_{dj} at known operating parameters: temperature and load $2\sigma_a$. In these conditions the element is subjected to n_j cycles. Knowing the value of stresses arising during the operation of the power unit elements, it is possible to determine their load cycles and, based on that, to read from the chart or, if the characteristics of wear due to low-cycle fatigue are defined by equations, to calculate the time to destruction for individual cycles and the load share in the loss of life using formula (1).

2.2. Wear degree due to creep

Creep-related fatigue is calculated using the time-to-destruction dependence on loads in operating temperatures higher than the limit point for the steel grade the element is made of. A relevant chart for 10CrMo9-10 steel is shown in Fig. 8.

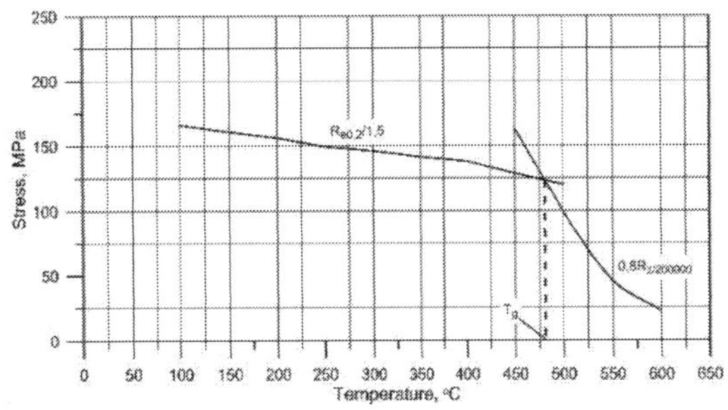

Fig. 8. Determination of limit temperature T_g = 480 °C for 10CrMo9-10 steel, data to find the yield point at elevated temperatures $R^t_{e0.2}$ and creep strength over time $R_{tz/200000}$; developed based on [10]

If creep-related destruction is calculated according to the recommendations of Standard EN-12952-4 [11], in the case of elements with the wall thickness similar to the walls of the analysed component, i.e. 100 mm, using plane stresses calculated in the same manner as for thin-walled vessels, at the design pressure value of 16.2 MPa, gives stress values lower than 50 MPa. Elements operating in temperatures close to 540°C and under loads smaller than 50 MPa can operate without any time limits. This is demonstrated by the chart presented in Fig. 7 illustrating creep strength depending on the time to rupture for 10CrMo9-10 steel.

Considering the real time of a pressure element operation t_k and using the creep strength chart for a given steel grade to find time t_{dk}, after which the specimen is ruptured, the element wear degree caused by creep is determined using the following relation:

$$D_p = \sum_{k=1}^{N_p} \frac{t_k}{t_{dk}} \tag{2}$$

where N_p – number of time intervals.
Utilizing stress monitoring systems, it is possible to determine creep-related wear more precisely using calculated hoop stresses (cf. Fig. 7). For the final steam outlet header of the OP-650 boiler, creep-related wear for the presented time of operation totals 0.0117; the wear degree due to low-cycle fatigue is 0.002.

The design remnant life is determined using the Palmgren-Miner relation:

$$\sum_{j=1}^{N_z}\left(\frac{n}{N_d}\right)_j + \sum_{k=1}^{N_p}\left(\frac{t}{t_d}\right)_k \le D \tag{3}$$

where:
D – total damage due to fatigue and creep, n – number of cycles at j[th] loading conditions, N_d – number of allowable cycles of changes in loads at j[th] loading conditions, t – time of the load duration at k[th] conditions, t_d – allowable time of static load duration at k[th] loading conditions.

According to the Palmgren-Miner hypothesis, the element total wear due to low-cycle fatigue and creep is thus equal to 0.0137. The time of operation in creep conditions, i.e. in temperatures higher than 400°C, is 273,000 hours. For the temperature higher than the limit point T_g = 480°C, the calculations are carried out using momentary creep strength for 10CrMo9-10 steel reduced for safety reasons by 20%; for the temperature lower than T_g – the yield point reduced by the safety factor of 1.5.

If operation was continued in the same conditions as in the period under analysis, i.e. if the temperature level allowable for the steel was not exceeded and, considering the long start-up times, thermal stresses were low, the element could operate for another 37.8 years. Taking account of 0.6t_r (t_r – remnant resistance to creep-related wear) to determine component service life [12], i.e. taking account of the time corresponding to the end of the secondary creep stage and the wear degree due to low-cycle fatigue, the time of further and safe operation is about 22 years. This is about 192,200 hours of operation. Because the estimation concerns the time of safe operation of an element weakened by holes, taking account a safety factor of 30%, the time of safe operation totals 147,850 hours. This time, assuming the boiler appropriate operation and considering such a low degree of wear, is close to the expected value because for new shell-and-hole elements designers set it at 140,000-240,00 hours.

3. ASSESSMENT OF THE FEASIBILITY OF THE HEADER OPERATION DURING ACCELERATED START-UPS

The assessment of the feasibility of the header operation during accelerated start-ups was performed within collaboration between the Cracow University of Technology and a Polish power plant. As no operating characteristics had been provided by the boiler manufacturer, the authors developed reference start-up curves for the cold, warm and hot start-up and their counterparts for the start-up time shortened by 20% and 40%. The developed curves define allowable changes in live steam temperature in the outlet header and in pressure in the boiler drum as a function of time. The start-up curves are presented in Fig. 9. The start-up of approx. 300 minutes (5 hours) is adopted as the reference start-up procedure. Each developed curve takes account of the necessary time of 10 minutes for the steam turbine preparation at the beginning of the procedure. For each proposed start-up curve, a transient thermal and structural FEM-based analysis is performed using the Ansys Mechanical APDL software.

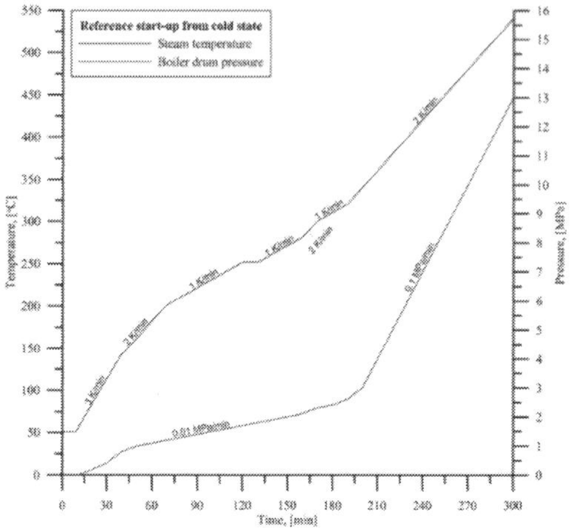

Fig. 9. Cold start-up curves for the OP-650 power unit

The numerical model is developed based on the outlet header technical documentation and the final number of finite cells satisfies the mesh sensitivity analysis. Fig. 10 illustrates the transient temperature distribution within the component during the heating process. Temperature gradients such as the ones observed in Fig. 10 cause high thermal stresses, especially in stress concentration zones on the inner surface of connector pipes

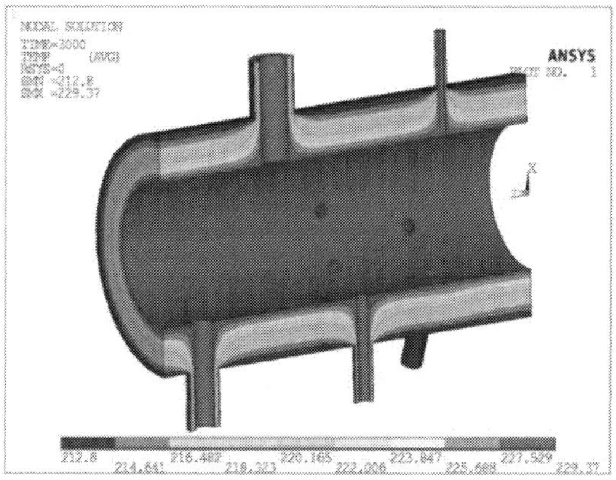

Fig. 10. Transient temperature field within the outlet header during the start-up procedure calculated using the FEM-based ANSYS Mechanical APDL program (heating rate: 3.33 K/min)

The finite element method enables an analysis of total stresses occurring within the header wall. Changes in the hoop (circumferential) and axial stresses on the header inner surface during cold start-up sequences are presented in Fig. 11. Analysing Fig. 11, it can be seen that a decrease in the start-up time causes significantly higher thermal stresses at the beginning of the heating process. At the end of the start-up procedure, the stress values are similar because temperature gradients approach zero and stresses are only induced mechanically by internal pressure. Higher stress values can be observed on the edge of the hole created by the connector pipe. Generally, headers embedded in the steam boiler are equipped with many connector pipes which deliver superheated (or reheated) steam from the superheater subsequent stages.

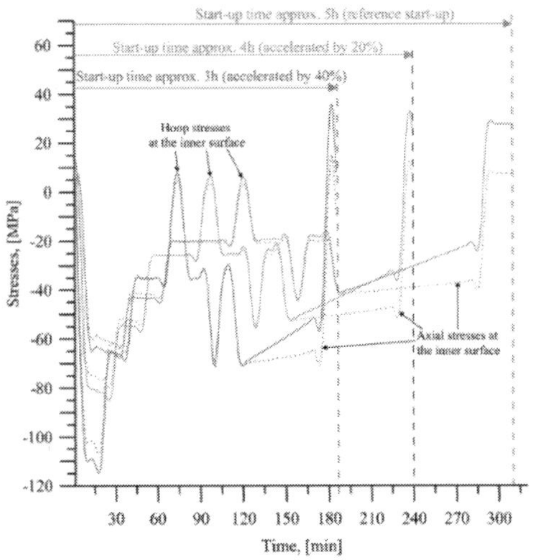

Fig. 11. Changes in hoop and axial stresses on the inner surface not weakened by a connector pipe

In the analysed case, the most common connector pipe has the outer diameter of 45.5 mm and the wall thickness of 8 mm. The changes in hoop and axial stresses determined during the start-up sequences are presented in Fig. 11 and Fig. 12, respectively. Fig. 11 and Fig. 12 indicate that the stress values are significantly higher compared to those on the inner surface, and axial stresses take slightly higher values. Typical stress values vary between -100 and 80 MPa for the reference start-up, and between -130/80 and -170/80 MPa for start-ups shortened by 20% and 40%, respectively.

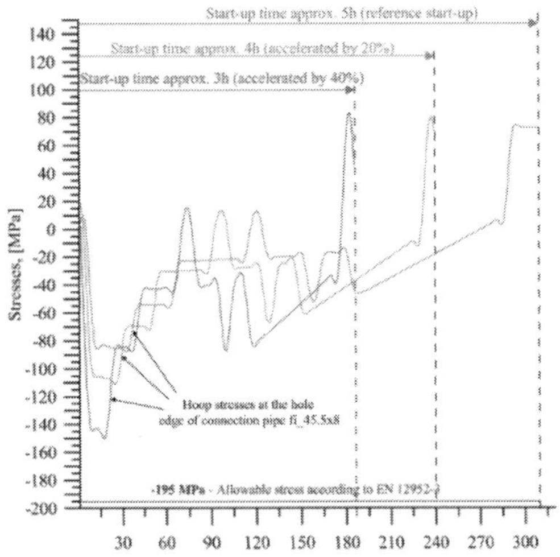

Fig. 12. Changes in hoop stress values on the edge of the hole of a ∅45.5x8 connector pipe according to the proposed start-up curves

None of the values exceed the allowable stress determined for the case under consideration (outlet header & ∅45.5x8 connector pipe) based on the EN standard [13]. The analysis results indicate that there is a margin for shortening the reference start-up sequences, at least from the point of view of thermal stresses. Very similar results were obtained for another critical component, i.e. the boiler drum.

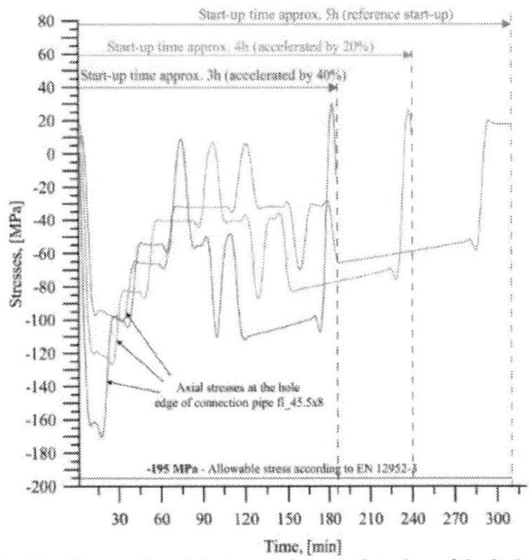

Fig. 13. Changes in axial stress values on the edge of the hole of a ∅45.5x8 connector pipe according to the proposed start-up curves

However, their presentation is beyond the scope of this paper. Although the analysis results can be valuable for boiler operators, the final effect in the form of an improvement in the steam boiler dynamics (i.e. accelerated start-ups) depends on the technical condition of the boiler itself and the auxiliary components (coal mills, air fans, feed pumps, etc.).

4. Conclusions

Assessing the time of further safe operation of elements, it is necessary to take account of their wear degree. It should also be noted that damage to pressure elements is not only caused by creep and low-cycle loads. The thermal shocks that affect horizontal thick-walled pressure elements of power boilers cause not only local cracks in the area of connector pipe holes but also bending in the horizontal axis. The thermal shocks arising during the boiler hot start-up can be an effect of the pumping of the cooler condensate accumulated in superheater coils, bad dewatering or water leaks from the valves of steam attemperators. Power stations and combined heat and power plants must be safe, reliable and efficient, and their operating costs should be as low as possible. It is therefore extremely important that areas which are especially prone to damage should be identified correctly. It is also essential to monitor stresses arising in critical elements during the boiler operation. The issues are topical not only because of the frequent changes in loads and the shutdowns and start-ups of power boilers but also due to the current trend aiming to lower the boiler technical minimum. It is therefore of the utmost importance to ensure a high level of operating safety, a measure of which can be the technical risk related to the operation of a power unit or facility. The presented issues can be of help in the operation of boilers and turbines of power units which are already in service and in the design of new coal-fired supercritical power units.

References

1. Rao, K.R. *Wind Energy for Power Generation, Meeting the Challenge of Practical Implementation* Springer, 2019
2. Kato T. *Prediction of wind power generation output and network operation*, chapter 5 *Integration of Distributed Energy Resources in Power Systems*, Edited by: Funabashi, ScienceDirect, 2016
3. Taler J., Węglowski B., Cebula A., An assessment of Polish power, Modern Power Systems, vol. 27 No. 5, pp. 13-17, May 2007.
4. *European Network of Transmission System Operators for Electricity ENTSOE*, Power in Transition, R&I Roadmap 2017–2026, 2017.
5. Taler, J., Węgłowski, B., Taler, D., Sobota T., Dzierwa, P., Trojan,M., Madejski, P., Pilarczyk M.: *Determination of start-up curves for a boiler with natural circulation based on the analysis of stress distribution in critical pressure components*. Energy, vol. 92, 2015, str. 153-159.
6. Pilarczyk M., Węglowski B., *Determination and validation of transient temperature fields within a cylindrical element using the inverse heat conduction method,* Applied Thermal Engineering 150 (2019) pp.1224-1232
7. Węglowski B., Pilarczyk M., *Experimental and numerical verification of transient spatial temperature distribution in thick-walled pressure components*, Journal of Mechanical Science and Technology 32 (3) (2018) pp. 1087-1098
8. Taler J., Dzierwa P., Jaremkiewicz M., Taler D., Kaczmarski K., Trojan M., Węglowski B., Sobota T., *Monitoring of transient 3D temperature distribution and thermal stress in pressure elements based on the wall temperature measurement*, Journal of Thermal Stresses, 2019, Vol. 42, No. 6, 698–724
9. Murakami Y., *The Rainflow Method in Fatigue*, Butterworth-Heinemann, 1992.
10. EN 10216-2, *Seamless steel tubes for pressure purposes – Technical delivery conditions* – Part 2: *Non-alloy and alloy steel tubes with specified elevated temperature properties*.
11. EN 12952-4, *Water-tube boilers and auxiliary installations* – Part 4: *In-service boiler life expectancy calculations*.
12. Dobrzański J., *Materials science interpretation of the life of steels for power plants*, Open Access Library, Volume 3, 2011.
13. EN 12952-3 : 2011 *Water-Tube Boilers and Auxiliary Installations* – PART 3: *Design and Calculation for Pressure Parts of the Boiler*

RESEARCH & DEVELOPMENT IN POWER ENGINEERING, 2019

Sizing of biomass-fired ORC cogeneration unit within coal-fired municipal heating plant back fitting project

Jacek Kalina[1,*], *Mateusz* Świerzewski[1]

[1] Silesian University of Technology, Akademicka 2A, 44-100 Gliwice, jacek.kalina@polsl.pl, Poland

Abstract. The problem discussed in this paper is optimal sizing of biomass-fired ORC cogeneration units into existing coal-fired district heating plants under given site-specific technical, economic and ecological constraints. In this paper the municipal heating plant in Krosno (Poland) is taken into account as the reference case. Basing on the operational experiences from this unit an optimisation study has been performed in order to examine the influence of current economic and legal conditions on the optimal design characteristics of the plant. Different electricity, biomass and coal prices are taken into account as well as the influence of the EUA (European Emission Allowance) price is examined. There are taken into account thermal energy storage and sale of electricity on balancing market. It has been found that in the studied case the implementation of hot water storage tank moves the optimal electric power output slightly towards higher values. On the other hand only a small improvement of financial performance has been gained. The results reveal importance of the optimisation of design parameters as well as the dependence of the plant's size and structure on local economic conditions.

1 Introduction

A possible way for deployment of Renewable Energy Sources (RES) technologies into the energy market is modernisation and retrofitting of existing fossil-fuel-fired production facilities [1]. In the case of biomass, there have been usually initiated cogeneration projects in locations where a heat market had existed. Therefore, most of biomass-fired ORC systems have been installed in district heating systems [2]. The biomass-fired ORC modules are usually being added to existing fossil-fuel-fired heating plants within backfitting projects focused on using locally available feedstock, making profits out of electricity generation and CO_2 emission reduction. Such projects result in change of a given heating plant into a dual-fuel cogeneration plant (CHP). The critical issue of a project of this type is optimal selection of the biomass-fired block size and its integration with the existing heat only boilers. The design task also includes elaboration of operational strategy and design of control algorithms as the system will run under variable load and price conditions. Current experience shows that cogeneration modules have been traditionally sized for either base or medium heating load of district heating network [3, 4]. However, for best effects the optimisation task must be properly formulated under site-specific heating load profile, economic (including financial support mechanism) and legal conditions [5], including different strategies to sell electric energy generated in cogeneration.

There have been presented in the literature several studies on optimisation of ORC cogeneration systems [4 - 9]. The critical problem of such task is selection of proper objective function and constraints. In industrial practice optimization is usually carried out using an economic objective function, expressed by local financial effect [10-11].

In this work, there is presented simulation based optimisation of ORC system size and technological structure within backfitting projects of district heating plants. The optimisation task has been solved by searching the area of allowable solutions and therefore impact of key design variables on objective function has been depicted. The proposed methodology has been based on operational experiences of a real case. The municipal heating plant in Krosno (Poland) has been taken into account as the reference system. In 2013 there was commissioned biomass biomass-fired cogeneration system with the Turboden T14 CHP SPLIT unit of 1255 kW net electric power. The optimization study has been performed in order to examine the impact of current economic and legal conditions on the optimal design characteristics of the ORC system. The main decision variables have been power output of the ORC unit, presence of thermal energy storage tank (TES) and its volume. Different electricity, biomass and coal prices are taken into account as well as the influence of the EUA (European Emission Allowance) price is examined. The results reveal importance of the optimisation of design parameters as well as the dependence of the plant's size and structure on local economic conditions.

2 Reference system

In the city of Krosno the biomass-fired CHP block has been added to existing coal-fired heating plant. At present, the technological system consists of two subsystems, namely the coal-fired heat-only boiler (HOB) subsystem and the biomass-fired CHP

 https://doi.org/10.1051/e3sconf/201913701042

subsystem. Schematic diagram of the integrated plant has been depicted in Fig. 1. Duration curves of heating load, network water temperature and flow have been shown in Fig. 2 and 3. The peak heating load of the system is around 30 MW and the maximum grid forward water temperature is around 125°C.

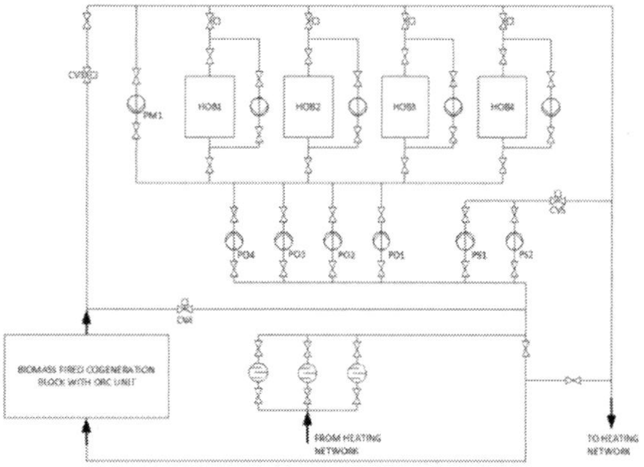

Fig. 1. Schematic process diagram of the CHP plant in Krosno

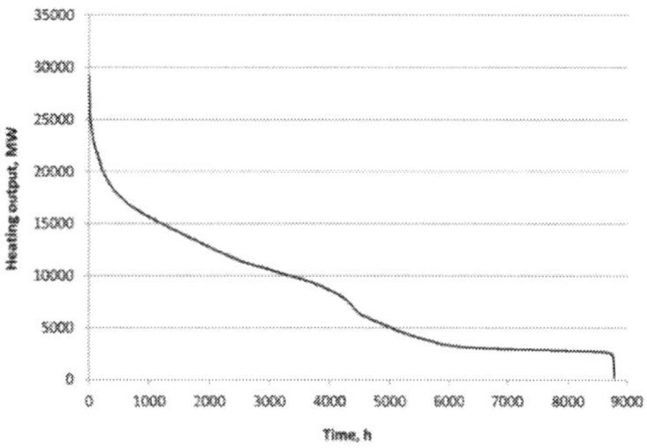

Fig. 2. Heat load duration curve

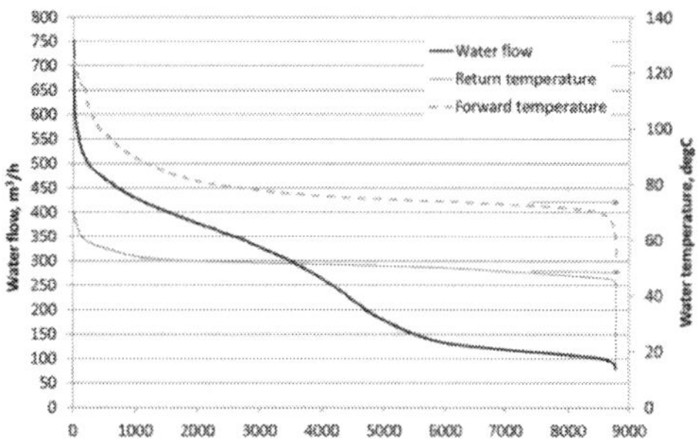

Fig. 3. Water temperature and flow duration curves

Within the heat only block there are 4 WR type forced circulation coal-fired water-tube boilers with mechanical grate (HOB1-HOB4). This is one of the most popular boiler constructions in the Polish heating sector. The nominal heating output of the WR10 boiler is 10 MWth and of the WR4.8 is 4.8 MWth. Design thermal efficiency of each boiler is at the level of 83% (ratio of heat output to fuel LHV chemical energy input). The fuel is hard coal of the lower heating value (average weighted): LHV = 23.2 MJ/kg.

Technological structure of the biomass-fired ORC block in Krosno has been depicted in Fig. 4.

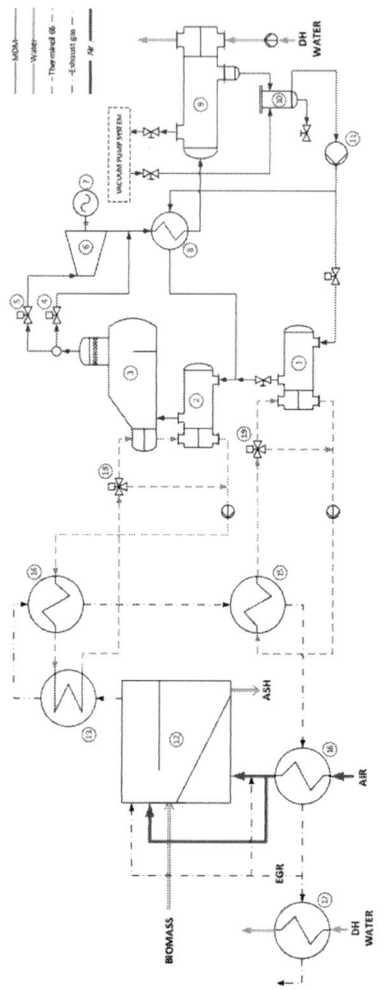

Fig. 4. Process diagram of biomass-fired ORC reference block (legend: 1 – SPLIT heat exchanger/MDM preheater; 2 – MDM heater; 3 – MDM evaporator; 4 turbine by-pass valve; 5 – turbine inlet valve; 6 – turbine; 7 – electric generator; 8 – working fluid regenerative heat exchanger; 9 – MDM condenser/DH network water heater; 10 – filter; 11 – pump; 12 – biomass combustion furnace; 13 – spiral type thermal oil heater; 14 – high temperature economiser; 15 - low temperature economiser; 16 – recuperator/combustion air heater; 17 – exhaust gas latent heat recovery unit (condensation) 18,19 – thermal oil three-way admission valves; EGR – exhaust has recirculation)

The ORC unit was sized for the base heating load of the district heating network. The rated thermal output of the condenser is 5341 kW and the maximum output is 5350 kW. The ORC module' generator has the rated electrical power output of 1400 kW. The rated power of the biomass-fired furnace is 8529 kW. The furnace is adjustable in the power range of 30-100%. The rated heat output of the flue gas-thermal oil heat exchangers is: 6715 kW. The energy efficiency of the boiler declared by the manufacturer is 85-88%, which has been confirmed by measurements. The fuel delivered into the biomass boiler has variable composition and water content of up to 55%. Its lower heating value is between 8 – 12 MJ/kg. The quality of fuel varies both seasonally and daily. Thermal oil system consists of two loops: high temperature HT (310/250°C) and low temperature LT (250/130°C). The thermal oil transfers heat to the evaporator and the working fluid preheater of the Turboden 14 CHP SPLIT ORC module. The working fluid within the ORC is MDM (Octamethyltrisiloxane). After expansion in the ORC turbine the working fluid goes through the regenerative heat exchanger to the condenser 9 where heat is transferred to the heating network. Operational experiences of this system have been discussed in [12].

3 Optimisation problem

The problem in this work is to optimally design new biomass-fired ORC cogeneration block for existing coal-fired heating plant. The plant is located in Poland where the system of incentives has been established for cogeneration and renewable energy projects in the form of market oriented Quota Obligation Systems [1]. The ORC unit will be installed at a plant of the fuel chemical energy input higher than 20 MW, thus being subject to European CO_2 Trading System (ETS). Therefore, another financial support instrument for the investment project is the trade of European Emission Allowances (EUA). Moreover, the electricity is partly used to cover own demand of the company and partly traded on the balancing market at variable prices. The project does not affect the amount and the parameters of the generated network heat.

There have to be decided values of key decision variables to achieve the best value of a selected quality indicator (objective function) under given constraints. These are: load profiles, feedstock characteristics, parameters of district heating water, energy and substance balances, equations of thermodynamic processes, equations of state, stoichiometry of combustion reactions, heat transfer principles, process dynamics, characteristics of equipment, properties and limitations of substances (working fluid, heat transfer oil, combustion gasses), etc. Eventually, the task is:

$$\min J\,(x_1, x_2, \dots, x_n) \qquad (1)$$

s.t.

$$G_i(x_1, x_2, \dots, x_n) = 0;\ i=1,2,..k \qquad (2)$$

$$H_j(x_1, x_2, \dots, x_n) > 0;\ j=1,2,..m \qquad (3)$$

where x_i are independent decision variables.

Traditional economic indices of investment projects, i.e. Net Present Value (NPV) and Internal Rate of Return (IRR) within given time horizon have been assumed to be alternative objective functions:

$$J = NPV \text{ or } J = IRR \qquad (4)$$

$$NPV = \sum_{t=1}^{N} \frac{\Delta CF_t}{(1+r)^t} - CAPEX \qquad (5)$$

$$\sum_{t=1}^{N} \frac{\Delta CF_t}{(1+IRR)^t} - CAPEX = 0 \qquad (6)$$

Where: CAPEX- capital expenditures [PLN]; CF – cash flow [PLN]; N – project duration (years); r- discounted cash flow rate; t – year number,

Although many system parameters need to be selected at the design stage [6-9] in this work the number of independent decision variables has been limited to the nominal power of the ORC generator and the volume of TES. The search limits have been set within the range of 200 – 5000 kW and 200 – 5000 m^3 for TES volume respectively. Instead of detailed modelling of the thermodynamic cycle the nominal electrical efficiency of the ORC unit has been determined using market data for biomass-fired ORC cogeneration units with MDM as working fluid and 60/80°C cooling water temperature [15]:

$$\eta_{el,nom} = 0.0122 \ln \dot{P}_{el,gen,nom} + 0.1004 \qquad (7)$$

Total capital expenditures (CAPEX) for biomass-fired ORC CHP systems have been assessed using data published by IFC [14]. These data have been fitted with the curve (in EUR/m^3):

$$CAPEX_{CHP} = 5584{,}4\left(\dot{P}_{el,gen,nom}\right)^{-0{,}266} \qquad (8)$$

In the case of TES the investment cost equation has been elaborated using vendors data (in PLN/m^3):

$$CAPEX_{TES} = -1409 \ln V_{TES} + 14797 \qquad (9)$$

Although accuracy of this initial model is limited it represents design nonlinear trend curves for ORC systems. The methodology is typical for prescreening of possible solutions and identification of potential location of the optimum. In the region of optimum an additional sensitivity analysis would be required in a real business case.

The most important component of the objective function is the differential annual cash flow ΔCF_t within the operational phase, which results from comparison of cash flows after and before the cogeneration project. It takes into account the incomes resulting from the support mechanisms, thus in Polish conditions it is:

$$\Delta CF_{t=1\div N} = \int_{\tau_0}^{\tau_a} \left[\dot{E}_{el,exp} p_{el,exp}(\tau) + \dot{E}_{el,self} sc_{el}(\tau) + \dot{E}_{el,gen} \sum_{sup} f_{\sup}(\tau) - \dot{m}_b sc_b - \Delta\dot{m}_c sc_c\right] d\tau - \Delta m_{env,t} sc_{env,t} - \Delta T_t \qquad (10)$$

where: $\dot{E}_{el,exp}$ - electricity exported [kW]; $p_{el,exp}(\tau)$ – electricity price at time [PLN/kWh]; $\dot{E}_{el,self}$ – self consumption of electricity [kW]; $sc_{el}(\tau)$ – specific cost of electricity at time τ [PLN/kWh]; $\dot{E}_{el,gen}$ – generator power output [kW]; $\sum_{sup} f_{\sup}(\tau)$ – total income from support instruments [PLN/kWh]; $\dot{m}_b$ – biomass flow [kg/s], sc_b – specific cost of biomass [PLN/kg]; $\Delta\dot{m}_c$ – reduction of coal consumption [kg/s]; sc_c – specific cost of coal [PLN/kg]; $\Delta m_{env,t}$ – environmental impact in year t [kg]; $sc_{env,t}$ – specific costs due to use of environment [PLN/kg]; ΔT_t – change of tax in year t

$$\Delta m_{env,t} sc_{env} = \Delta m_{f,i} \zeta_{env,j,i} sc_{env,j} + \Delta m_{f,i} \zeta_{CO2} p_{EUA}(\tau) + \Delta m_a sc_a \quad (11)$$

Where: $\Delta m_{f,i}$ – change of i-th fuel consumption [kg], $\zeta_{env,j,i}$ – specific emission index of pollutant j from fuel i [kg/kg]; $sc_{env,j}$ – specific cost for pollutant j [PLN/kg]; ζ_{CO2} – emission index for CO_2 [kg/kg], $p_{EUA}(\tau)$ – specific prices of EUA at time τ [PLN/kg].

Equation (10) takes into account only key income and cost components. Changes of other financial items after the project, such cost of materials, cost of labour etc. have been assumed to be negligible.

The main equality constraint G_i is the overall heat balance of the district heating plant:

$$\sum_k \dot{Q}_{d,k} \pm \dot{Q}_{TES} = \dot{Q}_{DHS} + \dot{Q}_{plant} \quad (12)$$

Where: $\dot{Q}_{d,k}$ – heat flux delivered by device k [kW], $\dot{Q}_{TES}$ – heat flux to/from storage [kW]; $\dot{Q}_{DHS}$ – heat delivered to the district heating grid, [kW]; $\dot{Q}_{plant}$ – heat consumed at the plant [kW]

It has been assumed that in order to minimize heat losses from TES the initial and final states of charge are the same. This assumption gives the following constraint:

$$\Delta Q_{TES}|_0^{24} = \sum_t \left(Q_{TES} - Q_{i,TES}\right)_t = 0 \quad (13)$$

The state of charge has been calculated assuming no heat storage in connecting pipes and no losses:

$$\frac{\delta(Q_{TES})}{\delta\tau} = \pm\dot{Q}_{TES} \quad (14)$$

Sales of electricity on balancing market result from momentary balance of electric power:

$$\dot{E}_{el,exp} = \dot{E}_{el,gen}(1-\alpha) - \dot{E}_{el,plant} - \sum_{cp} \dot{E}_{el,cp} \quad (15)$$

Where $\dot{E}_{el,cp}$ – self consumption of electricity in other consumption points that the plant itself.

Fuel consumption is calculated using efficiency characteristics:

$$\dot{m}_c = \frac{\dot{Q}_{HOB}}{\eta_{HOB} LHV_c} \quad (16)$$

$$\dot{m}_b = \frac{\dot{Q}_{ORC}}{(\sigma\eta_{el}\eta_{TO}\eta_{BB}) LHV_b} \quad (17)$$

Nominal energy efficiency of coal boilers has been determined at $\eta_{HOB,nom} = 0.835$. Nominal efficiency of biomass boiler has been assessed as $\eta_{BB,nom} = 0.854$.

As the annual financial effect of system operation depends on part load characteristics of equipment relevant curves have been developed. Part load energy efficiency curve of coal boilers have been adopted from design documentation of the WR type boilers. It takes the form:

$$\frac{\eta_{HOB}(\tau)}{\eta_{HOB,nom}} = a\left(\frac{\dot{Q}_{HOB}(\tau)}{\dot{Q}_{HOB,nom}}\right)^2 + b\left(\frac{\dot{Q}_{HOB}(\tau)}{\dot{Q}_{HOB,nom}}\right) \quad (18)$$

Where a=-1.1056 and b=2.1056. Part load energy efficiency characteristics of biomass boiler has been developed using long term historical measurement data of the Krosno system in real operational conditions, which are available within the archive of the plant's SCADA. It has been applied in the normalised form:

$$\frac{\eta_{BB}(\tau)}{\eta_{BB,nom}} = a\left(\frac{\dot{Q}_{ORC}(\tau)}{\dot{Q}_{ORC,nom}}\right)^2 + b\left(\frac{\dot{Q}_{ORC}(\tau)}{\dot{Q}_{ORC,nom}}\right) + c \quad (19)$$

Where a=-0.1142, b=0.1315 and c=0.9828. Performance characteristics of the ORC unit have been also developed and presented in [13]. These parameters have been normalised and applied in dimensionless form. In the first step, for an assumed nominal heating output of the ORC unit $\dot{Q}_{ORC,nom}$, momentary heating load $\dot{Q}_{ORC}(\tau)$ and water outlet temperature t_{w2} the equivalent heating output of the T14 ORC unit is calculated:

$$\dot{Q}_{eqv} = \left(\frac{\dot{Q}_{ORC}(\tau)}{\dot{Q}_{ORC,nom}}\right) \cdot \dot{Q}_{T14,nom} \quad (20)$$

Then the ORC's generator momentary power output is calculated using formula:

$$\dot{E}_{el}(\tau) = \frac{\dot{E}_{el,nom}}{\dot{E}_{T14,nom}}\left[a + b\left(\frac{t_{oil}(\tau)}{t_{w2}(\tau)}\right) + c\dot{Q}_{eqv} + d\left(\frac{t_{oil}(\tau)}{t_{w2}(\tau)}\right)^2 + e\left(\frac{t_{oil}(\tau)}{t_{w2}(\tau)}\right)\dot{Q}_{eqv} + f\left(\dot{Q}_{eqv}\right)^2\right] \quad (21)$$

Where: a=-183.3; b=87.49; c=0.3291; d=14.21; e=-0.03988; f=4.998e-06. Electrical efficiency of the ORC unit is:

$$\eta_{el}(\tau) = \frac{\eta_{el,nom}}{\eta_{T14,nom}}\left[a + b\left(\frac{t_{oil}(\tau)}{t_{w2}(\tau)}\right) + c\left(\frac{t_{oil}(\tau)}{t_{w2}(\tau)}\right)^2 + d\dot{Q}_{eqv}\right] \quad (22)$$

Where: a=0.02492; b=0.0839; c=0.3291; d=2.88e-06 [13]. Therminol 66 heat transfer oil temperature at the ORC evaporator inlet is:

$$t_{oil}(\tau) = a + bt_{w2,i} + c\dot{Q}_{eqv} + dt_{w2}(\tau)\dot{Q}_{eqv} + e\dot{Q}_{eqv} \quad (23)$$

where: a=155.7; b=0.8288; c=0.023; d=-0.81e-05; e=1.95e-07 (Kalina, Świerzewski, 2019).

Inequality constraints result from allowable load range of each heating device k:

$$\dot{Q}_{k,min} \le \dot{Q}_k \le \dot{Q}_{k,max} \quad (24)$$

According to measurements, minimum load of the CHP unit has been assumed $0.3\dot{Q}_{ORC,nom}$ whereas minimum load of coal boilers is $0.2\dot{Q}_{HOB,nom}$ and the maximum is $1.2\dot{Q}_{HOB,nom}$.

The inequality constraint of TES charging and discharging processes is the charging/discharging rate, which results from the water flow rate in connecting pipelines:

$$\dot{Q}_{TES,min} \le \dot{Q}_{TES} \le \dot{Q}_{TES,max} \quad (25)$$

$$w_{min} \le w \le w_{max} \quad (26)$$

In addition the state of charge should not be higher than the maximum one, which results from the given volume of the storage tank:

$$Q_{TES} \leq Q_{TES,max} = V_{max}c_{p,w} \cdot \Delta\bar{T}_w \cdot \rho_w \quad (27)$$

The annual time of CHP operation results from its availability. Basing on experiences of the Krosno system, the maximum availability has been assumed at 95%, what gives 8322 hours per year of the ORC unit operation. There have been distinguished two modes of operation: summer and winter [13]. In the case with TES, the winter mode is realised in periods when daily average DHS heating load is in the range:

$$\overline{\dot{Q}_{DHS}} \geq \dot{Q}_{ORC,max} \cup \overline{\dot{Q}_{DHS}} < \dot{Q}_{ORC,min} \quad (28)$$

In this mode the TES loading and unloading is controlled to minimize coal consumption:

$$m_c LHV_c = \int_{\tau=0}^{\tau=24} \left(\frac{\dot{Q}_{HOB}}{\eta_{HOB}}\right) d\tau \rightarrow \min \quad (29)$$

The summer mode is realised in periods when daily average DHS heating output is in the range:

$$\overline{\dot{Q}_{DHS}} < \dot{Q}_{ORC,max} \cap \overline{\dot{Q}_{DHS}} \geq \dot{Q}_{ORC,min} \quad (30)$$

Then the local objective function is income from the sale of electricity on balancing market:

$$\int_{\tau=0}^{\tau=24} \dot{E}_{el,exp} p_{el,exp}(\tau) d\tau \rightarrow \max \quad (31)$$

The task can be classified as a Mixed Integer Non Linear Programming (MINLP) problem, in which the constrains vary with time and the set of decision variables may be discontinuous. There can also occur local extremes, which have been identified by searching the area of trial solutions.

4 Results and discussion

In order to examine the influence of design assumptions on financial performance of projects the optimisation task has been solved for Krosno DHS by searching the area of acceptable solutions. For a given trial size of the ORC system the simulation of annual operation was performed. Then the results of annual mass and energy balance have been used for financial calculation within 15 years of project economic lifetime. Prices have been valid for Polish conditions at the beginning of the year 2019. Coal price is 14.50 PLN/GJ, biomass price is 18.22 PLN/GJ, EUA price is 103.20 PLN/Mg, price of certificate of electricity origin from RES is 130.0 PLN/MWh, electricity price for own consumption is 370 PLN/MWh, discounting cash flow rate is r = 0.05, possible subsidies for investment project are at the level of 0.3CAPEX.

Figures 5 and 6 depict operation of ORC units of different size and HOBs within the DHS in the case there is no TES. The annual operating time of larger ORC systems is shorter due to the fact that their minimum permissible heating load is higher than the load of the grid.

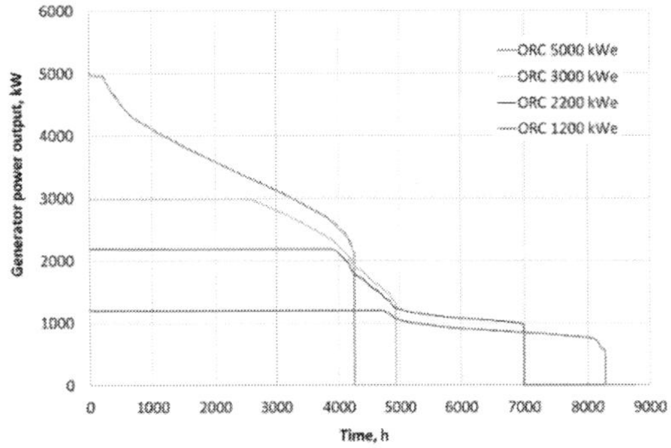

Fig. 5. ORC power duration curve

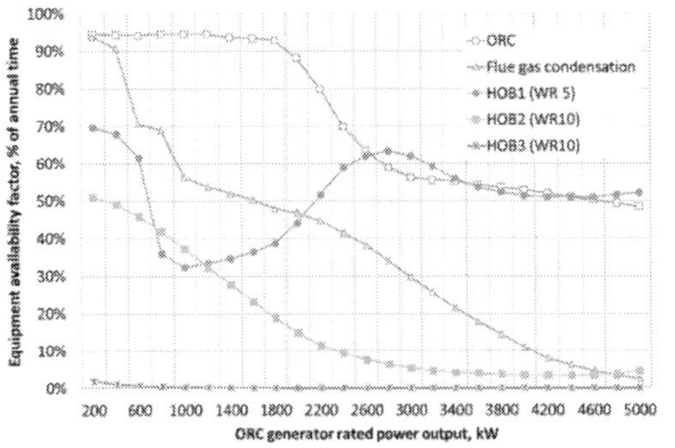

Fig. 6. Annual time of equipment operation

NPV and IRR of the project under different assumptions on financing have been depicted in Fig. 7 to 10 for the system without and with TES respectively. Adding TES to the system moves optimum size of the ORC unit towards larger values. However, under current economic conditions the TES only slightly changes values of objective functions NPV and IRR.

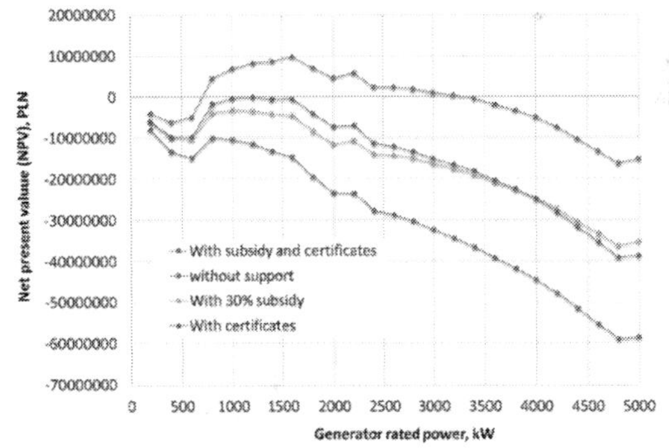

Fig. 7. NPV under different financing options (system without TES)

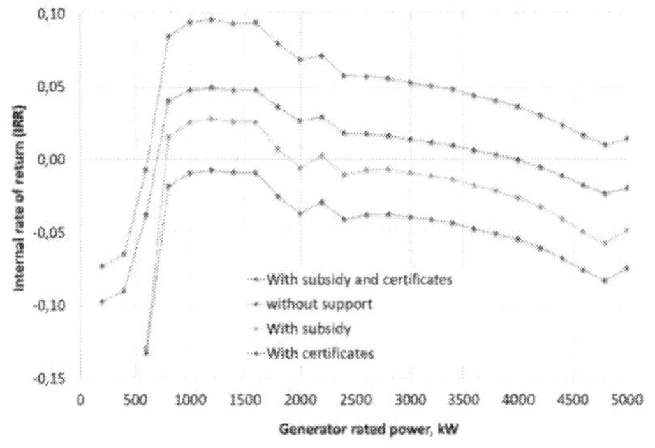

Fig. 8. IRR under different financing options (system without TES)

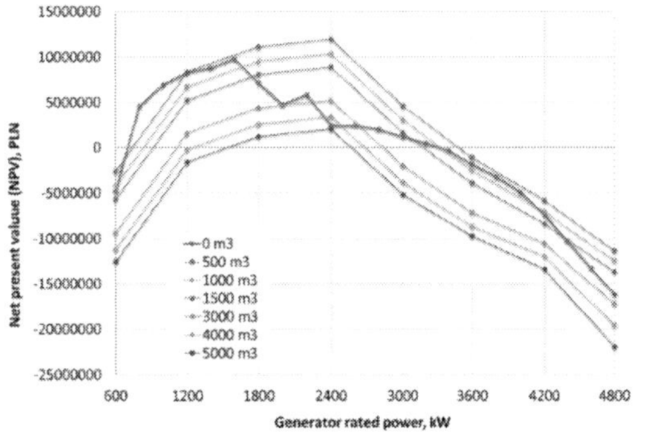

Fig. 9. IRR under different financing options (system with TES)

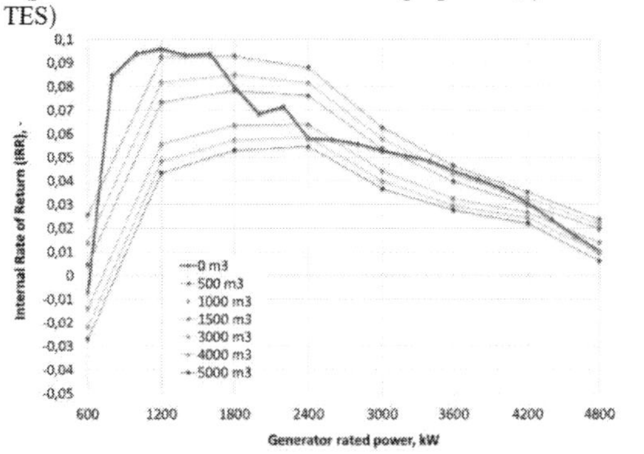

Fig. 10. IRR under different financing options (system with TES)

Results revealed that although the ORC technology has been intensively developed for last two decades biomass-fired cogeneration projects are still heavily dependent on financial support. Without support profitability of the project can't be reached. Figure 11 depicts structure of income of the best case without TES. It can be noted at current prices the support mechanism, which mostly influences the profitability is emission trading system. As this mechanism is marked based and connected with significant uncertainty and risk it does not encourage investment projects to be undertaken.

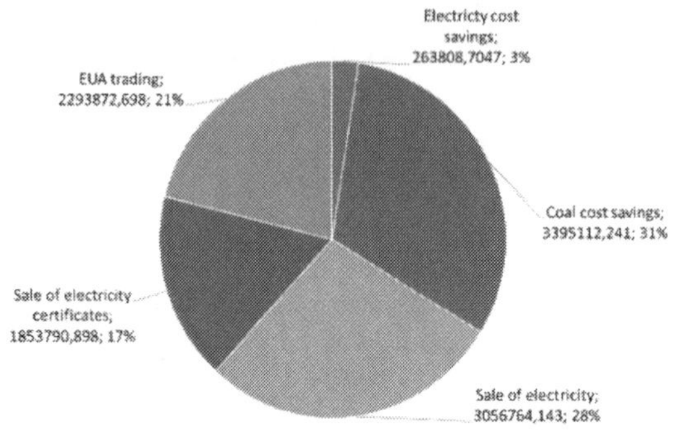

Fig. 11. Income structure for the best system of 2400 kW power output with 500 m³ TES

After taking into consideration current conditions of electricity and EUA prices there had been assumed within the next simulation price variability paths. For CO_2 several forecasts have been taken into account and average value has been estimated. In the case of electricity the price path has been developed on the basis of the report provided by the Jagiellonian Institute [16]. Both paths are depicted in Fig. 12.

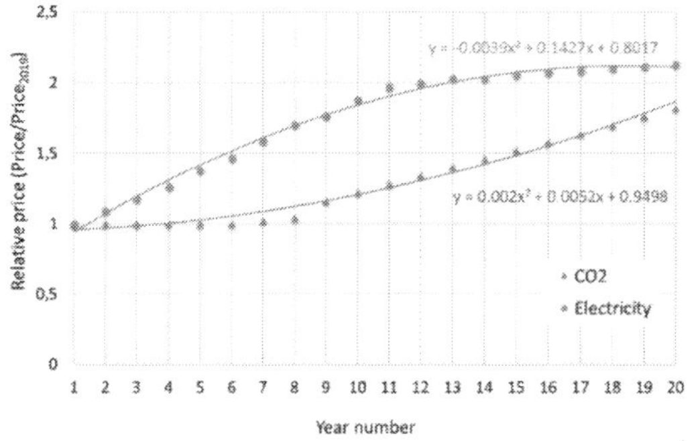

Fig. 12. Adopted price paths for electricity and CO_2 (year 1 is 2019)

Figures 13 and 14 depict NPV and IRR for different ORC generator rated power output of the system without heat storage. It can be concluded that qualitative results are the same for both studies. However, in the case of variable prices the NPV curve gets flattened. This is because losses due to shorter annual time of operation in the case of larger generators are compensated by higher incomes. Quantitatively the increase of electricity and EUA prices significantly improves financial performance of the project. Discounted payback period (DBP) for the optimal system in the case of variable electricity prices shortens from 12 to around 8.5 years.

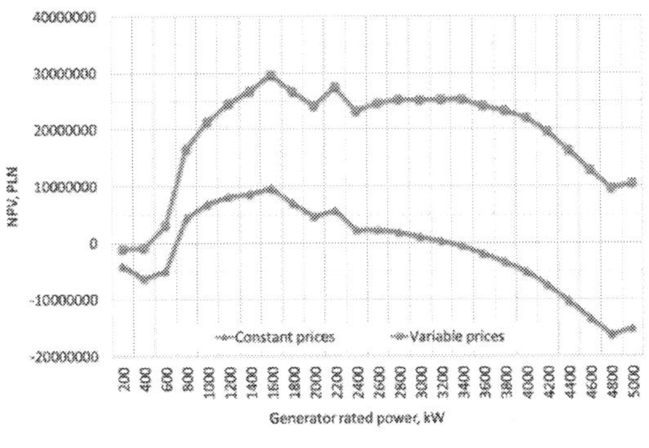

Fig. 13. NPV under constant (2019) and variable prices of electricity and EUA (system without TES)

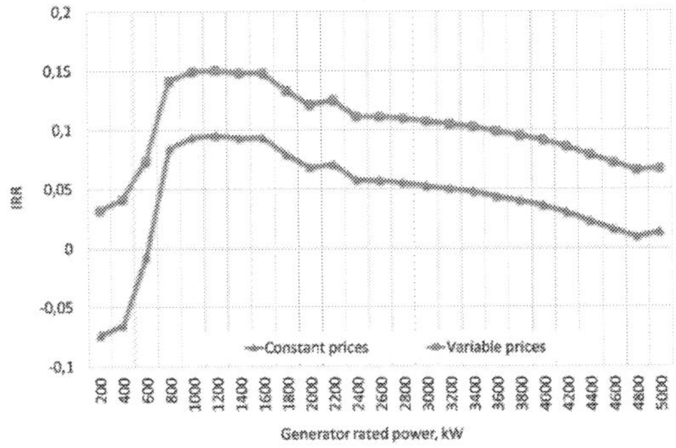

Fig. 14. IRR of projects under constant (2019) and variable prices of electricity and EUA (system without TES)

Results for ORC cogeneration system with 500 m^3 TES are depicted in Fig. 15 and 16. In the case of NPV the shape of curve has changed significantly strongly indicating the location of the optimal generator rated power, which is the same as in the case of constant prices. In the case of IRR the optimal value has slightly move towards higher values.

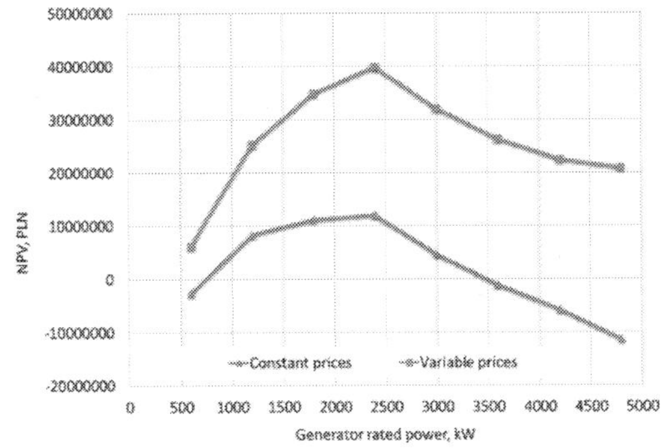

Fig. 15. NPV under constant (2019) and variable prices of electricity and EUA (system with 500 m^3 TES)

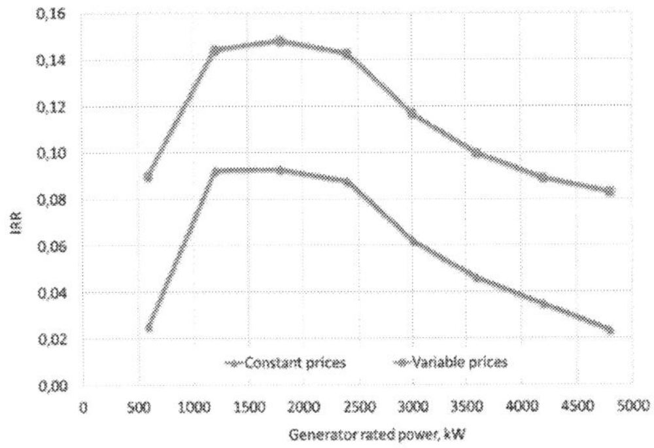

Fig. 16. IRR of projects under constant (2019) and variable prices of electricity and EUA (system with 500 m^3 TES)

5 Conclusions

The problem of optimal sizing and selection of technological structure of biomass-fired ORC cogeneration system for backfitting of fossil fuel-fired district heating plant has been presented in the paper. In Polish conditions, there are many DHS where such projects could be potentially executed. Results of this study revealed that feasible projects are possible if the ORC system is optimally sized for given site conditions. On the other hand profitability of investment is still relatively low and dependent on unstable market conditions and financial support. Discounted payback period for a feasible project can be expected within the range of 12 years. This value doesn't suggest significant investments will be triggered in the near future in the field of biomass fired cogeneration based on the ORC technology. Adding thermal energy storage and selling electricity on balancing market have improved financial performance only to a small extent.

An important conclusion is that economic optima are relatively flat and the range of system sizes, for which the value of objective function is close to the optimum is quite wide. Recommended solutions are the ones with relatively long annual time of operation. The analysis proved that sizing of ORC cogeneration unit should aim at covering base heating load of the district heating network. In case of the Krosno project the size of the implemented ORC system appeared to be close to the optimum.

This work was carried out in the project IntBioCHP titled: System integration of biomass fired cogeneration plants. The project is financed by German Federal Ministry of Education and Research and Polish Ministry of Science and Higher Education within the framework of the Polish - German Sustainability Research Programme STAIR.

References

1. Kalina, J.: Equipment sizing in a coal-fired municipal heating plant modernisation project with support for renewable energy and cogeneration technologies. En. Conv. and Man., **Vol. 86** (2014), p. 1050-1058.

 https://doi.org/10.1051/e3sconf/201913701042

2. Tartière T.: ORC Market: A World Overview. Web project available at http://orc-world-map.org/analysis.html
3. Tańczuk, M., Ulbrich, R.: Implementation of a biomass-fired co-generation plant supplied with an ORC (Organic Rankine Cycle) as a heat source for small scale heat distribution system - A comparative analysis under Polish and German conditions. Energy, **Vol. 62** (2013), p. 132-141.
4. Franco, A., Bellina F.: Methods for optimized design and management of CHP systems for district heating networks (DHN). En. Conv. and Man., **Vol. 172** (2018), p. 21–31.
5. Duvia, A., Guercio, A., Rossi, C.: Technical and economic aspects of biomass fuelled CHP plants based on ORC turbogenerators feeding existing district heating networks. *Proceedings of the 17th European Biomass Conference*, Hamburg, Germany, June. 2009, p. 2030–2037.
6. Elsido, C., Mian, A., Martelli, E., A systematic methodology for techno-economic optimization of Organic Rankine Cycles. Energy Proceedia, **Vol. 129** (2017), p. 26-33.
7. Martelli, E., Federico Capra, F., Consonni S.: Numerical optimization of Combined Heat and Power Organic Rankine Cycles - Part A: Design optimization. Energy, **Vol. 90** (2015), p. 310-328.
8. Capra, F., Martelli, E.: Numerical optimization of combined heat and power Organic Rankine Cycles - Part B: Simultaneous design & part-load optimization. Energy. **Vol. 90** (2015), p. 329-343.
9. Taljan, G., Verbic, G., Pantos, M., Sakulin, M., Fickert, L.: Optimal sizing of biomass-fired Organic Rankine Cycle CHP system with heat storage. Energy, **Vol. 41** (2012), p. 29-38.
10. Baños, R., Manzano-Agugliaro, F., Montoya, F., G., Gil, C., Alcayde, A., Gómez, J.: Optimization methods applied to renewable and sustainable energy: A review. Ren. and Sust. En. Rev., **Vol. 15**, (2011), p. 1753–1766.
11. Østergaard, P., A.: Reviewing optimisation criteria for energy systems analyses of renewable energy integration. Energy, **Vol. 34** (2009), p. 1236–1245.
12. Kalina, J., Świerzewski, M., Strzałka, R.: Operational experiences of municipal heating plants with biomass-fired ORC cogeneration units. En. Conv. and Man., **Vol. 181** (2019), p. 544-561.
13. Kalina, J., Świerzewski M.: Identification of ORC unit operation in biomass-fired cogeneration system. Ren. En.. **Vol. 142** (2019), p. 400-414.
14. International Finance Corporation (IFC) in partnership with Austrian Federal of Ministry of Finance, 2017, Converting Biomass to Energy – a Guide for Developers and Investors. Washington, D.C. June 2017. ifc.org
15. Turboden srl. ORC standard units. Datasheet. www.turboden.com.
16. Schnell Ch., Roszkowski M.: Trwały wzrost cen energii zmieni dopiero modyfikacja polskiego miksu energetycznego. Czy stać nas na marnowanie czasu ? Report of the Jagiellonian Institute: Ceny energii. Warsaw, October 2018. http://jagiellonski.pl/files/other/2018.10_Instytut_JagielloYski_-Ceny_energii_ANALIZA.pdf.

E3S Web of Conferences 137, 01043 (2019)
RDPE 2019
https://doi.org/10.1051/e3sconf/201913701043

RESEARCH & DEVELOPMENT IN POWER ENGINEERING, 2019

Influence of unsteady pressure-flow conditions on strength of steel pipelines with volumetric defects reinforced by composite sleeves

Ferdinand Uilhoorn* and Maciej Witek

Warsaw University of Technology, Department of Heating and Gas Systems, Nowowiejska 20, 00-653 Warsaw, Poland
E-mail addresses: ferdinand.uilhoorn@pw.edu.pl, maciej.witek@pw.edu.pl

Abstract. Structural integrity and risk management have a wide interest because of its practical applications, such as oil and gas pipelines, piping systems under pressure in power stations, urban water, and heating networks. The main goal of this paper is twofold. Firstly, to estimate the unsteady pressure-flow variations in a gas transmission grid within the framework of sequential data assimilation. This technique enables to determine accurately the maximum pressure at the localized defect on the pipeline by merging measurements that contain random errors into the inexact numerical flow model. For this purpose, a particle filter is used. The semi-discretization approach is applied to convert the nonisothermal flow model into an initial value problem of ordinary differential equations. The spatial discretization is based on a five-point, fourth-order finite difference approximation and the time marching was done using a diagonally implicit Runge-Kutta scheme. Secondly, to study the strength of steel tubes reinforced with composite sleeves containing localized part-wall thickness losses caused by corrosion while taking into consideration a safe operating pressure. For a steel thin-walled cylinder containing a wrap of fiberglass with epoxy resin, the burst pressure and sleeve thickness are determined. Finally, the repaired pipeline with a fiber-reinforced composite sleeve is investigated. The results enable operators to handle problems of corroded steel pipelines and develop effective repair activities during operation. For this reason, current research is important for the maintenance of underground steel networks.

1 Introduction

Magnetic flux leakage (MFL) technology using axial excitation of magnetic field lines is the most effective and widespread in-line inspection method for long-distance high-pressure underground pipelines [1]. The data gives insight into tube wall losses along the pipeline. Defect assessment after periodic diagnostics plays an important role for safe operation of natural gas grids [2]. It enables to evaluate the severity of flaws and to select the anomalies to be repaired with polymer sleeves. A three-step methodology was employed in [3] as a useful tool for operators to evaluate the severity of part-wall metal losses and aging process of underground steel pipelines that transport natural gas. Time-dependent reliability of corroded steel pipelines with correlations of random variables was presented inter alia in [4].

The role of pressure is utmost important in estimating the tube structural integrity. In previous work [5] a steady-state flow model was used to determine the internal pressure as a main load of the pipe wall. This steady-state approach is adequate if the system is not subject to significant changes in flow and pressure. In this research, a more realistic situation is investigated, namely unsteady flow conditions. Moreover, these transients are considered as realizations of a stochastic process. As far as the authors are aware, such an approach has a poor representation in the literature.

Flow models are inexact and contain uncertainties that are related to initial and boundary conditions, fluctuations in gas composition, ambient temperature, soil properties and so on. As a result, the estimates of state variables such as pressure, temperature and flow rate might too much deviate from the reality. To handle this problem, an ensemble data assimilation approach is implemented. In particular, the particle filter [6,7,8] which provides a framework for solving estimation problems that is not limited to Gaussianity and has attractive convergence properties.

The aim of this study is to investigate the strength of pipelines containing volumetric defects due to general and pitting corrosion observed by in-line inspection. The decrease in the tube wall thickness caused by electrochemical corrosion is analysed considering the limit state of plastic collapse and corresponding maximum operating pressure of the remaining pipe wall residual thickness. This value was compared to the estimated internal pressure obtained from the data assimilation approach at the point of the localized defect. Thickness of multi-layer fiber-reinforced composite sleeve was determined in order to reach the minimum strength of the pipeline.

Structural integrity and risk management have a wide interest because of its practical applications, such as oil and gas steel pipelines [5,9], piping systems under pressure in power stations, urban water and heating

https://doi.org/10.1051/e3sconf/201913701043

networks. This research enables network operators of steel pipelines subjected to corrosion to handle capacity limitations based on transient pressure calculations.

2 Nonisothermal gas flow model

The compressible gas flow model constitutes a nonhomogeneous hyperbolic system of partial differential equations (PDEs) derived from the conservation principles of mass, momentum and energy. Using a primitive formulation, the governing equations can be written as [10].

$$\frac{\partial \mathbf{W}}{\partial t} + \mathbf{F}(\mathbf{W})\frac{\partial \mathbf{W}}{\partial x} + \mathbf{S}(\mathbf{W}) = 0, \tag{1}$$

with $\forall x \in \mathbb{R}, t \in \mathbb{R}^+$, $\mathbf{W} = \begin{bmatrix} p & \dot{m} & T \end{bmatrix}^\mathrm{T}$ and

$$\mathbf{F}(\mathbf{W}) = \begin{bmatrix} -\frac{\dot{m}\left(a_s^2\alpha_2 - RTz\right)}{Ap} & \frac{a_s^2}{A} & \frac{a_s^2\dot{m}\alpha_1}{AT} \\ A - \frac{a_s^2\alpha_2^2 c_p \dot{m}^2 - Ra_s^2\alpha_1^2\alpha_2\dot{m}^2 z}{Ac_p p^2} & \frac{\dot{m}\left(\alpha_2 c_p a_s^2 - Rza_s^2\alpha_1^2 + RTc_p z\right)}{Ac_p p} & \frac{a_s^2\alpha_1\dot{m}^2\left(\alpha_2 c_p - R\alpha_1^2 z\right)}{ATc_p p} \\ -\frac{RTa_s^2\alpha_1\alpha_2\dot{m}z}{Ac_p p^2} & \frac{RTa_s^2\alpha_1 z}{Ac_p p} & \frac{R\dot{m}z\left(a_s^2\alpha_1^2 + Tc_p\right)}{Ac_p p} \end{bmatrix}, \tag{2}$$

$$\mathbf{S}(\mathbf{W}) = \begin{bmatrix} -\frac{a_s^2\alpha_1\left(\Omega Ap + RT\dot{m}wz\right)}{A^2Tc_p p} \\ w + \frac{pAg}{zTR}\sin\theta \\ -\frac{a_s^2\alpha_2\left(\Omega Ap + RT\dot{m}wz\right)}{A^2c_p p^2} \end{bmatrix}, \tag{3}$$

with pressure p, mass flow rate $\dot{m}$, temperature T, cross-sectional area A, frictional force w, gravitational acceleration g, angle of inclination θ, rate of heat transfer Ω, specific gas constant R, compressibility factor z and specific heat at constant pressure c_p. The parameters α_1 and α_2 are calculated as follows

$$\alpha_1 = 1 + \frac{T}{z}\left(\frac{\partial z}{\partial T}\right)_p, \quad \alpha_2 = 1 - \frac{p}{z}\left(\frac{\partial z}{\partial p}\right)_T. \tag{4}$$

The isentropic wave speed $(\partial p/\partial \rho)_s^{1/2}$ is denoted as a_s and obtained from the expression

$$\left(\frac{\partial p}{\partial \rho}\right)_s = \left[\frac{\rho}{p}\left(1 - \frac{p}{z}\left(\frac{\partial z}{\partial p}\right)_T\right. - \frac{p}{\rho c_p T}\left(1 + \frac{T}{z}\left(\frac{\partial z}{\partial T}\right)_p\right)^2\right]^{-1}, \tag{5}$$

where ρ is the density. The frictional force per unit length is defined as

$$w = \frac{1}{8} f_r \rho v\left|v\right|\pi D_i, \tag{6}$$

with velocity v and internal diameter D_i. The friction factor f_r is obtained from the Colebrook-White equation [11]

$$\frac{1}{\sqrt{f_r}} = -2\log\left(\frac{\varepsilon}{3.7D_i} + \frac{2.51}{\mathrm{Re}\sqrt{f_r}}\right), \tag{7}$$

where ε denotes the roughness and Re refers to Reynolds number. The steady heat transfer between the fluid and soil per unit length and time is defined as

$$\Omega = -\pi dU\left(T - T_s\right), \tag{8}$$

where U is the total heat transfer coefficient and T_s the soil temperature. Finally, to the flow model, the following initial and boundary conditions are imposed:

$$p(x,0) = p_0(x), \quad T(x,0) = T_0(x), \quad \dot{m}(x,0) = \text{const}. \tag{9}$$

$$p(0,t) = p_0, \quad \dot{m}(\ell,t) = \phi(t), \quad T(0,t) = T_0, \tag{10}$$

respectively. The pipeline length is denoted as ℓ.

The numerical approximation is done using the semi-discretization approach, which converts the flow model into ordinary differential equations (ODEs) via discretization of the spatial derivatives. The sparse system of ODEs reads

$$\frac{\mathrm{d}\mathbf{w}(t)}{\mathrm{d}t} = -\mathbf{F}(\mathbf{w})\mathbf{T}\mathbf{w}(t) - \mathbf{S}(\mathbf{w}(t)) = f(\mathbf{w}(t)), \tag{11}$$

with $\mathbf{w}(t_0) = \mathbf{w}_0$ and

$$\mathbf{w}(t) = \left[p_1(t), \cdots, p_i(t), \cdots, p_{n_x}(t), \dot{m}_1(t), \cdots, \dot{m}_i(t), \cdots, \dot{m}_{n_x}(t), T_1(t), \cdots, T_i(t), \cdots, T_{n_x}(t)\right]^\mathrm{T}, \tag{12}$$

whereas the discretization is done uniformly on the interval $\left[0, \ell\right]$ with n_x nodes. In Eq. (11), the parameter $\mathbf{T}$ represents a block-diagonal matrix defined as $\mathbf{T} = \sum_{j=1}^{n} \mathbf{I}_j \otimes \mathbf{T}^{(j)}$. Since $n = 3$ and the sub-matrices are identical, $\mathbf{T}^{(j)} = \mathbf{T}^{(1)}\ \forall j$ where

$$\mathbf{T}^{(1)} = \frac{1}{2\Delta x}\begin{bmatrix} -3 & 4 & -1 & 0 & \cdots & 0 \\ -1 & 0 & 1 & 0 & \cdots & 0 \\ 0 & -1 & 0 & 1 & \cdots & 0 \\ \vdots & \ddots & \ddots & \ddots & \ddots & \vdots \\ 0 & \cdots & 0 & -1 & 0 & 1 \\ 0 & \cdots & 0 & 1 & -4 & 3 \end{bmatrix}. \quad (13)$$

The coefficients in the sub-matrix refer to the classical three-point finite difference approximation.

For the time stepping, a second order diagonally implicit Runge-Kutta (DIRK) method [12,13] is used. This scheme starts with the trapezoidal rule (TR) followed by the second order backward-differentiation formula (BDF2) and showed to be robust with good accuracy and efficiency properties, if the tolerance is not set too tight [14]. Within the framework of particle filtering, the large set of ODEs is sampled many times, therefore in order to reduce the computation time, the sparsity of the system is exploited.

3 Particle filter

The idea of data assimilation is to combine noisy measurements with the numerical flow model. The heart of the approach is based on Bayes theorem that states that the model pdf can be updated via measurements. Mathematically, it reads:

$$\zeta_k = g(\zeta_{k-1}) + v_{k-1}, \quad (14)$$

$$y_k = \zeta_k + n_k, \quad (15)$$

where $\zeta_k \in \mathbb{R}^{n_\zeta}$ refers to the states, $y_k \in \mathbb{R}^{n_y}$ denote the measurements, $v_k \in \mathbb{R}^{n_v}$ and $n_k \in \mathbb{R}^{n_n}$ are i.i.d. random variables that represent the model and measurement noise, respectively. The noise statistics are not unimportant, if wrongly selected filter divergence might occur. On the other hand, a priori knowledge about these statistics is often based on trial-and-error or tuning [15]. Here $g : \mathbb{R}^{n_\zeta} \times \mathbb{R}^{n_v} \mapsto \mathbb{R}^{n_\zeta}$ refers to the finite dimensional model. In Eq. (15) it can be seen that the state variables of interest, namely, pressure, temperature and mass flow rate are directly measured. However, the state variables contain noise that we seek to reduce by data assimilation. A technique that is used here is particle filtering where the aim is to approximate the posterior pdf $\pi(\zeta_{0:k} \mid y_{1:k})$ where $\zeta_{0:k} = \{\zeta_0, \zeta_1, ..., \zeta_k\}$ and $y_{1:k} = \{y_1, y_2, ..., y_k\}$. The advantage of particle filters is that they are not limited by linearity or Gaussianity. This technique is tracking a set of states $\zeta_{0:k}$ up to time *k* by updating a random measure $\{\zeta^i_{0:k}, \omega^i_k\}^{N_p}_{i=1}$ that approximates $\pi(\zeta_{0:k} \mid y_{1:k})$. Here $\{\zeta^i_{0:k}, i = 0, ..., N_p\}$ is a set of support points and associated weights $\{\omega^i_k, i = 1, ..., N_p\}$. The particle filter starts with sampling N_p particles ζ^i_0 from the initial model probability density $\pi(\zeta_0)$ followed by integrating all particles in time up to the moment measurements become available, i.e., $\zeta^i_k \sim \pi(\zeta_k \mid \zeta^i_{k-1})\ \forall i$. In this work, the importance density $\pi(\zeta_k \mid \zeta^i_{k-1}, y_k)$ is set equal to the prior $\pi(\zeta_k \mid \zeta^i_{k-1})$. The next step is to compute the weights $\omega^i_k = \pi(y_k \mid \zeta^i_k) / \sum^{N_p}_{i=1} \pi(y_k \mid \zeta^i_k)$ followed by resampling of particles such that the weights are equal to N_p^{-1}. This sequence is repeated for all measurements. In the resampling step particles with low weights are excluded, whereas particles with high weight are kept for the posterior pdf. The reader is referred to [6,7,8] for details.

It is known, that particle filters might suffer from the curse of dimensionality, this means that the ensemble size to generate a few samples from the high-probability areas of the posterior is growing exponentially with the dimension of the state vector [16,17]. However, in the numerical experiments a coarse interval is used, and the sampling is done only at the inlet and outlet node of the spatial discretization. For this reason, it is reasonable to assume that the curse of dimensionality does not apply here.

4 Strength of defected tube

The Det Norske Veritas Recommended Practice methodology [18] was selected as it is considered adequate for electrochemical corrosion defects and steel grade investigated in this work. The maximum allowable operating pressure of the defected pipeline with longitudinal oriented single metal-loss (see Fig. 1) is defined by [18]

$$MAOP_{\mathrm{DNV}} = \gamma_m \frac{2\delta f_u \left(1 - \gamma_d \left(\frac{d}{\delta}\right)^*\right)}{\left(D_o - \delta\right)\left(1 - \frac{\gamma_d \left(\frac{d}{\delta}\right)^*}{Q}\right)}, \quad (16)$$

where

$$\left(\frac{d}{\delta}\right)^* = \left(\frac{d}{\delta}\right)_{\mathrm{means}} + \varepsilon_d StD\left(\frac{d}{\delta}\right), \quad (17)$$

$$Q = \sqrt{1 + 0.31\left(\frac{L^2}{D_o \delta}\right)}, \quad (18)$$

with outside diameter D_o, ultimate tensile strength of the pipe f_u to be used in the design, single defect depth *d* derived from diagnostics, defect length *L* derived from in-line inspection, wall nominal thickness δ, measured (relative) single defect depth $(d/\delta)_{means}$ obtained from the inspection results, and standard deviation of the measured ratio $StD(d/\delta)$ derived from the specification of MFL tool by the assumption of normal distribution of random variable d/δ [18].

The safety factors γ_m, γ_d and ε_d in Eq. (16) and (17) are determined by: (i) pipeline safety class, usually derived from design, (ii) testing method, relative or absolute and (iii) inspection accuracy and a confidence level. According to the Polish regulation, the degree of area urbanization concerning the location of high-pressure gas pipelines is classified as follows:

- Class 1 (safety class high corresponding to the pipeline segment design factor $f \leq 0.4$) covers an area built over with single- or multi-family collective dwelling houses and public utility facilities, with intensive vehicular traffic and developed underground infrastructure, such as water networks, sewage systems, heat distribution networks, gas grids, power and telecommunication networks, as well as streets, roads and mining sites;
- Class 2 (safety class normal corresponding to the pipeline segment design factor $0.4 < f \leq 0.6$) comprises an area with single-family and farmstead housing and individual leisure buildings, along with their necessary infrastructure;
- Class 3 (safety class low corresponding to the pipeline segment design factor $0.6 < f \leq 0.72$) includes an undeveloped area and an area in which only isolated single-family houses and farm and livestock buildings can be situated, along with their necessary infrastructure.

The safety factor γ_m is obtained from in-line inspections that are based on relative feature depth measurements such as the MFL method. Its value equals $\gamma_m = 0.79$ for class location low. The partial safety factor for corrosion depth is $\gamma_d = 1.156$. The fractile value for corrosion $\varepsilon_d = 0$. The latter two factors depend on the safety class and standard deviation and selected according to [18].

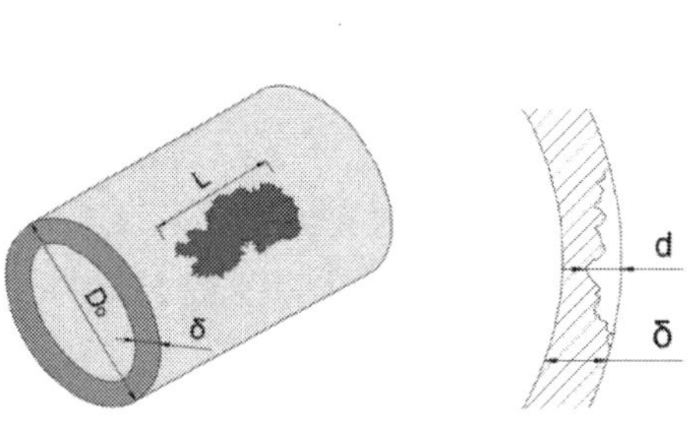

Fig. 1. Geometry of the axially oriented tube wall material loss.

5 Pipe strength repaired by composite sleeve

The components of a multi-layer polymer repair system applied to a steel pipe with volumetric defect in current research as well as in publications [3,20] is shown in Fig. 2.

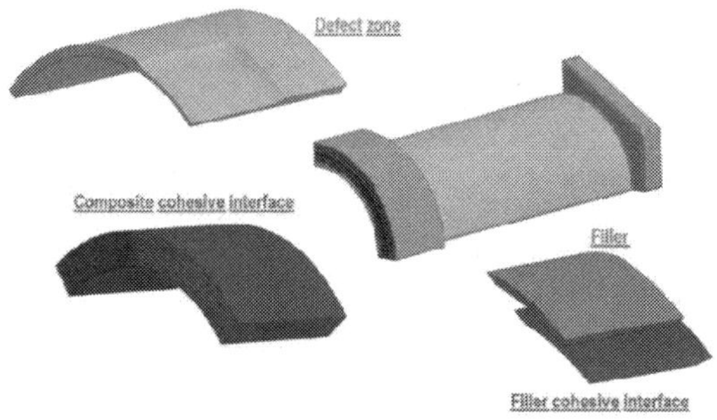

Fig. 2. Layered system of composite repair used for defected steel pipe reinforcement.

The strength calculations applied in this paper are reasonable for estimating the failure pressure of a corroded pipe, reinforced with a multi-layer composite wrap [19]. Assuming an elastic behavior of both steel tube and polymer sleeve, the failure pressure of a straight, thin-walled cylinder with volumetric defect reinforced by composite wrap can be calculated as follows [3, 19]

$$P_{\max}^{\text{th}} = \frac{f_u\left(r_o - r_i\right)}{\alpha_\theta\left(r_i - \eta r_o\right)}, \tag{19}$$

where

$$\alpha_\theta = \left[\frac{1-\left(\frac{d}{\delta}\right)\left(\frac{1}{Q}\right)}{1-\left(\frac{d}{\delta}\right)}\right], \tag{20}$$

$$\eta = \left(\frac{r_o^2 E_{\text{pipe}}\left(r_o - r_i\right)}{r_i^2 E_{\text{sleeve}}\left(r_e - r_o\right)} + \frac{r_o}{r_i}\right)^{-1}, \tag{21}$$

with failure pressure of a straight pipe with volumetric defect reinforced by composite wrap $P_{\max}^{\text{th}}$, external radius of the steel tube plus thickness of the sleeve r_e, inner radius of the steel pipe r_i, external radius of the steel pipe r_o, elasticity modulus of steel E_{pipe} and elasticity modulus of the composite sleeve E_{sleeve}.

The elasticity modulus of a sleeve depends on the polymer and other components of the layered system used for the pipeline reinforcement. However, elasticity modulus of the sleeve in fiber-reinforced composite wraps depends on the fiber volume. According to [3], the elasticity modules of the fiberglass sleeve with 50% of fiber equals to 44.2 GPa. Analyses of fiberglass reinforcement systems such as applied in [20] confirm the assumption that the resin volume in a sleeve is negligible and an averaged elasticity modulus of the polymer in the hoop direction is approximately equal to a mean of an axial and a transverse value. These calculation assumptions were applied in [3]. The elasticity moduli used in this work are shown in Table 1.

Table 1. Elasticity modulus of steel, fiberglass sleeve and epoxy filler [3].

Elasticity modulus of steel E_{pipe}/GPa	Axial modulus of fiberglass sleeve E_{Gx}/GPa	Transverse modulus of fiberglass sleeve E_{Gy}/GPa	Axial modulus of epoxy resin E_{Rx}/GPa	Transverse modulus of epoxy resin E_{Ry}/GPa
206	44.2	44.2	3.4	3.4

The acceptability condition of tube wall metal loss without reinforcement can be written as follows

$$P(x)_{est} \le MAOP_{DNV}, \qquad (22)$$

with estimated pressure from the flow model $P(x)_{est}$ at the metal loss coordinate x along the pipeline, maximum allowable operating pressure of the defected thin-walled cylinder with longitudinal oriented single metal-loss $MAOP_{DNV}$ calculated according to [18].

The thickness of the multi-layer composite sleeve ($k = r_e - r_o$) has to be determined in order to reach the minimum strength of the reinforced tube. This to resist the estimated operating pressure at the point of the defect occurrence, i.e.,

$$P(x)_{est} \le f\, P_{max}^{th}, \qquad (23)$$

with design factor f depending on the pipeline segment safety class.

The condition for maximum admissible operating pressure of the pipeline with volumetric surface flaw reinforced with composite sleeve can be expressed by the inequality:

$$MAOP^{th} \le f\, P_{max}^{th}, \qquad (24)$$

with maximum admissible operating pressure of the thin-walled cylinder containing volumetric surface defect reinforced with multi-layer composite sleeve $MAOP^{th}$.

If as usual in steel underground structures maintenance, many pipe wall metal losses were recognized as a result of in-line inspection, the conditions (23) and (24) have to be fulfilled for each volumetric surface defect.

6 Case study

In the first part of the case study, the flow dynamics are estimated using the particle filter. From the spatial-temporal evolution of the pressure estimates, the maximum pressure is determined at the defect location. This point is obtained from hypothetical MFL diagnostics. In the second part, the maximum allowable operating pressure of the tube with part-wall metal loss is computed and compared with the estimated internal fluid pressure obtained from the flow model. Based on these calculations the strength of the defected pipe, reinforced with fiber-based polymer sleeve is analyzed in order to determine the appropriate thickness of multi-layered repair wrap.

6.1 Pressure estimates

For the investigation, a section of the Yamal transit pipeline on Polish territory between two compressor stations was selected with the following parameters: ℓ = 177 km, D_i = 1.380 m, ε = 0.0015 mm, T_s = 12 °C and U = 1.66 W m^{-2} K^{-1}. The molar fraction of the natural gas is: 98.3455 CH_4, 0.6104 C_2H_6, 0.1572 C_3H_8, 0.0299 i-C_4H_{10}, 0.0253 n-C_4H_{10}, 0.0055 i-C_5H_{12}, 0.0040 n-C_5H_{12}, 0.0303 N_2 and 0.7918 CO_2. The thermodynamic and transport properties required in flow model (1) were fitted to the calculated values of GERG-2008. This equation of state is highly accurate, but computationally intensive and therefore, less practically simulating flow transients. Therefore, the following multivariate regression model is used:

$$\kappa(p,T) = \sum_{i=1}^{n} \iota_i p^{\mathcal{I}_{i,1}} T^{\mathcal{I}_{i,2}}, \qquad (25)$$

with $\mathcal{I}$ as the power and ι_i refer to the corresponding coefficients and 5 degrees freedom. The simulations were done on the interval t_F/h ∈ [0,24] and with $p(0,t)$ = 8.4 MPa, $T(0,t)$ = 308.15 K and $\dot{m}(\ell,t) = f(t)$. The latter boundary condition is depicted in Fig. 3.

In the absence of real data for the state variables, it is common to generate synthetic data and to avoid inverse crime, the numerical model is solved on a coarser grid (n_x = 101) than the one that is used to generate the data (n_x = 1001). The measurements were generated by adding a random Gaussian noise to the model solutions with a fine grid, i.e., $n_{k,p} \sim \mathcal{N}(0, 0.04^2)$, $n_{k,T} \sim \mathcal{N}(0, 1.5^2)$ and $n_{k,\dot{m}} \sim \mathcal{N}(0, 2.5^2)$ with variance in MPa^2, K^2 and (kg s^{-1})^2, respectively. It is assumed that the model noise is based on Gamma random variables, i.e., $v_{k-1,p} \sim \text{Gamma}(9, 0.5)$, $v_{k-1,T} \sim \text{Gamma}(5, 1)$ and $v_{k-1,\dot{m}} \sim \text{Gamma}(3, 2)$. The Gaussian and Gamma random variables were generated using Matlab functions randn and gamrnd. The ensemble size N_p is set to 100. The measurement sampling time is set to 10 min but in practice, it can vary from one to several tens of minutes. The spatial-temporal evolution of the filtered estimates for all state variables is shown in Fig. 4.

It is assumed that the part-wall volumetric defect is located at 60 km from the up-stream compressor station. Since, the distance does not coincide with a grid point after spatial discretization, the pressure values are obtained from its nearest values using piecewise cubic interpolation. The maximum pressure at x=60 km is P_{est} (60) = 8.37 MPa after 16 h from the beginning of the simulation (Fig. 5).

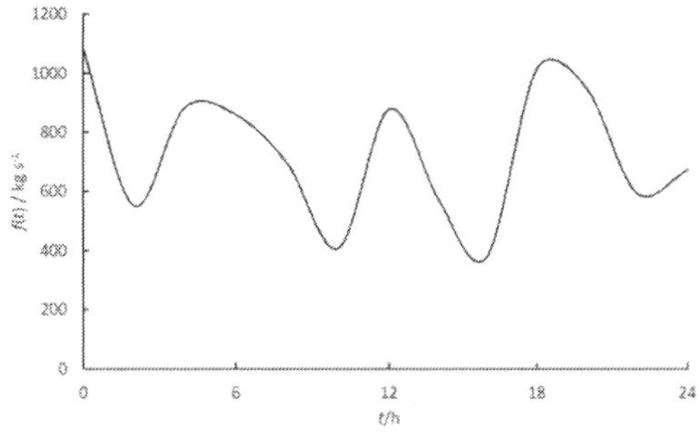

Fig. 3. Mass flow rate boundary condition.

6.2 Strength of steel tube repair

In the second part of this case study, the allowable pressure is calculated based on the following pipeline parameters [21]: f_u = 570.0 MPa, f_y = 485 MPa, E_{pipe} = 206 GPa, D_o = 1422 mm, δ = 21 mm, r_o = 711 mm and r_i = 690 mm. Dimensions of the longitudinal oriented corrosion area shown in Fig. 1 are derived from hypothetical in-line inspection results. The maximum defect depth as a percentage of nominal tube wall thickness is $(d/\delta)_{\text{means}}$ = 43% and the axial length L of the corroded area is 1000 mm.

Based on the feature dimensions obtained from hypothetical MFL diagnostics, the maximum allowable operating pressure of the defected pipe without reinforcement calculated from the formula (16) equals to $MAOP_{\text{DNV}}$ = 7.98 MPa. The maximum pressure obtained from the data assimilation approach at the defect location equals to $P_{est}(60)$ = 8.37 MPa. This means that condition (22) is not fulfilled and therefore the pipeline must be immediately excavated, and the external surface flaw needs to be directly assessed as it was described in [2]. If the inspection tool defect sizing is confirmed, the volumetric metal loss feature has to be reinforced by composite wrap in order to reach minimum required strength of the tube. Determination results of multi-layer polymer sleeve thickness k to reach the strength of the defected pipe corresponding to the maximum pressure at the point of the specific defect occurrence is presented below. For the number of layers of the composite wrap k = 16, the corresponding burst pressure of the repaired tube equals to $P^{\text{th}}_{\text{max}}$ = 11.63 MPa. From Eq. (24), the maximum admissible operating pressure while considering a low safety class for the pipeline that contains a volumetric surface defect reinforced with a composite sleeve equals to $MAOP^{\text{th}}$= 8.37 MPa. Hence, for the specific defect dimensions and k=16 obtained from Eqs (19-21), condition (23) is satisfied.

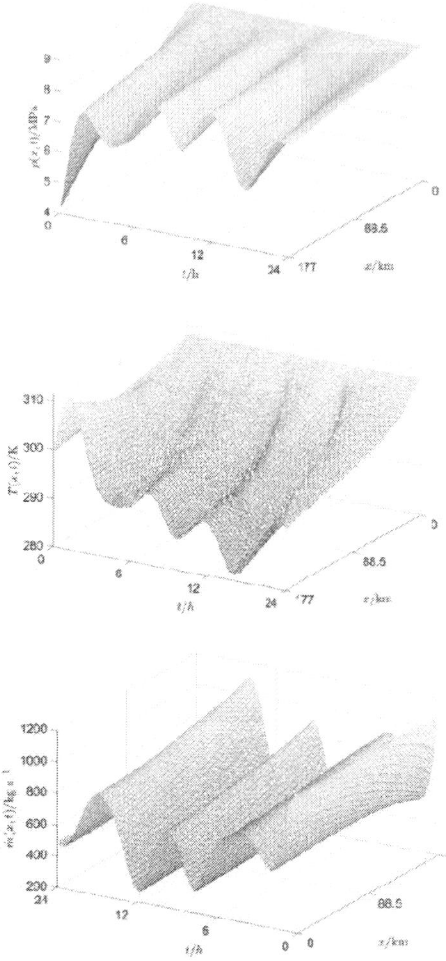

Fig. 4. Evolution of state variables.

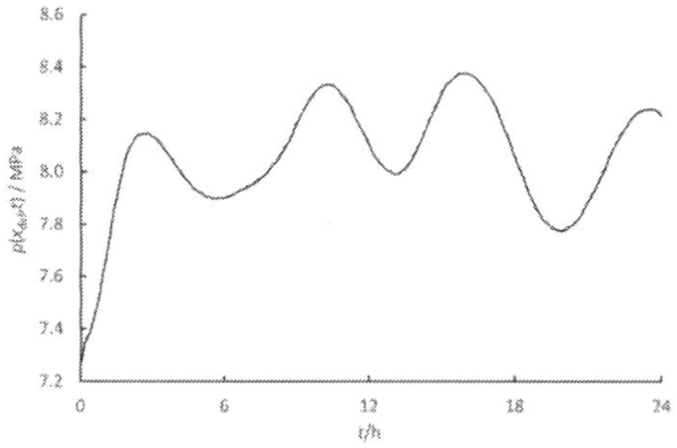

Fig. 5. Pressure conditions at the volumetric defect location.

7 Conclusions

In this work, a data assimilation technique was used to estimate the pressure under transient flow conditions at the point where cylinder wall defect exists in order to evaluate the underground pipeline integrity. This approach is more accurate and realistic compared to a situation where the pressure at the certain coordinate does not change in time, which is nowadays assumed in thin-walled tubes strength analysis.

Repairs of part-wall surface metal losses derived from periodic in-line inspection by composite sleeves avoid unnecessary capacity limitations of steel network subjected to corrosion due to local reinforcement of pipe in order to maintain design-operating pressure of the infrastructure. The analytical strength computations applied in the current paper is reasonable to estimate the failure pressure of corroded tubes reinforced with polymer sleeves. A more accurate strength analysis requires a numerical model, which enables to simulate a multi-layered repair system. This approach is more complicated but recommended for further research.

References

1. Bouledroua O., Zelmati D., Hassai M., Inspections, statistical and reliability assessment study of corroded pipeline, *Engineering Failure Analysis* Vol. 100, pp. 1-10, (2019).
2. Witek M., Validation of in-line inspection data quality and impact on steel pipeline diagnostic intervals, *Journal of Natural Gas Science and Engineering*, Vol. pp. 121-133, (2018).
3. Witek M., Gas transmission failure probability estimation and defect repairs activities based on in-line inspection data, *Engineering Failure Analysis*, Vol. 70, pp. 255-272, (2016).
4. Zhou W., System reliability of corroding pipelines, *International Journal of Pressure Vessels and Piping*, Vol. 87, pp. 587–595, (2010).
5. Witek M., Batura A., Orynyak I., Borodii M., An integrated risk assessment of onshore gas transmission pipelines based on defect population, *Engineering Structures*, Vol. 173, pp. 150-165, (2018).
6. Arulampam S., Maskell S., Gordon N., Clapp T., A tutorial on particle filters for on-line non-linear/non-Gaussian Bayesian tracking, *Signal Processing, IEEE Transactions on* Vol. 50 (2), pp. 174, 2002.
7. Doucet A., On sequential Monte Carlo methods for Bayesian filtering. *Tech. rep., Tech. Rep., Univ. Cambridge, Department of Engineering*, UK, (1998).
8. Gordon N., Salmond D., Smith A., Novel approach to nonlinear/non-Gaussian Bayesian state estimation, *Radar and Signal Processing, IEE Proceedings F* Vol. 140(2), pp. 107-113, (1993).
9. Han Z.Y., Weng W.G., An integrated quantitative risk analysis method for natural gas pipeline network, *Journal of Loss Prevention in the Process Industries*, Vol. 23, pp. 428-436, (2010).
10. Uilhoorn F.E., Comparison of Bayesian estimation methods for modeling flow transients in gas pipelines, *Journal of Natural Gas Science and Engineering*, Vol. 38, pp. 159-170, (2017).
11. Colebrook C., Turbulent flow in pipes, with particular reference to the transition region between the smooth and rough pipe laws, *J. Inst. Civil Eng.* Vol. 11, pp. 133–156, (1939).
12. Bank R.E., Coughran, Jr. W.M., Fichtner W., Grosse E. H., Rose D. J., Smith R. K., Transient Simulation of Silicon Devices and Circuits, *IEEE Transactions on Computer-Aided Design CAD, Vol* 4(4), pp. 436-451 (1985).
13. Shampine L., Hosea M., Analysis and implementation of TR-BDF2, *Applied Numerical Mathematics*, Vol. 20(1-2), pp. 21-37, (1996).
14. Uilhoorn F.E., A comparison of numerical integration schemes for particle filter-based estimation of gas flow dynamics, *Physica Scripta*, Vol. 93 (12), (2018).
15. Uilhoorn F.E., A multiobjective optimization approach to filter tuning applied to coupled hyperbolic PDEs describing gas flow dynamics, *Optimal Control Applications and Methods*, Vol. 40 (4), pp. 750-763, (2019).
16. Bickel P., Li B., Bengtsson T., Sharp failure rates for the bootstrap particle filter in high dimensions, IMS Collections: *Pushing the Limits of Contemporary Statistics: Contributions*, pp. 318-329, (2018).
17. Snyder C., Bengtsson T., Bickel T., Anderson J., Obstacles to high-dimensional particle filtering, *Monthly Weather Review* Vol. 136, pp. 4629-4640, (2008).
18. DNV-RP-F101 Det Norske Veritas Recommended practice Corroded pipelines (2010).
19. Da Costa Mattos H.S., Reis J. M. L, Paim L. M., Silva da M. L., Lopes Junior R., Perrut V.A., Failure analysis of corroded pipelines reinforced with composite repair systems, *Engineering Failure Analysis*, 59, pp. 223-236, (2016).
20. Mazurkiewicz Ł., Tomaszewski M., Małachowski J., Sybilski K., Chebakov M., Witek M., Yukhymets P., Dmitrienko R., Experimental and numerical study of steel pipe with part-wall defect reinforced with fibre glass sleeve, *International Journal of Pressure Vessels and Piping*, Vol. 149, pp. 108-119, (2017).
21. Gerwatowski, D., Analiza czynników wpływających na przepustowość układu przesyłowego gazu ziemnego, PhD Thesis, Warsaw University of Technology (2011).

RESEARCH & DEVELOPMENT IN POWER ENGINEERING, 2019

Virtual power plants - types and development opportunities

Arkadiusz Przychodzień[1]

[1] Institute of Heat Engineering, Warsaw University of Technology, Power Generation Strategic Analysis Department, PKN ORLEN, arkadiusz.przychodzien@orlen.pl, Poland

Abstract. Modern power engineering meets new challenges. With the development of new energy production and storage technologies, creates new demands for energy services. To support this development, it is necessary to implement new teleinformatic systems that will allow for resource management. Such systems are called Virtual Power Plants (VPP). There are many definitions of this type of solutions due to the very wide range of possible applications. VPPs can be developed by many types of entities, e.g. distribution system operators, electricity generators, energy clusters. The ability to build a system based on modules allows you to customize the system to user's needs. An opportunity for the development of VPP will be a package "Clean energy for all Europeans" (so called "Winter package") that introduces regulations that allow for the development of renewable energy sources, including prosumers, and enables an active participation in the energy market for energy consumers. In addition, more stringent requirements for balancing production and energy consumption are introduced, requiring greater balancing accuracy.

1 Introduction

We are witnessing a great breakthrough in the energy sector in which we can observe changes in the shape of power systems and market models. Nowadays, power systems are dominated by large generation units owned by big power companies, while trends are moving towards a distributed generation model, consisting of a large number of smaller generation units owned by households and small enterprises. The development of small generation installations allowed many consumers to produce energy for their own needs, making them prosumers. The potential escape of prosumers from current energy suppliers has made them adapt their business models to retain customers by offering them new services and giving them the opportunity to cooperate.

Virtual Power Plants are one of the innovative solutions that fit the trend of distributed energy development. They are created to maximize the economic effect of units, but also to effectively balance the network. In VPPs an aggregator combines distributed units and manages them so that they are one unit selling energy on the market and offering system services. For this purpose aggregator optimizes VPP in real time, through the load forecasting, net-metering and automatic generation control. Ability to generate additional benefit occurs through the proper balancing of the production and consumption of energy in the system, while predicting the behavior of the market and the factors affecting the efficiency of power units. Obtaining such a result requires a high level of flexibility of production balancing and energy consumption using IT systems.

Further development of uncontrollable renewable energy sources will cause that generation flexibility and demand control capability will gain significant value. This is one of the reasons why energy concerns are interested in the concept of virtual power plants. The idea of VPP gives the possibility of combining many technologies, such as conventional power plants, demand side response (DSR), prosumer installations, as well as energy storage. This combination allows generation of additional profits thanks to the synergy effect of cooperating technologies. VPP are an example of the "Internet of Things" using existing networks and IT systems to create services in the field of electricity generation, while avoiding large capital expenditures that would require the construction of new flexible peak sources. The use of a combination of existing resources gives many new opportunities, such as balancing deviations between the forecasted and actual generation of wind sources under VPP. This limits the business risk, reduces balancing costs and increases the network©operational safety. [1]

The development of virtual power plants has many challenges on its way. One of them is a large diversity of market models and legal regulations in different countries. This makes impossible to copy ready-made VPP solutions developed in other countries. VPP must be tailor-made for a given system and energy market. This situation may change due to the entry into force of the package "Clean energy for all Europeans" also called the "Winter Package". It contains a number of regulations enabling the development of renewable sources of energy, prosumer installations, increasing the rights of energy consumers, more flexible sales, combine and unify the European energy markets in order to create a common market for EU countries. Regulatory changes may create an appropriate environment for the development of VPP, allowing their operators to use the generating units even more efficiently and offer new services on the electricity market, e.g. on interconnectors.

The development of the VPP will also affect changes in the markets for electricity and energy resources, changes in the demand profile for electricity (e.g.. as a result of the development of electromobility) forcing an increase in production efficiency. An important impulse for development may also be further progress in the field of electricity storage technologies, increasing the possibilities of optimal management of generating assets. The development of appropriate models of functioning suited to the regulatory and market environment as well as the needs of the system operators will be key for the development of VPPs, because other objectives will have an enterprise with large production units, and another group of cooperating prosumers.

2 Virtual Power Plant definition

There is no one clear definition of the Virtual Power Plant and the use of the term depends on the authors of the described solution. This definition is applied to a very wide range of terms, because it can only define the organizational structure connecting entities [2], in most definitions it refers to the IT system connecting entities, while in other definitions it is defined as a set of units together with the system. Generally, the Virtual Power Plant can be described as a system of generating, consuming and storing electricity units connected by means of an ICT system that accomplishes the assumed goals using IT systems.

3 The purposes of creating Virtual Power Plants

The shape and functionality of VPP depends on the purpose for which it was created. This purpose depends on the range of activities that the VPP's owner performs and is different due to the different role of the entity in the operation of the power system.

The objectives of the VPP's operation created by the energy producer:

• Increasing revenues from sales of energy generated by units through optimization of load. The use of flexibility of production in order to offer on the SPOT market and balancing electricity that meet the demand of customers.

• Reducing the risk of operations, e.g. imbalance of generation units on the energy market with the use of advanced forecasting modules and cooperation of many units. The possibility of quick and optimal reaction on unplanned events.

• Decreasing operating costs by unifying and integrating systems responsible for forecasting, optimization, control and reporting. No need to maintain separate systems and infrastructure, as well as less work to support the system.

• Ability to offer new system services for the needs of distribution network operators possible to implement thanks to the management of distributed sources.

The objectives of the VPP operation created by the distribution system operator:

• Increasing the security of the system by reducing the risk of rapid changes in units production or customers demand. VPP also provides the ability to predict network threatening situations, and in case of emergency situations, flexible network balancing while maintaining appropriate priorities.

• Reduction of operating costs of the distribution network through the optimal use of available resources, as well as the unification and integration of operator systems with source control systems that allows to limit the number of systems maintained.

• The use of power units for the system services

4 Virtual Power Plant division

We can divide them both due to the location of sources, the type of system control and its organization.

We can divide VPPs in terms of dispersion of units:

• Focused - units are located within one connection to the external power grid, e.g. a wind farm in which individual turbines play the role of elements managed by VPP.

• Distributed - units have a lot of connections to the grid, e.g. VPP managing a group of wind turbines connected in many places in the network

Distributed VPPs can also be divided in terms of the location of units:

• Regional - are located in a limited area and are connected to the same distribution network, e.g. local energy clusters, where the synergy effect is also achieved thanks to the possibility of optimizing the loads in the distribution network and reducing the related costs.

• Supra-regional - can be located in the whole power system using the model of "copper plate", the synergy effect is achieved by the operation of units on one energy market or through the delivery of distributed system services

According to [3] VPP, we can divide the type of control :

• Centralized Controlled Virtual Power Plant - management takes place only in one central control unit responsible for all aspects of the VPP operation, which is connected directly to each DER (Distributed Energy Resource) managed by it

• Distributed Controlled Virtual Power Plant - management takes place in a multi-level manner through the cooperation of control units of various levels, i.e. the central control unit issues commands to lower-order control units and they issue commands to the control units of the next order or to DERs

• Fully Distributed Controlled Virtual Power Plant - DERs independently implement the control on the basis of data transferred by the VPP system, which has no direct impact on the functioning of generating units.

VPP can include many types of power units [4]:

• Uncontrollable Energy Sources - mainly Renewable Energy Sources, can occur both in the form of large generation units and smaller units requiring coordination.

• Controllable Energy Sources - they can exist occur as large generation units (coal fired units or CCGT), as well as smaller units, e.g. small cogeneration for the needs of a district heating system, and also controllable renewable energy sources, such as biogas, can be included in this group.

• Energy storage - they can occur both as large energy stores built to balance the network by the operator or large RES producers, as well as small energy stores installed in homes or smaller companies to balance and ensure the security of energy supply by its customers.

• Energy consumer - they can appear both as large industrial recipients, small and medium enterprises, local government units, public institutions and households.

In addition, consumers can be divided into the following categories:

• Passive - they are only consumers of electricity, they do not carry out any VPP management center commands, they only provide data on demand plans and measurement data.

• Active - having the tools and opportunities to change the volume of charged electricity:

• Prosumers - having their own generation sources and producing a part of energy for their own needs, as well as having the opportunity to resell energy to the power grid (at this point they are included in the VPP balance sheet as an energy source).

• DSR (Demand Side Response) - have ability to temporary decrease charged electricity by using own generation, energy stores or limiting of consumption.

5 Modules of the management centre

The universality of VPP applications means that the construction of the management center should meet the various needs of the aggregator and be able to change or expand depending on the needs. Therefore, the best solution in the construction of this type of systems is the use of a modular structure. The modules communicate with each other send information in a unified form. Additionally, thanks to the separation of individual modules, it is possible to update one of them, not the entire system. Similarly in the case of failures or errors, the protection at the module boundaries prevents them from spreading throughout the system.

The division into modules can have a different shape due to the needs of the users. An example of a division to cover basic VPP functionalities:

• Balancing module - responsible for scheduling production and consumption of electricity, participation in the Balancing Market, ongoing monitoring of production and energy consumption, balancing units within VPP.

• Forecasting module - forecasting the electricity demand and production of renewable sources, as well as the market situation on the basis of analyzes and external data, e.g. meteorological data.

• Production optimization module - calculation of optimal load distributed for each unit using economic and technical criteria.

• Trade module - trading electricity on the market, offering on the stock exchange, settlements of contracts.

• Analytical module - a wide range of analyzes of both internal factors such as technical capabilities of devices as well as external factors related to the market and power system.

• System services module - offering distribution and transmission system operators services services related to power reserves or regulation of system operation, performance of obligations resulting from capacity market contracts.

• Reporting module - fulfilment of reporting obligations in the scope of electricity generation and trading resulting from REMIT and other regulations.

6 An example of the operation of VPP

6.1 Integration of large and distributed power units

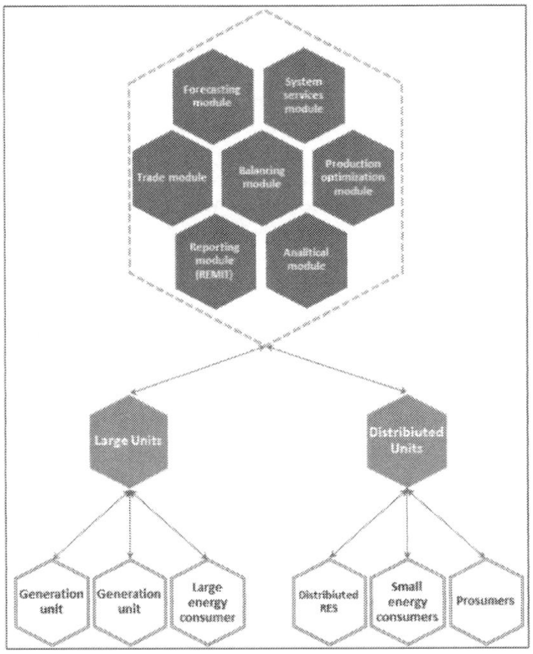

Fig. 1. Integration of large and distributed power units.

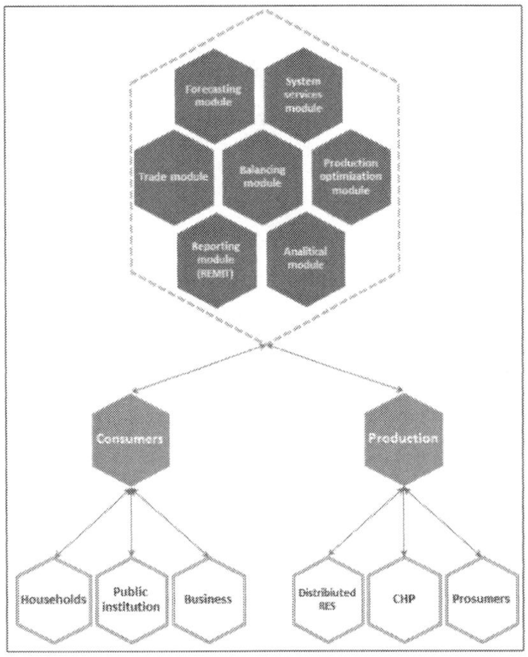

Fig. 2. Integration of energy units operating in a limited area

The first example of the use of VPP is the integration of units with a diverse work character and the volume of produced and consumed electricity.

Such VPPs may include:

• Large generation units and large electricity consumers:

- Generation units - conventional flexible power stations, electricity is produced continuously to cover the demand of industrial customers or the market demand

- Large consumption units - industrial plants consuming large amounts of electricity, characterized by continuity of demand and requiring power reliability due to production processes

• Distributed units:

- Many small electricity consumers - a network dispersed over a large area, characterized by high volatility of electricity demand and small volume of individual consumers

- Distributed renewable energy sources - a network of small and micro installations generating electricity, e.g. photovoltaic or wind turbines dispersed over a large area, are characterized by high variability and the inability to control the level of electricity production.

- Prosumers - units being both the consumer and the generator of electricity, part or all of the energy generated by the source is consumed on-side and in the case of excess or deficit of energy is balanced by the external network.

6.2 Integration of energy units operating in a limited area (clusters)

The second example of the use of VPP is the integration of units operating within the so-called "energy clusters", being a form of energy cooperative operating in a limited area that gathers local generation units and consumers for the purpose of regional development of renewable energy sources, as well as reducing energy and distribution costs.

Such VPPs may include:

• Electric energy consumers:

- Households - recipient with an obligatory approved tariff for electricity, characterized by high volatility and low volume of energy consumed by individual consumer, depending on the type of the cluster could be the main group of consumers,

- Public institutions - this group includes offices, schools, local government institutions, characterized by high volatility and moderate volume of energy consumed by individual consumer

- Companies - depending on the type of their business activity can be characterized by both stable and large volume or variable and low level of the volume of consumed energy

• Production units:

- Distributed renewable energy sources - a network of small and micro installations generating electricity, e.g. photovoltaic or wind turbines distributed over a large area, are characterized by high variability and the inability to control the level of electricity production.

- Local heat and power plants - generating units producing electricity and heat for the needs of the local heating network, depending on the technological system there is a possibility to change the level of energy production,

- Prosumers - units being both the consumer and the generator of electricity, part or all of the energy generated by the source is consumed on-side and in the case of excess or deficit of energy is balanced by the external network.

7 Regulatory and market environment affecting the development of Virtual Power Plants

7.1 Winter Package

On November 30, 2016, a package of legislative proposals was published. "Clean energy for all Europeans" (so-called " Winter Package ") containing a set of directives and regulations that together create a new model of the European electricity market. These documents define the rules support energy renewable, functioning common electricity market in the EU, define the principles of energy security and risk management, promoting the development of energy efficiency and functioning of regulatory authorities.

The entire Winter Package consists of 8 documents, but for the functioning of VPPs, the most important are 3 of them concerning common rules of the electricity market and renewable energy sources:

• Directive on the promotion of the use of energy from renewable sources

• Regulation on the internal electricity market

• Directive on common rules for the internal market in electricity

Their importance for the development of VPP results from the establishment of clear rules enabling the functioning of such systems in all EU countries, as well as through support for renewable energy sources (including largely prosumer sources) will increase the demand for such systems balancing and managing production sources.

7.2 Regulation and Directive on common rules for the internal market in electricity

The most important documents of the "Winter Package" are the Regulation and the Directive on common rules for the internal market in electricity. They contain articles that will have the strongest impact on the shape of the new common electricity market in the European Union. These documents concern the functioning of wholesale energy markets, network balancing, the operation of generation capacity mechanisms, as well as end-user service in concluding contracts with it, energy bills and the possibility of their active participation in the market.

Significant changes will be introduced in the functioning of SPOT and balancing markets. Due to the fact that these markets make it possible to participate in a larger number of flexible resources. It is necessary to implement an effective pricing mechanism so that it reflects power shortages and encourages market participants to actively respond to market signals. Thanks to the possibility of recovering the costs incurred in the moments of unbalancing the network, flexible resources will be available when the market needs them most without the need to implement additional power mechanisms. Therefore, all market participants should be financially responsible for imbalance that they cause in the system, while imbalance prices should reflect the value of energy in real time. It is also necessary to change electricity price limits on wholesale markets, so that they are flexible and automatically adjust to the market situation reflecting the maximum value of non-delivered energy. Additionally, until January 1, 2021, the imbalance settlement period must be 15 minutes in all graphic areas, unless regulatory authorities granted a derogation from this rule (by 31 December 2024 at the latest). All these changes introduce greater risk to the activity of energy market participants, and require them to have higher than previously quality forecasting of production and consumption levels. They also give a chance to obtain better economic results of units by using the available flexibility of units. All these circumstances create a space for the development of VPPs that will optimally manage and balance all energy resources, both reducing the risk of imbalance and giving the opportunity for additional revenues.

An important part of the energy market will be consumers. Until now, they were passive and most often bought electricity at regulated prices. This makes the offer for them unrelated to the market. On the one hand, it protects them from the risk of a sudden increase of prices, but on the other hand it does not give full opportunity to optimize their consumption, as would be in the case with dynamic prices. The introduced changes give each consumer the right to active participation in the energy market, directly or through the aggregator. In addition, the obligation to change the seller within 24 hours will be introduced, as well as the possibility of having several energy purchase contracts from many suppliers. These changes will positively affect the competitiveness of energy sales offers, and by simplifying and making the processes more flexible, they will activate the consumers to change the seller. The greater demand for such services will mean that entities offering such contracts will need tools such as VPPs that will allow flexible management of the constantly changing portfolio of consumers.

Another important element of the new common electricity market will be its full digitization. During the introduction of smart metering systems, many models of managing these data were created. It is necessary to introduce transparent rules of access to data on non-discriminatory terms and providing cyber security. These systems will enable active participation of consumers in the electricity markets if they are interoperable and able to generate the data needed for the proper functioning of consumer energy management systems. For this purpose, appropriate standards implementing the best practices in this area should be applied, so that market participants can easily adapt their IT systems to uniform technical requirements. The introduction of uniform principles of smart metering will facilitate the development of VPP through the possibility of applying the same standards in the areas of various operators and various energy markets. This will improve the offer of IT solutions in this area, because it will be possible to easily transfer the best solutions used in this field in other systems and will eliminate the need to develop many standards adapted to different energy markets. [5] [6]

7.3 Directive on the promotion of the use of energy from renewable sources

The RES Directive establishes a common framework for the promotion of energy from renewable sources. It sets a EU overall target for the total share of energy from renewable sources in gross final energy consumption in the Union in 2030 at 32%. It also establishes rules on financial support for electricity from renewable sources and prosumption.

One of the main goals of the Winter Package is to make the electricity market customer-friendly. In the RES Directive, it is implemented by introducing rules facilitating the operation of renewable energy prosumers. The Directive ensures the right to become a prosumer of renewable energy, while maintaining the rights and obligations of the final consumer of energy. Prosumers can produce renewable energy, also for their own needs, store and sell surpluses of their renewable electricity produced through electricity suppliers or through peer-to-peer arrangements. At the same time, they are not subject to discriminatory or disproportionate procedures, as well as to any fees that do not reflect real costs. They also have the right to participate in support schemes that cover energy from renewable energy entering the network.

The Winter Package also provides end users with the opportunity to participate in energy communities operating in the field of renewable energy while maintaining their rights and obligations as consumers, provided that their participation is not the main economic activity. Energy communities will have the right to produce, use for their own needs, store and sell renewable energy as well as access to all their respective energy markets. Importantly, the operator of a given distribution system should cooperate with such a community, and fees related to the use of the system should reflect the real level of costs, without inadequate fees

Such a shape of RES support regulation generates significant potential for the development of VPP systems, especially in the case of prosumers and energy communities. VPP is an ideal solution for energy communities that bring together renewable energy producers as well as energy consumers located in a small area. Unified basis for the functioning of such organizations, adapting the system to individual needs will allow to maximize the economic effect of these communities and allow better use of available resources to ensure the security of energy supply. [7]

8 Summary

Virtual Power Plants are an ideal tool that can meet the needs of modern energy sector in the management of a group of diversified energy units. The versatility of solutions and the ability to adjust the shape of VPPs to the needs makes that in the near future systems of this type will be used by an increasing number of entities. The regulatory environment will also be conducive to the development of VPP. The increasing amount of distributed renewable energy sources, and on the other hand, the ever-increasing balancing requirements will make the implementation of such systems necessary for conducting economically justified activities in this area. The new regulations also give opportunities for better use and giving more value to the flexibility of generating units and energy storage. All these factors in the near future will make VPP the most important tool for entities operating on the electricity market.

References

1. Bayar, T. (2013). Virtual Power Plants: A New Model for Renewables Integration. *Renewable Energy World*, Vol 16(5), 2013.
2. Ropuszynska-Surma, E., & Weglarz, M. (2018). A Virtual Power Plant as a Cooperation Network. *Marketing and Management of Innovations*, 4, 136-149.
3. Nikonowicz Ł., Milewski J., Virtual Power Plants – general review: structure, application and optimization, *Journal of Power Technologies* 92 (3) (2012)
4. Rafał Magulski, Koncepcja funkcjonowania wirtualnej elektrowni integrującej rozproszone źródła energii jako dostawcy usług systemowych, *Zeszyty Naukowe Wydziału Elektrotechniki i Automatyki PG*, Nr 53/2017 (2017)
5. *Regulation (EU) 2019/943 of the European Parliament and of the Council of 5 June 2019 on the internal market for electricity.*
6. *Directive (EU) 2019/944 of the European Parliament and of the Council of 5 June 2019 on common rules for the internal market for electricity and amending Directive 2012/27/EU.*
7. *Directive (EU) 2018/2001 of the European Parliament and of the Council of 11 December 2018 on the promotion of the use of energy from renewable sources.*

 https://doi.org/10.1051/e3sconf/201913701045

RESEARCH & DEVELOPMENT IN POWER ENGINEERING, 2019

Quantity and Quality of Dual Fuel IC Marine Engine Exhaust Sensible Waste Heat Considering Different Sulphur Content in Fuels

Mateusz Przybyła[1], *Andrzej* Adamkiewicz[2]

[1] mateusz.przybyla@wp.pl, Poland
[2] Maritime University in Szczecin, a.adamkiewicz@am.szczecin.pl, Poland

Abstract. Maritime transport is facing a significant set of technical challenges due to planned by the International Maritime Organization implementation of ecological criterions on 01 January 2020 and 2021 regarding the emission of sulphur oxides, carbon dioxide and nitrogen oxides. The advantageous properties of natural gas (NG) as fuel in conjunction with dual fuel (DF) internal combustion (IC) engines potentially enables the fulfilment of all planned criterions. However, if the methane emission of DF IC is taken into consideration in CO2 emission balance it completely devaluates the advantages arising from high hydrogen to carbon ratio of NG. On the other hand, the planned global sulphur cap in combination with its low content in NG potentially enables to recover higher rates of waste heat and exergy of combustion products without the risk of low temperature corrosion (compared to liquid fuel). In this study the influence of sulphur content in NG and pilot fuel on the sulphuric acid condensation temperature was investigated in order to determine the rate of waste heat (quantity) and exergy (quality) of four stroke DF IC engine combustion products. In ideal process scenario (exergy based) the methane slip effect compensation was achieved only for a 0,5 engine load.

1 Introduction

The decision made by the International Maritime Organization (IMO) on the simultaneous implementation of regulations on 01 January 2020 regarding the global sulphur content in marine fuel and the energy efficiency of ships poses a new, complex set of problems for maritime transport. In the case of vessels operated within the Baltic and North Sea areas this set of problems is extended by the implications resulting from the decision to classify them on 01 January 2021 as Emission Control Area (ECA) [1].

Using low-sulphur fuel (fuel with sulphur mass fraction not exceeding 0.5%) in order to meet the first of these criteria leaves the problem of excessive emission of nitrogen oxides unsolved. Meeting the Tier III criterion implies the need to process combustion products in the Selective Catalytic Reduction (SCR) system. On the other hand, the SCR system significantly limits the possibility of exhaust waste heat conversion due to the required minimum temperature determining the efficiency of the process [2]. In addition, the amount of available waste heat flux of liquid fuel exhaust gases and the efficiency of its conversion is limited mainly by the condensation temperature of the sulphuric acid [3].

Liquified Natural Gas (LNG) is an alternative marine fuel that reduces the problems described above. The advantageous properties of natural gas (NG) enabling the fulfilment of the IMO ecological criteria are the main reason for significant changes in the configuration of LNG tanker power systems (LNGC - Liquified Natural Gas Carriers). The quantitative shares of ship power systems with steam turbines (ST - Steam Turbine) recognized on the basis of [4] had amounted to 100% in the LNGC population until 2004. Due to the higher effective efficiency of dual fuel (DF) piston engines and favorable ecological indicators, the current shares of ST power systems are only about 50% (Figure 1).

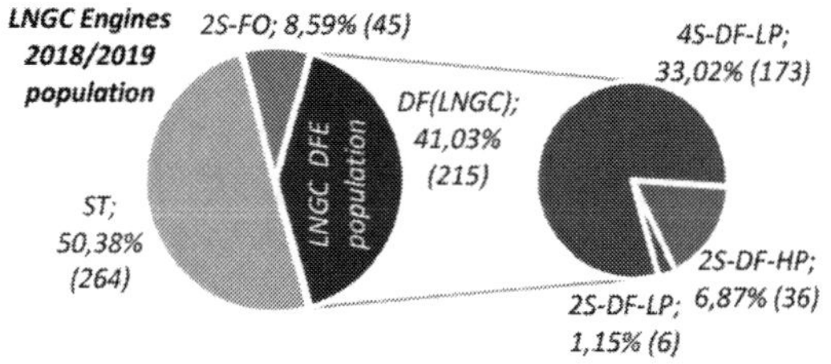

Fig. 1. Quantitative shares of engines of the LNGC population.

Global population of marine DF heat engines (Figure 2) was created using the LNGC population of dual fuel engines (Figure 1) and the population of Liquified Natural Gas Fueled Ships (LNGFS) identified on the basis of [5].

The research problem studied in this work was separated from the identified set of problems on DF heat engines population based on the criterion of its position in the hierarchy of current problems and the number of objects it concerns. The group, constituting 71% of the global population of marine DF heat engines, are four stroke DF IC engines which uses liquid fuel (Common rail pilot-fuel injection [2]) to ignite natural gas-air lean mixture ($\lambda > 1$). One of the most important problems related to this group of engines is the significant emission of methane to the atmosphere [6, 7].

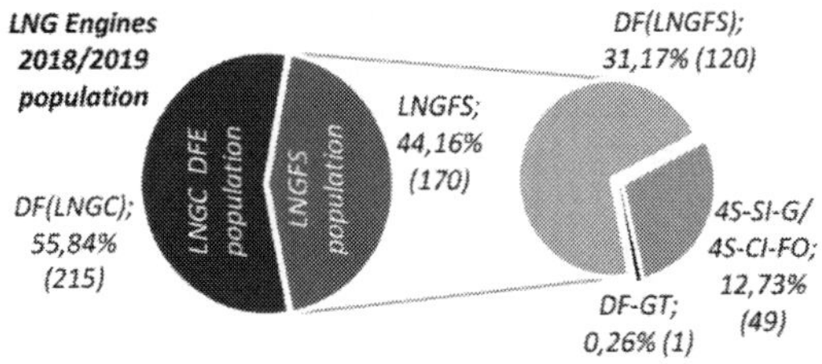

Fig. 2. Quantitative shares of engines of the LNG population.

Research that has been conducted and described in [6] points out the negative effect of methane slip resulting in excessive greenhouse gases emission. On the other hand, LNG properties potentially limit the described difficulties of waste heat conversion of liquid fuel exhaust gases [3, 8, 9, 10].

The reason to consider the possibility of reducing the methane slip effect by increasing the rate of recovered waste heat of dual-fuel engine exhaust were the following observations:

- possibility of significant reduction of the minimum exhaust gas temperature due to the insignificant sulphur content in LNG [9, 10],
- higher temperatures of natural gas exhaust in the turbocharger outlet section than liquid fuel exhaust temperatures [2],
- the possibility of utilizing waste heat at low temperatures in cycles with an organic working medium [3, 8, 9, 10],
- evolution of Organic Rankine Cycle (ORC) marine WHRS, which is proved by the physical implementation of such a system described in [11].

A slightly different position regarding observation a) is presented in publications [12, 13] warning about the observed sulphur content in natural gas that may lead to condensation of sulphuric acid on the heat transfer surface. In the case of the considered group of engines, the sulphur present in the combustion process is also fed in pilot fuel in the amount resulting from the IMO regulation. That is why this study considers the influence of sulphur content in LNG and pilot fuel on:

a) the sulphuric acid condensation temperature,
b) amount of available exhaust waste heat (quantity),
c) exergy of combustion products (quality),
d) the potential to compensate the methane slip effect.

2 Thermodynamic analysis

As the subject of this study the combustion products of marine dual fuel four stroke internal combustion engine (MAN B&W 51/60DF [2]) were considered. Thermodynamic analysis was carried assuming:

- excess air coefficient $\lambda = idem = 2{,}2$,
- homogeneous mixture of NG, air and liquid fuel,
- engine maximum effective power $\dot{W}_{e(M)} = 1\ MW$,
- air parameters – temperature $T_0 = 298{,}15\ K$, pressure $p_0 = 101{,}325\ kPa$ and relative humidity $\varphi = 0{,}7$,
- range of sulphur mass fraction in liquid fuel $0{,}001 \leq x_{S(FO)} \leq 0{,}005$ according to IMO [1],
- range of sulphur mole fraction in natural gas $1 \cdot 10^{-6} \leq y_{S(NG)} \leq 20 \cdot 10^{-6}$ according to [12, 13].

The molar flow rates balance of NG, FO and air constituents was considered as for an open steady-state thermodynamic system (Figure 3). Combustion products parameters based on abovementioned balance were used to calculate: the condensation temperature of sulphuric acid, exhaust specific heat as a function of temperature later used to establish the time rate of waste heat transfer. The quality of the waste heat source was assessed using exergy and the waste energy quality index. Based on the calculated time rate of exhaust exergy flow the maximum reduction potential of carbon dioxide emission due to ideal energy conversion in WHRS was determined. The methane slip effect was included in the carbon dioxide balance by the methane specific emission and its global warming potential (GWP) indicator.

2.1 Molar flow rates balance of DF IC engine

The balance (1) of incoming and outgoing fluids crossing the thermodynamic system boundary (Figure 3) is based on matter conservation principle. The molar flow rates entering the control volume are natural gas $\dot{n}_{NG}\ [mol/s]$, liquid fuel $\dot{n}_{FO}\ [mol/s]$ and air $\dot{n}_a\ [mol/s]$.

$$\sum_{i=1}^{m} \dot{n}_i + \sum_{j=1}^{n} \dot{n}_j + \dot{n}_a = \sum_{k=1}^{n} \dot{n}_k \qquad (1)$$

$$\dot{n}_{NG} + \dot{n}_{FO} + \dot{n}_a = \dot{n}_{ex} \qquad (2)$$

The outgoing exhaust molar flow rate $\dot{n}_{ex}\ [mol/s]$ was determined as the sum of incoming molar flow rates.

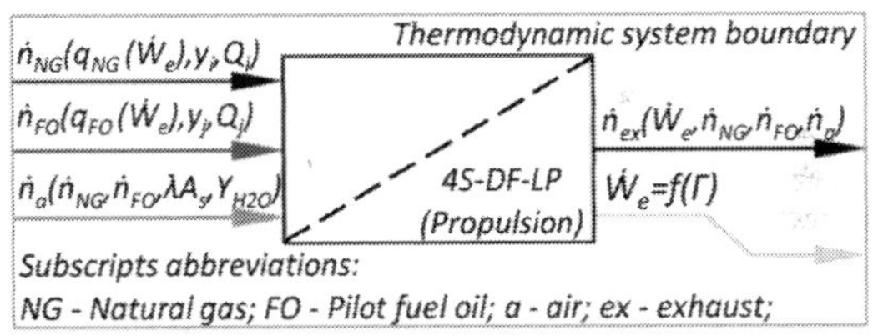

Fig. 3. Diagram of parameters relations in studied system.

Molar flow rates of both fuels $\dot{n}_{FO} + \dot{n}_{NG}\ [mol/s]$ were determined basing on: the required specific heats of natural gas $\dot{q}_{NG}\ [J/Ws]$ and liquid fuel $\dot{q}_{FO}\ [J/Ws]$ published in [17], components mole fractions y_i, y_j of fuels and their heat of combustion $\tilde{Q}_i, \tilde{Q}_j\ [J/mol]$ taken from [4, 12, 13].

$$\dot{n}_{C_\alpha H_\beta O_\gamma S_\delta} = \dot{n}_{FO} + \dot{n}_{NG} = \dot{W}_e \left(\frac{\dot{q}_{NG}}{\sum_{i=1}^{m} y_i \tilde{Q}_i} + \frac{\dot{q}_{FO}}{\sum_{j=1}^{n} y_j \tilde{Q}_j} \right) \qquad (3)$$

The determined molar flow rates of fuels were used to calculate by equation (4) their mole fractions as a function of engine load $y_{NG}, y_{FO} = f(\Gamma = \dot{W}_e / \dot{W}_{e(M)})$.

$$y_{NG} = \frac{\dot{n}_{NG}}{\dot{n}_{C_\alpha H_\beta O_\gamma S_\delta}},\ y_{FO} = \frac{\dot{n}_{FO}}{\dot{n}_{C_\alpha H_\beta O_\gamma S_\delta}} \qquad (4)$$

Due to a significant number of NG and FO constituents the analysis of the combustion process was based on the stoichiometric combustion balance of quasi-molecule $C_\alpha H_\beta O_\gamma S_\delta$ of fuel given by (5).

$$\dot{n}_{C_\alpha H_\beta O_\gamma S_\delta}\left[1 + \lambda \tilde{A}_s\left(1 + 3{,}76 + Y_{H_2O}\right)\right] = \dot{n}_{C_\alpha H_\beta O_\gamma S_\delta}\{\alpha + \delta + \tfrac{\beta}{2} + \lambda \tilde{A}_s[Y_{H_2O} + (1 - \tfrac{1}{\lambda}) + 3{,}76] \quad (5)$$

The stoichiometric mole ratio of oxygen and fuel $\tilde{A}_s\ [mol(O_2)/1\ mol]$ was determined basing on eq. (6) and the number of carbon α, hydrogen β, oxygen γ and sulphur δ atoms identified by the system of equations (7).

$$\tilde{A}_s = \alpha + \frac{\beta}{4} - \frac{\gamma}{2} + \delta \quad (6)$$

$$\begin{cases} \alpha = y_{NG} \sum_{i=1}^{m} y_i\, \alpha_i + y_{FO} \sum_{j=1}^{n} y_j\, \alpha_j \\ \beta = y_{NG} \sum_{i=1}^{m} y_i\, \beta_i + y_{FO} \sum_{j=1}^{n} y_j\, \beta_j \\ \gamma = y_{NG} \sum_{i=1}^{m} y_i\, \gamma_i + y_{FO} \sum_{j=1}^{n} y_j\, \gamma_j \\ \delta = y_{NG} \sum_{i=1}^{m} y_i\, \delta_i + y_{FO} \sum_{j=1}^{n} y_j\, \delta_j \end{cases} \quad (7)$$

The mole ratio of water vapor contained in the air to oxygen $Y_{H_2O}[mol(H_2O)/mol(O_2)]$ was calculated using equation (8).

$$Y_{H_2O} = \frac{n_{H_2O}}{n_{O_2}} \equiv \frac{p_{H_2O}}{p_{O_2}} = \frac{\varphi p_{H_2O(s)}}{y_{O_2}(p_a - \varphi p_{H_2O(s)})} \quad (8)$$

The saturation vapor pressure as a function of temperature $p_{H_2O(s)} = f(T)$ was determined using the Antoine equation with the values of coefficients according to [14].

$$p_{H_2O(s)} = e^{\left(A - \frac{B}{T+C}\right)} = e^{\left(16{,}262 - \frac{3799{,}89}{T - 46{,}38}\right)} \quad (9)$$

Partial pressures of combustion products components $p_k\ [Pa]$ were calculated using Dalton's law given by eq. (10) and molar flow rates balance (1) and (2) solved on the basis of presented equations (3)-(9).

$$p_k = y_k p_{ex} = \frac{\dot{n}_k}{\sum_{k=1}^{n} \dot{n}_k} p_{ex} = \frac{\dot{n}_k}{\dot{n}_{ex}} p_{ex} \quad (10)$$

2.2 Condensation temperature of H_2SO_4

The sulphuric acid condensation temperature was determined using the Verhoff-Banchero empirical relationship published in [15] as a following function $T_{H_2SO_4(c)} = f\left(p_{H_2O}, p_{H_2SO_4}\right)$. Partial pressure of water vapor at equilibrium state $p_{H_2O}\left(p_{H_2O(\alpha)}; p_{H_2O(\beta)}\right)$ was determined using eq. (1)-(10). It was assumed that water vapor is composed of H_2O molecules introduced with ambient air and H_2O molecules formed in the hydrogen oxidation process. Partial pressure of sulfuric acid $p_{H_2SO_4}$ is determined by the efficiency of the oxidation process of sulphur dioxide to sulphur trioxide (assumed $\eta_{SO_3} = 0{,}05$ [16]) and the efficiency of the gaseous SO_3 and H_2O conversion process to gaseous H_2SO_4 (assumed $\eta_{H_2SO_4} = 1$ [14]). The determined H_2SO_4 condensation temperature was assumed as the minimum temperature of exhaust gases ($T_{H_2SO_4(c)} = T_{ex(1)}$) leaving the WHRS.

2.3 The rate of exhaust waste heat and exergy

The rate of exhaust sensible waste heat $\dot{Q}_{ex}$ was identified on the basis of solved balance (11), exhaust gas temperature leaving the turbocharger $T_{ex(0)}$ [2] and calculated condensation temperature of sulphuric acid assuming that $T_{H_2SO_4(c)} = T_{ex(1)}$.

$$\dot{Q}_{ex} = \dot{n}_{ex} \Delta \tilde{h}_{ex} = \dot{n}_{ex} \int_{T_{ex(0)}}^{T_{H_2SO_4(c)}} \tilde{c}_p(T) dT \quad (11)$$

Molar specific heat was calculated using eq. (12) as the sum of products – mole fractions of exhaust components y_k and their molar specific heats as a function of temperature $\tilde{c}_{pk} = f(T)$ [14, 17].

$$\tilde{c}_p(T) = \sum_{k=1}^{5} y_k \tilde{c}_{pk}(T) = aT^3 + bT^2 + cT + d \quad (12)$$

Equation (13) was obtained substituting the polynomial (12) into the relation (11) and solving the integral within the limits of the identified exhaust temperatures.

$$\dot{Q}_{ex} = \dot{n}_{ex} \Delta \tilde{h}_{ex} = \dot{n}_{ex} \int_{T_{ex(0)}}^{T_{H_2SO_4(c)}} \tilde{c}_p(T) dT$$
$$\dot{n}_{ex}\left(\frac{a}{4}\Delta T^4 + \frac{b}{3}\Delta T^3 + \frac{c}{2}\Delta T^2 + d\Delta T\right) \quad (13)$$

Exergy of combustion products indicates the maximum theoretical work that can be obtained in the clock-wise reversible thermodynamic cycle realizing ideal energy conversion, i.e. $B_{ex} \equiv W_{ex(id)}\ [J]$. From this it results that the time rate of exhaust exergy flow determines the maximum power that potentially can be generated in the clockwise reversible thermodynamic cycle (maximum time rate of work that can be produced in ideal conditions using exhaust gases as a heat source of reversible cycle), i.e. $\dot{B}_{ex} \equiv \dot{W}_{ex(id)}\ [J/s = W]$.

$$\dot{B}_{ex} = \dot{Q}_{ex} \psi_{ex} = \dot{n}_{ex} \Delta \tilde{h}_{ex}\left(\tilde{b}_{ex}/\Delta \tilde{h}_{ex}\right) = \dot{n}_{ex} \tilde{b}_{ex} \quad (14)$$

The rate of exhaust gases exergy flow $\dot{B}_{ex}$ was calculated using function of logarithmic mean temperature $\bar{T}_{ex}$ (assumed as only variable parameter) represented by the eq. (15) – waste heat quality index ψ_{ex} [8]. Sea water was introduced to the thermodynamic cycle as cooling medium of condenser. Assumed fixed cold sink temperature was $T_0 = 298{,}15\ K = idem$.

$$\psi_{ex} = 1 - \frac{T_0}{\bar{T}_{ex}} = 1 - \frac{T_0}{\frac{T_{ex(0)} - T_{H_2SO_4(c)}}{\ln\left(T_{ex(0)}/T_{H_2SO_4(c)}\right)}} \quad (15)$$

Considered waste heat quality index ψ_{ex} in this paper takes into account only the quality of thermal energy, neglecting the energy of exhaust pressure.

2.4 Emission balance of CO_2 considering CH_4

The total mass flow rate of emitted carbon dioxide $\dot{m}_{\Sigma CO_2}[g/s]$ is composed of CO_2 molecules formed in combustion process of fuel $\dot{m}_{CO_2}$ and the so-called equivalent CO_2 mass flow rate $\dot{m}_{CO_2(eq)}^{CH_4}$. The equivalent mass flow rate of carbon dioxide takes into account the impact of methane on the greenhouse effect through the

Global Warming Potential (GWP). GWP indicates the amount of heat absorbed of certain mass of CH_4 relative to the same mass of CO_2. Assuming that the life cycle of CH_4 molecule in the atmosphere is hundred years, the GWP of methane is $GWP_{CH_4}^{100} = 21$ [6].

$$\dot{m}_{\Sigma CO_2} = \dot{m}_{CO_2} + \dot{m}_{CO_2(eq)}^{CH_4} = \dot{W}_e\left(\bar{e}_{CO_2} + \bar{e}_{CH_4}GWP\right) \quad (16)$$

The average specific emission of carbon dioxide $\bar{e}_{CO_2} = 444{,}2\ g/kWh$ and methane $\bar{e}_{CH_4} = 6{,}9\ g/kWh$ was taken for the considered engine type according to [7].

The maximum reduction of carbon dioxide emission $\Delta\dot{m}_{CO_2(id)}$ that can be obtained in the ideal process of waste heat conversion of exhaust gases was determined basing on the eq. (14) – the definition of exergy flow rate ($\dot{B}_{ex} \equiv \dot{W}_{ex(id)}\ [W]$).

$$\Delta\dot{m}_{CO_2(id)} = \dot{W}_{ex(id)}\left(\bar{e}_{CO_2} + \bar{e}_{CH_4}GWP\right) \quad (17)$$

The potential of the methane slip effect reduction due to ideal conversion of exhaust sensible waste heat into mechanical work was assessed basing on the eq. (18)

.

$$\Delta\dot{m}_{CO_2(CH_4)} = \dot{m}_{CO_2(eq)}^{CH_4} - \Delta\dot{m}_{CO_2(id)} \quad (18)$$

Fulfilling the inequality $\Delta\dot{m}_{CO_2(CH_4)} > 0$ means that there is no compensation for the methane slip effect, if the inequality $\Delta\dot{m}_{CO_2(CH_4)} < 0$ is fulfilled, the effect is fully compensated.

3 Analysis results and discussion

The solution to the problems formulated in the introduction was carried out for two sets of operational parameters of a selected engine determined by the set of engine loads $\Gamma = idem = \{0{,}5; 0{,}85\}$. The lowest load value $\Gamma = 0{,}5$ was chosen based on the highest sulphur mole fraction criterion (Figure 4), the design load value $\Gamma = 0{,}85$ was chosen based on the highest frequency of occurrence in operational profile. The diagrams in Figures 4 and 5 were calculated for two different values of sulphur mass fraction in FO.

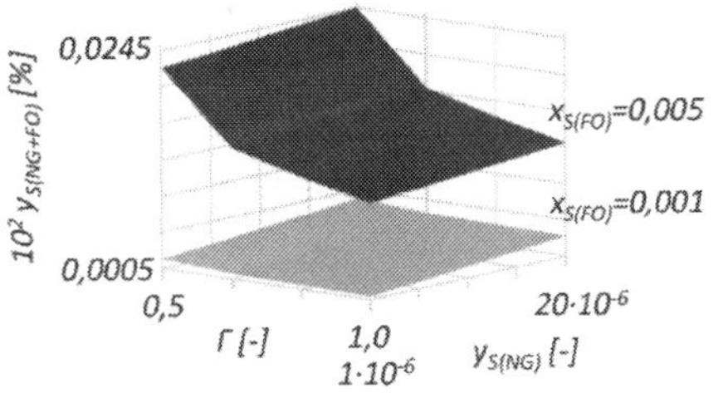

Fig. 4. Total sulphur mole fraction in fuel as a function of engine load and sulphur mole fraction in NG.

In order to solve selected set of problems the following were created, for: problem a) fig. 6, problem b) fig. 7, 8, 9, problem c) fig. 10, 11, 12, problem d) fig. 13.

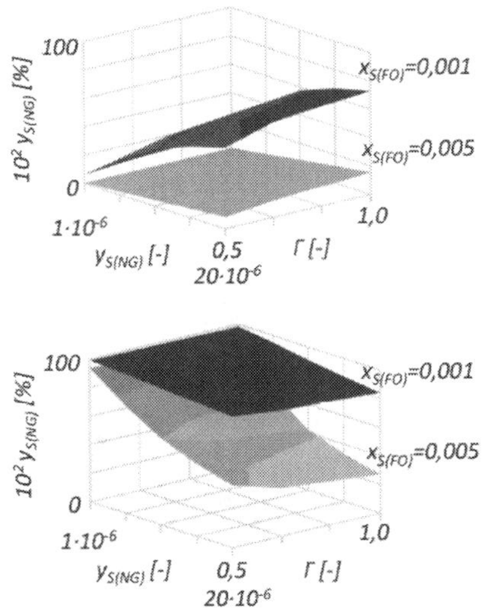

Fig. 5. Sulphur mole fraction of NG and FO as a function of engine load and sulphur mole fraction in NG.

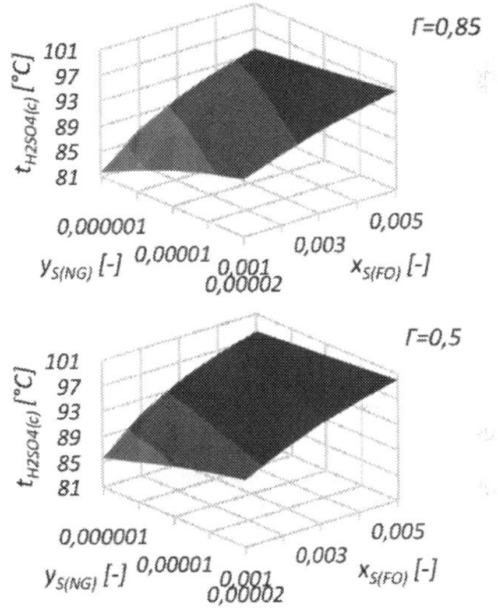

Fig. 6. Condensation temperature of sulphuric acid as a function of sulphur mole fraction in NG and sulphur mass fraction in FO.

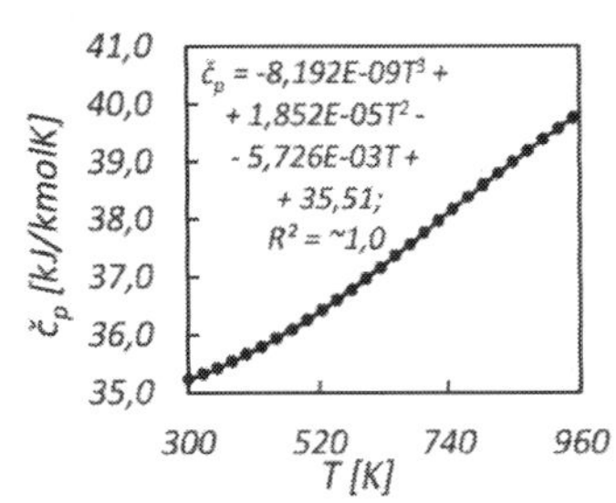

Fig. 7. Molar specific heat of isobaric process of exhaust gases as a function of absolute temperature.

By substituting the coefficients presented in Figure 7 into eq. (13), the relationship (19) describing the rate of waste heat flux of exhaust gases $\dot{Q}_{ex}$ as a function of temperature difference was determined.

$$\dot{Q}_{ex} = \dot{n}_{ex} \int_{T_{ex(0)}}^{T_{H_2SO_4(c)}} \tilde{c}_p(T)dT = \dot{n}_{ex} \ (\underbrace{a/4}_{-2{,}048\cdot 10^{-9}} \ \Delta T^4 + \underbrace{b/3}_{6{,}173\cdot 10^{-6}} \ \Delta T^3 + \underbrace{c/2}_{-2{,}863\cdot 10^{-3}} \ \Delta T^2 + \underbrace{d}_{35{,}51} \ \Delta T) \quad (19)$$

The rates of waste heat of combustion products presented in Figure 9 were calculated using eq. (19) and the temperature differences presented in Figure 8.

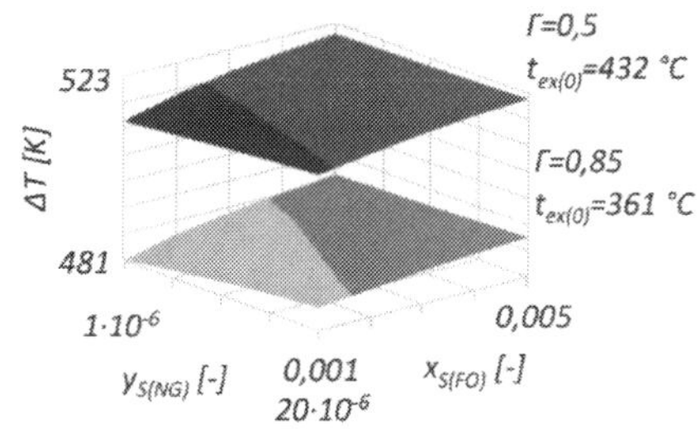

Fig. 8. Temperature differences of combustion products as a function of sulphur fractions – mole in NG and mass in FO.

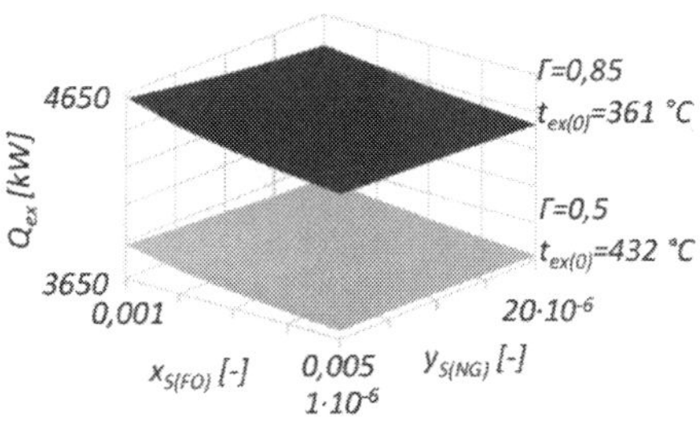

Fig. 9. The rate of combustion products waste heat as a function of sulphur mole fraction in NG and sulphur mass fraction in FO.

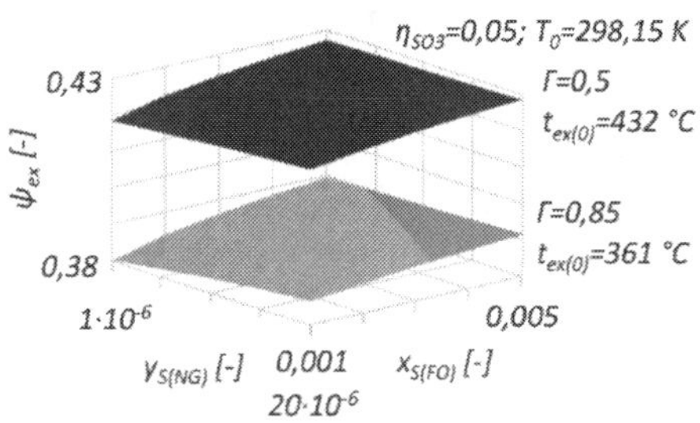

Fig. 10. Waste heat quality index of combustion products as a function of sulphur fractions – mole in NG and mass in FO.

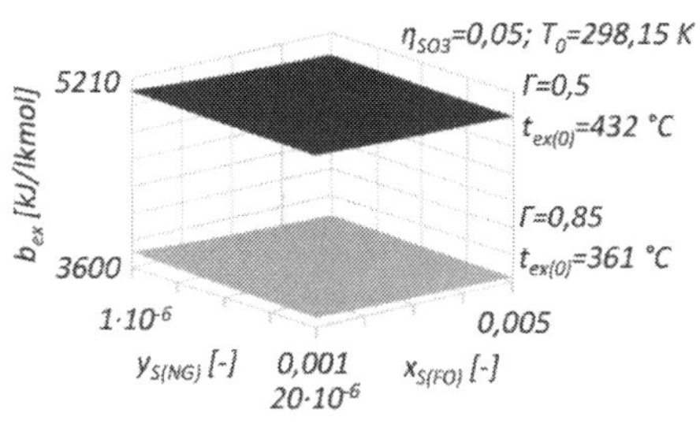

Fig. 11. Molar specific exergy of exhaust gases as a function of sulphur mole fraction in NG and sulphur mass fraction in FO.

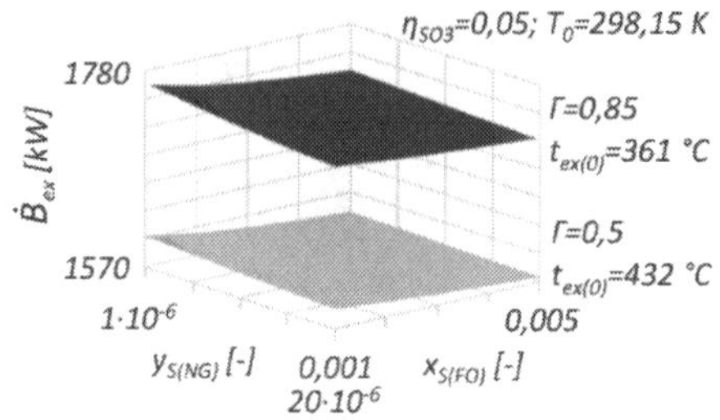

Fig. 12. Exergy flow rate of exhaust gases as a function of sulphur mole fraction in NG and sulphur mass fraction in FO.

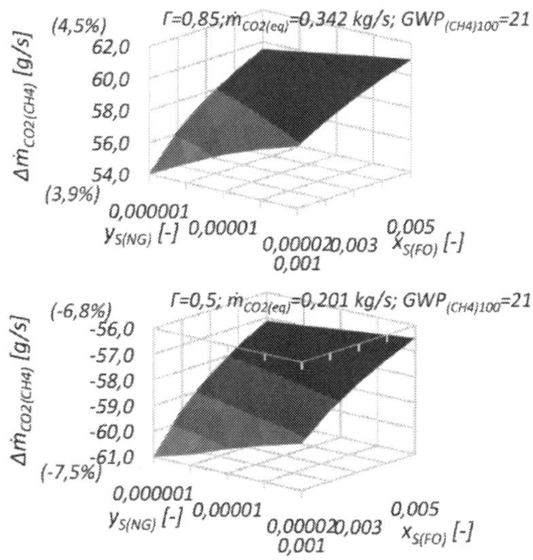

Fig. 13. Equivalent carbon dioxide mass flow rate as a function of sulphur mole fraction in NG and sulphur mass fraction in FO.

According to thermodynamic analysis results the following was observed (notation the same as for identified set of problems):

a) sulphuric acid condensation temperature is influenced by engine load and sulphur content in fuels (Fig. 6). The highest sulphuric acid condensation temperature occurred at engine load of $\Gamma = 0{,}5$ which corresponded with the highest sulphur mole fraction in fuels (Fig. 4).
b) The highest rate of waste heat of combustion products (Fig. 8) was observed at the load of $\Gamma = 0{,}85$ which results from higher molar flow rate, despite lower exhaust temperature differences (Fig. 8). These observations have a significant impact on the quality of waste heat of exhaust.
c) More favorable indicators of the waste heat quality of combustion products were observed for the lower engine load of $\Gamma = 0{,}5$ (Fig. 10). The reason for this is the following temperature relation $T_{ex(0)}(\Gamma = 0{,}5) > T_{ex(0)}(\Gamma = 0{,}85)$ (with the assumption of constant temperature of the environment). An analogous relationship occurs for molar specific exergies presented in Figure 11. Despite observed differences of analyzed engine loads parameters, the exergy flow rates are comparable (Fig. 12).

d) The potential to compensate the methane slip effect was observed in Fig. 13 only for the engine load of $\Gamma = 0{,}5$ which satisfied the inequality $\Delta\dot{m}_{CO_2(CH_4)} > 0$. For the engine load of $\Gamma = 0{,}85$ observed equivalent mass flow rate of carbon dioxide satisfied the inequality $\Delta\dot{m}_{CO_2(CH_4)} < 0$ meaning that the methane slip effect was not compensated.

4 Conclusions and future research

The present study has examined the influence of future IMO global, and actual (ECA), regulation of sulphur mass fraction in marine fuels, as well as possible volume fraction of sulphur in NG, on the sulphuric acid condensation temperature of four stroke DF IC engine combustion products, thus on the rate of its waste heat and exergy. The particular attention has been paid to the possibility of methane emission compensation by conversion of waste heat in WHRS into effective power. Above has been assessed only for combustion products using exergy to determine effective power generation limits. Thermodynamic analysis has been conducted for two engine loads – nominal (or design) load ($\Gamma = 0{,}85$) and low load ($\Gamma = 0{,}8$). Using values determined in analysis and reported in the literature – the mean specific emission of CO_2 and CH_4 of four stroke DF IC engine and the Global Warming Potential factor of CH_4, the emission rates of CO_2, equivalent CO_2 (eq. 16) and maximum possible reduction of CO_2 emission in the ideal exhaust waste heat conversion process (eq. 17) were determined. Negative values of balanced CO_2 emission rates which are indicating the potential to compensate the methane slip effect were observed only for low engine load.

The results of this study allowed following conclusions to be made:

1. Despite relatively low sulphur volume fractions in NG and mass fraction in pilot FO the condensation of sulphuric acid can occur at high exhaust temperatures which are also related with engine load. It indicates the importance of H_2SO_4 condensation analysis in design process of DF IC engine exhaust sensible waste heat recovery systems on ships in order to utilize maximum energy avoiding at the same time low temperature corrosion of heat transfer surfaces.
2. The main limitation of conducted analysis is the idealization of waste energy conversion process whose assumptions cannot be met in real conditions. However, determined boundary values under these assumptions led to another conclusion.
3. Compensation of four stroke DF IC marine engine methane emission and its influence on global warming by means of exhaust sensible waste heat conversion needs to be supported in real conditions by other CO_2 reduction technologies.

The combination of above conclusions and availability of acid-proof engineering materials presented for example in [18] led to the decision on future research in the field of combustion products latent waste heat recovery by introducing condensing heat exchangers to marine power plants.

References

1. Lloyd's Register, Future IMO and ILO Legislation, (2019).
2. MAN B&W, MAN 51/60DF Project Guide, (2017).
3. Wu Sh.-Y., Li Ch., Xiao L., Li Y.-R., Liu Ch., The role of outlet temperature of flue gas in organic Rankine cycle considering low temperature corrosion, *Journal of Mechanical Science and Technology*, Vol. 12, pp. 5213-5219, (2014).
4. International Gas Union, World LNG Report, (2019).
5. DNV-GL, LNG fueled fleet statistics, (2019), available at: http.://afi.dnvgl.com/Statistics, (accessed 22.09.2019).
6. Attah E. E., Bucknall R., An analysis of the energy efficiency of LNG ships powering options using the EEDI, *Ocean Engineering*, Vol. 110, pp. 62-74, (2015).
7. Stenersen D., Thonstad O., GHG and NOx emissions from gas fueled engines, *SINTEF Ocean AS*, (2017).
8. Adamkiewicz A., Michalski R., Zeńczak W., Wybrane problemy technologii konwersji energii w okrętowych systemach energetycznych, *Kaprint*, (2012).
9. Mondejar M. E., Andreasen J. G., Pierobon L., Larsen U., Thern M., Haglind F., A review of the use of organic Rankine cycle power systems for maritime applications, *Renewable and Sustainable Energy Reviews*, Vol. 91, pp. 126-15, (2018).
10. Singh D. V., Pedersen E., A review of waste heat recovery technologies for maritime applications. *Energy Conversion and Managment*, Vol. 111, pp. 315-328, (2016).
11. Haglind F., Mondejar M. E., Andreasen J. G., Pierobon L., Meroni A., Organic Rankine cycle unit for waste heat recovery on ships (Pilot ORC), *Danish Technical University*, (2017).
12. CIMAC, Impact of Gas Quality on Gas Engine Performance, (2015).
13. The European Association of Internal Combustion Engine Manufacturers, Total Sulphur Levels in Natural Gas with Special Consideration of IC Engines, (2012).
14. Szarawara J., Termodynamika chemiczna, *Wydawnictwa Naukowo-Techniczne*, (1985).
15. Verhoff F. H., Banchero J. T., Predicting dew points of flue gases, *Chemical Engineering Progress*, Vol. 8, pp. 71-72, (1974).
16. Cordtz R., Schramm J., Rabe R.: Investigating SO_3 formation from the Combustion of Heavy Fuel Oil in a Four-Stroke Medium-Speed Test Engine, *Energy & Fuels*, Vol. 10, pp. 6279-6286, (2013).
17. Burcat A., Ruscic B., Third Millennium Ideal Gas and Condensed Phase Thermochemical Database for Combustion with updates from Active Thermochemical Tables, *Argonne National Laboratory*, (2005).
18. Xiong Y., Tan H., Wang Y., W., Mikulčić H., Duić N., Pilot-scale study on water and latent heat recovery from flue gas using fluorine plastic heat exchangers. *Journal of Cleaner Production*, Vol. 161, pp. 1416-1422, (2017).

RESEARCH & DEVELOPMENT IN POWER ENGINEERING, 2019

Fuel-Water emulsion impact on miniature gas turbine pollutant emission

Maciej Chmielewski[1,*], *Marian* Gieras[1], and *Paweł* Niszczota[1]

[1] Warsaw University of Technology, Institute of Heat Engineering, Nowowiejska 21/25, 00-665 Warsaw, Poland

Abstract. The innovative use of the Fuel-Water emulsion in a small gas turbine for distributed energy generation is proposed. The FWE in this situation is considered as a nonhomogeneous mix of water and fuel, where water is a dispersed phase in the continuous fuel phase with an addition of surfactants. The Fuel-Water emulsion has a great mainly due to two mechanisms: temperature reduction due to heat absorption by the water phase and enhanced homogeneity of the fuel-air mixture due to micro-explosion of the superheated water phase inside the emulsion droplet. Proposed paper presents theoretical background on Fuel-Water emulsion combustion mechanism. Finally initial results of numerical research of fuel-water emulsion injection to miniature gas turbine are presented. Theoretical predictions of NO_X pollutant emissions are compared with accuracy of the gas analyzer planned to be used during experimental research.

1 Introduction

Historically water injection into reciprocating engines was used as a mean to cool down an air/fuel mixture temperature and prevent knocking, which was the main limitation for increasing a compression ratio [1]. Modern reciprocating engines water delivery systems includes: direct water injection into the engine cylinder, water injection through the intake port, water injection into the exhaust pipe, oil-water emulsification and combustible gas humidification [2-4]. Currently water addition to reciprocating engines is used mainly for engine pollutant and noise reduction [5, 6]. The water absorbs part of the heat generated in the cylinder which reduces the flame peak temperature. As the oxides of nitrogen (NO_X) formation is exponentially dependent on the temperature [7], reduction in temperature helps to meet stringent emission regulations [8, 9].

Mechanism described above works also inside gas turbine, which can explain the popularity of this method in the case of stationary gas turbines. In case of direct water injection into combustion chamber, water is delivered directly through separate fuel nozzle circuit or through dedicated water nozzles located on combustor domeplate [10]. Water injection upstream combustor helps to atomize water by small droplets detachment from trailing edge of the swirler vanes [11]. Water or steam injection into compressor is also popular, as it provides more homogeneous water-air mixture. On the other hand it generates also some losses, because only part of the compressed air reaches combustor primary zone [12]. It has been proven that water is approx. 60% more effective than steam to achieve the same NO_X reduction levels [13].

Despite clear benefits, traditional water and steam injection systems into gas turbines have many disadvantages. It is clear that additional water/steam supply systems have to be installed, which is connected with increased capital and maintenance cost of the gas turbine. There is also high risk for hot section components corrosion. In order to prevent that a water has to purified using reverse osmosis and de-ionization phenomena. Finally, as the combustion temperature is reduced, increase in the carbon oxides (CO) and unburned hydrocarbons (UHC) is expected.

The use of Fuel-Water Emulsion (FWE) inside small gas turbines, for distributed energy generation, proposed in this article might be an interesting alternative to traditional systems of water injection. The system provides all of the benefits of the traditional water injection systems and eliminates most of the their drawbacks. The FWE is considered as a nonhomogeneous mix of water and fuel, where water is a dispersed phase in the continuous fuel phase. Most commonly FWE contains between 3% to 15% of water with about 1% to 2% of surfactant (fatty acid esters). The surfactant role in the FWE is to improve emulsion stability by forming a boundary surface around dispersed water droplets and preventing them to join back together [14]. Since fuel-water mixture is prepared in advance, therefore no separate water tanks or delivery system is required. In addition, in the FWE water droplets are surrounded by the fuel, which prevents direct contact of the hot section components with the water, thus reducing corrosion risk.

Another notable advantage of the Fuel-Water Emulsion use are puffing and micro-explosion phenomena observed in number of research [15, 16]. It has been noticed that during the temperature increase in fuel-water emulsion injected into combustion chamber, causes rapid expansion of the vapour inside the droplet. This can lead to either to partial breakup of the droplet

* Corresponding author: maciej.chmielewski@pw.edu.pl

caused by vapor blowing out of droplet surface called "puffing" or to total disintegration of the droplet called "micro-explosion". Both phenomena lead to better atomization and intensified mixing of the fuel with the air inside combustion chamber [17]. Number of studies of droplet micro-explosion show that the size and distribution of emulsion water droplets inside the parent diesel droplet has a great impact on micro-explosion [18-24].

Impact of the FWE use in the turbine engine is investigated with the use of the polish production small gas turbine "GTM-120". The kerosene propelled gas turbine operating parameters are documented in [25, 26]. In this research paper a Fuel-Water Emulsion containing 3% of water is studied. The emulsion used in the study was developed at Warsaw University of Technology (WUT). The composition of the emulsion was precisely developed and studied. Microscope picture of the actual emulsion is shown in Figure 1. The full stability period of 24 hours of undisturbed emulsion and up to 2 weeks stability after emulsion stirring have been achieved.

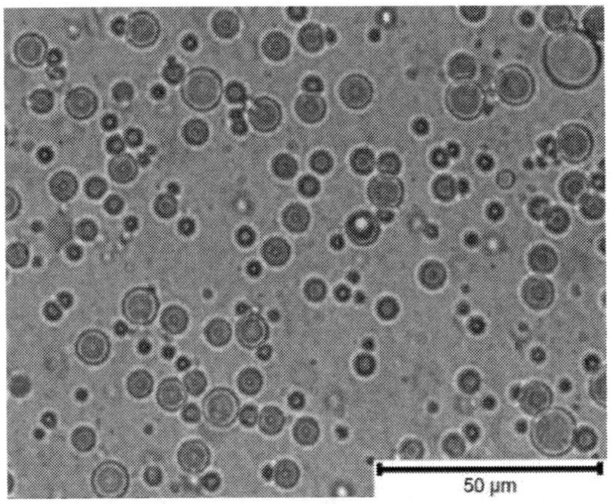

Fig. 1. Fuel-water emulsion microscope picture at 40x magnification.

The Computational Fluid Dynamics (CFD) model of reactive flow through the turbine is used in order to investigate the expected NO_X reduction levels. Finally, the comparison between numerical and baseline experimental data for the engine fueled with kerosene is discussed.

2 CFD model of reactive flow

The CFD models of the combustor chamber, fueled with standard fuel – kerosene and proposed fuel - emulsion containing 3% of water, were developed in order to find out what level of NO_X reduction is expected. This information was needed prior to experimental bench testing of the engine running on FWE to make sure that the gas analyzer accuracy allows to measure statistical difference in NO_X emissions.

The commercial CFD ANSYS Fluent code is used. Steady state engine operating conditions corresponding to 80k rpm are simulated. The geometry selected for numerical simulation consists of compressor diffuser, combustion chamber and nozzle. Tetrahedral mesh with 2.5M cells is used. Sensitivity study for the proposed mesh was performed in [27]. Boundary condition were taken from previous research [26]. The mass flow boundary condition was set at diffuser inlet. The values of mass flow for the simulated engine operating conditions were taken from experimental results [25]. The direction of air flow was determined by the solution of compressor velocity triangles for given rotational speed. Gauge pressure and total temperature at the inlet were adjusted iteratively to match experimental results for the static pressure and total temperature at diffuser outlet. The static pressure and total temperature were adjusted iteratively to match experimental results. Target mass flow and radial equilibrium pressure distribution were assumed at the test stand outlet, which was modeled with the use of pressure outlet boundary condition. The k-ε turbulence model with Scalable wall function was used. Convective heat transfer was calculated with the use of wall function. The Discrete Ordinates model was used for radiative heat transfer. Discrete Phase Model (DPM) was applied for modeling the injection, tracking and evaporation of fuel and water (only for emulsion case) droplets. Fuel-water emulsion parameters are taken directly from the microscopic study of the emulsion developed at WUT. Combustion was modeled as a non-premixed combustion model. The only difference between numeric models of kerosene combustion and emulsion combustion is water addition as separate DPM source to the combustion chamber through the fuel inlet for emulsion combustion model. The proposed emulsion combustion model aims only to account for an impact of water addition without simulating puffing or micro-explosion of the emulsion droplet. Such model is expected to provide the NO_X reduction, expected from water quenching phenomena. The obtained results are going to be compared against the actual NO_X emissions from the engine running on kerosene. That information will be used to decide if the gas analyzer accuracy is appropriate to show statistical difference of NO_X reduction between both fuels prior to running actual engine trials.

Thermal NO_X formation model was used to predict NO_X emissions from combustion chamber. It has to be stated that considered reactive flow is highly turbulent. In the highly turbulent flows significant fluctuations of the temperature and species concentration occurs. Since the relationships between temperature, species concentrations and NO_X formation rate are highly nonlinear, the time-average calculations of NO_X concentrations results in significant errors [28]. Much better results are obtained with Flamelet model which accounts for moderate chemical equilibrium. Simulation of the combustion process inside GE LM1600 gas turbine showed that Non-Premixed PDF model NO concentration relative error checks up to 80% comparing to approximately 10% when using the Flamelet model [29]. The best numerical tool for accurate NO_X formation modeling incorporated in ANSYS FLUENT is Decoupled Detailed Chemistry (DDC) models which decouples NO_X from flame calculations [28].

Authors understand that there is a significant room for numeric model of quantitative NO_X predictions

 https://doi.org/10.1051/e3sconf/201913701046

improvement. However, the proposed numerical model is not going to be used directly to report quantitative results of NO_X productions. Results of both kerosene and emulsion combustion models are to be compared on "back-to-back" basis in order to find expected reduction of NO_X quantities. In addition, general conclusions about combustion with water addition are to be drawn. Future plans assumes Unsteady DDC model implementation to improve the quantitative results of NO_X concentrations towards experimental data.

Results of the numerical combustion numerical modeling are show in Figures 2-4. One can notice that in case of baseline kerosene combustion model, heat up of the kerosene liquid occurs inside the evaporating tubes (Figure 2a). Full evaporation and subsequent formation of the combustible mixture occurs at the evaporating tubes outlet. Mixture is ignited at the moment in which it mixes with hot combustion products in the combustor primary zone. On the other hand for the FEW combustion case the fuel evaporates inside evaporating tube (Figure 2b) and ignites almost immediately due to contact with hot surface of the tubes. The difference in the combustion beginning between the models can be clearly noticed on Figure 3.

The high temperature of the combustion process impacts the NO production rate and mole fraction which can be observed in Figure 4. One can easily observe that NO high concentration regions of the combustion chamber corresponds with the highest temperature area. The high NO mol fraction regions have been significantly reduced inside combustion chamber fueled with FEW (based on comparison of Figures 4a and 4b).

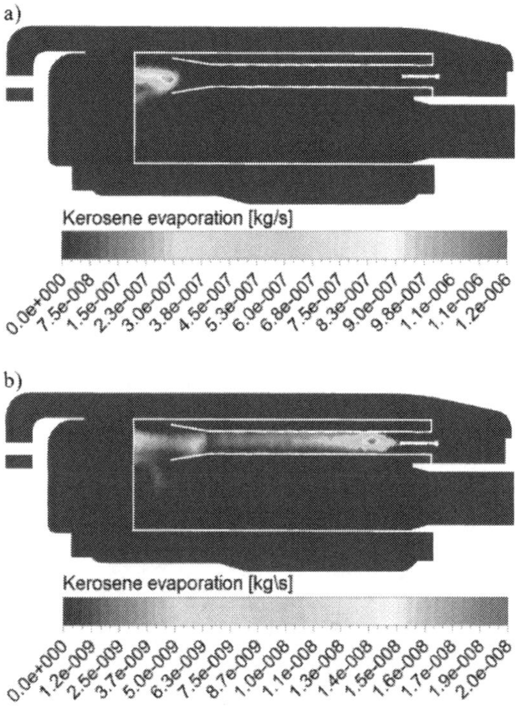

Fig. 2. Kerosene evaporation rate: a) engine running on kerosene, b) engine running on FEW.

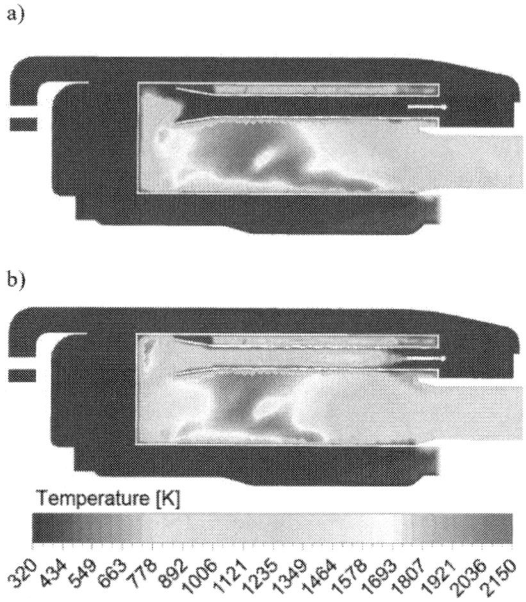

Fig. 3. Temperature: a) engine running on kerosene, b) engine running on FWE.

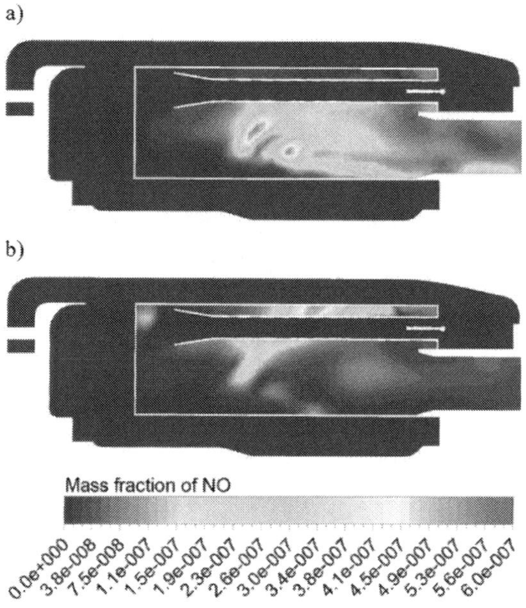

Fig. 4. Mass fraction of NO: a) engine running on kerosene, b) engine running on FWE.

Finally based on the numerical results the NO concentration (which is the main NO_X component in hot combustion gasses) produced by the engine running on kerosene checks 256 ppm compared to 186 ppm produced by the engine running on the FWE. Clearly the quantitative results of NO production are overestimated when comparing to experimental data given in [25, 26]. As explained above, qualitative results are not the aim of the proposed numerical model. Proposed data shows that

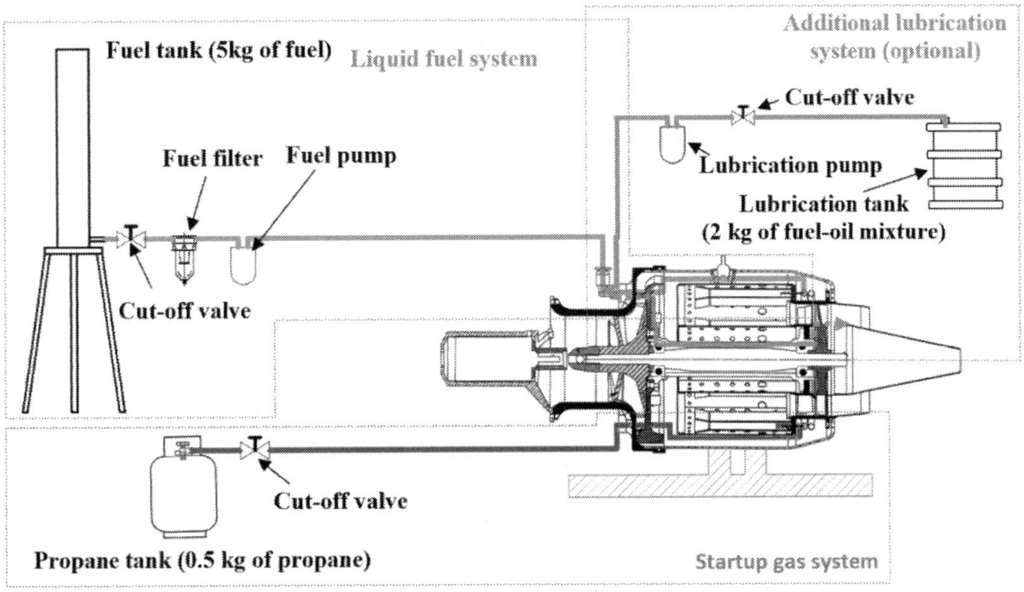

Fig. 5. Engine test bench supporting systems.

there is 27% of the NO emission reduction due to addition of the water. Based on the experimental data mean values of overall NO_X production levels are between 16 ppm (for 40k rpm) up to 27 ppm (for 120k rpm) with approximate linear dependence on the engine rotational speed. Assuming 27% of NO_X reduction due to water addition, the experimental results of the engine running on FWE containing 3% of water should show between 12 ppm up to 20 ppm of NO_X concentration, for 40k rpm and 120k rpm respectively, measured at the engine exhaust by the emission analyzer.

Above results shows that between 4 ppm to 7 ppm of NO_X concentration reduction is expected which, for engine rotational speeds above 40k rpm, is more than ±5 ppm of NO_X measurement accuracy by gas analyzer according to [30]. Based on above the proposed measuring technique and equipment is suitable to prove statistical difference between NO_X concentrations produced by engine operating on kerosene and FWE above 40k rpm of engine rotational speed.

3 Experimental research

Experimental research work to confirm results of the numerical predictions is to be carried on the GTM-120 miniature engine test bench developed in Institute of Heat Engineering, Warsaw University of Technology [25]. Test bench (Figure 5) consists of GTM-120 small jet turbine with ECU mounted on engine mount, fuel tank, exhaust gas analyzer, data acquisition card and PC with test bench software. Pressure and temperature sensors have been placed in all major engine sections. In the inlet section of the engine differential pressure sensor with static pressure sensor and type K thermocouple are installed. One static pressure sensor and one type K thermocouple are mounted at compressor and combustor outlet sections. At the engine outlet a type K thermocouple and gas analyzer probe are installed. Test bench is equipped with strain gauge Mavin NS-6 with WOBit WDT-1 strain gauge amplifier used to measure thrust force. Fuel tank is equipped with differential pressure sensor Dudek P400U used to measure fuel consumption. All sensors are connected to the National Instruments USB-6259 BNC card connected to PC with test bench software programmed in LabVIEW. Portable emission analyzer Testo 350 is used to measure engine exhaust gas composition.

In addition to test bench components listed above, a separate lubrication system was developed (Figure 5). The main advantage of the GTM-120 engine is trust/weight ratio. A total-loss oil system is used in default configuration of the GTM-120 to reduce weight of the engine. In the original oil system a fuel with 5% of oil addition is used to lubricate and cool down two sets of ball bearings installed in the engine. Corrosive properties of water contained in FWE delivered to hot bearings were primary reason to design standalone lubrication system delivering an oil during engine operation. Additional lubrication system consists of lubrication pomp, separate lubrication liquid tank and flow restrictor (Figure 5). Flow restrictor assures that approximately 2.5% of main fuel flow is delivered to engine bearings, which is the same amount as for the default engine.

Predictions of experimental data based on [26, 27] and numerical model are shown in Figure 6. Experimental data was scaled down by 27%.

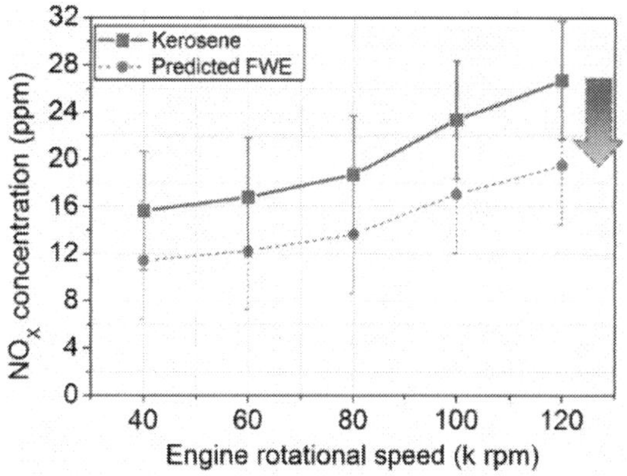

Fig. 6. Predicted NO_X emissions from engine running on FWE.

4 Conclusions

Proposed, simple PDF numerical model of the kerosene and FWE combustion shows up to 27% of reduction in terms of NO concentration in hot combustion products at the small gas turbine combustor outlet. Large potential for the emissions reduction from miniature gas turbine can be explained by poor combustion efficiencies inside small combustion chambers. The results obtained from the model are used for "back-to-back" comparison of the kerosene and FEW combustion and to draw general conclusions on the water addition impact on combustion. More sophisticated, non-equilibrium, combustion models have to be incorporated in order to obtain more accurate absolute values of NO_X emission results.

Predicted NO_X emission reductions (up to 7 ppm) allows to claim that the gas analyzer accuracy (up to ±5ppm) which is going to be use in the experimental research allows to show statistical difference between both fuels. In addition the model shows that evaporation of liquid kerosene and ignition of the kerosene-air mixture accelerates when the water addition is used. According to model, the combustion can start in the evaporating tubes when FWE is used, whereas the evaporation and subsequent ignition of the combustible mixture takes place at the outlet of the evaporating tube for pure kerosene. The proposed equilibrium model does not account for chemical non-equilibrium due to aerodynamic straining of the flame by the turbulence. Base on above ore accurate model together with experimental data is needed to confirm the phenomena. Currently a Flamelet combustion model is developed to improve numerical NO_X predictions and ongoing experimental research will be used to anchor numerical model and confirm whether combustion occurs inside evaporating tubes.

Numerical study proves that there is a high potential to reduce pollutant emissions from miniature gas turbines especially used in distributed energy generation systems where additional mass of the fuel does not adversely impact performance of the entire system. It is proposed to confirm the conclusions from the numerical model by using more sophisticated combustion modelling an investigating phenomena occurring inside the evaporating tubes during engine operation on FWE.

References

1. Launius R.D, *Innovation and the development of flight*, Vol. 14, pp. 117-118, (1999)
2. Chenyao W., Zhang F., Wang E., et al, *Energy Procedia*, vol. 159, pp. 5735-5740, (2019)
3. Sipeng Z., Hu, B., Akehurst S., et al, *Energy Conversion and Management*, vol. 184, pp. 139-158, (2019)
4. Wei M., Nguyen T.S., Turkson R.F., et al, *Journal of the Energy Institute*, vol. 90, no. 2, pp. 285-299, (2017)
5. Bharathiraja M, Venkatachalam R, Tiruvenkadam N., *Transportation Research Part D Transport & Environment*, vol. 49, pp. 291-300, (2016)
6. Kim N.H, *Applied Thermal Engineering*, vol. 128, pp.1502-1509, (2017)
7. Zel'dovich Y.B, *Acta Physicochimica*, vol. 21, pp. 577–628, (1946)
8. Francqueville L.D., Michel J.B, *SAE International Journal of Engines*, vol. 7, no. 4, pp. 1808-1823, (2014)
9. Li B., Wei M., *Small Internal Combustion Engine and Motorcycle*, vol. 41, no. 6, pp. 87-90, (2012)
10. Davis, L.B., and Washam, R.M, *ASME Paper 89-GT-255*, (1989)
11. Lefebvre A.H., Ballal D.R, *Gas Turbine Combustion: Alternative Fuels and Emissions*, Third Edition, CRC Press, (2010)
12. Hilt, M.B., Waslo, J., *Journal of Engineering for Gas Turbines and Power*, vol. 106, pp. 825–832, (1984)
13. Schorr, M.M, *Turbomachinery International*, Nov./Dec.1991, pp. 28–36, (1991)
14. Rodzewicz R., Gieras M., *Archivum Combustionis*, vol. 38, no. 1.pp. 1-10, (2018)
15. Law C.K, *Progress in Energy and Combustion Science*, vol. 8, no. 3, pp. 171-201, (1982)
16. Botero M.L. , Haung Y. , Zhu D.L, et al, *Fuel*, vol. 94, pp. 342-347, (2012)
17. Shingo J., Xia J., Ganippa L.C, et al, *Physics of Fluids*, vol. 26, no. 10, (2014)
18. Mura E., Massoli P., Josset C., et al, *Experimental Thermal and Fluid Science*, vol. 43, pp. 63-70, (2012)
19. Mura E., Calabria R., Califano V., et al, *Experimental Thermal and Fluid Science*, vol. 56, pp. 69-74, (2014)
20. Califano V., Calabria R., Massoli P., *Fuel*, vol. 117, pp. 87-94, (2014)
21. Mura E., Josset C., Loubar K., et al, *Atomization and Sprays*, vol. 20, no. 9, pp. 791-799, (2010)
22. Kadota T., Yamasaki H., *Progress in Energy and Combustion Science*, vol. 28, pp. 385-404, (2002)
23. Hirotatsu W., Yohsuke M., Hideyuki A., et al, *Combustion and Flame*, vol. 157, no. 5, pp. 839-852, (2010)

24. Hirotatsu W., Yoshiyuki S., Takuji H., et al, *Energy*, vol. 35, pp. 806-813, (2010)
25. Chmielewski M., Gieras M., *Journal of KONES*, vol. 22, pp. 47–54, (2015)
26. Chmielewski M., Gieras M., *Journal of the Energy Institute*, vol. 90, pp. 257–264, (2017)
27. Chmielewski M., *Impact of variable geometry combustion chamber on combustion efficiency and pollutant emissions from miniature gas turbine*, PhD Thesis, Warsaw University of Technology, (2016)
28. ANSYS Inc., *ANSYS Fluent Theory Guide*, 275 Technology Drive Canonsburg, PA 15317, (2013)
29. Baukal C.E., Gershtein V.Y., Li X., *Computational Fluid Dynamics in Industrial Combustion*, Fist Edition, CRC Press, (2000)
30. Testo AG, *Testo 350 Combustion & Emission Analyzer Instruction Manual*, (2011)

E3S Web of Conferences **137**, 01047 (2019) https://doi.org/10.1051/e3sconf/201913701047
RDPE 2019

RESEARCH & DEVELOPMENT IN POWER ENGINEERING, 2019

Development and implementation of mathematical models of working mediums for gas part of combined cycle gas turbine system in Python programming environment

Paweł Trawiński

Warsaw University of Technology, Institute of Heat Engineering, Nowowiejska 21/25, 00-665 Warsaw, ptraw@itc.pw.edu.pl

Abstract. The working mediums in the gas turbine systems are: atmospheric air, natural gas and exhaust gases. For the detailed analysis of thermodynamic performance and values at characteristic points of the cycle it is necessary to know the relations defining: specific volume, specific isobaric and isochoric heat capacity, isentropic exponent, specific enthalpy and specific entropy. Mathematical models of thermodynamic parameters for the mentioned mediums were developed based on dependencies for mixtures of ideal and semi-ideal gases. The functions obtained for semi-ideal gas mixtures were extended by pressure correction factors derived from the Redlich-Kwong and Peng-Robinson equations of state. The thermodynamic parameters of the working mediums were dependent on the mass fractions of the components, temperature and pressure. Developed models approximated the behaviour and parameters of real gas mixtures. All calculation algorithms were implemented and optimized using appropriate numerical methods in the Python programming environment. As a result, mathematical models of working mediums for the gas part of the combined cycle gas turbine system were obtained.

1 Introduction

Modelling thermal-flow systems should be started with precise identification of working mediums of the cycle. For the gas turbine system it will be: atmospheric humid air, natural gas and hot exhaust gases. The developed mathematical description should contain the functions of thermodynamic parameters necessary to determine unambiguously the state of the medium at characteristic points and to analyse the performance of the entire system.

In many engineering applications models are based on dependencies describing the performance of ideal and semi-ideal gas. The assumptions of the physical model for ideal gas are limiting the range of its applicability, particularly for conditions that significantly deviate from standard pressure and temperature. The area in which the Ideal Gas Law may be an equitable approximation is indicated in the range up to 2-3 MPa and 1200 °C [1]. More advanced and complex physical models for working mediums should be used to achieve better accuracy and more precise results. These models are usually based on the real gas equations of state [2-4]. Most of them originate from the van der Waals equation of state, which was the first equation that successfully took into account corrections for molecular volume and intermolecular interactions.

In this work the Redlich-Kwong and Peng-Robinson equations of state were used. This required the determination of additional physic-chemical parameters for pure compounds (critical conditions and acentric factors [3, 5-8], binary interaction parameters [9, 10]), properties of mixtures [2, 3, 9, 10] and analytical or numerical dependencies resulting from the mentioned equations [1, 4].

2 Working mediums

2.1 Steam

The values of thermodynamic parameters of steam were implemented on the basis of the publications of the International Association for the Properties of Water and Steam [11-15]. They include sets of equations approximating thermodynamic parameters of steam in the range: 0-800 °C for pressure up to 100 MPa and 800-2000 °C for pressure up to 50 MPa.

Steam was considered as one of the components of mixtures of humid atmospheric air and exhaust gases. The obtained thermodynamic functions concerning the properties of steam were used to create mathematical models of the discussed working mediums.

2.2 Atmospheric air

Dry atmospheric air was modelled as a mixture of nitrogen, oxygen and argon [16]. The physical model of humid atmospheric air additionally introduces the fraction of steam. The values of thermodynamic parameters for humid air were determined as for a mixture of dry air and steam with appropriate fractions, partial pressures and temperature.

https://doi.org/10.1051/e3sconf/201913701047

2.3 Natural gas

Natural gas was considered as a mixture of methane, nitrogen, carbon dioxide and higher-order hydrocarbons [17]. The composition of natural gas in general includes a variety of other compounds. The content of individual components of a mixture depends not only on the type (high-methane, nitrated) but also on the supplier and location. The content of higher-order hydrocarbons (e.g. propane and butane) was included in the molar fraction of ethane.

2.4 Exhaust gases

Exhaust gases are a mixture of carbon dioxide, nitrogen, oxygen, argon and steam [1]. The composition of the exhaust gases depends on the reagent ratio and the combustion process. Only the complete combustion reactions of methane and ethane were considered. The fractions of particular components of the mixture were determined on the basis of mass balance and stoichiometry.

3 Equations of state

From a wide range of available equations of state, the Ideal Gas Law and two real gas equations of state were chosen: Redlich-Kwong and Peng-Robinson EoS. Selected equations can be implemented in undemanding way in calculation algorithms. Moreover, number of additional parameters is limited (e.g. in comparison to Beattie-Bridgeman and Benedict-Webb-Rubin EoS). Calculations based on these relationships have relatively low computing costs and better precision of results (e.g. in comparison to van der Waals and Berthelot EoS). The choice of implemented real gas equations of state is therefore a compromise between their complexity and accuracy [2-4].

3.1 Ideal Gas Law

The Clapeyron equation binds three parameters of state – pressure p, temperature T and molar volume V_m of ideal gas [1, 3]:

$$pV_m = BT \tag{1}$$

where B is the universal molar gas constant.

For the purposes of the calculation program, specific volume ϑ was determined by converting the given relation using molar mass M:

$$\vartheta = \frac{BT}{pM} \tag{2}$$

3.2 Redlich-Kwong equation of state

The Redlich-Kwong EoS is considered to be the first significant improvement of the van der Waals equation. It introduces an empirical dependence of the cohesive pressure on the gas temperature [3, 4]:

$$\left(p + \frac{a}{\sqrt{T}V_m(V_m+b)}\right)(V_m - b) = BT \tag{3}$$

where:
a – correction constant for molecular interactions (cohesive pressure)
b – correction constant for volume of molecules (co-volume).

The characteristic constants for the Redlich-Kwong equation were estimated experimentally in relation to the critical parameters of the gas [3, 4]:

$$a = 0.427480\frac{B^2{T_c}^{2.5}}{p_c} \tag{4}$$

$$b = 0.086640\frac{BT_c}{p_c} \tag{5}$$

where:
T_c, p_c – critical temperature and critical pressure.

3.3 Peng-Robinson equation of state

The Peng-Robinson equation of state is another modification of the van der Waals equation. It introduces the dependence of the cohesive pressure on the gas temperature (similarly to the Redlich-Kwong EoS) and the acentric factor of molecules [2]:

$$\left(p + \frac{\alpha a}{{V_m}^2+2bV_m-b^2}\right)(V_m - b) = BT \tag{6}$$

where:
a – correction constant for molecular interactions (cohesive pressure)
b – correction constant for volume of molecules (co-volume)
α – correction coefficient for acentric factor of molecules.

The characteristic constants for the Peng-Robinson equation were estimated experimentally in relation to the critical parameters of the gas [2]:

$$a = 0.45724\frac{B^2{T_c}^{2.5}}{p_c} \tag{7}$$

$$b = 0.07780\frac{BT_c}{p_c} \tag{8}$$

The α coefficient in the Peng-Robinson equation of state is defined as:

$$\alpha = \left[1 + f(\omega)\left(1 - \sqrt{\frac{T}{T_c}}\right)\right]^2 \tag{9}$$

where:

$$f(\omega) = 0.37464 + 1.54226\omega - 0.2699\omega^2 \tag{10}$$

where ω is the acentric factor of molecules.

3.4 Mixing and combining rules for real gas equations of state

Appropriate rules were established for the calculation of equivalent coefficients for multicomponent mixtures. For the equations of state based on the van der Waals equation it is recommended to use a combination of linear (coefficient b) and geometric (coefficients a and α) approach [2, 3, 9, 10]:

 https://doi.org/10.1051/e3sconf/201913701047

$$a = \sum_i \sum_j (1 - k_{ij}) n_i n_j \sqrt{a_i a_j} \quad (11)$$

$$b = \sum_i n_i b_i \quad (12)$$

$$\alpha = \sum_i \sum_j (1 - k_{ij}) n_i n_j \sqrt{\alpha_i \alpha_j} \quad (13)$$

or:

$$(\alpha a) = \sum_i \sum_j (1 - k_{ij}) n_i n_j \sqrt{\alpha_i a_i \alpha_j a_j} \quad (14)$$

where:
$n_{i,j}$ – molar fractions of components
$a_{i,j}$ – characteristic coefficients a for particular components
b_i – characteristic coefficients b for particular components
k_{ij} – binary interaction parameter for a pair of components
$\alpha_{i,j}$ – coefficients α for particular components.

Geometric approach was also used to calculate pseudocritical temperature of the mixture [3]:

$$T_c = \sum_i \sum_j (1 - k_{ij}) n_i n_j \sqrt{T_{c_i} T_{c_j}} \quad (15)$$

where:
$T_{c_{i,j}}$ – critical temperature of particular components.

4 Thermodynamic parameters of semi-ideal and real gases

The thermodynamic parameters of a real gas can be expressed as a sum of the components for a semi-ideal gas and a pressure correction factor [1]. Correction factors for pressure were determined with the Redlich-Kwong or Peng-Robinson equations of state. For multicomponent systems the dependencies used for semi-ideal gas mixtures were applied [18, 19]. The values of derivatives and integrals were determined analytically (if possible) or by numerical methods. Detailed presentation of the relations and derivations of thermodynamic parameters of real gas can be found in the literature [1, 3, 4].

4.1 Specific isobaric heat capacity

The specific isobaric heat capacity of a semi-ideal gas is a function of temperature. The general form of the relationship can be approximated as [1]:

$$c_p{}^{id}(T) = \sum_{i=0}^{n} a_i T^i \quad (16)$$

where:
a_i – coefficients of the approximation polynomial.

There are many models available in the literature that approximate specific isobaric heat capacities of pure compounds [1, 3, 5, 6, 7]. They vary in accuracy and temperature range of application. Dependencies used in this paper can be found in thermodynamic tables [3, 6]. As a result implemented models covered the temperature range from 50 K to 3000 K (1500 K for ethane) as it is presented in the figures below.

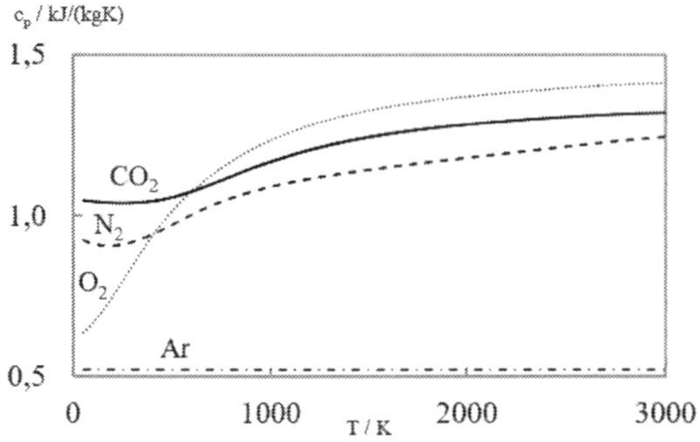

Fig. 1. Specific isobaric heat capacities of nitrogen, oxygen, carbon dioxide and argon as for semi-ideal gases. Author®own study based on [3, 6].

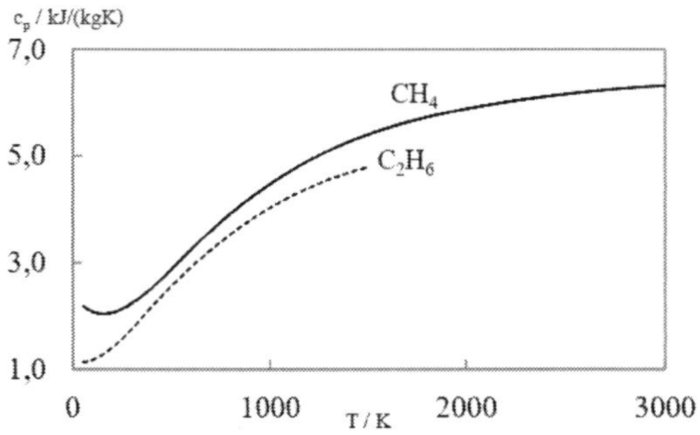

Fig. 2. Specific isobaric heat capacities of methane and ethane as for semi-ideal gases. Author®own study based on [3, 6].

The specific isobaric heat capacity of a real gas can be expressed as the sum of the components for a semi-ideal gas $c_p{}^{id}(T)$ and a correction factor for pressure $\Delta c_p(p,T)$:

$$c_p(p,T) = c_p{}^{id}(T) + \Delta c_p(p,T) \quad (17)$$

Eventually, the relation can be presented as [1]:

$$c_p(p,T) = c_p{}^{id}(T) - T \int_0^p \left(\frac{\partial^2 \vartheta}{\partial T^2}\right)_{p=idem} dp \quad (18)$$

4.2 Specific isochoric heat capacity

The specific isochoric heat capacity of semi-ideal gas can be found using the relation [18, 19]:

$$c_\vartheta{}^{id}(T) = c_p{}^{id}(T) - B \quad (19)$$

The specific isochoric heat capacity of the real gas is given in the form of [3]:

$$c_\vartheta(p,T) = c_p{}^{id}(T) + T \int_0^p \left[\frac{\left(\frac{\partial p}{\partial T}\right)^2_{\vartheta=idem}}{\left(\frac{\partial p}{\partial \vartheta}\right)_{T=idem}}\right] dp \quad (20)$$

In this case the partial derivatives were calculated analytically directly from the equations of state.
For the Redlich-Kwong equation of state:

$$\left(\frac{\partial p}{\partial T}\right)_{\vartheta=idem} = \frac{B}{V_m - b} + \frac{a}{2V_m(V_m+b)\sqrt{T^3}} \quad (21)$$

$$\left(\frac{\partial p}{\partial \vartheta}\right)_{T=idem} = -\frac{BT}{(V_m-b)^2} + \frac{a(V_m+b)}{\sqrt{T}V_m{}^2(V_m+b)^2} \quad (22)$$

For the Peng-Robinson equation of state:

$$\left(\frac{\partial p}{\partial T}\right)_{\vartheta=idem} = \frac{B}{V_m-b} + \frac{a}{V_m{}^2+2bV_m-b^2}\left(\frac{f^2(\omega)}{T_{cr}} - \frac{f^2(\omega)+f(\omega)}{\sqrt{T_{cr}T}}\right) \quad (23)$$

$$\left(\frac{\partial p}{\partial \vartheta}\right)_{T=idem} = -\frac{BT}{(V_m-b)^2} + \alpha a\frac{2V_m+2b}{(V_m{}^2+2bV_m-b^2)^2} \quad (24)$$

4.3 Isentropic exponent

Isentropic exponent is the ratio of specific heat capacity at constant pressure to specific heat capacity at constant volume [18, 19]. For real gas its value can be calculated as:

$$\kappa(p,T) = \frac{c_p(p,T)}{c_\vartheta(p,T)} \quad (25)$$

4.4 Specific enthalpy

Specific enthalpy of semi-ideal gas can be calculated on the basis of the relation [18, 19]:

$$i^{id}(T) = \int_0^T c_p(T)dT \quad (26)$$

Taking into account the reference value of specific enthalpy $i(T_0)$ at a given temperature T_0 [5-7]:

$$i^{id}(T) = i_0 + \int_{T_0}^T c_p(T)dT \quad (27)$$

The specific enthalpy of a real gas can be presented as [1]:

$$i(p,T) = i^{id}(T) + \int_0^p \left[v - T\left(\frac{\partial \vartheta}{\partial T}\right)_{p=idem}\right] dp \quad (28)$$

4.5 Specific entropy

In a similar way the specific entropy of semi-ideal gas [18, 19] can be determined in relation to the reference specific entropy s_0 in the reference conditions p_0, T_0 [5-7]:

$$s^{id}(T) = s_0 + \int_{T_0}^T \frac{c_p(T)}{T} dT - B\ln\frac{p}{p_0} \quad (29)$$

Eventually, the specific entropy of a real gas can be calculated as the sum of components:

$$s(p,T) = s^{id}(T) - \int_0^p \left(\frac{\partial \vartheta}{\partial T}\right)_{p=idem} dp \quad (30)$$

5 Implementation of numerical methods in the mathematical model in Python

All dependencies determining thermodynamic parameters were implemented in the object-oriented Python programming language. It is a general-purpose language that allows to create complex applications and integrate with other programming environments. Moreover, it is independent of the operating system and has a significant number of accessible libraries. Comprehensive overview of the implemented numerical methods can be found in the literature [20, 21].

5.1 Analytical solution of third-degree equations

In order to solve the real gas equations of state a calculation algorithm was implemented. It examines the existence of real solutions on the basis of the value of canonical form discriminant. The created function takes coefficients of unknown variables as input data and returns value of the gas specific volume.

5.2 Numerical solution of non-linear equations

Backward functions (calculating temperature on the basis of pressure and specific enthalpy or entropy) were built upon algorithms solving non-linear equations. The methods of bisection and Fibonacci (golden-section search) were chosen.

5.3 Numerical integration

Numerical integration was performed with the use of rectangular or trapezoidal methods. Integration step for each iteration for the temperature components of specific enthalpy and entropy was set to 0.1 K. Integration step for the pressure components of specific enthalpy and entropy was dependent on temperature and pressure (narrower intervals for the low-temperature range due to the greater variability of the specific heat capacity in that area).

5.4 Numerical differentiation

Numerical estimation of the values for the first and second derivative of function at a given point was based on the expansion of the function into Taylor's series. Algorithms used central differences: four-point for the first derivative and five-point for the second derivative.

5.5 Discretisation of intervals

Additionally, the functions discretising intervals (creating lists of nodes) and determining the step of discretisation depending on the enforced precision of division were implemented.

6 Results

Mathematical models of all working mediums were implemented in such a way that thermodynamic parameters were a function of: mass/molar fractions of components (various humidity content of atmospheric air, composition of natural gas, air–fuel equivalence ratio in the combustion process), temperature and pressure. One of the input data to the functions was also the chosen equation of state which was used to perform calculations.

Presented results were chosen in order to illustrate the possibilities of created calculation program on substantive examples. These are demonstrative combinations of chosen thermodynamic parameters of working mediums as functions of input parameters with use of implemented equations of state. The program allows to test any combination of the above. Selected results are presented in the figures below.

6.1 Specific isobaric heat capacity of dry air as a function of temperature and pressure

The influence of pressure on the specific isobaric heat capacity of dry air is significant for the low-temperature range. Decreasing distances between molecules result in increasing role of intermolecular interactions. As the temperature rises, the differences for the respective pressure values fade. Intermolecular interactions increase due to the growth of kinetic energy of the molecules. For this reason, an additional increase in pressure does not cause a further significant increase in intermolecular interactions and specific isobaric heat capacity.

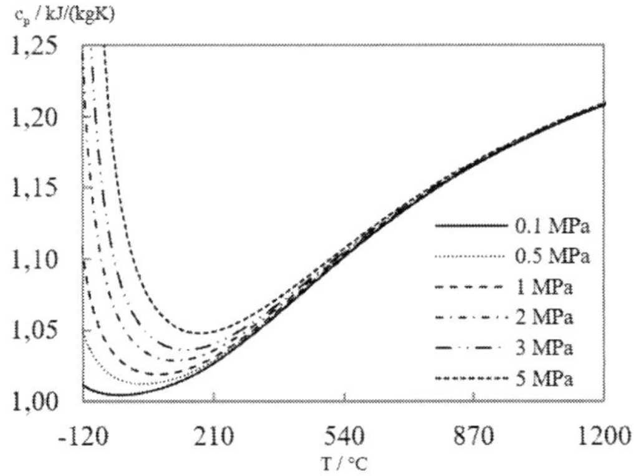

Fig. 3. Specific isobaric heat capacity of dry air as a function of temperature and pressure (Redlich-Kwong EoS).

6.2 Specific isobaric heat capacity of air as a function of temperature and relative humidity

The influence of relative humidity on the specific isobaric heat capacity of atmospheric humid air increases with the temperature. The progression of function for respective levels of relative humidity is illustrated in the diagram. Steam is characterised by a much higher specific isobaric heat capacity than dry air. Increase of water vapour content in the humid air mixture directly affects the increase in specific isobaric heat capacity of the whole mixture.

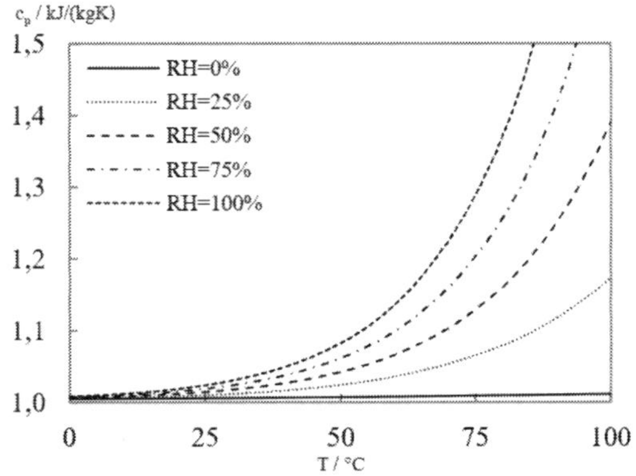

Fig. 4. Specific isobaric heat capacity of air as a function of temperature and relative humidity (Redlich-Kwong EoS).

6.3 Specific enthalpy of natural gas as a function of temperature and pressure

The influence of pressure on the specific enthalpy of natural gas is visible in the low-temperature region. The pressure impact decreases with temperature. Specific enthalpy can be considered as the sum of internal energy of the system and work done on the environment to create space for this system [18]. From this relation it results that an increase in pressure leads to a decrease in its specific enthalpy at the constant temperature.

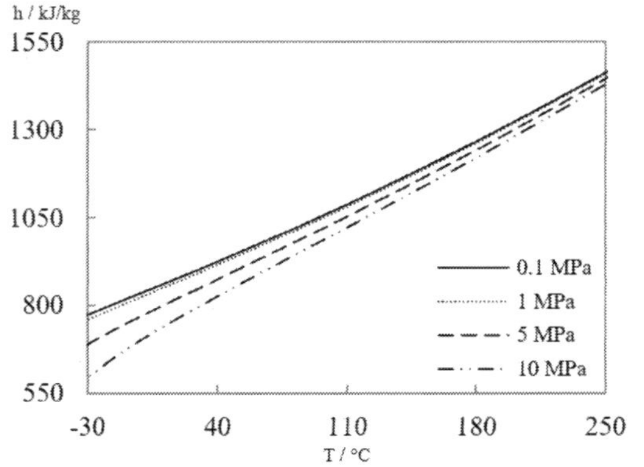

Fig. 5. Specific enthalpy of natural gas as a function of temperature and pressure (Peng-Robinson EoS).

6.4 Specific entropy of exhaust gases as a function of temperature and air–fuel equivalence ratio

The progression of specific entropy as a function of air-fuel equivalence ratio and temperature is illustrated in the diagram. Specific entropy increases with temperature due to increasing kinetic energy of molecules that leads to growth of disorder and randomness of the system. Moreover, increasing air-fuel equivalent ratio leads to dilution of exhaust gases, decrease fractions of compounds generated in combustion process and approximate its composition and properties to the atmospheric air mixture.

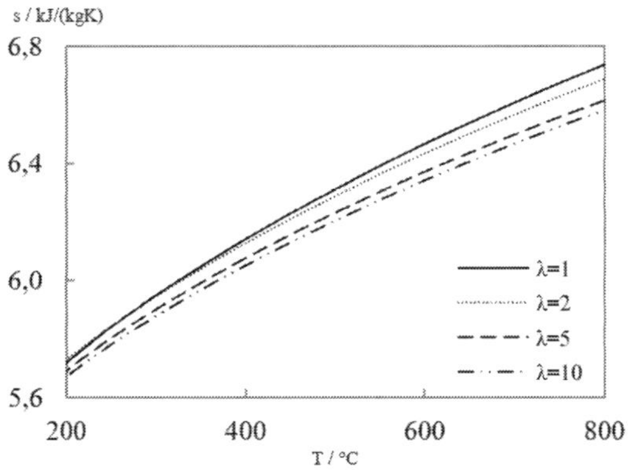

Fig. 6. Specific entropy of exhaust gases as a function of temperature and air–fuel equivalence ratio (Redlich-Kwong EoS, 2 MPa).

7 Summary and conclusions

Individual computational modules were developed for each working medium. They consisted of functions and calculation algorithms, which on the basis of input data (fractions of individual components, pressure, temperature, equation of state) determined the values of thermodynamic parameters. In areas where it was impossible to determine analytical solutions, numerical methods were used. In this case it is necessary to point out the operations of integration, differentiation, interval discretization, solving non-linear equations and optimization (e.g. with use of bisection method and golden-section search). Algorithms of necessary numerical methods were implemented in a separate calculation module.

The semi-ideal gas model was extended by applying pressure correction factors based on the Redlich-Kwong and Peng-Robinson equations of state. The obtained mathematical models of working mediums allowed for more accurate calculation of thermodynamic parameters in a wide range of temperature and pressure.

The designed calculation modules can be used for thermodynamic analysis of gas turbine performance and other combined heat and power systems based on natural gas combustion. Furthermore, the way of implementing computational algorithms allows to extend the program with new pure compounds and more advanced real gas equations of the state.

References

1. Chmielniak T. *Zagadnienia cieplnych maszyn przepływowych : Czynniki robocze. Obiegi parowe i gazowe* (Skrypty Uczelniane - Politechnika Śląska Nr 695), Gliwice: Politechn. Śląska, (1977).
2. Peng D., Robinson D., A New Two-Constant Equation of State, *Industrial & Engineering Chemistry Fundamentals*, Vol. 15, pp. 59-64, (1976)
3. Poling B., Prausnitz J., O'Connell J., *The properties of gases and liquids* (5th, International ed.), McGraw-Hill, (2007).
4. Sado J., Jednowymiarowy przepływ gazu rzeczywistego, *Biuletyn Instytutu Techniki Cieplnej Politechniki Warszawskiej*, Vol. 86, pp. 3-37, (1999).
5. Borgnakke C., Sonntag R., *Thermodynamic and transport properties*, John Wiley, (1997).
6. Chhabra R., *CRC handbook of thermal engineering* (Second ed., Mechanical and Aerospace Engineering), CRC Press/Taylor&Francis Group, (2018).
7. Wark K., *Tables and figures to accompany thermodynamics* (Mechanical Engineering Series), MacGraw-Hill Book Comp., (1989) .
8. Speight J., Lange N., *Lange's handbook of chemistry* (16th ed., McGraw-Hill Standard Handbooks), McGraw-Hill, (2005).
9. Fateen S., Khalil M., Elnabawy A., Semi-empirical correlation for binary interaction parameters of the Peng-Robinson equation of state with the van der Waals mixing rules for the prediction of high-pressure vapour-liquid equilibrium, *Journal of Advanced Research*, Vol. 4, pp. 137-145, (2013).
10. Stoll J., Vrabec J., Hasse H., Vapor-Liquid Equilibria of Mixtures Containing Nitrogen, Oxygen, Carbon Dioxide, and Ethane, *American Institute of Chemical Engineers. AIChE Journal*, Vol. 49, pp. 2187-2198, (2003).
11. IAPWS, *Revised Release on the IAPWS Industrial Formulation 1997 for the Thermodynamic Properties of Water and Steam*, (2007).
12. IAPWS, *Revised Supplementary Release on Backward Equations for Pressure as a Function of Enthalpy and Entropy p(h,s) for Regions 1 and 2 of the IAPWS Industrial Formulation 1997 for the Thermodynamic Properties of Water and Steam*, (2014).
13. IAPWS, *Revised Supplementary Release on Backward Equations for the Functions T(p,h), v(p,h) and T(p,s), v(p,s) for Region 3 of the IAPWS Industrial Formulation 1997 for the Thermodynamic Properties of Water and Steam,* (2014).
14. IAPWS, *Revised Supplementary Release on Backward Equations p(h,s) for Region 3, Equations as a Function of h and s for the Region Boundaries, and an Equation Tsat(h,s) for Region 4 of the IAPWS Industrial Formulation 1997 for the Thermodynamic Properties of Water and Steam,* (2014).
15. IAPWS, *Revised Supplementary Release on Backward Equations for Specific Volume as a Function of Pressure and Temperature v(p,T) for Region 3 of the IAPWS Industrial Formulation 1997 for the Thermodynamic Properties of Water and Steam,* (2016).
16. Oleśkowicz-Popiel C., Wojtkowiak J., *Właściwości termofizyczne powietrza i wody przeznaczone do obliczeń przepływów i wymiany ciepła*, Wydawnictwo Politechniki Poznańskiej, (2010).
17. http://pgnig.pl/dla-domu/poradnik/czym-jest-gaz-ziemny (on-line: 16.10.2019).
18. Staniszewski B., *Termodynamika* (Wyd. 4. ed.), Państ. Wydaw. Nauk., (1986).
19. Litke B., Właściwości gazów doskonałych i półdoskonałych, *Problemy Nauk Stosowanych*, Vol. 4, pp. 95-106, (2016).
20. Łukaszewicz J., Warmus M., *Metody numeryczne i graficzne. Cz. 1*, Biblioteka Główna Politechniki Warszawskiej, (2010).
21. Fortuna Z., *Metody numeryczne* (7th ed.), Wydawnictwo Naukowe PWN, (2017).

https://doi.org/10.1051/e3sconf/201913701048

RESEARCH & DEVELOPMENT IN POWER ENGINEERING, 2019

Oily water emulsification as the method of its pre-treatment for utilization via co-combustion

Aleksandra Dzido[1,*], Piotr Krawczyk[1], Michalina Kurkus-Gruszecka[1] and Krzysztof Badyda[1]

[1] Institute of Heat Engineering, Warsaw University of Technology, Nowowiejska 21/25, 00-665 Warsaw, aleksandra.dzido.dokt@pw.edu.pl, Poland

Abstract. Oily water can be the side product of many processes, like crude oil processing and transportation leakages, industries such as steel industry, machine industry, petroleum refinery, petrochemical industry, textile industry, metal industry, pyrolysis etc. It seems to be the difficult to dispose waste. Although several treatment methods have been developed, many times additional water disposal installation is required. When the main process consumes natural gas, to simplify oily water utilization, the method of its direct injection to the natural gas burner is suggested by the paper authors. To provide the proper combustion and the mixture homogenization, emulsification was proposed. The aim of the research was to find the optimal recipe from the stability point of view, taking into account technical and economic conditions. The article covers the variety of surfactants examination for the water-hydrocarbons mixture emulsification, including the surfactants concentration optimization.

1 Oily water disposal

Oily water seems to be the larger waste product of oil and gas production, although the others sectors as steel industry, machine industry, textile industry, metal industry, pyrolysis also contribute to this kind of waste world production. Nowadays challenge is to propose economic viable way of the oily water purification, as it can go back to the natural environment without any damage for the biotope and inanimate nature. The clean-up process focuses mainly on the oil and water separation. Significant amount of researchers have reported interesting results in this field [1], [2], [3], [4], [5], [6], [7]. Among the most common purification methods the following can be enumerated [8], [9]:

Flotation – it is the process in which the tiny air bubbles are introduced to the polluted water. Due to the fact that the floating density of oil is less than that of water, oil droplets via adhesion go up with the bubbles. Finally the scum layer has to be separated from water fraction.

Coagulation – the process in which the emulsified or dissolved oil, with the presence of coagulant – typically metallic salts, forms a greater structure characterized by the irregular shape. In this form it is significantly easier to remove the oil from water.

Biological treatment – the process bases on the contaminant consumption by the microorganisms. The phenomena occur usually in the sludge, which consist of biological matter. In the following step, water is separated from the sludge.

Membrane separation technology – bases generally on the pressure driven processes with the help of the porous membrane. Membrane separation is a kind of physical removal, among the others: microfiltration, ultrafiltration, nanofiltration and reverse osmosis can be enumerated.

As justified above the oily water is a difficult dispose waste, but several purification technologies were developed. The authors of this paper propose another method of its utilization. It can be noticed that the oily water many times is generated in the industry in which the natural gas or petrol is used. In the paper the idea of the oily water emulsification and co-combustion with the natural gas is presented as the way of its utilization.

2 Emulsions

2.1 Emulsions types

An emulsion by the definition is a mixture of two or more liquids that are normally immiscible and can be classified as a kind of colloid, where both phases are liquid. Emulsions can be also classified as thermodynamically unstable multiphase systems [10], that is why to increase its stability some chemicals called surfactants (discussed deeply in the further part of the article) can be added. Typical dispersed liquid droplets size for emulsions can take values from 0.001 to 0.05 mm. Various types of emulsions can be enumerated: water in oil (w/o), oil in water (o/w) and multiple emulsions [11] as shown in the fig. 1. Emulsion does not have to consist of only two liquids. Multi-component double emulsions are described in [12] [13]. Another emulsion type are emulsions with extremely high internal volume fraction [14]. Their internal structure takes the shape of hexagons, which enable to obtain inner phase concentration greater than 99%.

https://doi.org/10.1051/e3sconf/201913701048

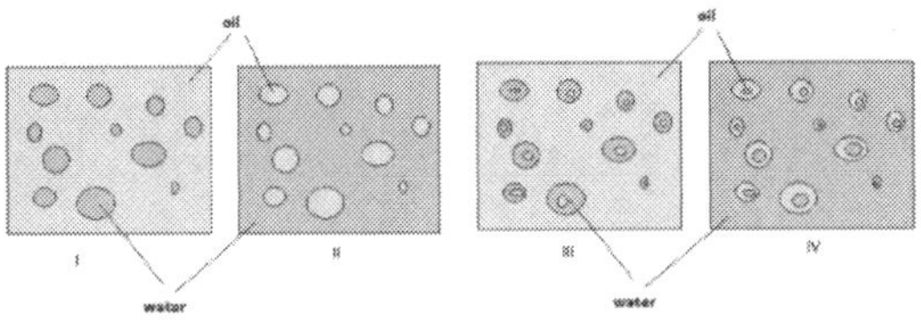

Fig. 1. Emulsions types: I- water in oil, II – oil in water, III – multiple emulsion w/o/w, IV – multiple emulsion o/w/o.

2.2 Emulsions generation methods

2.2.1 Mechanical [15]

One of the most common emulsification techniques is mechanical mixing. The rotation of the stirrer causes turbulences, which lead to the fluid elements motion, and finally to the phase mixing resulting in the breakage of droplets and the emulsion formation. The process of emulsification with a mechanical agitator consists of two stages: initial emulsification and proper emulsification. In the first stage, water, organic phase and emulsifier (surfactant) are mixed. The emulsifier can be added first to only one phase or to both phases simultaneously. As a result of the initial emulsification, a primary emulsion is formed, from which the specific emulsion is obtained in the second stage. The second emulsification stage involves the deformation of large droplets and their decay.

2.2.2 High pressure homogenization [16], [17]

High pressure homogenizers pumps the liquid, through a narrow gap formed by stationary parts. The high flow velocity, which in the cross-section of the gap can reach up to 200 m/s, causes hydraulic shear and a considerable pressure drop, so that high disintegration of the dispersed phase into droplets of about 1 μm or less can be obtained. The high pressure homogenizer is supplied by the pre-emulsion, which can be characterized by relatively large droplet diameter, which are dispersed into the smaller ones while shear and cavitation forces act on the flow in the described device. Values of these forces are regulated by the size of the gap in the homogenizing valve. Usually two-stage high-pressure emulsification is used. In the first stage, the emulsion of the pre-emulsion is ground up. The second step is to stabilize the emulsion, which consists in breaking up the disperse phase droplets formed after the first stage. Homogenizers of this type are used to create emulsions characterized by increased stability obtained due to the high grinding of the emulsion particles. Industrial pressure homogenizers usually work in the pressure ranges from 3 to 20 MPa. There is a linear relationship between the size of the applied pressure and the size of the obtained droplets diameter of the dispersed emulsion phase. Emulsification under higher pressure conditions results in greater disintegration of the dispersed phase.

2.2.3 Ultrasound emulsification [18]

In this type of devices, capillary interfacial waves are responsible for the emulsification processes, in addition to the cavitation. These waves propagate at the interface, without penetrating into the center. When increasing the intensity of ultrasound, the amplitude of the capillary wave increases to such a value at which the droplets of liquid break off the wave spines and are ejected into the scattering phase (fig. 2, step I). Then the initial droplet can be divided into smaller ones via the same mechanism (fig. 2, step II).The most important process parameters affecting the efficiency of ultrasonic emulsification are: power, duration of emulsification and frequency of ultrasonic waves. In ultrasonic homogenizers, emulsions with a very high disintegration of the dispersed phase can be obtained. The process of emulsion production in the ultrasonic field is very complex and depends not only on the physical conditions of ultrasound, but also on the type, physicochemical properties and acoustic parameters of immiscible liquids.

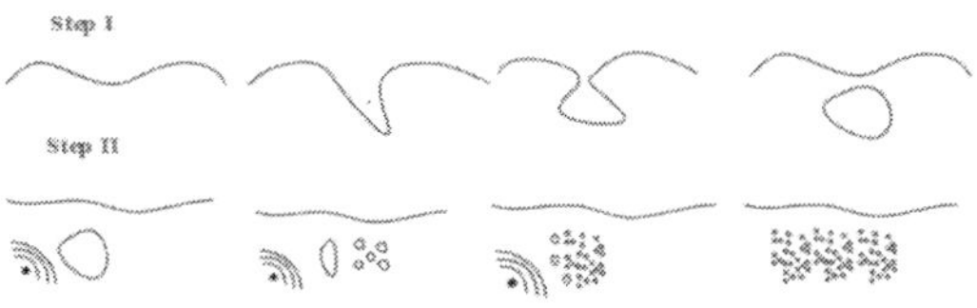

Fig. 2. Ultrasound emulsification: droplet formation and breakup.

2.2.4 Membrane emulsification [19], [20], [21]

In the emulsification process using membranes made of ceramic or porous glass, the dispersed phase is forced through the pores of the membrane into a second liquid which is a continuous phase (fig. 3). The continuous phase flows along the inner surface of the membrane and entrains the forming drops at the pore outlets. During the continuous phase squeezing through the tubular membranes with the same pore diameter, low pressures at about 200 kPa are used. Diaphragm homogenizers can work in a serial or parallel system. When emulsifying with porous membranes, the emulsion drops are formed by direct dispersion and there is no effect of breaking forces of the emulsion particles to create the smaller ones - as occur in the others emulsification methods. The supplied energy is consumed to overcome the porous structure of the membranes and to detach the forming droplet phase dispersed by the flow of the continuous phase. The membranes applied in the process can by characterized by pores of a various diameters to obtain emulsions of a wide range of droplet sizes. The membranes should be made of materials resistant to cracks under the pressure applied during emulsification. The most common materials are glass or ceramic. Emulsions with a higher degree of disintegration are obtained by set of membranes application. Each set consists of several membrane layers. Polarity is also a very important feature. The membranes should not be wetted by the dispersed phase. Hydrophobic membranes are used in the preparation of w/o emulsions, and hydrophilic in the preparation of o/w type. The particle size of the disperse phase depends on the pore size of the membrane, the emulsifier and its concentration, the type of membrane surface and the flow velocity of the continuous phase but also of the transmembrane pressure.

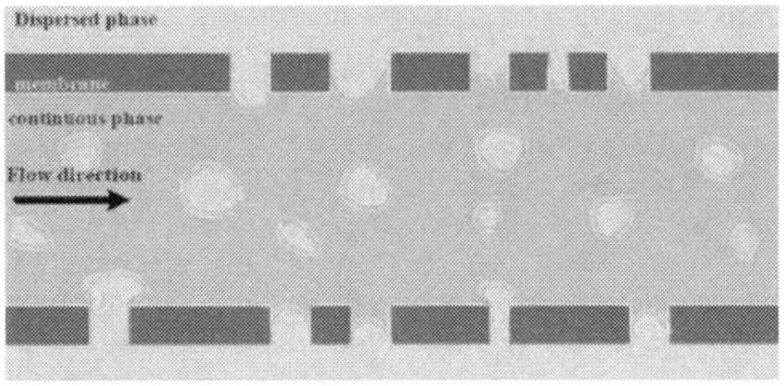

Fig. 3. Membrane emulsification.

2.3 Emulsions stability

As mentioned before, emulsions as a colloid systems are thermodynamically unstable [22], although their kinetic stability can be achieved. In the two immiscible liquids system minimum energy corresponds with the two layer system (only one contact surface) [23]. In emulsions droplets at the scale of microns are considered. This is relatively large scale, so due to the significant droplet area, the excess Gibbs energy is high and not easily compensable by the entropy contributions like in case of microemulsions. Surface forces working in the emulsion system were described in [24]. Due to the interfacial surface tension, the emulsion can evaluate via various mechanisms leading to its dehomogenization. Main mechanisms are presented in the fig. 4.

Creaming – is the process in which the emulsion droplets concentrate above the main emulsion fraction [25]. Finally the majority of the suspended droplets is located at the top. The creaming mechanism can be realized in different ways, depending on i.e. droplet concentration, polydispersity or inter-particle interaction.

Sedimentation – is the emulsion destabilization mechanism in which the dispersed droplets concentrate at the bottom of the vessel. Initial droplet size has an influence on the sedimentation time (the larger the droplet the faster goes down). Sedimentation occurs i.e. in the w/o emulsions in which the density of the dispersed phase is greater than the continuous phase density [26].

Flocculation is the process basing on the droplets agglomeration with no rupture of the stabilization layer [25]. Sometimes flocculation contributes to the creaming, because of the fact that agglomerates are greater than a single particle so can move faster in the continuous phase. Flocculation can also lead to coalescence.

Coalescence is the process of droplet merging. After coalescence the dispersed phase droplets have significant greater dimensions [27], which can contribute to force other emulsion destabilization mechanisms like creaming. The coalescence mechanism is deeply described in [28].

Emulsion breaking is the process consisting in the phase separation into two layers. In some cases the coalescence and creaming can cause an inadvisable emulsion break, which means going back to the thermodynamic equilibrium. In several industries emulsion break can be desired, i.e. when the phase separation is required. For this purpose electrochemical techniques [29] or microfiltration membranes [30] can be applied.

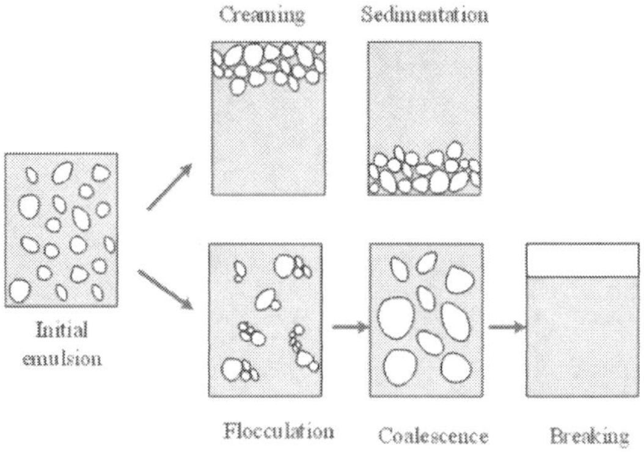

Fig. 4. Emulsion instability mechanisms.

3 Surfactants

Generally, surfactants are compounds that's aim is to decrease the surface tension (or interfacial tension) between two phases i.e. liquids. Surfactants are widely used to grow up the emulsion stability [31] and prevent the emulsion break. The surfactant molecule consists of two main parts: hydrophilic and hydrophobic. Surfactant particles concentrate in the region of phase boundaries, and orient in the way that the hydrophilic part is submerged in water phase while the hydrophobic – in the organic phase. Hereby the surfactant particle works as a phase connector by decreasing the surface tension.
According to [32] surfactants can be classified as:

Anionic

- Sulfate Esters
- Sulfonic Acid Salts
- Carboxylate Soaps and Detergents
- Phosphoric Acid Esters and Related Surfactants

Cationic

Nonionic Surfactants

- Polyoxyethylene-Based Surfactants
- Derivatives of Polyglycerols and Other Polyols
- Block Copolymer Nonionic Surfactants
- Miscellaneous Nonionic Surfactants

Amphoteric Surfactants

- Imidazoline Derivatives
- Surface-Active Betaines and Sulfobetaines
- Phosphatides and Related Amphoteric Surfactants

The scientists have addressed the problem in the surfactant selection so the concept of the hydrophilic-lipophilic balance (HLB) parameter were created. The HLB is an experimental conception to describe the hydrophilic or lipophilic properties of the surfactant particle [33]. This parameter was accepted by the wide range of researchers and industrial engineers. The HLB is determined basing on the hydrophilic and lipophilic group strength and number of each groups per particle. Surfactant application depending on the HLB value is presented in the Table 1.

https://doi.org/10.1051/e3sconf/201913701048

Table 1. Surfactant application for various HLB values.

HLB	application
1.5-3	anti-foaming
3-6	w/o emulsions
7-9	wetting agents
8-12	o/w emulsions
15-20	solubilizer or hydrotrope

While introducing more than one emulsifier to the system, the HLB value can be calculated as a sum of each HLB weighted by added mass of each of them [34]

4 Methodology

The aim of the research was to determine the proper surfactant for the water emulsification with the light gasoline fractions addiction. As reported in [35], typical surfactant concentration vary from 1-8 %, that is why in the paper the concentration was assumed at the level of 4.6%. A mixture consisting of the following ingredients was created:

- 1 ml of heavy oil
- 3 ml of light gasoline fraction
- 6 ml of water
- 0.25 ml Rokwinol 80 (stabilizer)
- 0.5 ml of surfactant (provided by PCC Rokita, listed in the Table 2)

The organic fraction (heavy oil and gasoline) was mixed with the surfactant and stabilizer. Water was added to the thus formed mixture and stirred for 120 s on a magnetic stirrer (1100 rpm). The emulsion was then poured into the test tube and its properties were evaluated at selected time steps.

5 Results

The results are presented as a water phase fraction change at time (fig. 5). One hour period was taken into account. Conducted research aim was to select surfactant for the further experiment, in which the authors focused on the surfactant concentration optimisation. Basing on the obtained results two surfactants were chosen to be taken into account during the further research: Roksol EMB2 and Rokanol O3.

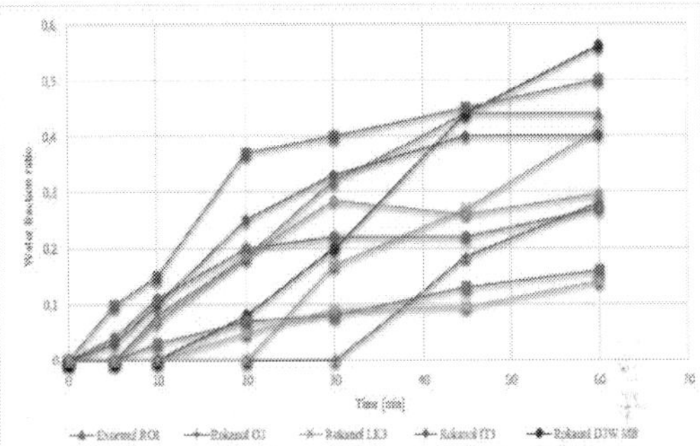

Fig. 5. Water phase ratio as a function of time.

Table 2. Examined surfactants.

No.	Commercial name	Composition	HLB	Solubility in water
1	Rokanol RZ4P11	Alcohols, C16-18 (even), ethoxylated, propoxylated	3.3	low, creates emulsions
2	Exoemul RO1	mix	6.9	low
3	Rokanol O3	(Z) -9-Octadecen-1-ol, ethoxylated	7.1	low
4	Rokanol LK3	Alcohols, C12-14, ethoxylated	7.8	low, creates turbid solutions
5	Rokanol IT3	C13 alcohols, branched, ethoxylated	8	low, creates emulsions
6	Rokanol D3W MB	C10 alcohol, ethoxylated	8.6	low, creates emulsions
7	Rokanol NL6	Alcohols, C9-11, ethoxylated	12.5	good
8	Rokanol IT8	Alcohols, C13, branched, ethoxylated	12.8	good
9	Rokwinol 80	sorbitan monooleate, ethoxylated	15	good
10	Rokwinol 20	Sorbitan monolaurate, ethoxylated	16.7	good
11	Roksol EMB2	2,2-iminodiethanol	no data	low, creates turbid solutions

6 Concentration optimization

To optimize the surfactant concentration the emulsions consisting of the following ingredients were prepared:

- 1 ml of heavy oil
- 3 ml of light gasoline fraction
- 6 ml of water
- 0,1 ml Rokwinol 80 (stabilizer)
- Surfactant

The procedure was the same as described in Chapter 4. The concentration of the surfactant vary from 2-10%. From the economic point of view, the lower surfactant consumption the more viable process. Comparing Roksol EMB 2 with Rokanol O3 it can be claimed that the second one is characterised by longer stability time while shifting to the lower surfactant concentration (fig. 6-7). The emulsion is stable for minutes, which seems to fulfil the requirement for the process needs. Taking into account the fact, that continuous mixing of the emulsion prevent its dissection Rokanol O3 at the lowers considered concentration seems to be the best solution for the thermal utilisation of the oily water purpose.

The paper authors considered also the emulsion in which the total sum of the surfactant and stabilizer equals to 1%. Two mixing times were taken into account: 120 s and 720 s. The obtained mixtures at the selected times are presented in the fig. 8-9. Comparing this two products, it can be noticed that increasing the mixing time can help to extend the stability time. Although the obtained product was not fully emulsified, so the surfactant concentration was too low.

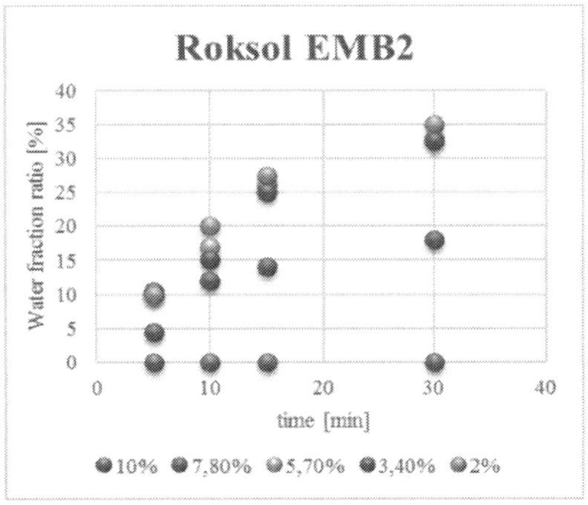

Fig. 6. Water fraction ratio change in time, various surfactant concentration, Roksol EMB2.

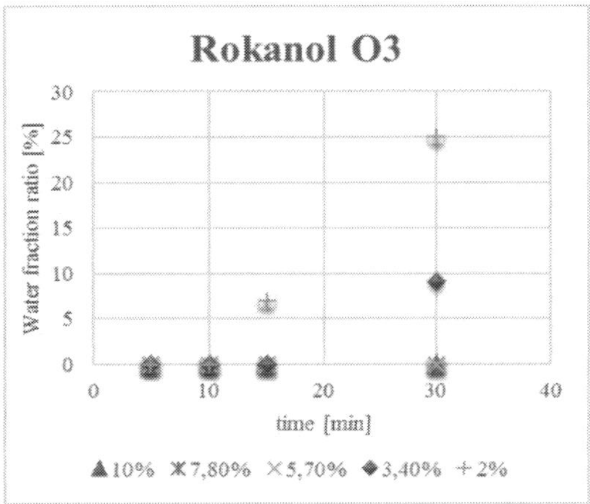

Fig. 7. Water fraction ratio change in time, various surfactant concentration, Rokanol O3.

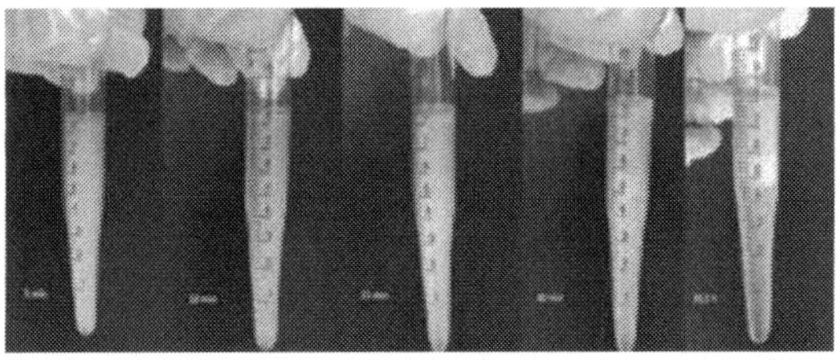

Fig. 8. Rokanol O3+Rokwinol 80, concentration 1 %, mixing time 120 s.

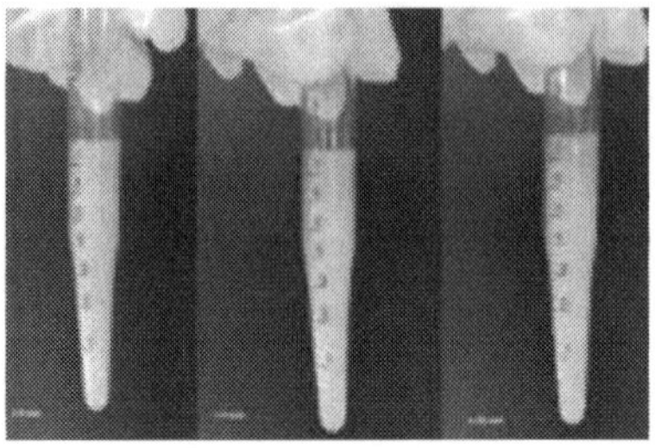

Fig. 9. Rokanol O3+Rokwinol 80, concentration 1 %, mixing time 720 s.

7 Outcomes

Oily water is the side product of wide range of industries and its disposal is still challenging. The authors proposed the way of its utilisation by emulsification with the light gasoline fraction and combustion. The aim of this paper was to examine various surfactants and optimise its concentration. The Rokanol O3 (HLB=7.1) with Rokwinol 80 (HLB=15) were selected as a best mixture of emulsifiers for the described purpose. Our results are in good agreement with the theory of the HLB factor (o/w emulsion stabilisation corresponds with HLB from 8 to 12). First emulsion combustion tests were conducted, but further research in this area is needed.

References

1. Shui Y, Yan L, Bao C, Jiang L. Treatment of oily wastewater by organic – inorganic composite tubular ultrafiltration (UF) membranes, 2006; 196: 76–83.
2. Soltani S, Mowla D, Vossoughi M, Hesampour M. Experimental investigation of oily water treatment by membrane bioreactor. DES, 2010; 250(2): 598–600.
3. Beall GW. The use of organo-clays in water treatment, 2003; 24: 11–20.
4. Secerov Sokolovic R, Sokolovic S, Sevic S. Oily water treatment using a new steady-state fiber-bed coalescer, 2009; 162: 410–5.
5. Syed S, Alhazzaa MI, Asif M. Treatment of oily water using hydrophobic nano-silica. Chem Eng J, 2011; 167(1): 99–103.
6. Cheryan M, Rajagopalan N. Membrane processing of oily streams Wastewater treatment and waste reduction, 1998; 151.
7. Bande RM, Prasad B, Mishra IM, Wasewar KL. Oil field effluent water treatment for safe disposal by electroflotation, 2008; 137(August 2006): 503–9.
8. Ahmadun F, Pendashteh A, Abdullah LC, Biak DRA, Madaeni SS, Abidin ZZ. Review of technologies for oil and gas produced water treatment. J Hazard Mater, 2009; 170: 530–51.
9. Yu L, Han M, He F. A review of treating oily wastewater. Arab J Chem, 2017.
10. Haeberle S, Zengerle R, Ducre J. Centrifugal generation and manipulation of droplet emulsions. Microfluid Nanoflui, 2007: 65–75.

11. Wong SF, Lim JS, Dol SS. Crude oil emulsion : A review on formation , classification and stability of water-in-oil emulsions. J Pet Sci Eng, 2015; 135: 498–504.
12. Matter S, Adams LLA, Kodger TE, Kim S, Shum C, Weitz DA. Single step emulsification for the generation of multi-component double, 2012: 10719–24.
13. Sauret A, Cheung Shum H. Forced generation of simple and double emulsions in all-aqueous systems, 2013; 154106(November 2011): 1–5.
14. Schuler F, Paust N, Zengerle R, Stetten F Von. Centrifugal Step Emulsification can Produce Water in Oil Emulsions with Extremely High Internal Volume Fractions, 2015: 1180–8.
15. Dalmazzone C. The Mechanical Generation of Emulsions. Lubr Sci, 2005(February).
16. Floury J, Desrumaux A, Lardieres J. Effect of high-pressure homogenization on droplet size distributions and rheological properties of model oil-in-water emulsions. Innov Food Sci Emerg Technol, 2000: 127–34.
17. Schultz BS, Wagner G, Urban K, Ulrich J. Review High-Pressure Homogenization as a Process for Emulsion Formation, 2004(4): 361–8.
18. Canselier JP, Delmas H, Wilhelm AM, Abismaïl B. Ultrasound Emulsification — An Overview, 2010; 2691(2002).
19. Holdich RG, Dragosavac MM, Vladisavljevic GT, Kosvintsev SR. Membrane Emulsification with Oscillating and Stationary Membranes. Ind Eng Chem Res, 2010: 3810–7.
20. Joscelyne SM, Trägårdh G. Membrane emulsification — a literature review. J Memb Sci, 2000; 169(October 1999): 107–17.
21. Charcosset C, Limayem I, Fessi H. The membrane emulsification process — a review, 2004; 218(May 2003): 209–18.
22. Florence AT, Rogers JA. Emulsion Stabilization by non-ionic surfactants: experiment and theory, 1971(1950): 233–51.
23. Goff HD. Instability and Partial Coalescence in Whippable Dairy Emulsions. J Dairy Sci, 1997; 80(10): 2620–30.
24. Sjoblom J. Emulsions and emulsion stability., 2005.
25. Boyd J, Parkinson C, Sherman P. Factors Affecting Emulsion Stability, and the HLB Concept, 1972; 41(2).
26. Rousseau D, Hodge SM. Stabilization of water-in-oil emulsions with continuous phase crystals. Colloids and Surfaces, 2005; 260: 229–37.
27. Yarranton HW, Urrutia P, Sztukowski DM. Effect of interfacial rheology on model emulsion coalescence II Emulsion coalescence, 2007; 310: 253–9.
28. Saito H, Kawagishi A, Tanaka M, Tanimoto T, Okada S. Coalescence of Lipid Emulsions in Floating and Freeze – Thawing Processes : Examination of the Coalescence Transition State Theory, 1999; 134: 129–34.
29. Canizares P, Martinez F, Lobato J, Rodrigo MA. Break-up of oil-in-water emulsions by electrochemical techniques. J Hazard Mater, 2007; 145: 233–40.
30. Hlavacek M. Break-up of oil-in-water emulsions induced by permeation through a microfiltration membrane, 1995; 102: 1–7.
31. Han G, Wit JS De, Chung T. Water reclamation from emulsi fi ed oily wastewater via effective forward osmosis hollow fi ber membranes under the PRO mode. Water Res, 2015; 81: 54–63.
32. Attwood D, Florence AT. Surfactant systems Their Chemistry, Pharmacy and Biology., 1983.
33. Laughlin RG. HLB, from a thermodynamic perspective. J Soc Cosmet Chem, 1981; 392: 371–92.
34. Magdassi S, Frenkel M, Garti N, Kasan R. Multiple Emulsions II : HLB Shift Caused by Emulsifier Migration to External Interface, 1984; 97(2): 374–9.
35. Wasan DT, Nikolov AD, Aimetti F. Texture and stability of emulsions and suspensions : role of oscillatory structural forces. Adv Colloid Interface Sci, 2004; 109: 187–95.

E3S Web of Conferences 137, 01049 (2019) https://doi.org/10.1051/e3sconf/201913701049
RDPE 2019

RESEARCH & DEVELOPMENT IN POWER ENGINEERING, 2019

Analysis of offshore wind farm located on Baltic Sea

Anna Sobotka[1], Kajetan Chmielewski[2], Marcin Rowicki[3], Justyna Dudzińska[4], Przemysław Janiak[5], Krzysztof Badyda[6]

[1] Institute of Heat Engineering, Warsaw University of Technology, Power Generation Strategic Analysis Department, PKN ORLEN, anna.sobotka@orlen.pl, Poland
[2] Power Generation Strategic Analysis Department, PKN ORLEN, kajetan.chmielewski@orlen.pl, Poland
[3] Power Generation Strategic Analysis Department, PKN ORLEN, marcin.rowicki@orlen.pl, Poland
[4] Power Generation Strategic Analysis Department, PKN ORLEN, justyna.dudzinska@orlen.pl, Poland
[5] Power Generation Strategic Analysis Department, PKN ORLEN, przemyslaw.janiak@orlen.pl, Poland
[6] Institute of Heat Engineering, Warsaw University of Technology, krzysztof.badyda@itc.pw.edu.pl, Poland

Abstract. Poland is currently at the beginning of the energy transformation. Nowadays, most of the electricity generated in Poland comes from coal combustion. However, in accordance to the European Union policy of reducing the emission of carbon dioxide to the atmosphere, there are already plans to switch to low-emission energy sources in Poland, one of which are offshore wind farms. The article presents the current regulatory environment of the offshore wind energy in Poland, along with a reference to Polish and European decarbonisation plans. In the further part of the article, the methods of determining the kinetic energy of wind and the power curve of a wind turbine are discussed. Then, on the basis of historical data of wind speeds collected in the area of the Baltic Sea, calculations are carried out leading to obtain statistical distributions of power that could be generated by an exemplary wind farm with a power capacity of 400 MW, located at the place of wind measurements. On their basis, statistical differences in the wind power generation between years, months of the year and hours of the day are analysed.

1 Introduction

Electricity production in Poland amounted to 165.3 TWh in 2018, of which approximately 79.5% was generated by coal-fired power plants [1]. In this regard, Poland is a black spot on the map of Europe. Due to the environmental requirements related to the climate policy of the European Union, which have been getting stricter for years, Poland faces the requirement to diversify its energy mix and transform its economy towards climate neutrality. On the other hand, it creates an opportunity for the development of other technologies, mainly Renewable Energy Sources (hereinafter referred to as RES). The example of the offshore wind farms fits into the policy of energy changes in the world. Their main advantages are the zero emission of pollutants and greenhouse gases to the atmosphere and the lack of waste generated in the process of their exploitation. Compared to photovoltaics or windmills on land, offshore wind farms have much higher coefficient of capacity utilisation per year. Furthermore, there are social benefits, such as support for the local labour market and entrepreneurs. From an economic point of view, the development of technology and additional revenues to the budgets of local governments can be remarkable. Cooperation with Polish companies in the area of supply chain management also drives the national economy. According to Polish Wind Energy Association (PWEA) estimations [2], by 2030 Poland will have installed additional 6 GW of generation capacity in offshore wind farms. This is a milestone that will allow to diversify generation sources to a large extent. In the Offshore Wind Farms (hereinafter referred to as OWF) construction projects are involved such companies as PGE, Polenergia and PKN Orlen, which proves the relevant potential of the Baltic Sea. The development plan for maritime areas lists about 2.3 thousand square kilometres of the sea area designated for wind farms, which translates into a potential of 10-12 GW [2]. These are the generation capacities which cannot be passed indifferently, not only due to the ongoing energy transformation forced by the global warming, but also due to the termination of lifetime of ageing conventional units, which will have to be replaced in the near future.

In the article, selected regulatory aspects that appear in discussions about the RES development, including the OWF, have been developed. The methodology of wind power analysis has been discussed, along with the results for an exemplary wind farm located in the Baltic Sea.

The aim of the article is to discuss the wind potential of the Exclusive Economic Zone of Poland and to present recommendations concerning the OWF that could be installed there.

2 Regulatory and market environment

At present, in Poland there are no generation units located at sea. The Act on renewable energy sources [3], which exists in Polish law, regulate the issue of the OWF only to a very limited extent, which causes uncertainty among investors. There is a plan of adoption of a completely new act on the development of the OWF, which would take probably about a year to implement.

However, there are regulations that impose restrictions as to the locations where the OWF could be established in the future. In the Act on maritime areas of the Republic of Poland and maritime administration [4] Article 23 paragraph 1a is stated that the construction and use of wind power plants in the internal waters and territorial sea is prohibited. This area includes a strip of waters which is 12 nautical miles wide from the coastline. This effectively results in the fact that an Exclusive Economic Zone remains the only free space for wind farm locations.

Another document limiting the available areas is the Spatial Development Plan of the Polish Maritime Areas [5]. Only 8 of the 93 sea areas mentioned in the document are intended for the development of renewable technologies. The total area of these water bodies is 2342 square kilometres. Additional restrictions result from taking into account the impact of the OWF on the environment in the area of the Exclusive Economic Zone. The key issue here is finding the solution to raise constructions in way that would not endanger the ecological functions of spawning grounds and survival of early stages of fish (roe and larvae) of commecrial species. For each investment, before issuing a building permit, it is recommended to develop rules of conducting fishing in the area of offshore wind farms. For some of the 8 areas, it is recommended to plan RES investments to enable the marking of a transport corridor for units up to 150 meters in length. All this means that the assumed area of approximately 2.3 thousand square kilometres may be reduced by several hundred square kilometres, which will directly translate into the reduction of the potential OWF capacity in Poland. In the Baltic Sea, at the time of writing the article, permits to build artificial islands were issued for the construction of farms with a maximum capacity of approx. 7,1 [6] GW, mainly in the areas of Ławica Słupska and Ławica Środkowa. This results in the need to plan the development of wind farms and their integration into the national power system so that they bring the greatest possible benefits to the Polish power system and the whole economy.

The OWF management requires forecasting of their operation in advance, as wind sources are uncontrolled. They generate electricity depending on wind changing conditions and the generation from conventional and controllable sources (coal, gas or water) must be adjusted to them in order to balance the supply and demand for energy.

In Europe, according to a report of WindEurope, at the end of 2018 about 189 GW was installed in wind farms (170 GW in onshore and 19 GW in offshore), which produced 14% of the electricity consumed in the European Union [7]. According to the data of WindEurope, more than 2/3 of wind power is installed in 5 countries: Germany, Spain, Great Britain, France and Italy. In Poland, according to Polskie Sieci Elektroenergetyczne (PSE) data, at the end of 2018 the installed capacity of wind farms was about 5.8 GW [8], of which onshore wind farms represented 100% . However, their development was inhibited by the so-called « distance act », which prohibits the construction of a wind turbine at a distance smaller than 10 km from the nearest residential buildings. Newertheless, the development of RES, including wind farms, is suggested by published domestic and foreign documents. According to both the draft of National Energy Policy until 2040 published in November 2019 [9] and the draft of National Energy and Climate Plan for 2021-2030 [10], sent to the European Commission in January 2019, a dynamic development of RES is projected in Poland. These plans include onshore and offshore wind farms. Both documents, despite the discrepancies in data (which should be unified in their final versions), indicate a clear trend to be followed by the Polish power industry in the coming years. Development works are also progressing in the companies holding concessions for the construction of the OWF„ even though there is still regulatory uncertainty.

3 Analysis of wind power potential

In this chapter the formulas for calculating the kinetic energy of wind have been derived, which can be easily translated into power generated by the turbine. It is also explained why a power curve was used for the analysis presented in the article. Finally, the potential of wind power for an exemplary wind farm located in the Baltic Sea has been discussed.

The source of wind speed data was FINO database [11], which provides free of charge measurements from the station located in the Baltic Sea. Data are collected at an altitude of 102 metres, which enables a realistic representation of the potential capacity of a wind turbine installed at this location. The frequency of data recording is 10 minutes, which introduces a measurement error due to fluctuations of wind power in a given range. Prepared calculation model uses the hourly value of speed, which is the average of measurements in a given hour. The wind power analysis was based on data from January 1, 2010 to June 30, 2019.

The kinetic energy carried by wind can be written with a following formula, in which „m" is the mass and „V" is its speed.

$$\varepsilon_K = \frac{1}{2} \cdot m \cdot V^2 \qquad (1)$$

By substituting „m" with the expression

$$m = \rho \cdot v \qquad (2)$$

where „ρ" is the density, while „v" is the volume of air, equation (1) takes the form of

$$\varepsilon_K = \frac{1}{2} \cdot \rho \cdot v \cdot V^2 \qquad (3)$$

By substituting „v" with the expression

$$v = V \cdot A$$

where „A" is the rotor cross-section, equation (3) takes the form of

$$P = \frac{1}{2} \cdot \rho \cdot A \cdot V^3 \qquad (4)$$

 https://doi.org/10.1051/e3sconf/201913701049

This formula allows to determine the power curve of a wind turbine, assuming that there are no losses resulting from the conversion of wind kinetic energy into electricity. It also assumes that a turbine generates electricity even at a wind speed not much higher than zero, which does not translate into reality. Turbine manufacturers provide their own power curves for each type of the wind turbine. This allows to determine the level of power generated with better accuracy depending on the speed of the wind stream flowing through the rotor blades. For the purpose of the article, a turbine power curve based on V164 - 8 MW model data from Vestas [12] was used. Data of power output in dependance of wind speed was presented in Table 1.

Tab.1. Power curve based on V164 - 8 MW Vestas [8]

Wind speed [m/s]	Power [kW]
4	110
4,5	350
5	600
5,5	850
6	1140
6,5	1490
7	1900
7,5	2370
8	2900
8,5	3500
9	4155
9,5	4870
10	5630
10,5	6420
11	7150
11,5	7610
12	7865
12,5	7940
13	7970
13,5	8000
14	8000
14,5	8000
15	8000
15,5	8000
16	8000
16,5	8000
17	8000
17,5	8000
18	8000
18,5	8000
19	8000
19,5	8000
20	8000
20,5	8000
21	8000
21,5	8000
22	8000
22,5	8000
23	8000
23,5	8000
24	8000
24,5	8000
25	8000

In order to conduct an in-depth analysis of wind data, an example of a 400 MW wind farm consisting of 50 wind turbines was taken into account. The calculation model assumes transmission losses at the level of 4% of the generated power, 100% of availability, lack of losses caused by the possible turbulent wind flow and lack of wake losses.

The results of the analysis are presented in the form of average power generated by the farm in hourly and monthly cross-section. Besides there are calculated percentiles referring to the percentage of time while the amount of generated power did not fall below a given percentage of the sample.

4 Results

On the basis of the calculations carried out, statistical power distributions are obtained for a 400 MW offshore wind farm located in the Baltic Sea at the wind speed measurement site. This chapter presents selected results.Conditional formatting in the form of a gradient of three colours: red, yellow and green has been applied to facilitate the reception of the analysis results presented in the following tables. If the value of the OWF power is relaively small, it is displayed on a green background, if the value is large, the background is red.

Tab.2. Average power generation of 400 MW OWF in a given hour for the years 2010-2019

Hour	Power [MW]
00:00	255,92
01:00	202,73
02:00	206,56
03:00	205,40
04:00	205,61
05:00	207,10
06:00	205,88
07:00	204,60
08:00	202,94
09:00	202,45
10:00	201,72
11:00	201,12
12:00	204,31
13:00	205,37
14:00	207,46
15:00	209,31
16:00	210,00
17:00	213,45
18:00	215,90
19:00	215,88
20:00	217,14
21:00	214,39
22:00	216,59
23:00	212,04

Tab.3. Average power generation of 400 MW OWF in a given month for the years 2010-2019

Month	Power [MW]
January	243,95
February	235,21
March	228,02
April	214,46
May	182,33
June	173,39
July	163,12
August	166,84
September	191,37
October	247,31
November	223,70
December	257,43

Table 2. shows that the average capacity of the calculated farm for the data from the whole period under consideration is higher in the afternoon and night hours. What is interesting, however, these are not so significant differences - they are within the range of less than 14% of the installed capacity. The differences in months, shown in Table 3, are much clearer, and the difference between the minimum (July) and maximum (December) is as much as 24% of the installed capacity.

The coefficient of installed capacity utilisation „C_p" has also been calculated. It was calculated as the quotient of the produced energy „E" in a given period and the product of the installed power „P_N" and the examined period „T". It allows to compare the power of the OWF with other RES or conventional ones.

$$C_p = \frac{E}{P_n \cdot T} \tag{5}$$

For the whole analysed period, the coefficient of installed capacity utilisation amounted to 48.98%, which is comparable to the data available in the literature [2]. Compared to other RES : onshore wind farms and photovoltaics, calculated coefficient of installed capacity utilisation is much higher. For modern onshore wind farms the coefficient is around 40%, while for photovoltaics it is only about 10% [2]. The partial results for individual years are presented in Table 4. They show how much the wind power differs from one year to the next. This is very important especially because it is a value over which the investor or operator has no influence.

Tab.4. Coefficient of utilisation of installed capacity in a given year for the full years 2010-2018

Year	Capacity factor [%]
2010	52,67%
2011	53,69%
2012	52,23%
2013	50,49%
2014	52,74%
2015	49,85%
2016	39,41%
2017	48,64%
2018	40,08%

In case of division of the generated power in 24 hours into individual percentiles of occurrence, as shown in Table 5, it can be clearly noticed that (according to the methodology described earlier and the wind data used) for 20% of time in a year the OWF works with 100% power, while for 10% of the time the OWF does not produce anything. It is also worth noting that hours without electricity production occur at any moment of the day, which clearly shows how important is to secure generation in controllable power plants for the efficient and uninterrupted operation of the electricity system.

Tab.5. Percentiles of power distribution in individual hours on average for the years 2010-2019

Hour	Percentile																		
	0,95	0,90	0,85	0,80	0,75	0,70	0,65	0,60	0,55	0,50	0,45	0,40	0,35	0,30	0,25	0,20	0,15	0,10	0,05
00:00	384	384	384	384	380	367	340	295	248	205	168	131	100	74	55	36	20	0	0
01:00	384	384	384	384	379	364	335	296	249	199	160	128	95	72	53	35	18	0	0
02:00	384	384	384	384	379	364	335	283	241	200	160	123	92	69	52	35	18	0	0
03:00	384	384	384	384	380	364	333	290	242	198	157	121	94	70	50	33	16	0	0
04:00	384	384	384	384	377	362	331	292	247	200	162	120	91	68	50	33	16	0	0
05:00	384	384	384	384	378	364	339	295	246	200	161	122	94	69	50	33	16	0	0
06:00	384	384	384	384	378	362	336	295	243	199	156	122	92	68	48	31	16	0	0
07:00	384	384	384	384	380	363	332	290	239	198	153	119	91	67	46	32	16	0	0
08:00	384	384	384	384	380	364	334	291	242	194	151	113	86	64	45	29	14	0	0
09:00	384	384	384	384	380	366	339	287	238	184	141	109	86	62	44	30	11	0	0
10:00	384	384	384	384	380	365	337	289	235	189	147	112	84	62	42	26	9	0	0
11:00	384	384	384	384	380	366	336	289	239	192	149	110	81	61	40	22	4	0	0
12:00	384	384	384	384	382	367	341	288	245	194	149	116	85	60	40	25	6	0	0
13:00	384	384	384	384	382	367	338	291	245	202	159	120	86	62	42	26	8	0	0
14:00	384	384	384	384	380	368	346	304	252	204	158	120	90	65	45	27	6	0	0
15:00	384	384	384	384	384	371	350	308	257	211	161	121	92	67	45	28	7	0	0
16:00	384	384	384	384	384	373	356	315	262	210	166	119	89	66	46	31	9	0	0
17:00	384	384	384	384	384	374	355	320	267	217	167	130	99	72	51	32	14	0	0
18:00	384	384	384	384	384	373	354	313	274	223	173	138	105	78	55	35	17	0	0
19:00	384	384	384	384	384	374	356	319	270	228	180	143	109	80	57	39	18	0	0
20:00	384	384	384	384	384	372	353	317	278	226	181	141	112	81	58	38	20	0	0
21:00	384	384	384	384	383	372	353	318	273	228	185	142	106	80	56	39	22	0	0
22:00	384	384	384	384	383	369	353	311	263	218	183	141	106	80	59	38	21	0	0
23:00	384	384	384	384	382	368	346	304	257	216	178	140	105	79	56	39	21	0	0

Based on the power factor coefficients in Table 4, simple calculations can be made to show the scale of changes in power generation for the most and least windy years. In the best case(2011), when C_p = 53.69%, power generation could be etsimated at about 4.7 TWh, while in the worst case (2016) C_p was only 39.41%, which translates into about 3.45 TWh. These fluctuations have impact directly on the generated profits.Unfortunately, outside the scope of the authors© knowledge is the answer to the question whether these are random fluctuations or the announcement of wider changes in the wind power on the Baltic Sea caused by e.g. climate changes. The described trend is also visible in Fig.1., which presents an ordered diagram of the generation of the example OWF per year.

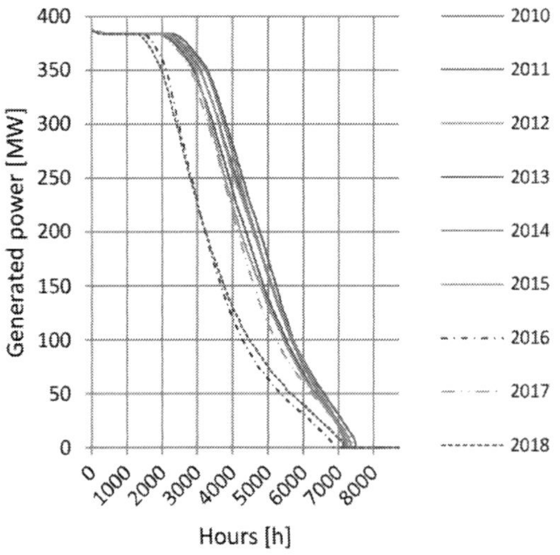

Fig.1. Structured diagram of wind generation in particular years of the analysed period

5 Summary

Considerations of selected regulatory aspects that are important in the context of the OWF development in Poland, presented in the article, indicate the lack of clearly defined borders of the areas in the Exclusive Economic Zone of Poland, where the OWF could be established. Possible reduction of this area will translate into a significant reduction of wind power potential. Additional concerns arise from the lack of adequate legal conditions that would regulate the offshore branch of industry. Nevertheless, the companies holding concessions for the construction of the OWF are progressing with the offshore projects development.

The publicly available wind measurements and the wind turbine power curve enabled performing considerations of power generated by a hypothetical OWF located in the Baltic Sea and changes in the level of this power at different periods of time in the year and during the day.

On the basis of the calculations performed, statistical power distributions were determined. The coefficient of capacity utilisation for the whole analysed period was 48.98%. Analysis of individual years showed that the most windy year was 2011, when total generation amounted to 4 703 244 MWh, and the least windy year was 2016 with generation of 3 452 316 MWh. These factors have a significant impact on revenue which the OWF could generate each year. It is worth noting, that it is out of a farm's owner control. The distribution of the percentiles clearly illustrates the situation which must be faced. In the analysed period, for 20% of time, the example OWF operates at 100% capacity, while for 10% of time generation is 0%. This clearly shows, that backup with the conventional power plants is necessary to ensure safe and uninterrupted work of the electricity system.

Compared to onshore wind farms and photovoltaics , the OWF generate more electricity from the same installed power. The coefficient of capacity utilisation for offshore wind farms is nearly 25% higher than the coefficient for the onshore wind farms, and five times higher than for the photovoltaics [2]. This also means, that the OWF generate electricity more steadily than mentioned renewable technologies.

All of this, along with the fact, that the OWF do not disrupt rural areas, truly shows the potential of Baltic Sea in generating clean energy.

References

[1] Polskie Sieci Elektroenergetyczne S.A. (2018). *Zestawienie danych ilościowych dotyczących funkcjonowania KSE w 2018 roku.* Retrieved from https://www.pse.pl/dane-systemowe/funkcjonowanie-rb/raporty-roczne-z-funkcjonowania-kse-za-rok/raporty-za-rok-2018

[2] Polskie Stowarzyszenie Energetyki Wiatrowej (2019). *Przyszłość morskiej energetyki wiatrowej w Polsce.* Retrieved from http://psew.pl/wp-content/uploads/2019/06/Przysz%C5%82o%C5%9B%C4%87-morskiej-energetyki-wiatrowej-w-Polsce-raport.pdf

[3] Ustawa z dnia 20 lutego 2015 r. *o odnawialnych źródłach energii*, Dz.U. 2015 poz. 478

[4] Ustawa z dnia 21 marca 1991 r. *o obszarach morskich Rzeczypospolitej Polskiej i administracji morskiej*, Dz. U. 2019.0.2169

[5] Rozporządzenie Ministra Gospodarki Morskiej i Żeglugi śródlądowej oraz Ministra Inwestycji i Rozwoju *Plan zagospodarowania przestrzennego morskich wód wewnętrznych, morza terytorialnego i wyłącznej strefy ekonomicznej (Projekt z dnia 22.07.2019 r.)*

[6] Polskie Sieci Elektroenergetyczne S.A. (2019). *Wykaz Podmiotów ubiegających się o przyłączenie do Krajowej Sieci Przesyłowej.* Retrieved from https://www.pse.pl/obszary-dzialalnosci/krajowy-system-

elektroenergetyczny/wykaz-podmiotow-ubiegajacych-sie-o-przylaczenie

[7] WindEurope (2019). *Wind energy in Europe in 2018. Trends and statistics.* Retrived from *https://windeurope.org/wp-content/uploads/files/about-wind/statistics/WindEurope-Annual-Statistics-2018.pdf*

[8] Polskie Sieci Elektroenergetyczne S.A. (2019). *Informacja o dostępności mocy przyłączeniowej do sieci przesyłowej*. Retrieved from *https://www.pse.pl/dane-systemowe/praca-kse/informacje-ogolne/informacja-o-dostepnosci-mocy-przylaczeniowej*

[9] Ministerstwo Energii (2019). *Projekt Polityki energetycznej Polski do 2040 roku.* Retrived from *https://www.gov.pl/web/aktywa-panstwowe/polityka-energetyczna-polski-do-2040-r-zapraszamy-do-konsultacji1*

[10] Ministerstwo Energii (2019). *Projekt Krajowego planu na rzecz energii i klimatu na lata 2021-2030.* Retrived from *https://www.gov.pl/attachment/01f5256f-582a-47a6-b41a-42215e4a76e0*

[11] FINO-Datenbank (2019). *FINO-Datenbank.* Retrieved from *http://fino.bsh.de/*

[12] Desmond C., Murphy J., Blonk L., Haans W., Description of an 8 MW reference wind turbine, *Journal of Physics : Conference Series* 753:092013, (2016)

E3S Web of Conferences 137, 01050 (2019) https://doi.org/10.1051/e3sconf/201913701050
RDPE 2019

RESEARCH & DEVELOPMENT IN POWER ENGINEERING, 2019

A comparative study on electric and gas engine heat pump

Konrad Kuryło[1], Adam Ruciński[1]

[1] Warsaw University of Technology, Warsaw University of Technology, The Faculty of Power and Aeronautical Engineering The Institute of Heat Engineering, Nowowiejska Str. 21/25, 00-665 Warsaw, email: adam.rucinski@pw.du.pl, Poland

Abstract. The paper compares heat pumps driven by electric motors (EHP) with heat pumps driven by gas engines (GEHP). GEHPs are still a novelty on the Polish HVACR market - therefore, the subject of the study is to indicate whether in Polish climatic conditions their use is profitable. A thorough analysis of the energy consumption of selected devices was carried out due to the consumption of utilities needed for their propulsion and the related costs. This has been done by calculating seasonal efficiency coefficients and using an innovative method of comparative modifiers allowing for unification of the performance of EHP and GEHP pumps. The results obtained include average energy efficiency coefficients, operating costs and payback times. Discussion of the calculation results proved that under certain assumptions GEHP pumps may be competitive with EHP heat pumps due to the possibility of managing waste heat at high temperatures.

1 Introduction

Gas engine driven heat pump (GEHP) is a relatively new technology, the first industrial application get back to 1985. GEHP is mainly used in industry, office buildings, warehouses and hotels. It is because of narrow range of products designed for individual customers. The equipment is dominated by devices with a capacity many times greater than the heat demand of a residential house. On the other hand, heat pumps driven by an electric motor have been used since the 1920s. So there is wide range of capacity comfortable for different customer needs. The share of heat pumps on the market of new detached houses in Poland currently reaches 12.5%.

Heat pumps driven by electric and gas motors operate a left-handed Linde refrigeration cycle. The described devices cooperate with lower and upper heat sources. The lower sources are air, water and soil. In the chemical, food and paper industry - these are high temperature waste heat streams. The upper source can be central heating and/or domestic hot water installations. For central heating, heat pumps can be operated in water, air or direct evaporation systems [1].

2 Technology

2.1 Engines

2.1.1 The electric engine

Modern EHP are equipped with high-performance motors with inverter which controls capacity from 30% to 100%. They are characterized by high energy efficiency, flexibility of work, low starting current, precise temperature control and long compressor life. The efficiency of high power electric motors is above 95%, for motors with small power the efficiency may decrease even up to 80%. It is due to the use of wires with a smaller cross-section and worse electromagnetic field distribution inside the machine.

2.1.2 The gas engine

The gas engine work is based on the Otto cycle. Devices with a power output of less than 60 kW are automotive engines adapted to this technology. More powerful motors are slow-running motors with a long service life specially designed for GEHP demands. The efficiency of the gas engine is much lower (from 30% to 45%), but the possibility of recovering waste heat generated during engine operation gives more efficiency [2, 3]. So, the technology can compete with other heating systems, including EHP.

In his report [4], Gasterra states, that internal combustion engines have an efficiency between 25% and 38%. The efficiency depends on the brand, engine type and power. More powerful engines usually have higher efficiency.

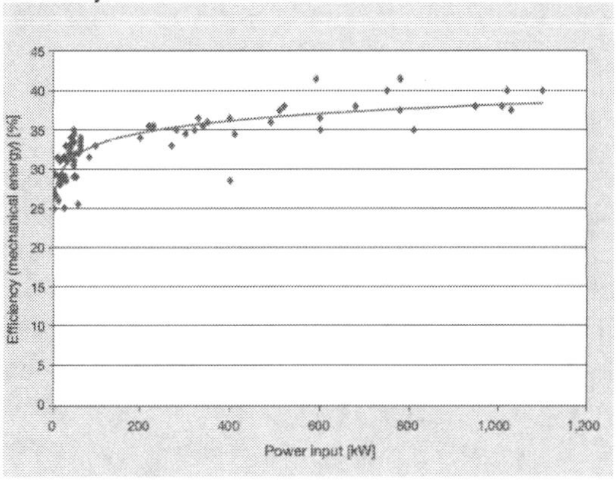

Fig. 1. Gas engine efficiency, Gasterra, 2010, [4]

In combustion engines there are parts the most exposed to wear which are replaceable. The service life of

https://doi.org/10.1051/e3sconf/201913701050

the engines driving heat pump compressors is longer than that of car engines. This is because of a smaller number of starts and stops, which lead to alternate cooling and heating of engine parts [4]. More frequent regular inspections are recommended for large installations or installations with old motors. The current engine generation requires less frequent inspections which include: valve adjustment, spark plug, oil and air filter inspection, fuel-to-air ratio inspection, and compressor measurement on all cylinders.

As far as regular maintenance is concerned, it is advisable to replace the oil filter, spark plugs and belt drive every 10,000 operating hours and to adjust the valves. However, an oil change is advisable every 30,000 operating hours.

New generation engines are equipped with a larger oil collector, modified ignition, special spark plugs and full synthetic oil [4]. For example, the manufacturer Aisin Toyota recommends that filters, spark plugs, V-belts and oil replenishment in the engine will be replaced every 5 years or 10,000 hours worked, while the oil will be replaced every 30,000 hours [5]. Assuming an annual operation of 3,120 hours (12 hours a day, 5 days a week), the inspections may be performed once every 3.2 years.

2.2 Compressors

In the case of heat pumps with power up to 100 kW, scroll compressors are used. In devices above 100 kW screw compressors are used.

2.3 Power system issues

Choosing GEHP instead of EHP leads to a reduction in electricity consumption, what could be important during summer peak demand for air conditioning drive. one favours the country's power system [6]. Then lower fluctuations in demand and lower maximum demand reduce the investment needs in peak and intermediate load power plants.

2.4 Specific applications

In addition to the advantages discussed above, GEHP has an additional benefit over EHP in situations where [7]:

- EHP connection would require a conversion of the voltage switchgear,

- the electricity consumption of the building must be reduced or part of the consumption must be allowed to supply other appliances,

- no possibility of connecting three-phase power supply,

- there is a constant high demand for hot water (swimming pools, spas, hotels, industrial processes),

- in countries where electricity prices significantly exceed gas prices.

3 Energy efficiency

We have 2 basic factors describing energy efficiency of heat pumps: EER and SEER in cooling and heating mode. EER is for examination of energy efficiency of heat pump in specific time, SEER is for whole heating and cooling season (Tab. 1).

Table 1. Conditions for heating and cooling modes of heat pumps

	Heating	Cooling
Outside temperature [°C]	7	35
Inside temperature [°C]	20	27

3.1 EER and SEER

The EER is the ratio of the useful energy output from the appliance to the energy supplied to the appliance to ensure its operation. It is the ratio that can be determined experimentally for a device under strictly defined temperature conditions. It describes the operation of the device in only one measuring point.

Table 2. Definitions of efficiency factor EER

Mode	EHP	GEHP
Heating (EER_{heat})	ratio of heating power to electric power consumed by the electric motor	ratio of heating power to the gas fuel heating power increased by electric power to drive auxiliary equipment
Cooling (EER_{cool})	ratio of cooling power to electric power consumed by the electric motor	ratio of cooling power to the gas fuel heating power increased by electric power to drive auxiliary equipment

Table 3. Examples of nominal EER values for the analysed heat pumps

Mode	EHP	GEHP
Heating (EER_{heat})	4.38	1.49
Cooling (EER_{cool})	4.58	1.18

Here, it is assumed that $SEER_{HEAT}$ and $SEER_{COOL}$ coefficients are not measured coefficients. They are estimated by using EER_{HEAT} and EER_{COOL} respectively, as provided by the manufacturer plus the conditions under which they were calculated.

The method of estimating the $SEER_{HEAT}$ and $SEER_{COOL}$ coefficients for GEHP is described in the European Standard DIN EN 16905 in Chapter 5. For EHP, the $SEER_{HEAT}$ and $SEER_{COOL}$ coefficients can be estimated using EN 14825.

A different way of estimating the above mentioned coefficients is presented below. In his report [7], Zaltash presents method of calculating the EER. It is evaluated for any temperature based on experimental measurements. In this way, it is possible to determine the EER for each temperature reached during the heating and cooling period. So, here SEER is determined as follows:

$$SEER = \sum_i t_i(T_i) \cdot EER_i(T_i) \quad (1)$$

where t_i is the time of occurrence of T_i temperature and EER_i is the EER calculated on the basis of Table 2.

4 GEHP calculation methodology

The heat pump driven by a gas engine at the Oak Ridge National Heat Pump Laboratory [8] was taken into account for the analyses. Laboratory measurements of the efficiency at 1650, 2000 and 2400 rpm for gas engine revolutions and operating parameters were made. On this basis, a mathematical model of the device operation was prepared.

For the current analyses, GEHP with a heating power of 35 kW was taken into account. Simulated demand for heating power of the building is shown in Fig. 3. On the basis of the calculation model the characteristics of work were built depending on the external temperature assumed from -22°C to +16°C (Fig. 2.).

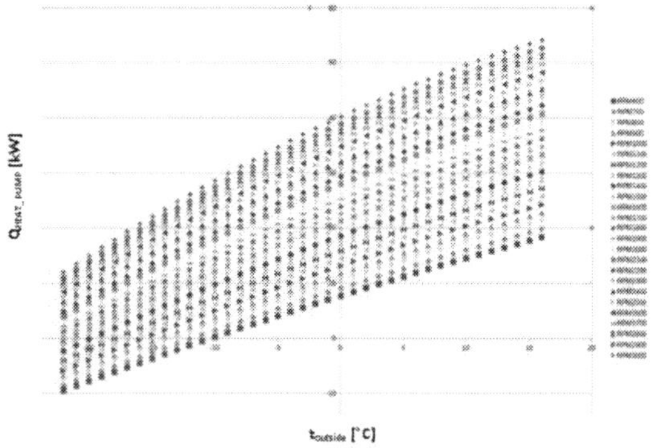

Fig. 2. Performance characteristics of the unit for different motor speeds as a function of outside temperature

Then the maximum and minimum pump capacity at a given temperature was calculated. The graph shows the ranges of smooth regulation of pump operation and building heat load (Fig. 3). The design temperature was assumed to be -10°C.

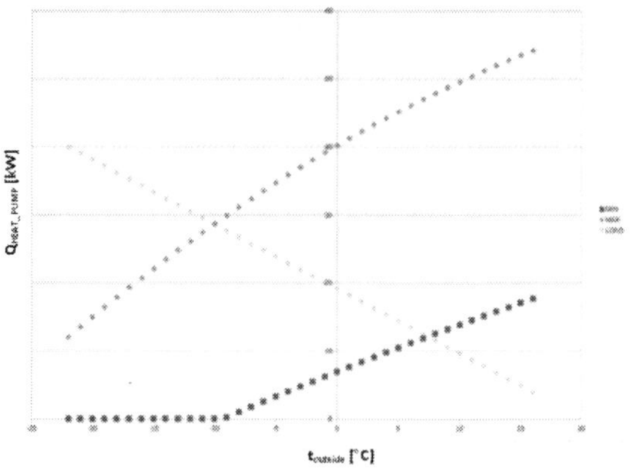

Fig. 3. Minimum and maximum operating range of GEHP taking into account the thermal load of the building

The characteristics of the heat pump indicate that it can operate continuously in the temperature range from -10°C to +7°C. Below -10°C, the GEHP must be supported by another emergency heating system to cover the building's needs.

From +7 to +16°C, the heat pump must be in ON-OFF mode. Operation in this mode is characterised by switching the motor off and in order to adjust the minimum heating capacity of the unit to the current building load. Such operation of the pump has a negative impact on the operation of the motor. Due to the frequency of occurrence of these temperatures in the heating period, an analysis of a system consisting of two heat pumps has been proposed. The diagram of the range of smooth work regulation is presented below.

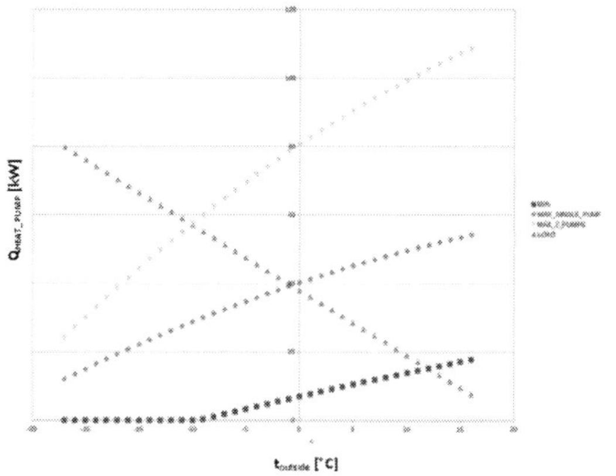

Fig. 4. Minimum and maximum operating range of two pump batteries combined with the heat load of the building in heating mode

The temperature's range of continuous operation of a single pump is from 0° to +11°C, while from -10° to -1°C is the range of simultaneous operation of two pumps. Below -10°C pump operation is supported by a peak heating system, while the range from +12° to +16°C is the operation of one of the pumps in ON-OFF mode.

To simplify further calculations, the thermal load of the building and the values directly related to it are given for continuous operation of one pump.

Further analysis allows us to create a computational model for determining the EER capacity ratio and the SEER seasonal capacity ratio.

The inverse characteristic of the EER_{HEAT} capacity coefficient is shown below (Fig. 5.)

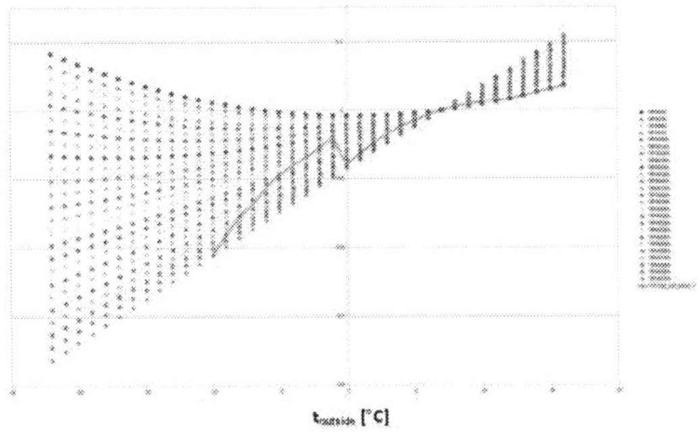

Fig. 5. Inversion of EER_{HEAT} of GEHP in heating mode

After calculating the $SEER_{HEAT}$ of 1.6, it was possible to estimate the total heating cost of a building for one heating period. This cost is 4560 EURO, of which 3,025 EURO was a cost associated with burning gas and 1,535 EURO was a component associated with the purchase of electricity to power auxiliary equipment.

Taking into account the achievable heat output from the cooling of the engine, the $SEER_{HEAT}$ factor has increased to 2.33. This significantly increases the economic benefits of using GEHP during the heating season. The pump is able to heat 1,538 m^3 of water (up to 45°C), which leads to savings of 20,806.85 PLN (approx. 4,624 EURO).

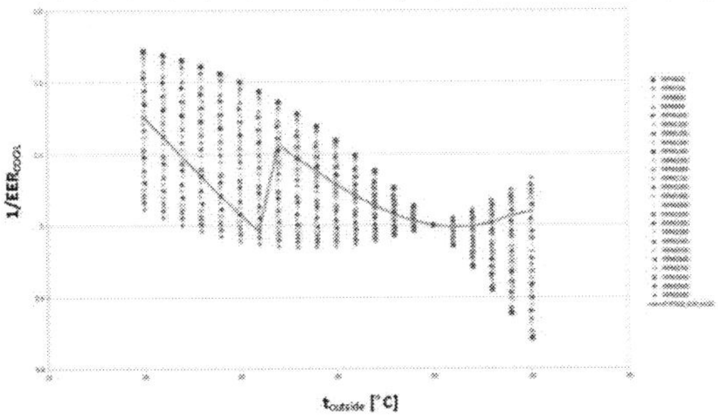

Fig. 6. Inversion of EER_{COOL} of GEHP in cooling mode

The same analysis was done for GEHP used in cooling mode (Fig. 6, 7). The $SEER_{COOL}$ coefficient was determined equals to 1.03. Then the total cost of GEHP work in cooling mode was 2,197 EURO, of which 1,887 EURO was related to the purchase of gas and 310 EURO to the purchase of electricity.

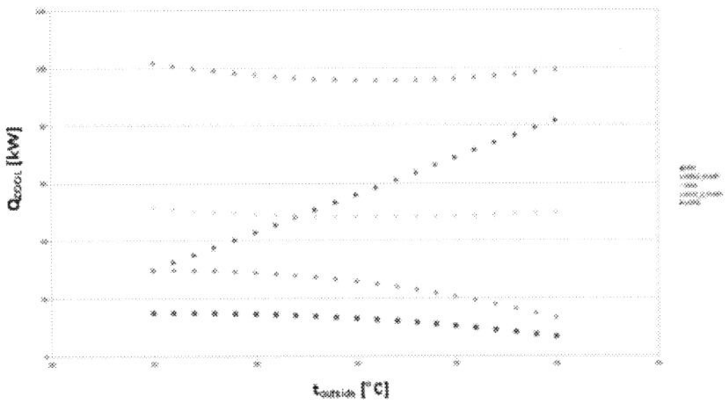

Fig. 7. Minimum and maximum operating range of two pump batteries combined with the heat load of the building in cooling mode

In the range from +24 to +26°C only one GEHP works. From +27 to +40, the need for building cooling can be satisfied by the simultaneous operation of two pumps. Also, as in the case of heating mode, the heat load of the building and the values directly related to it are given in relation to the continuous operation of one pump.

Taking into account the achievable heat from cooling of the engine, the $SEER_{COOL}$ factor has increased to 1.58. In the summer season, the pump is able to heat 809 m^3 of water, which leads to savings of 10,942 PLN or 2,431 EURO.

5 EHP calculation methodology

The following calculations have been made for the available data: heat load of the building in winter (Fig. 8) and summer (Fig. 10), hourly frequency of a given outdoor temperature, data on heat pumps driven by an electric motor. Characteristics of pump operation as a function of temperature were built, coefficients of heating and cooling capacity were determined. The SEER seasonal coefficient for heating and cooling was calculated, which allowed to determine the total cost of use of the pump in winter and summer.

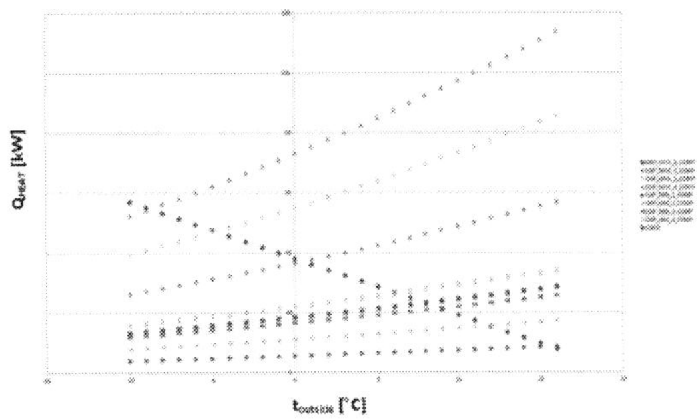

Fig. 8. Minimum and maximum operating range of two EHP heat pump batteries combined with the heat load of the building in heating mode

The power control range of a single compressor is from 30% to 100% of heating power. According to the

https://doi.org/10.1051/e3sconf/201913701050

fact that dual compressor pumps are the subject of analysis, the minimum efficiency is considered to be 15% of the heating power. In the range from +1 to +15°C, one EHP pump will be sufficient. At +16°C, one pump will operate in the ON/OFF mode. In the range from -8 to 0°C both pumps will operate simultaneously, while in the range from -10 to -9°C the operation of EHP pumps will be supported by an additional heating device.

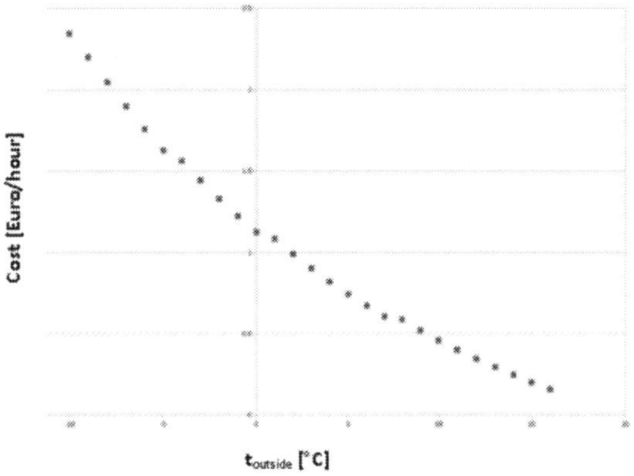

Fig. 9. Cost of EHP operation in the heating period as a function of the outside temperature

The total cost of the heating season was calculated as 4,379 EURO.

Next the cooling mode in summer was considered. The characteristics were built for work of EHP in cooling mode (Fig. 10)

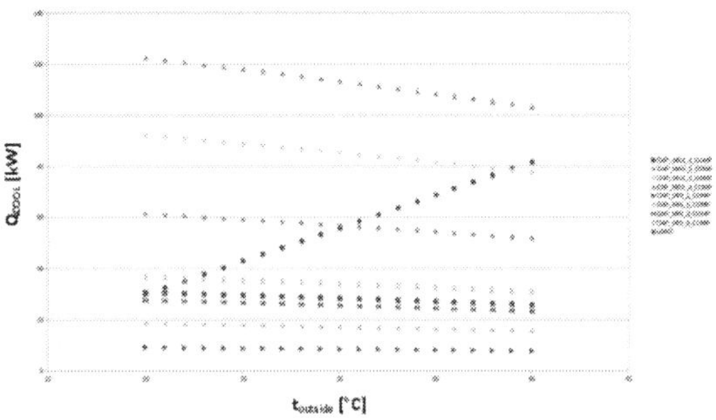

Fig. 10. Minimum and maximum operating range of two EHP heat pump batteries combined with the heat load of the building in cooling mode

The whole cost of exploitation of EHP for summer season was estimated upon partial costs (Fig. 11) as 1,070 EURO ($SEER_{COOL}$ = 4.63)

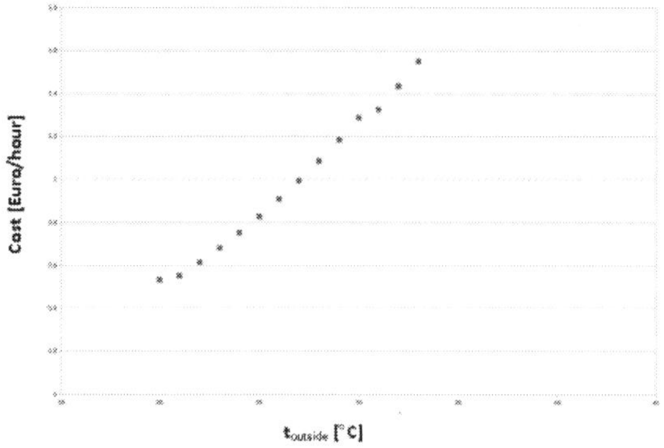

Fig. 11. Cost of EHP operation in the cooling period as a function of the outside temperature

Investment costs were estimated on the basis of available data [9]. Ultimately, the cost of installing GEHP was established as 30,000 EURO while EHP respectively as 25,000 EUR. Taking into account the savings due to the production of domestic hot water (GEHP) the operating costs of GEHP and EHP are estimated over a period of 10 years. This cost for GEHP is 65,500 EUR, compared to 90,200 EUR in case of EHP. Without the production of domestic hot water, GEHP is much worse. The operating costs for this configuration can be estimated as 122,000 EUR (Fig. 12).

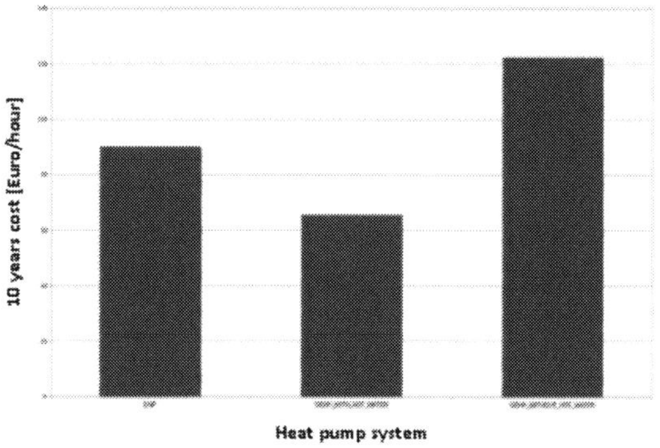

Fig. 12. Comparison of operating costs of GEHP and EHP systems

The operating costs of EHP and GEHP may differ from those estimated below due to annual fluctuations in electricity and gas prices.

6 Conclusions

The paper presents a comparison of the profitability of using heat pumps driven by a gas engine and conventional electric heat pumps. New calculation method of any gas and electric pump operating point allows to present characteristics of efficiency in heating and cooling mode, seasonal efficiency coefficients SEER and total costs of using the devices in heating and cooling season. Next, an economic analysis is made and the projected operating costs of the equipment for 10 years are presented. Conducted analysis allows to draw the following conclusions:

1. GEHP technologies are competitive in comparison with EHP when used in buildings with an energy demand above 30 kW. Usage of GEHP is economically justified in commercial applications because of good ratio of the gas to electricity price. However, still relatively low efficiency of the gas engine must be taken into account.

2. The choice of heat pumps driven by gas engines instead of electric pumps, apart from economic reasons, also leads to a smoothing of the electricity demand peak in the summer season, lower carbon dioxide emissions and higher primary energy utilisation rate.

3. In uncommon applications where the cost of the electrical connection is very high or there is no continuous electricity supply GEHP technology seems to be better choice than EHP.

4. The characteristics of heat pump operation can be reduced to regression curve equations which describe efficiency, energy and waste heat. Therefore, there is a standardized way of representing the work of both technologies.

5. In the scientific literature collected there are coefficients for calculating the regression curves for both electric and gas engine heat pumps. These models are based on experimental measurements in all possible conditions of temperature and heat load.

6. Reliable data on gas consumption, electric current consumption, efficiency and waste heat values are measured experimentally by placing the device tested in a thermostatic chamber. A number of measurements are made according to the ARI 340/360-2004 standard.

7. With experimental data, regression curve coefficients, heat load characteristics of the building and temperature data, it is possible to present operating characteristics of heat pumps and to estimate the costs of heating and cooling of the building using the given technology.

8. From the analysis it can be concluded that taking into account only heating and cooling of the building, electric pumps are almost 1.5 times cheaper solution than gas pumps (during 10 years). On the other hand the analysis considers the use of waste heat from the water cooling of the gas engine. If the gas pump system is adapted to produce domestic hot water, the benefits of it should be included in the analysis. The money savings will represent a significant percentage of the annual costs and will be a decisive item to choose GEHP instead of EHP.

References

1. Ruciński A., Rusowicz A., Grzebielec A.: *Gas engine driven haat pump – characteristics, analysis of applications in buildings energy systems*. 9th International Conference. Environmental Engineering,Wilno, (2014).
2. Elgendy E., Schmidt J., Optimum utilization ofr recovered heat of a gas engine heat pump used for water heating at low air temperature, *Energy and Buildings*, Vol. 80, pp. 375-383, (2014)
3. Hepbasli A., Erbay Z. et al., A review of gas engine driven heat pumps (GEHPs) for residential and industrial applications, *Renewable and Sustainable Energy Reviews*, Vol. 13, pp. 85-99, (2009)
4. Western Cooling Efficiency Center, Gas Engine Heat Pump Technology Final Report, 2015.
5. Rosato A., Sibilio S., Scorpio M., A Review of Gas Engine Driven Heat Pumps for Residential Applications, Conference paper: *CLIMA 2013 - 11th REHVA World Congress and the 8th International Conference on Indoor Air Quality, Ventilation and Energy Conservation in Buildings, At Prague, Czech Republic*, (2013)
6. Wan X., Cai L. et al., Power management strategy for a parallel hybrid-power gas engine heat pump system, *Applied Thermal Energy*, Vol. 110, pp. 234-243, (2017)
7. Zaltash A., Performance of gas engine driven heat pump unit, *Final summary technical report by Blue Montain Energy*, (2009)
8. Mahderekal I., Experimental evaluation and modeling of engine driven heat pump, *UNLV Theses, Dissertations, Professional Papers, and Capstones*, (2009)
9. Abuheiba A., Mahderekal I., Momen A., Vineyard E., Challenges and opportunities of Gas Engine Heat Pumps - Two Case Studies. *12th IEA Heat Pump Conference*, (2017).

RESEARCH & DEVELOPMENT IN POWER ENGINEERING, 2019

Close loop optimisation of large CHP based on approximation model

Łukasz Śladewski[1], Rafał Wereszczyński[2], Jerzy Majchrzak[2], Krzysztof Możejko[2], Wojciech Bujalski[3], Konrad Świrski[3] and Maks Sędzielski[1]

[1]Transition Technologies – Advanced Solutions Sp. Z O.O.., Pawia St. 55, 01-030 Warsaw, Poland
[2]Polski Koncern Naftowy ORLEN S.A., Chemików St. 7, 09-411 Plock, Poland
[3]Institute of Heat Engineering, Warsaw University of Technology, Nowowiejska St. 21/25, 00-665 Warsaw, Poland

Abstract. Complex optimization of CHP plants becomes a very important issue in research and implementation, particularly in the context of increasing environmental requirements. The process in industrial CHP plants could be decomposed into several subprocesses, which could be optimized individually using dedicated solutions. The article presents the results of work of complex, multi-modular optimization project of one CHP plant located in petrochemical and refinery plant in Poland. The scope of the project is economical load dispatch optimizer aimed to increase economical profit of CHP operation, combustion optimization for boiler efficiency increase and NOX emission reduction, steam temperature advanced control for improved control quality, sootblowing optimization for reduction of steam demand for sootblowing process. The solution includes also measurement validation and correction system, which is based on data reconciliation algorithm and on-line performance monitoring system.

1 Introduction

The fossil fuel heat and power generation industry faces nowadays demanding environmental regulations as well as accent on economical profit. Both conditions request investments in modernization of exiting plants in terms of low emission and emission reduction technologies like new burners or air staging, flue gas desulfurization and selective catalytic or no-catalytic reduction of NOX emission. From economical profit point of view steam generation boilers and steam turbines are modernized in the way to reduce losses and to improve efficiency of the process. Once either industrial CHPs or municipal CHPs reach a high level of modernization with best available technologies new solutions for better utilization of the plants are applied. In most model based advanced control and optimization software solutions are used.

This paper presents solution and implementation results of model based advanced control and optimization in PKN Orlen CHP plant located in refinery and petrochemical plant in Plock, Poland. Design of the solution responds to design of the CHP plant and, in most, conditions of process steam supply. The process in the plat had been decomposed into subprocesses, where a benefit could be achieved, so the solution integrates the economical load dispatch optimizer, combustion optimizer, steam temperature advanced control, sootblowing optimizer, measurement validation and reconciliation system, performance monitoring system.

All systems in the solution operate in on-line mode. The economical load dispatch optimizer [0], [2], [3] is a closed-loop system, that based on the thermodynamic and economical model of the plant calculates load demand individually for boilers and turbines, to improve economical profit of the plant. In other words the system maximize electricity generation and minimize fuel consumption. The combustion optimization system [4], [5] is a closed-loop system that, based on the combustion empirical model calculates proper amount of air and air distribution. Goal of the system is to maximize boiler efficiency and minimize NOX emission, while respecting constraints of key process parameters like steam temperatures and O2 content in flue gases. The steam temperature advanced control [6] is to improve control quality of steam temperatures, what means minimize difference of steam temperature and its setpoint. The sootblowing optimizer [7], [8] operates in advisory mode and the goal of the optimization is to save steam for sootblowing process, what consequently saves fuel energy. The measurements validation and reconciliation system [9] is the system, that monitors mass and energy flow and correct the measurements in the way to have balanced mass and energy of the plant as well as individual units like boilers, turbines pressure reduction stations etc. The performance monitoring system is on-line system, that based on measurements calculates key performance parameters of the process like boiler and turbine efficiency, boiler losses, turbine performance coefficients etc.

2 The CHP plant description

The Orlen CHP plant is located in refinery and petrochemical plant in Plock, Poland. It is composed of eight oil and gas boilers and seven steam turbine generators.

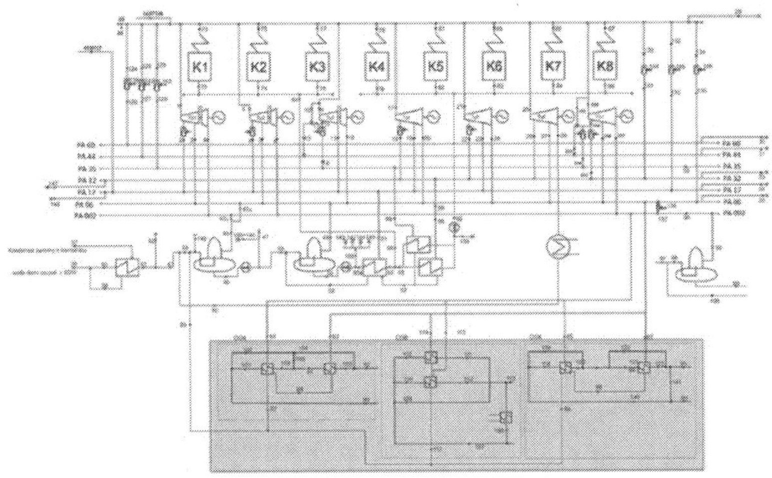

Fig. 1 - The CHP plant model

Three boilers are with the capacity of 320 t/h and five boilers are with the capacity of 420 t/h of steam generation. All boilers are able to fire simultaneously different types of oil and gas fuels.

The boilers are equipped with twelve burners, installed in there vertical levels on all four walls. Most burners are able to fire either oil or gas fuel. The steam temperature is controlled by attemperators installed on left and right side before second and third superheaters of each boiler. Boilers are equipped with sootblowing system and sootblowers are installed on steam superheaters, SCR, economizer and air heaters.

The are 6 extraction-backpressure turbines and one extraction-condensing turbine with capacities ranging from 320 t/h of inlet steam flow and 55 MW to 420 t/h of inlet flow and 70 MW. All turbines are able to control MW. Turbines 1, 2, 4 and 5 are able to control pressure on second extraction and on outlet. Turbines 3, 6, and 7 are able to control pressure on first and second extraction and on outlet. Simplified model of the CHP plant is presented in the **Fig. 1**.

The model presents boilers, process steam collectors and pressure of each, turbines extractions pressure and feedwater system.

The main purpose of the CHP plant is to satisfy demand of process steam. Number of boilers, turbines and steam reduction stations in operation depends on process steam demand but considering ability of responding of rapid changes of process steam demand. In other words the CHP plant must be ready, at any time to reduce or increase load according to demand of the process steam.

3 The economical load dispatch optimization

The economical load dispatch optimizer is aimed to control the CHP plant in closed-loop mode to maximize the economic profit of the plant operation. The system composes of three main components: thermodynamic and economic model of the plant, calculation engine for searching optimal solution graphical user interface for system tuning and monitoring. The model is a combination of first-order thermodynamic model, where mass and energy balance is modelled and empirical models of efficiency of boilers and turbines. Efficiency of boilers and fuel demand is calculated with the following formulas:

$$\eta = f(t, m) \qquad (1)$$

$$Q_b = \frac{m(i_{steam} - i_{water})}{3600} \quad [MW] \qquad (2)$$

$$Q_{fuel} = \frac{Q_b}{\eta} \quad [MW] \qquad (3)$$

where:

η – boiler efficiency [-],
t – ambient temperature [°C],
m – steam generation [t/h],
Q_b – boiler heat demand [MW],
i_{steam} – enthalpy of steam generated [kJ/kg*K],
i_{water} – enthalpy of feed water [kJ/kg*K],
Q_{fuel} – fuel demand [MW].

Turbine performance coefficients and turbine power are calculated with the following formulas:

$$u1 = f(m_{in}, m_{out}, i_{in}, m_{u2}) \qquad (4)$$

$$u2 = f(m_{in}, m_{out}, i_{in}, m_{u2}) \qquad (5)$$

$$out = f(m_{in}, m_{out}, i_{in}, m_{u2}) \qquad (6)$$

$$eff = f(m_{in}, m_{out}, i_{in}, m_{u2}) \qquad (7)$$

$$P_{el} = \left(\frac{m_{u1}}{u1} + \frac{m_{u2}}{u2} + \frac{m_{out}}{out}\right) * eff \qquad (8)$$

where:

m_{u1} – steam flow at first extraction [t/h],
m_{u2} – steam flow at second extraction [t/h],
m_{out} – steam flow at outlet [t/h],
m_{in} – steam flow at turbine inlet [t/h],
i_{in} – enthalpy of steam at inlet [kJ/kg*K],
$u1$ – performance parameter at first extraction [t/MW],
$u2$ – performance parameter at second extraction [t/MW],
out – performance parameter at outlet [t/MW],

eff – turbine mechanical efficiency [-],
P_{el} – power generation [MW].

The economical model is expressed by the following formula:

$$Income = \sum_i P_{ee} \cdot P_{el}^i - \sum_j P_{fuel} \cdot Q_{fuel}^j \quad (9)$$

where:
P_{ee} – electricity price [PLN/MWh],
P_{el}^i – energy production of i-th turbine [MWh],
P_{fuel} – fuel price [PLN/MWh],
Q_{fuel}^j – energy in fuel demand of j-th boiler [MWh].

Every optimization step the calculation engine searches for a solution that on one hand satisfies the demand of process steam and on the other hand is most effective from economical point of view. The calculation engine is based on constrained optimization by linear approximation – COBYLA. The COBYLA algorithm is numerical optimization solution for the problems that have constraints and derivative of the goal function is not known. The algorithm is designed to solve non-linear problems with linear approximation. The goal function of the economical load dispatch is non-linear because the empirical models of boiler efficiency are second degree polynomials and models of turbine performance coefficients are third degree polynomials. Optimization is executed every 60 seconds. A solution for certain optimization step is a vector of load individually boilers and turbine's extractions and outlets. The solution must fulfill certain conditions like:

1. To satisfy the demand for process steam,
2. To satisfy minimum and maximum constraints for boiler load,
3. To satisfy minimum and maximum constraints for turbine extractions and outlet flow – distinguishing between extractions/outlets in pressure control mode and extractions/outlets in standard mode, turbine inlet flow and turbine MW.
4. To satisfy maximum gas temperature at SCR inlet for each boiler.
5. To satisfy maximum pressure difference in live steam manifold.
6. To satisfy minimum and maximum load of power supply sections.

Once a solution is found control algorithm calculates values of boiler master control, turbine load and turbine valve positions in order to reach the optimum point. The control algorithm is based on PID algorithm. The control algorithm is significantly slower than tuning of PID controller of pressure in process steam collectors implemented originally in DCS logics. This design allows to reach optimal point without influencing process steam pressure significantly.

4 The combustion optimization

The task of the combustion optimization system is to perform an on-line optimization of a current process operating point. The optimizer is implemented above a base control layer, which is DCS. The system calculates setpoints or setpoints corrections for controllers that operate in a base control layer. Control systems of processes in CHP plants are based on PI (Proportional-Integral) controllers. These controllers controls sub-processes that have an influence on a main optimized MIMO (Multi Input Multi Output) process (i.e. combustion process).

In specific the combustion optimization system is aimed to calculate proper amount of air for the combustion and air distribution. The goal for the system is to increase boiler efficiency and reduce NOX emission while considering a number of constraints. The system utilizes artificial intelligence algorithm, which is artificial immune system. The algorithm has self-learning and auto-adaptive abilities. The system is equipped with several algorithms but there are two main: learning algorithm and optimization algorithm. The learning algorithm is responsible for monitoring the combustion process and collecting knowledge about the process at certain operating point. This allows the optimization algorithm to work with the most updated knowledge. The optimization algorithm uses the knowledge database calculates linear model of the combustion process. Once a boiler changes operating point, for example load or fuel mix, the optimization algorithm automatically searches for new knowledge units that represent new operating point and basing on these calculate new model.

The optimisation module allows to define the goal function with the following formula. For each optimization signal the formula consists of two main parts linear and square. This feature allows to change optimization priority once one of optimization goals differ from the setpoint significantly. In other words, if value of one of optimization signals exceeds square insensibility level, the penalty of the signal increases over penalties of other signals significantly so the optimizer changes the priority order.

$$J = \sum_{k=1}^{n_m} \begin{bmatrix} \alpha_k\left(|\breve{m}_k^c - \tilde{m}_k^c| - \tau_k^{lm}\right)_+ \\ +\beta_k((|\breve{m}_k^c - \tilde{m}_k^c| - \tau_k^{sm})_+)^2 \end{bmatrix} + \\ + \sum_{k=1}^{n_y} \begin{bmatrix} \gamma_k\left(|\breve{y}_k - \tilde{y}_k| - \tau_k^{ly}\right)_+ \\ +\delta_k\left(\left(|\breve{y}_k - \tilde{y}_k| - \tau_k^{sy}\right)_+\right)^2 \end{bmatrix} \quad (10)$$

where:
α_k – linear penalty coefficient for k-th manipulated variable,
β_k – square penalty coefficient for k-th manipulated variable,
γ_k – linear penalty coefficient for k-th monitored process output,
δ_k – square penalty coefficient for k-th monitored process output,
τ_k^{lm} – insensibility zone for linear penalty for k-th manipulated variable,
τ_k^{sm} – insensibility zone for square penalty for k-th manipulated variable,
τ_k^{ly} – insensibility zone for linear penalty for k-th monitored process output ,
τ_k^{sy} – insensibility zone for square penalty for k-th monitored process output,

$(\cdot)_+$ – "positive" operator $(x)_+ = \frac{1}{2}(x + |x|)$,
$\tilde{m}_k^c$ – current value for k-th manipulated variable,
$\tilde{y}_k$ – current value for k-th monitored process output,
$\breve{m}_k^c$ – demand for k-th manipulated variable,
$\breve{y}_k$ – demand value for k-th o monitored process output.

The **Fig. 2** presents visual representation of example penalty function of NOX emission. The green zone is the range, within which no penalty is applied, so the parameter is within optimum point. The blue zone is linear penalty zone, which means the emission is either too low or too high, so optimizer calculates setpoint changes in the way to reduce the penalty The red zone is square penalty zone, which elevates priority of certain signal automatically, so the optimizer focuses on reducing this particular penalty.

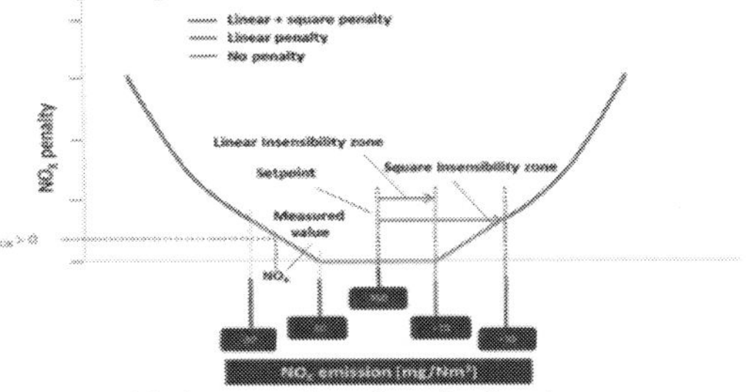

Fig. 2 - Graphical presentation of penalty function

During optimization the optimization algorithm is searching for a solution with the following formula:

$$\min_{\Delta m^d}\left\{\sum_{k=1}^{n_m}\begin{bmatrix}\alpha_k\left(|\breve{m}_k^c + \Delta m^d - \tilde{m}_k^c| - \tau_k^{lm}\right)_+ \\ +\beta_k((|\breve{m}_k^c + \Delta m^d - \tilde{m}_k^c| - \tau_k^{sm})_+)^2\end{bmatrix} + \sum_{k=1}^{n_y}\begin{bmatrix}\gamma_k\left(|\breve{y}_k + \Delta m^d K_k - \tilde{y}_k| - \tau_k^{ly}\right)_+ + \\ \delta_k\left(\left(|\breve{y}_k + \Delta m^d K_k - \tilde{y}_k| - \tau_k^{sy}\right)_+\right)^2\end{bmatrix}\right\} \quad (11)$$

where:
Δm^d – optimal change vector of manipulated variables,
K_k – matrix of the automatically identified input-output gains of the process.

The K_k matrix represents static gains in mathematical model. It is automatically identified at each optimization period basing on the most recent knowledge gathered in the knowledge database.

The calculated change of manipulated variables must satisfy the following constraints:

$$\Delta m_{low}^d \leq \Delta m^d \leq \Delta m_{hi}^d \quad (12)$$

$$m_{low}^c \leq \tilde{m}^c + \Delta m^d \leq m_{hi}^c \quad (13)$$

where:
Δm_{low}^d – low constrain vector of control variables in single optimization step,
Δm_{hi}^d – high constrain vector of control variables in single optimization step,
$\tilde{m}^c$ – current values vector of control variables,
m_{low}^d – low constrain vector of absolute values of control variables,
m_{hi}^d – high constrain vector of absolute values of control variables.

First constrain formula refers to the allowable range of decision variable change in single optimization step and the second refers to absolute range of decision variable.

5 The steam temperature advanced control

The steam temperature advanced control system is a MPC solution, that is aimed to improve control quality of steam temperature. A significant change of controlled value could be caused by a change of disturbance signal – boiler load. The steam temperature advanced control is to minimize the side effect of changing boiler load. The system basing on dynamic model of the process calculates correction of setpoints of steam temperature control in on-line and closed-loop mode.

Standard control of steam temperature is based on two PID controllers working in cascade mode. The steam temperature advanced control calculates a correction of setpoint to the lower PID. The correction is calculated with the following formula:

$$\Delta u = -\frac{\sum_{i=1}^{N} G_{1,i}\Delta v_i}{G_2} = -\sum_{i=1}^{N}\frac{G_{1,i}}{G_2}\Delta v_i \quad (14)$$

where:
Δu – correction of setpoint,
G_1 – transfer function of disturbance change e.g. load change,
G_2 – transfer function of control change e.g. PID setpoint,
Δv – disturbance change e.g. load change.

The general effect of the steam temperature advance control is improved control quality of steam temperatures. The system calculates the control trajectory in the most efficient way to minimize control error. The PID is to balance inaccuracy of the model. Second goal for the system is to balance load of steam attemperators between two levels of steam control. The algorithm monitors if the second level of steam control is out of range, e.g. spray valves are closed and modifies the setpoint for first level of steam control, e.g. increases the temperature after the first attemperators. This feature of the system allows for better operation of the first algorithm. The **Fig. 3** presents design of base control change with steam temperature advance control signals – STO.

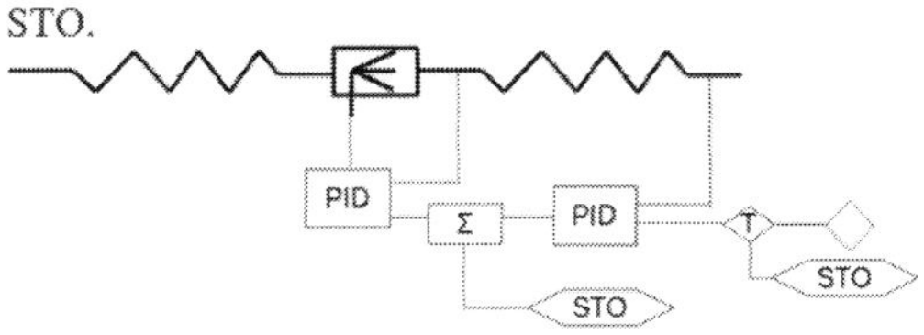

Fig. 3 - Base control modification with steam temperature advanced control

One signal is a setpoint correction of lower PID and second signal is a replacement of setpoint of upper PID, which corresponds of two algorithms described above.

6 The sootblowing optimizer

The sootblowing optimization system is an advisory mode system that monitors the sootblowing performance and activates the sootblowing according to certain optimization goals. In a standard procedure, the sootblowing is activated every fixed period with all sootblowers. The main feature of the system is calculation and monitoring of heat fluxes of the boiler heat exchange sections and decide, when and which sootblowers to activate. The optimization goal is to maintain cleanliness of boiler heat exchange sections with lowest steam consumption for the sootblowing. The cleanliness factor is calculates as a ratio of current heat flux and the theoretical heat flux at current operating conditions and clean boiler sections. The current heat flux is calculated with standard thermodynamics lows basing on steam flow, temperature and pressure measurements. The theoretical heat flux is empirical models, that depends on boiler load, enthalpy of steam in to section and type of fuel.

$$CF = \frac{Q_{current}}{Q_{theoretical}} \quad (15)$$

$$Q_{current} = \dot{m}(H_{out} - H_{in}) \quad (16)$$

$$Q_{th} = f(\dot{m}, H_{in}, \frac{oil}{gas} ratio, oil\ type, gas\ type) \quad (17)$$

where:
$\dot{m}$ – steam flow kg/s,
H_{wy} – enthalpy of steam out of section [kJ/kg],
H_{we} – enthalpy of steam in to section [kJ/kg].

The sootblowing optimizer calculates the time for activation of two groups of sootblowers – steam supereheater, SCR and economizer group of sootblowers and air preheaters group of sootblowers. The algorithm that calculates the time considers following conditions:

1. Minimize steam consumption for the sootblowing.
2. Maintain cleanliness of boiler heat exchange sections.
3. Minimize flue gas temperature at SCR.
4. Maintain general cleanliness of the boiler to not exceed limits for SCR temperature in case a rapid increase load of is needed.

The time is displayed on the process graphic, so operators are informed when to activate certain group of sootblowers. The time is calculated in on-line mode and could be adjusted if operating conditions are changing, e.g. boiler load change or fuel mix changed.

7 The measurement validation and reconciliation system

The measurement validation and reconciliation system is aimed to monitor and to correct measurements considering mass and energy balances, thermodynamics lows etc. Each measurement has an error. When considering a complex design of a plant with number of measurements a proper mass and energy balance is a challenge. This could be solved with the reconciliation algorithm.

The reconciliation algorithm calculates corrections of measurements or not measured values that, with the highest probability, are representing real values. The algorithm is basing on mathematical and thermodynamic model of the plant. Each measurement is described with a parameter that represents the accuracy of the measurements. The specific value of accuracy coefficient is defined considering quality of measurement and experience, but a good estimation would provide analysis of the standard deviation and the variance of the measurement. When estimating the accuracy the measurements are divided into three groups: top accuracy measurements which are MW measurements or steam flow measurements at CHP/plant connections, standard accuracy measurements which are superheated steam or water measurements, which accuracy is relatively high, low accuracy measurements which are wet steam measurements or the measurements with low reliability.

The reconciliation algorithm searches for a minimum of the following goal function:

$$J(\hat{y}, \hat{z}) = (y - \hat{y})^T V^{-1}(y - \hat{y}) \quad (18)$$

$$F(\hat{y}, \hat{z}) = 0 \quad (19)$$

where:
y – raw measurements,
$\hat{y}$ – corrected measurements,
$\hat{z}$ – not measured values,
V – covariance of measurements,
F – conditions matrix – mass and energy equations.

Additionally, the system is equipped with algorithm for correcting raw measurements according to density. The density correction factor is calculated with the following formula:

$$c = \sqrt{\frac{\rho}{\rho_{cor}}} \quad (20)$$

where:
ρ_{cor} – density at standard conditions,
ρ – density at current conditions.

8 Results

The general goal of the project was to improve the efficiency/economical profit of operation of CHP plant with closed-loop and advisory mode with artificial intelligence, machine learning, model based IT solutions. Specific goals for specific systems were defined if the following way:

- Economical load dispatch:
  6 100 MWh increased electricity generation,
  56 000 GJ reduced fuel consumption.
- Combustion optimization with steam temperature advanced control:
  0.2% increased boiler efficiency,
  5-15% reduced NOX emission.
- Sootblowing optimization:
  Boiler K2 – 8800 GJ reduced fuel consumption,
  Boiler K5 – 5900 GJ reduced fuel consumption,
  Boiler K7 – 5900 GJ reduced fuel consumption,

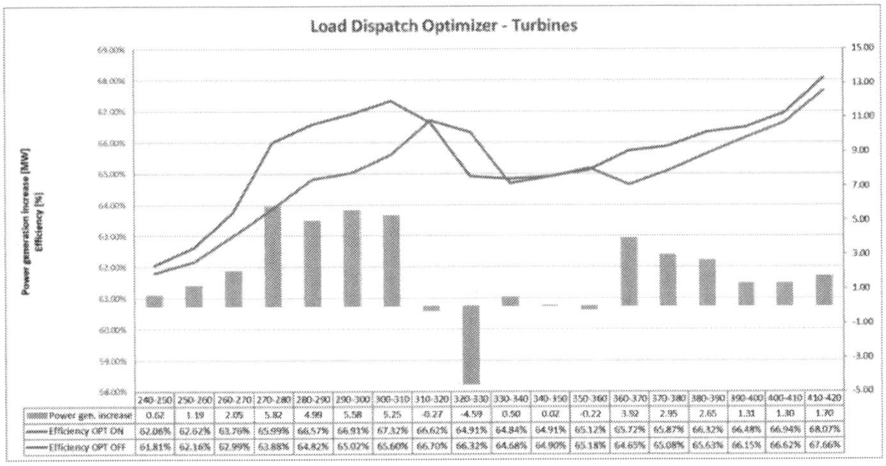

Fig. 4 - Economical load dispatch optimizer – turbines

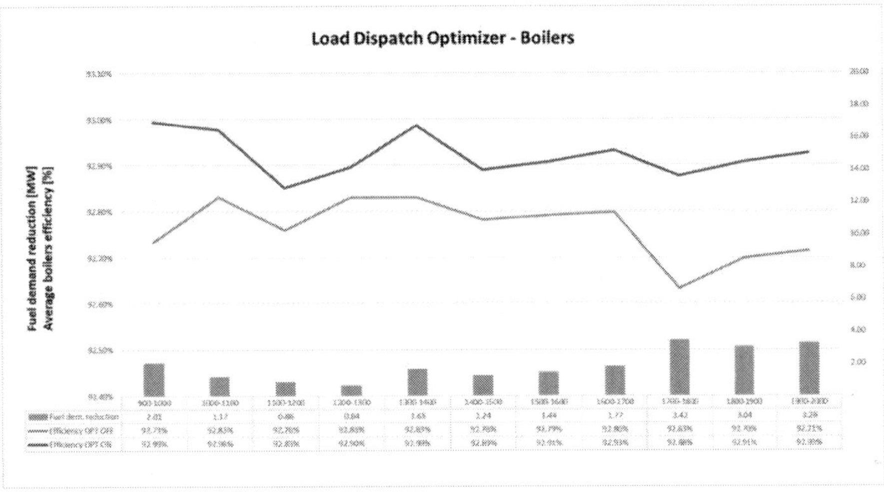

Fig. 5 - Economical load dispatch optimizer – boilers

8.1 The economical load dispatch optimization - turbines

The effects of economical load dispatch optimization are calculated based on historical data from central SCADA system for the period of January 2019 – June 2019. The **Fig. 4** presents effect of the optimization on the turbines.

The optimization effect is calculated basing on analysis of average isentropic efficiency of all turbines T1-T7. The average isentropic efficiency is a ratio of total MW and total isentropic MW. The isentropic MW is a total of isentropic expansion and steam flow for each turbine's extraction and outlet. The formula is presented below:

$$\eta_{isentr} = \frac{P_{CHP}}{P_{isentr}} \cdot 100\% \quad (21)$$

$$P_{CHP} = \sum_{i=1}^{7} P_{MW_i} \quad (22)$$

$$P_{isentr} = [\dot{m}_{PA44} \cdot (i_{PA136} - i_{PA44_S}) + \dot{m}_{PA32} \cdot (i_{PA136} - i_{PA32_S}) + \dot{m}_{PA17} \cdot (i_{PA136} - i_{PA17_S}) + \dot{m}_{PA06} (i_{PA136} - i_{PA06_S}) + \dot{m}_{PA002} \cdot (i_{PA136} - i_{PA002_S})]/3600 \quad (23)$$

where:

η_{isentr} – average isentropic efficiency [%],

P_{CHP}. – total CHP MW production [MW],

P_{isentr} – total isentropic MW production [MW],

P_{MW_i} – MW production of i-th turbine [MW],

$\dot{m}_{PA44}$, $\dot{m}_{PA32}$, $\dot{m}_{PA17}$, $\dot{m}_{PA06}$, $\dot{m}_{PA002}$ – stem flow to certain process steam collectors PA44, PA32, PA17, PA06 and PA002 [t/h],

i_{PA136} – enthalpy of live steam [kJ/kg*K],

i_{PA44_S}, i_{PA32_S}, i_{PA17_S}, i_{PA06_S}, i_{PA002_S} – enthalpy of steam of theoretical isentropic expansion PA44, PA32, PA17, PA06, PA002 [kJ/kg*K].

As a result of optimization the average increase of isentropic efficiency increase 0.621%, what results 1.93 MW of higher MW production. The total increase of electricity generation annually is 16 245 MWh.

 https://doi.org/10.1051/e3sconf/201913701051

The highest result has been recorded for lower load of the CHP plant, but lower results are for within minimum and maximum range of total load of the CHP. This is influenced mostly by constraints, which limits optimization decisions for minimum and maximum load of the CHP.

8.2 The economical load dispatch optimization - boilers

The effects of economical load dispatch optimization for boilers are calculated based on historical data from central SCADA system for the period of January 2018 – June 2019. The presents **Fig. 5** effect of the optimization on the boilers.

The optimization effect is calculated basing on analysis of average efficiency of all boilers. The efficiency is calculated basing on polish and European standards – PN-EN 12952-15. The average boiler efficiency of the CHP plant is a weighted average of boiler load, calculated with the following formulas:

$$\eta_B = \frac{1-\sum S_i^*}{1+\frac{\dot{Q}_{rad}+\dot{Q}_{sootbl}-\dot{Q}_{XL}\sum S_i^*}{\dot{Q}_{use}}} \quad (24)$$

$$\eta_{avg} = \frac{\sum_{i=1}^{8} \dot{m}_{Bi} \cdot \eta_{Bi}}{\sum_{i=1}^{8} \dot{m}_{Bi}} \quad (25)$$

where:
η_B – boiler efficiency[%],
$\sum S_i^*$ – total boiler losses [kJ/kg*K],
$\dot{Q}_{rad}$ – heat loss of the radiation [kJ/kg*K],
$\dot{Q}_{sootbl}$ – heat loss for sootblowing [kJ/kg*K],
$\dot{Q}_{XL}$ – heat to steam air heaters [kJ/kg*K],
$\dot{Q}_{use}$ – useable heat [kJ/kg*K],
η_{avg} – weighted average of steam generation [%],
$\dot{m}_{Bi}$ – steam generation of i-th boiler [t/h],
η_{Bi} – boiler efficiency [%].

As a result of the optimization average increase of total boiler efficiency is 0.158% what results in 1.86 MW reduction of fuel demand. The total reduction fuel consumption annually is 56 421 GJ.

8.3 Combustion optimization

The effects of combustion optimisation for boilers are calculated based on historical data from central SCADA system for the period of January 2018 – June 2019.

The **Fig. 6b** presents effect on boiler efficiency and **Fig. 6bBłąd! Nie można odnaleźć źródła odwołania.** presents effect on NOX emission of the combustion optimization for boiler K2.

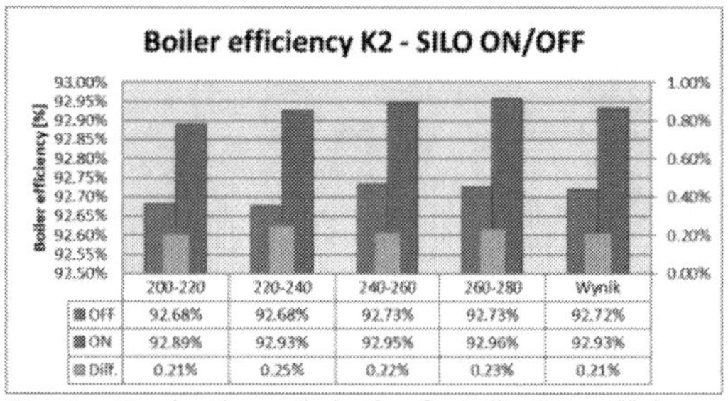

Fig. 6a - Combustion optimization for boiler K2 – efficiency

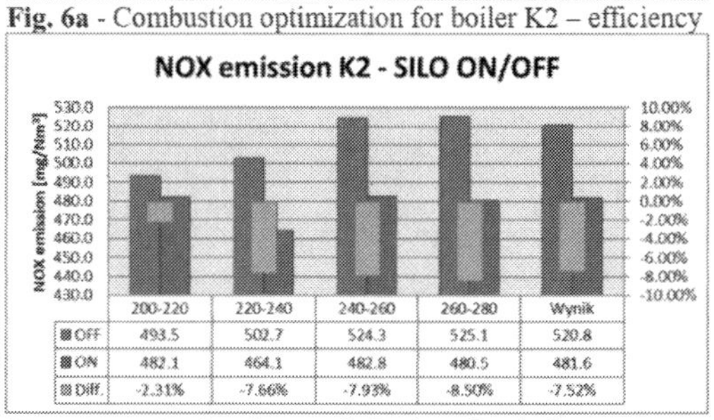

Fig. 6b - Combustion optimization for boiler K2 –NOX emission

The average increase of boiler efficiency is 0.21 % reduction of NOX emission is 7.52%. The main factor is reduction of excess air and air distribution with respect of constraints like minimum of steam temperatures – 530 °C, minimum of O2 in flue gas – 1.85% or maximum of CO emission – 15 ppm. The lowest efficiency gain is recorded for lower boiler load, which is caused by lower steam temperatures, which require higher excess air.

The **Fig. 7b** presents effect on boiler efficiency and **Fig. 7b** presents effect on NOX emission of the combustion optimization for boiler K5.

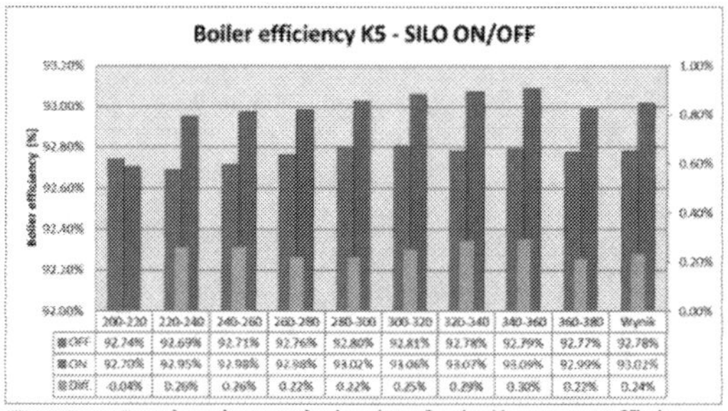

Fig. 7a - Combustion optimization for boiler K5 – efficiency

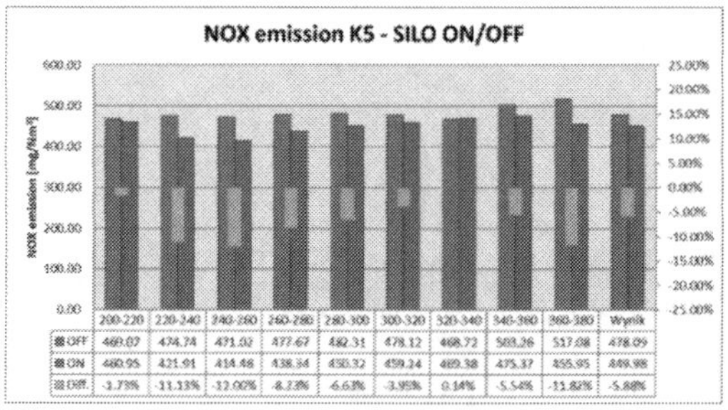

Fig. 7b - Combustion optimization for boiler K5 – NOX emission

 https://doi.org/10.1051/e3sconf/201913701051

The average boiler efficiency increase is 0.24 % and reduction of NOX emission is 5.88%. The main factor is reduction of excess air and air distribution with respect of constraints like steam temperatures, O2 in flue gas or CO emission. The lowest efficiency gain is recorded for lower boiler load, which is caused by lower steam temperatures, what requires higher excess air. An insignificant drop of efficiency within minimum load rage is a result of changing order optimization priorities in the way to increase steam temperatures.

The **Fig. 8b** presents effect on boiler efficiency and **Fig. 8b** presents effect on NOX emission of the combustion optimization for boiler K7.

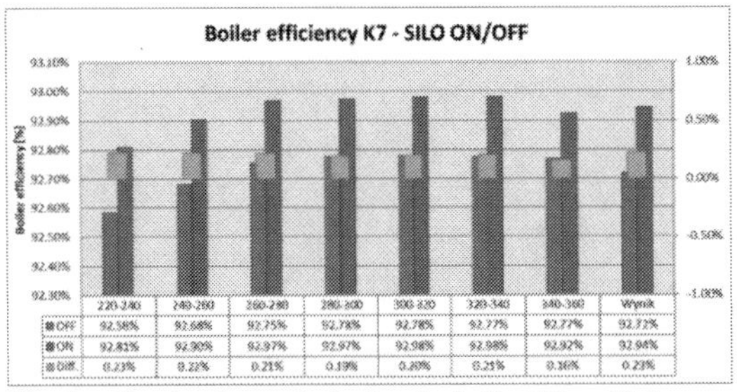

Fig. 8a - Combustion optimization for boiler K7 – efficiency

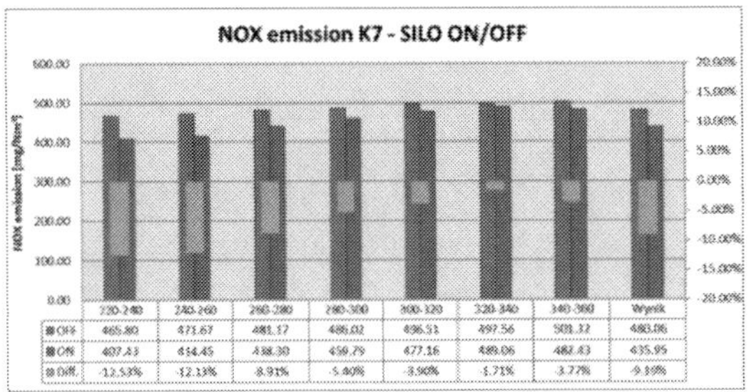

Fig. 8b - Combustion optimization for boiler K7 – NOX emission

The average boiler efficiency increase is 0.23 % and reduction of NOX emission is 9.19%. The main factor is reduction of excess air and air distribution with respect of constraints like steam temperatures, O2 in flue gas or CO emission. The lowest efficiency gain is recorded for lower boiler load, which is caused by lower steam temperatures, what requires higher excess air.

8.4 The sootblowing optimization

The sootblowing optimization results in sootblowing steam reduction of 4311 t for boiler K2 and 3498 t for boiler K5 and K7 individually. This allows for annual reduction of fuel consumption respectively 11662 GJ for boiler K2 and 8421 GJ for boilers K5 and K7 individually. The optimization effects are presented on **Fig. 9b** and **Fig. 9b**.

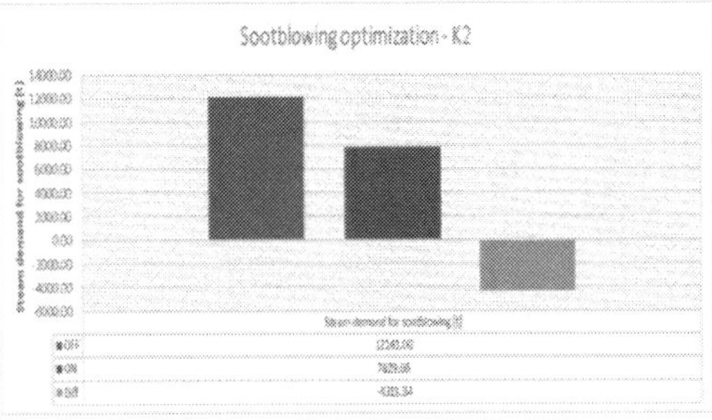

Fig. 9a - Sootblowing optimization of boiler K2

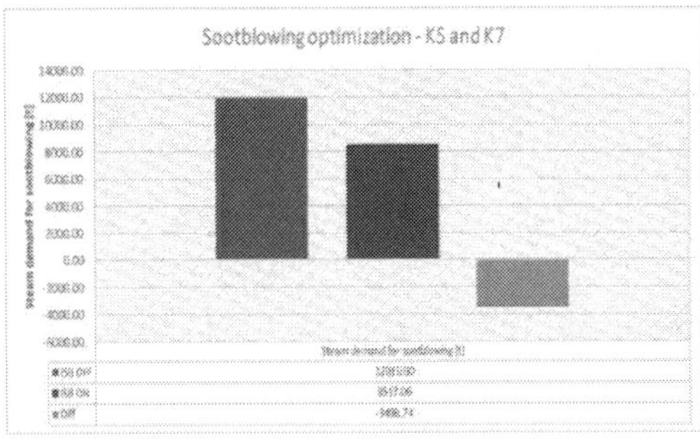

Fig. 9b - Sootblowing optimization of boiler K5 and K7

9 Conclusions

The project has a positive influence on the process efficiency and economical profit of the CHP plant. The model based optimization and control allows for more efficient calculation of process control setpoints and control trajectory. The system respects constraints what allows to maintain the necessary range for base process control. The constraints management feature combined with balanced mass and energy brings a new quality. Depending on the solution systems operate with 1-5 minutes resolution supporting process operators with their duty. In other words the systems support operators to work in more efficient way and generally, the plant operates respecting standard procedures, defined by CHP management board.

https://doi.org/10.1051/e3sconf/201913701051

Bibliography

1. Kler A.M.; Stepanova E.L.; Maximov A.S., The optimization problems of CHP operation, *Journal of Physics: Conference Series*, Vol.891(1), (2017).
2. Badyda K., Milewski J., Bujalski W., Warchoł M., Heat Accumulator in Large District Heating Systems – Simulation and Optimisation, *Proceedings of ASME Turbo Expo 2010: Power for Land, Sea and Air GT2010*, Glasgow, Scotland, UK. ASME Paper GT2010 -22032, (2010).
3. Badyda K., Bujalski W., Niewinski G., Warchol M., Selected issues related to heat storage tank modelling and optimisation aimed at forecasting its operation, *Archives of Thermodynamics*, Vol.32(3), p.3, (2011).
4. Śladewski Ł., Wojdan K., Świrski K., Janda T., Nabagło D., Chachuła J., Optimization of combustion process in coal-fired power plant with utilization of acoustic system for in-furnace temperature measurement, *Applied Thermal Engineering*, Vol. 123, pp. 711–720, (2017).
5. Wojdan K., Świrski K., Chomiak T., Immune Inspired System for Chemical Process Optimization using the example of a Combustion Process in a Power Boiler, *IEEE Proc. of 14th International Conference on Intelligent System Applications to Power Systems -ISAP 2007*, Kaohsiung, Tajwan, (2007).
6. Sanchez-Lopez A., Arroyo-Figueroa G., Villavicencio-Ramirez A., Advanced control algorithms for steam temperature regulation of thermal power plants, *International Journal of Electrical Power and Energy Systems*, Vol.26(10), pp.779-785, (2004).
7. Custer, J. ; Fuller, T.A. P ower Engineering, February 2017, Vol.121(2)
8. Pattanayak L., Ayyagari S., Sahu J., *Clean Technologies and Environmental Policy*, Vol.17(7), pp.1897-1906, (2015).
9. Szargut J., Rachunek wyrównawczy w technice cieplnej, *Wrocław: Zakład Narodowy im. Ossolińskich*, (1984).

RESEARCH & DEVELOPMENT IN POWER ENGINEERING, 2019

Fuel processing in a swirl flow: numerical modelling of combustion and gasification

Robert Zarzycki, Justyna Jędras, Rafał Kobyłecki

Czestochowa University of Technology, ul. Dąbrowskiego 69, 42-201 Częstochowa, Poland; E-mail: robert.zarzycki@pcz.pl

Abstract. This study presents the concept of a cyclone furnace for coal dust combustion and gasification under conditions of oxy-fuel combustion. A two-chamber design of the cyclone furnace allows for the separation of the process of heating, drying and devolatilization of fuel from processes of its combustion and gasification of carbon residue. The choice of process parameters helps control fuel gasification. Supplying the driving gas with a specific composition (O_2, CO_2) to the chamber PC1 ensures the control of temperature and composition of combustible gases obtained through gasification. These gases can be used as a fuel for power boilers and, consequently, allow for utilization of the oxy-fuel combustion technology in new or existing power boilers.

1 Introduction

The burning of fossil fuels in many countries is the main source of electricity and heat. The processes of fossil fuel combustion in conventional boilers (stoker boilers, direct flow boilers, or fluidized bed boilers) are already well-recognized and no further significant development of these technologies is expected. Due to their negative impact on the natural environment, the use of fossil fuels in the power industry (CO_2, SO_x, NO_x, Hg, HCl and dust emission) will be reduced in the coming years. The recently rising costs of CO_2 emission allowances will limit the development of energy based on burning fossil fuels. However, it is planned to replace the processes of direct coal combustion with the processes of its gasification [1, 2]. After purification, the gas produced during coal gasification can be used as a substrate for various chemical processes or as a fuel.

2 Cyclone furnace

Cyclone furnaces are power-generating devices where fuel combustion and gasification processes occur in a strong swirl flow of gas [3 - 5]. With effective mixing and a relatively small combustion chamber in cyclone furnaces, it is possible to ensure combustion of solid, liquid and gaseous fuels. These fuels may also have low calorific value, with high ash and water contents. With a relatively high heat load in the combustion chamber, it is possible to implement the process of combustion and gasification, e.g. fragmented hard coal and black coal. After purification and preparation, gas obtained during gasification can be used as a substrate for various industrial purposes or for combustion in chambers of power boilers by replacing it by e.g. ignition fuel. Therefore, the gasification process in the atmosphere of air, oxygen, recirculated flue gas (CO_2) or water steam can be performed in a cyclone furnace. The present study discusses the results of computations for the process of coal dust gasification in the O_2/CO_2 atmosphere.

The cyclone furnace (Fig. 1) consists of two upper (PC2) and lower (PC1) chambers. The chamber PC2 is cylinder-shaped. Its upper part features tangentially designed channels used for pneumatic supply of fragmented fuel (coal dust). The upper part is equipped with a plunger used to limit the transport of fine coal dust grains outside the chamber PC2 to the chamber PC3 for combustion of gasification products. The lower chamber PC1 is comprised of several steps with reduced diameters. Each step features tangential nozzles that allow for the supply of the "driving" gas with the specific composition (O_2, CO_2) to the chamber PC1. Both chambers (PC1 and PC2) are connected with a channel where an internal plunger allows for separation of the flow of flue gas from the chamber PC1 to PC2 and the fuel flow (carbon residue) from the chamber PC2 to PC1 (Fig. 1). The lower part of the chamber PC1 features an axially located nozzle for the supply of gases with specific composition to the chamber PC1.

The discussed design of a cyclone furnace containing two chambers allows for:

- heating, drying and gasification of fuel in the chamber PC2;
- coal gasification process in the atmosphere of recirculated flue gas (CO_2 and O_2) or (CO_2, O_2 and H_2O) in order to maximize production of CO and H_2, as well as combustion of a certain part of carbon residue in order to generate the necessary amount of heat for maintaining and control over endothermic reactions.

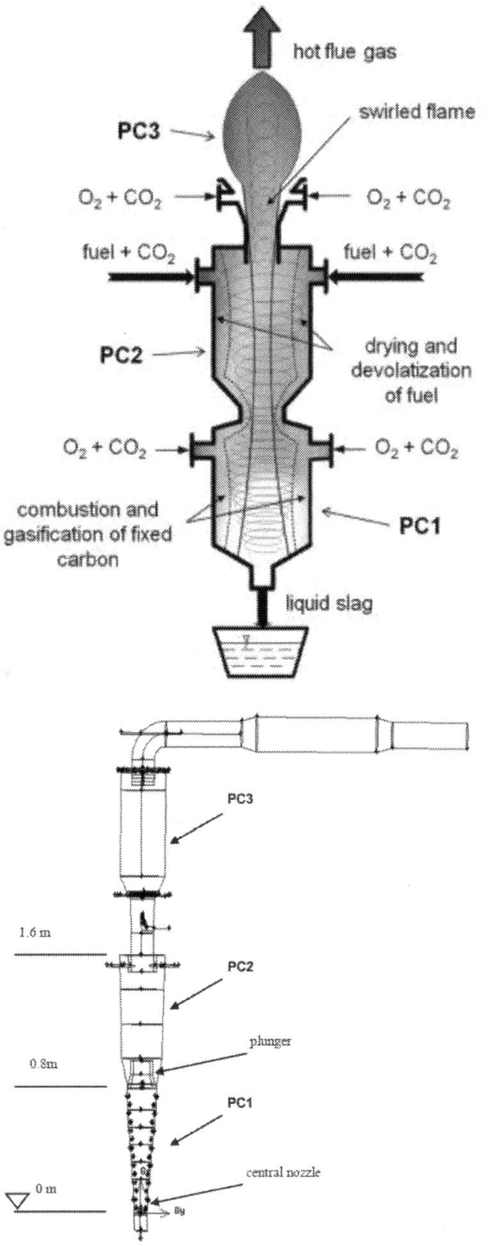

Fig. 1. The design and view of the geometry of a cyclone furnace.

3 Modelling of the process of coal combustion and gasification in a cyclone furnace

ANSYS FLUENT 14 software was used for calculations of the flow of coal dust and the combustion and gasification process. The numerical model of the process allows for the computation of coal dust gasification in the atmosphere of O_2, CO_2 through flue gas recirculation.

The lower part of the cyclone furnace (the chamber PC1) (Fig. 1) features nozzles that allow for implementation of the driving medium with oxidant (O_2, CO_2), whereas the upper part of the chamber PC2 contains inlets for coal dust through which the dust is supplied to a cyclone furnace pneumatically, using the stream of CO_2. The central nozzle in the lower part of the chamber PC1 is used for supply the gas mixture to the chamber PC1 (e.g. CH_4/O_2 that allows for stabilization of the processes of gasification and combustion of carbon residue.

Calculations of gas and fuel flow were performed using the *Reynolds Stress*turbulence model [4,5], which can also be successfully used for strong swirl flow. Modelling of the flow of coal grains was based on the *Discrete Phase Model*, whereas coal dust combustion was performed using the *Species Transport* model, which allows for modelling chemical reactions both in the solid and gaseous phases [6-10]. The calculations were based on the radiation model termed *Discrete Ordinate (DO)*. Calculations of the process of coal dust combustion and gasification with replacement diameter of 500 μm were carried out for the fuel with physicochemical parameters as presented in Table 1. For simplification purposes, it was adopted that the fuel does not contain sulphur. Reaction rate constants were derived from the studies [6, 7].

Table 1. Result of technical and elemental analysis of the fuel used for simulations (dry state).

Proximate analysis	[-]	Ultimate analysis	[-]
VM	0.45	C	0.85
FC	0.45	H	0.1
A	0.05	O	0.04
M	0.05	N	0.01

Calculations for combustion and gasification of coal dust was described with seven reactions:

Reaction of volatile matter combustion

$$x_1\,VM + x_2\,O_2 = y_1\,CO + y_2\,H_2O + y_3\,N_2 \quad (1)$$

where: x_1, x_1, y_1, y_2, y_3 are stoichiometric coefficients.

Reaction of oxidation of carbon oxide

$$CO + 0.5\,O_2 = CO_2 \quad (2)$$

Reaction of oxidation of fixed carbon (FC)

$$C_{(s)} + 0.5\,O_2 = CO \quad (3)$$

Boudouard©reaction

$$C_{(s)} + CO_2 = 2\,CO \quad (4)$$

Synthesis of water gas

$$C_{(s)} + H_2O = CO + H_2 \quad (5)$$

Hydrogen oxidation reaction

$$H_2 + 0.5\,O_2 = H_2O \quad (6)$$

Methane oxidation reaction

$$CH_4 + 1.5\,O_2 = CO + 2H_2O \quad (7)$$

Conditions of coal dust gasification process, which are the subject of numerical modelling were as follows: flow of coal dust with mean grain size of 500 μm of $1 \cdot 10^{-3}$ kg/s supplied tangentially with two inlets located at the opposite sides of the chamber PC2 (Fig. 1) in the CO_2 stream at the rate of 3 m/s. Following a helical line, the fuel moves downwards in the chamber PC2 as it is heated, dried and devolatilized through the effect of hot flue gas that flows in the furnace axis and contact with hot walls of the chamber PC2. After devolatilization of the fuel, the process of partial burning of fixed carbon (FC) occurs in the chamber PC1. Through the set of feed nozzles for PC1 chamber, the "driving" gas was tangentially fed to the one forming the "driving" gas at the rate of 3 m/s and temperature of 600 K with molar fraction of O_2/CO_2 40/60%. Introduction of the driving gas supplied tangentially from the nozzles in the chamber PC1 is necessary to maintain proper conditions of the swirl flow of the coal dust. The lower part of the chamber PC1 features an additional central nozzle (located axially), from which the mixture of gases (80% O_2, 20% CH_4) at a temperature of 350 K was discharged at a constant rate of 10 m/s.

The numerical model also takes into consideration the heat transfer through furnace walls to its surroundings. Calculation of heat loss took into consideration the actual thickness of individual components of the laboratory furnace [4, 5] with coefficients of heat penetration and conduction for the materials used. It was assumed that air temperature around the furnace is 300 K.

4 Analysis of the process of coal dust combustion and gasification

Analysis of the distribution of selected process parameters in two perpendicular cross-sections of the cyclone furnace. Distributions of fuel concentration on internal walls of the cyclone furnace are presented in Fig. 2a, whereas vertical cross-sections in two perpendicular planes are presented in Fig. 2b. As expected, the results show that the fuel, which was supplied in the upper part of the chamber PC2 as a result of strong swirl flow and effect of the centrifugal force, moves only near the internal walls of the furnace (Fig. 2b). For the adopted process conditions, the highest concentration of fuel in the upper part of the PC1 chamber occurs in the vicinity of its walls.

Fig. 2c shows the distribution of the combustion rate for the devolatilized fuel (fly ash). It is noticeable that the area of the highest combustion rates is located in the upper part of the PC1 chamber (near the symmetry axis) and coincides with the area of decreasing fuel concentration in Figs. 2a and 2b.

The consequence of the high fuel combustion rate is temperature distributions shown in Fig. 2d near the inner walls of the furnace and in Fig. 2e in the vertical cross-section of the cyclone furnace. Analysis of Figs. 2d and 2e reveals a high-temperature region exceeding 2000 K in the upper part of the PC1 chamber. It is directly related to the burning of fuel in the form of fly-ash. In the lower part of the PC1 chamber (Fig. 2e), the area of high temperatures exceeding 2000 K is also observed, coinciding with the location of combustion of the CH_4/O_2 mixture supplied from the central nozzle.

The combustion or gasification process depends on the presence of fuel and oxygen concentration. Fig. 2f presents distributions of molar fractions of oxygen in the vertical cross-section of the cyclone furnace. In order to reduce the loss of incomplete combustion at the design stage, it was assumed that the lower part of the PC1 chamber should have the highest possible oxygen concentration. This can be achieved by supplying excess oxygen in the O_2/CH_4 mixture through a burner installed in the lower part of the PC1 chamber. Analysis of the distributions of oxygen concentration presented in Fig. 2f reveals that the oxygen concentration in the lower part of the PC1 chamber exceeds 30% (for the accepted process conditions). These conditions allow for minimizing the loss of incomplete combustion. The upper part of the PC1 chamber shows an oxygen concentration close to zero. The area in which there is a rapid decline in oxygen concentration coincides with the location of the highest temperatures (Fig. 2e), high rates of fly ash combustion (Fig. 2c), which indicates the location of the front of fuel combustion in this place. In this area, significant amounts of heat are released, which results in a large surface and volume load to the PC1 chamber. Due to the properties of the materials used in the construction of the fireplace, it is not recommended to further increase the temperature. It is necessary to control the temperature, which can be achieved e.g. by reducing the fuel flow and heat generation. Temperature can also be controlled by means of an endothermic Boudouard reaction. In order to ensure the effectiveness of the gasification process in accordance with the Boudouard reaction, it is necessary to meet several conditions of the process. This process takes place at sufficiently high temperatures. At temperature above 900 K, the balance of Boudouard© reaction is shifted towards carbon monoxide formation. The Boudouard reaction is only possible with the presence of reactive fixed carbon (devolatilized fuel i.e. fly ash) and sufficient time for the fixed carbon to react with carbon dioxide, and the most intensive contact of fixed carbon and carbon dioxide. These conditions are ensured by the design of the PC1 chamber of the furnace in which a stream of strong swirl flow of coal dust devolatilized in the PC2 chamber, transported through individual steps of the PC1 chamber, allows for virtually infinite contact of fuel with carbon dioxide, which is supplied as one of the components of the "driving" gas (flue gas recirculation) and is a product of fixed carbon combustion. In this case, it is only important to ensure a sufficient temperature to initiate and then continue the Boudouard reaction.

The distributions of CO_2 and CO concentrations in the vertical cross-sections of the furnace are shown in Figures 2g and 2f, respectively. It is noticeable that in the lower part of the PC1 chamber, the CO_2 concentration is ca. 60%, which is related to the supply of the "driving" gas with the assumed composition of O_2/CO_2 of 40/60 %. In the central part of the PC1 chamber, the CO_2 concentrations increase locally, which is associated with the combustion of fly ash (location of the combustion front) and then it declines quite rapidly. Furthermore, the analysis of the CO concentrations presented in Figure 2h reveals concentrations of this component in the lower part of the PC1 chamber close to zero (due to the excess oxygen occurring there, see Fig. 2f), while a rapid increase in CO concentration occurs above the place identified as the fly ash combustion front. This increase is caused by the presence of devolatilized fuel (Figs. 2a and 2b), high temperatures above 2000 K (Figures 2d and 2e), low oxygen concentration (Fig. 2f) and high concentration of CO_2 (Figures 2g). Compliance with these conditions allows for fuel gasification in accordance with the Boudouard reaction (4).

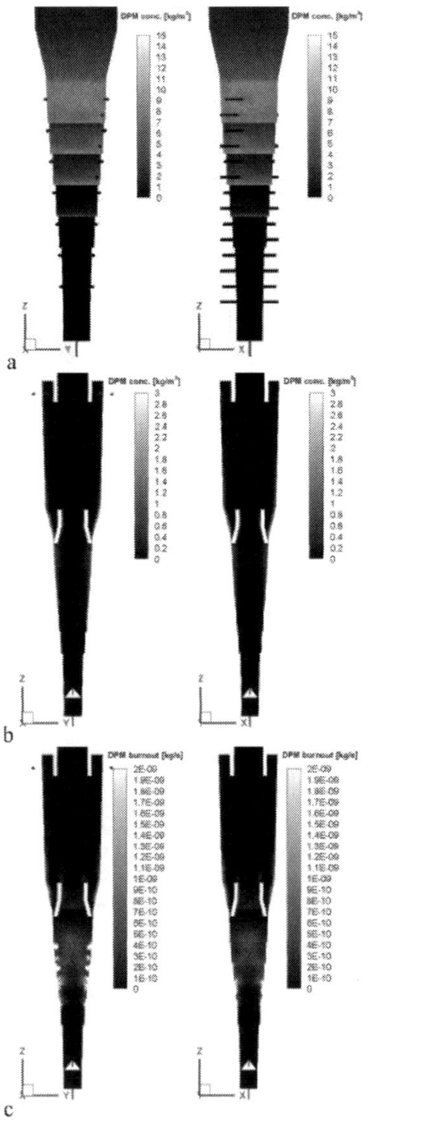

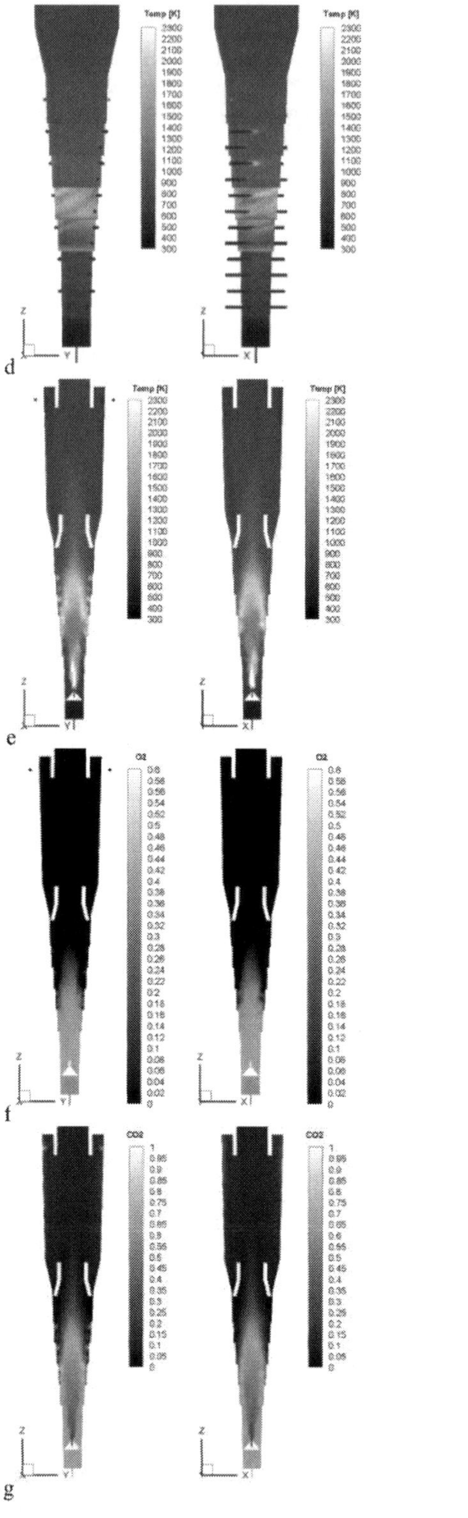

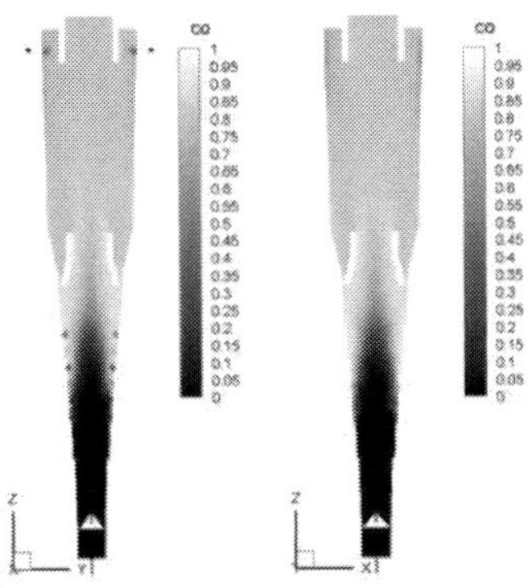

h

Fig. 2. Distributions of selected values for gas supplied by "driving" nozzles of the PC1 chamber with composition of 40% O_2, 60% CO_2: a) fuel concentration on the furnace walls, b) fuel concentration in the vertical section of the furnace, c) combustion rate for the devolatilized fuel (fly ash), d) temperature near the furnace walls, e) temperature in the vertical section of the furnace, f) molar fractions of O_2 in the vertical section, g) molar fractions of CO_2 in the vertical section of the furnace, h) molar fractions of CO in the vertical section.

Table 2 presents selected mean process parameters at the outlet from PC1 and PC2 chambers. The temperature drop between PC1 and PC2 cross-sections can be observed. It results from the process of heating and degassing of the fuel in the PC2 chamber and supplying of a CO_2 stream into the PC2 chamber, transporting pneumatically the fuels at a temperature of 300 K. For the assumed process parameters, the O_2 concentration at the outlet from the PC1 and PC2 chambers is close to zero. Analysis of the CO_2 and CO concentrations at the outlet from the PC1 chamber reveals the effect of the coal gasification process. At the outlet from the PC1 chamber, the CO concentration of 0.859 and the CO_2 concentration of 0.071 are obtained for the adopted process conditions. Minor changes in concentrations are noticeable at the outlet from the PC2 chamber and result mainly from the dilution of the gas generated in the process of gasification with carbon dioxide used for pneumatic fuel supply to the PC2 chamber. The function of the PC2 chamber as a fuel degassing area is confirmed by the values of volatile matter (VM) concentrations. Despite the dilution of the gases in the PC2 chamber with CO_2 supplied with the fuel, the VM concentration increases to 0.064.

Table 2. Selected process parameters at the outlet from the PC1 and PC2 chambers.

	PC1	PC2
Temp [K]	978.2	767.2
O_2 [-]	4.22E-05	1.05E-19
CO_2 [-]	0.071	0.114
CO [-]	0.859	0.805
VM [-]	0.053	0.064

5 Conclusions

The results of numerical modelling of coal dust gasification and combustion in an innovative cyclone furnace lead to the following conclusions:

- The configuration of the geometry of the lower part of the furnace and distribution of the "driving" nozzles at individual levels helps control the time of fuel grains (fly ash) remaining so that the maximal level of its reaction is achieved.
- The process of fuel devolatilization and its gasification as a result of the Boudouard reaction allows for controlling the temperature inside the PC1 chamber.
- For the adopted conditions of cyclone furnace operation, the CO concentrations at the outlet from the chamber PC2 were ca. 80%.

The results presented in the paper were obtained in a project co-financed by NCBR within the contract No. BIOSTRATEGG3/345940/7/ NCBR/2017 titled Water in soil: satellite monitoring to improve water retention using biochar (project acronym: SoilAqChar). The paper was co-financed by BS/PB-400-301/19.

References

1. Zhao Y., Feng D., Sun S., Luan J., Che H., Characteristics of rice husk gasification in cyclone pyrolysis-suspended combustion system, *Thermal Science*, vol. 22, Suppl. 2, pp. S439-S447, (2018).
2. Osborne D., The Coal Handbook: Towards Cleaner Production: Volume 2: Coal Utilisation, Woodhead Publishing, (2013).
3. Guo J., Zhang L., Cang D., Zhang C., Wang D., Yin H., Cold experimental studies of flow field in the swirling coal combustion device, *IOP Conf. Series: Earth and Environmental Science*, vol. 94, pp.012-021, (2017).
4. Zarzycki R., Bis Z., Kobyłecki R., The Concept of Coal Burning in a Cyclone Furnace, *Procedia Engineering*, Vol. 157, pp. 472-479, (2016).
5. Zarzycki R., Bis Z., Modelling of the process of coal dust combustion in a cyclone furnace, *Journal of Thermal Science*, vol. 26, Issue 2, pp 192–198, (2017).
6. Toporov D., Bocian P., Heil P., Kellermann A., Stadler H., Tschunko S., Förster M., Kneer R., Detailed investigation of a pulverized fuel swirl flame in CO2/O2 atmosphere. *Combustion and Flame*, vol. 155, pp. 605-618, (2008).
7. Vascellari M., Cau G., Numerical simulation of pulverized coal oxycombustion with exhaust gas recirculation. *Proceeding of CCT2009 Fourth International Conference on Clean Coal Technologies*. Dresden, Germany, (2009).
8. Chen L., Gazzino M., Ghoniem A.F., Characteristics of pressurized oxy-coal combustion under increasing swirl number. *35th International Technical Conference on Coal Utilization & Fuel Systems*. Clearwater, Florida, (2010).
9. Warzecha, P. ,Bogusławski, A., Symulacje numeryczne spalania pyłu węglowego w atmosferze

https://doi.org/10.1051/e3sconf/201913701052

O2-CO2. *Archiwum Spalania*, vol. 12, No. 3, pp. 145-151, (2012).
10. ANSYS Fluent Theory Guide, (2011).

AUTHOR INDEX

AUTHOR INDEX